# Fundamentals and Applications of Microfluidics

## Third Edition

For a listing of recent titles in the
*Artech House Integrated Microsystems Series*,
turn to the back of this book.

# Fundamentals and Applications of Microfluidics

**Third Edition**

Nam-Trung Nguyen
Steven Wereley
Seyed Ali Mousavi Shaegh

ARTECH HOUSE
BOSTON | LONDON
artechhouse.com

**Library of Congress Cataloging-in-Publication Data**
A catalog record for this book is available from the U.S. Library of Congress.

**British Library Cataloguing in Publication Data**
A catalog record for this book is available from the British Library.

ISBN-13: 978-1-63081-364-2

**Cover design by John Gomes**


**685 Canton Street**
**Norwood, MA 02062**



10 9 8 7 6 5 4 3 2 1

*To my wife, Jill—my friend, my companion, and the love of my life*
*—Nam-Trung Nguyen*

*To my wife Kristina Bross and daughters Katie and Gracie, who have been so understanding and patient with the demands of academic pursuits*
*—Steve Wereley*

*To my mother and my father, Alliyeh and Seyed Jafar, my first teachers, who nourished my soul to never stop learning*
*—Seyed Ali Mousavi Shaegh*

# Contents

# Preface to the Third Edition

It has been more than 10 years since the second edition of this book was published, and which has provided readers an overview of fundamentals and applications of microfluidics. In the last decade, microfluidics has experienced extensive growth. Currently, the technology is at an advanced and mature level that can support the commercialization of various microfluidic devices. Detailed understanding of microfluidic phenomena, and the development of microfabrication techniques suitable for rapid prototyping of polymeric microdevices, have been the main motives for the adaptation of microfluidics for various applications. In particular, microfluidics has opened new horizons for analytical tasks such as diagnostics and field-deployable portable detections, drug-related studies such as disease modeling, drug discovery, and drug delivery, as well as regenerative medicine and tissue engineering. The third edition of the book maintains the core structure of the previous edition that covers a wide spectrum of the microfluidics literature. Since significant advancements have been achieved in the field of microfluidics, the current edition mainly presents the major and revolutionizing developments. More than 10 figures/tables were updated while more than 50 new multipanel figures were added to the third edition of the book. Similar to the second edition, the book offers information for upper-level undergraduates and graduate students in an introductory course in microelectromechanical systems (MEMS), bio-MEMS, or microfluidics. The structure for the book is suitable for classroom use. In general, the book is divided into a *fundamentals* section and an *applications* section.

*Fundamentals*:

Chapter 1 introduces the field of microfluidics including its definition and commercial and scientific aspects.

Chapter 2 discusses when to expect changes in fluid behavior as the length scale of a flow is reduced to microscopic sizes. New microfluidic phenomena such as micromagnetofluidics, optofluidics and microacoustofluidics are introduced in the new edition.

Chapter 3 provides the technology fundamentals required for making microfluidic devices, ranging from silicon-based microfabrication to alternative nonbatch techniques appropriate for small-scale production and prototyping. In the new edition, fabrication methods and rapid prototyping techniques for making polymeric and paper-based devices are presented. In addition, detailed discussion on the use of additive manufacturing (3-D printing) methods is provided.

Chapter 4 presents experimental characterization techniques for microfluidic devices with a concentration on full-field optical techniques.

*Applications*:

Chapter 5 describes the design of microdevices for sensing and controlling macroscopic flow phenomena such as velocity sensors, shear stress sensors, microflaps, microballoons, microsynthetic jets, and microair vehicles.

Chapters 6, 7, and 8 present design rules and solutions for microvalves, micropumps, and microflow sensors, respectively. New designs of microvalves, micropumps and microflow sensors are also discussed.

Chapters 9 to 13 discuss a number of tools and devices for the emerging fields of life sciences and chemical analysis in microscale, such as needles, mixers, dispensers, separators, and reactors. Major advancements occurred for each category of tools have been added to the corresponding chapters.

# Acknowledgments

The authors would like to express our gratitude to the many faculty colleagues, research staff, and graduate students who helped this work come together. Conversations with Carl Meinhart of the University of California Santa Barbara, Juan Santiago at Stanford University, Ali Beskok at Texas A&M University, Kenny Breuer at Brown University, and Ron Adrian at the University of Illinois were especially helpful, both in this book as well as in developing microfluidic diagnostic techniques.

Nam-Trung Nguyen is grateful to Hans-Peter Trah of Robert Bosch GmbH (Reutlingen, Germany), who introduced him to the field of microfluidics almost 30 years ago. Dr. Nguyen is indebted to Wolfram Dötzel, who served as his advisor during his time at Chemnitz University of Technology, Germany, and continues to be his guide in the academic world. He is thankful to his colleagues at Nanyang Technological University, Singapore. He expresses his love and gratitude to his wife Thuy-Mai and his three children Thuy-Linh, Nam-Tri and Nam-An for their love, support, patience, and sacrifice.

Steve Wereley is particularly indebted to Richard Lueptow at Northwestern University, who served as his dissertation advisor and continues to offer valuable advice on many decisions, both large and small. He would also like to thank his wife and fellow traveler in the academic experience, Kristina Bross. This book project has exacted many sacrifices from both of them. Certain portions of Chapter 2 were written at the hospital while Kristina was laboring to deliver our second child. Finally, he would like to thank his parents, who nurtured, supported, and instilled in him the curiosity and motivation to complete this project.

Seyed Ali Mousavi Shaegh is indebted to Nam-Trung Nguyen who served as his amazing dissertation advisor during his PhD program at Nanyang Technological University, Singapore, and continues to be his mentor in his academic career. Importantly, Ali would like to thank Nam-Trung Nguyen for giving him such a great opportunity to contribute to the preparation of the book's third edition. He is also thankful to Ali Khademhosseini as his postdoc mentor at Harvard University. His inspirations encouraged Seyed Ali to define a new horizon in his academic career. He would like to thank his parents and wife for their love, curiosity and support to finish the book project.

# Chapter 1

# Introduction

## 1.1 WHAT IS MICROFLUIDICS?

**mi·cro·flu·id·ics** *(mī′krō flōō id′iks)* n. The science and engineering of systems in which fluid behavior differs from conventional flow theory primarily due to the small length scale of the system.

### 1.1.1 Relationships Among MEMS, Nanotechnology, and Microfluidics

Since Richard Feynman's thought-provoking 1959 speech "There's Plenty of Room at the Bottom" [1], humanity has witnessed the most rapid technology development in its history—the miniaturization of electronic devices. Microelectronics was the most significant enabling technology of the last century. With integrated circuits and progress in information processing, microelectronics has changed the way we work, discover, and invent. From its inception through the late 1990s, miniaturization in microelectronics followed Moore's law [2], doubling integration density every 18 months. Presently poised at the limit of photolithography technology (having a structure size less than 100 nm), this pace is expected to slow down to doubling integration density every 24 to 36 months [3]. Until recently, the development of miniaturized nonelectronic devices lagged behind this miniaturization trend in microelectronics. In the late 1970s, silicon technology was extended to machining mechanical microdevices [4]—which later came to be known as microelectromechanical systems (MEMS). However, it is inappropriate, though common, to use MEMS as the term for the microtechnology in use today. With fluidic and optical components in microdevices, microsystem technology (MST) is a more accurate description. The development of microvalves, micropumps and microflow sensors in the late 1980s dominated the early stage of microfluidics. However, the field has been seriously and rapidly advanced since the introduction by Manz et al. at the Fifth International Conference on Solid-State Sensors and Actuators (*Transducers '89*), which indicated that life sciences and chemistry are the main application areas of microfluidics [5]. Several competing terms, such as "microfluids," "MEMS-fluidics," or "bio-MEMS," and "microfluidics" appeared as the name for the new research discipline dealing with transport phenomena and fluid-based devices at microscopic length scales [6]. With the emergence of nanotechnology, terms such as "nanofluidics" and "nanoflows" have also been growing in popularity. With all these different terms for basically the same thing (i.e., flows at small scales), it makes sense to try to converge on an accepted terminology by asking a few fundamental questions, such as:

- What does the "micro" refer to in microfluidics?
- Is microfluidics defined by the device size or the fluid quantity that it can handle?
- How does the length scale at which continuum assumptions break down fit in?

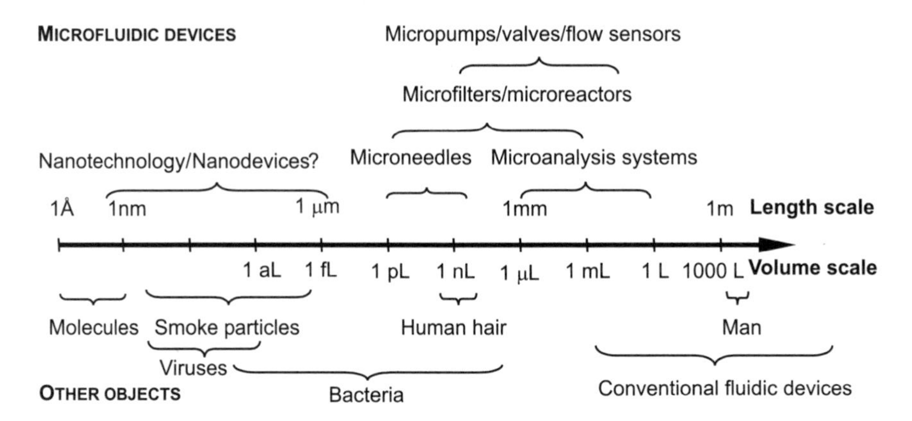


**Figure 1.1** Size characteristics of microfluidic devices.

While for MEMS, it may be reasonable to say that the device size should be smaller than a millimeter (after all, the first "M" stands for *micro*), the important length scale for microfluidics is not the overall device size but rather the length scale that determines flow behavior. In the other words, microfluidics can be defined as the science and technology of manipulating small amount of fluids ($10^{-9}$-$10^{-18}$ liters) in channels with dimensions of tens to hundreds of micrometers [7]. The main advantage of microfluidics is utilizing scaling laws and continuum breakdown for new effects and better performance. These advantages are derived from the microscopic amount of fluid a microfluidic device can handle. Regardless of the size of the surrounding instrumentation and the material of which the device is made, only the space where the fluid is processed has to be miniaturized. The miniaturization of the entire system, while often beneficial, is not a requirement of a microfluidic system. However, the development of microfluidic-based systems for portable and field-deployable applications such as point-of-care testing and diagnostics necessitates the miniaturization of the whole device. The microscopic quantity of fluid is the key issue in microfluidics. The term "microfluidics" is used here not to link the fluid mechanics to any particular length scale, such as the micron, but rather to refer in general to situations in which small-size scale causes changes in fluid behavior. This use of "microfluidics" is analogous to the way that "microscope" is used to refer both to low magnification stereo microscopes that have spatial resolutions of 100 μm, as well as to transmission electron microscopes that can resolve individual atoms. Hence, in this text "microfluidics" is a generic term referring to fluid phenomena at small length scales, and while nanometer-scale and even molecular-scale flow phenomena are certainly discussed at length in this text, the term "nanofluidics" is not used. The working definition of microfluidics used in this text is given above.

These differing points of view regarding device size and fluid quantity are to be expected considering the multidisciplinary nature of the microfluidics field and those working in it. Electrical and mechanical engineers came to microfluidics with their enabling microtechnologies. Their common approach was shrinking down the device size, leading to the idea that microfluidics should be defined by device size. Analytical chemists, biochemists, and chemical engineers, for years working in the field of surface science, came to microfluidics to take advantage of the new effects and better performance. They are interested primarily in shrinking down the pathway the chemicals take, leading to the idea that microscopic fluid quantities should define microfluidics. Biologists and bioengineers started to develop microfluidic systems to take advantage of the small dimensions of microfluidic devices for easy handing of cells and various biological components and mimicking cellular microenvironment under in vitro conditions. To put these length scales in context, Figure 1.1 shows the size characteristics of typical microfluidic devices compared to other common objects.

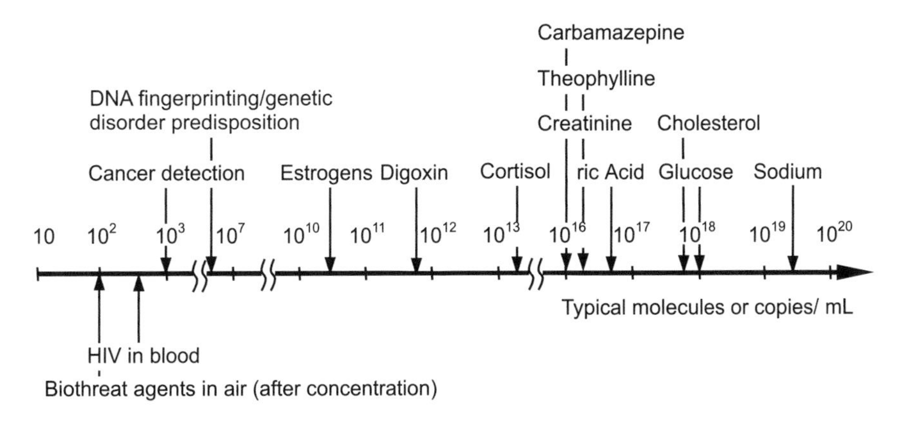


**Figure 1.2** Concentrations of typical diagnostic analytes in human blood or other samples. (*After*: [8].)

When beginning in the field of microfluidics, one must ask whether working at microscopic length scales is really beneficial. For example, the maximum sensitivity that a sensor can have is limited by the analyte concentration in a sample. The relation between the sample volume *V* and the analyte concentration $A_i$ is given by [5]:

$$V = \frac{1}{\eta_{\mathrm{s}} N_{\mathrm{A}} A_i} \tag{1.1}$$

where $\eta_{\mathrm{s}}$ is the sensor efficiency ($0 < \eta_{\mathrm{s}} < 1$), $N_{\mathrm{A}}$ is the Avogadro number, and $A_i$ is the concentration of analyte *i*. Equation (1.1) demonstrates that the sample volume or the size of the microfluidic device is determined by the concentration of the desired analyte. Figure 1.2 illustrates the concentrations of typical diagnostic analytes in human blood or other samples of interest. Concentration determines how many target molecules are present for a particular sample volume. Sample volumes that are too small may not contain any target molecules and thus will be useless for detection purposes. This concept is illustrated in Figure 1.3. Common human clinical chemistry assays require analyte concentrations between $10^{14}$ and $10^{21}$ copies per milliliter. The concentration range for a typical immunoassay is from $10^8$ to $10^{18}$ copies per milliliter. Deoxyribonucleic acid (DNA) probe assays for genomic molecules, infective bacteria, or virus particles require a concentration range from $10^2$ to $10^7$ copies per milliliter [5]. Clinical chemistry with a relatively high analyte concentration allows shrinking the sample volume down to femtoliter range or 1 $\mu\mathrm{m}^3$ (see Figure 1.1 and Table 1.1).

Immunoassays with their lower analyte concentrations require sample volumes on the order of nanoliters. Nonpreconcentrated analysis of the DNA present in human blood requires a sample volume on the order of a milliliter [8]. Some types of samples, like libraries for drug discovery, have relatively high concentrations.

**Table 1.1**
Unit Prefixes

| *Atto* | *Femto* | *Pico* | *Nano* | *Micro* | *Milli* | *Centi* | *Deka* | *Hecto* | *Kilo* | *Mega* |
|---|---|---|---|---|---|---|---|---|---|---|
| $10^{-18}$ | $10^{-15}$ | $10^{-12}$ | $10^{-9}$ | $10^{-6}$ | $10^{-3}$ | $10^{-2}$ | 10 | $10^2$ | $10^3$ | $10^6$ |

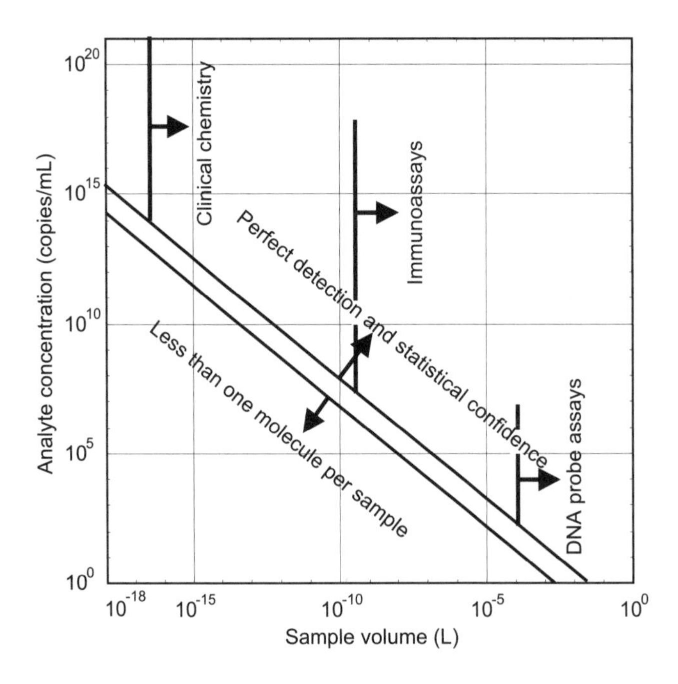


**Figure 1.3** The required analyte concentration/sample volume ratio for clinical chemistry assays, immunoassays, and DNA probe assays. (*After*: [8].)

### 1.1.2 Commercial Aspects

With the achievements in the Human Genome Project and the huge potential of both biotechnology and nanotechnology, microfluidic devices promise to be a huge commercial success. Microfluidic devices are tools that enable novel applications unrealizable with conventional equipment. The apparent interest and participation of the industry in microfluidics research and development show the commercial values of microfluidic devices for practical applications. With this commercial potential, microfluidics is poised to become the most dynamic segment of the MEMS technology thrust and an enabling technology for nanotechnology and biotechnology. From its beginnings with the now-traditional microfluidic devices, such as inkjet print heads and pressure sensors, a much broader microfluidics market is now emerging.

Figure 1.4 shows the estimated sales of microfluidic devices until 2020. The estimation predicts a $6 billion (U.S.) market for microfluidic devices [9]. The major market share belongs to devices associated with healthcare and life sciences such as drug discovery, drug delivery, and diagnostics that are considered "killer applications" of microfluidics. The major impact of such systems and devices is predicted for analytical laboratory instrumentation market. Microfluidics, and in particular lab-on-a-chip, will help to address the high cost of drug development and the pressure to reduce the drug development cycle time. Also, portable and desktop analytical devices for rapid testing and detection, telemedicine, and low-cost point-of-care diagnostics using single-use and disposable test chips hold a considerable share of the emerging market.

Microfluidics can have a revolutionizing impact on chemical analysis and synthesis, similar to the impact of integrated circuits on computers and electronics. Microfluidic devices could change the way instrument companies do business. Instead of selling a few expensive systems, companies could have a mass market of cheap, disposable devices. Making analysis instruments, tailored drugs, and disposable drug dispensers available for everyone will secure a huge market similar to that of computers today.

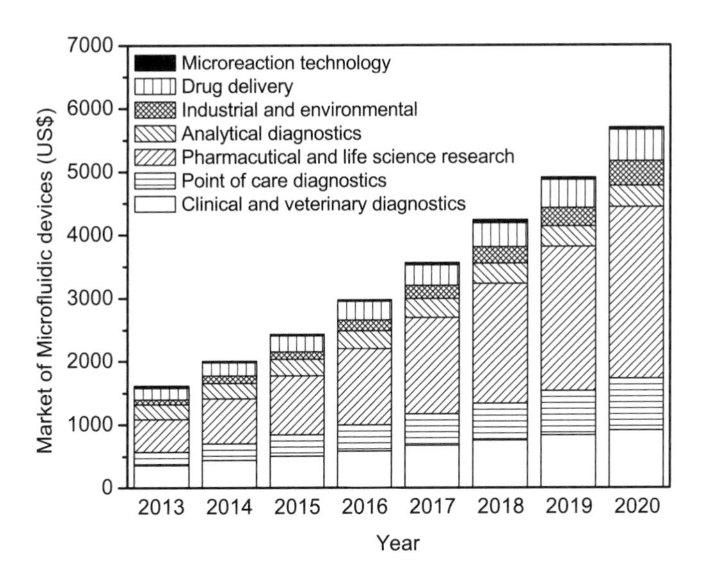


**Figure 1.4** Estimated sales of major microfluidic devices until 2020. (*After*: [9].)

Consider the analogy of the tremendous calculating labors required of hundreds of people (known as computers) necessary for even a relatively simple finite element analysis at the beginning of the twentieth century, compared to the fraction of a second that a personal computer needs today. Computing power is improved from generation to generation by higher operation frequency as well as parallel architecture. Exactly in the same way, microfluidics revolutionizes chemical screening power. Furthermore, microfluidics will allow the pharmaceutical industry to screen combinatorial libraries with high throughput—not previously possible with manual, bench-top experiments. Fast analysis is enabled by the smaller quantities of materials in assays. Massively parallel analysis on the same microfluidic chip allows higher screening throughput. While modern computers only have about 20 parallel processes, a microfluidic assay can have several hundred to several hundred thousand parallel processes. This high performance is extremely important for DNA-based diagnostics in pharmaceutical and healthcare applications.

### 1.1.3 Scientific Aspects

In response to the commercial potential and better-funded environments, microfluidics quickly attracted the interest of the scientific community. Scientists from almost all traditional engineering and science disciplines have begun pursuing microfluidics research, making it a truly multidisciplinary field representative of the new economy of the twenty-first century. Electrical and mechanical engineers contribute novel enabling technologies to microfluidics. Initially, microfluidics developed as a part of MEMS technology, which in turn used the established technologies and infrastructure of microelectronics. Fluid mechanics researchers are interested in the new fluids phenomena possible at the microscale. In contrast to the continuum-based hypotheses of conventional macroscale flows, flow physics in microfluidic devices is governed by a transitional regime between the continuum and molecular-dominated regimes. Besides new analytical and computational models, microfluidics has enabled a new class of fluid measurements for microscale flows using in situ microinstruments. Life scientists and chemists also find novel, useful tools in microfluidics. Microfluidic tools allow

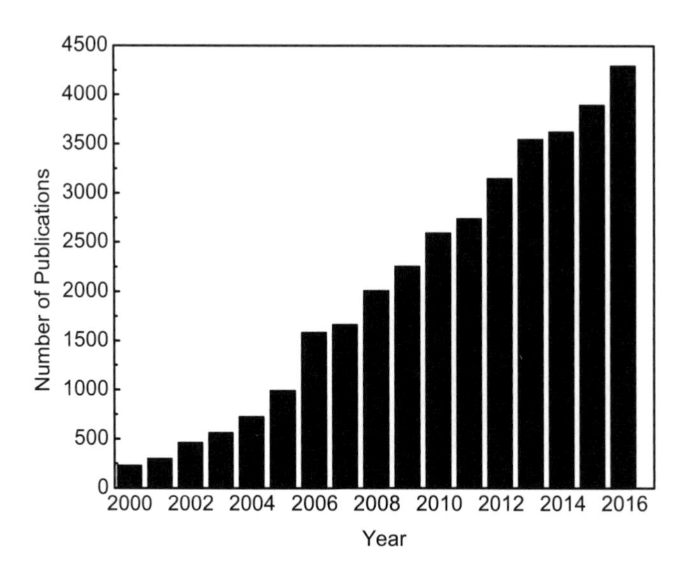


**Figure 1.5** Trend of publications with topic of microfluidics obtained from *Web of Science*.

them to explore new effects not possible in traditional devices. These new effects, new chemical reactions, and new microinstruments lead to new applications in chemistry and bioengineering. These reasons explain the enormous interest of research disciplines in microfluidics. Nowadays, beside technical conferences organized for various disciplines of microfluidics such as microTAS, almost all conferences of professional societies, such as the Institute of Electrical and Electronic Engineers (IEEE), American Society of Mechanical Engineers (ASME), International Society for Optical Engineering (SPIE), and American Institute of Chemical Engineers (AIChe), have technical sessions for microfluidics. Figure 1.5 shows a steady increase of publications on microfluidics in the last two decades.

## 1.2 MILESTONES OF MICROFLUIDICS

At the time of this writing, microfluidics was considered a mature technology that explores applications and commercialization. With the worldwide effort in microfluidics research, major developments have been made in the field that can be considered as discovery and detailed understanding of various microscale fluidic phenomena, applications-driven development of microdevices, and the development of fabrication technologies.

### 1.2.1 Microscale Fluidic Phenomena

*New physics of microscale flow.* Development and understanding of fluid flow behavior at microscale have been a major portion of investigations in the field of microfluidics. Currently, detailed understanding of various phenomena, required for on-chip flow manipulation, such as electrokinetic, mixing and pumping, microscale chemical reactions and synthesis, is available. Over the last decade, novel microscale phenomena such as ion concentration polarization (ICP) at the interface of a nanochannel and microchannel, optomicrofluidic, micromagnetofluidics, micro acoustofluidics, centrifugal microflows, microfluidic electrochemistry, droplets, and multiphase systems in microchannels have been investigated in details. Taking advantage of these phenomena, various microdevices and components including concentrators, separators, sensors and detectors, micro power generators, drug delivery systems, and drug carriers have been realized. Continual advancement of our understanding from microscale fluid flows allows for further development of novel microdevices.

### 1.2.2 Device Development

*Miniaturization approach.* With silicon micromachining as the enabling technology, researchers have been developing silicon microfluidic devices. The first approach for making miniaturized devices was shrinking down conventional principles. This approach is representative of the research conducted in the 1980s through the mid-1990s. In this phase of microfluidics development, a number of silicon microvalves, micropumps, and microflow sensors were developed and investigated [10].

Two general observations of scaling laws can be made in this development stage: the power limit and the size limit of the devices. Assuming that the energy density of actuators is independent of their size, scaling down the size will decrease the power of the device by the length scale cubed. This means that we cannot expect micropumps and microvalves to deliver the same power level as conventional devices. The surface-to-volume ratio varies as the inverse of the length scale. Large surface area means large viscous forces, which in turn requires powerful actuators to be overcome. Often, integrated microactuators cannot deliver enough power, force, or displacement to drive a microfluidics device, so an external actuator is the only option for microvalves and micropumps. The use of external actuators limits the size of those microfluidic devices, which can range from millimeters to centimeters.

*Exploration of new effects.* Since the mid-1990s, development has been shifted to the exploration of new actuating schemes for microfluidics. Because of the power and size constraints discussed above, for some applications research efforts were concentrated on actuators with no moving parts and nonmechanical pumping principles. Electrokinetic pumping, surface tension-driven flows, electromagnetic forces, and acoustic streaming are effects that usually have negligible effect at macroscopic length scales. However, at the microscale they offer particular advantages over mechanical principles. New concepts, which mimic the way cells and molecules function, are the current developmental stage of microfluidics. With this move, microfluidics is entering the era of nanotechnology and biotechnology.

*Application developments.* Concurrent with the exploration of new effects, microfluidics today is looking for further application fields beyond conventional fields, such as flow control, chemical analysis, biomedical diagnostics, and drug discovery. Owing to the fact that microfluidic technologies enable rapid processing and precise control of liquids, a major portion of investigations have been dedicated to diagnostic methods and biology research [11]. In particular, microfluidics is able to manipulate cells or create cellular microenvironment at in vitro conditions. To this end, organ-on-chip, as a new class of microfluidic technology, was developed to mimic the behavior of tissues for disease modeling and drug testing [12]. Chemical production and synthesis of new materials, particularly in pharmaceutical industries, using microreactors make new products possible. The large-scale production can be realized easily by running multiple identical micro reactors in parallel. Scalability is inherent in microfluidics and can be approached from the point of view of "numbering up" rather than scaling up. In this regard, the microreactor concept mirrors potential of nanotechnology, in which technology imitates nature by using many small parallel processors rather than a single larger reactor.

### 1.2.3 Technology Development

Similar to the trends in device development, the technology of making microfluidic devices has also seen a paradigm shift. Starting with silicon micromachining as the enabling technology, a number of microfluidic devices with integrated sensors and actuators were made in silicon. However, unlike microelectronics, which manipulates electrons in integrated circuits, microfluidics must transport molecules and fluids in larger channels, due to the relative size difference between electrons and the more complicated molecules that comprise common fluids. This size disparity leads to much larger microfluidic devices. Significantly fewer microfluidic devices than electronic devices can be placed on a silicon wafer. Adding material cost, processing cost, and the yield rate, microfluidic devices based on silicon technology are too expensive to be accepted by the commercial market, especially as disposable devices.

Since mid 1990, with chemists joining the field, microfabrication technology has been moving to plastic micromachining. With the philosophy of functionality above miniaturization and simplicity above complexity, microfluidic devices have been kept simple, sometimes only with a passive system of microchannels. The actuating and sensing devices are not necessarily integrated into the microfluidic devices. These microdevices are incorporated as replaceable elements in benchtop and handheld tools.

Some of such microdevices are results of the integration of miniaturized components, such as microchannels and microvalves, on a single substrate for applications where high-throughput or complex assays with various on-chip functions are required. The actuation of microvalves for required flow manipulations can be achieved using external sources such as vacuum, pressure, or a magnetic field.

Batch fabrication of microfluidic devices is required for commercialization aspects. To this end, thermoplastics have gained popularity. For the mass production of plastic microfluidic devices, fabrication methods for thermoplastic devices with integrated microvalves and micropumps have been developed [13, 14]. Large-scale production of plastic-based microfluidic devices can be easily realized using microinjection molding and hot embossing. The master (stamp or mold) for replication can be fabricated with traditional silicon-based micromachining methods or conventional high-precision machining technologies. Complex microfluidic devices based on plastic microfabrication could be expected in the near future with further achievements of plastic-based microelectronics. One example of this migration from silicon to plastic fabrication is the i-STAT point-of-care blood chemistry diagnostic system in which only a very small fraction of the device is made using silicon microfabrication technologies. The bulk of the device is fabricated from two pieces of plastic that are taped together. Three-dimensional (3-D) printing methods [15], also known as additive manufacturing techniques, have created a very distinct paradigm shift for the fabrication of microfluidic devices. Additive manufacturing methods enable the rapid prototyping of microdevices having complex and 3-D architectures without any need to use a master mold in a single step. Further developments, both in the fabrication process and material development, are still required for the implementation of functional elements and actuators for printing a microdevice.

With new applications featuring highly corrosive chemicals, microfluidic devices fabricated in materials such as stainless steel or ceramics are desired. Microcutting, laser machining, microelectro discharge machining, and laminating are a few examples of these alternative fabrication techniques. The relatively large-scale microfluidic devices and the freedom of material choice make these techniques serious competitors for silicon micromachining.

## 1.3 ORGANIZATION OF THE BOOK

This book is divided into 13 chapters. The topic of each chapter follows.

This chapter introduces the field of microfluidics, including its definition and commercial and scientific aspects. It also addresses the historical development of this relatively new research field and provides important resources for microfluidics research.

Chapter 2 discusses when to expect changes in fluid behavior as the length scale of a flow is shrunk to microscopic sizes. The appropriate means for analytically treating as well as computationally simulating microflows is addressed. The theoretical fundamentals that form the basis for design and optimization of microfluidic devices are addressed. Several microfluidic-based phenomena such as micromagnetofluidics, optofluidics, and microacoustofluidics along with multiphysics couplings are discussed, including thermofluid and electrofluid.

Chapter 3 provides the technology fundamentals required for making microfluidic devices. Conventional MEMS technology, such as bulk micromachining and surface micromachining, are discussed. Varieties of plastic micromachining such as molding, laser micromachining, and micromilling techniques are presented in this chapter. Fabrication methods for production of paper-based microfluidic devices such as wax printing and inkjet printing are presented. Alternative non-batch techniques and assembly methods, which are interesting for small-scale production and prototyping, are also discussed in this chapter.

Chapter 4 analyzes various experimental characterization techniques for microfluidic devices. The chapter presents a number of novel diagnostic techniques developed for investigation of fluid flow in microscale. Chapters 2 through 4 consider the fundamentals of the typical development process of microfluidic devices—from analysis to fabrication to device characterization.

Chapters 5 through 13 present design examples of microfluidic devices, which were the objects of the worldwide microfluidics research community in the last two decades. The examples are categorized in their application fields, such as external flow control, internal flow control, and microfluidics for life sciences and chemistry.

Chapter 5 describes the design of microdevices for controlling macroscopic flow phenomena, such as velocity sensors, shear stress sensors, microflaps, microballoon, microsynthetic jets, and microair vehicles.

Chapters 6, 7, and 8 present design rules and solutions for microvalves, micropumps, and microflow sensors, respectively. The chapters analyze in detail the operation principles, design considerations, and fabrication techniques of these microfluidic devices.

Chapters 9 through 13 list a number of tools and devices for the emerging fields of life sciences and chemical analysis in microscale. Typical devices are filters, separators, needles, mixers, reactors, heat exchangers, dispensers, and separators. Special examples on the use of such microtools for chemical and biological applications, including particle and cell separators, synthesis of drug carriers using droplet-based microfluidics, and bioreactors for cell culture, were presented.

The field of microfluidics has progressed remarkably in the last two decades. However, many topics, such as new fluid effects and multiphysics effects at the microscale, and novel methods of fabrication are still under intensive research. Importantly, in recent years, a larger number of investigations has been directed to the use of microfluidics in healthcare, life sciences, and chemistry. Therefore, while the materials presented in this book serve to bring the reader to a complete understanding of the present state of the art, the examples and the references listed at the end of each chapter will allow the reader to grow beyond the subject matter presented here. This book is intended to serve as a source of reference on microfluidics.

## References

[1] Feynman, R. P., "There's Plenty of Room at the Bottom," *Journal of Microelectromechanical Systems*, Vol. 1, No. 1, 1992, pp. 60–66.

[2] Moore, G., "VLSI, What Does the Future Hold," *Electron. Aust.*, Vol. 42, No. 14, 1980.

[3] Chang, C. Y., and Sze, S. M., *ULSI Devices*, New York: Wiley, 2000.

[4] Petersen, K. E., "Silicon as Mechanical Material," *Proceedings of the IEEE*, Vol. 70, No. 5, 1982, pp. 420–457.

[5] Manz, A., Graber, N., and Widmer, H. M., "Miniaturized Total Chemical Analysis Systems: A Novel Concept for Chemical Sensing," *Sensors and Actuators B*, Vol. 1, 1990, pp. 244–248.

[6] Gravesen, P., Branebjerg, J., and Jensen, O. S., "Microfluidics—A Review," *Journal of Micromechanics and Microengineering*, Vol. 3, 1993, pp. 168–182.

[7] Whitesides, G., "The Origins and the Future of Microfluidics," *Nature*, Vol. 422, 2006, pp. 369–373.

[8] Petersen, K. E., et al., "Toward Next Generation Clinical Diagnostics Instruments: Scaling and New Processing Paradigms," *Journal of Biomedical Microdevices*, Vol. 2, No. 1, 1999, pp. 71–79.

[9] Yole Microfluidic Applications in the Pharmaceutical, Life Sciences, In-Vitro Diagnostic and Medical Device, Markets 2015 report, http://www.yole.fr.

[10] Shoji, S., and Esashi, M., "Microflow Devices and Systems," *Journal of Micromechanics and Microengineering*, Vol. 4, No. 4, 1994, pp. 157–171.

[11] Sackmann, E. K., et al., "The Present and Future Role of Microfluidics in Biomedical Research," *Nature*, Vol. 507, No. 7491, 2014, pp. 181–189.

[12] Bhatia, S. N., and Ingber, D., E., "Microfluidic Organs-on-Chips," *Nature Biotechnology*, Vol. 32, No. 8, 2014, pp. 760–772.

[13] Ren, Kangning, et al., "Whole-Teflon microfluidic chips," *PNAS*, Vol. 108, No. 20, 2011, pp. 8162–8166.

[14] Mousavi Shaegh, S. A., et al., "Rapid Prototyping of Whole-Thermoplastic Microfluidics with Built-In Microvalves Using Laser Ablation and Thermal Fusion Bonding," *Sensors and Actuators B: Chemical*, Vol. 255, 2018, pp. 100-109.

[15] Ho, C. M., et al., "3D Printed Microfluidics for Biological Applications," *Lab Chip*, Vol. 15, No. 18, 2015, pp. 3627–3637.

# Chapter 2

# Fluid Mechanics Theory

## 2.1 INTRODUCTION

Although everyone has an intuitive sense of what a fluid is, rigorously defining just what fluids are is more troublesome. According to *Merriam-Webster's Collegiate Dictionary* [1], a fluid is: "a substance (as a liquid or gas) tending to flow or conform to the outline of its container."

While this definition gives us a sense of what a fluid is, it is far removed from a technical definition. It also begs the question: What is a liquid or a gas? Again, according to [1], a liquid is: "a fluid (as water) that has no independent shape but has a definite volume and does not expand indefinitely and that is only slightly compressible," while a gas is: "a fluid (as air) that has neither independent shape nor volume but tends to expand indefinitely."

These definitions are circular and ultimately rely on examples such as water and air to explain what a fluid is. Clearly we must look further for a good definition of what a fluid is.

According to one of the leading undergraduate textbooks in fluid mechanics [2], the definition of a fluid is: "a substance that deforms continuously under the application of shear (tangential) stress, no matter how small that stress may be." This definition proves to be a suitable working definition that we can use to determine whether some material that is not air or water is a fluid.

Consider a thought experiment wherein both a solid block of an elastic material (such as aluminum) and a layer of fluid are subjected to the same shearing force. Figure 2.1 depicts just this situation. The block of material in Figure 2.1(a), when subjected to the shearing force, will deform from its equilibrium shape, indicated by the vertical solid lines, to its deformed shape, indicated by the dashed inclined lines. If this force is released the block will return to its original, equilibrium position. As long as the elastic limit of the solid is not exceeded, it will always behave like this. A fluid subject to a constant shearing force in Figure 2.1(b) will behave very differently. The fluid is deformed from its original position, indicated by the vertical solid lines, to another position, indicated by the first set of dashed inclined lines, by the shearing force. If the force is removed, the fluid will remain in this position, its new equilibrium position. If the force is reapplied, the fluid will deform still further to the shape indicated by the second set of dashed lines. As long as the shear force persists, the fluid will continue deforming. As soon as the shear force is removed, the fluid will cease deforming and remain in its present position. For true fluids, this process will continue indefinitely and for any size of shearing force. When the shear stress (shearing force/area fluid contact) is directly proportional to the rate of strain (typically $\partial u/\partial y$) within the fluid, the fluid is said to be Newtonian. This simple demonstration also alludes to the no-slip boundary condition, which is of fundamental importance in fluid mechanics and needs to be considered on a case-by-case basis in microscopic domains.

Flowing fluids can be characterized by the properties of both the fluid and the flow. These can be organized into four main categories [3]:

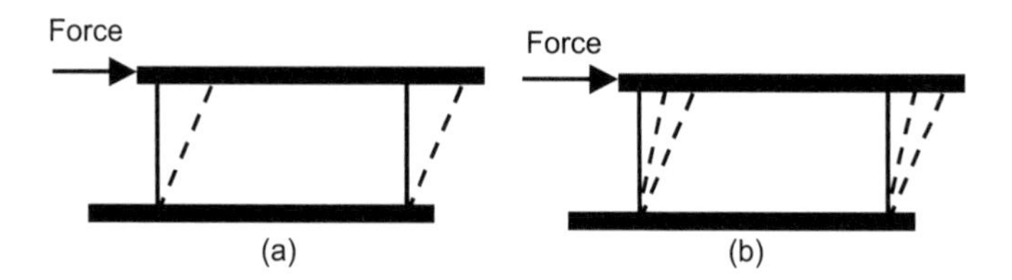


**Figure 2.1** (a) A block of solid material, and (b) a fluid contained between two plates are subjected to a shearing force. Both materials are shown in an original position (solid lines) and deformed positions (dashes). When the force is removed from the solid material, it returns from its deformed position to its original or equilibrium position. The fluid remains deformed upon removal of the force.

- Kinematic properties such as linear and angular velocity, vorticity, acceleration, and strain rate;
- Transport properties such as viscosity, thermal conductivity, and diffusivity;
- Thermodynamic properties such as pressure, temperature, and density;
- Miscellaneous properties such as surface tension, vapor pressure, and surface accommodation coefficients.

Knowledge of these properties is necessary for quantifying the fluid's response to some set of operating conditions. The kinematic properties are actually properties of the flow, but they usually depend on the properties of the fluid. The transport properties and thermodynamic properties are generally properties of the fluid, but they may depend on the properties of the flow. The miscellaneous properties are properties that may depend on the interaction of fluid and the vessel in which it is flowing such as the surface tension or the surface accommodation coefficient. They can also be constitutive properties of the fluid (or any other hard-to-classify properties of the fluid/flow).

A fluid can be modeled in one of two ways: as a collection of individual, interacting molecules or as a continuum in which properties are defined to be continuously defined throughout space. The former approach is addressed in the introductory chapters of many fluid mechanics texts [3–5] and then ignored as either impractical or unnecessary. However, as the length scale of a system decreases in size, the question of whether to treat the fluid as a collection of molecules or as a continuum acquires critical significance. To use a continuum approach in a situation where a molecular approach is necessary would certainly produce incorrect results. In this chapter, these two distinctly different approaches will be explored.

### 2.1.1 Intermolecular Forces

The behavior of all states of matter—solids, liquids, and gases—as well as the interaction among the different states, depends on the forces between the molecules that comprise the matter. An accurate model of the interaction of two simple, nonionized, nonreacting molecules is given by the Lennard-Jones potential, $V_{ij}$:

$$V_{ij}(r) = 4\varepsilon \left[ c_{ij} \left(\frac{r}{\sigma}\right)^{-12} - d_{ij} \left(\frac{r}{\sigma}\right)^{-6} \right] \tag{2.1}$$

where $r$ is the distance separating the molecules $i$ and $j$, $c_{ij}$ and $d_{ij}$ are parameters particular to the pair of interacting molecules, $\varepsilon$ is a characteristic energy scale, and $\sigma$ is a characteristic length scale. The term with the $r^{-12}$ dependence is a phenomenological model of the pairwise repulsion that exists between two molecules when they are brought very close together. The term with the $r^{-6}$ dependence is a mildly attractive potential due to the van der Waals force between any two molecules. The van der Waals forces are analytically derivable and reflect the contribution from several phenomena. These are:

- Dipole/dipole interactions (Keesom theory);

**Table 2.1**

A Selection of Lennard-Jones Constants Derived from Viscosity Data [9]. *Note*: The Boltzmann constant $K$ is equal to $1.381 \times 10^{-23} J/K$.

| *Fluid* | $\frac{\varepsilon}{K}$ *(K)* | $\sigma$ *(nm)* |
|---|---|---|
| Air | 97 | 0.362 |
| $N_2$ | 91.5 | 0.368 |
| $CO_2$ | 190 | 0.400 |
| $O_2$ | 113 | 0.343 |
| Ar | 124 | 0.342 |

- Induced dipole interactions (Debye theory);
- Fundamental electrodynamic interactions (London theory).

Since (2.1) represents the potential energy between two interacting molecules, the force between those two molecules is given by its derivative as:

$$F_{ij}(r) = \frac{\partial V_{ij}(r)}{\partial r} = \frac{48\varepsilon}{\sigma}\left[c_{ij}\left(\frac{r}{\sigma}\right)^{-13} - \frac{d_{ij}}{2}\left(\frac{r}{\sigma}\right)^{-7}\right] \tag{2.2}$$

With an appropriate choice of the parameters $c_{ij}$ and $d_{ij}$, this equation can represent the force between any two molecules. The characteristic time scale for molecular interactions for which Lennard-Jones is a good model is given by:

$$\tau = \sigma\sqrt{\frac{m}{\varepsilon}} \tag{2.3}$$

which is the period of oscillation about the minimum in the Lennard-Jones potential. The $m$ is the mass of a single molecule. The Lennard-Jones potential has been shown to provide accurate results for liquid argon using $\varepsilon/K = 120$K, $\sigma = 0.34$ nm, and $c = d = 1$. For these parameters, the time scale works out to $\tau = 2.2 \times 10^{-12}$ sec—a very short time indeed [6]. Several recent articles provide a more in-depth discussion of the Lennard-Jones potential and its various uses [6–8].

The values of the various parameters in (2.1) through (2.3) are available in the literature [7] for a limited number of molecules. These values are summarized in Table 2.1. The parameters $c_{ij}$ and peclet should be taken as unity when using the parameters from Table 2.1. These parameters should be used only to calculate the potential between two fluid molecules of the same species. It is also acceptable to use these parameters between molecules of the same species even if they are in different phases (e.g., liquid and solid). However, these constants should not be used to calculate the interaction of a fluid molecule with a flow boundary of different molecules.

These parameter values can be used to calculate the potential energy and force between two interacting molecules. The effect of the constants in Table 2.1 is to change the magnitude of the potential energy and force, but not its general shape. Figure 2.2 shows this general shape by plotting potential energy scaled by $\frac{1}{4\varepsilon}$ and force scaled by $\frac{\sigma}{48\varepsilon}$ versus dimensionless separation distance $r/\sigma$. Were this a dimensional figure, it would be stretched or compressed in both the horizontal and vertical directions, but the overall shape would appear the same. One important observation concerning this force plot is that, although its magnitude decreases very quickly with distances beyond the location of the minimum (e.g., it drops from $-0.05$ at a distance of 1.2 to nearly nothing at 2.0), it never reaches zero. This has important implications for molecular dynamics simulations, as we will see in Section 2.3.1.

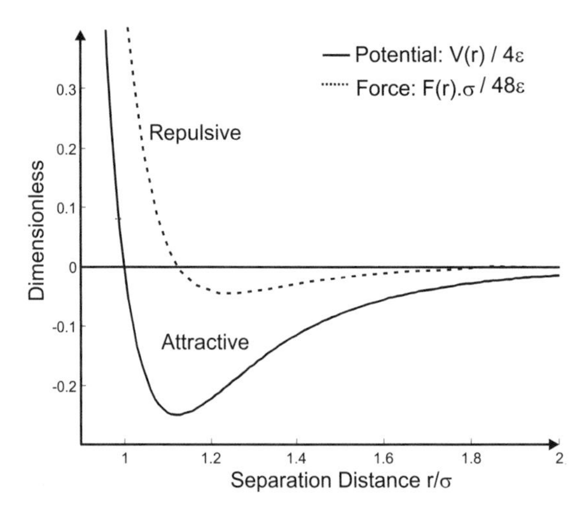


**Figure 2.2** Generalized plot of intermolecular potential energy and force according to the Lennard-Jones model.

### 2.1.2 The Three States of Matter

All three states of matter—solids, liquids, and gases—are comprised of molecules that are all interacting through the Lennard-Jones force discussed earlier. The molecules in a solid are densely packed and rigidly fixed within some particular molecular arrangement. The molecules are in constant contact with one another with a mean intermolecular spacing of about $\sigma$. The molecules of the solid interact with all of their neighbors through the Lennard-Jones force. Each molecule is held in place by the large repulsive forces that it would experience if it were to move closer to one of its neighbors. Hence, for a molecule to leave its particular molecular neighborhood and join a nearby arrangement of molecules requires a significant amount of energy—an amount that it is unlikely to receive when the solid is held below its melting temperature. As the solid is heated up to and beyond its melting temperature, the average molecular thermal energy becomes high enough that the molecules are able to vibrate freely from one set of neighbors to another. The material is then called a liquid. The molecules of a liquid are still relatively close together (still approximately $\sigma$), which agrees with our observations that the density of a liquid just above the melting temperature is only slightly lower than a solid just below the melting temperature. Water is the most common exception to this rule and is a few percent more dense as a liquid near the melting temperature than as a solid. If the temperature of the liquid is raised, the vibration of the molecules increases still further. Eventually, the amplitude of vibration is great enough that, at the boiling temperature, the molecules jump energetically away from each other and assume a mean spacing of approximately $10\sigma$ (at standard conditions). The material is now called a gas. The molecules interact strictly through brief, highly energetic collisions with their neighbors.

They are no longer in continuous contact with each other. In fact, only during these brief collisions are the molecules close enough together that the Lennard-Jones forces become significant when compared with their kinetic energy. The molecules comprising the gas will expand to fill whatever vessel contains them. These attributes of solids, liquids, and gases are summarized in Table 2.2.

### 2.1.3 Continuum Assumption

The study of fluid mechanics (at conventional, macroscopic length scales) generally proceeds from the assumption that the fluid can be treated as a continuum. All quantities of interest such as density, velocity, and pressure are assumed to be defined everywhere in space and to vary continuously from

**Table 2.2**
Summary of Solid, Liquid, and Gas Intermolecular Relationships (*After*: [4].)

| *Phases* | *Intermolecular Forces* | *Ratio of Thermal Vibration Amplitude Compared to* $\sigma$ | *Approach Needed* |
|---|---|---|---|
| Solid | Strong | $\ll 1$ | Quantum |
| Liquid | Moderate | $\sim 1$ | Quantum/classical |
| Gas | Weak | $\gg 1$ | Classical |

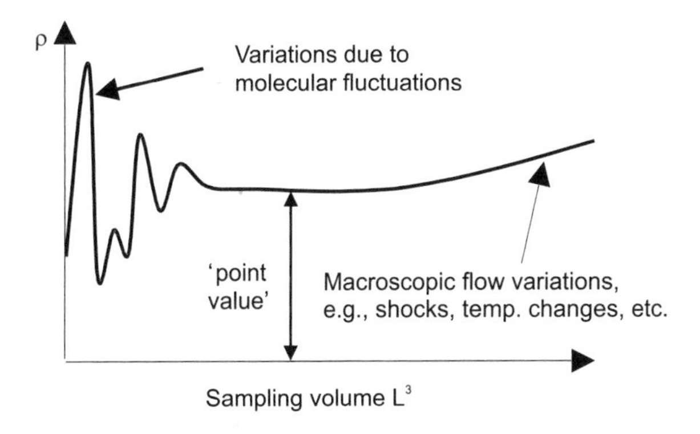


**Figure 2.3** Continuum assumption in fluids illustrated by thought experiment for measuring density. (*After*: [4].)

point to point within a flow. We have just seen that all matter is actually comprised of quanta of mass—discrete atoms and molecules—not a uniformly distributed featureless substance. Whether assuming continuity of mass and other properties is reasonable at microscopic length scales depends greatly on the particular situation being studied. If the molecules of a fluid are closely packed relative to the length scale of the flow, the continuum assumption is probably valid. If the molecules are sparsely distributed relative to the length scale of the flow, assuming continuity of fluid and flow properties is probably a dangerous approach. However, even at microscopic length scales, there can still be many thousands of molecules within a length scale significant to the flow. For example, a 10-$\mu$m channel will have approximately 30,000 water molecules spanning it—certainly enough molecules to consider the flow as continuous.

To illustrate this point of the molecular versus continuum nature of fluids, let us consider a thought experiment in which we measure the density of a fluid at a point, as is shown in Figure 2.3. Most graduate-level fluid mechanics textbooks will have an argument similar to that which follows [4, 5, 10]. This is not a point in the geometric sense, but rather, a small sampling volume of space surrounding the geometric point in which we are interested. To make this argument more straightforward, let us consider that the molecules are frozen in space. The argument is easily extensible to moving molecules, but this simplification separates the temporal and spatial variations of the flow. If we consider measuring the average density of the fluid within the sampling volume, we could do it by counting the number of molecules within the sampling volume, multiplying by the molecular mass of each molecule, and dividing by the volume of the sampling volume according to:

$$\rho = \frac{N \cdot m}{L^3} \tag{2.4}$$

where the sampling region is assumed to be a cube measuring $L$ on a side, and $N$ is the number of molecules that were found in it. The mass of a single molecule is $m$.

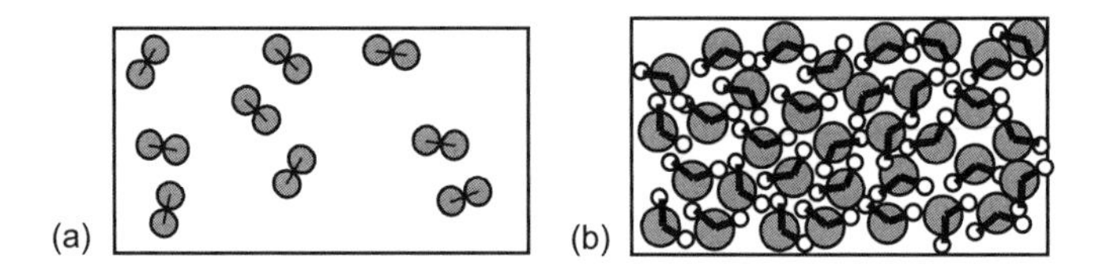


**Figure 2.4** Sketches of (a) a gas, such as $N_2$, at standard conditions, and (b) a liquid, such as $H_2O$.

**Table 2.3**
Properties of a Typical Gas and Liquid at Standard Conditions (*After*: [10].)

| *Property* | *Gas* ($N_2$) | *Liquid* ($H_2O$) |
|---|---|---|
| Molecular diameter | 0.3 nm | 0.3 nm |
| Number density | $3 \times 10^{25}\text{m}^{-3}$ | $2 \times 10^{28}\text{m}^{-3}$ |
| Intermolecular spacing | 3 nm | 0.4 nm |
| Displacement distance | 100 nm | 0.001 nm |
| Molecular velocity | 500 m/s | 1,000 m/s |

When the size of the sampling volume is very small, it will at any given time contain only a few molecules. Gradually expanding the sampling volume will at times add a new molecule to the sampling volume and at other times merely increase the volume of the sampling volume. Hence, we would expect that, when the number of molecules in the sampling volume is small, the calculated density would vary rapidly, but as the volume becomes larger and larger, the calculated density would approach some relatively constant value. That value is called the point value in continuum fluid mechanics. If the size of the sampling volume is increased beyond that necessary to achieve a point value, spatial variations in the flow will begin to get averaged out. In the case of density, spatial variations could be due to a shock wave or to local heating of the flow.

To make this distinction between continuous and molecular behavior, it is useful to consider concrete examples. Figure 2.4 shows two sketches, one of a gas, which for the sake of concreteness we will take to be diatomic nitrogen, $N_2$, at standard conditions, as well as a liquid, water, also at standard conditions. Table 2.3 compares the properties of the gas and the liquid.

In order that a fluid can be modeled as a continuum, all of its properties must be continuous [10]. There are different length scales for different types of properties. A fluid's kinematic properties, such as velocity and acceleration, as well as its thermodynamic properties, such as pressure and density, can be treated as the point quantities in the thought experiment above. According to random process theory, in order to get reasonably stationary statistics, less than 1% statistical variations, $10^4$ molecules must be used to compute an average value. Consequently, the point quantities can be thought of as continuous if the sampling volume is a cube that measures:

$$L_{\text{gas,pt}} = \sqrt[3]{\frac{10^4}{3 \times 10^{25}\text{m}^{-3}}} = 70 \times 10^{-9}\ \text{m} \tag{2.5}$$

and

$$L_{\text{liquid,pt}} = \sqrt[3]{\frac{10^4}{2 \times 10^{25}\text{m}^{-3}}} = 8 \times 10^{-9}\ \text{m} \tag{2.6}$$

The transport quantities such as viscosity and diffusivity must also be continuous in order for the fluid to be treated as a continuum. The analysis of the transport quantities is somewhat different from that of the point quantities where a certain number of molecules were required in order for the property to be treated as continuous. For the transport quantities to behave continuously, it is important that the fluid molecules interact much more often with themselves than with flow boundaries. As a somewhat arbitrary criterion, we can choose the measurement point to be a cube whose sides are 10 times as large as the molecules' interaction length scale. For a gas, the best estimate of an interaction length scale is the displacement distance, also known as the mean free path, which is on the order of 100 nm. For a liquid, the molecules are essentially in a continual state of collision or interaction, so their displacement distance is not a good estimate of how many interactions will be present in some cube of space. Their molecular diameter is a much better estimate. Hence, the transport quantities will be continuous in a cube measuring:

$$L_{\text{gas,tr}} = \sqrt[3]{10^3} \times 100 \text{ nm} = 10^{-6}\text{m} \tag{2.7}$$

and

$$L_{\text{liquid,tr}} = \sqrt[3]{\frac{10^3}{2 \times 10^{25}\text{m}^{-3}}} = 4 \times 10^{-9}\text{m} \tag{2.8}$$

In order to be able to treat a flow as continuous, both its point quantities and its transport quantities must be continuous. Hence, taking the greater of the two length scales, the length scale at which continuous behavior can be expected is:

$$L_{\text{gas}} = 1 \text{ µm } (10^{-6}\text{m}) \tag{2.9}$$

and

$$L_{\text{liquid}} = 10 \text{ nm } (10^{-8}\text{m}) \tag{2.10}$$

This analysis is very approximate and is meant merely to help the reader get a feel for when continuum behavior can be expected and when the flow must be treated as an ensemble of individual interacting molecules.

The ultimately quantized nature of matter combined with the diminutive length scales that are the subject of this book beg the question of why we should bother with the continuum approach at all if the downside is an incorrect analysis. Scientists and engineers had been developing the field of fluid mechanics since long before the atomic nature of matter was conclusively demonstrated in the late nineteenth century. The governing continuum fluid mechanics equations have existed in their present form, called the Navier-Stokes equations, when certain reasonable approximations are used, for more than a century. Consequently, if the continuum approach is valid in a given microfluidic situation, it should be used because of the tremendous body of work available for immediate application to the flow situation. Also, describing fluids by continuously varying fields greatly simplifies modeling the fluid. Using the continuum approach, many flows can be computed analytically with no more equipment necessary than a pencil. Molecular approaches to solving fluid mechanics problems generally consist of knowing the state (position and velocity) of each molecule of the fluid and then evolving that state forward in time for every single molecule. While the evolution equations might be particularly simple, for example, Newton's second law plus the Lennard-Jones force, it might need to be repeated excessively. Because most microflows are relatively large compared to molecular length scales, they necessarily contain a moderate (several thousands) to large (billions or more) number of molecules. Two of the main molecular fluids techniques are discussed later in this chapter. These are the molecular dynamics (MD) and direct simulation Monte Carlo (DSMC) techniques. The remaining sections in this chapter explore:

- Continuum approaches to fluid mechanics at small scales (Section 2.2);
- Molecular approaches to fluid mechanics (Section 2.3);
- Electrokinetics (Section 2.4).

## 2.2 CONTINUUM FLUID MECHANICS AT SMALL SCALES

Even in very small devices and geometries, fluids, especially liquids, can be considered to be continuous. Hence, well-established continuum approaches for analyzing flows can be used. These approaches have been treated in great detail in a variety of fluid mechanics textbooks [2–5] and hence will only be summarized here. The three primary conservation laws that are used to model thermofluid dynamics problems are conservation of mass, momentum, and energy. In partial differential equation form, these are:

$$\frac{\partial \rho}{\partial t} + \frac{\partial \rho}{\partial x_i}(\rho u_i) = 0 \tag{2.11}$$

$$\frac{\partial}{\partial t}(\rho u_i) + \frac{\partial}{\partial x_j}(\rho u_j u_i) = \rho \mathbf{F}_i - \frac{\partial p}{\partial x_i} + \frac{\partial}{\partial x_j}\tau_{ji} \tag{2.12}$$

$$\frac{\partial}{\partial t}(\rho e) + \frac{\partial}{\partial x_i}(\rho u_i e) = -p\frac{\partial u_i}{\partial x_i} + \tau_{ji}\frac{\partial u_i}{\partial x_j} + \frac{\partial q_i}{\partial x_i} \tag{2.13}$$

where repeated indices in any single term indicate a summation according to the accepted practice in index notation (Einstein summation convention). In these equations, $u_i$ represents the flow velocity, $\rho$ is the local density, $p$ is the pressure, $\tau$ is the stress tensor, $e$ is the internal energy, $\mathbf{F}$ is body force, and $q$ is the heat flux. Most often at large length scales, the body force $\mathbf{F}$ is taken to be gravity. However, because of cube-square scaling, the effect of gravity in small systems is usually negligible. In order to overpower the cube-square scaling, the body force needs to be some force that can be increased to large values—such as a magnetic body force, which can be several orders of magnitude larger than the body force generated by gravity. These equations comprise five partial differential equations (one for mass, three for momentum, and one for energy) in 17 unknowns: $\rho$, $u_i$, $\tau_{ji}$, $e$, and $q_i$. Thus, they do not represent a closed system of equations. To close these equations, a necessary but not sufficient condition for their solution, we must look to constitutive relationships among these unknowns—specifically to relationships between the stress tensor and the velocity field, heat flux and temperature field, and constitutive thermodynamic properties of the fluid, such as the ideal gas law. These relationships can be very different for gases and liquids, so they will be considered separately in the next sections.

One note of commonality between the continuum approach to solving fluid mechanics problems in gases and liquids is that computational fluid dynamics (CFD) methods have been developed to solve problems in both gases and liquids, and these methods have been very successful in large-scale problems. As long as the continuum assumption holds, there should be no particular problems with adapting CFD to solve microscale problems. Several commercial packages presently are available which do just that.

### 2.2.1 Gas Flows

#### 2.2.1.1 Kinetic Gas Theory

As explained in Section 2.1.2, gases consist of mostly space with a few molecules colliding only infrequently. Consequently, they can be described quite well by the kinetic gas theory, in which a gas molecule is considered to move in a straight line at a constant speed until it strikes another molecule—which it does infrequently. Many of the conclusions of kinetic gas theory are presented in this section. The equation of state for a dilute gas is the ideal gas law, which can have several equivalent forms. Two of these forms are:

$$p = \rho R T \tag{2.14}$$

and

$$p = nKT \tag{2.15}$$

where $p$ is the pressure, $\rho$ is the density of the gas, $R$ is the specific gas constant for the gas being evaluated, $n$ is the number density of the gas, $K$ is Boltzmann's constant ($K = 1.3805 \times 10^{-23} J/K$), and $T$ is the absolute temperature. The specific gas constant $R$ can be found from:

$$R = \frac{\overline{R}}{M} \text{ where } \overline{R} = 8.3185 \frac{\text{kJ}}{\text{kmol} \cdot \text{K}} \tag{2.16}$$

where $\overline{R}$ is the universal gas constant and $M$ is the molar mass of the gas.

Using the second form of the ideal gas law, it is possible to calculate that at standard conditions $(273.15K; 101,625Pa)$, the number density of any gas is $n = 2.70 \times 10^{25} \text{m}^{-3}$. The mean molecular spacing will be important in the following analysis. Since the molecules in the gas are scattered randomly throughout whatever vessel contains them, the mean molecular spacing $\delta$ can be estimated as $\delta = n^{-1/3} = 3.3 \times 10^{-9}$m (at standard conditions).

Comparing this value to the diameter of a typical gas molecule, say, $N_2$ from above

$$\frac{\delta}{d} = \frac{3.3 \times 10^{-9} \text{ m}}{3 \times 10^{-10} \text{ m}} \approx 10 \gg 1 \tag{2.17}$$

gives a measure of effective density. Gases for which $\delta/d \gg 1$ are said to be dilute gases [11], while those not meeting this condition are said to be dense gases. For dilute gases, the most common mode of intermolecular interaction is binary collisions. Simultaneous multiple molecule collisions are unlikely. Generally, values of $\delta/d$ greater than 7 are considered to be dilute.

Several transport quantities are important in gas dynamics. These are the mean free path $\lambda$, the mean-square molecular speed $\overline{c}$, and the speed of sound $c_s$. The mean free path is the distance that the average molecule will travel before experiencing a collision. For a simple gas of hard spheres at thermodynamic equilibrium, Bird [11] gave the equation:

$$\lambda = \frac{1}{\sqrt{2} \cdot \pi d^2 n} \tag{2.18}$$

The mean-square molecular speed was given by Vincenti and Kruger [12] as:

$$\overline{c} = \sqrt{3RT} \tag{2.19}$$

where $R$ is the gas constant specific to the gas under consideration. The speed at which sound (infinitesimal pressure waves) travels through a gas is important in determining the type of flow that will develop in the gas. The speed of sound $c_s$ is given by:

$$c_s = \sqrt{kRT} \tag{2.20}$$

where $k$ is the ratio of specific heats, defined to be the ratio of the specific heat at a constant pressure $c_p$ to the specific heat at a constant volume $c_v$. For common gases (air, $N_2$, $O_2$), $k$ is generally about 1.4 but can depart significantly from that value for more complicated molecules.

One last quantity of importance in gas dynamics is the viscosity of the gas. According to the kinetic theory of gases, the viscosity is given by:

$$\nu = \frac{\eta}{\rho} = \frac{1}{2} \lambda \overline{c} \tag{2.21}$$

where $\nu$ is the kinematic viscosity and $\eta$ is the dynamic viscosity. Both $\lambda$ and $\bar{c}$ are given as above. This equation thus relates viscosity to temperature and pressure through the mean free path and mean molecular speed. If we solve (2.21) for the dynamic viscosity $\eta$, we can find that the dynamic viscosity is not a function of pressure, because density is proportional to pressure while the mean free path is inversely proportional to it.

In addition to these parameters, there are several dimensionless groups of parameters that are very important in assessing the state of a fluid in motion. These are the Mach number $\mathrm{Ma}$, the Knudsen number $\mathrm{Kn}$, and the Reynolds number $\mathrm{Re}$. The Mach number is the ratio between the flow velocity $u$ and the speed of sound $c_\mathrm{s}$, and is given by:

$$\mathrm{Ma} = \frac{u}{c_\mathrm{s}} \tag{2.22}$$

The Mach number is a measure of the compressibility of a gas and can be thought of as the ratio of inertial forces to elastic forces. Flows for which $\mathrm{Ma} < 1$ are called subsonic and flows for which $\mathrm{Ma} > 1$ are called supersonic. When $\mathrm{Ma} = 1$, the flow is said to be sonic. The Mach number can either be a local measure of the speed of the flow if $u$ is a local velocity, or it can be a global measure of the flow if the mean flow velocity $u_\mathrm{avg}$ is used. Under certain conditions, the density of a gas will not change significantly while it is flowing through a system. The flow is then considered to be incompressible even though fluid, a gas, is still considered compressible. This distinction is an important one in fluid mechanics. The analysis of an incompressible gas flow is greatly simplified, since it can be treated with the same versions of the governing equations that will be derived for liquid flows in Section 2.2.2. If the Mach number of a gas flow is greater than 0.3, the flow must be treated as a compressible flow. If the Mach number of a flow is less than 0.3, the flow may be treated as incompressible provided some further conditions are satisfied. The Mach number of a flow being less than 0.3 is a necessary but not sufficient condition for the flow to be treated as incompressible [8]. Gas flows in which the wall is heated locally can cause large changes in the density of the gas even at low Mach numbers. Similarly, gas flows in long, thin channels will experience self-heating due to the viscous dissipation of the fluid, which can cause significant changes in the gas density. For a further discussion of these effects, see [8]. Absent effects like these, when the Mach number is less than 0.3, gas flows may be treated as incompressible.

The Knudsen number has tremendous importance in gas dynamics. It provides a measure of how rarefied a flow is, or how low the density is, relative to the length scale of the flow. The Knudsen number is given by:

$$\mathrm{Kn} = \frac{\lambda}{L} \tag{2.23}$$

where $\lambda$ is the mean free path given in (2.18) and $L$ is some length scale characteristic of the flow.

The Reynolds number is given by:

$$\mathrm{Re} = \frac{\rho u L}{\eta} = \frac{uL}{\nu} \tag{2.24}$$

where $u$ is some velocity characteristic of the flow, $L$ is a length scale characteristic of the flow, $\rho$ is the density, $\eta$ is the dynamic viscosity, and $\nu$ is the kinematic viscosity. The physical significance of the Reynolds number is that it is a measure of the ratio between inertial forces and viscous forces in a particular flow. The different regimes of behavior are:

- $\mathrm{Re} \ll 1$ viscous effects dominate inertial effects (completely laminar flows);
- $\mathrm{Re} \approx 1$ viscous effect comparable to inertial effects (vortices begin to appear);
- $\mathrm{Re} \gg 1$ inertial effects dominate viscous effects (turbulence occurs).

Flow patterns will typically be functions of the Reynolds number. In channel flows, Reynolds numbers smaller than about 1,500 typically indicate laminar flow, while flows with Reynolds numbers greater than 1,500 are increasingly likely to be turbulent. These dynamics are discussed in greater detail in Section 2.2.4.3. The three dimensionless groups just presented are all related by the expression:

$$\mathrm{Kn} = \sqrt{\frac{k\pi}{2}}\frac{\mathrm{Ma}}{L} \tag{2.25}$$

where $k$ is the ratio of specific heats. The Knudsen number is one of the main tools used to classify how rarefied a flow is—or how far apart the molecules of the flow are relative to the scale of the flow. There are four main regimes of fluid rarefaction. These are:

- $\mathrm{Kn} < 10^{-3}$ Navier-Stokes equation, no-slip boundary conditions;
- $10^{-3} < \mathrm{Kn} < 10^{-1}$ Navier-Stokes equation, slip boundary conditions;
- $10^{-1} < \mathrm{Kn} < 10^{1}$ Transitional flow regime;
- $\mathrm{Kn} > 10^{1}$ Free molecular flow.

**Example 2.1: Gas Flow Calculation**

Calculate all the significant parameters to describe a flow of diatomic nitrogen $N_2$ at 350K and 200 kPa at a speed of 100 m/s through a channel measuring 10 μm in diameter.

*First step*: Look up properties for diatomic nitrogen gas:

- Molar mass: $M = 28.013$ kg/kmol
- Molecular diameter [9]: $d = 3.75 \times 10^{-10}$m

*Second step*: Calculate the dimensional parameters for the flow:

$$\begin{aligned}
R &= \frac{8.3145\ \text{kJ/kmol}\cdot\text{K}}{28.013\ \text{kg/kmol}} = 0.2968\frac{\text{kJ}}{\text{kg}\cdot\text{K}}\\
n &= \frac{2\times 10^{5}\ \text{N/m}^2}{1.3805\times 10^{-23}\text{J/K}\times 350\text{K}}\cdot\frac{\text{J}}{\text{Nm}} = 4.14\times 10^{25}\ \text{m}^{-3}\\
\delta &= n^{-1/3} = 2.89\times 10^{-9}\text{m}\\
\lambda &= \left(\sqrt{2}\pi d^2 n\right)^{-1} = \left[\sqrt{2}\pi\left(3.75\times 10^{-10}\ \text{m}\right)^2\cdot 4.14\times 10^{25}\ \text{m}^{-3}\right]^{-1} = 3.9\times 10^{-8}\text{m}\\
\bar{c} &= \sqrt{3\times 0.2968\frac{\text{kJ}}{\text{kgK}}\times 350\text{K}\times 1,000\frac{\text{kg}\,\text{m}^2/\text{s}^2}{\text{kJ}}} = 558\ \text{m/s}\\
c_s &= \sqrt{1.4\times 0.2968\frac{\text{kJ}}{\text{kgK}}\times 350\text{K}\times 1,000\frac{\text{kg}\,\text{m}^2/\text{s}^2}{\text{kJ}}} = 381\ \text{m/s}\\
\nu &= \frac{1}{2}\lambda\bar{c} = \frac{1}{2}\times 3.9\times 10^{-8}\text{m}\times 558\ \text{m/s} = 1.09\times 10^{-5}\text{m}^2/\text{s}
\end{aligned} \tag{2.26}$$

*Third step*: Calculate the dimensionless parameters for the flow and draw reasonable conclusions:

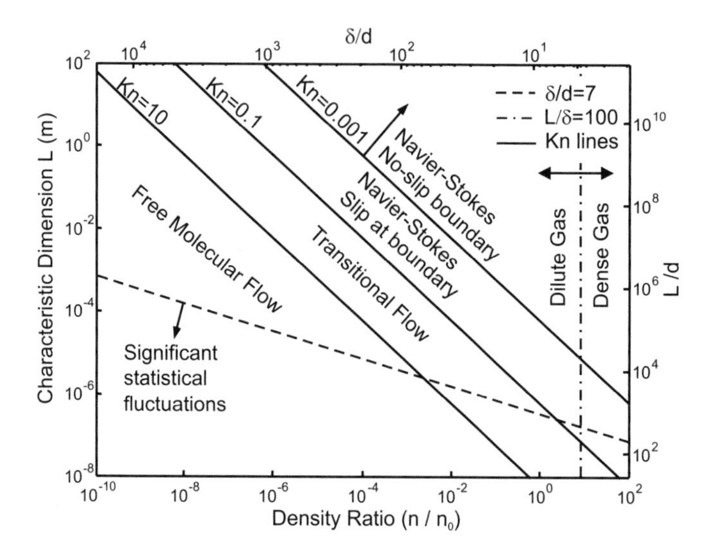


**Figure 2.5** Graphical representation of the relationship among the important dimensionless quantities.

| *Quantity* | *Comparison* | *Conclusion* |
|---|---|---|
| $\frac{\delta}{d} = \frac{2.89\times10^{-9}\text{m}}{3.75\times10^{-10}\text{m}} = 7.7$ | $\delta/d \gg 1$ | Dilute gas |
| $\text{Ma} = \frac{u}{c_\text{C}} = \frac{100\text{m}}{381\text{ m/s}} = 0.26$ | $\text{Ma} < 0.3$ | Incompressible |
| $\text{Kn} = \frac{\lambda}{L} = \frac{3.9\times10^{-8}\text{m}}{10^{-5}\text{m}} = 0.004$ | $10^{-3} < \text{Kn} < 10^{-1}$ | N-S with slip BC |
| $\text{Re} = \frac{uL}{\nu} = \frac{100\text{ m/s}\times10^{-5}\text{m}}{1.09\times10^{-3}\text{m}} = 92$ | $\text{Re} < 1000$ | Not turbulent |

dUse an applied electric field to induce movement

Once these dimensionless parameters are known for a particular gas flow, it is possible to categorize the flow and determine the best way to solve for the flow. Figure 2.5 shows the relationship among the three important dimensionless parameters for gas flows. The dilute gas assumption is valid to the left of the vertical dotted line. The continuum assumption is valid above the dashed line. Below the dashed line, in the area labeled "significant statistical fluctuations," point quantities are ill-defined and fluctuate rapidly due to an insufficient number of molecules in the sampling region. The three Knudsen number lines quantify how rarefied the gas is, as well as demark regions where the Navier-Stokes equations can be used, both with and without slip boundary conditions, as well as areas where other approaches must be used to treat the transitional flow and free molecular flow regimes.

### 2.2.1.2 Governing Equations for Gas Flows

Assuming the flow of a compressible gas governed by the ideal gas law in which the fluid can be considered Newtonian and isotropic (i.e., the same in all directions), to exhibit heat conduction according to Fourier's law of heat conduction, the shear stress tensor and the flux vector can be stated as follows:

$$\tau_{ji} = -p\delta_{ji} + \eta\left(\frac{\partial u_i}{\partial x_j} + \frac{\partial u_j}{\partial x_i}\right) + \lambda\frac{\partial u_k}{\partial x_k}\delta_{ji} \tag{2.27}$$

$$q_i = -\kappa \frac{\partial T}{\partial x_i} \tag{2.28}$$

These relationships can be substituted into the mass, momentum, and energy equations to yield the closed conservation equations:

$$\frac{\partial \rho}{\partial t} + \frac{\partial}{\partial x_i}(\rho u_i) = 0 \tag{2.29}$$

$$\rho \left( \frac{\partial u_i}{\partial t} + \frac{\partial u_j u_i}{\partial x_j} \right) = \rho \mathbf{F} - \frac{\partial p}{\partial x_i} + \frac{\partial}{\partial x_i} \left[ \eta \left( \frac{\partial u_i}{\partial x_j} + \frac{\partial u_j}{\partial x_i} \right) + \lambda \frac{\partial u_k}{\partial x_k} \delta_{ji} \right] \tag{2.30}$$

$$\rho c_v \left( \frac{\partial T}{\partial t} + u_i \frac{\partial T}{\partial x_i} \right) = -p \frac{\partial u_i}{\partial x_i} + \phi + \frac{\partial}{\partial} \left( \kappa \frac{\partial T}{\partial x_i} \right) \tag{2.31}$$

$$\phi = \frac{1}{2} \eta \left( \frac{\partial u_i}{\partial x_j} + \frac{\partial u_j}{\partial x_i} \right)^2 + \lambda \left( \frac{\partial u_k}{\partial x_k} \right)^2 \tag{2.32}$$

This set of equations is now closed, and represents a system of equations that may be solved for the velocity and thermodynamic state of a compressible gas—or a liquid for that matter—but the equations can be simplified further for an incompressible liquid. These equations are accurate for Knudsen numbers less than 0.1. For Knudsen numbers greater than 0.1, alternate methods must be used to arrive at a solution for a flow. One of these techniques is the direct simulation Monte Carlo technique and is described further in Section 2.3.2.

### 2.2.2 Liquid Flows

#### 2.2.2.1 Behavior of Liquids

As discussed in Section 2.1.2, liquid molecules exist in a state of continual collision. Their behavior is completely different from that of gases and is significantly more complex. Consequently, the molecular theory of liquids is not as well developed as that for gases. There are no parameters such as the Knudsen number to help in determining the behavioral regime in which a liquid might be. Turning to experimental results yields no help either. Researchers have found that the apparent viscosity of a liquid flowing through a small channel is either higher than, the same as, or lower than the same liquid flowing through a channel of the same material but at a larger length scale. Clearly, more work needs to be done on both the theoretical and experimental sides of this issue. For a good review of the experiments to date, the reader is encouraged to refer to Gad-el-Hak's recent review article [8].

As discussed in the gas dynamics case in Section 2.2.1.1, several constitutive relationships are needed to close the conservation equations governing liquid flows. The two primary assumptions are that the fluid is Newtonian (i.e., stress is directly proportional to strain rate) and Fourier's law of heat conduction (i.e., heat flow is directly proportional to temperature gradient). For liquids comprising of a single type of simple molecules (such as water), these assumptions are quite effective. Of course, there are many deviations from these assumptions for liquid mixtures, suspensions, and slurries, each of which must be treated on an individual basis. Because of the generality of the Newtonian and Fourier relationships, those are the two that will be used exclusively here.

As mentioned above, the theory and experimental data are of little help in determining when to expect a flow to behave continuously and when to expect it to behave as a collection of individual molecules. The argument outlined in Section 2.1.3 suggests a guideline of when to treat a liquid, that is, the material itself, as a continuum—whenever the length scale is larger than 10 nm. However, this guideline is no help in determining if the flow should be treated as continuous. There are certain dynamic situations that may cause an otherwise continuous flow to behave discontinuously. An example in gas flow would be the occurrence of a shock wave, where the properties of the flow (such as velocity or acceleration) might vary discontinuously even though there are more than enough molecules present to treat the gas as a continuum. As mentioned above, the paucity of theoretical work and the contradictory nature of the experimental work in this area provide few examples to consider.

One important situation during which a flowing liquid changes behavior due to its molecular nature is when it is sheared too much. Loose and Hess [13] showed that proportionality between stress and strain rate (the Newtonian assumption) breaks down when the strain rate exceeds approximately twice the molecular interaction frequency:

$$\dot{\gamma} = \frac{\partial u}{\partial y} \geq \frac{2}{\tau} \tag{2.33}$$

where $\dot{\gamma}$ is the strain rate, and the molecular interaction frequency is the inverse of the characteristic time scale given in (2.3). To give the reader an idea of the magnitude of the shear, if liquid argon (properties given above) is contained between two parallel plates separated by 1 μm, one of the plates fixed and one moving at a constant speed, the plate would have to be moving at $10^6$ m/s—a relativistic speed much higher than that found in most mechanical systems. The conclusion is that in most microflows, liquids can be regarded as continuous.

#### 2.2.2.2 Governing Equations

When considering the flow of an incompressible, Newtonian, isotropic, Fourier conducting fluid, the previous equations can be simplified considerably to:

$$\frac{\partial u_i}{\partial x_i} = 0 \tag{2.34}$$

$$\rho\left(\frac{\partial u_i}{\partial t} + u_i \frac{\partial u_i}{\partial x_j}\right) = \rho \mathbf{F}_i - \frac{\partial p}{\partial x_i} + \frac{\partial}{\partial x_i}\left[\eta\left(\frac{\partial u_i}{\partial xj} + \frac{\partial u_j}{\partial x_i}\right)\right] \tag{2.35}$$

$$\rho c_v\left(\frac{\partial T}{\partial t} + u_i \frac{\partial T}{\partial x_i}\right) = \frac{\partial}{\partial x_i}\left(\kappa \frac{\partial T}{\partial x_i}\right) \tag{2.36}$$

To close this set of equations, a constitutive relationship between viscosity and temperature is needed. This type of information is available in table and equation forms from many fluid mechanics textbooks [2].

Assuming that the flow remains at a constant temperature (a reasonable assumption in a microscopic surface-dominated flow, provided no reacting species are present) or that the viscosity does not depend on temperature, still further simplification is possible. The energy equation can be eliminated altogether (or at least decoupled from conservation of mass and momentum), and the conservation of mass and momentum equations can be simplified to:

$$\frac{\partial u_i}{\partial x_i} = 0 \tag{2.37}$$

$$\rho\left(\frac{\partial u_i}{\partial t} + u_i \frac{\partial u_i}{\partial x_j}\right) = \rho \mathbf{F}_i - \frac{\partial p}{\partial x_i} + \eta \frac{\partial^2 u_j}{\partial x_j^2} \tag{2.38}$$

### 2.2.3 Boundary Conditions

The proper boundary condition (BC) between a flowing fluid and the solid that bounds it has been an issue of debate since the inception of modern fluid mechanics more than 150 years ago. Generally, the complete thermomechanical state of the fluid must be known at the boundary. This means temperature and velocity for an incompressible liquid or temperature, pressure, and velocity for a compressible gas. The following discussion will focus on the velocity boundary condition, since it has received the most attention, but the other boundary conditions are important as well. Goldstein [14] provided a thorough discussion of the history of fluid boundary conditions (among other topics) by the great names in fluid mechanics: Navier, Newton, Prandtl, Rayleigh, and Stokes—among many others. Beginning in the late 1800s through to the present day, the accepted practice at macroscopic length scales has been to assume a no-slip velocity condition at the fluid–solid boundary (i.e., the fluid has the same speed as the boundary). Further, the temperature of the fluid immediately adjacent to the wall is assumed to be the same as that of the wall itself. This boundary condition is usually called the no-temperature-jump boundary condition. The pressure is treated similarly. In the direction normal to the wall, this velocity boundary condition amounts to a no-penetration boundary condition in which flow is prevented from flowing through a solid boundary. In the direction tangential to the wall, this boundary condition is called the no-slip boundary condition because the fluid is assumed to stick to the wall. This no-slip boundary condition has been widely used for more than 100 years because it is simple and generates results in agreement with macroscopic experiments. The emergence of microfluidics with its wall-dominated flows raises the question of what is the most appropriate fluid–solid boundary condition at small length scales. There have been a large number of computational, analytical, and experimental studies of the fluid–solid boundary condition in recent years. The reader is referred to a review of these recent activities [15], rather than summarizing all of them here. Several different types of fluid slip are identified in the review article. In the interest of developing an accepted terminology for these different types of fluid slip, they are presented here. They are [15]:

- *Phenomenon of slip*: Any flow situation where the tangential fluid velocity component appears to be different from that of the solid surface immediately in contact with it.
- *Molecular slip* (also *intrinsic slip*): The flow situation where liquid molecules immediately adjacent to a solid boundary "slip" past the solid molecules. Such a concept necessarily involves large forces (see Section 2.2.3.2) [15, 16].
- *Apparent slip*: Flow situation where no-slip is enforced on a small length scale but is not evident on a large length scale. Examples of apparent slip include electrokinetics (see Section 2.3.3), acoustic streaming (see Section 7.1.1.8), and a liquid flowing over a gas layer.
- *Effective slip*: The situation where molecular or apparent slip is estimated from a macroscopic (at least in the molecular sense) measurement or calculation. This is called effective slip because the source of the slip cannot be determined from this measurement alone.

One of the earliest works proposing a fluid–solid boundary condition was that of Navier [17], in which he proposed that the relative velocity at which the fluid slips along the wall $\Delta u_{\text{wall}}$ is directly proportional to the velocity gradient at the wall—and hence the shear stress. The constant of proportionality is called the slip length $L_s$ because it has dimensions of length. In equation form, this boundary condition is

$$\Delta u_{\text{wall}} = u_{\text{fluid}}(y \to \text{wall}) - u_{\text{wall}} = L_{\text{s}} \frac{\partial u_{\text{fluid}}(y)}{\partial y}\bigg|_{\text{wall}} \tag{2.39}$$

where $u_{\text{fluid}}$ is the velocity field in the fluid that depends on the height above the wall $y$, and $u_{\text{wall}}$ is the velocity of the wall. It is important to note here that the velocity field in the fluid will be discontinuous at $y = 0$, so the limit of $y \to$wall must be evaluated from the fluid side of the wall, not the solid side. The significance of the slip length $L_s$ can be seen in Figure 2.6, which shows both no-slip and partial-slip Couette flows. In the case of this exact shear flow, $L_s$ can be interpreted

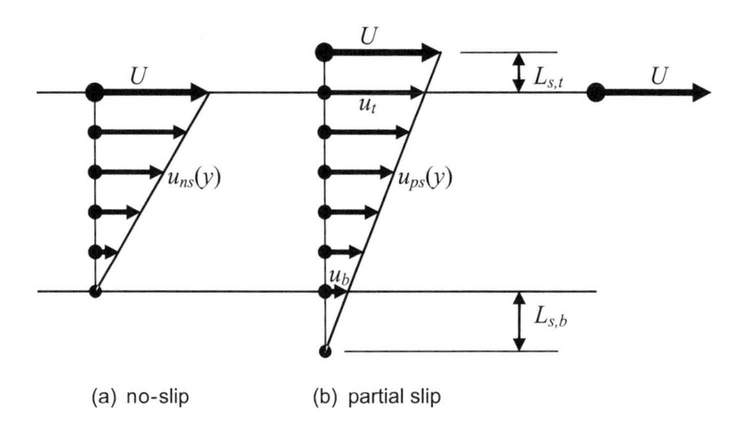


**Figure 2.6** Comparison of (a) no-slip and (b) partial-slip Couette velocity profiles.

as the hypothetical distance within the top and bottom walls where the no-slip boundary condition would be satisfied. In the no-slip case, the flow speed exactly matches the wall speed, 0 at the bottom wall and $U$ at the top wall. Both $L_{s,t}$ and $L_{s,b}$ would be zero. In the partial slip case, $L_{s,t}$ is the distance above the top interface where the flow speed would reach $U$, and $L_{s,b}$ is the distance below the bottom interface where the flow speed would reach $U$. In a flow that has curvature, for example, pressure-driven flow that has a parabolic flow profile, the interpretation of $L_s$ explained above is approximate. The degree of approximation is generally quite good because $L_s$ is usually small compared to the length scale of the system.

#### 2.2.3.1 Gas Flows

According to kinetic gas theory, the no-slip and no-temperature-jump conditions are valid for Knudsen numbers less than $10^{-3}$. In fact, as will be seen below, these boundary conditions are the Kn→0 limit for the more general case boundary conditions. For larger Knudsen numbers, the boundary conditions need to be modified.

Equation (2.39) is based on general empirical and dimensional arguments. The slip length $L_s$ must be given a particular form in order to have any value in fluid mechanics. Maxwell [18] and von Smoluchowski [19] did some of the foundational work in gas dynamics, and derived the velocity slip and temperature jump conditions as:

$$u_{\text{gas}} - u_{\text{wall}} = \lambda \frac{2-\sigma_{\text{v}}}{\sigma_{\text{v}}} \frac{\partial u}{\partial y}\bigg|_{\text{wall}} + \frac{3}{4} \frac{\eta}{\rho T_{\text{gas}}} \frac{\partial T}{\partial x}\bigg|_{\text{wall}} \tag{2.40}$$

$$T_{\text{gas}} - T_{\text{wall}} = \frac{2-\sigma_{\text{T}}}{\sigma_{\text{T}}} = \frac{2-\sigma_{\text{T}}}{\sigma_{\text{T}}} \frac{2k}{k+1} \frac{\lambda}{\text{Pr}} \frac{\partial T}{\partial y}\bigg|_{\text{wall}} \tag{2.41}$$

where $\sigma_{\text{v}}$ and $\sigma_{\text{T}}$ are the tangential momentum and temperature accommodation coefficients, respectively, and Pr is the Prandtl number. The accommodation coefficients are defined according to:

$$\sigma_{\text{v}} = \frac{\tau_i - \tau_r}{\tau_i - \tau_w} \quad \text{and} \quad \sigma_{\text{T}} = \frac{\text{d}E_i - \text{d}E_r}{\text{d}E_i - \text{d}E_w} \tag{2.42}$$

where $\tau_{\text{i}}$ and $\tau_{\text{r}}$ are the tangential momentum of the incoming and reflected molecules and $\tau_{\text{w}}$ is the tangential momentum of reemitted molecules, that is, those that may have been adsorbed to the surface, and is equal to zero for stationary walls. Similarly for $\sigma_{\text{T}}$, $\text{d}E_{\text{i}}$ and $\text{d}E_{\text{r}}$ are the energy fluxes of incoming and reflected molecules per unit time, respectively, and $\text{d}E_{\text{w}}$ is the energy flux

**Table 2.4**

Tangential Momentum and Thermal Accommodation Coefficients
for Various Gases Flowing Through Channels of Various Materials [20, 21]

| *Gas* | *Metal* | $\sigma_\mathrm{v}$ | $\sigma_\mathrm{T}$ |
|---|---|---|---|
| Air | Aluminum | 0.87–0.97 | 0.87–0.97 |
| Air | Iron | 0.87–0.96 | 0.87–0.93 |
| Air | Bronze | 0.88–0.95 | NA |
| Ar | Silicon | 0.80–0.90 | NA |
| $N_2$ | Silicon | 0.80–0.85 | NA |

of reemitted molecules corresponding to the surface temperature $T_\mathrm{wall}$. The Prandtl number is given by:

$$\mathrm{Pr} = \frac{c_p \eta}{\kappa} \tag{2.43}$$

where $c_p$ is the specific heat at constant pressure, $\eta$ is the viscosity, and $\kappa$ is the thermal conductivity.

The first term on the right-hand side of both (2.40) and (2.41) has the expected form insofar as wall-normal gradients in the velocity and temperature field result in velocity and temperature jumps at the wall. The second term on the right-hand side of (2.40) is somewhat unexpected. If the gas in a channel is stationary with no imposed external forces, but there is a temperature gradient along the wall, the gas at the wall will experience a slip velocity in the direction of the temperature gradient, that is, toward the higher temperature end of the channel. This phenomenon is called *thermal creep*, and when used to move gases from one reservoir to another is called a *Knudsen pump*. For low-speed flows or large temperature changes, the effect of thermal creep can be significant.

The accommodation coefficients have physical interpretations that make their significance apparent. When $\sigma_\mathrm{v} = 0$, the tangential momentum of all incident molecules equals that of the reflected molecules. Hence, the gas collisions impart no momentum to the wall. Effectively, the flow can be considered inviscid for $\sigma_\mathrm{v} = 0$. This mode of molecular reflection is called *specular reflection*. When $\sigma_\mathrm{v} = 1$, the net reflected tangential momentum is equal to zero. Hence, the gas molecules impart all of their tangential momentum to the wall.

This mode of reflection is called *diffuse reflection*. For the case of perfect energy exchange between the wall and the gas, $\sigma_\mathrm{T} = 1$. Table 2.4 gives the accommodation coefficients for air flowing through some common metals. These values are meant to be representational only. In general, these accommodation coefficients will depend on the temperature, pressure, and velocity of the flow, as well as the surface finish and surface roughness of the material. Accommodation coefficients as low as 0.2 have been observed. The only recent (i.e., MEMS-era) measurements of accommodation coefficients are those of Arkilic [21].

In order to assess relative size of the terms that contribute to the slip velocity and temperature jump, it is necessary to nondimensionalize (2.40) and (2.41), using a characteristic velocity $u_0$ and temperature $T_0$ (see Section 2.2.5 for more details about nondimensionalization). The resulting boundary conditions are:

$$u^*_\mathrm{gas} - u^*_\mathrm{wall} = \mathrm{Kn}\frac{2-\sigma_\mathrm{v}}{\sigma_\mathrm{v}}\frac{\partial u^*}{\partial y^*}\bigg|_\mathrm{wall} + \frac{3}{2\pi}\frac{k-1}{k}\frac{\mathrm{Kn}^2\mathrm{Re}}{\mathrm{Ec}}\frac{\partial T^*}{\partial x^*}\bigg|_\mathrm{wall} \tag{2.44}$$

$$= \mathrm{Kn}\frac{2-\sigma_\mathrm{v}}{\sigma_\mathrm{v}}\frac{\partial u^*}{\partial y^*}\bigg|_\mathrm{wall} + \frac{3}{4}\frac{\Delta T}{T_0}\frac{1}{\mathrm{Re}}\frac{\partial T^*}{\partial x^*}\bigg|_\mathrm{wall} \tag{2.45}$$

$$T^*_\mathrm{gas} - T^*_\mathrm{wall} = \frac{2-\sigma_\mathrm{T}}{\sigma_\mathrm{T}}\frac{2k}{k+1}\frac{\mathrm{Kn}}{\mathrm{Pr}}\frac{\partial T^*}{\partial y^*}\bigg|_\mathrm{wall} \tag{2.46}$$

where the superscript * denotes dimensionless quantities. The quantity Ec is the Eckert number given by:

$$\mathrm{Ec} = \frac{u_0^2}{c_p \Delta T} = (k-1)\frac{T_0}{\Delta T}\mathrm{Ma}^2 \tag{2.47}$$

where $\Delta T = T_{\text{gas}} - T_0$. The first version of (2.45) shows that thermal creep should become significant at larger Knudsen numbers, while the second version shows that thermal creep will be more prevalent for slower flows (smaller Reynolds number) or when the relative temperature gradient is large.

These boundary conditions are all the boundary conditions that are needed to cover the entire range for which the Navier-Stokes equations are valid. The boundary conditions in (2.45) and (2.46) are applicable for any Knudsen number less than 0.1. Of course, when the Knudsen number is smaller than 0.001, these equations will degenerate to the no-slip and no-temperature-jump conditions.

#### 2.2.3.2 Liquid Flows

For most microscale liquid flows, the no-slip and no-temperature-jump conditions will hold. In equation form, this is:

$$\begin{aligned} u_{\text{wall}} &= u_{\text{liquid}}\Big|_{\text{wall}} \\ T_{\text{wall}} &= T_{\text{liquid}}\Big|_{\text{wall}} \end{aligned} \tag{2.48}$$

where $u_{\text{wall}}$ is the speed of the wall, and $u_{\text{liquid}}$ is the speed of the liquid in contact with the wall. A similar interpretation holds for the temperature field. These boundary conditions are just the large length scale (when measured in molecular diameters) approximations of the linear Navier boundary condition introduced in Section 2.2.3.1:

$$\Delta u_{\text{wall}} = u_{\text{fluid}}(y \to \text{wall}) - u_{\text{wall}} = L_S \frac{\partial u_{\text{fluid}}(y)}{\partial y}\Big|_{\text{wall}} \tag{2.49}$$

in which the velocity slip at the wall $\Delta u_{\text{wall}}$ is approximately zero, due to the low shear rates found in most microdevices and the relatively large length scale. The slip length $L_S$ is a constant for any liquid-solid pairing but can vary significantly for different pairs of liquids and solids. Determining this slip length is difficult and must be done through experiments or molecular dynamics simulations. Presently, there is little data available of this type. One such source of data is the molecular dynamics simulations of Thompson and Troian [16], in which they simulated the motion of a liquid comprised of simple molecules between atomically flat plates comprised of simple atoms in a (111) plane of a face-centered cubic arrangement. One of the plates was moved at a constant velocity relative to the other—a typical Couette arrangement—in which the shear rate could be defined and controlled well. When the spacing between the atoms comprising the wall was commensurate with the mean molecular spacing in the fluid, the slip length was found to be zero (i.e., the no-slip condition governs the flow when the liquid and solid are matched in molecular spacing). However, when the lattice spacing of the solid was four times the mean molecular spacing of the liquid, the slip length had increased to 16.8 times the mean molecular spacing in the liquid. These results are expected to be general for any choice of simple fluids and simple solids in the (111) plane of a face-centered cubic arrangement. Unfortunately, water is not such a simple liquid, and therefore these results should be used with caution when considering water as the liquid.

The most important conclusion of Thompson and Troian's [16] work is that the linear Navier boundary condition, which should be independent of the velocity field and hence the shear rate, does break down at large shear rates. In fact, the breakdown is quite catastrophic. They observed the slip

**Table 2.5**

Slip Length Properties from Thompson and Troian's Data [16]. The characteristic energy scale $\varepsilon^{wf}$ and length scale $\sigma^{wf}$ of the Lennard-Jones potential between wall atoms and the fluid molecules are given in terms of the fluid's properties, as is the density of the wall material $\rho^{w}$. The constant slip length at low shear rates is $L_{\text{S}}^{0}$, while the critical shear rate is $\dot{\gamma}_{\text{c}}$.

| $\varepsilon^{wf}$ | $\sigma^{wf}$ | $\rho^{w}$ | $L_{\text{S}}^{0}/\sigma$ | $\dot{\gamma}_{\text{c}}\tau$ |
|---|---|---|---|---|
| 0.6 | 1.0 | 1 | 0.0 | NA |
| 0.1 | 1.0 | 1 | 1.9 | 0.36 |
| 0.6 | 0.75 | 4 | 4.5 | 0.14 |
| 0.4 | 0.75 | 4 | 8.2 | 0.10 |
| 0.2 | 0.75 | 4 | 16.8 | 0.06 |

length to change at large values of the shear rate according to the relation:

$$L_{\text{S}} = L_{\text{S}}^{0}\left(1 - \frac{\dot{\gamma}}{\dot{\gamma}_{\text{c}}}\right)^{-\frac{1}{2}} \tag{2.50}$$

where $L_{\text{S}}$ is the slip length as a function of the shear rate $\dot{\gamma}$, the shear rate at which the slip length breaks down $\dot{\gamma}_{\text{c}}$, and $L_{\text{S}}^{0}$ is the constant slip length found at low values of the shear rate. Another important conclusion is that this breakdown of the Navier boundary condition happens well below the breakdown of Newtonian behavior in the bulk of the fluid—see Section 2.2.2.1—at $\dot{\gamma}_{\text{c}} \geq 2/\tau$. The last row of Table 2.5 shows that the critical shear rate occurs at approximately 3% of the value of the breakdown of the Newtonian assumption (0.06/2 = 0.03). Significant deviations from the Navier boundary condition occur at an order of magnitude lower than the critical shear rate, where the slip length has already increased by 5% when compared with the low shear rate value of the slip length. Deviations from the Navier boundary condition are significant at a shear rate that measures 0.3% of that at which breakdown of Newtonian behavior occurs. In terms of the example used above to illustrate when to expect a breakdown of Newtonian behavior, if liquid argon (properties given above) is contained between two parallel plates separated by 1 μm, one of the plates fixed and one moving at a constant speed, the plate would have to be moving at 3,000 m/s—still a high speed, but one which could potentially be encountered in actual microflows.

Another exception to the Navier boundary condition is moving contact lines that require a relaxation of the no-slip assumption in order to avoid unbounded shear stresses at the point where a solid, liquid, and gas all meet. Since a liquid will pull itself into and through a hydrophilic channel, the point at which the solid, liquid, and gas meet must move along the wall of the channel, violating the no-slip assumption. The solution to this problem lies in the molecular reordering very near the point where the three phases come together, and thus is an example of effects at the molecular level causing macroscopic effects. This phenomenon will be discussed further in Section 2.2.7.

### 2.2.4 Parallel Flows

Many microdevices have long straight passages of constant cross sections in them. Flow through such channels is thus an important flow phenomenon at small length scales. In addition, many of these long straight passages have cross-sectional geometries that are considerably different from the usual macroscopic circular pipe. As a simplification, let us consider an isothermal flow of an incompressible, isotropic, Newtonian liquid through a long channel of a constant arbitrary cross section, as shown in Figure 2.7. Also consider that the body force is absent. Because this is a liquid flow, we will take the boundary conditions to be no-slip at the wall in the direction parallel to the flow $(x)$ and no penetration through the wall in the directions normal to the flow $(y, z)$.

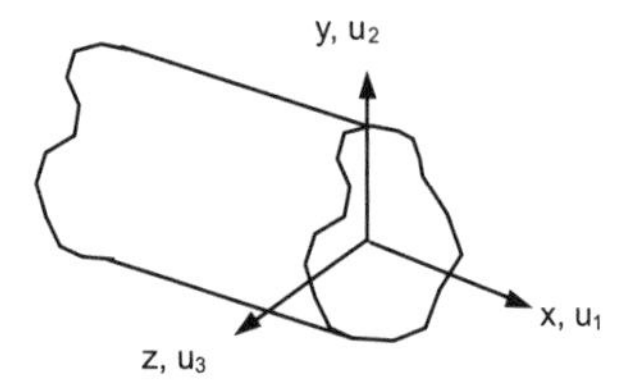


**Figure 2.7** Parallel flow geometry through a channel of an arbitrary cross section.

For laminar flows in parallel geometries, the flow is purely axial—only the $u_1$ component is nonzero—and $u_1$ is a function of $y$ and $z$ only. Under these assumptions, (2.34) and (2.35) simplify to:

$$0 = -\frac{\partial p}{\partial x} + \eta\left(\frac{\partial^2 u_1}{\partial y^2} + \frac{\partial^2 u_1}{\partial z^2}\right) \tag{2.51}$$

$$0 = \frac{\partial p}{\partial y} = \frac{\partial p}{\partial z} \tag{2.52}$$

Equation (2.52) indicates that the pressure is strictly a function of $x$; that is, it is independent of $y$ and $z$. Since the pressure is strictly a function of $x$ and (2.51) requires that its derivative equal a function of $y$ and $z$ only, not $x$, must be a constant. Further, it stands to reason that the flow must be in the direction of decreasing pressure, so $\partial p/\partial x$ must be a negative constant.

#### 2.2.4.1 Analytical Solutions

To proceed to a solution, we need only fix the geometry of the cross section. Solutions for several common shapes (and many unusual ones) are available from Berker [22] and Shah and London [23]. The details of the solutions presented here are left to the reader. The isosceles triangle is included in the solutions presented here because an anisotropic wet etch in single crystal (100) silicon, if allowed to self-terminate, will form an isosceles triangle cross section.

*Circular cross section:*

$$u_1 = \frac{-dp/dx}{4\eta}(r_0^2 - r^2) \tag{2.53}$$

$$\dot{Q} = \frac{\pi r_0^4}{8\eta}\left(\frac{\mathrm{d}p}{\mathrm{d}x}\right) \tag{2.54}$$

*Rectangular cross section:*

$$u_1(y,z) = \frac{16a^2}{\eta\pi^3}\left(-\frac{\mathrm{d}p}{\mathrm{d}x}\right)\sum_{i=1,3,5\ldots}^{\infty}(-1)^{(i-1)/2}\left[1 - \frac{\cosh(i\pi z/2a)}{\cosh(i\pi b/2a)}\right]\frac{\cos(i\pi y/2a)}{i^3} \tag{2.55}$$

$$\dot{Q} = \frac{4ba^3}{3\eta}\left(-\frac{\mathrm{d}p}{\mathrm{d}x}\right)\left[1 - \frac{192a}{\pi^5 b}\sum_{i=1,3,5,\ldots}^{\infty}\frac{\tanh(i\pi b/2a)}{i^5}\right] \tag{2.56}$$

*Triangular (isosceles) cross section:*

$$u_1(y,z) = \frac{1}{\eta}\left(-\frac{\mathrm{d}p}{\mathrm{d}x}\right)\frac{y^2 - z^2\tan^2\phi}{1 - \tan^2\phi}\left[\left(\frac{z}{2b}\right)^{B-2} - 1\right] \tag{2.57}$$

$$\dot{Q} = \frac{4ab^3}{3\eta}\left(-\frac{\mathrm{d}p}{\mathrm{d}x}\right)\frac{(B-2)\tan^2\phi}{(B+2)(1-\tan^2\phi)} \tag{2.58}$$

where

$$B = \sqrt{4 + \frac{5}{2}\left(\frac{1}{\tan^2\phi} - 1\right)} \tag{2.59}$$

These equations provide the velocity field as a function of location within the cross section of a channel and as a function of applied pressure gradient. They also provide the integral of the velocity profile across the cross section—the volume flow rate—as a function of pressure drop. When designing microdevices that contain long channels, such as heat exchangers and chromatography columns, these relations will provide the pressure drop for a given flow rate, or, given the pressure drop, will allow calculating the flow rate.

Equation (2.54) for flow through a channel of circular cross section can be rearranged into an especially useful form for calculating pressure drop through a channel of length $L$ and diameter $D$ as:

$$\Delta p = \frac{8\eta \dot{Q} L}{\pi r_0^4} = f\frac{L}{D}\rho\frac{u^2}{2} = \mathrm{Re} f \frac{\eta L}{2D^2} u \tag{2.60}$$

where $f$ is the Darcy friction factor and $u = \dot{Q}/A$ is the average velocity in the channel. For reference, a competing version of the friction factor, called the Fanning friction factor, is one-fourth the Darcy friction factor. For laminar flows, the Fanning friction factor can be found by inspection of (2.60) to be:

$$f = 64/\mathrm{Re} \tag{2.61}$$

This is a particularly useful equation for calculating the pressure drop through a long channel. The form of the friction factor varies depending on channel cross section shape, whether the flow is laminar or turbulent, and whether or not the flow is fully developed. For a further discussion of this useful expression, see [24].

#### 2.2.4.2 Hydraulic Diameter

Clearly, there are several shapes missing from this list that can be significant in microfluidics, as well as many others that the reader may chance to encounter less frequently. Two examples of these shapes are the trapezoidal cross section created by anisotropic wet etches in silicon, and the rectangular cross section with rounded corners often created by isotropic wet etches of amorphous materials. These shapes are shown in Figure 2.8. Of course, there are many other shapes that could also be significant, which are left off this list. Addressing the flow through these geometries is often difficult because of their irregular shapes. One method for approximating the flows through these geometries is using a concept known as the hydraulic diameter $D_\mathrm{h}$. The hydraulic diameter is given by:

$$D_\mathrm{h} = \frac{4 \times \text{cross section area}}{\text{wetted perimeter}} = \frac{4A}{P_\mathrm{wet}} \tag{2.62}$$

The concept of the hydraulic diameter can be used to assess flows through both completely filled channels, as well as those that are only partially filled. The term *wetted perimeter* refers to the perimeter of the channel that is in direct contact with the flow, while the area refers to the area through which the flow is occurring. For channels that are completely filled, the wetted perimeter is simply the perimeter of the channel and the area is the cross-sectional area of the channel.

The utility of the hydraulic diameter can be demonstrated by considering the example of flow through a passage of circular cross section. In that case,

$$D_\mathrm{h} = \frac{4 \times \frac{\pi}{4} \times d^2}{\pi d} = d \tag{2.63}$$

so the hydraulic diameter reduces to simply the diameter of the passage.

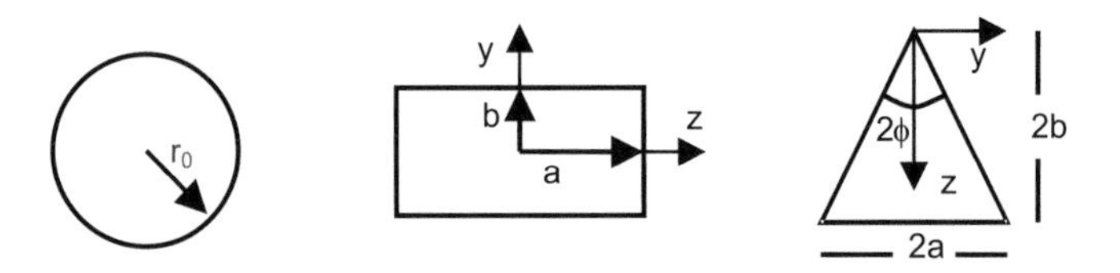


**Figure 2.8** Channel cross sections with practical applications in microsystems. (*After*: [13].)

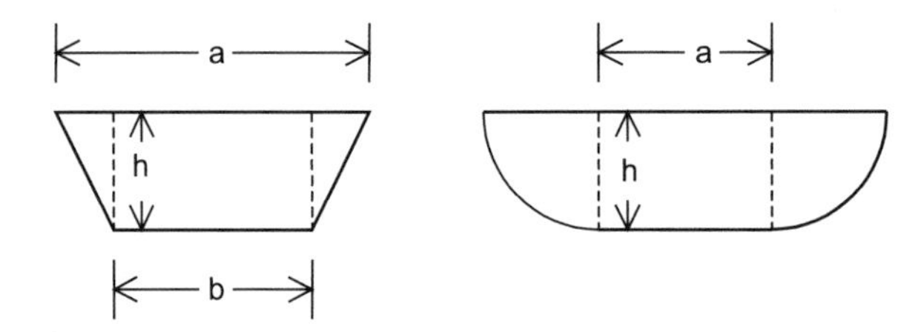


**Figure 2.9** Common channel shapes produced by microfabrication.

### Example 2.2: Hydraulic Diameter

Calculate the hydraulic diameter for the shapes in Figure 2.9, assuming the shapes to exhibit left-right reflection symmetry.

*Trapezoid:*

$$\text{Area} = \frac{h}{2}(a+b)$$

$$\begin{aligned}\text{Wetted perimeter} &= a+b+2\sqrt{h^2+\left(\frac{a-b}{2}\right)^2} \\ &= a+b+2h\sqrt{1+\left(\frac{a-b}{2h}\right)^2}\end{aligned}$$

$$D_\mathrm{h} = \frac{4\left[\frac{h}{2}(a+b)\right]}{a+b+2h\sqrt{1+\left(\frac{a-b}{2h}\right)^2}} = \frac{2h}{1+\frac{2h}{a+b}\sqrt{1+\left(\frac{a-b}{2h}\right)^2}}$$

*Rounded rectangle:*

$$\text{Area} = ah + \frac{1}{2}\pi h^2$$

$$\text{Wetted perimeter} = 2a + \pi h + 2h$$

$$D_\mathrm{h} = \frac{4\left(ah+\frac{1}{2}\pi h^2\right)}{2a+\pi h+2h} = 2h\frac{\frac{a}{h}+\frac{\pi}{2}}{\frac{a}{h}+\frac{\pi}{2}+1}$$

Once these hydraulic diameters are known, they may be used in the expressions for flow through a circular cross section [(2.54) or (2.60)] to estimate the volume flow rate as a function of pressure gradient or conversely to estimate the expected pressure drop for a given volume flow rate.

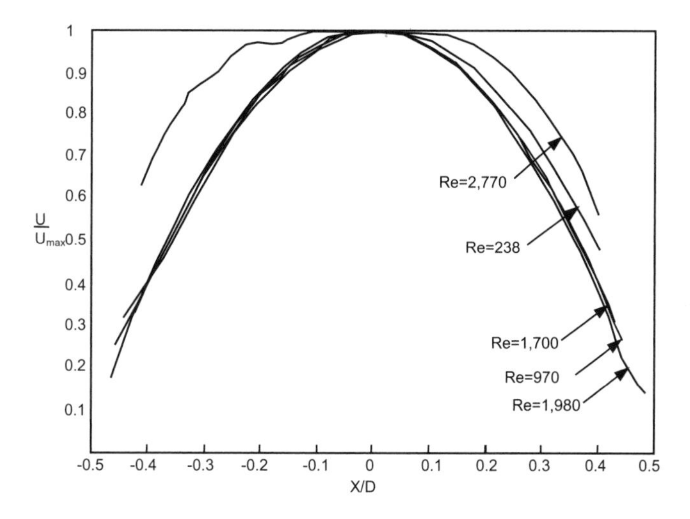


**Figure 2.10** Fully developed velocity profiles for flow through a channel with rectangular cross section ($D_h = 327$ mm). The highest Reynolds number exhibits the broadened profile characteristic of turbulent flow while the others agree to the parabolic profile within experimental uncertainty.

### 2.2.4.3 Transition to Turbulence in Microchannels

For flows in long straight channels of constant cross section as discussed above, the Reynolds number remains a good predictor of when turbulence will occur (see Section 2.2.5 for further explanation of the Reynolds number), even at microscopic length scales. The Reynolds number is given by:

$$\mathrm{Re} = \frac{\rho D_h u}{\eta} = \frac{D_h u}{\nu} \tag{2.64}$$

where $D_h$ is the hydraulic diameter (as discussed above), $u$ is some measure of the characteristic velocity in the channel, $r$ is the density of the fluid, $\eta$ is the dynamic viscosity of the fluid, and $\nu$ is the kinematic viscosity of the fluid. In channels of circular cross section [24] as well as rectangular cross section [25], recent experiments have demonstrated no significant deviation of the transitional Reynolds number—the Reynolds number below which the flow is laminar and above which the flow is turbulent—from that typically encountered at macroscopic length scales. The transitional Reynolds number has been found to be in the range of 1,500 to 2,500. Figure 2.10 shows fully developed velocity profiles for five different Reynolds numbers ranging from 238 to 2,770. The scatter between the velocity profiles at the lower Reynolds numbers is fairly small, while the fifth profile is considerably wider and blunter—indicating that the transition to turbulent flow has begun somewhere between Reynolds numbers of 1,980 and 2,770. The particular Reynolds number at which transition will occur is a function of many parameters, such as the channel shape, aspect ratio, and surface roughness, but is expected to be in the range of 1,000 to 2,000 for most situations. This conclusion is at odds with some other researchers' findings [26, 27] in which the transitional Reynolds number has been reported to be as low as several hundred. However, those findings were largely based on global pressure drop measurements and not confirmed with a close inspection of the velocity profile as a function of Reynolds number. Sharp [24] provided an excellent discussion of this topic.

### 2.2.5 Low Reynolds Number Flows

A great many microfluidic devices operate in regimes where the flow moves slowly—at least by macroscopic standards. Biomedical microdevices are a good example where samples are carried from one reaction chamber or sensor to the next by a flowing fluid that may be driven by an external syringe pump. To determine whether a flow is slow relative to its length scale, we need to scale the original dimensional variables to determine their relative size. Consider a flow in some geometry whose characteristic size is represented by $D$ and average velocity $u$. We can scale the spatial coordinates with $D$ and the velocity field with $u$ according to the relations:

$$x_i^* = \frac{x_i}{D} \quad \text{and} \quad u_i^* = \frac{u_i}{u} \tag{2.65}$$

with the inverse scaling:

$$x_i = Dx_i^* \quad \text{and} \quad u_i = uu_i^* \tag{2.66}$$

The starred quantities are then dimensionless quantities whose size varies from zero to order one. These inverse scalings can be substituted into the dimensional forms of the governing equations given in (2.34) and (2.35) to arrive at a dimensionless set of governing equations. Assuming an isothermal flow of a Newtonian, isotropic fluid, the conservation of mass and momentum equations can be simplified to:

$$\frac{\partial u_i^*}{\partial x_i^*} = 0 \tag{2.67}$$

$$\frac{\rho u D}{\eta}\frac{D}{u}\frac{\partial u_i^*}{\partial t} + \frac{\rho u D}{\eta}u_j^*\frac{\partial u_i^*}{\partial x_j^*} = \frac{\rho D^2}{u\eta}\mathbf{F}_i - \frac{D}{u\eta}\frac{\partial p}{\partial x_i^*} + \frac{\partial^2 u_i^*}{\partial x_j^{*2}} \tag{2.68}$$

This transformation is incomplete because some of the variables in (2.68) are dimensionless while others remain in dimensional form. Considering the constants that multiply each term of (2.68), the following scalings suggest themselves:

$$t^* = \frac{t}{D/u} \quad \text{and} \quad p^* = \frac{p}{\eta u/D} \tag{2.69}$$

Upon substituting these expressions into (2.68), the following convenient, dimensionless versions of the governing equations emerge:

$$\frac{\partial u_i^*}{\partial x_i^*} = 0 \tag{2.70}$$

$$\mathrm{Re}\left(\frac{\partial u_i^*}{\partial t^*} + u_j^*\frac{\partial u_i^*}{\partial x_j^*} - \frac{\mathbf{F}_i D}{u^2}\right) = -\frac{\partial p^*}{\partial x_i^*} + \frac{\partial^2 u_i^*}{\partial x_j^{*2}} \tag{2.71}$$

in which all of the variables ($u_i$, $x_i$, $t$, $p$) appear in dimensionless form, all of the terms are dimensionless, and the Reynolds number emerges as a key parameter. When the body force $\mathbf{F}_i$ is gravity, the group of parameters $\mathbf{F}_i D/u_2$ is the inverse of the Froude number [2]. Because of the nondimensionalization process, each of the groups of variables, such as $\partial u_i^*/\partial t^*$, $\partial p^*/\partial x_i^*$, or $\partial u_i^*/\partial t^*$, has a size of order one, while the groups of parameters, namely the Reynolds number and $\mathbf{F}_i D/u_2$, have a size determined by flow parameters. If the Reynolds number is very small, $\mathrm{Re} \ll 1$, the entire left side of (2.71) becomes negligible, leaving only:

$$\frac{\partial u_i^*}{\partial x_i^*} = 0 \tag{2.72}$$

$$0 = -\frac{\partial p^*}{\partial x_i^*} + \frac{\partial^2 u_i^*}{\partial x_j^{*2}} \tag{2.73}$$

as the governing equations. Although this equation requires Re $\ll 1$ to be true, the agreement with experiments is usually acceptable up to Re $\sim 1$. The momentum equation assumes the form of the Poisson equation.

*Observations About (2.72) and (2.73)*:

- The equation is *linear* and thus relatively easy to solve. Tabulated solutions for many common geometries are available in standard fluid mechanics texts [3, 4].
- If the pressure gradient is increased by some constant factor A, the entire flow field is increased by the same factor; that is, the velocity at every point in the flow is multiplied by A. The shape (*streamlines*) of the flow does not change. Thus, an experiment done at *one* Reynolds number is valid for predicting all low Reynolds number behaviors of a system.
- Time does not appear explicitly in the equation. Consequently, low Reynolds number flows are completely reversible (except for diffusion effects that are not included in the equation). Since many microsystems contain periodically moving boundaries such as membranes, any motion the membrane imparts to the fluid in the first half of the membrane's period will be exactly reversed in the second half of the period, and sum to zero net motion.

**Example 2.3: Low Reynolds Number Systems**

To get an idea of what kind of systems might exhibit low Reynolds behavior, it is useful to perform a sample calculation. A typical biomedical microdevice might exhibit the following behavior:

| | |
|---|---|
| Fluid properties similar to water: | $\eta = 10^{-3}$ kg/(s m), $\rho = 10^3$ kg/m$^3$ |
| Length scale: | 10 μm $= 10^{-5}$ m |
| Velocity scale: | 1 mm/s $= 10^{-3}$ m/s |
| Then: | Re $= 10^{-2}$ |

Clearly low Reynolds number behavior.

As another example, consider the flow in a microchannel heat exchanger:

| | |
|---|---|
| Fluid properties similar to water: | $\mu = 10^{-3}$ kg/(s m), $\rho = 10^3$ kg/m$^3$ |
| Length scale: | 100 $\mu$m $= 10^{-4}$m |
| Velocity scale: | 10 m/s |
| Then: | Re $= 10^3$ |

Clearly *not* low Reynolds number behavior—borderline turbulent behavior.

### 2.2.6 Entrance Effects

Sometimes the differences between microscale flows and larger flows are due to changes in fundamental physics, such as working at size scales where the length at the wall becomes important or there is a large electrical field, and sometimes they are due to more practical differences. A great many of the fabrication processes that are used to produce small-scale systems (see Chapter 3) are planar techniques in which some pattern having an $x$ and $y$ dependence is effectively extruded in the $z$ direction to produce a three-dimensional geometry. Since the geometry through which the flow is happening is independent of the $z$ position, it might be tempting to consider the flow to be two-dimensional. However, the aspect ratio (ratio between the height of channels and the lateral feature size) is usually near 1, meaning that the flow must definitely be considered three-dimensional.

One example where this situation can be important is in considering the entrance length effect. When a fluid flows from a large vessel or reservoir into a smaller one of a constant cross section (a common situation in both large- and small-scale fluid mechanics), the flow profile requires a certain distance along the flow direction to develop or adjust to the presence of the walls of the channel. Beyond the entrance length, the flow is considered fully developed and the velocity profile does not change any further. The point at which flow adjusts to the presence of the walls most slowly is at

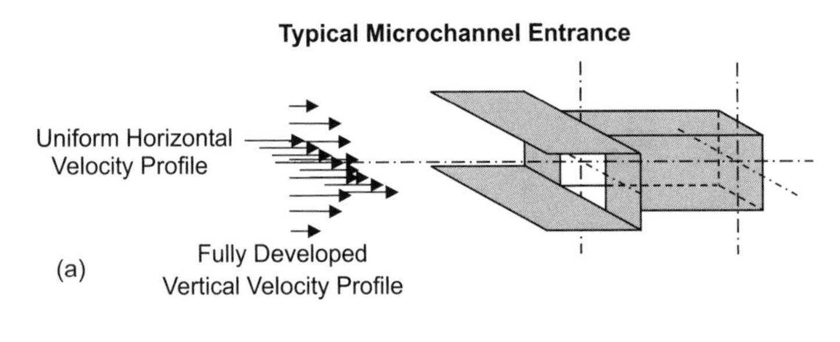


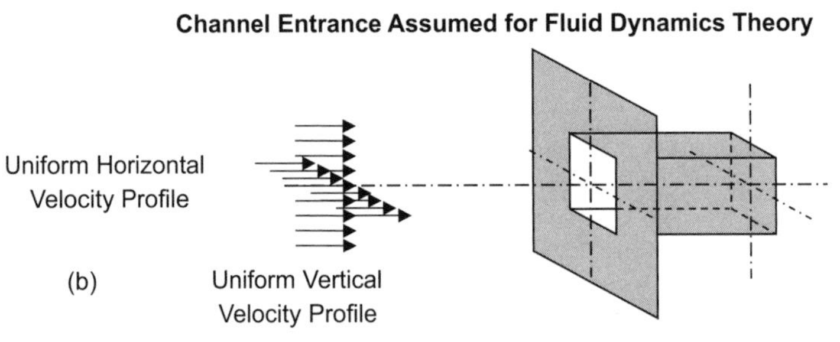


**Figure 2.11** (a) Typical macroscopic entrance geometry versus (b) microscopic entrance geometry as a consequence of planar microfabrication.

its furthest distance from the walls—its centerline. Intuition leads us to expect that the centerline velocity in a very long pipe would initially rapidly approach its fully developed value, and that farther and farther downstream, the centerline velocity would continue to increase, albeit ever more slowly. Consequently, an arbitrary standard needs to be set beyond which the flow is considered close enough to fully developed to be considered fully developed. Shah and London [23] defined the entrance length $L_\mathrm{e}$ as the point where the developing centerline velocity is equal to 0.99 times its fully developed value. A large-scale example of such a flow would be the flow out of a large reservoir through a pipe. A corresponding small-scale flow might be the flow in a microchannel heat exchanger where the flow proceeds from a plenum (or large chamber) into the microchannel. This type of geometry is most often made by a planar microfabrication technique, so that while the plenum may be large compared to the microchannel, they are both the same height. Figure 2.11 depicts this situation.

As a consequence of the plenum and the microchannel having the same height in the planar microfabricated geometry, the flow in the plenum is uniform in the direction parallel to the top and bottom walls, but parabolic in the direction normal to the top and bottom walls, whereas in the macroscopic reservoir, the flow in the reservoir is everywhere uniform. The implication is that since the flow is already partially developed in the microscopic case, the entrance length should be decreased in the planar geometry when compared to the macroscopic situation.

For macroscopic flows where the Reynolds number is usually assumed to be relatively high, the entrance length $L_\mathrm{e}$ can be accurately predicted by:

$$\frac{L_\mathrm{e}}{D_\mathrm{h}} \approx 0.06\mathrm{Re}_{D_\mathrm{h}} \tag{2.74}$$

where $D_\mathrm{h}$ is the hydraulic diameter and $\mathrm{Re}_{D_\mathrm{h}}$ is the Reynolds number based on the hydraulic diameter [2]. Since the Reynolds number scales with the length scale of a system, small-scale systems often exhibit low Reynolds number behavior. As the Reynolds number tends towards zero, the entrance length predicted by (2.74) also tends towards zero, in disagreement with low Reynolds

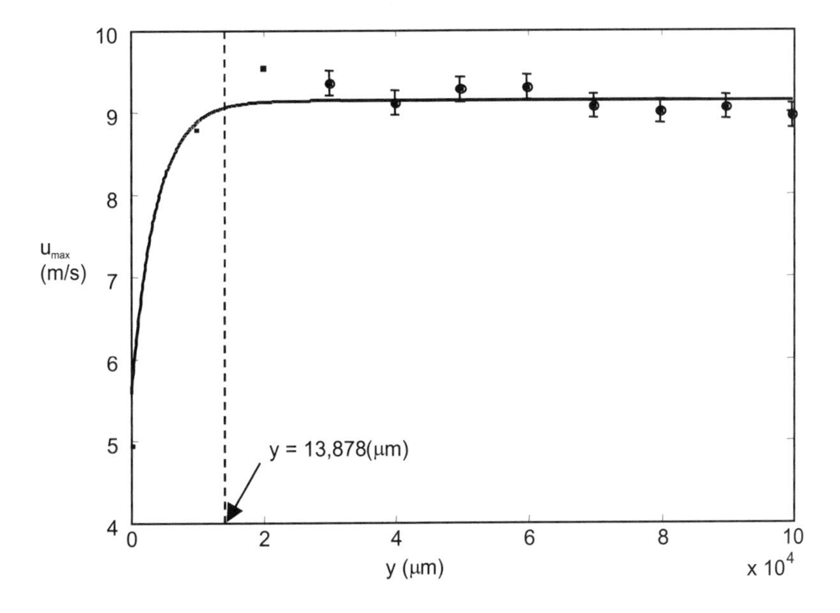


**Figure 2.12** Development of flow entering rectangular channel. The flow becomes fully developed 1.4 cm downstream from the entrance, whereas (2.75) predicts 3.6 cm—a significant deviation. The square boxes are data points and the solid curve is an exponential curve fit.

number experiments. A better expression for low Reynolds numbers is given by:

$$\frac{L_\mathrm{e}}{D_\mathrm{h}} \approx \frac{0.6}{1 + 0.035\mathrm{Re}_{D_\mathrm{h}}} + 0.056\mathrm{Re}_{D_\mathrm{h}} \tag{2.75}$$

according to Shah and London [23]. This expression predicts that even in the limit of the Reynolds number approaching zero, the entrance length is approximately 0.6 hydraulic diameters long.

Using (2.75) to predict the entrance length provides more reasonable results at low Reynolds numbers. However, the problem of the different macroscopic and microscopic geometries still remains. Developing a better expression for entrance length in planar geometries will require many sets of spatially resolved measurements at various Reynolds numbers. Preliminary experiments demonstrating that there is a difference in the entrance length have been conducted in the microfluidics laboratory at Purdue University [25]. Using the microparticle image velocimetry (μPIV) technique explained further in Chapter 4, velocity profiles were measured in a rectangular channel (hydraulic diameter of 327 $\mu$m) being fed by a planar plenum of the same height (600 $\mu$m). For a Reynolds number of 1,980, the differences between the experimentally observed entrance length and that predicted by (2.75) are quite obvious. Figure 2.12 demonstrates this developing behavior and clearly shows that the flow develops in approximately one-half the distance predicted by (2.75). Given that the convective heat transfer behavior in the developing region differs from that in the fully developed region, and given that many microchannel heat exchangers are shorter than the entrance length of the flow, this change in the entrance length could prove significant in many cases. The shear stress at the wall in the developing region is less than that in the fully developed region, which has implications for biomedical microdevices.

### 2.2.7 Surface Tension

Consider again the Lennard-Jones potential represented by (2.1) and shown in Figure 2.2. The Lennard-Jones potential models accurately the interaction between any two simple molecules. As

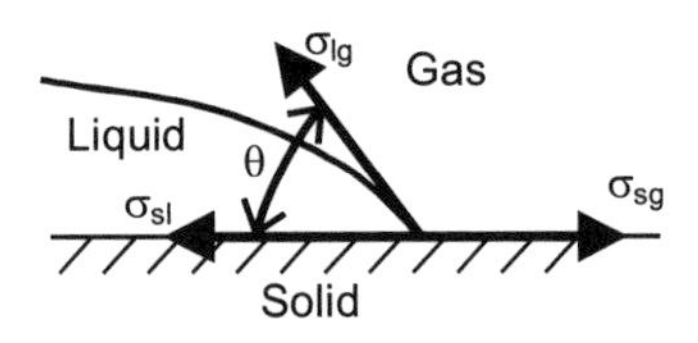


**Figure 2.13** Typical contact line arrangement. The contact angle is labeled $\theta$, and the interfacial tensions $\sigma_{\mathrm{sl}}$, $\sigma_{\mathrm{lg}}$, and $\sigma_{\mathrm{sg}}$ are the solid–liquid, liquid–gas, and solid–gas, respectively.

discussed in Section 2.1.2, molecules of a liquid have a mean spacing, such that they vibrate around the zero crossing location of the force versus distance plot in Figure 2.2. The behavior of a liquid molecule is independent of its location within the liquid as long as that location is far from any interfaces when measured in molecular units. Consider the oscillation of a liquid molecule that is exactly on the interface between a liquid and a gas. As it vibrates away from its equilibrium position toward the gas phase, it experiences the attractive van der Waals forces (the $r^{-6}$ term in the LJ potential) of all of its neighboring liquid molecules. Recall that gas molecules interact through infrequent energetic collisions and are not in a state of continual interaction with their neighbors, as are liquids. Consequently, the gas will not exert any attractive force on the liquid molecule that would tend to pull it free from the bulk of the liquid. Hence, the molecule will experience a net attractive force toward the bulk of the fluid. This behavior is generally the case at liquid-gas interfaces.

The liquid-solid interface behavior is more difficult to predict because the molecules of the liquid on the interface are continually interacting with liquid molecules in one half-space and solid molecules in the other. Whether the liquid molecules on the interface are more attracted to the solid molecules on the other side of the interface or the liquid molecules on their side of the interface depends on the particular form of the Lennard-Jones potential for liquid-liquid interactions versus that for liquid-solid interactions. When all three phases come together at a line, the phenomenon is called a *contact line*.

Surface tension has vexed microsystem designers since they began building devices that are meant to be filled with liquid, as well as devices that are manufactured using liquid-based processing. For example, removing a sacrificial silicon dioxide layer from under a layer of silicon (see Section 3.2) should be a simple wet etch. However, once the etch is complete, the area where the sacrificial layer used to be is filled with a liquid. The surface tension of the liquid makes it difficult to remove. It may even break the silicon surface because of the high pressures that can be generated by interfaces with small radii of curvature. Clearly, it is worthwhile to know more about this phenomenon.

Figure 2.13 shows a typical contact line phenomenon where a solid, liquid, and gas all meet along a line of contact. Because the solid is rigid, the contact line can only move left or right along the surface. It cannot be displaced in a vertical direction. Hence, the energy balance must be:

$$\sigma_{\mathrm{sl}} + \sigma_{\mathrm{lg}} \cos\theta = \sigma_{\mathrm{sg}} \tag{2.76}$$

where $\theta$ is called the contact angle. When the contact angle is smaller than 90° (as shown in the figure), the liquid is said to wet the surface. When the contact angle is greater than 90°, the liquid is said to be nonwetting. A surface that a liquid can wet is said to be hydrophilic, and a surface that a liquid cannot wet is said to be hydrophobic. Because of the tension along the interfaces, there will be a net increase or decrease in pressure upon crossing a curved interface. This pressure change $\Delta p$ is given by:

$$\Delta p = \sigma\left(\frac{1}{R_1} + \frac{1}{R_2}\right) \tag{2.77}$$

where $R_1$ and $R_2$ are the two radii of curvature of the interface surface [5].

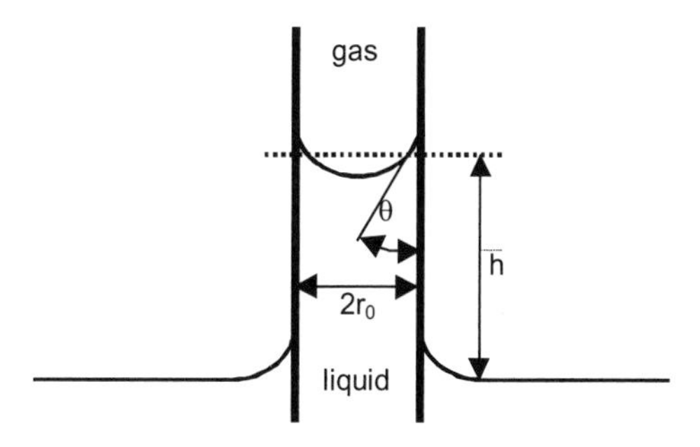


**Figure 2.14** Capillarity of liquids filling small channels.

The most commonly used liquid in microflows is water. The surface tension of water is predicted well by:

$$\sigma_{\mathrm{lg}} = (0.076 - 0.00017T \cdot \mathrm{K}^{-1})\ \mathrm{N/m} \tag{2.78}$$

where $T$ is the temperature of the system. This expression is accurate to within 1% of tabulated values over the range of 0°C to 374°C [3]. The surface tension of a liquid depends strongly on the presence of contaminants such as surfactants or dirt. These contaminants can greatly change the observed surface tension. The contact angle of water on glass is approximately 0°. For further information about surface tension, including tables of surface energies for many liquids, please see [28].

Equations (2.76) and (2.77) can be combined to solve simple but important contact line problems. First, consider the case of a glass capillary of diameter $2r_0$ inserted vertically into a reservoir of water, as in Figure 2.14. The water climbs up the inside of the capillary to some equilibrium height $h$ that is given by:

$$\rho g h = \Delta p = \frac{2\sigma_{\mathrm{lg}} \cos\theta}{r_0} \tag{2.79}$$

where the hydrostatic pressure $\rho g h$ is equated to that encountered by crossing the air-water interface. The tendency of liquids to draw themselves into small channels comprised of materials that they wet is known as *capillarity*. Capillarity makes it difficult to fill hydrophobic channels and difficult to empty hydrophilic channels.

## 2.3 MOLECULAR APPROACHES

As we saw in Section 2.2, the continuum assumption will fail for some range of parameters—for gas flows when the Knudsen number is larger than 0.1, and for liquid flows when the fluid is sheared at a rate faster than twice the characteristic molecular interaction frequency $(2/\tau)$. When the continuum assumption fails, the continuum approach to modeling flows must also fail and a new means for modeling flows must be used. Two approaches have emerged in recent years based on radically different methods—MD and DSMC. There have been several thorough explorations of these techniques [6–8] to which the reader is directed for a thorough discussion of these issues. The next two sections will provide an introduction to these techniques. The reader is encouraged to consult the works cited above for a more thorough coverage of these techniques.

### 2.3.1 MD

The MD technique is in principle a straightforward application of Newton's second law, in which the product of mass and acceleration for each molecule comprising the flow is equated to the forces on the molecule that are computed according to a model—quite often the Lennard-Jones model. Koplik and Banavar [6] provided a thorough, accessible introduction to the technique, along with the latest research results in their recent review article. The technique begins with a collection of molecules distributed in space. Each molecule has a random velocity assigned to it, such that the distribution of velocities obeys a Boltzmann distribution corresponding to the temperature of the distribution. The molecular velocities are integrated forward in time to arrive at new molecular positions. The intermolecular forces at the new time step are computed and used to evolve the particles forward in time again. As the user desires, the molecular positions and velocities can be sampled, averaged, and used to compute flow quantities, such as velocity, density, and viscosity. Although the algorithm is relatively simple to explain, it is computationally intensive to implement because of the very short time steps needed for the integration and the extremely large number of molecules in even modest volumes of space. Theoretically, MD can be used for any range of flow parameters from slow liquid flows to highly rarefied, hypersonic gas flows. However, the MD technique is used primarily for liquid flows because the molecules are in a continual collision state. MD is very inefficient for gas flows because collisions are infrequent. DSMC (see Section 2.3.2) is often used for gas flows.

#### 2.3.1.1 Difficulties with MD Implementation

Even though the algorithm can be outlined in a single paragraph as above, the MD technique is quite complicated to implement. Chief among these complications is the sheer volume of calculations required to simulate a reasonably sized volume of space for a reasonable duration. In this instance, "reasonable" is used to mean large enough that continuum behavior can be observed. MD simulations typically consider volumes measuring 10 nm on a side—meaning approximately $30^3 = 27,000$ molecules. The characteristic molecular interaction time is $2.2 \times 10^{-12}$ seconds and perhaps a total of $10^{-9}$ seconds would be used for the entire simulation—just enough time for continuum behavior to set in.

A second problem is constructing a reasonable pairwise molecular interaction potential. As discussed previously, the Lennard-Jones potential:

$$V_{ij}(r) = 4\varepsilon \left[ c_{ij} \left( \frac{r}{\sigma} \right)^{-12} - d_{ij} \left( \frac{r}{\sigma} \right)^{-6} \right] \tag{2.80}$$

where $\varepsilon$ and $\sigma$ are characteristic energy and length scales, respectively, is widely used for this purpose. The force between interacting molecules is given by the derivative of the potential as:

$$F_{ij}(r) = -\frac{\partial V_{ij}(r)}{\partial r} = \frac{48\varepsilon}{\sigma} \left[ c_{ij} \left( \frac{r}{\sigma} \right)^{-13} - \frac{d_{ij}}{2} \left( \frac{r}{\sigma} \right)^{-7} \right] \tag{2.81}$$

With an appropriate choice of the parameters $c_{ij}$ and $d_{ij}$, this equation can represent the force between any two simple molecules. With appropriate modifications, it can even represent the force between more complicated molecules, like water, by considering two hydrogen atoms tethered to an oxygen atom through another potential. The characteristic time scale for molecular interactions, for which Lennard-Jones is a good model, is given by:

$$\tau = \sigma \sqrt{\frac{m}{\varepsilon}} \tag{2.82}$$

which is the period of oscillation about the minimum in the Lennard-Jones potential. The Lennard-Jones potential has been shown to provide accurate results for liquid argon using $\varepsilon/k_\mathrm{B} = 120$ K, $\sigma = 0.34$ nm, and $c = d = 1$. For these parameters, the time scale is $\tau = 2.2 \times 10^{-12}$ seconds [6].

The Lennard-Jones potential (and the resulting force) has the inconvenient property of never quite reaching zero, even for large distances. Consequently, every molecule in a flow will exert a force on every other molecule. If there are $N$ molecules in a flow, then there are $N$ factorial possible pairwise interactions between molecules. The number of possible interactions scales approximately with $N^2$. One solution to this problem is to shift the potential by a function linear in $r$, such that the force at some cutoff distance $r_\mathrm{c}$ is zero. Beyond that cutoff distance, the force is set to zero identically. One such modified Lennard-Jones force is:

$$F_{ij,\mathrm{TRUNC}}(r) = \begin{cases} F_{ij}(r) - F_{ij}(r_\mathrm{c}) & r < r_\mathrm{c} \\ 0 & r > r_\mathrm{c} \end{cases} \tag{2.83}$$

The potential is modified similarly but instead of being shifted by a constant value, it is shifted by an amount linear in $r$. These modified potentials are plotted in Figure 2.15 along with the difference between the modified and original Lennard-Jones expressions. The differences are multiplied by 100 to make them apparent. In the dimensionless units of the figure, the Lennard-Jones force is shifted upward by a constant amount, approximately 2% of the maximum attractive force. Once the form of the pairwise potential is decided, the state (position and velocity) of all the molecules is evolved forward in time according Newton's second law in the form:

$$m_i \frac{\mathrm{d}^2\mathbf{r}_i}{\mathrm{d}t} = -\frac{\partial}{\partial \mathbf{r}_i} \sum_{i=j} V_{ij}(|\mathbf{r}_i - \mathbf{r}_j|) \tag{2.84}$$

where the mass of the $i$th molecule is given by $m_i$ and its vector position is $r_i$. Any suitable finite difference method may be used for this evolution. Macroscopic flow quantities are calculated by averages over small volumes of space. For instance, the vector velocity $\mathbf{u}(\mathbf{x})$ in some neighborhood $B$ would be calculated by:

$$\mathbf{u}(\mathbf{x}) = \frac{1}{N}\left\langle \sum_{\mathbf{x}_i \in B} \frac{\mathrm{d}\mathbf{x}_i}{\mathrm{d}t} \right\rangle \tag{2.85}$$

where $N$ is the number of molecules in the neighborhood and $\mathbf{x}_i$ is their position.

Molecular dynamics has been used to investigate a great many flows, such as the investigation into the Navier boundary condition described in Section 2.2.3.2. In practice, it is a great deal more difficult than described in this brief introduction. The reader is referred to Koplik and Banavar's [6] review article on the subject for a more thorough discussion, as well as a description of significant results achieved by the technique.

### 2.3.2 DSMC Technique

The DSMC technique was developed by Bird [29] nearly 30 years ago, and has been used successfully by many researchers for computing gas flows. Several comprehensive review articles concerning the technique are available [7, 30, 31]. In Section 2.2, we saw that for Knudsen numbers larger than 0.1, the continuum assumption will fail. Necessarily, the analytical techniques based on the continuum assumption will also fail for Kn $> 0.1$. For flows with higher Knudsen numbers, there are several alternatives for obtaining solutions. One is the Burnett equation [32], a continuum approach that uses a higher order stress-strain rate model and is applicable to higher Knudsen numbers than the Navier-Stokes equation. Unfortunately, the Burnett equation is unwieldy and difficult to solve. Solutions are generally reached computationally.

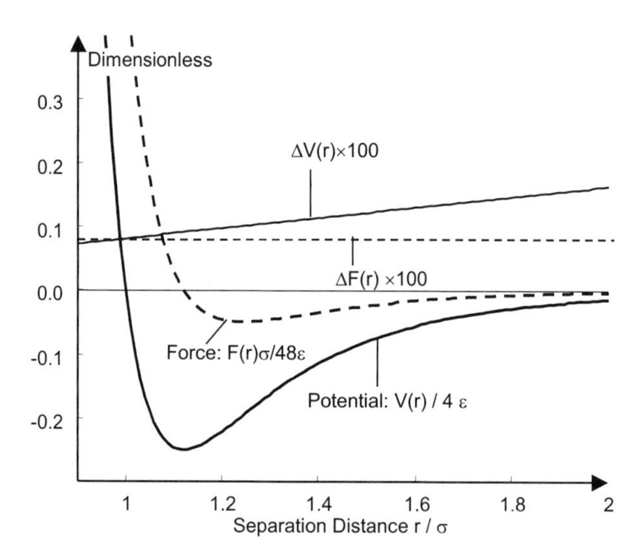


**Figure 2.15** Difference between truncated and untruncated Lennard-Jones potential and force.

Another approach that has been used successfully is the DSMC technique, which is a simulation technique based on kinetic theory. DSMC tracks the motion of hundreds of thousands or millions of "particles," each of which represents a large number of molecules. DSMC uncouples molecular motions from intermolecular collisions. The molecular motions are treated deterministically, while the collisions are treated statistically. DSMC is a very efficient method (compared to its competitors) to compute flows in the transitional flow regime, $0.1 < \text{Kn} < 10$, although technically it can be used in any dilute gas flow, $\delta/d > 10$—see (2.17). This means that there is an overlap where both the Navier-Stokes equations can be used ($\text{Kn} < 0.1$) and where the DSMC technique can be used ($\delta/d > 10$) to solve the same problem.

#### 2.3.2.1 Overview of the DSMC Procedure

DSMC consists of four main steps [30, 31], which are repeated at each time step of the simulation. These are:

- Moving the particles;
- Indexing and cross-referencing the particles;
- Simulating particle collisions;
- Sampling macroscopic properties of the flow.

These steps will be discussed in greater detail below.

The first step in DSMC consists of evolving the position of all the particles forward in time according to their velocity at the beginning of the time step. The length of the time step is chosen to be smaller than the mean collision time $\Delta t_c$, which is the ratio of the mean free path $\lambda$ to the rms molecular speed $\bar{c}$. When $\Delta t_c$ is calculated, it is usually found to be on the order of $10^{-10}$ seconds for most flows. Once the particles are advanced in space, some of them will have left the computational domain through inlets and exits, and some will be inside the boundaries of the flow. Hence, the boundary conditions must be applied at this point. Particles that are found to be inside walls will be subjected to wall collisions according to the appropriate boundary conditions (2.42).

The second process is indexing and tracking the particles. For more efficient computations, the volume of the flow being simulated is divided into cells, each of which measures approximately one-third of the mean free path of the gas. During the first step, particles may leave one cell and enter another cell. Because potentially millions of particles are being tracked, it is important that this step is efficient.

The third step consists of modeling the collisions of particles within each cell. Since each particle represents thousands of molecules, a statistical collision model must be used. The no-time-counter technique of Bird [11] is commonly used, in which collision rates are calculated within the cells and collision pairs are selected within subcells, improving the accuracy of the collision model by enforcing that particles collide with their nearest neighbors.

The fourth and last step is calculating macroscopic flow quantities (e.g., velocity within the cell) from the microscopic quantities (e.g., individual particle velocity) within each cell. This can be done in a process similar to that used in MD (2.81). At this point, the simulation can conclude or can loop back to the first step. Some amount of iteration is required, even with steady flows, to reach a steady-state after the initial start-up transience.

#### 2.3.2.2 Drawbacks and Limitations of DSMC

Beskok [30] and Oran et al. [31] identified several possible sources of error and limitations of the DSMC technique. These are:

*Finite cell size:* DSMC cell sizes should be chosen to be approximately one-third of the local mean free path. Cells larger than this size can result in artificial increases in gas viscosity, reduced sensitivity to macroscopic spatial gradients in the flow, and larger errors. Cells smaller than this size will have too few particles in them for statistically reliable collision results.

*Finite time step:* The time step must be smaller than the mean collision time $\Delta t_c$, or a significant fraction of the collisions in the flow will be overlooked when the particles travel through more than one cell during the time $\Delta t_c$.

*Ratio of simulated particles to real molecules:* When each of the simulated particles represents too many real molecules, large statistical scatter results. Typically, each particle represents $10^{14}$ to $10^{18}$ molecules for a three-dimensional calculation. This ratio is too large, especially if there are complicated dynamics beyond simple collisions happening in each cell.

*Boundary condition uncertainties:* It can be challenging to provide the correct boundary conditions for DSMC flows. For a subsonic flow, boundary conditions are usually specified at the entrance to a channel flow as number density of molecules, temperature, and mean macroscopic velocity. At the exit, number density and temperature can be specified. The mass flow rate through the pipe will develop according to these boundary conditions. It is difficult to specify the mass flow rate directly.

*Uncertainties in the physical parameters:* These effects include the inaccuracies of simulating collisions of real molecules, like $N_2$, with the variable hard sphere (VHS) and variable soft sphere (VSS) models frequently used in DSMC.

Despite these limitations, DSMC has proved to be very useful in computing rarefied gas dynamics. It has been used successfully to model the flow between hard drive read/write heads and the magnetic media—a lubrication-type flow in which the head "flies" on a cushion of air on the order of 10 nm thick. DSMC has also been used for simple, constant cross-section microchannel flows for verifying high Knudsen number boundary condition models [30] and in complicated microgeometries, such as diffusers [33] and backward-facing steps [34].

## 2.4 ELECTROKINETICS

Electrokinetic pumping and particle manipulation techniques are widely used to move liquids and particles at small length scales because they are implemented through surface forces, which scale

**Table 2.6**
Classification of Main Electrokinetic Phenomena (*After*: [35].)

| *Name* | *Type of Movement* | *Electrokinetic Coupling* |
|---|---|---|
| Electrophoresis | Charged surface moves relative to a stationary liquid | Use an applied electric field to induce movement |
| Electro-osmosis | Liquid moves relative to a stationary charged surface | |
| Streaming potential | Liquid moves relative to a stationary charged surface | Use movement to create an electric field |
| Sedimentation potential | Charged surface moves relative to a stationary liquid | |

well when length scales are reduced. Electrokinetic techniques also have the advantage of being easily integrable into microfluidic systems when compared to external systems such as syringe pumps.

This section will provide a brief overview of the field of electrokinetics. For a thorough discussion of this topic, the reader is referred to several comprehensive texts on the subject [35, 36]. Several overviews of the topic have also recently appeared [24, 37] that discuss the topic in somewhat more detail than that presented here but are more accessible than the first two sources cited. In addition, Devasenathipathy and Santiago [37] provided a thorough discussion of experimental diagnostic techniques that can be applied to electrokinetic flows. Probstein [35] followed Shaw's [38] classification of electrokinetic phenomena into four main types. These are listed in Table 2.6. The discussion here will concentrate on electrophoresis and electro-osmosis—the two electrokinetic phenomena that use an applied electric field to induce motion. The remaining two phenomena, the streaming potential and the sedimentation potential, have the opposite electrokinetic coupling in that they use motion to produce an electric field.

### 2.4.1 Electro-Osmosis

Electro-osmosis is a good place to begin discussions of electrokinetic effects because the geometries involved can be simplified to considering a liquid in contact with a planar wall. When polar liquid, such as water, and the solid are brought into contact, the surface of the solid acquires an electric charge. The surface charge then influences the migration of charges within the liquid near the wall. Ions in the liquid are strongly drawn toward the surface and form a very thin layer called the *Stern layer* in which the ions in the liquid are paired with the charges on the surface. The Stern layer then influences the charge distribution deeper in the fluid creating a thicker layer of excess charges of the same sign, as those in the Stern layer called the *diffuse* or *Gouy-Chapman layer.* Together these two layers are called the *electric double layer* (EDL). Because of the proximity of charges, the Stern layer is fixed in place while the diffuse layer can be moved. In particular, the diffuse layer has a net charge and can be moved with an electric field. Consequently, the boundary between the Stern layer and the diffuse layer is called the *shear surface* because of the relative motion across it. The potential at the wall is called the *wall potential* $\phi_W$ and the potential at the shear plane is called the *zeta potential* $\zeta$. This situation is shown in Figure 2.16 and is typical of the charge distributions observed in many microfluidic devices. Both glass [36] and polymer-based [39] microfluidic devices tend to have negatively charged or deprotonated surface chemistries, which means that the EDL is positively charged.

After assuming a Boltzmann distribution of the charge in the EDL and the Poisson relationship:

$$\nabla^2\phi = -\frac{\rho_E}{\varepsilon} \tag{2.86}$$

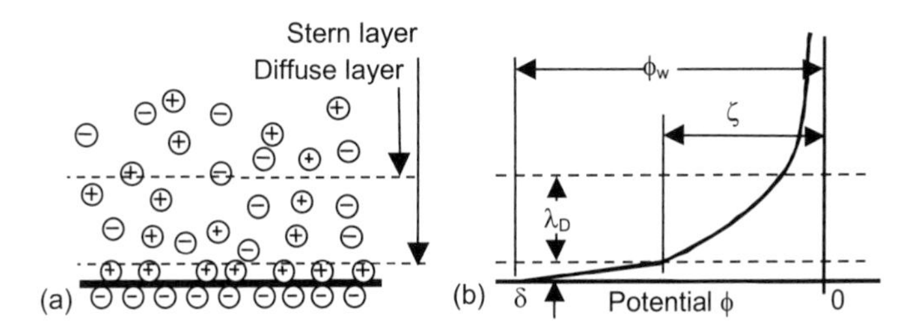


**Figure 2.16** Sketch of the electric double layer showing (a) the Stern layer, and the diffuse layer and (b) the resulting potential.

between the charge density $\rho_E$ and the potential $\phi$, and that the electrolyte is symmetric (e.g., $Na^+$, $Cl^-$), the governing equation for the potential is found to be the Poisson-Boltzmann equation:

$$\frac{d^2\phi}{dy^2} = \frac{2Fzc_\infty}{\varepsilon} \sinh\left(\frac{zF\phi}{KT}\right) \tag{2.87}$$

where $c_\infty$ is the concentration of ions far from the surface, $z$ is the charge number (valence) of each ion, $\varepsilon = \varepsilon_r\varepsilon_0$ is the dielectric constant, and $F$ is Faraday's constant. Faraday's constant is equal to the charge of 1 mole of singly-ionized molecules—$F = 9.65 \times 10^4$ C mol$^{-1}$. This equation is clearly nonlinear and difficult to solve. However, the relative thickness of the EDL is usually small enough in micron-sized systems, that the hyperbolic sine term can be replaced by the first term in its Taylor series—just its argument. This approximation is called the Debye-Hückel limit of thin EDLs, and it greatly simplifies (2.87) to:

$$\frac{d^2\phi}{dy^2} = \frac{\phi}{\lambda_D^2} \text{where} \lambda_D^2 = \sqrt{\frac{\varepsilon KT}{2z^2F^2c_\infty}} \tag{2.88}$$

where $\lambda_D$ is called the Debye length of the electrolyte. The solution to this ordinary differential equation is quite straightforward and found to be:

$$\phi = \phi_w \exp\left(-\frac{y}{\lambda_D}\right) \tag{2.89}$$

Hence, the Debye length represents the decay distance for the potential as well as the electric field at low potentials.

This potential can be added into the Navier-Stokes equation from Section 2.2.2 to calculate the flow produced by the electro-osmotic effect. Consider the geometry shown in Figure 2.17 where the electro-osmotic flow is established in a flow of constant cross sections. This geometry is reminiscent of the parallel flow geometry of Section 2.2.4. Hence, we can use the simplified form of the Navier-Stokes equation given in (2.51). Combining these two equations, we get:

$$0 = -\nabla p + \eta\nabla^2 u + \rho_E E_{el} = -\nabla p + \eta\nabla^2 u - \varepsilon\nabla^2\phi \tag{2.90}$$

where in the first equation, the force generated by the electrical field is given by the product of the charge density and the field strength. In the second equation, the electrical force is written in terms of the potential using Poisson's equation. In microfluidic systems, the imposed pressure is usually a condition set by the designer as is the potential. The goal of this analysis is to calculate the velocity field that these two input conditions generate. Fortunately, (2.90) is linear, implying that the velocity

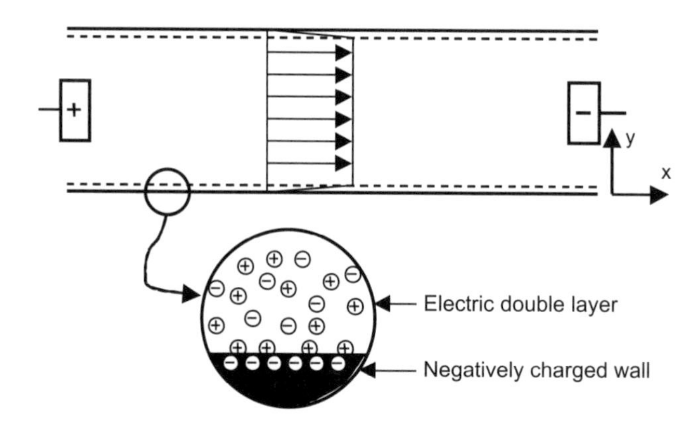


**Figure 2.17** Schematic representation of the flow in a constant cross-section channel subject to electro-osmosis.

field generated by the imposed pressure field can be added to the velocity field generated by the electrical field. Since the pressure, generated velocity field was considered above, only the effect of the electrical field will be considered here. The simplified equation is then:

$$\eta \nabla^2 u_{\text{eof}} = \varepsilon E_{\text{el}} \nabla^2 \phi \tag{2.91}$$

where the component of the flow due to electro-osmosis is denoted $u_{\text{eof}}$. If, as in the parallel flow case, we consider the flow to be fully developed, the only gradients that will be present are those in the $y$ direction. So the equation becomes:

$$\eta \frac{\mathrm{d}^2 u_{\text{eof}}}{\mathrm{d}y^2} = \varepsilon E_{\text{el}} \frac{\mathrm{d}^2 \phi^2}{\mathrm{d}y^2} \tag{2.92}$$

This differential equation can be solved by two integrations with respect to $y$ and application of appropriate boundary conditions. The boundary conditions are that at the edge of the EDL, the gradients in the fluid velocity and the potential should both be zero (i.e., that outside the double layer, both the potential and the velocity are constant). At the wall location, the potential is set equal to the potential of the zeta potential of the wall. The final version of this equation becomes:

$$u_{\text{eof}} = \frac{\varepsilon E_{\text{el}} \zeta}{\eta} \tag{2.93}$$

This equation is known as the Helmholtz-Smoluchowski equation, and is subject to the restriction that the Debye layer is thin relative to the channel dimension. This velocity field can be directly added to that obtained by imposing a pressure gradient on the flow to find the combined result of the two forces. Obtaining solutions for the flow when the Debye length is large generally requires resorting to numerical solutions because the Debye-Hückel approximation is not valid when the Debye layer is an appreciable fraction of the channel size.

#### 2.4.1.1 Fluid Velocity/Electric Field Similitude

Equation (2.93) can be used to find approximate electro-osmosis solutions in more complicated channel geometries [37, 40]. The equation can be generalized from flow in a constant cross-section channel with a field aligned with the channel axis to any imposed electrical field, provided a few constraints are satisfied according to:

$$u_{\text{eof}}(x, y, x, t) = \frac{\varepsilon \zeta}{\eta} E_{\text{el}}(x, y, z, t) \tag{2.94}$$

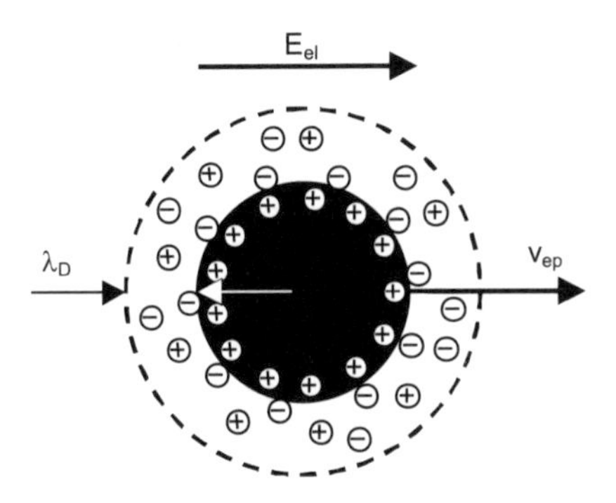


**Figure 2.18** Sketch of charge distribution around an electrophoretic particle.

where the electrical field dependence on space and time is explicitly included. The velocity field is assumed to be proportional to the electrical field. This expression provides reasonably accurate results provided:

- The zeta potential is uniform everywhere;
- The Debye length is small compared to channel geometry;
- There is no applied pressure gradient.

For the purpose of comparing the effectiveness of several different electro-osmotic channel/solution combinations, the electro-osmotic mobility $\mu_{eo}$ is defined as:

$$\mu_{eo} = \frac{u_{eo}}{E_{el}} \tag{2.95}$$

The electro-osmotic mobility is a useful, empirical quantity that aids in predicting flow velocities expected for different imposed electrical fields. In the absence of appreciable Joule heating, the proportionality is very good. The zeta potential $\zeta$ that is necessary in calculating the electro-osmotic flow is an approximate quantity. It is a quantity that should be a function of wall material and surface chemistry. However, presently there is no method to independently determine the zeta potential other than to conduct electro-osmosis experiments, which presupposes the correctness of the EDL theory.

### 2.4.2 Electrophoresis

Electrophoresis is the manipulation of particles or molecules through the use of an electrical field, as in Figure 2.18. Capillary electrophoresis (CE) is commonly used to fractionate or separate DNA molecules according to length—a technique that is being used widely in police forensics (e.g., the O. J. Simpson trial), as well as genetic research such as the human genome project. This phenomenon is closely related to the electro-osmosis phenomenon discussed in Section 2.4.1. Consequently, the analysis presented in this section will be abbreviated. The reader is directed to Section 2.4.1 for a more in-depth discussion.

The analysis of particles moving in fluids necessarily includes some drag model to account for the effect of the fluid on the particle. Because the particles tend to be small and slow-moving (at least relatively speaking), a Reynolds number based on particle diameter and relative fluid speed would be a small number, probably significantly less than unity. Given this line of reasoning, it is reasonable to balance the viscous Stokes drag (low Reynolds number) with the electrical force for a particle moving at a constant speed. When solving problems using the Stokes drag assumption, it is always advisable after working the problem to check that the Reynolds number is small. The force balance on the particle then becomes:

$$q_s E_{el} = 6\pi\eta u_{ep} r_0 \tag{2.96}$$

where $u_{ep}$ is the electrophoretic velocity of the particle relative to the fluid, $r_0$ is the radius of the particle, and all other parameters are defined as above. The particle is assumed to be nonconducting. This is a reasonable assumption, even for materials that would normally be conducting, because in many cases conducting particles will become polarized by the applied field and behave as nonconductors. Working in spherical coordinates, the relationship between the surface charge on the particle $q_s$ and the zeta potential $\zeta$ can be found to be:

$$q_s = -\varepsilon \left(\frac{\partial \phi}{\partial r}\right)_{r=r_0} = \varepsilon\zeta \left(\frac{1}{r_0} + \frac{1}{\lambda_D}\right) = \frac{\varepsilon\zeta}{\lambda_D}\left(\frac{\lambda_D}{r_0} + 1\right) = \frac{\varepsilon\zeta}{\lambda_D}\left(1 + \frac{r_0}{\lambda_D}\right) \tag{2.97}$$

where $\lambda_D$ is defined as in the electro-osmosis case. It is immediately obvious that there are two interesting limits to this expression—Debye length *small* compared to the particle radius, and Debye length *large* compared to the particle radius. These expressions now must be related to the net charge in the EDL according to:

$$\begin{aligned} q &= 4\pi\varepsilon r_0^2 \left(\frac{\partial \phi}{\partial r}\right)_{r=r_0} = 4\pi\varepsilon r_0^2 \zeta \left(\frac{1}{r_0} + \frac{1}{\lambda_D}\right) \\ &= \frac{4\pi\varepsilon r_0^2 \zeta}{\lambda_D}\left(\frac{\lambda_D}{r_0} + 1\right) = 4\pi\varepsilon r_0 \zeta \left(1 + \frac{r_0}{\lambda_D}\right) \end{aligned} \tag{2.98}$$

The electrophoretic motion of molecules often meets the limit of Debye length large compared to the effective size of the molecule simply because molecules can be very small. In addition, with the emergence of gold and titania nanoparticles, and fullerenes, this limit becomes a very important one for nanotechnology. The expression for the electrophoretic velocity $u_{ep}$ becomes:

$$u_{ep} = \frac{q_s E_{el}}{6\pi\eta r_0} = \frac{2}{3}\frac{\varepsilon\zeta E_{el}}{\eta} \tag{2.99}$$

where the first form of the equation is well suited to molecules in which the total charge $q = q_s$ of the molecule may be known (valence number) rather than some distributed surface charge. The second form of the equation is more appropriate for very small particles for which the zeta potential $\zeta$ might be known. This form of the equation is called the Hückel equation.

The limit of small Debye length compared to particle radius is an appropriate limit to consider for particles in excess of 100 nm. Examples of these types of particles include polystyrene latex spheres used to "tag" biomolecules, as well as single-cell organisms, which tend to have diameters measured in microns. When the Debye length is small compared to particle radius, the EDL dynamics are approximately reduced to the flat plate scenario discussed in the case of electro-osmosis. Hence, the equation of motion becomes:

$$u_{ep} = \frac{\varepsilon\zeta E_{el}}{\eta} \tag{2.100}$$

which is simply the Helmholtz-Smoluchowski equation from the electro-osmosis phenomenon. One interesting thing to note about (2.99) and (2.100) is that even though they are developed for opposite limiting cases, they differ only by the constant factor 2/3. When the Debye length is neither large nor small relative to the particle radius, the dynamics of the particle motion are significantly more difficult to calculate—beyond the scope of this text; however, (2.99) is still a reasonable estimate of particle velocity.

As with the electro-osmosis case, the effectiveness of electro-phoresis is quantified using an electrophoretic mobility parameter defined as:

$$\mu_{\text{ep}} = \frac{u_{\text{ep}}}{E_{\text{el}}} \tag{2.101}$$

where $\mu_{\text{ep}}$ can be thought of as motion produced per unit field.

### 2.4.3 Dielectrophoresis

Dielectrophoresis (DEP) refers to the motion of polarizable particles (or cells) suspended in an electrolyte and subjected to a nonuniform electric field [41, 42]. Since the particles or cells are moving in a liquid electrolyte, the DEP force is balanced by the viscous drag on the particle. Following the notation of Ramos [42], the DEP force for a homogeneous dielectric particle in an aqueous solution can be written as:

$$\bar{\mathbf{F}}_{\text{DEP}} = \frac{1}{2}\Re\left[\left(\bar{\mathbf{m}}(\omega) \bullet \nabla \bar{E}^*_{\text{el}}\right)\right] \tag{2.102}$$

where $\Re$ represents the real component, $\bar{E}^*_{\text{el}}$ is the complex conjugate of the electrical field $\bar{E}_{\text{el}}$, and $\bar{\mathbf{m}}(\omega)$ is the dipole moment that can be written as:

$$\bar{\mathbf{m}}(\omega) = 4\pi\varepsilon_{\text{m}} r_0^3 K(\omega)\bar{E} \tag{2.103}$$

for a spherical particle of radius $r_0$ and a solution of permittivity $\varepsilon_{\text{m}}$, with $\omega$ being the angular field frequency (radians per second). The parameter $K(\omega)$ is the Clausius-Mossotti factor, which is given by:

$$K(\omega) = \frac{\tilde{\varepsilon}_p - \tilde{\varepsilon}_m}{\tilde{\varepsilon}_p + 2\tilde{\varepsilon}_m} \tag{2.104}$$

where $\tilde{\varepsilon}_p$ and $\tilde{\varepsilon}_m$ are the complex permittivities of the particle and solution, respectively. For an isotropic homogeneous dielectric, the complex permittivity is:

$$\tilde{\varepsilon} = \varepsilon - \frac{k_{\text{el}}}{\omega} i \tag{2.105}$$

where $\varepsilon$ is the permittivity and $k_{\text{el}}$ is the conductivity of the dielectric, and $i$ is the square root of $-1$ (also denoted by $j$, depending on the source). For real electrical fields [i.e., $\bar{E}_{\text{el}} = \Re(\bar{E}_{\text{el}})$], the DEP force can be found by substituting (2.103) through (2.105) into (2.102), averaging in time, and simplifying to get:

$$\langle \bar{\mathbf{F}}_{\text{DEP}}(t) \rangle = 2\pi\varepsilon_m r_0^3 \Re[K(\omega)] \nabla |\bar{E}_{\text{rms}}|^2 \tag{2.106}$$

where $\nabla|\bar{E}_{\text{rms}}|^2$ is the gradient of the square of the root-mean-square (rms) electrical field. The dependence of the DEP force on the various parameters is clear except in the case of the $K(\omega)$ parameter. The limits on this parameter are:

$$-\frac{1}{2} < \Re[K(\omega)] < 1 \tag{2.107}$$

The exact value of the parameter varies with the frequency of the applied field and the complex permittivity of the medium. The DEP force is classified as either positive or negative depending on the sign of the real part of $K$. The classifications are given in Table 2.7.

**Table 2.7**
Positive and Negative DEP Force Classifications and Particle Behavior

| | | |
|---|---|---|
| Positive DEP $\Re[K(\omega)] > 0$ | DEP force is toward higher electrical fields | Particles collect at electrode edges |
| Negative DEP $\Re[K(\omega)] < 0$ | DEP force is toward lower electrical fields | Particles are repelled from electrode edges |

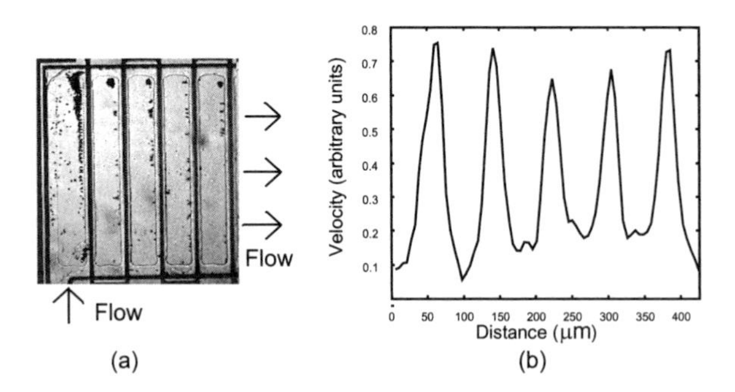


**Figure 2.19** (a) Image of particles captured by DEP force while flowing through microfluidic device, and (b) velocity fluctuations (arbitrary units) as the particles flow into and out of regions of high DEP force at a voltage lower than that needed for particle capture.

The effect of the DEP force is shown in Figure 2.19, which is based on recent work at Purdue University [43]. Figure 2.19(a) shows particles being collected at the edges of electrodes—positive DEP—as they flow through a chamber with interdigitated electrodes patterned on its bottom. There are five fingers in each electrode, with each finger measuring 70 $\mu$m wide and 10 $\mu$m from its neighboring electrode. The beads are polystyrene, 2.38 $\mu$m in diameter, and are captured with a field produced by a sinusoidal voltage of 10-V amplitude, and a frequency of 1 MHz. Figure 2.19(b) shows the mean velocity (arbitrary units) of the particles as they flow through the device along a horizontal line from left to right at a voltage slightly lower than that needed to capture the particles. The velocities were measured using the μPIV technique described in Chapter 4. The wavelength of oscillations in the speed directly correspond to the pitch of the electrodes in the device.

## 2.5 MICROMAGNETOFLUIDICS

Micromagnetofluidics is the science and technology of using magnetism in combination with microfluidics for implementing various fluidic functions such as pumping, mixing, magnetowetting and motion control of magnetic particles [44]. Magnetic manipulation of fluids and particles has various advantages. Compared to an electrical field, an external magnetic field, that does not come into contact with the fluid, can be employed for fluidic control in micromagnetofluidics. Magnetic particles and molecules can be adhered to magnetic particles for sorting, separation, and detection applications. Also, magnetic control is not affected by surface charge, pH and ionic concentration of the fluid [44].

According to the type of working fluid and the associated phenomenon, micromagnetofluidics is categorized into four established fields as magnetohydrodynamics (MHD), ferrohydrodynamics

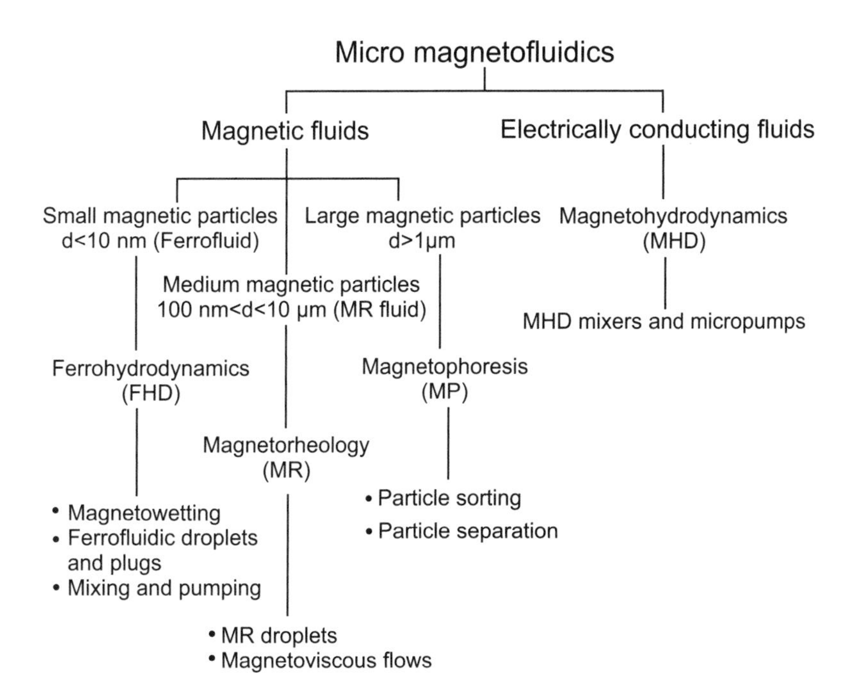


**Figure 2.20** Micromagnetofluidics, its subfields, properties and examples of applications. (*After* [44].)

(FHD), magnetorheology (MR), and magnetophoresis (MP), Figure 2.20 [44]. In addition, in a similar analogy with conventional microfluidics, micromagnetofluidics can be studied in continuous and digital fields [45].

MHD studies the interaction of magnetic field and the flow of electrically conducting fluids. This is a well established field that has been explored for the last hundred years. FHD, MR, and MP cover the effect of a magnetic field on magnetic fluids. A magnetic fluid is composed from a carrier fluid and a suspension of magnetic particles. Size of the magnetic particles determine the behavior of the magnetic fluid and the associated microscale fluidic phenomena at the presence of a magnetic field.

### 2.5.1 Ferrohydrodynamics

Ferrohydrodynamics (FHD) is based on fluids containing magnetic particles with dimensions smaller than about 10 nm. Thus, particles are dispersed well and form a ferrofluid that is a special class of nanofluid. In this way, the fluid is treated as a continuum and behaves as a paramagnetic liquid. Ferrohydrodynamic behavior can be predicted by Navier-Stokes equation with an additional term of magnetic force:

$$\rho\frac{D\mathbf{u}}{Dt} = -\nabla p + \eta\nabla^2\mathbf{u} + (\mathbf{M}\bullet\nabla)\mathbf{B} \tag{2.108}$$

where $\mathbf{M}$ is local magnetization in A/m, $\mathbf{B}$ is flux density in Tesla (T), and $\eta$ is the dynamic viscosity of the fluid. The above equation shows that to move a fluid against a pressure gradient, either a gradient of the magnetic field or a gradient of magnetization field of the ferrofluid is required. Using FHD principles, a number of microfluidic devices for pumping and valving have been realized [46, 47, 48]. Ferrofluidic micropumping will be further discussed in Section 7.3.

A gradient of magnetization can induce movement of ferrofluids in a microchannel. The magnetization degrades through increasing temperature, and the net magnetization of particles are lost beyond Curie temperature. Thus, temperature gradient can induce magnetization gradient that leads to a gradient of magnetic body force necessary for driving a ferrofluid in a microchannel. This phenomenon was called the *magnetocaloric effect* and the experienced body force by the liquid is expressed as [49]:

$$\mathbf{f}_{\mathrm{mc}} = \frac{1}{2}(\mathbf{B} \bullet \frac{\partial \mathbf{M}}{\partial T})\nabla T \tag{2.109}$$

where $T$ is temperature. In addition, formation of ferrofluid droplets has been realized using T-junction or flow-focusing junction. Since ferrofluid is a subclass of nanofluids, droplet formation of ferrofluid is affected by nanoparticles and their surfactant. In addition, FHD provide unique platform for digital microfluidics due to its ability for controlling single stand-alone magnetic droplets.

### 2.5.2 Magnetorheology

A magnetorheological fluid contains magnetic particles on the order of 100 nm to 10 $\mu$m. Thus, the large size of particles does not allow for keeping them suspended by Brownian motion. Upon applying a magnetic field, magnetic particles align under dipole interaction to shape chains and *supraparticle structures* (SPS) [44]. These chains hinder the fluid motion perpendicular to the magnetic field and results in the increase of the apparent fluid viscosity that is known as the magnetoviscous phenomenon. The magnetoreological fluid can perform as a viscoelastic solid once the strength of the applied magnetic field is strong enough. The controllable viscosity of MR fluids has been employed for separation of large molecules in a microchannel, however, there are opportunities to perform more researches on the use of MR fluids for various microfluidic applications.

### 2.5.3 Magnetophoresis

Magnetophoresis mainly concerns with the movement of magnetic particles driven by a magnetic field gradient. Experienced magnetic force by a magnetic particle ($\mathbf{F}_m$) in a carrier fluid is given by:

$$\mathbf{F}_{\mathrm{m}} = \frac{V(\chi_p - \chi_f)}{\mu_0}(\mathbf{B} \bullet \nabla)\mathbf{B} \tag{2.110}$$

where $V$ is the volume of the particle, $\chi_p$ and $\chi_f$ refer to particle and fluid susceptibility, and $\mu_0$ is the permeability of vacuum. Magnetophoretic velocity of a particle is obtained by balancing the friction force of the particle:

$$\mathbf{u} = \frac{d^2(\chi_p - \chi_f)(\mathbf{B} \bullet \nabla)\mathbf{B}}{18\mu_0\eta} = \frac{1}{\mu_0}\zeta(\mathbf{B} \bullet \nabla)\mathbf{B} \tag{2.111}$$

where $\zeta$ is called the magnetophoretic mobility. Owing to the fact that the magnetophoretic velocity is a function of size and magnetic susceptibility, separation of magnetic and diamagentic particles, and also magnetic particles in different diameters is feasible using magnetophoresis. This has enabled the development of various microfluidic devices for the sorting, trapping, and separation of the particles [50, 51, 52].

## 2.6 OPTOFLUIDICS

Optofluidics is known as the field of integrating photonics with microfluidics [53]. Optofluidic devices take advantage of the fluid properties available at small spatial scales to manipulate light traveling through the fluid. To this end, optofluidics changes the optical property of the fluid medium in a microfluidic device by replacing one fluid with another, creating optically smooth interfaces between two immiscible fluids, and producing gradients in optical properties by controlled diffusion between multiple streams of immiscible fluids [54]. Thus, optofluidic devices can be categorized into three major groups: whole fluid-based systems that only optical properties of fluids are important, hybrid structures of solid-liquid where optical properties of both media are critical, and colloid-based systems where the operation of the devices relies on employing unique properties of a colloidal solution or manipulation of solid particles in fluid [54].

Taking advantage of optofluidics, various optical devices such as lenses, prisms, lasers, switches, and waveguides in microfluidic systems have been realized [53]. The main goal of optofluidics was to build optical components within a microfluidic systems using fluidic features to increase the portability and maneuverability of on-chip processes. However, in recent years, a major focus has been occurring on the use of microfluidics for image capture, known as optofluidic imaging, such as 2-D imagery and 3-D tomography [55]. In this section, examples of optofluidic lenses are discussed to illustrate the basics of optofluidics.

### 2.6.1 Optofluidic Lenses

The operation of optofluidic lenses relies on changing refractive index or geometry of the fluids. By regulating the refractive index, the amplitude, the phase, and the polarization of the light passing through the liquid lens is modulated [56].

Controlled diffusion between colaminar flows results in a concentration gradient within a microfluidic channel. Concentration gradient creates a gradient of refractive index in the optical liquid. Figure 2.21(a) shows an optofluidic lens realized by controlled diffusion of calcium chloride ($CaCl_2$), as core flow, to DI water as cladding streams. 3.5 M $CaCl_2$ has a refractive index of 1.41 while the refractive index of water is 1.33. Thus, through diffusion of $CaCl_2$ to DI water, the refractive index decreases gradually from 1.41 to 1.33.

The concentration gradient can be adjusted by the flow rate of the cladding streams and the core stream. Peclet number is used to calculate the relative ratio between the advection transport and molecular diffusion:

$$\mathrm{Pe} = \frac{UL_{\mathrm{ch}}}{D} \tag{2.112}$$

where $U$ is the mean flow velocity, $L_{ch}$ is considered as the characteristic diffusion length, and $D$ is the coefficient of molecular diffusion. Once the flow rate is increased, the advective transport dominates over molecular diffusion. Thus, a larger gradient of concentration is produced once two or multiple colaminar streams flow in a microchanel. This results in a larger gradient of refractive index that enables a sharper bend of light beams and a shorter focal point in the direction of the flow. Also, the focal point can be modulated in the perpendicular direction to the flow.

Figure 2.21(b) shows that increasing the flow rate ratio of cladding streams pushed up the focal point. In this way, the concentration of $CaCl_2$ is higher in lower half of the channel, allowing the light to swing up.

Another method to establish a lens is to create a curved liquid-air interface. Such pneumatically tunable in-plane liquid lens using liquid-gas interface can be produced either in two separate chambers or in a single compartment with direct liquid-air contact. Figure 2.21(c) shows a lens in a one-compartment configuration. The air pressure is adjusted according to the liquid flow rate to keep the liquid lens in place as a stable curved interface. There is a force balance between the liquid pressure, air pressure and the interfacial tension.

In addition, tuning of the lens can be achieved using dielectrophoresis [Figure 2.21(d)]. A dielectric liquid, once located in a gradient of electric field, experiences a net body-force. The dielectric force density is expressed as:

$$\mathbf{f} = \frac{\epsilon_0}{2}\nabla[(\epsilon_1 - \epsilon_2)|\mathbf{E}|^2] \tag{2.113}$$

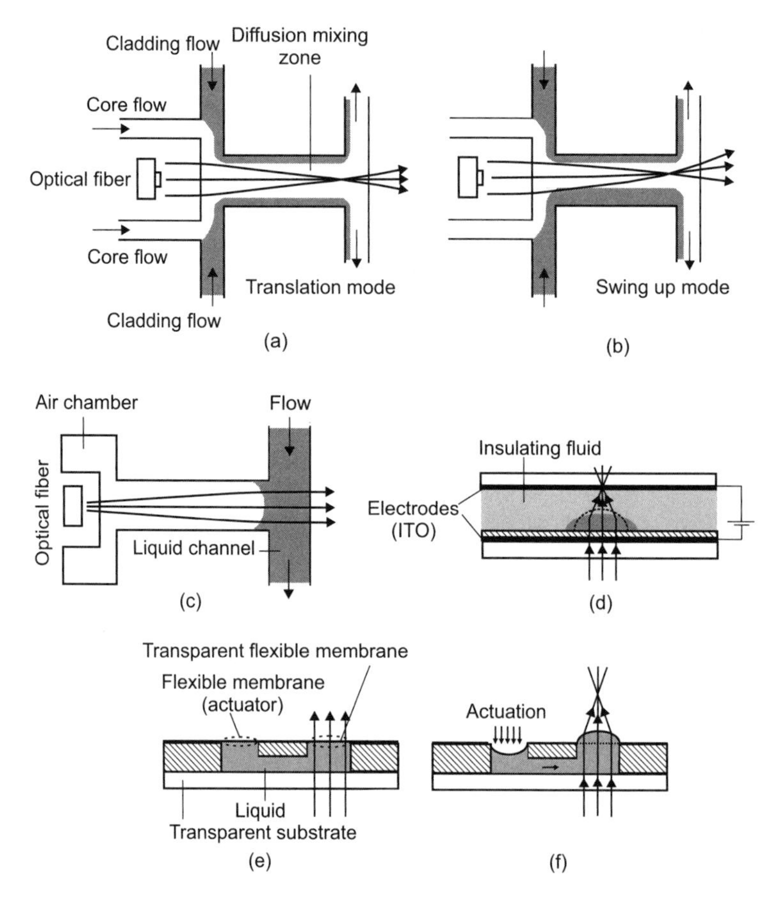


**Figure 2.21** Micro-optofluidic lenses: (a) a micro-optofluidic device to generate a gradient of refractive index at translation mode; (b) operation of (a) at swing up mode; (c) pneumatically tunable in-plane planoconcave liquid lens; (d) tunable liquid lens using dielectrophoresis; (e) a two-chamber tunable liquid lens with implemented actuator; and (f) device shown in (e) upon actuation. (*After:* [56].)

where $\epsilon_0$ is the dielectric constant of vacuum and $\epsilon_1$ and $\epsilon_2$ are the relative dielectric constants of the insulating fluid and the lens fluid, respectively. **E** is the strength of the electric field across the liquid/liquid interface.

Pneumatically tunable out-of-plane liquid lenses use external pressure supply to tune a lens shape. The lenses consist of a circular liquid chamber covered by a flexible membrane. The membrane radius of curvature and the lens focal point is adjusted upon applying pressure change in the liquid chamber [Figure 2.21(e)]. The regulation of liquid pressure can be achieved using another flexible membrane deformed by an actuation scheme. Electroactive polymers (EAP), electromagnetic field, and pneumatics have been employed to generate required actuation pressure to deform the lens membrane [Figure 2.21(f)].

Hydrodynamic forces also can be employed to form and control a liquid-liquid interface. Once a liquid core flow is sandwiched between liquid cladding flows, a hydrodynamic focusing for an optofluidic system is formed, Figure 2.22(a).

Equation (2.114) represents the relationship between the widths of the core and cladding streams and the corresponding flow rate ratio:

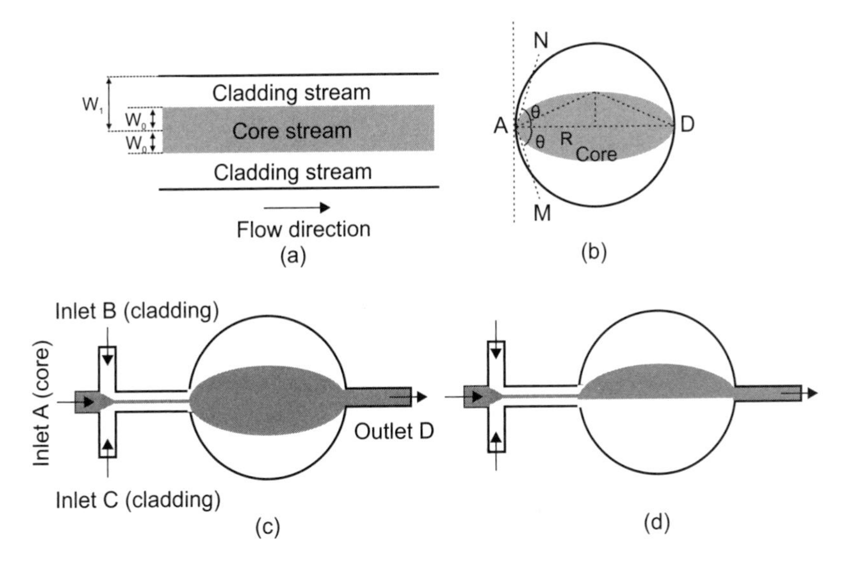


**Figure 2.22** Hydrodynamic focusing for optofluidic lens: (a) straight channel; (b) circular chamber; and (c), (d) in-plane micro-optofluidic lens formed in a circular chamber with hydrodynamic tuning. (*After:* [56].)

$$\frac{\mu_{\text{core}}}{\mu_{\text{cladding}}}\frac{Q_{\text{core}}}{Q_{\text{cladding}}}=\frac{2W_0}{W_1-W_0} \tag{2.114}$$

where $\mu$ represents the viscosity of the fluid and $Q$ is the flow rate. The width of the streams are shown in Figure 2.22(a).

In addition to straight channels, a circular chamber can be used for flow focusing [Figure 2.22(b)]. In this flow configuration, the streamlines spread from point A and flow back to point D as a sink with circular bounded domain. The radius of the liquid-liquid interface and its shape can be tuned by modulating the flow rates [Figures 2.22(c), (d)].

Similar to straight channels, the opening angles of the core and cladding streams are determined by flow rate ratio and viscosity ratio as expressed in the following equation:

$$\frac{\mu_{\text{core}}}{\mu_{\text{cladding}}}\frac{Q_{\text{core}}}{Q_{\text{cladding}}}=\frac{2\theta}{\pi/2-\theta} \tag{2.115}$$

The radius of the interface can be analytically described as:

$$\frac{Q_{\text{core}}}{Q_{\text{cladding}}}=\frac{\mu_{\text{cladding}}}{\mu_{\text{core}}}\frac{2\,arctan(r/R-\sqrt{r^2-R^2}/R)}{\pi/4-arctan(r/R-\sqrt{r^2-R^2}/R)} \tag{2.116}$$

where $r$ is the radius of the lens interface.

Once the radius of curvature is known, the relationship between the flow rate of core and cladding streams and the focal length is explained as [57]:

$$\frac{Q_{\text{core}}}{Q_{\text{cladding}}}=\frac{\mu_{\text{cladding}}}{\mu_{\text{core}}}\frac{2\,arctan[(2(n-1)f-\sqrt{4(n-1)^2f^2-R^2})/R]}{\pi/4-arctan[(2(n-1)f-\sqrt{4(n-1)^2f^2-R^2})/R]} \tag{2.117}$$

where $f$ is the focal length and $n$ represents the relative index of core stream. Equation (2.117) shows that the focal length is a function of flow rate ratio and the diameter of the lens chamber.

Also, there are few other schemes to form and manipulate liquid lens that can be studied in [56].

## 2.7 MICROACOUSTOFLUIDICS

Microacoustofluidics is the use of acoustic waves in microfluidic devices for manipulating fluids and particles at microscale. It mainly relies on the use of acoustic shock waves traveling on the surface of a solid substrate [58], known as *surface acoustic waves* (SAW). SAWs are categorized into *traveling surface acoustic waves* (TSAWs) and *standing surface acoustic waves* (SSAWs). In general, SAWs are mechanical waves with typical amplitudes of few nanometers [58]. The generation of surface acoustic waves is mainly achieved through implementation of interdigital electrodes (IDT) on top of a piezoelectric layer [58] [Figure 2.23(a)]. Strong SAWs can induce acoustic pressure over droplets to actuate and move them once located over the hydrophobic surface of the chip [58].

SAW has relatively slow speed in comparison with the bulk wave speed of sound in the medium itself [59]. It causes the entrapment of wave and its energy on the surface within a few wavelengths [59]. The acoustic shock waves moving on the surface can translate a liquid plug over the substrate or propel it through an enclosed channel implemented over the substrate [59, 60] [Figures 2.23(b), (c)].

In practice, the shock waves induce a stream on the solid-liquid interface that results in the movement of the drop [Figure 2.23(d)]. To predict the effect of acoustic streaming on liquid behavior, the motion of piezoelectric substrate along with the equations describing fluid motion (Navier-Stokes equations) should be solved. Detailed explanations about the flow behavior of acoustofluidics can be found in references [61, 62].

The maximum induced vibration velocity is of the order of 1 m/s; however, extreme accelerations on the order of at least $10^8 \mathrm{m}/s^2$ can be achieved. Such high accelerations are the outcome of the frequency and the vibrating velocity. Taking advantage of the acceleration induced by TSAW, astonishing unique inertial behavior such as jetting and atomization from fluids can be obtained. Jetting occurs when the streaming in a drop is restricted and directed toward the drop free surface with adequate inertia. To this end, the streaming overwhelms the capillary stress and deforms the drop interface. Hence, an extrusion of a thin liquid jet from a drop is formed.

At higher powers of actuation, the acoustic streaming can destabilize the drop surface with its subsequent breakup to form 1-$\mu$m-order aerosol droplets. This phenomenon is known as atomization that happens at powers typically of the order of 1W [59, 60]. Vibrating forces of SAW induces an internal liquid recirculation inside a droplet that has enabled on-chip SAW mixing and microcentrifugation schemes [59].

Standing surface acoustic waves are produced once identical traveling SAWs propagate at opposite directions [63]. Figure 2.24(a) shows a one-dimensional SSAW field that can be realized by implementing a pair of IDTs on a single substrate. Two-dimensional SSAW field can be formed by producing SAWs using two pairs of IDTs implemented on the substrate in an orthogonal arrangement. Once exposed to a SSAW field, immersed particles in a liquid experience primary radiation forces that push them to pressure nodes or pressure anti-nodes.

Figure 2.24(b) shown focusing of beads in a single line [64]. In this case, a SSAW was established across the fluidic channel with a pressure node situated at the center of the channel. Beads are pushed to the center of the microchannel by radiation acoustic forces once exposed to the SSAWs. The primary acoustic radiation forces have been extensively employed for various applications including focusing a flow of particles into a single line, differentiating particles into multiple streams based on their properties, manipulating single cell/particle or a group of cells/particles in a stagnant fluid, and protein manipulation [63].

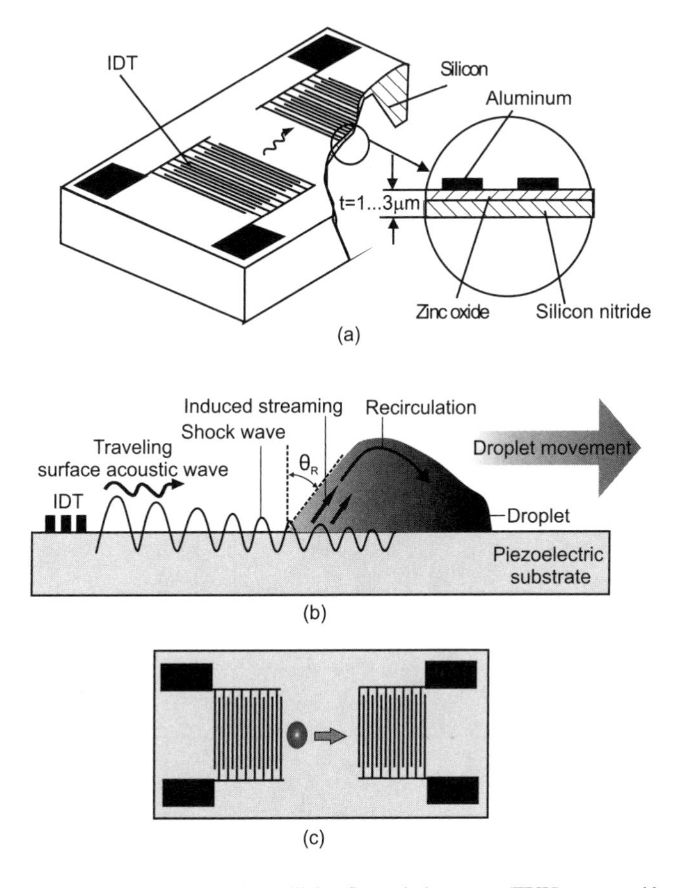


**Figure 2.23** Microacoustofluidics: (a) a chip design utilizing flexural plate wave (FPW) generated by IDTs; (b) schematics of traveling surface acoustic wave (TSAW) propagation to translate liquid droplet over the substrate; (c) top view of (b) for actuation of liquid droplet by TSAWs. (*After:* [58].)

## 2.8 CONCLUSION

Gases and liquids, although very different in nature, together comprise the class of matter called fluids. In this chapter we have explored the range of parameters for which various types of fluid flow analysis are accurate. For both gas and liquid flows of moderate size and flow parameters, the standard continuum analysis as taught in most fluid mechanics courses is sufficient. However, with both gases and liquids in microscopic domains, flow behavior can often deviate from that expected according to macroscopic intuition. When and in what manner gases and liquids deviate from typical macroscopic behavior depends strongly on whether the fluid is a gas or a liquid. The behavior for gases and liquids can be summarized as follows.

For gases, the Knudsen number is the parameter that predicts how the gas will flow when acted upon by some external force.

- When the Knudsen number is smaller than 0.001, a traditional continuum approach yields sufficient accuracy for most situations. The assumption that the flow sticks to the wall (i.e., the no-slip condition) is reasonably accurate.
- When the Knudsen number is larger than 0.001 but smaller than 0.1, the gas still behaves as a continuum away from the flow boundaries. Near the flow boundaries, the molecular nature of the gas becomes apparent as the molecules begin to slip along the surface of the flow boundary

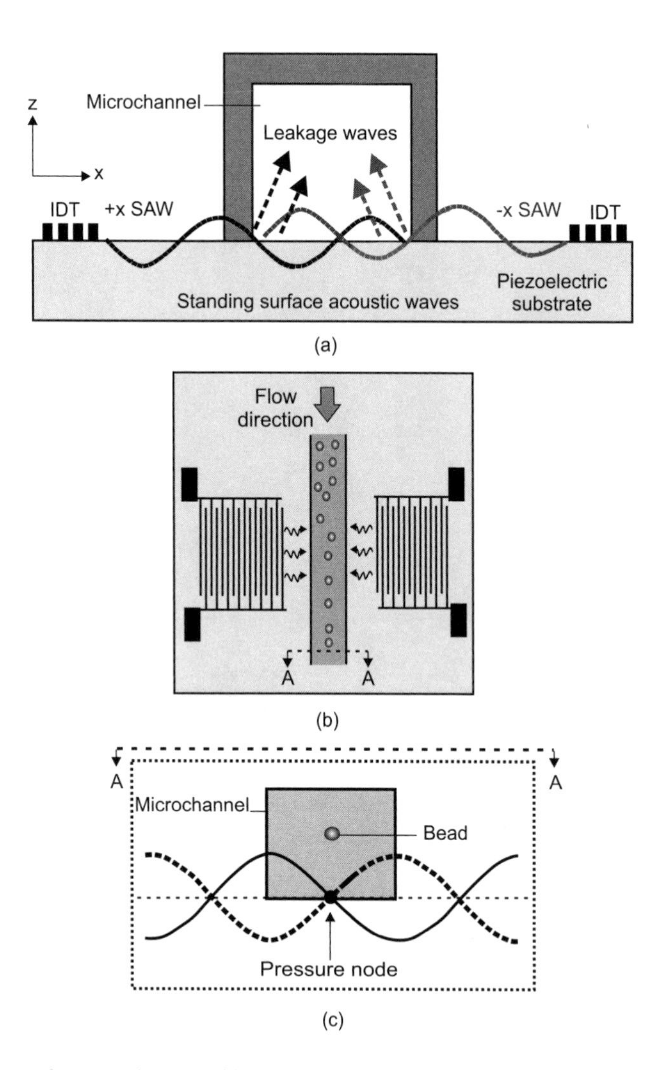


**Figure 2.24** Standing surface acoustic waves (SSAWs): (a) chip design with a microchannel and two IDTs for generation of SAWs at opposite directions; (b) on-chip beads focusing using TSAWs; and (c) cross section of channel shown in (b) representing pressure node located in the microchannel. [(a) *After:* [63], (b) and (c) *After:* [64].]

and no longer assume the temperature of the boundary. A model is presented that relates the degree of velocity slip and temperature jump at the wall to the velocity gradient, mean free path, accommodation coefficients, and temperature gradients. With this boundary condition modification, the traditional continuum analysis can be used for these Knudsen numbers.

- When the Knudsen number is larger than 0.1 and smaller than 10, the gas is in a transitional regime where it does not behave as a continuum, even in the bulk of the gas, and it may still have too many collisions to be regarded as a free molecular flow. For this range of parameters, the flow can be analyzed by a computational method called the DSMC technique.

- When the Knudsen number is larger than 10, the gas can be considered to be a free molecular flow in which the collision of two gas molecules is rare. The solution of these types of problems is not addressed in this chapter.

For liquids, the situation is somewhat more complicated. There is no parameter for liquids comparable to the Knudsen number for gases to use as a basis for determining how the liquid will respond to external forces.

- In most cases, a liquid will behave as a continuum and observe no-slip and no-temperature-jump at a boundary because of the tight spacing of the liquid's molecules. The typical continuum analysis can be used except when special circumstances exist as described in the next few points.
- An argument is presented based on having a population of molecules sufficiently large to prevent noticeable statistical variations in point quantities. When the length scale of a flow of simple molecules is larger than about 10 nm, the liquid should behave as a continuum. For more complicated molecules, the argument should be modified based on the types of molecules being considered.
- When the liquid is sheared especially vigorously—at shear rates greater than twice the molecular interaction frequency—it will cease to behave as a Newtonian fluid. At that point, only computational means can be used to analyze the flow.
- MD simulations show that the correct boundary condition for liquid flows is the constant slip length Navier boundary condition, where the slip length, measured in molecular diameters, is too small to be noticed (i.e., on the scales of microns).
- The MD simulations also show that the Navier boundary condition will begin to break down and allow slip lengths at the wall as the shear rate is increased beyond a value about two orders of magnitude smaller than that at which the Newtonian approximation breaks down.

Small-scale domains also have plenty of advantages. Liquid flows can be manipulated by electrical and magnetic fields and surface acoustic waves. Once liquid and channel have the correct chemical composition, an EDL at the wall of the channel is formed. EDL can be manipulated by an applied electric field to accomplish electro-osmotic pumping. A similar phenomenon occurs for particles immersed in liquids when the particle and the liquid have the correct chemical relationship. They can be moved using the EDL and an electrical field in a process called electrophoresis. Even if the surface chemistry for a particular combination of liquid and particle is not favorable for the formation of an EDL, if the particles are polarizable, they can be manipulated by an ac electric field in a process called dielectrophoresis. As for micromagnetofluidics, the flow behavior is manipulated based on the size and magnetic properties of immersed particles within the magnetic fluid. Once the magnetic particles are at the order of 10 nm, ferrofluid is formed that can be easily manipulated as bulk liquid in microchannels or as individual microdroplets. If the size of magnetic particles are located within the range of 100 nm to 1 $\mu$m, magnetoviscous fluid is produced whose viscosity is changed according to the strength of the magnetic field. If the dimension of the magnetic particle is about or beyond 1 $\mu$m, they can be manipulated individually once exposed to a magnetic field gradient. This phenomenon is known as magnetophoresis that has been widely employed for sorting, separation, and detection purposes.

Micro-optofluidics is the fusion of microfluidic technology with optics to fabricate optical elements using unique spatial fluid characteristics at microscale. In particular, optofluidic lenses have been widely developed for integration with microfluidic chips for on-chip optical applications. Attention has been paid to employ optofluidic principles for high-resolution 2-D and 3-D imaging.

Acoustofluidics is mainly based on the manipulation of acoustic waves at the interface of a solid substrate in contact with a liquid plug or drop. The energy of the acoustic waves creates streaming within a liquid plug located over the acoustofluidic substrate. The transferred energy is used to actuate the liquid plug and manipulate the suspended particles within the plug.

## Problems

**2.1** At what separation distance can one molecule influence another molecule? Provide proof of your statement. Plot the intermolecular potential and force versus separation distance for $N_2$.

**2.2** Calculate all the significant parameters to describe a flow of diatomic nitrogen $N_2$ at 350K and 200 kPa at a speed of 100 m/s through a two-dimensional slot measuring 100 mm in height and infinite in depth. Classify the type of flow that you will have based on these parameters and how you would solve this problem.

**2.3** Develop a spreadsheet containing the formulae used in Problem 2.2. Using this spreadsheet, repeat the analysis in Problem 2.2 for channel dimensions of 10 mm, 1 mm, 100 nm, 10 nm, and 1 nm. Classify the types of flows encountered for these operating parameters.

**2.4** Consider a flow of diatomic nitrogen $N_2$ at 350K and 200 kPa at a speed of 100 m/s through a two-dimensional slot measuring 1 mm in height and infinite in depth. Calculate the slip length at the wall.

**2.5** Consider that the flow in the slot from Problem 2.4 is not moving—the pressure gradient has been set to zero. Instantaneously, the wall temperature is modified so that it varies from 100K to 1,000K within the space of 1 mm. What is the slip velocity at the wall at the instant the flow begins moving?

**2.6** Compute the velocity profiles for the conditions in Problem 2.4 given the two channel heights of 100 mm and 1 mm. What is the slip velocity at the wall in each case?

**2.7** Consider a simple liquid such as water. At what shear rates do the no-slip condition and the Newtonian assumption break down? Working in terms of variables, develop expressions for the electrophoretic particle velocity of both a small and large particle in a channel that exhibits electro-osmotic flow. Assume that the channel and particle can be made of different materials.

**2.8** Develop an expression for the velocity profile in an electro-osmotic flow in a long, two-dimensional slot in which both ends are blocked. Assume that you are looking at the flow far from the blocked ends. Hint: The blocked ends imply the existence of a pressure gradient that causes the net flow across any cross section to be zero.

## References

[1] Merriam-Webster Dictionary.

[2] Fox, R. W., and McDonald, A. T., *Introduction to Fluid Mechanics*, New York: Wiley, 1999.

[3] White, F. M., *Viscous Fluid Flow*, Boston, MA: McGraw-Hill, 1991.

[4] Batchelor, G. K., *An Introduction to Fluid Dynamics*, Cambridge, U.K.: Cambridge University Press, 2000.

[5] Panton, R. L., *Incompressible Flow*, New York: Wiley, 1996.

[6] Koplik, P. J., and Banavar, J. R., "Continuum Deductions from Molecular Hydrodynamics," *Annu. Rev. Fluid Mech. 1995*, Vol. 27, pp. 257-292.

[7] Karniadakis, G. E., and Beskok, A., *Microflows*, New York: Springer, 2002.

[8] Gad-el-Hak, M., "Flow Physics", in M. Gad-el-Hak (ed.), *The MEMS Handbook*, Boca Raton, FL: CRC Press, 2002.

[9] Hirschfelder, J., Gurtiss, C., and Bird, R., *Molecular Theory of Gases and Liquids*, New York: Wiley, 1954.

[10] Deen, W. M., *An Analysis of Transport Phenomena*, Oxford, U.K.: Oxford University Press, 1998.

[11] Bird, G. A., *Molecular Gas Dynamics and the Direct Simulation of Gas Flows*, Oxford, U.K.: Clarendon Press, 1994.

[12] Vincenti, W., and Kruger, C., *Introduction to Physical Gas Dynamics*, Huntington, New York: Robert E. Krieger Publishing Company, 1977.

[13] Loose, W., and Hess, S., "Rheology of Dense Fluids via Nonequilibrium Molecular Hydrodynamics: Shear Thinning and Ordering Transition," *Rheologica Acta*, Vol. 28, 1989, pp. 91–101.

[14] Goldstein, S., "Fluid Mechanics in First Half of This Century" *Ann. Rev. Fluid Mech.*, Vol. 1, 1969, pp. 1-28.

[15] Lauga, E., Brenner, M. P., and Stone, H. A., "Microfluidics: The No-Slip Boundary Condition," in Tropea, C., Yarin, A. L., Foss, J. F., (eds) *Handbook of Experimental Fluid Dynamics*, Berlin: Springer, 2007.

[16] Thompson, P. A., and Troian, S. M., "A General Boundary Condition for Liquid Flow at Solid Surfaces," *Nature*, Vol. 389, 1997, pp. 360–362.

[17] Navier, C. L. M. H, "Memoire sur les lois du Mouvement des Fluides," *Mem. Acad. Roy. Sci. Inst. France*, Vol. 1, 1823, pp. 389-440.

[18] Maxwell, J. C., "On Stresses in Rarefied Gases Arising from Inequalities of Temperature," *Philosophical Transactions of the Royal Society Part 1*, Vol. 170, 1879, pp. 231–256.

[19] Smoluchowski, von M., "Über Wärmeleitung in verdünnten Gasen," *Annalen der Physik und Chemie*, Vol. 64, 1898, pp. 101–130.

[20] Seidl, M., and Steinheil, E., "Measurement of Momentum Accomodation Coefficients on Surfaces Characterized by Auger Spectroscopy, SIMS and LEED," *Proceedings of the Ninth International Symposium on Rarefied Gas Dynamics*, 1974, pp. E9.1–E9.2.

[21] Arkilic, E., "Measurement of the Mass Flow and Tangential Momentum Accommodation Coefficient in Silicon Micromachined Channels," Ph.D. thesis, Massachusetts Institute of Technology, 1997.

[22] Berker, A. R., "Intégration des Équations du Movement d'un Fluide Visqueux Incompressible," in S. Flgge (ed.), *Encyclopedia of Physics*, Vol. 8, No. 2, Berlin: Springer, 1963, pp. 1–384.

[23] Shah, R. K., and London, A. L., *Laminar Flow Forced Convection in Ducts*, New York: Academic, 1978.

[24] Sharp, K. V., et al., "Liquid Flows in Microchannels," in M. Gad-el-Hak (ed.), *The MEMS Handbook*, Boca Raton, FL: CRC, 2001.

[25] Lee, S. Y., et al., "Microchannel Flow Measurement Using Micro-PIV," *Proceedings of IMECE2002 ASME International Mechanical Engineering Congress & Exposition*, New Orleans, Louisiana, November 17–22, 2002, IMECE2002-33682.

[26] Wu, P., and Little, W. A., "Measurement of Friction Factors for the Flow of Gases in Very Fine Channels Used for Microminiature Joule-Thomson Refrigerators," *Cryogenics*, Vol. 23, 1983, pp. 273-277.

[27] Peng, X. F., Peterson, G. P., and Wang, B. X., "Frictional Flow Characteristics of Water Flowing Through Rectangular Microchannels," *Exp. Heat Transfer*, Vol. 7, 1994, pp. 249-264.

[28] Israelachvili, J., *Intermolecular & Surface Forces*, London: Academic Press, 1992.

[29] Bird, G., *Molecular Gas Dynamics*, Oxford, U.K.: Clarendon Press, 1976.

[30] Beskok, A., "Molecular-Based Microfluidic Simulation Models," in Gad-el-Hak (ed.), *The MEMS Handbook*, Boca Raton, FL: CRC Press, 2002.

[31] Oran, E. S., Oh, C. K., and Cybyk, B. Z., "Direct Simulation Monte Carlo: Recent Advances and Applications," *Annu. Rev. Fluid Mech.*, Vol. 30, 1998, pp. 403–441.

[32] Agarwal, R. K., and Yun, K. Y., "Burnett Simulations of Flows in Microdevices," in Gad-el-Hak (ed.), *The MEMS Handbook*, Boca Raton, FL: CRC Press, 2002.

[33] Piekos, E. S., and Breuer, K. S., "Numerical Modeling of Micromechanical Devices Using the Direct Simulation Monte Carlo Method," *J. Fluids Eng.*, Vol. 118, 1996, pp. 464–469.

[34] Beskok, A., and Karniadakis, G.E., "Modeling Separation in Rarefied Gas Flows," *28th AIAA Fluid Dynamics Conference*, AIAA 97-1883, Snowmass Village, CO, June 29–July 2, 1997.

[35] Probstein, R. F., *Physicochemical Hydrodynamics: An Introduction*, 2nd ed., New York: Wiley, 1994.

[36] Hunter, R. J., *Zeta Potential in Colloid Science*, London: Academic Press, 1981.

[37] Devasenathipathy, S., and Santiago, J. G., "Diagnostics for Electro-Osmotic Flows," in Breuer, K. (ed.), *Microflow Diagnostics*, New York: Springer, 2002.

[38] Shaw, D. J., *Introduction to Colloid and Surface Chemistry*, 3rd ed., London: Butterworth's, 1980.

[39] Roberts, M. A., et al., "UV Laser Machined Polymer Substrates for the Development of Microdiagnostic Systems," *Analytical Chemistry*, Vol. 69, 1997, pp. 2035–2042.

[40] Cummings, E. B., et al., "Conditions for Similitude Between the Fluid Velocity and Electric Field in Electro-Osmotic Flow," *Anal. Chem.*, Vol. 72, 2000, pp. 2526–2532.

[41] Pohl, H., *Dielectrophoresis*, Cambridge, U.K.: Cambridge University Press, 1978.

[42] Ramos, A., et al., "AC Electrokinetics: A Review of Forces in Microelectrode Structures," *J. Phys. D: Appl. Phys.*, Vol. 31, 1998, pp. 2338–2353.

[43] Gomez, R. et al., "Microfabricated Device for Impedance-Based Detection of Bacterial Metabolism," *Materials Research Society Spring 2002 Meeting*, San Francisco, CA, April 1-5, 2002.

[44] Nguyen, N. T., "Micro-Magnetofluidics: Interactions Between Magnetism and Fluid Flow on the Microscale," *Microfluidics and Nanofluidics*, Vol. 12, 2012, pp. 1-16.

[45] Zhang, Y., Nguyen, N. T., "Magnetic Digital Microfluidics: A Review," *Lab Chip*, Vol. 17, No. 6, 2017, pp. 251-256.

[46] Ando, B., et al., "The One Drop Ferrofluidic Pump With Analog Control," *Sens Actuators A: Phys*, Vol. 156, No. 1, 2009, pp. 251-256.

[47] Ando, B., et al., "Ferrofluidic Pumps: a Valuable Implementation without Moving Parts," *IEEE Trans Instrum Meas*, Vol. 58, No. 9, 2009, pp. 3232-3237.

[48] Nguyen, N. T., et al., "A stepper Micropump for Ferrofluid Driven Microfluidic Systems," *Micro Nanosyst*, Vol. 1, No. 1, 2009, pp. 3232-3237.

[49] Love, L. J., et al., "A magnetocaloric Pump for Microfluidic Applications," *IEEE Transactions on NanoBioscience*, Vol. 3, 2004, pp. 1-16.

[50] Pamme, N., "On-chip Bioanalysis with Magnetic Particles," *Current Opinion in Chemical Biology*, Vol. 16, 2012, pp. 1-16.

[51] Rikken, R. S. M., et al., "Manipulation of Micro- and Nanostructure Motion with Magnetic Fields," *Soft Matter*, Vol. 10, 2014, pp. 1295-1308.

[52] Hejazian, M., et al., "Lab on a Chip for Continuous-Flow Magnetic Cell Separation," *Lab Chip*, Vol. 15, 2015, pp. 959-970.

[53] Pang, L., et al., "Optofluidic Devices and Applications in Photonics, Sensing and Imaging," *Lab Chip*, Vol. 12, 2012, pp. 3543-3551.

[54] Psaltis, D., et al.,"Developing Optofluidic Technology Through the Fusion of Microfluidics and Optic," *Nature*, Vol. 442, 2006, pp. 381-386.

[55] Zhao, Y., et al., "Optofluidic Imaging: Now and Beyond," *Lab Chip*, Vol. 13, 2013, pp. 17-24.

[56] Nguyen, N. T., "Micro-optofluidic Lenses: A review," *Biomicrofluidics*, Vol. 4, 2010, pp. 1-15.

[57] Song, C., et al., "Modelling and Optimization of Micro Optofluidic Lenses," *Lab Chip*, Vol. 9, 2009, pp. 1178-1184.

[58] Mark, D., et al., "Microfluidic lab-on-a-chip Platforms: Requirements, Characteristics and Applications," *Chem. Soc. Rev.*, Vol. 39, 2010, pp. 1153-1182.

[59] Yeo, L. Y., and Friend, J. R., "Surface Acoustic Wave Microfluidics," *Annual Review of Fluid Mechanics*, Vol. 46, 2014, pp. 379-406.

[60] Yeo, L. Y., and Friend, J. R., "Ultrafast Microfluidics Using Surface Acoustic Waves," *Biomicrofluidics*, Vol. 3, 2009, pp. 1-23.

[61] Bruus, H., "Acoustofluidics 1: Governing Equations in Microfluidics," *Lab Chip*, Vol. 11, 2011, pp. 3742-3751.

[62] Tan, M. K., et al., "Capillary Wave Motion Excited by High Frequency Surface Acoustic Waves," *Physics of Fluids*, Vol. 22, 2010, pp. 1-22.

[63] Ding, X., et al., "Surface Acoustic Wave Microfluidics," *Lab Chip*, Vol. 13, 2013, pp. 3626-3649.

[64] Shi, J., et al., "Focusing Microparticles in a Microfluidic Channel with Standing Surface Acoustic Waves (SSAW)," *Lab Chip*, Vol. 8, 2008, pp. 221-223.

# Chapter 3

# Fabrication Techniques for Microfluidics

## 3.1 BASIC MICROTECHNIQUES

This section aims to provide the readers an overview of basic microtechniques, which are not only used for microfluidics but also for microelectronics and MEMS. Refer to [1] for an excellent reference describing all these techniques in detail. It is essential to mention that some of the fabrication techniques for microfluidics are performed in a clean room with controlled particle number and particle size to prevent their adverse effect on the fabrication quality. In a clean room, other potential contaminations such as chemical vapors and particular light wavelengths are also kept away from the microfabrication process.

### 3.1.1 Photolithography

Lithography is the most important technique for fabricating microscale structures. Depending on the type of energy beam, lithography techniques can be further divided into photolithography, electron lithography, X-ray lithography, and ion lithography [2]. Photolithography and X-ray lithography for LIGA[1]-technique (see Section 3.3.1.1) are the most relevant techniques for the fabrication of microfluidic devices. The patterning process with photolithography is limited to two-dimensional (2-D), lateral structures. This technique uses a photosensitive emulsion layer called resist, which transfers a desired pattern from a transparent mask to the substrate. The mask is a transparent glass plate with metal (chromium) patterns on it. For microfluidic applications with relatively large structures, a mask printed on a plastic transparency film by a high-resolution imagesetter is an option for low-cost and fast prototyping. Photolithography consists of three process steps:

- *Positioning process:* Lateral positioning of the mask and the substrate, which is coated with a resist, adjusting the distance between the mask and the substrate.
- *Exposure process:* Optical or X-ray exposure of the resist layer, transferring patterns to the photoresist layer by changing properties of exposed area.
- *Development process:* Dissolution (for negative resist) or etching (for positive resist) of the resist pattern in a developer solution.

Generally, photolithography is categorized as contact printing, proximity printing, and projection printing. In the first two techniques, the mask is brought close to the substrate. Contact printing lets the mask even touch the photoresist layer. The resolution $b$ depends on the wavelength $\lambda$ and the distance $s$ between the mask and the photoresist layer [2]:

$$b = 1.5\sqrt{\lambda s} \tag{3.1}$$

1 German acronym of "Lithographie, Galvanoformung, Abformung."

**Table 3.1**
Spectrum of Mercury Lamps

| *Types* | *I-line* | *H-line* | *G-line* | *E-line* | — | — | — |
|---|---|---|---|---|---|---|---|
| Wavelength (nm) | 365.0 | 404.7 | 435.8 | 546.1 | 577.0 | 579.1 | 623.4 |

**Example 3.1: Resolution of Proximity Photolithography**
A resist layer at the bottom of a 5-mm-deep channel and a 20-mm-deep channel is to be patterned. The photoresist is exposed to ultraviolet (UV) light of a 400-nm wavelength. Compare the resolutions at the bottom of the two channels.
**Solution.** Following (3.1), the resolutions at the bottom of the two channels can be estimated as:

$$b_1 = 1.5\sqrt{\lambda s_1} = 1.5\sqrt{0.4 \times 5} = 2.1 \ \mu\text{m}$$
$$b_2 = 1.5\sqrt{\lambda s_2} = 1.5\sqrt{0.4 \times 20} = 4.2 \ \mu\text{m}$$

Deeper channels cause larger distances and "blurrier" images.

The resolution of projection printing system is estimated as:

$$b = \frac{\lambda}{2\text{NA}} \tag{3.2}$$

where NA is the numerical aperture of the imaging lens system. While contact printing and projection printing offer a resolution on the order of 1 μm, proximity printing has a lower resolution on the order of several microns. Depth of focus of projection printing (0.1 to 0.25 μm) is poor compared to proximity printing (around 5 μm). Most photolithography systems use a mercury lamp as a light source, which has a spectrum as listed in Table 3.1.

Photolithography of thick resists has a big impact in the fabrication of microfluidic devices. Microfluidic devices use thick resists directly as functional material or as a template for polymer molding as well as electroplating of metals. In this case, the lithography process should offer a high aspect ratio, which is the ratio of resist thickness and the smallest structure dimension.

Because of the thickness and the required high aspect ratio, special resists such as SU-8 or high-energy beam such as X-ray should be used. Section 3.3 on polymeric micromachining will discuss these techniques in detail. Thick resist layers can be achieved with multiple spin-coating steps or with viscous resist at a slower spinning speed. Because the best depth of focus offered by proximity printing is on the order of 5 μm, a resist layer thicker than this depth may degrade the resolution. Three is the rule-of-thumb number for the aspect ratio between resist thickness and resolution.

The resolution of photolithography is limited by the wavelength. For the fabrication of nanostructures with a size on the order of tens of nanometers, the interferometric effect can be utilized. This technique is called *interferometric lithography* (IL). IL is a simple process where two coherent light beams interfere to generate a standing wave. The gratings have a spatial period of one-half the wavelength of interfering light. If ultraviolet laser is used, structures smaller than 100 nm can be created by IL.

### 3.1.2 Additive Techniques

#### 3.1.2.1 Chemical Vapor Deposition

Chemical vapor deposition (CVD) is an important technique for creating material films on a substrate. In a CVD process, gaseous reactants are introduced into a reaction chamber. Reactions occur on heated substrate surfaces, resulting in the deposition of the solid product. Other gaseous reaction products leave the chamber. Depending on the reaction conditions, CVD processes are categorized as:

- Atmospheric pressure chemical vapor deposition (APCVD);
- Low-pressure chemical vapor deposition (LPCVD);
- Plasma-enhanced chemical vapor deposition (PECVD).

APCVD and LPCVD involve elevated temperatures ranging from 500°C to 800°C. These temperatures are too high for metals with low eutectic temperature with silicon, such as gold (380°C) or aluminum (577°C). For APCVD or LPCVD processes, some metals such as tungsten are suitable. PECVD processes have a part of their energy in the plasma; thus, lower substrate temperature is needed, typically on the order of 100°C to 300°C.

#### 3.1.2.2 Thermal Oxidation

Although silicon dioxide can be deposited with CVD, thermal oxidation is the simplest technique to create a silicon dioxide layer on silicon. In silicon-based microfluidic devices, thermal oxidation can be used for adjusting gaps such as filter pores or channel width with submicron accuracy. Based on the type of oxidizer, thermal oxidation is categorized as dry oxidation or wet oxidation.

In dry oxidation, pure oxygen reacts with silicon at high temperatures from about 800°C to 1,200°C:

$$\mathrm{Si} + \mathrm{O_2} \rightarrow \mathrm{SiO_2} \tag{3.3}$$

In wet oxidation, water vapor reacts with silicon at high temperatures:

$$\mathrm{Si} + 2\mathrm{H_2O} \rightarrow \mathrm{SiO_2} + \mathrm{H_2} \uparrow \tag{3.4}$$

Based on the densities of silicon and silicon dioxide and their molecular masses, one can estimate the amount of consumed silicon for a certain oxide film, as illustrated in the following example.

**Example 3.2: Thickness of Oxide Layer of Thermal Oxidation**

The density of silicon and silicon dioxide are 2,330 kg/m$^3$ and 2,200 kg/m$^3$, respectively. Molecular masses of silicon and oxygen are 28.09 kg/kmol and 15.99 kg/kmol, respectively. Determine the consumed silicon thickness for a silicon dioxide film of thickness $d$.

**Solution.** For 1 kmol silicon, one will get 1 kmol silicon dioxide. For the same surface area, the ratio of the thickness is equal to the ratio of volume:

$$\frac{d_{\mathrm{Si}}}{d_{\mathrm{SiO_2}}} = \frac{V_{\mathrm{Si}}}{V_{\mathrm{SiO_2}}} = \frac{M_{\mathrm{Si}}/\rho_{\mathrm{Si}}}{d_{\mathrm{SiO_2}}/\rho_{\mathrm{SiO_2}}} = \frac{28.09/2,330}{(28.09 + 2 \times 15.99)/2,200} = 0.44$$

Thus, for growing an oxide thickness $d$, a silicon layer of $0.44 \times d$ is consumed.

#### 3.1.2.3 Physical Vapor Deposition

Physical vapor deposition (PVD) creates material films on a substrate directly from a source. PVD is typically used for deposition of electrically conducting layers such as metals or silicides. The technique covers two physically different methods: evaporation and sputtering.

*Evaporation* deposits thin film on a substrate by sublimation of a heated source material in a vacuum. The vapor flux from the source coats the substrate surface. Based on the heating methods, evaporation can be categorized as vacuum thermal evaporation (VTA), electron beam evaporation (EBE), molecular beam epitaxy (MBE), or reactive evaporation (RE). The first two methods are the most common. VTA uses resistive heating, laser heating, or magnetic induction to elevate the source temperature. In EBE, the electron beam is focused on the target material, which locally melts. Alloys can be deposited with evaporation using two or more material sources. However, evaporation of alloys is not as stable as sputtering.

*Sputtering* uses flux of atoms, which are released from a target material by bombarding it with chemically inert atoms, such as argon atoms. In a strong electromagnetic field, argon gas is ionized and becomes plasma. The positively charged argon atoms are accelerated and bombard the cathode surface, which is the target material. The bombardment knocks out target atoms, which are then condensed on the substrate surface as a thin film. Sputtering can deposit all kinds of materials, such as alloys, insulators, or piezoelectric ceramics. Sputtering is less directional then EBE but has higher deposition rates.

#### 3.1.2.4 Sol-Gel Deposition

In sol-gel deposition processes, solid particles of a polymer compound dissolved in a solvent are spin-coated on the substrate surface. The process forms a gelatinous network on the substrate surface. Subsequent removal of the solvent solidifies the gel, resulting in a solid film. This technique can also deposit various ceramics such as lead zirconate titanate (PZT).

#### 3.1.2.5 Spin Coating

Spin coating is the simplest method to fabricate a film on a substrate. This method is used for coating resists for photolithography. Spin coating is typically used for coating of polymers or chemical precursors to a polymer called prepolymers. A solvent dissolves the material to be deposited. Subsequently, the solution is dripped on the substrate surface. The wafer is spun at high speed on the order of 5,000 rpm. Centrifugal forces caused by the spinning speed, surface tension, and viscosity of the solution determine the thickness of the coated film. A part of the solvent is evaporated during the spinning process, and the rest of it is removed by subsequent baking at temperatures from about 100°C to 200°C. Spin coating results in a relatively planar surface, even for a nonplanar substrate surface. Therefore, this technique can be used for planarization purposes. Besides spin coating, dry lamination, dip coating, spray coating, and electrodeposition can also be used for transferring a resist layer to the substrate surface.

#### 3.1.2.6 Doping

*Ion Implantation.* Ion implantation is one of the most important techniques for microelectronics. The technique adds impurities to semiconductors such as silicon. Adding dopant atoms with three valence electrons, such as boron in silicon (with four valence electrons), creates positively charged carriers called *holes*. Silicon of this type is called p-type or p-Si. Adding dopant atoms with five valance electrons such as phosphorus creates negatively charged carriers called *electrons*. Silicon of this type is called n-type or n-Si.

Ion implantation can be used to fabricate an insulating layer such as silicon dioxide buried in the substrate. At high temperatures, oxygen ion implantation creates an oxide layer with depths ranging from 0.1 to 1 μm from the surface. The crystal defects in silicon above this layer are repaired automatically at high temperatures. The technique is called *separation by implantation of oxygen*

(SIMOX). The single crystalline silicon layer is called *silicon on insulator* (SOI), which has large impacts on both microelectronics and MEMS.

*Drive-In Diffusion.* After ion implantation, dopants are distributed in a layer on the silicon surface. Subsequent annealing redistributes the dopant atoms. This process is based on the diffusion of dopants and is called drive-in diffusion.

#### 3.1.2.7 Silicon on Insulator

For applications in MEMS and microfluidics, a much thicker SOI layer than that resulting from SIMOX is needed. Most of SOI wafers used in MEMS are fabricated with the bonded etched-back silicon on insulator (BESOI) technique. This technique uses two polished silicon wafers with an oxide layer on each. The two wafers are then bonded together using fusion bonding. One wafer is then thinned to the desired thickness using chemical-mechanical polishing (CMP). The major advantages of SOI fabricated with this technique are the following.

- The thickness of SOI is adjustable and allows a thicker structural layer of the device.
- The thickness of silicon dioxide is adjustable.
- The SOI layer has the same quality of the bulk substrate, while SOI from SIMOX may have crystal defects caused by ion implantation.

### 3.1.3 Subtractive Techniques

#### 3.1.3.1 Wet Etching

Wet etching is referred to as etching processes of solid materials in a chemical solution. During these processes, the substrate is dipped in the solution or the solution is sprayed on the substrate. Wet etching processes in microelectronics are mostly isotropic, independent of crystalline orientation. Anisotropic etching is discussed in Section 3.3.1 on bulk micromachining.

Because of the underetching effect, isotropic etching has drawbacks in designing lateral structures. If the etch solution is well stirred, the isotropic etch front has a spherical form. In the case of a microchannel, the channel width also depends on the channel depth [Figure 3.1(a)]. For example, if a microchannel is to be fabricated with isotropic wet etching in glass, the etch rate, channel width, and channel depth should be considered in the mask design. The major advantages of wet etching are:

- High selectivity;
- Relatively planar etching surface;
- High repeatability;
- The etch rate is controllable with etchant concentration.

Isotropic wet etching is often used for removing thin layers or thinning a film. Stirring has a big impact on etching results, because reaction products may work as a barrier between etchants and material surface [Figure 3.1(b)]. Table 3.2 lists some common recipes of wet etching.

#### 3.1.3.2 Dry Etching

*Physical Dry Etching.* Physical dry etching utilizes beams of ions, electrons, or photons to bombard the material surface. The kinetic energy of ions knocks out atoms from the substrate surface. The high beam energy then evaporates the knocked-out material. Almost all materials can be etched by this method. However, the drawbacks are:

- Slow etch rates;

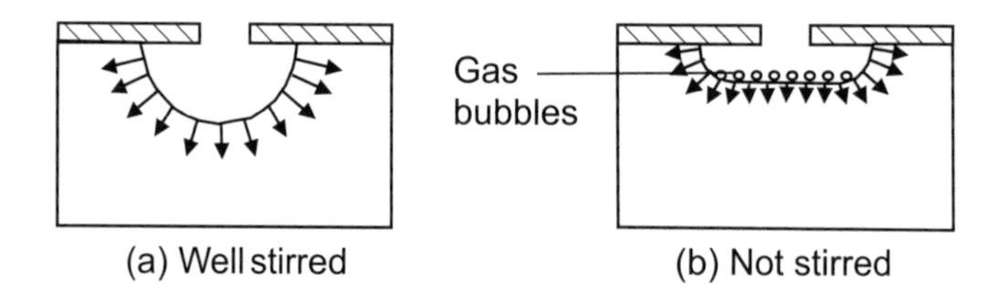


**Figure 3.1** Profile of wet etching: (a) well stirred and (b) not stirred; the lengths of arrows represent the etch rates.

**Table 3.2**
Examples of Wet Etchant Recipes for Thin Films of Functional Materials (*After*: [3].)

| *Material* | *Etchants* | *Selective To* |
|---|---|---|
| Si | HF, $HNO_3$, $CH_3COOH$ | $SiO_2$ |
| Si | KOH | $SiO_2$ |
| $SiO_2$ | $NH_4$, HF | Si |
| $SiO_2$ | HF, $NHO_3$, $H_2O$ | Si |
| $SiO_2$ | $H_3PO_4$, $NHO_3$, $H_2O$ | Si |
| $Si_3N_4$ | $H_3PO_4$ | $SiO_2$ |
| Al | $H_3PO_4$, $HNO_3$, $H_2O$ | $SiO_2$ |

- Low selectivity, because ions attack all materials;
- Trench effects caused by reflected ions.

*Chemical Dry Etching.* Chemical dry etching uses a chemical reaction between etchant gases to attack material surface. Gaseous products are conditions for chemical dry etching because deposition of reaction products will stop the etching process. Chemical dry etching is isotropic. This technique is similar to wet etching and exhibits relatively high selectivity. Etchant gases either can be excited in an RF field to plasma or react directly with the etched material.

Chemical dry etching can be used for cleaning wafers. For instance, photoresist and other organic layers can be removed with oxygen plasma. Table 3.3 lists some typical recipes of dry etchant gases.

*Physical-Chemical Etching.* Dry etching is actually referred to as *physical-chemical etching*. Physical-chemical etching is categorized as reactive ion etching (RIE), anodic plasma etching (APE), magnetically enhanced reactive ion etching (MERIE), triode reactive ion etching (TRIE),

**Table 3.3**
Recipes of Dry Etchant Gases for Thin Films of Functional Materials (*After*: [3].)

| *Material* | *Etchant Gases* | *Selective To* |
|---|---|---|
| Si | $BCl_3/Cl_2$, $BCl_3/CF_4$, $BCl_3/CHF_3$, $Cl_2/CF_4$, $Cl_2/He$, $Cl_2/CHF_3$, HBr, $HBr/Cl_2/He/O_2$, HBr /$NFl_3/He/O_2$, $HBr/SiF_4/NF_3$, HCl, $CF_4$ | $SiO_2$ |
| SiO2 | $CF_4/H_2$, $C_2F_6$, $C_3F_8$, $CHF_3$, $CHF_3/O_2$, $CHF_3/CF_4$, ($CF_4/O_2$) | Si (Al) |
| $Si_3N_4$ | $CF_4/H_2$, ($CF_4/CHF_3/He$, $CHF_3$, $C_2F_6$) | Si ($SiO_2$) |
| Al | $BCl_3$, $BCl_3/Cl_2$, $BCl_3/Cl_2/He$, $BCl_3/Cl_2/CHF_3/O_2$, HBr, $HBr/Cl_2$, HJ, $SiCl_4$, $SiCl/Cl_2$, $Cl_2/He$ | $SiO_2$ |
| Organics | $O_2$, $O_2/CF_4$, $O_2/SF_6$ | — |

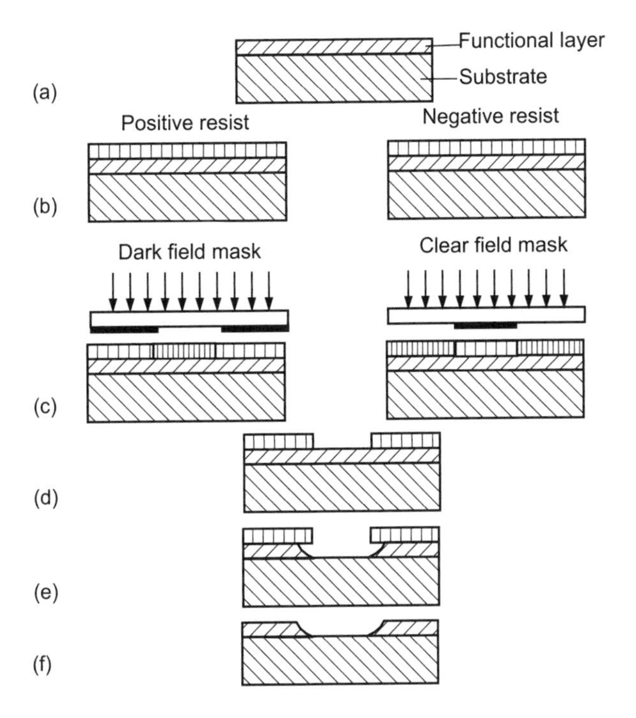


**Figure 3.2** Pattern transfer with additive technique, lithography, and subtractive technique: (a) deposition of functional layer; (b) coating photoresist (negative or positive); (c) photolithography (dark field mask or clear field mask); (d) developing photoresist; (e) selective etching of functional layer (photoresist is not attacked); and (f) structure is transferred to functional layer.

and transmission coupled plasma etching (TCPE) [3]. RIE is the most important technique for micromachining. Reactant gases are excited to ions. Under low pressures and a strong electrical field, ions meet the substrate surface almost perpendicularly. Therefore, this method can achieve relatively high aspect ratios. The etch rates range between those of physical etching and chemical etching. Etch techniques with extremely high aspect ratios are called *deep reactive ion etching* (DRIE). DRIE will be discussed in more detail in Section 3.3.1.2.

### 3.1.4 Pattern Transfer Techniques

#### 3.1.4.1 Subtractive Transfer

Selective etching with a photolithography mask can transfer structures from a mask to a functional material. Figure 3.2 illustrates a typical pattern transfer process. To start with, a film of functional material is deposited [Figure 3.2(a)]. After spin coating the photoresist layer and taking the right combination of resist type (negative or positive) and mask type (dark-field or clear-field), the pattern is transferred from the mask to the resist layer [Figure 3.2(d)]. A subsequent selective etching process transfers the structure further to the functional layer [Figure 3.2(e)]. Washing away the photoresist reveals the desired structure in the functional layer [Figure 3.2(f)].

#### 3.1.4.2 Additive Transfer

Lift-off technique is the common additive technique to transfer a pattern [Figure 3.3]. The functional material is deposited directly on a patterned photoresist layer [Figure 3.3(d)]. After resolving photoresist with acetone, the deposited material on resist is removed. The remaining material is

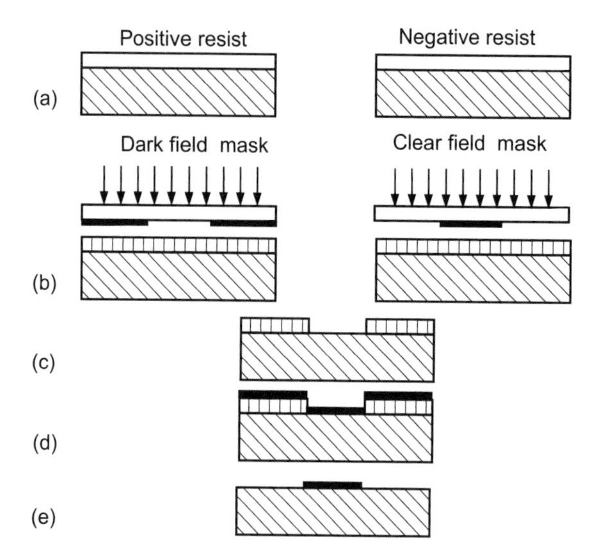


**Figure 3.3** Pattern transfer with lift-off technique: (a) coating photoresist (negative or positive); (b) photolithography (dark field mask or clear field mask); (c) developing photoresist; (d) deposition of functional layer; and (e) washing away photoresist with material on top; transferred structure remains.

the transferred structure. Lift-off technique is an important process for microfluidics. For example, polymeric membranes and catalytic metals such as platinum (Pt) and palladium (Pd) in biochemical sensors avoid direct wet etching, which may change the properties needed for sensing applications. The presence of photoresist keeps the temperature limit of the subsequent deposition process below about 300°C.

The other additive technique uses selective electroplating to transfer structures. The technique starts with a deposition of a metal seed layer on a substrate. After spin coating and developing a thick photoresist layer, the functional material is deposited on the seed layer by electroplating. Electroplating, or electrodeposition, is an electrochemical process in which ions in a solution are deposited onto a conductive substrate by an applied electric current. The seed layer mentioned above acts as an electrode or the conductive substrate. The advantages of electroplating over sputtering are the low cost and the fast deposition rate. However, sputtered or evaporated films are smoother than those which are electroplated. Since the deposition rate depends on the current density, which in turn depends on the electric field, it is difficult to control the uniformity of an electroplated structure. Electroplating can be used for making high aspect ratio structures for microfluidics, if thick-film photoresists are used. In the LIGA process, which is explained later in Section 3.4.1.1, the resist layer is patterned with high aspect ratio X-ray lithography. The electroplated structures can be used as molds for making polymeric microfluidic devices such as microchannels, microvalves, and micropumps (see Chapters 6 and 7).

## 3.2 FUNCTIONAL MATERIALS

### 3.2.1 Materials Related to Silicon Technology

#### 3.2.1.1 Single Crystalline Silicon

Because of the historical background of MEMS, which has emerged from microelectronics, silicon remains as the most important electrical and mechanical material in MEMS and microfluidics. The

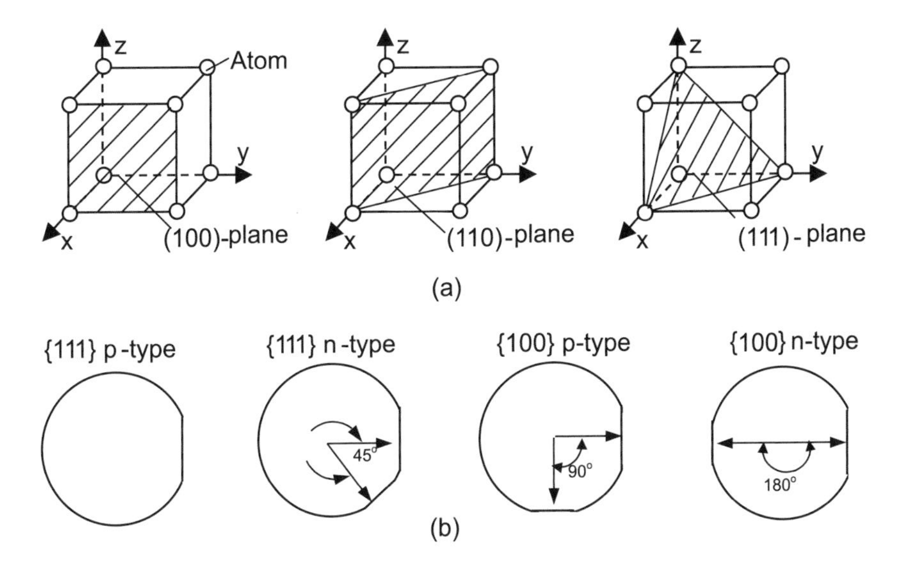


**Figure 3.4** Single crystalline silicon: (a) different crystal planes in a cubic lattice of silicon atoms; and (b) flat orientations and the corresponding silicon wafer types.

technology is established, and single crystalline silicon wafers with high purity are commercially available at a relatively low cost.

Single crystalline silicon wafers are classified by the crystalline orientation of their surfaces. The classification is based on the Miller indices, which are shown in Figure 3.4(a). A particular direction is described with square brackets, such as [100]. Because of the symmetry, there are a number of [100]-directions. The set of equivalent directions is described with angle brackets, such as <100>. If this direction is the normal vector of a plane, the plane is denoted with parentheses such as (100). The set of equivalent planes is described with braces, such as {100}.

Single crystalline silicon is mostly fabricated with the Czocharalski method (CZ-method). A small seed crystal with a given orientation is dipped into a highly purified silicon melt. The seed is slowly pulled out of the melt while the crucible containing the melt is rotated. Silicon crystals are grown along the selected orientation of the seed to a rod. The other method of fabricating silicon crystals is called the *floating zone method* (FZ-method). A polysilicon rod is used as the starting material. A seed crystal at the end of the rod defines the orientation. A radio-frequency heater locally melts the polysilicon rod. Crystal growth starts from the end with the seed. The silicon rod is then sawed into wafers, whose surface is subsequently polished readily for use. Figure 3.4(b) shows the typical flat position for identifying the orientation and type of a silicon wafer.

The thickness of silicon wafers is an important parameter for designing microfluidic devices. Commercially available wafers have standardized thickness, which depends on the wafer size. With the trends of larger wafers for higher productivity, the wafer thickness also increases to maintain the mechanical stability (see Table 3.4).

#### 3.2.1.2 Epitaxial Silicon

Epitaxy is a single crystalline layer grown from another single crystalline substrate. The most important technique for epitaxy growth is CVD. Table 3.5 lists the common CVD reactions using silane $SiH_4$ and dichlorosilane $SiH_2Cl_2$ at high temperatures on the order of 1,200°C. The epitaxial layer can be doped if dopant gases, such as diborane $B_2H_6$ for p-type or phosphine $PH_3$ for n-type, are mixed during the CVD process.

**Table 3.4**
Dimensions of Silicon Wafers in Different Technology Periods

| *Period* | *Wafer Diameter (mm)* | *Wafer Thickness (μm)* |
|---|---|---|
| 1970 to 1975 | 76 | 375 |
| 1975 to 1980 | 100 | 450/500* |
| 1980 to 1985 | 125 | 525 |
| 1985 to 1990 | 150 | 675 |
| 1990 to 1995 | 200 | 725 |
| 1995 to 2000 | 300 | 770 |

* Typical for MEMS.

Epitaxy can be grown by MBE. The process is similar to an evaporation process using silicon melt in a crucible. MBE is carried out under ultrahigh vacuum and temperatures between 400°C and 800°C. The growth rate is on the order of 0.1 μm/min [4].

Highly doped silicon epitaxial layers are used in MEMS and microfluidics for fabricating membrane with a precise thickness. The etch stop occurs on the p-n junction (see Section 3.3.1.1).

#### 3.2.1.3 Polysilicon

Polycrystalline silicon is referred toas polysilicon, which is deposited during a LPCVD process with silane (see Table 3.5). The deposition temperatures range from 575°C to 650°C. At temperatures below 575°C, the silicon layer is amorphous. Above 650°C, polycrystalline has a columnar structure. The grain size is typically between 0.03 and 0.3 μm. After annealing at 900°C to 1,000°C for several minutes, crystallization and grain growth occur. The grain size is then on the order of 1 μm. Polysilicon can be doped in situ with the same gases used for epitaxial silicon. The deposition rates range from 10 to 20 nm/min [5].

Polysilicon layers are generally conformal. In surface micromachining, polysilicon is used directly as mechanical material. In microfluidics, polysilicon can be used for making channel walls and sealing etched channel structures. The annealing process described above increases grain size and decreases intrinsic stress in polysilicon one order of magnitude, from several hundreds of megapascals to several tens of megapascals. The low intrinsic stress is required for applications in MEMS and microfluidics.

#### 3.2.1.4 Silicon Dioxide

Silicon dioxide can be grown with thermal oxidation, as described in Section 3.1.2.2. Because thermal oxidation relies entirely on the diffusion of oxygen, the growth rate of thermal oxidation decreases significantly with thicker oxide layers. In addition, thermal oxidation requires a silicon substrate. CVD can make thicker silicon dioxide layers and does not need a silicon substrate. Table 3.5 lists the major reactions used for deposition of silicon dioxide.

Low-temperature oxide (LTO) can be deposited at temperatures below 350°C. Thus, it can be used as insulation coating for aluminum. Since silicon dioxide is used as a sacrificial layer in surface micromachining and other related techniques, the step coverage quality is an important parameter. The low deposition temperature causes bad step coverage. The uniformity improves with higher temperature and lower pressure. Silane oxide CVD has the worst conformity. LTO-CVD and PECVD give an average conformity. Tetra-ethyl-ortho-silicate (TEOS) CVD and high-temperature oxide (HTO) CVD have the best step coverage characteristics.

The selectivity of silicon dioxide to many silicon etchants makes it a good mask material for self-aligned etching processes. In combination with silicon nitride, multistep etching processes of

**Table 3.5**

Chemical Reactions Used in CVD for Different Material Films (*After*: [3].)

| *Material* | *Chemical Reactions* | *Techniques* |
|---|---|---|
| Silicon | $SiH_4 \rightarrow Si + 2H_2 \uparrow$ | Silane CVD |
| | $SiH_2Cl_2 \rightarrow SiCl_2 + 2H_2 \uparrow$ | Dichlorosilane CVD |
| | $SiCl_2 + H_2 \rightarrow Si + 2HCl \uparrow$ | |
| Polysilicon | $SiH_4 \overset{630^\circ C, 60Pa}{\longrightarrow} Si + 2H_2 \uparrow$ | Low-pressure CVD (LPCVD) |
| Silicon dioxide | $SiH_4 + O_2 \overset{430^\circ C, 1bar}{\longrightarrow} SiO_2 + 2H_2 \uparrow$ | Silane oxide CVD |
| | $SiH_4 + O_2 \overset{430^\circ C, 40Pa}{\longrightarrow} SiO_2 + 2H_2 \uparrow$ | Low-temperature oxide (LTO) CVD |
| | $Si(OC_2H_5)_4 \overset{700^\circ C, 40Pa}{\longrightarrow} SiO_2 + Gas \uparrow$ | Tetra-ethyl-ortho-silicate (TEOS) CVD |
| | $Si(OC_2H_5)_4 + O_2 \overset{400^\circ C, 0.5bar}{\longrightarrow} SiO_2 + Gas \uparrow$ | Subatmospheric CVD (ACVD) |
| | $SiH_2Cl_2 + 2N_2O \overset{900^\circ C, 40Pa}{\longrightarrow} SiO_2 + Gas \uparrow$ | High-temperature oxide (HTO) CVD |
| | $SiH_4 + 4N_2O \overset{350^\circ C, plasma, 40Pa}{\longrightarrow} SiO_2 + Gas \uparrow$ | Plasma-enhanced CVD (PECVD) |
| Silicon nitride | $SiH_2Cl_2 + 4NH_3 \overset{750^\circ C, 30Pa}{\longrightarrow} Si_3N_4 + Gas \uparrow$ | Low-pressure CVD (LPCVD) |
| | $3SiH_4 + 4NH_3 \overset{700^\circ C, plasma, 30Pa}{\longrightarrow} Si_3N_4 + Gas \uparrow$ | Plasma-enhanced CVD (PECVD) |
| | $3Si + 4NH_3 \overset{300^\circ C, plasma, 30Pa}{\longrightarrow} Si_3N_4 + 6H_2 \uparrow$ | Plasma-enhanced CVD (PECVD) |
| Silicide | $4SiH_4 + 2WF_6 \overset{400^\circ C, 30Pa}{\longrightarrow} 2WSi_2 + 12HF \uparrow + 2H_2 \uparrow$ | |
| | $4SiH_2Cl_2 + 2TaCl_5 \overset{600^\circ C, 60Pa}{\longrightarrow} 2TaSi_2 + 18HCl \uparrow$ | |
| | $2SiH_4 + TiCl_4 \overset{450^\circ C, plasma, 30Pa}{\longrightarrow} 2TiSi_2 + 4HCl \uparrow + 2H_2 \uparrow$ | |

three-dimensional (3-D) structures are possible. Similar to polysilicon, silicon dioxide can be used to seal microchannels. Its insulating properties make silicon dioxide a good coating layer for channels in microfluidics. The electric properties of CVD oxide are worse than those of thermal oxide. The growth rates of CVD oxide increase with process temperature; they are about 0.5 to 1 μm/min for silane oxide, 5 to 100 nm/min for LTO, and 5 to 50 nm/min for TEOS [5]. Silicon dioxide in the form of glasses is a familiar material for chemical analysis (see Table 3.6).

#### 3.2.1.5 Phosphorous Silicate Glass

Phosphorous silicate glass (PSG) is phosphorous silicon dioxide. The mass analysis of phosphorous ranges from 2% to 10%. PSG can be deposited with the CVD processes for silicon dioxide described

**Table 3.6**

Composition (in Mass Analysis) of Common Glasses Used for Chemical Analysis [6]

| *Glass* | $SiO_2$ | $Al_2O_3$ | $Na_2O$ | $CaO$ | $MgO$ | $B_2O_3$ | $BaO$ |
|---|---|---|---|---|---|---|---|
| Soda glass | 68 | 3 | 15 | 6 | 4 | 2 | 2 |
| Boronsilicate glass | 81 | 2 | 4 | $< 0.1$ | $< 0.1$ | 13 | $< 0.1$ |
| Fused silica | 100 | $< 0.02$ | $< 0.02$ | $< 0.02$ | $< 0.02$ | $< 0.02$ | $< 0.02$ |

above. In addition, silicon dioxide is doped with phosphorous by mixing phosphine $PH_3$ or $POCl_3$ in the reactant gas. After annealing at about 1,000°C, phosphorous-doped silicon dioxide flows to PSG. The PSG layer is conformal and is a good sacrificial material for surface micromachining, which is explained in Section 3.3.2.1.

#### 3.2.1.6 Silicon Nitride

Silicon nitride is a good insulator and acts as a barrier against all kinds of diffusion. Table 3.5 lists the most important reactions for creating silicon nitride. Thermal growth by exposing silicon to ammonia can be achieved at high temperatures. However, due to the limitation of diffusion, the nitride film is very thin. Silicon nitride is deposited as a reaction product between ammonia ($NH_4$) and silane ($SiH_4$) or dichlorosilane ($SiH_2Cl_2$).

Because of its low thermal conductivity, silicon nitride is often used in microfluidic devices as a thermal insulator. For instance, heater structures can be suspended on silicon nitride membrane or flexures. Similar to other CVD films, silicon nitride exhibits a high intrinsic stress. Films thicker than 200 nm may crack because of the tensile stress [5]. The tensile stress on the order of 1 GPa can be decreased by silicon-rich deposition. Instead of the stoichiometric flow rate of silane or dichlorsilane, excess amounts are used. Silicon-rich nitride films with tensile stress on the order of 100 MPa are acceptable for microfluidic applications such as membrane filters.

#### 3.2.1.7 Silicide

Silicides are alloys between silicon and other metals. Silicides are used in microelectronics as high-temperature electrical interconnections. The most common silicides are $MoSi_2$, $WSi_2$, $TaSi_2$, $TiSi_2$ (for interconnections), PtSi, and $PdSi_2$ (for contacts). There are three major methods for fabrication of silicides: sputtering with silicon and metal targets, sputtering with metal target and subsequent bombardment with Si-ions, and CVD. Silicides can withstand temperatures above 1,000°C and can replace metals in high-temperature applications.

#### 3.2.1.8 Metals

With PVD, CVD, and electroplating, almost all metals can be deposited. The choice of a metal layer depends on the actual application. While aluminum is the common material for electrical interconnections, other metals are used for sensing and actuating purposes. Because of its availability, aluminum can also be used as a sacrificial layer for surface micromachining.

In microfluidics, thermoelectrical properties of metals are used for temperature sensing. Thermomechanical properties are used for actuating. Permalloy, an iron nickel alloy, can be deposited for magnetic sensing and actuating. Some metals such as platinum and palladium have catalytic properties, and are useful in chemical sensors and microreactors. Table 3.7 compares typical mechanical properties of the materials discussed above.

### 3.2.2 Polymers

In contrast to many other MEMS applications, microfluidic devices are relatively large, due to the long microchannels and the required sample volume, which cannot be too small. Therefore, the cost of the substrate material plays an important role for large-scale production. Figure 3.5 compares the prices between polymers and common glasses used for microfluidics, such as boro-float glass, boro-silicate glass, and photo-structurable glass. For the same area and optical transparency, a glass substrate may cost 10 to 100 times more than a polymer substrate. Besides the cost, a wide range of polymers with different surface chemistries exist. Thus, the material choice can be tailored by applications. As low-cost materials, polymers can be used directly as mechanical materials. Their electrical and chemical properties are interesting for physical, chemical, and biochemical sensing

**Table 3.7**

Mechanical Properties of Typical Functional Materials

($T_m$: Melting Temperature, $E$: Young's Modulus, $\sigma_y$: Yield Strength, $\nu$: Poisson's Ratio, $\rho$: Density, $H$: Knoop Hardness)

| *Materials* | $T_m(°C)$ | *E (GPa)* | $\sigma_y(GPa)$ | $\nu$ | $\rho(kg/m^3)$ | *H (GPa)* |
|---|---|---|---|---|---|---|
| Bulk silicon | 1,415 | 160–200 | — | 0.22 | 2,330 | 5–13 |
| Polysilicon | 1,415 | 181–203 | — | — | — | 10–13 |
| Silicon dioxide | 1,700 | 70–75 | 8.4 | 0.17 | 2,200 | 15–18 |
| Pyrex glass | — | 64 | — | 0.2 | 2,230 | — |
| Silicon nitride | 1,800 | 210–380 | 14 | 0.25 | 3,100 | 8 |
| Silicon carbide | — | 300–400 | 21 | 0.19 | 3,210 | 24–27 |
| CVD-diamond | — | 800–1,100 | 0.07 | 3,530 | — | |
| Aluminum | 661 | 70 | 0.2 | 0.33 | 2,700 | — |
| Platinum | 1,772 | 170 | 0.137–0.170 | 0.38 | 21,440 | — |
| Gold | 1,065 | 80 | 0.120 | 0.38–0.42 | 19,280 | — |
| Stainless steel | — | 200 | 2.1 | 0.3 | 7,900 | 6.5 |

[7]. Polymer membranes and matrices are widely used in macroscale for the separation of DNA and proteins [8]. Some basic properties, advantages, and drawbacks of polymers in microfluidic applications are discussed below.

Polymers are organic materials consisting of macromolecules, which may have more than 1,000 monomeric units. The cross-linking process of the monomers is triggered chemically by an initiator substance, or physically by photons, pressure, or temperature. In a polymerization reaction, monomer units react to form linear chains or three-dimensional networks of polymer chains. If only one type of polymer is used, the material is called *homopolymer*. Polymerization of two or more monomer units results in a *copolymer*. Polymers containing specific additives are called *plastics*. Polymers exist in two basic forms: amorphous and microcrystalline. The macromolecules in a polymeric material have different lengths. Thus, there is no fixed melting temperature for polymers. Several temperature ranges exist in the melting process of a polymeric material. The two characteristic temperatures for a polymeric material are the glass transition temperature and the decomposition temperature. At the glass transition temperature, the material will lose the strength it used to have at lower temperatures, but still can keep its solid shape. Increasing the temperature further damages the bondage between the monomers, and the plastic will lose its solid shape. Above the glass transition temperature, a polymeric material becomes soft and can be machined by molding or hot embossing. This temperature can be adjusted by mixing a softener with the original polymeric material. Above the decomposition temperature, the polymeric material starts to degrade and ceases to function.

Based on their molding behavior, polymers can be categorized into three groups:

- Elastomeric materials;
- Duroplastic meterials;
- Thermoplastic materials.

Elastomeric materials have weakly cross-linked polymer chains. These polymer chains can be stretched under external stress, but regain their original state if the stress is removed. Elastomeric polymer does not melt before reaching decomposition temperature. Elastomeric materials are suitable for prototyping of microfluidic devices. The elastic property is ideal for sealing of the fluidic interfaces. In contrast to elastomeric materials, duroplastic materials have strong cross-linked polymer chains. Duroplastics do not soften much before decomposition temperature. They are strong and brittle. Thermoplastic materials are between the above two extremes. The material consists of

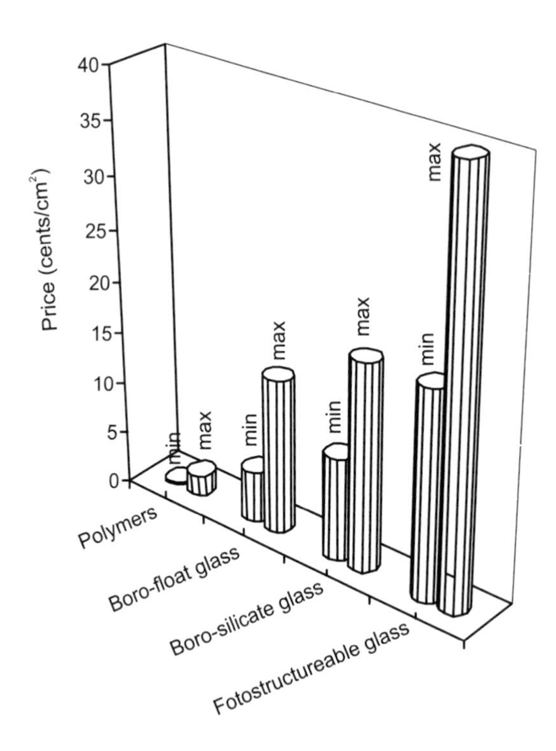


**Figure 3.5** Price comparison between polymers and common glasses used for microfluidics. (*From*: [6].)

weakly linked polymer chains. Thus, thermoplastics can be softened and structured at temperatures between the glass transition point and the decomposition point. Due to this characteristic, thermoplastic polymers are commonly used for micromolding.

Most microfluidic applications for chemical analysis and life sciences require an optically transparent material. Many polymers are self-fluorescent at low-excitation wavelengths. Self-fluorescence may affect the sensitivity of microfluidic applications with fluorescent detection. Another drawback of polymers is their poor chemical resistance to solvents. With applications in the chemical industry and drug discovery, polymeric microfluidic devices may need to handle a variety of solvents. Furthermore, aging, chemical resistance, and UV resistance may limit the use of polymers for some microfluidic applications.

Surface properties play an important role in microfluidic devices utilizing electro-osmotic pumping. A high charge density on the surface assures a stable and controllable electro-osmotic flow. Due to the lack of ionizable groups, most polymers have a lower surface charge density compared to glass. Thus, for applications with electro-osmotic flows such as CE separation, the surface of the polymeric substrate should be treated accordingly.

The biggest advantage of polymers compared to silicon-based or glass-based materials is their superior biocompatibility. Polymeric devices are best for DNA analysis, polymerase chain reactions, cell handling, and clinical diagnostics. Many polymers are compatible to blood and tissue. Micromachining of these materials may realize implantable microfluidic devices for applications such as drug delivery. Table 3.8 lists the properties of some typical polymers.

Besides their use as substrate materials, polymers can be spin-coated, laminated, or vapor-deposited on other substrates. In traditional microelectronics and micromachining, polymers are utilized as a photoresist or as a passivation layer. The layer thickness is typically on the order of a few microns. Section 3.4 discusses polymeric micromachining in detail.

**Table 3.8**

Properties of Typical Polymers

($T_g$: Glass Transition Temperature, $\rho$: Density, $\kappa$: Thermal Conductivity, $\gamma$: Thermal Expansion Coefficient)

| *Materials* | $T_g(^\circ C)$ | $\rho(kg/m^3)$ | $\kappa(W/K\text{-}m)$ | $\gamma \times 10^{-6} K^{-1}$ |
|---|---|---|---|---|
| Parylene-N | 410 | 1,100 | 0.13 | 69 |
| Parylene-C | 290 | 1,290 | 0.08 | 35 |
| Parylene-D | 380 | 1,418 | — | 30–80 |
| Polyamide 6(PA 6) | 60 | 1,130 | 0.29 | 80 |
| Polyamide 66 (PA 66) | 70 | 1,140 | 0.23 | 80 |
| Polycarbonate (PC) | 150 | 1,200 | 0.21 | 65 |
| Polymethylmethacrylate (PMMA) | 106 | 1,180–1,190 | 0.186 | 70–90 |
| Polyimide | — | 1,420 | 0.10–0.35 | 30–60 |
| Polystyrene (PS) | 80–100 | 1,050 | 0.18 | 70 |

## 3.3 SILICON-BASED MICROMACHINING TECHNIQUES

### 3.3.1 Silicon Bulk Micromachining

#### 3.3.1.1 Anisotropic Wet Etching

*Basic Etching Processes.* For single crystalline materials such as silicon, etch rates of anisotropic wet etching depend on crystal orientation. In an anisotropic wet etching process, hydroxides react with silicon in the following steps [9]:

$$\begin{aligned} \mathrm{Si} + 2\mathrm{OH}^- &\rightarrow \mathrm{Si(OH)}_2^{2+} + 4\mathrm{e}^- \\ 4\mathrm{H_2O} + 4\mathrm{e}^- &\rightarrow 4\mathrm{OH}^- + 2\mathrm{H_2} \\ \mathrm{Si(OH)}_2^{2+} + 4\mathrm{OH}^- &\rightarrow \mathrm{SiO_2OH}_2^{2-} + \mathrm{H_2O} \end{aligned} \tag{3.5}$$

The overall reaction is:

$$\mathrm{Si} + 2\mathrm{OH}^- + 2\mathrm{H_2O} \rightarrow \mathrm{Si(OH)}_2^{2+} + 2\mathrm{H_2} \tag{3.6}$$

In the steps of (3.5), four electrons are transferred from each silicon atom to the conduction band. The presence of electrons is important for the etching process. Manipulating the availability of electrons makes a controllable etch stop possible. Etch stop methods are discussed later in this section. Silicon etchants, which can provide hydroxide groups, are categorized as [9]:

- Alkali hydroxide etchants: KOH, NaOH, CsOH, RbOH, or LiOH;
- Ammonium hydroxide etchants: ammonium hydroxide $NH_4OH$, tetramethyl ammonium hydroxide (TMAH) $(CH_3)_4NOH$;
- Ethylenediamine pyrochatechol (EDP, which is hazardous and causes cancer, and should be accompanied by safety measures): a mixture of ethylenediamine $NH_2(CH_2)_2NH_2$, pyrochatechol $C_6H_4(OH)_2$, and water;
- Other etchants: hydrazine/water, amine gallate etchants.

Because of its crystalline structure, silicon atoms in $\{111\}$-planes have stronger binding forces, which make it more difficult to release electrons from this plane. Thus, etch rates at $\{111\}$-planes are the slowest. Anisotropy or orientation dependence is caused by the different etch rates in different crystal planes.

**Example 3.3: Trench Profile of Anisotropic Wet Etching of $\{100\}$-Silicon Wafer**

A microchannel is etched in a $\{100\}$-wafer with KOH solution. (a) Determine the angle between the channel wall and the front surface. (b) If the top channel width and the etch rate are 100 mm and 1 mm/min, respectively, what is the bottom channel width after 20 minutes of etching? (c) How long will it take until the etching process stops?

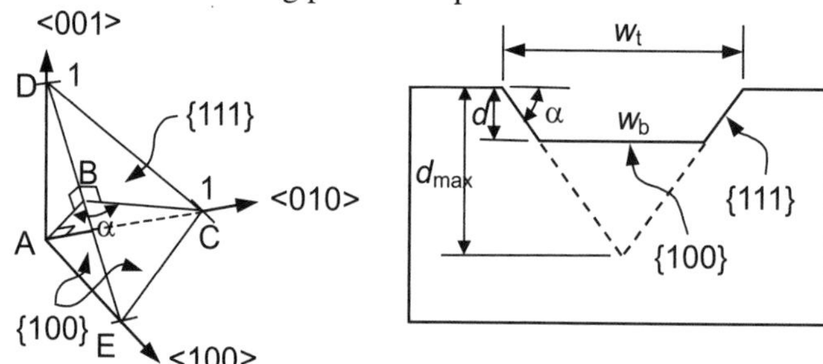


**Solution.** Based on the crystal structure (Miller indices), the angle between the fast etching planes $\{100\}$ and the slow etching planes $\{111\}$ is the angle $\angle \text{ABC} = \alpha$. In triangle $\triangle \text{DAE}$ with $\angle \text{DAE} = 90°$, the height AB is $\sqrt{2}/2$. In triangle $\triangle \text{ABC}$ with $\angle \text{BAC} = 90°$ and $\text{AC} = 1$, the angle $\angle \text{ABC}$ is:

$$\alpha = \arctan \frac{\text{AC}}{\text{AB}} = \arctan \frac{1}{\sqrt{2}/2} = \arctan \sqrt{2} = 54.74°$$

After 20 minutes of etching with an etch rate of 1 μm/min, the channel depth is:

$$d = 20 \times 1 = 20 \ \mu\text{m}$$

Thus, the bottom channel width is:

$$w_\text{b} = wt - 2 \times \frac{d}{\tan \alpha} = w_\text{t} - \sqrt{2}d = 100 - \sqrt{2} \times 20 = 71.7 \ \mu\text{m}$$

The etching process stops if the two $\{111\}$-planes cross each other, or $w_\text{b} = 0$:

$$w_\text{b} = wt - 2 \times \frac{d}{\tan \alpha} = 0 \rightarrow d_\text{max} = \frac{w_\text{t} \tan \alpha}{2} = \frac{w_\text{t}}{\sqrt{2}} = \frac{100}{\sqrt{2}} = 70.71 \ \mu\text{m}$$

The time until etch stop is:

$$t = \frac{d_\text{max}}{1 \ \mu\text{m/min}}$$

Table 3.9 [10, 20] compares the most important parameters of common anisotropic etchant solutions. KOH offers the best selectivity between the $\{100\}$-plane and the $\{111\}$-plane. However, KOH attacks aluminum structures on the wafer. TMAH etches faster in the $\{111\}$-plane but does not attack aluminum. All etchants are selective to silicon nitride and silicon dioxide. Thus, these two materials can be used as masks for anisotropic etching processes.

Figure 3.6 describes typical etch profiles for different wafer types. In Figure 3.6(c), the deep trench in the $\{111\}$-wafer is first etched with DRIE (see Section 3.3.1.2). Underetching is achieved with anisotropic wet etching [21].

**Example 3.4: Trench Profile of Anisotropic Wet Etching of the $\{110\}$-Silicon Wafer**

Determine the angle between a slanted (111)-surface and a vertical (111)-surface in Figure 3.5(b). What is the angle between the slanted (111)-surface and the wafer surface?

**Solution.** Use Miller indices in a crystal structure. The two nonparallel $\{111\}$-planes are depicted in the following figure.

**Table 3.9**
Characteristics of Different Anisotropic Wet Etchants

| *Characteristics* | *KOH* | *$NH_4OH$* | *TMAH* | *EDP* | *Hydrazine* | *Amine Gallate* |
|---|---|---|---|---|---|---|
| References | [10, 13] | [14, 15] | [16, 17] | [12] | [18, 19] | [20] |
| Concentration (weight %) | 40–50 | 1–18 | 10–40 | See [a] | See [b] | See [c] |
| Temperature (°C) | 80 | 75–90 | 90 | 70–97 | 100 | 118 |
| {111} etch rate (nm/min) | 2.5–5 | — | 20–60 | 5.7–17 | 2 | 17–34 |
| {100} etch rate (mm/min) | 1–2 | 0.1–0.5 | 0.5–1.5 | 0.2–0.6 | 2 | 1.7–2.3 |
| {110} etch rate (mm/min) | 1.5–3 | — | 0.1 | — | — | — |
| $Si_3N_4$ etch rate (nm/min) | 0.23 | — | 1–10 | 0.1 | — | — |
| $SiO_2$ etch rate (nm/min) | 1–10 | — | 0.05–0.25 | 0.2 | 0.17 | Slow |
| Al attack | Yes | No | No | Yes | — | Yes |

a. 1 L ethylene diamine $NH_2$-$CH_2$-$CH_2$-$NH_2$, 160 g pyrocatechol $C_6H_4(OH)_2$, 6 g pyrazine $C_4H_4N_2$, 133 mL $H_2O$
b. 100 mL $N_2H_4$, 100 mL $H_2O$ (explosive, very dangerous!)
c. 100 g gallic acid, 305 mL ethanolamine, 140 mL $H_2O$, 1.3 g pyrazine, 0.26 mL FC-129 surfactant

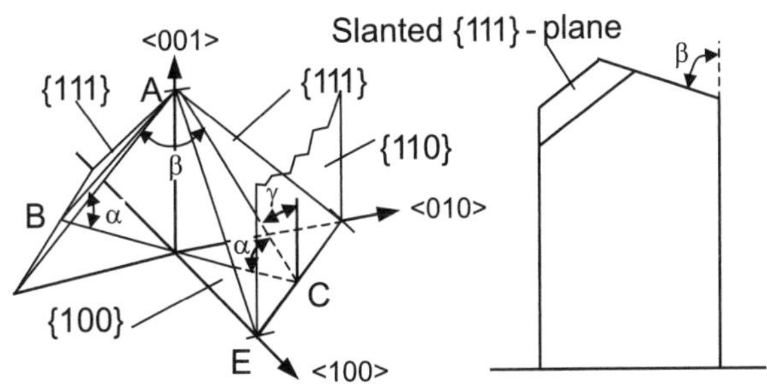


In triangle $\triangle ABC$, the two angles at A and at C are the angles between {111}- and {100}-planes $\alpha = 54.74°$. Thus, the angle between two slanted {111}-planes is:

$$\beta = 180° - 2\alpha = 180° - 2 \times 54.74° = 70.52°$$

The angle between the slanted (111)-plane and the wafer surface (110)-plane is:

$$\gamma = 90° - \alpha = 90° - 54.74° = 35.26°$$

*Etch-Rate Modulation.* Controlled etch-stop is an important technique for precise fabrication with anisotropic wet etching. Different methods to slow down or eliminate the etch rate are:

- Using selectivity of etchants, coating silicon surfaces with a protective layer, such as nitride or oxide;
- Using orientation dependency of etch rates;
- Controlled hole generation.

The first method is often used for selective etching with a layer of silicon dioxide and silicon nitride as a mask. By combining multiple silicon/nitride layers, structures with different depths can be realized. This passivation layer is used to protect the sidewall in the underetching process of a freestanding structure on a {111}-wafer [Figure 3.6(c)]. Since the etch rate of the {111}-plane is two orders of magnitude slower than those of {110}- and {100}-planes (see Table 3.7), the etch front stops at the {111}-plane. This unique property can be used to fabricate microchannels or underetched structures [Figure 3.6(c)].

Equation (3.5) shows that there should be enough electrons available for a successful wet etching process. Four electrons are needed for etching away one silicon atom. Releasing electrons

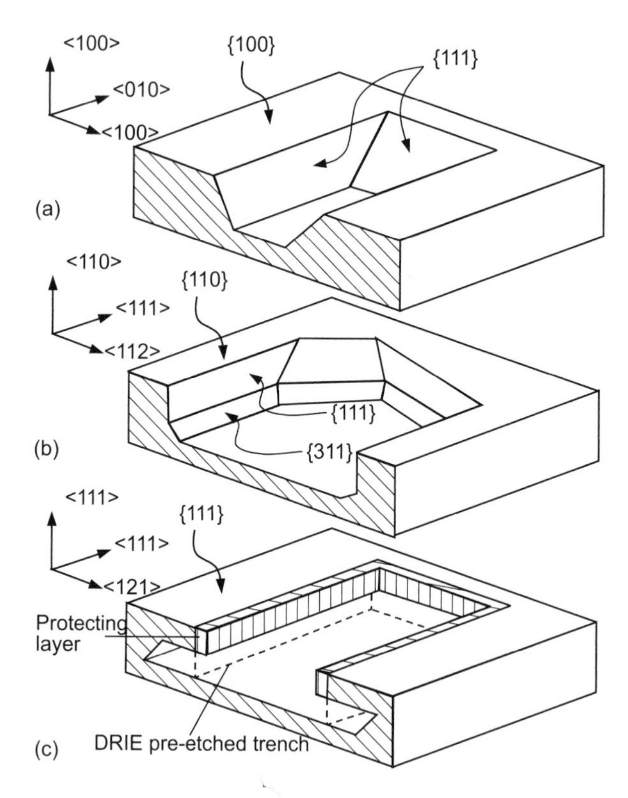


**Figure 3.6** Etch profile in different wafer types: (a) {100}; (b) {110}; and (c) {111}.

**Table 3.10**

Characteristics of Different Anisotropic Wet Etchants

| *Parameters* | *KOH* | *NaOH* | *TMAH* | *EDP* |
|---|---|---|---|---|
| Boron concentration ($cm^{-3}$) | $> 10^{20}$ | $> 3 \times 10^{20}$ | $> 10^{20}$ | $> 3 \times 10^{19}$ |
| Etch rate ratio $Si/Si^{++}$ | 20–500 | 10 | 40–100 | 10 |

generates holes, which in turn attract more hydroxide ions to the substrate surface and speed up the etching process. There are two ways of controlling the availability of holes: highly boron-doped p-silicon and electrochemical etching with a p-n junction.

The doping process is carried out with a solid or gaseous boron source. Silicon dioxide or silicon nitride can work as a diffusion barrier. The depth of the doped layer depends on the diffusion process and is limited by a maximum value on the order of 15 mm. Table 3.10 compares the etch rate reduction of different etchants in highly boron-doped silicon.

Etch rates can be controlled by the surface potential of the silicon substrate. If the silicon surface is biased with a positive potential relative to a platinum electrode, hydroxide ions are attracted to the substrate surface and speed up the etching process [Figure 3.7(a)]. Two critical potential values are:

- *Open circuit potential (OCP)* is the potential resulting to a zero current. The etching process works similar to the case without the circuit. OCP is on the order of 1.56V.
- *Passivation potential (PP)*: Decreasing the potential from OCP increases the current. The current reaches its maximum value and decreases again because of oxide formation, which

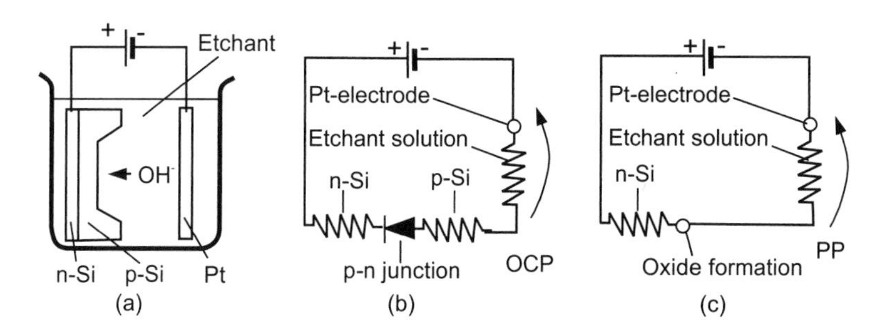


**Figure 3.7** Electrochemical etch-stop: (a) setup; (b) simplified circuit during etching; and (c) simplified circuit at etch-stop.

prevents further etching. The potential at which oxide formation is reached is called the *passivation potential.* PP is on the order of 1V.

The above potential characteristics can be used for controlling etch-stop with a p-n junction as described in Figure 3.7(a). Because the p-n junction is reverse-biased, most of the voltage drops at this junction. Thus, p-silicon is allowed to float at OCP and is etched away [Figure 3.7(b)]. If p-silicon is entirely etched, the p-n junction is destroyed. Consequently, the voltage across the two electrodes drops to a PP value. The formation of oxide at the n-silicon surface stops the etching process automatically [Figure 3.7(c)].

The other approach of generating holes is photon pumping. Holes are generated in positive bias n-silicon by illumination. This technique can also fabricate high-aspect-ratio structures [22].

#### 3.3.1.2 Anisotropic Dry Etching

Anisotropic dry etching allows high-aspect-ratio trenches to be etched in silicon. The process does not depend on crystal orientation of the wafers. Two major approaches of DRIE commercially available are:

- Etching assisted by cryogenic cooling;
- Alternate etching and chemical vapor deposition.

The common problem of physical-chemical dry etching (or RIE) used in microelectronics is that etch trenches are not vertical. The top sections of trench walls are exposed longer to etching plasma and ions. Consequently, the trench is wider on top. The wall should be protected during the dry etching process to keep trench walls parallel and to achieve a high aspect ratio.

With the first approach, the substrate is cooled with liquid nitrogen. The cryogenic temperatures allow reactant gas such as $SF_6$ or $O_2$ to condense on the trench surface. While the condensation film protects the sidewall from etching, it is removed at the bottom by ion bombardment. Since the trench bottom is not protected, it is etched further into the substrate [see Figure 3.8(a)].

The second approach uses chemical vapor deposition to protect sidewalls [23]. This technique was invented and patented by Robert Bosch GmbH in Reutlingen, Germany. Therefore, the technique is often called the Bosch process. The etch cycle consists of two steps: etching and deposition. In the etching step, silicon is etched by $SF_6$. The etching step lasts from 5 to 15 seconds, in which the etch front advances from 25 to 60 nm in silicon. In the deposition step, supply gas is switched to $C_4F_8$. A film of fluorocarbon polymer of about 10 nm is deposited on the trench wall. In the next cycle, the polymer film at the bottom surface is removed by ion bombardment, while the film at the sidewalls is intact and protects the sidewalls from etching. By this way, the etch front advances into the substrate at rates ranging from 1.5 to 4 μm/min [Figure 3.8(b)].

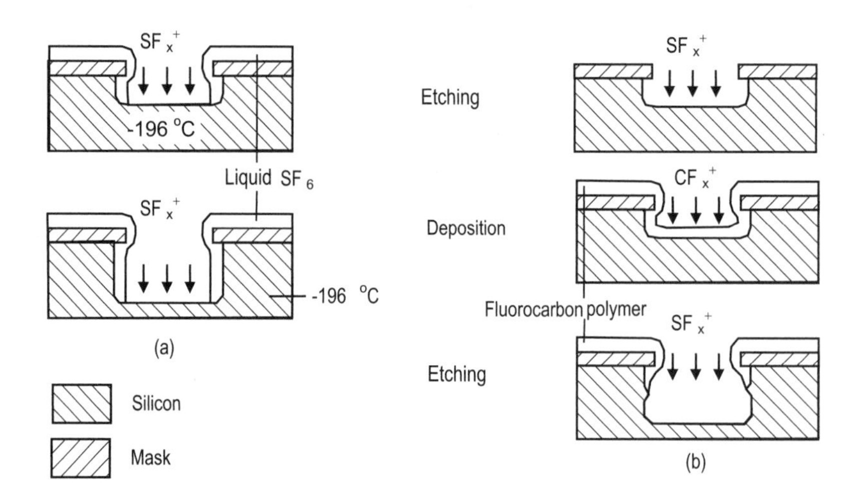


**Figure 3.8** Anisotropic dry etching: (a) cryogenic etching; and (b) alternate etching and deposition.

### 3.3.1.3 Bulk Micromachined Microchannels and Nanochannels

Although details on fabrication of different microfluidic devices are discussed in Chapters 5 to 13, examples in this section illustrate the fabrication of microchannels based on different micromachining techniques. Both isotropic and anisotropic etching can be used to fabricate microchannels in bulk material such as glass and silicon. A variety of cross-sectional channel shapes can be achieved by combining different micromachining techniques discussed above.

Since glass is a familiar material in chemistry and life sciences, microchannels in glass attract attention of applications in these fields [24, 28]. Glass consists mainly of silicon dioxide and therefore can be etched with oxide etchants listed in Tables 3.2 and 3.7. Channels can be etched in fused silica wafers using a polysilicon mask [24]. The two glass wafers are bonded together at high temperatures [Figure 3.9(a)]. Channels in other glass types can be etched in fluoride-based solutions [25, 26]. Photosensitive glasses such as Foturan can also be used for fabricating microchannels [27].

Isotropic etching in silicon results in channel shapes similar to those of glass etching. Figures 3.9(c, d) show channels formed by anisotropic etching of {100}- [27] and {110}-wafers [28] respectively. A glass wafer or a silicon wafer covers the channels using anodic bonding or silicon fusion bonding (see Section 3.6). By combining two etched wafers, different channel shapes can be achieved (Figure 3.9).

The microchannels described in Figure 3.9 have drawbacks of wafer-to-wafer bonding. For channels shown in Figures 3.9(e–h), misalignments and voids trapped during bonding processes can change the desired shapes and consequently the device function. Therefore, fabrication of covered channels in a single wafer is increasingly important. The general concept of these methods is to fabricate a buried channel in a single substrate and to cover the etch access with a subsequent deposition process.

Figure 3.10 describes the basic steps of making a buried channel in {100}-wafer. To start with, a highly boron-doped silicon layer with a doping concentration higher than $7 \times 10^{19}$ cm$^{-3}$ is used as mask layer for the subsequent wet etching process. Etch accesses are opened by RIE through the highly boron-doped layer. The buried channel is etched by EDP, which does not attack the boron-doped layer. After anisotropic etching, the access gaps are sealed by thermal oxidation. The final deposition of silicon nitride covers the entire structure [29].

If the channel wall in Figure 3.10(b) is further doped with boron and the bulk wafer is etched from the backside, microneedles can be fabricated with this technique [30].

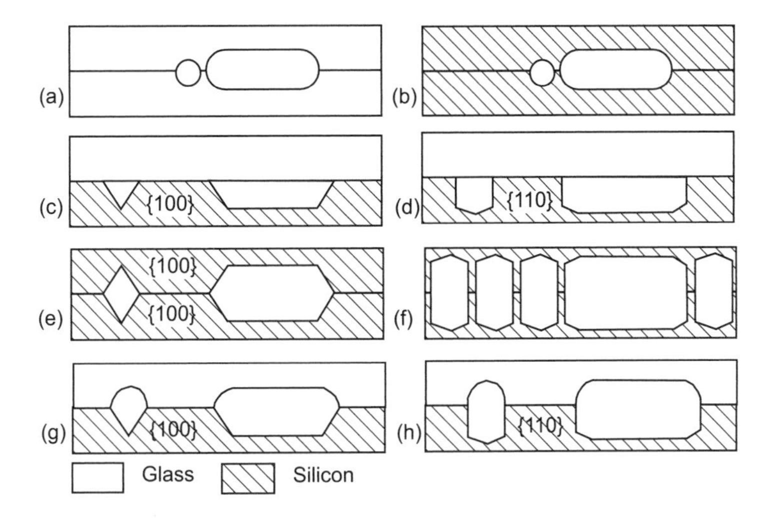


**Figure 3.9** Bulk micromachined channels: (a) glass-glass; (b) silicon-silicon; (c) glass-silicon; (d) glass-silicon; (e) silicon-silicon; (f) silicon-silicon; (g) glass-silicon; and (h) glass-silicon.

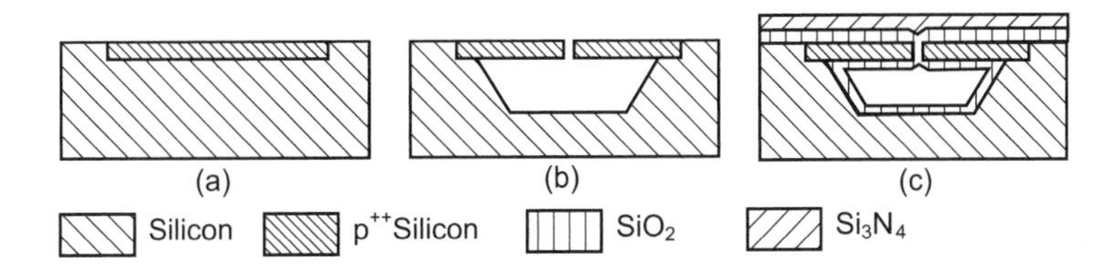


**Figure 3.10** Buried channel with highly boron-doped silicon layer as cover: (a) boron doping; (b) opening etch access, anisotropic wet etching; and (c) deposition of silicon oxide and silicon nitride.

The other method, called *buried channel technology* (BCT), creates microchannels by exploiting the concept of buried channels [31, 32]. The burying depth of the channel described above depends on the thickness of the highly boron-doped layer, which is a maximum of 5 mm due to limits of diffusion processes [29, 30]. BCT overcomes this problem by using deep trenches etched by DRIE. Figure 3.11 describes the concept of this technique.

The process starts with DRIE of a narrow trench [Figure 3.11(a)]. The depth of this trench defines the burying depth of the channel. In the next step, the trench wall is protected by deposition of silicon nitride or by thermal oxidation [Figure 3.11(b)]. The layer at trench bottom is then removed by RIE to create the etch access. Anisotropic or isotropic etching can be used to form the channel [Figure 3.11(c)]. After stripping the protecting layer, conformal LPCVD of silicon nitride seals the

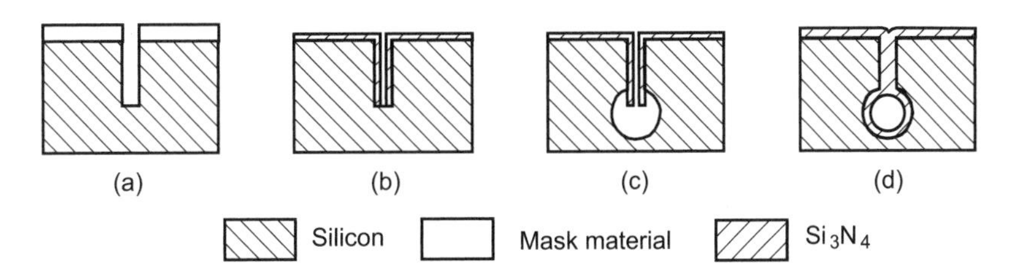


**Figure 3.11** Buried channel with silicon nitride sealing: (a) DRIE; (b) deposition of silicon nitride; (c) isotropic etching; and (d) deposition of silicon nitride.

channel [Figure 3.11(d)]. The advantage of BCT is that a network of channels can be fabricated at different depths in bulk silicon. A similar approach for a buried channel using crystal orientation as an etch stop in a {111}-wafer is depicted in Figure 3.7(c) [33].

Because of the wavelengths on the order of few hundred nanometers, optical lithography cannot create *nanochannels* with widths on the order of several tens of nanometers. As already mentioned in Section 3.1.1, interferometric lithography can be used for making structure size less than 100 nm. Some other techniques with high-energy beams can give a higher resolution. Direct writing with scanning electron beam lithography can create structures with 10-nm resolution. However, the throughput of this technique is extremely low due to its serial nature. X-ray lithography can deliver 50-nm resolution, but such a facility is expensive and impractical for mass production. Another technique for making nanostructures is called nanoimprint lithography or soft lithography. This technique uses a master, which is fabricated with the more expensive technologies such as electron beam writing. Resist patterns can be transferred by imprinting on a substrate surface. Open nanochannels can be fabricated by subsequent etching processes. The nanochannels can be covered with a deposition process. Cao et al. [34] used electron beam evaporation of a silicon oxide source on a tilted wafer with the nanochannels. The tilted wafer results in nonuniform deposition on the channel wall and subsequently seals it. After sealing, a flat and strong surface can be achieved with a straight deposition process.

### 3.3.2 Silicon Surface Micromachining

Silicon surface micromachining refers to different techniques using etching of sacrificial layers or underetching a substrate to form freely movable structures.

The term *surface micromachining* was originally defined for processes using sacrificial layers to form thin-film microstructures [35]. In this book, *silicon surface micromachining* describes the silicon-based techniques using deposited polysilicon and near-surface single crystalline silicon. *Polymeric surface micromachining* in Section 3.4 is referred to as a similar technique with polymers as functional and sacrificial materials.

#### 3.3.2.1 Polysilicon Surface Micromachining

The basic process of polysilicon surface micromachining is divided into four steps [35]:

- Substrate passivation and interconnect;
- Sacrificial layer deposition and patterning;
- Structural polysilicon deposition, doping, and stress annealing;
- Microstructure releasing, rinsing, and drying.

To start with, the silicon substrate is passivated with a double layer of thermal oxide and silicon nitride. Contact windows for the following interconnect layer are opened [Figure 3.12(a)]. The interconnect layer is n-polysilicon, which is deposited and structured to form electrical interconnects to the next mechanical structures [Figure 3.12(b)].

In the second step, a sacrificial layer, such as PSG, is deposited and patterned. The patterns define anchor positions for the coming mechanical structures [Figure 3.12(c)].

In the third step, structural polysilicon is deposited [Figure 3.12(d)]. This polysilicon layer is doped with phosphorous from both sides with the sacrificial PSG layer and an additional PSG layer on top. After annealing, the structural layer is doped symmetrically and does not have gradients in residual strain. The polysilicon layer is then etched to desired forms using RIE [Figure 3.12(e)].

In the last step, sacrificial PSG is etched in buffered hydrofluoric acid (HF). The released structures are rinsed in deionized water and undergo a special drying process to avoid the sticking effect caused by capillary forces [Figure 3.12(f)]. One common technique is supercritical drying,

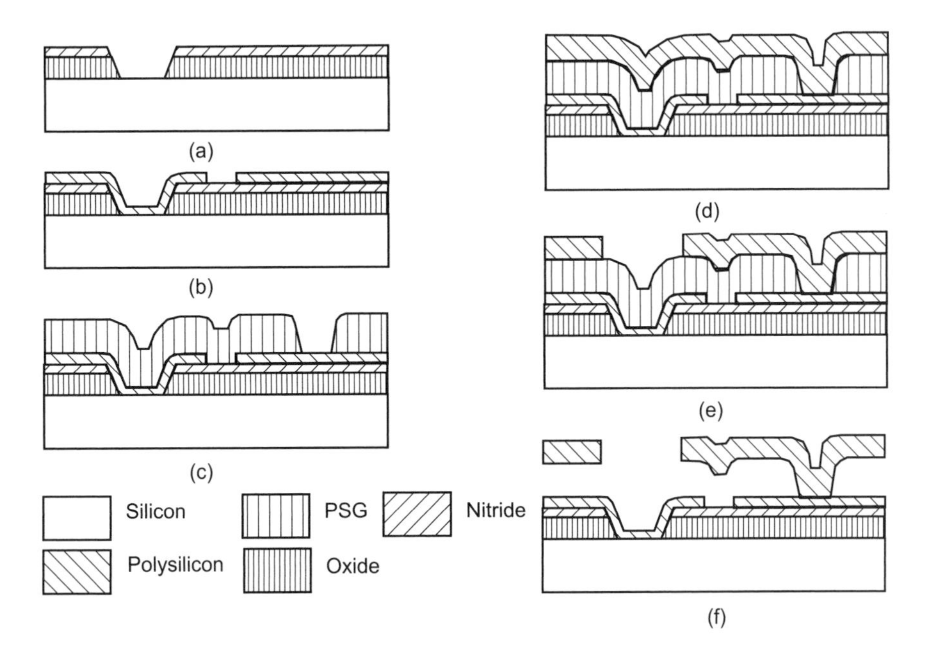


**Figure 3.12** Standard process steps of polysilicon surface micromachining: (a) opening the contact layer; (b) deposition of the interconnect layer; (c) deposition and structuring of the sacrificial layer; (d) deposition of the structural layer; (e) patterning of the structure; and (f) release etch. (*After*: [36].)

which uses a supercritical region of $CO_2$ to avoid the liquid phase. The other antistiction technique coats fluorocarbon polymers or self-assembling monolayer (SAM) to create a hydrophobic surface.

The aspect ratio of surface micromachining depends on the thickness of the functional layer. A thick structure would require unreasonably long deposition times. Using surface micromachining on nonplanar structures, such as high-aspect-ratio trenches, can fabricate three-dimensional polysilicon structures [37]. The technique is called by its author "HexSil" [38].

The HexSil process starts with plasma etching of deep trenches in a silicon wafer [Figure 3.13(a)]. The sidewall is then wet-etched isotropically to create a smooth surface. In the next step, a sacrificial layer, such as PSG, is deposited on the walls [Figure 3.13(b)]. Filling the trench with CVD polysilicon creates the first structural layer [Figure 3.13(c)]. After annealing, the first polysilicon layer is polished and prepared for the deposition of the second structural layer [Figure 3.13(d)]. The second layer is deposited and patterned in a way that it physically connects the structures of the first layer [Figure 3.13(e)]. The entire structure is released in buffered hydrofluoric acid [Figure 3.13(f)]. The silicon template can be reused if the above steps are repeated.

Almost all materials that can be deposited on the mold can utilize the above technique. Silicon carbide, for example, is a good material for high-temperature applications and aggressive conditions [39]. For molding silicon carbide, a silicon mold can be fabricated with DRIE [40]. Polycrystalline silicon carbide is then deposited in an APCVD process at growth rates ranging from 0.4 to 0.8 mm/min. The entire mold is used as the sacrificial material, which resolves completely in KOH to release the carbide structure. Silicon carbide devices with a thickness up to 400 μm can be fabricated with this technique. The technique is relevant for fabrication of high-temperature microreactors (see Section 13.2.1).

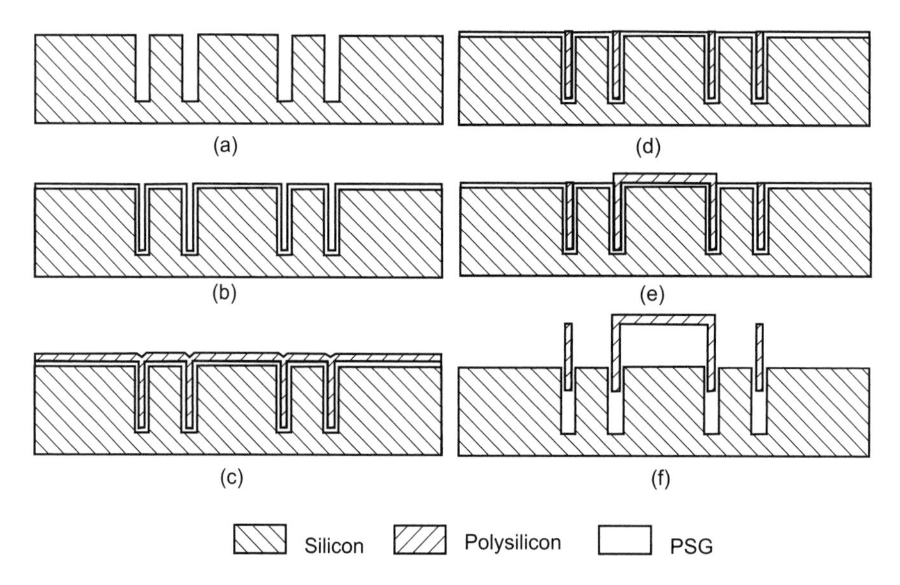


**Figure 3.13** HexSil-process: (a) DRIE; (b) deposition of the sacrificial layer; (c) deposition of the structural layer; (d) polishing; (e) deposition of linkage structures; and (f) release etch. (*After*: [37].)

### 3.3.2.2 Single Crystalline Silicon Surface Micromachining

Single crystalline silicon surface micromachining refers to machining of near-surface single crystalline silicon by etching an underlying sacrificial layer. All these techniques utilize DRIE to open high-aspect-ratio trenches, but differ in the type of sacrificial layers.

The techniques discussed here are: single crystal reactive etching and metalization (SCREAM), silicon micromachining by single step plasma etching (SIMPLE), silicon on insulator (SOI), black silicon method (BSM), porous silicon method (PSM), and surface/bulk micromachining (SBM) using anisotropic underetching of the $\{111\}$-wafer.

All of these processes are developed for lateral MEMS devices. However, they are well suited for planar microfluidic systems. The sacrificial etching process, with slight modifications, can be used for making buried microchannels.

*SCREAM.* SCREAM is a single mask process in single crystalline silicon. First, the functional structure is etched in silicon using standard RIE [41] or the more advanced DRIE. In the next step, the structure surface is coated with PECVD oxide. Further anisotropic RIE with $CF_4/O_2$ plasma removes the protecting oxide layer at the trench bottom and exposes silicon for the following release etch. Free moveable structures are released by a subsequent RIE with $SF_6$, which is selective to the protecting oxide layer. In the last step, a metal layer is sputtered and patterned on the structures to form the electrical interconnects. The final SCREAM structure is depicted in Figure 3.14(a).

*SIMPLE.* SIMPLE uses an epitaxial layer as the functional layer. The sacrificial layer is a highly n-doped region. The technique uses $Cl_2$-based plasma, which etches p-silicon and lightly n-doped silicon anisotropically, but highly n-doped silicon isotropically [42].

The process starts with the ion implantation of a highly n-doped silicon layer. This highly n-doped region is buried under an epitaxial layer, which is grown subsequently. Functional structures are etched in the epitaxial layer by RIE with PECVD oxide as a mask.

The RIE process has two steps. In the first step, epitaxial silicon and bulk silicon are etched anisotropically with $BCl_3$. In the second step, $BCl_3$ is mixed with $Cl_2$ and results in an isotropic

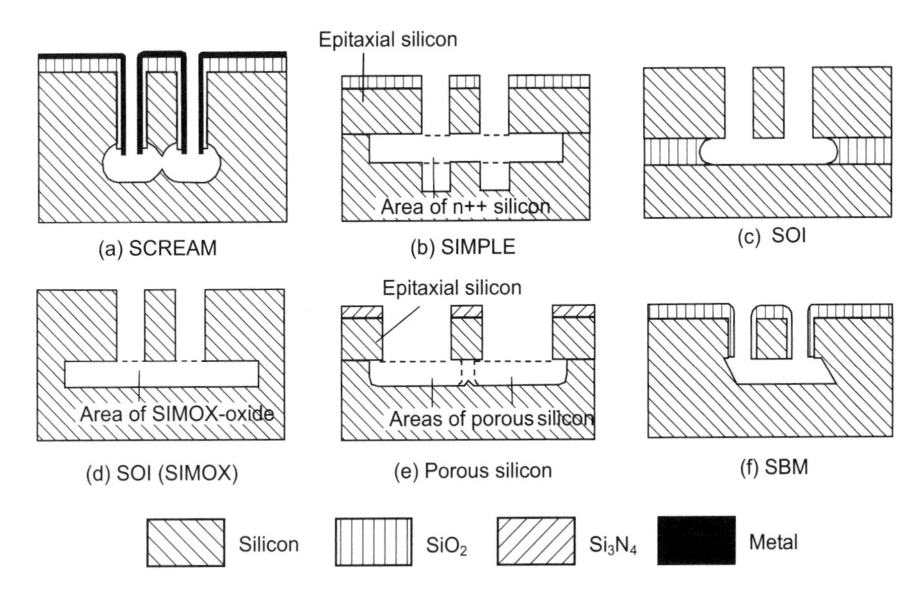


**Figure 3.14** Single crystalline silicon surface micromachining: (a) SCREAM; (b) SIMPLE; (c) SOI; (d) SOI (SIMOX); (e) porous silicon; and (f) SBM.

etchant, which attacks the highly n-doped silicon layer and frees the moveable epitaxial structure [Figure 3.14(b)].

*SOI.* SOI wafers are commercially available with a single crystalline silicon layer, which is separated from the handle wafer by an oxide layer. The single crystalline layer on top is the structural layer. The buried oxide layer can be used as sacrificial layer.

The mechanical structures are defined by DRIE, and the etched trenches should meet the buried sacrificial oxide layer. Etching the oxide layer releases the moveable structures [Figure 3.14(c)]. If the buried layer is fabricated by SIMOX, selective oxygen implantation and consequently selective sacrificial etching can be achieved [Figure 3.14(d)].

*BSM.* BSM uses $SF_6/O_2/CHF_3$ plasma [43, 44, 45]. Because of alternate deposition and etching characteristics of the plasma, small spikes are formed on the etched surface. These spikes keep incoming light in the area between them and make the silicon surface appear black, which gives the name to this method. Sacrificial etching processes of BSM are similar to SIMPLE and SOI. They are called by their authors BSM-SOI and BSM-SISI (silicon on insulator on silicon on insulator), respectively.

*PSM.* PSM uses the n-doped epitaxial silicon as the structural layer and porous silicon in the substrate as the sacrificial layer [46]. Porous silicon is created by electrochemical etching in HF solution.

The process begins with the growth of a lightly n-doped epitaxial layer on a p-silicon substrate. In the next steps, silicon nitride and silicon dioxide are deposited on the epitaxial layer. Silicon dioxide serves as a mask for subsequent reactive ion etching of nitride and epitaxial silicon. Silicon nitride is used as a self-aligned mask for the subsequent electrochemical etching process to form porous silicon in the bulk substrate. In the final step, KOH solution removes porous silicon from the bulk substrate and releases the free moveable structures [Figure 3.14(e)].

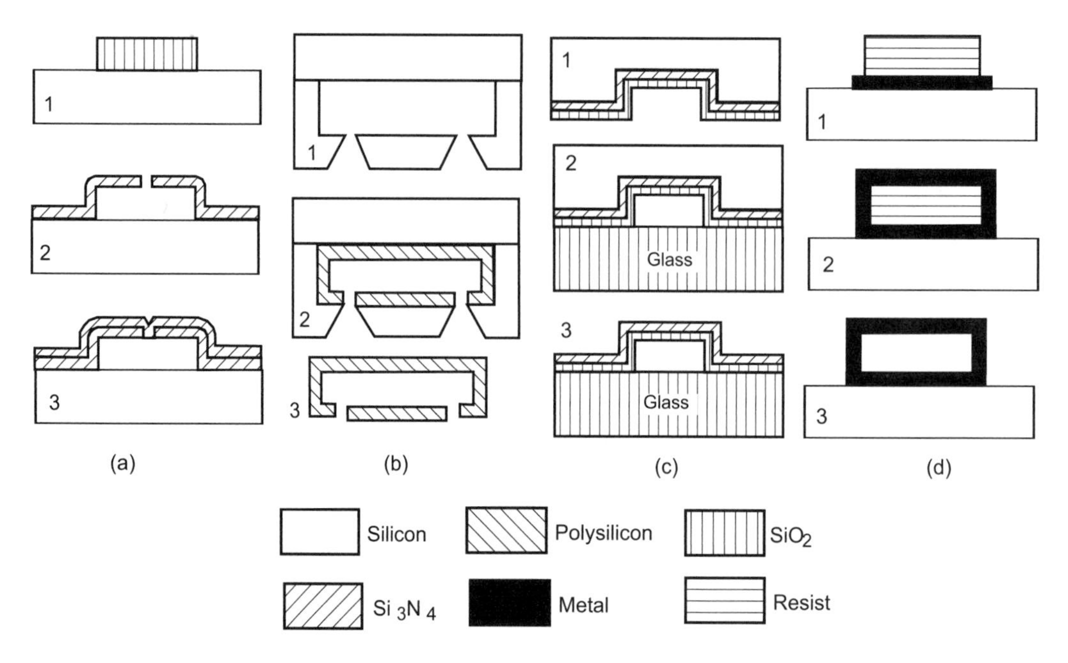


**Figure 3.15** Surface-micromachined channels: (a) polysilicon channel; (b) molded silicon channel; (c) oxide/nitride channel; and (d) metal channel.

*SBM.* SBM refers to a combination of DRIE and subsequent anisotropic underetching in an {111}-oriented silicon wafer [47]. The process starts with DRIE in a {111}-silicon wafer, which has a PECVD oxide layer as the mask. In the next step, PECVD oxide is deposited on the trench wall as the passivation layer for the subsequent wet etching step. The oxide layer at the bottom surface is then removed by RIE. Silicon at the trench bottom is exposed to a wet etchant. Finally, the entire wafer is dipped in the KOH solution, which etches silicon anisotropically in a lateral direction. The underetching process stops if all etch fronts reach {111}-planes [Figure 3.14(f)].

#### 3.3.2.3 Fabrication of Microchannels with Silicon Surface Micromachining

Surface micromachining is used to fabricate most of the microneedles discussed in Chapter 9. This section summarizes typical surface micromachining processes for microchannels.

A general surface micromachining process for microchannels starts with deposition of the sacrificial layer [Figure 3.15(a), part 1]. The channel material is then deposited over the structured sacrificial layer. After opening etch accesses through the channel wall, the sacrificial layer is etched to hollow the channel [Figure 3.15(a), part 2]. A subsequent deposition of channel material seals the hollow channel.

The channel in a microneedle can be fabricated with the above technique [48]. PSG is deposited as a sacrificial layer over a ground silicon nitride layer. The structured sacrificial layer is then encapsulated by LPCVD silicon nitride. Etch access is opened with RIE of the nitride wall. After removing the sacrificial PSG, a second LPCVD process seals the empty channel with silicon nitride. In a similar process, the channel is underetched resulting in a suspended nitride channel [49].

Using the HexSil method described in Section 3.3.2.1, polysilicon channels can be fabricated with a silicon mold [50]. The mold is fabricated in silicon using bulk micromachining [Figure 3.15(b), part 1]. Before depositing the channel material, a sacrificial oxide layer is deposited on the inner wall of the mold wafer. Polysilicon deposition defines the channel wall [Figure 3.15(b), part 2]. Etching away silicon dioxide releases the polysilicon channel [Figure 3.15(b), part 3]. The molding approach overcomes the thickness limitation of the standard surface micromachining process, where the channel height is constrained by the thickness of the sacrificial layer.

The molding approach can be further developed with the silicon substrate as sacrificial material [32]. Similar to the process of [51], the mold is fabricated in a handle wafer with bulk micromachining. The channel wall is defined by deposition of nitride/oxide double layer [Figure 3.15(c), part 1]. With silicon dioxide on top, the silicon wafer is bonded anodically to a glass wafer [Figure 3.15(c), part 2]. Etching away the silicon handle wafer releases the nitride/oxide channel on glass [Figure 3.15(c), part 3]. If the channel wall is too thin for certain applications, the surface of the structure on glass can be coated with a thick polymer layer [51].

Besides the above molding approaches, microchannels with reasonable heights can be fabricated with the process described in [52]. The process starts with deposition of a metal seed layer on the substrate [Figure 3.15(d)]. A subsequent electroplating process defines the bottom wall of the channel. Next, a thick-film photoresist such as AZ4620 is deposited and developed to form the sacrificial structure for the channel [Figure 3.15(d), part 1]. Gold is then sputtered on the resist structure as the second seed layer. Electroplating on this seed layer forms the side wall and top wall of the channel [Figure 3.15(d), part 2]. Etching the gold layer exposes the sacrificial photoresist. Removing photoresist with acetone creates a hollow metal channel [Figure 3.15(d), part 3]. A similar technique was used in [53] to fabricate more sophisticated microfluidic devices such as microvalves.

## 3.4 POLYMER-BASED MICROMACHINING TECHNIQUES

Polymeric micromachining uses polymers as structural material. The most well-known polymeric technique is LIGA, which includes thick-resist photolithography, electroplating, and micromolding. Besides the conventional X-ray lithography, there are a number of alternative polymeric techniques for high-aspect-ratio structures. This section discusses each of these techniques separately:

- Thick-resist lithography;
- Polymeric bulk micromachining;
- Polymeric surface micromachining;
- Microstereo lithography;
- Micromolding.

### 3.4.1 Thick Resist Lithography

#### 3.4.1.1 Polymethylmethacrylate (PMMA) Resist

Polymethylmethacrylate (PMMA) was originally used as a resist material for the LIGA technique [54]. The material is well known by a variety of trade names such as Acrylic, Lucite, Oroglas, Perspex, and Plexiglas.

PMMA can be applied on a substrate by different methods: multiple spin coating, prefabricated sheets, casting, and plasma polymerization. Multilayer spin coating causes high interfacial stresses, which lead to cracks in the resist layer. These cracks can be avoided by using a preformed PMMA sheet, which is bonded to the substrate [55]. Monomer MMA (methylmethacrylate) can be used as the adhesive material for the bonding process [56]. PMMA can also be polymerized in situ with casting resin [57] or with plasma [58].

Structuring PMMA requires collimated X-ray with wavelengths ranging from 0.2 to 2 nm, which are only available in synchrotron facilities. X-ray also requires special mask substrates such as beryllium and titanium, which further increases the cost of this technique. The beryllium mask with its higher Young's modulus and thickness is optimal for X-ray lithography. The absorber material

**Table 3.11**
Film Thickness of Different SU-8 Types at a Spin Speed of 1,000 rpm (*After*: [64, 65])

| *Type* | *Kinematic Viscosity* ($m^2/s$) | *Thickness* (μm) |
|---|---|---|
| SU-8 2 | $4.3 \times 10^{-5}$ | 5 |
| SU-8 5 | $29.3 \times 10^{-5}$ | 15 |
| SU-8 10 | $105 \times 10^{-5}$ | 30 |
| SU-8 25 | $252.5 \times 10^{-5}$ | 40 |
| SU-8 50 | $1,225 \times 10^{-5}$ | 100 |
| SU-8 100 | $5,150 \times 10^{-5}$ | 250 |

of an X-ray mask can be gold, tungsten, or tantalum. The thicker the absorber layer, the higher the X-ray energy, and the higher the aspect ratio in PMMA.

X-ray changes PMMA property in the exposed area, which is chemically etched in the development process. The developer consists of a mixture of 20 vol% tetrahydro-1,4-oxazine, 5 vol% 2-aminoethanol-1, 60 vol% 2-(2-butoxy-ethoxy) ethanol, and 15 vol% water [59]. The limited access and costs of a synchrotron facility are the main drawbacks of the LIGA technique in general and PMMA as polymeric structural material in particular. Thick-film resist such as SU-8 and the AZ-4000 series has the advantage of using low-cost UV exposure. However, structure heights and aspect ratios of UV exposure cannot meet those of PMMA with X-ray exposure.

#### 3.4.1.2 SU-8 Resist

SU-8 is a negative photoresist based on EPON SU-8 epoxy resin for the near-UV wavelengths from 365 to 436 nm. At these wavelengths the photoresist has very low optical absorption, which makes photolithography of thick films with high aspect ratios possible. This resist was developed by IBM [60, 61]. The material was adapted for MEMS applications during collaboration between EPFL-Institute of Microsystems and IBM-Zurich [62, 63]. Structure heights up to 2 mm with an aspect ratio better than 20 can be achieved with standard lithography equipment.

Photoresists such as SU-8 are based on epoxies, which are referred to as oxygen bridges between two atoms. Epoxy resins are molecules with one or more epoxy groups. During the curing process, epoxy resins are converted to a thermoset form or a 3-D network structure. SU-8 photoresist consists of three basic components:

- Epoxy resin, such as EPON SU-8;
- Solvent, called *gamma-Butyrolactone* (GBL);
- Photoinitiator, such as triarylium-sulfonium salts.

SU-8 photoresists are commercially available with different viscosities. Table 3.11 gives an example of the SU-8 series of MicroChem Inc. in Newton, Massachusetts. The higher product number indicates a higher viscosity and consequently a higher film thickness at the same spin speed. A standard SU-8 process consists of the following.

- *Spin coating.* The film thickness and consequently the structure height are determined by the viscosity of the photoresist and the spin speed. Higher viscosity or lower spin speed results in a thicker resist film.
- *Soft bake.* After spin coating, the film is soft-baked to evaporate the solvent. Soft bake can be carried out on a level hot plate or in a convection oven. Two-step temperature ramping between 65°C and 95°C is recommended [64, 65].
- *Exposure.* SU-8 can be exposed with I-line equipment, which uses a mercury lamp with near-UV wavelengths. Optical absorption of SU-8 increases sharply below 350 nm. Therefore,

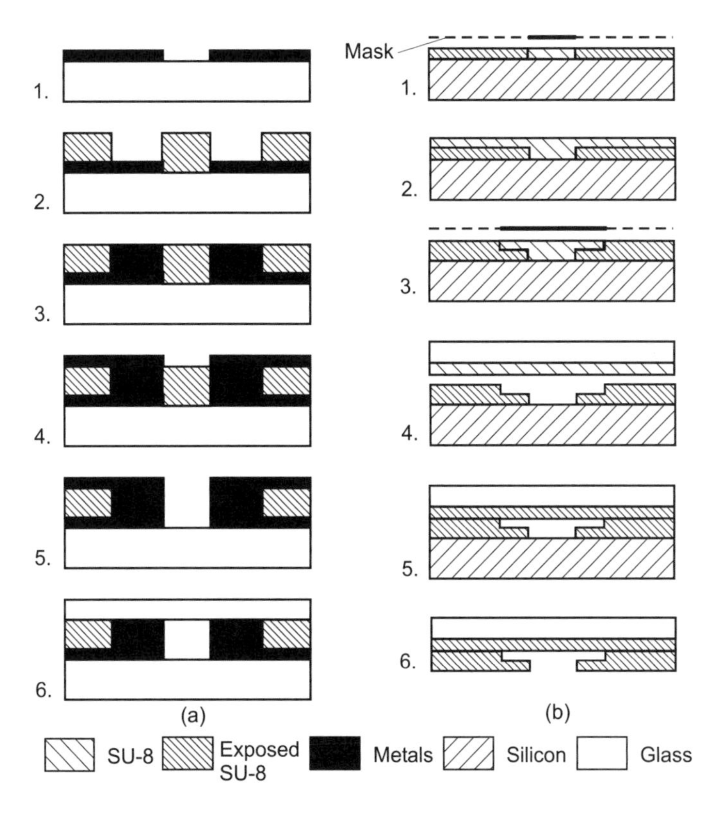


**Figure 3.16** Fabrication of microchannel with SU-8: (a) single layer combined with electroplating (*After*: [67]); and (b) multiple layer (*After*: [68]).

wavelengths higher than 350 nm should be used for the exposure. The thicker the film, the higher the exposure dose required.

- *Post-exposure bake (PEB).* The exposed area of SU-8 film is selectively cross-linked by a post-exposure bake. The cross-link process can cause high film stress, which damages the film with cracks. To avoid this problem, a two-step ramp between 65°C and 95°C [64, 65] or between 50°C and 100°C is recommended. Rapid cooling after PEB should be avoided.
- *Developing.* Immersion processes or spray processes can be used to develop the resist. Solvent-based developers, such as ethyl lactate and diacetone alcohol, dissolve areas that are not polymerized during exposure and PEB.
- *Hard bake.* If necessary, the developed structure can be hard-baked at elevated temperatures from 150°C to 200°C. However, hard baking can increase stress and cause cracks in structures [66].
- *Remove.* Removing a polymerized SU-8 film is the most difficult process, because SU-8 film becomes highly cross-linked after exposure and PEB. Etching with acid solutions, RIE, and laser ablation [67] are some of the methods for removing SU-8.

Because of its simple processes and the relatively good mechanical properties, SU-8 is used as structural material for many microfluidic applications. There are different ways to fabricate microchannels with SU-8: the use of SU-8 as spacer, whole SU-8 channel with embedded mask, whole SU-8 channel with selective photon writing, and whole SU-8 channel with a sacrificial layer.

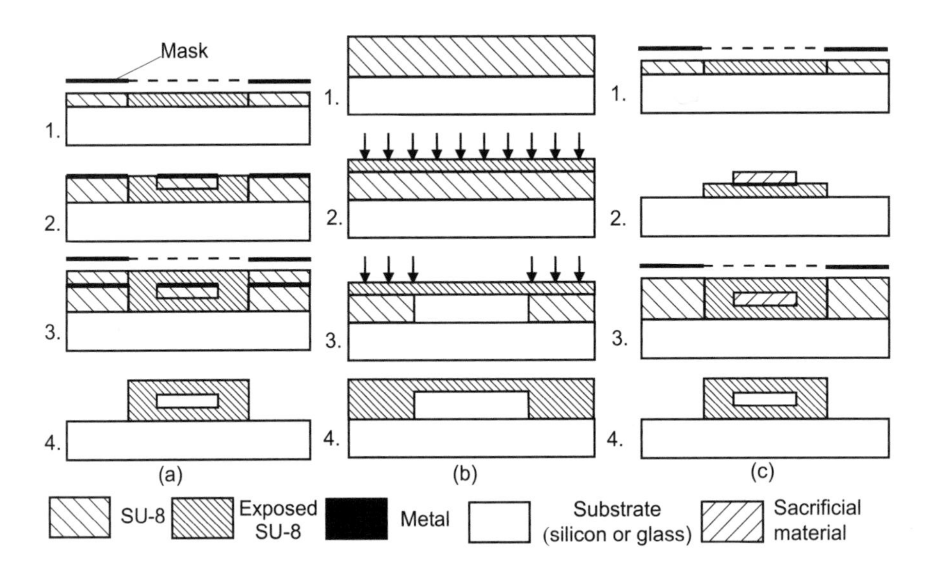


**Figure 3.17** Fabrication of covered channels with SU-8: (a) with embedded mask (*After*: [71]); (b) selective proton writing (*After*: [74]); and (c) with sacrificial layer (*After*: [72].)

*Fabrication Example with SU-8 as Spacer.* The simplest technique to form a microchannel with SU-8 is using the structured SU-8 film to define the channel's sidewall. The top wall and the bottom wall are made of other materials such as silicon and glass. A LIGA-like process is described in [68] [Figure 3.16(a)]. Starting with a glass wafer as substrate material, a metal seed layer is deposited [Figure 3.16(a), part 1]. SU-8 is spin-coated and structured [Figure 3.16(a), part 2]. The high-aspect-ratio SU-8 structure is used as a mold for electroplating of gold [Figure 3.16(a), part 3]. The actual microchannel is etched in SU-8 with oxygen plasma using an aluminum mask [Figure 3.16(a), parts 4 and 5]. Finally, a glass plate covers the structure using adhesive bonding.

In the above example, microchannels are etched in the exposed SU-8 film. Actually, microchannels can be directly formed by photolithography. The example shown in Figure 3.16(b) uses two SU-8 layers to form a 3-D structure [69]. The first layer is coated and exposed on a silicon substrate [Figure 3.16(b), part 1]. The next layer is spin-coated on top of the exposed first layer [Figure 3.16(b), part 2]. The mask for the second layer should cover completely the unexposed areas of the first layer to avoid double exposure [Figure 3.16(b), part 3]. After exposure of the second layer, the two layers are developed to form the channel structure. The channel is covered by a glass plate with a thin unexposed SU-8 layer on it [Figure 3.16(b), part 4]. The bonded surface is cross-linked by a blanket exposure through the glass plate. After PEB, a microchannel is formed between a glass substrate and a silicon substrate.

Optionally, the silicon substrate can be etched away to yield an optically transparent device. In many applications, a single SU-8 layer can be used to form microchannels and other microfluidic components [70]. The top and bottom of the channels can be sealed with SU8-coated glass plates and subsequent blanket exposure as described above.

*Closed SU-8 Channel with Embedded Mask.* Three-dimensional structures are formed with multi-layer exposure, as described in Figure 3.16(a). The only constraint is that the mask of the later layer should cover completely the previous layers to protect their unexposed areas. That means the direct fabrication of a closed structure, such as a covered channel, is not possible. One solution for the double-exposure problem is the use of an embedded mask [71, 72]. The process starts with exposure of the first SU-8 layer to form the bottom of the channel [Figure 3.17(a), part 1].

Subsequently, the second layer is coated. The next step is the fabrication of the embedded mask. A thin metal layer such as gold [71] is sputtered on the second SU-8 layer. This metal layer is then patterned by common photolithography and etching. The patterned metal layer is used as an embedded mask for the subsequent exposure of the second SU-8 layer [Figure 3.17(a), part 2]. A third SU-8 layer is spin-coated and exposed to fabricate the top wall of the channel [Figure 3.17(a), part 3]. In the final step, all three layers are developed in a single process, resulting in a covered microchannel. The embedded mask is washed away after the developing process [Figure 3.17(a), part 4]. Instead of the embedded metal mask, a antireflection film, such as CK-6020L resist (FujiFilm Olin Inc., Japan), can be used for making covered SU-8 microchannel [73]. The use of antireflection coating ensures that this coating and the structural SU-8 can be developed at the same time.

*Covered SU-8 Channel with Selective Proton Writing.* A covered channel can be fabricated with selective proton writing or proton beam micromachining [74]. Similar to near-UV exposure, a proton beam also causes polymerization in SU-8. The depth of the polymerized area depends on the proton beam energy. This feature is used to form a 3-D structure in SU-8. Figure 3.17(b) illustrates the relatively simple steps of this technique. To start with, the SU-8 layer is spin-coated on the substrate [Figure 3.17(b), part 1]. Writing with low energy forms the top of the channel. With low energy, a proton beam can only penetrate shallowly into the SU-8 layer. A thin polymerized layer is formed [Figure 3.17(b), part 2]. In the next step, the proton beam with higher energy polymerizes the sidewalls of the channel. The high energy allows the beam to penetrate through the SU-8 layer down to the substrate surface [Figure 3.17(b), part 3]. In the final step, the exposed SU-8 is developed, resulting in a covered microchannel [Figure 3.17(b), part 4].

*Whole SU-8 Channel with a Sacrificial Layer.* A further method of fabricating a closed SU-8 channel is using a sacrificial layer [72, 75]. The first SU-8 layer is coated, exposed, and developed to form the bottom of the channel [Figure 3.17(c), part 1]. Subsequently, a sacrificial structure is deposited and patterned [Figure 3.17(c), part 2]. The sacrificial material can be thermoplastics, waxes, epoxies [72], or positive photoresist [74]. The sidewalls and the channel ceiling are formed with a second SU-8 layer [Figure 3.17(c), part 3]. After developing the second layer, the sacrificial material inside the channel is removed, leaving a closed SU-8 microchannel [Figure 3.17(c), part 4].

#### 3.4.1.3 AZ4562 (Clariant)

AZ4562 (Clariant, Charlotte, North Carolina) is a positive photoresist that is commercially available. This resist belongs to the Novolak resist system, in common with most commercially available positive resists. Using multilayer spin-coating, thick-resist layers up to 100 μm can be achieved. This photoresist has no oxygen sensitivity, but a high resistance to plasma etching, good adhesion properties, and high-resolution capability [76]. This photoresist is typically used as a mold for subsequent metal electroplating [77, 78] or as master templates for micromolding. There are no reports of using AZ4562 directly as structural material. Table 3.12 gives a summary of the parameters of PMMA, SU-8, and AZ4562.

#### 3.4.1.4 Other Thick-Film Resists

AZ9260 is the other Novolak photoresist from Clariant. AZ9260 exhibits a better transparency compared to AZ4562, and therefore promises a better aspect ratio. Aspect ratios up to 15 are achieved with a film thickness of 100 μm [78]. A theoretical thickness of 150 μm is expected from this photoresist.

Ma-P 100 (Microresist Technology, Berlin, Germany) is the other photoresist that can give structure heights up to 100 μm. This photoresist has aspect ratios on the order of 5, poorer than that of the AZ-family [79].

**Table 3.12**
Comparison of Different Thick-Film Resists

| *Resist* | *PMMA* | *SU-8* | *AZ4562* |
|---|---|---|---|
| Exposure type | X-ray (0.2–2 nm) | UV (365, 405, 435 nm) | UV (365, 405, 435 nm) |
| Light source | Synchrotron facility | Mercury lamp | Mercury lamp |
| Mask substrate | Beryllium (100 μm) | Quartz (1.5–3 mm) | Glass (1.5-3 mm) |
| | Titanium (2 mm) | Glass (1.5–3 mm) | Quartz (1.5–3 mm) |
| Mask absorber | Gold (10–15 μm) | Chromium (0.5 μm) | Chromium (0.5 μm) |
| Maximum height | 1,000 μm | 250 μm | 100 μm |
| Aspect ratio | $\sim$ 500 | 20–25 | $\sim$ 10 |
| Young's modulus (GPa) | 2–3 | 4–5 | — |
| Poisson's ratio | — | 0.22 | — |
| Glass temperature (°C) | 100 | $>$ 200 | — |

### 3.4.2 Polymeric Bulk Micromachining

Similar to its silicon-based counterpart, polymeric bulk micromachining uses etching to transfer a pattern directly into the bulk substrate. Polymers and almost all organic material can be etched with an oxygen plasma. Since photoresists are organic, they cannot be used for the the etching process in oxygen plasma. Metals such as titanium [80] or aluminum [81] work well as the masking material for this purpose.

Polymeric bulk micromachining can form simple microchannels in a polymeric substrate. Moveable structures in a polymeric substrate can be fabricated using two basic methods: multiple etching steps with different masks and backside blanket-etching. In the first method, the first etching step defines the thickness of the moveable structure. In the second step, the masking layer is removed. The exposed area is etched, until the moveable structure is released [Figure 3.18(a)]. In the second method, the moveable structure is formed in the front side of the substrate using any polymeric technique, such as etching, molding, or hot embossing [82]. The substrate is then turned upside-down. Blanket-etching of the backside in oxygen plasma releases the moveable structure [Figure 3.18(b)].

Since most polymers have glass transition temperatures between 100°C and 200°C, movable structures fabricated by polymeric bulk micromachining can be assembled by thermal direct bonding at a relatively low temperature. Combining both micromachining technique and bonding technique would allow a great flexibility in making polymeric microfluidic components. Almost all existing design concepts for micropumps, microvalves, and microflow sensors can be transferred from silicon-based technology to polymeric technology.

### 3.4.3 Polymeric Surface Micromachining

Polymeric surface micromachining is similar to silicon surface micromachining. Polymers are used either as structural material or as sacrificial material. The following common polymers can be deposited and patterned in batch processes.

#### 3.4.3.1 SU-8

SU-8 is a thick-film resist, which can be structured using UV lithography (Section 3.4.1.2). With a Young's modulus of 4 to 5 GPa and a Poisson's ratio of 0.22, hardbacked SU-8 poses excellent mechanical properties and can be used for moveable parts. The sacrificial material for the release of the SU-8 part can be the silicon substrate, a metal layer, or a polymer layer. Figure 3.19 demonstrates this technology for the fabrication of a microcheck valve [83]. The valve was first structured on

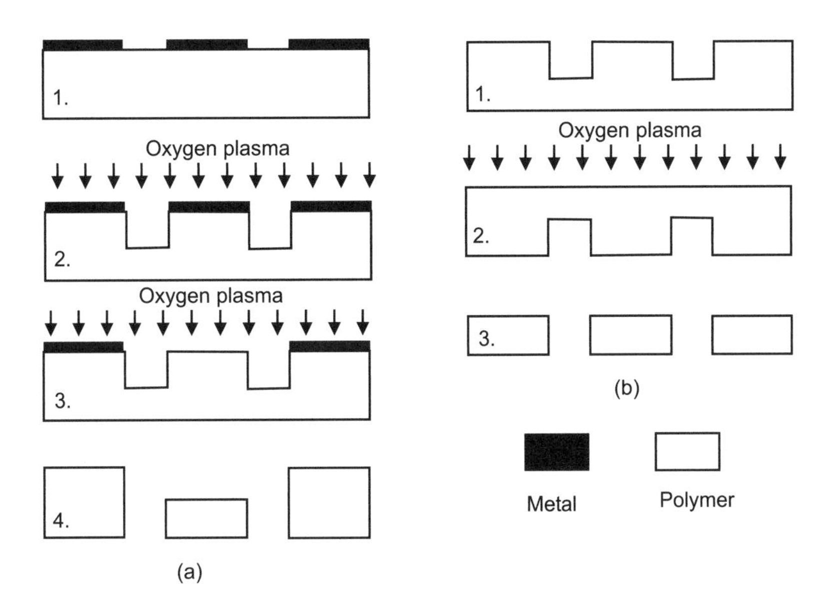


**Figure 3.18** Fabrication of movable structures using polymeric bulk micromachining: (a) multistep etching in oxygen plasma; and (b) backside blanket-etching in oxygen plasma.

silicon substrate with a two-layer process. Developing both layers results in a 3-D valve structure with spring beams, a valve disc, and a sealing ring. Underetching silicon with KOH releases the valve. Circular access holes were placed on the structure for faster release. The smooth contours of the design help to arrest surface stress and avoid cracks in the structure.

For many applications, a metal layer on the structural polymeric material is needed. The metal layer can be structured to form electrodes and heaters. Figure 3.20 shows the fabrication steps of a SU-8 microgripper, which has a thin metal layer on top acting as a heater [84]. Instead of silicon, a thin layer of polystyrene can be used as sacrificial layer. The polymeric sacrificial material can be dissolved by organic solvents such as toluene. In contrast to KOH for sacrificial silicon etching, solvent does not attack the thin metal layer.

### 3.4.3.2 Polyimide

Polyimide is commercially available as photoresists such as Proimide 348 or 349 (Ciba Geigy) or PI-2732 (DuPont). A single spin can result in a film thickness up to 40 μm. Photosensitive polyimide can be used for the same purpose as other thick resists described in the previous sections [85].

Fluorinated polyimide is an interesting material because of its optical transparency and simple machining. In RIE processes of this material, fluorine radicals are released from the fluorinated polyimide and act as etchants [86].

Polyimide is a good substrate material. Metals such as aluminum, titanium, and platinum can be sputtered on it using the lift-off technique [87]. Similar to other polymers, polyimide can be etched with RIE in oxygen plasma. Combining photolithography, RIE, and lamination, complex channel structures with metal electrodes can be fabricated in polyimide [88].

### 3.4.3.3 Parylene

Parylene is a polymer that can be deposited with CVD at room temperature. The CVD process allows coating a conformal film with a thickness ranging from several microns to several millimeters.

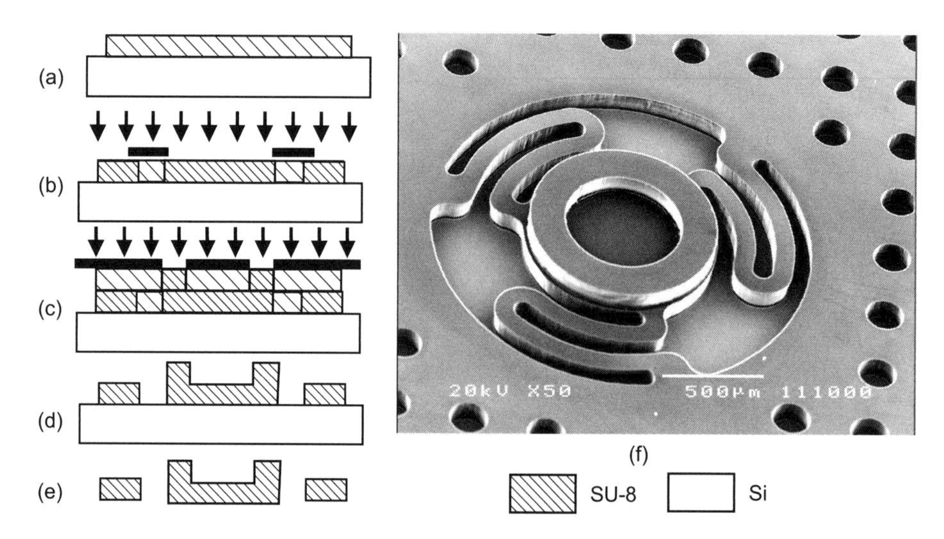


**Figure 3.19** Fabrication of a SU-8 micro check valve using polymeric surface micromachining: (a) coating the first SU-8 layer; (b) structuring the spring structure and the valve disc; (c) coating the second SU-8 layer and structuring the sealing ring; (d) developing SU-8; (e) underetching silicon to release the SU-8 structure; and (f) the fabricated check valve.

The basic type is parylene N, which is poly-paraxylylene. Parylene N is a good dielectric, exhibiting a very low dissipation factor, high dielectric strength, and a frequency-independent dielectric constant.

Parylene C is produced from the same monomer, modified only by the substitution of a chlorine atom for one of the aromatic hydrogens. Parylene C has a useful combination of electrical and physical properties as well as a very low permeability to moisture and other corrosive gases. Parylene C is also able to provide a conformal insulation.

Parylene D is modified from the same monomer by the substitution of the chlorine atom for two of the aromatic hydrogens. Parylene D is similar in properties to parylene C with the added ability to withstand higher temperatures. Figure 3.21 shows the chemical structures of these three parylene types. Deposition rates are fast, especially for parylene C, which is normally deposited at a rate of about 10 μm/min. The deposition rates of parylene N and parylene D are slower. Parylene can be used in microfluidic devices as a structural material, which offers low Young's modulus. Such a soft material is needed in microvalves and micropumps (see Chapters 6 and 7). Furthermore, parylene coating can improve the biocompatibility of a microfluidic device.

#### 3.4.3.4 Electrodepositable Photoresist

Eagle ED2100 and PEPR 2400 (Shipley Europe Ltd., England) are electrodepositable photoresists. These photoresists were originally developed for printed circuit boards. The photoresist is an aqueous emulsion consisting of polymer micelles. The resist is deposited on wafers by a cataphoretic electrodeposition process [76]. In an electric field, positively charged micelles move to the wafer, which works as a cathode. The polymer micelles coat the wafer until the film is so thick that deposition current approaches zero. The nominal resist thickness is about 3 to 10 μm.

The thickness and aspect ratios of electrodepositable photoresists are not relevant for thick-film application. Electrodepositable photoresist is suited for applications where it is difficult to spin on a resist after patterning the substrate surface. Due to the relatively conformal nature of the deposition process, electrodepositable photoresists are interesting sacrificial materials for polymeric surface micromachining.

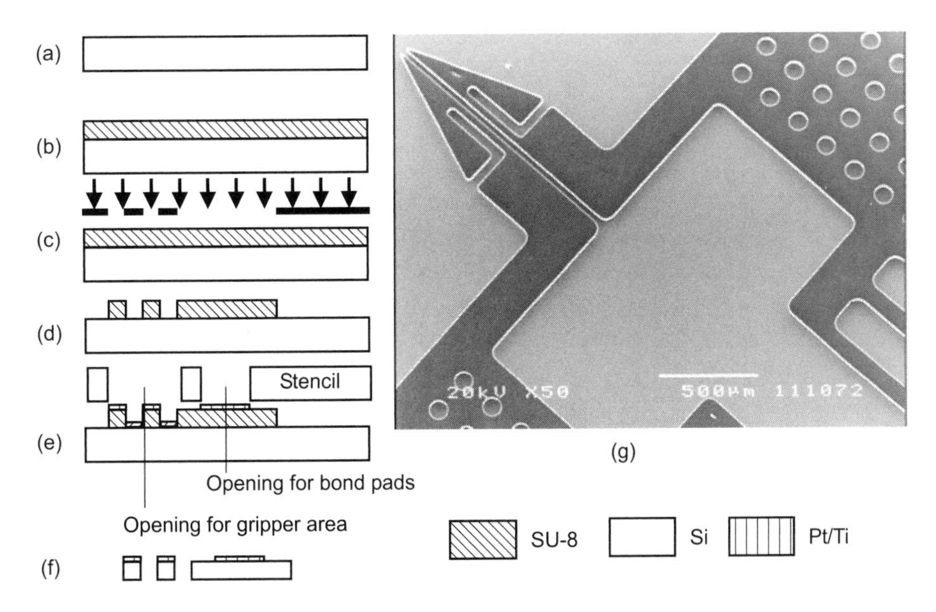


**Figure 3.20** Fabrication of an SU-8 microgripper using polymeric surface micromachining: (a) blank silicon substrate or coated with a polymeric sacrificial layer; (b) coating the SU-8 layer; (c) UV exposure; (d) developing the gripper structure; (e) coating the metal layer through a silicon stencil; (f) releasing the gripper; and (g) the fabricated microgripper.

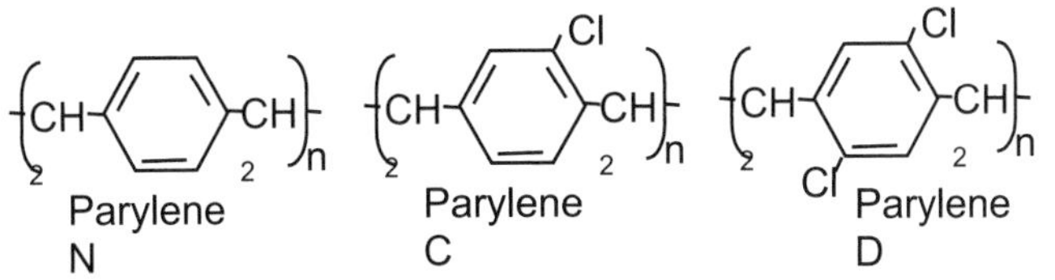

Parylene N  Parylene C  Parylene D

**Figure 3.21** Chemical structures of parylene.

### 3.4.3.5 Conductive Polymers

Conductive polymers or conjugated polymers are polymeric materials, which have received the growing attention of the MEMS community. Conjugated polymers have alternating single and double bonds between a carbon atom along the polymer backbone. The conjugation results in a bandgap and makes the polymers behave as semiconductors.

Conjugated polymers can be doped electrochemically, electrically, chemically, or physically with ion implantation. In the doped state, conjugated polymers are electrically conducting. These characteristics allow conductive polymers to be used as the material for electronic devices such as diodes, light emitting diodes, and transistors.

Furthermore, the doping level of polymers is reversible and controllable. In some conducting polymers, the change of doping level leads to volume change, which can be used as an actuator. The most common and well-researched conjugated polymer is polypyrrole (PPy).

The major deposition techniques for conjugated polymers are the following.

- *Spin-coating of polymers.* The polymer is dispersed or dissolved in a solvent. The resin can be spin-coated on a substrate. Melted polymers can also be applied directly on the substrate.
- *Spin-coating of precursor and subsequent polymerization.* Precursor polymers are dissolved in a solvent. The polymerization occurs at an elevated temperature. Polymerization can also be

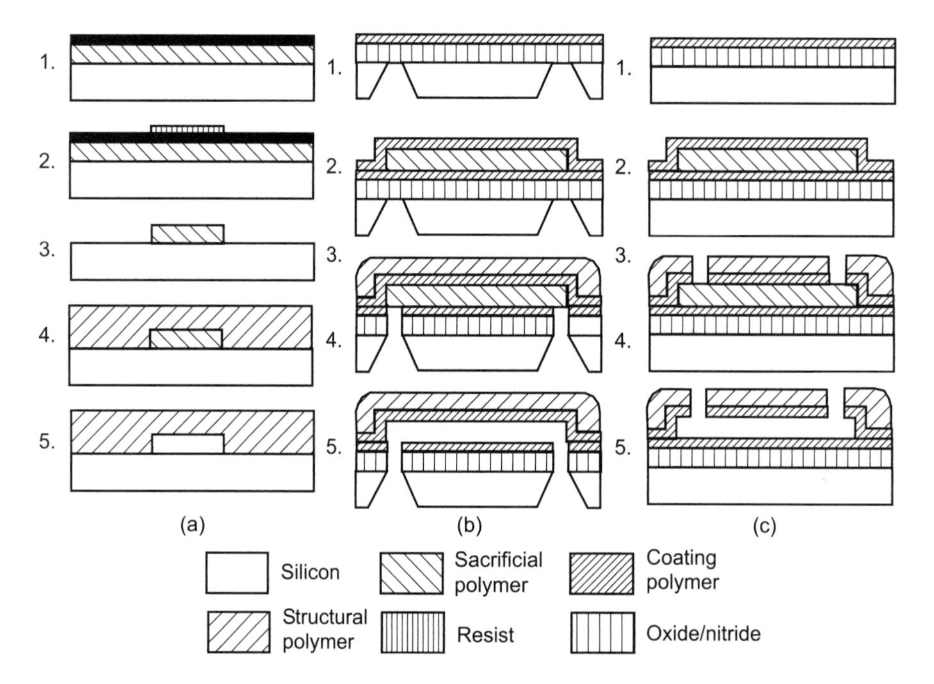


**Figure 3.22** Fabrication of microchannel with polymeric surface micromachining: (a) a simple channel (*After*: [90]); (b) and (c) microchannel with functional coating of inner wall (*After*: [113]).

achieved by a UV cross-link of the precursor with a photoinitiator, similar to the developing process of SU-8.

- *Chemical vapor deposition* is similar to the technique used for parylene.
- *Electrochemical deposition* is similar to electroplating of a metal layer [89].

### 3.4.3.6 Fabrication of Microchannels with Polymeric Surface Micromachining

Polymeric surface micromachining is perfectly suitable for closed microchannels. Both the structural layer and the sacrificial layer can be made of polymers. The typical fabrication process is shown in Figure 3.22(a). To start with, the sacrificial polymer is spin-coated on the substrate, which can be silicon or glass. The thickness of this layer determines the channel height. Similar to thick-film resists, the thickness of this polymer layer is controlled by the viscosity of the solution and the spin speed. Next, a metal layer is sputtered over the sacrificial layer as a mask [Figure 3.22(a), part 1]. Using photolithography, channel patterns are transferred to the mask [Figure 3.22(a), part 2]. The sacrificial layer is then structured by RIE with oxygen plasma [Figure 3.22(a), part 3]. After removing the resist and mask layers, the structural layer is deposited over the sacrificial structures [Figure 3.22(a), part 4]. At elevated temperatures, the sacrificial polymer decomposes into volatile products and leaves behind the desired microchannel [90].

Polynorbornene (PNB) is a good sacrificial polymer [91]. The decomposition temperatures of PNB are between 370°C and 425°C. Generally, silicon dioxide and silicon nitride are ideal encapsulation materials at these relatively high temperatures. If polymers as structural materials are required, they should have high glass transition temperatures and high thermal stability. Polyimides (Amoco Ultradel 7501, Dupont PI-2611, and Dupont PI-2734) are ideal for this purpose because of their glass transition temperature of over 400°C [91].

Polycarbonates, such as polyethylene carbonate (PEC) and polypropylene carbonate (PPC), offer relatively low decomposition temperature on the order of 200°C to 300°C [90]. The low decomposition temperature is needed for structural materials with less thermal stability. Inorganic glass, silicon dioxide, thermoplastic polymers, and thermoset polymers can be used as structural materials [90].

Because of the increasing number of applications in life sciences, polymeric microchannels may need a biocompatible coating for their inner walls. Figures 3.22(b,c) illustrate the fabrication processes of such microchannels [113]. The two channel types differ in the type of fluidic access. The biocompatible material used in this example is parylene C. First, parylene is vapor-deposited on a silicon substrate, which is covered by a nitride/oxide barrier layer [Figures 3.22(b), part 1; and 3.22(c), part 1]. Thick-film resist AZ4620 is used as the sacrificial material. After photolithography, developing, and hard bake of the resist structures, a second parylene layer is deposited [Figures 3.22(b), part 2; and 3.22(c), part 2]. After roughening the parylene surface with oxygen plasma, photosensitive polyimide is spin-coated as a structural layer on top of the second parylene layer. Next, polyimide is exposed and developed. In the case of fluidic access from the backside, via-holes to the sacrificial layer are opened by a combination of $O_2/CF_4$ and $O_2$ plasma etch [Figure 3.22(b), part 4]. In the case of fluidic access from the front side, the top parylene layer is etched in oxygen plasma with an aluminum mask [Figure 3.22(c), part 4]. In the last step, the sacrificial layer is removed with acetone. The resulting microchannels are optically transparent and hermetic.

### 3.4.4 Microstereo Lithography (MSL)

In many polymeric micromachining techniques discussed above, resins are cured by UV exposure. The polymerization reactions are generated by the absorption of photons. The most common absorption process used in photolithography is single-photon absorption [92]. Concurrently, if the combined energy of two photons matches the transition energy between the ground state of the excited state of a material, a nonlinear phenomenon called two-photon absorption occurs [93].

The rate of two-photon absorption is a square function of incident light intensity, while the rate of single-photon absorption is a linear function of intensity. For a two-photon absorption, the laser beam power should be high, which in turn requires a short pulse width on the order of femtoseconds.

#### 3.4.4.1 Polymerization with Single-Photon Absorption

Using a directed UV beam, two-dimensional "slides" can be formed on the focusing plane in a liquid resin. Stacking many such slides allows shaping a 3-D structure. Stereo lithography is used for rapid prototyping in macroscale. Based on the state of the surface of the liquid resin, there are two major conventional stereo configurations: constrained surface and free surface [94]. If the beam is focused at a point in the liquid, the entire liquid volume from the surface to the focusing point is polymerized because of the relatively high absorbance of most resin in the UV range. The only solution for this problem is to keep the polymerization at the liquid surface.

*Configurations with constrained surface* keep the UV beam focused on the liquid surface. The advantage of this configuration is that moving the transparent glass plate can precisely control the thickness of each structural "slide" [Figure 3.23(a)].

Similar to SU-8, prepolymers for stereo lithography are epoxy-based materials, which work as an adhesive at the contact surface. This feature can be used for moving the prototype, as shown in Figure 3.23(b), but causes problems in separating the prototype from the glass plate in Figure 3.23(a).

Figure 3.24 describes the typical setup of microstereo lithography with constrained surface. The most important part is the 3-D precision position control. The layout of the device is transferred from computer aided design (CAD) data to the control unit, and is scanned directly by the x-y stage and the z-stage. The optical lens and the constraining glass are kept at a fixed distance, so that the UV beam is always focused at the surface [94].

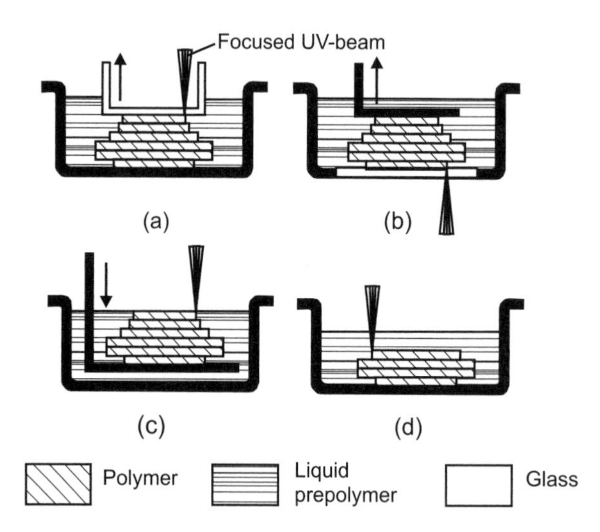


**Figure 3.23** Standard configurations of microstereo lithography: (a) and (b) free surface configuration; and (c) and (d) constrained surface configuration. (*After*: [94].)

*Configurations with free surface* can be realized by moving the prototype holder from the surface into the liquid, while keeping the UV beam focused on the free surface [Figure 3.23(c)]. With this technique, the free surface of the liquid is not stable and leads to difficulties in controlling the slide's thickness.

The configuration shown in Figure 3.23(d) is feasible with transparent liquid and a nonlinear response of polymerization to incident light intensity. First, the light itself should have higher wavelengths than the UV range to keep the resin transparent to the incident light. He-Cd blue laser ($\lambda = 422$ nm) is optimal for this purpose [95]. Second, keeping the laser intensity low avoids polymerization propagating. Under an optimum exposure, the resin only polymerizes in a spot around the focus and not in the out-of-focus region, due to thermal nonlinearity of a photopolymerization reaction [92]. This configuration has the advantage of true stereo lithography. Structures can be formed without the supporting base layer. By avoiding moving parts in the liquid and consequently the influence of viscosity and surface tension, this configuration can achieve a resolution better than 1 μm [95].

#### 3.4.4.2 Polymerization with Two-Photon Absorption

Polymerization with two-photon absorption is the second microstereo lithography technique that can use the setup depicted in Figure 3.25. This technique has several requirements for the liquid resin and the laser.

First, the liquid resin should be transparent to the laser, so that the focus point can move freely inside the liquid resin. The resin used in [93] is a mixture of urethane acrylate oligomers/monomers and photoinitiators. This resin has high absorbance in the UV range. The absorbance almost drops to zero after the spectrum of green wavelengths. Thus, the resin is transparent to near infrared wavelengths. Titanium-sapphire laser has a wavelength of 770 nm, which is optimal for this purpose. Optical transparency means that there is almost no single-photon absorption.

Second, the condition for two-photon absorption requires a high-energy laser with an extremely short pulse width. The titanium-sapphire laser used in [93] has a peak power of 3 kW at pulse width of 130 fs. With frequency doubling, a wavelength of 398 nm is used to form acrylic-ester-based Nopcocure 800 resin (San Nopco, Tokyo, Japan) [96].

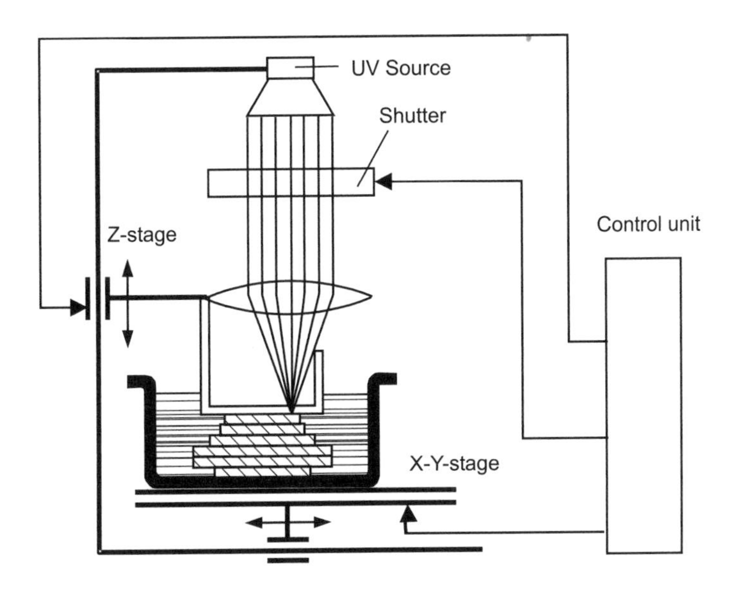


**Figure 3.24** System setup for microstereo lithography. (*After* [94].)

Third, the refractive index between polymerized resin and liquid resin should be small enough so that it does not affect the spatial accuracy of the scanning laser beam. At a wavelength of 770 nm, the resin described above has a refractive index of 0.02 [93].

The disadvantage of a prototyping approach can be solved with the setup shown in Figure 3.26. Instead of a single light beam, the source beam splits into several beams that are guided by optical fibers [97]. The number of optical fibers determines the number of devices, which can be fabricated in parallel.

#### 3.4.4.3 Polymerization with Layer-by-Layer Photolithography

The third approach of microstereo lithography combines the advantages of photolithography and stereo lithography [98]. Instead of scanning a 2-D pattern in each layer, the structures are exposed at once with a pattern generator. The pattern generator works as a mask and shapes the laser beam, so that it contains the image of the layer to be formed.

This process is similar to conventional photolithography. Selective polymerization of the liquid resin occurs in the exposed areas. This technique allows a much faster processing time than the scanning technique. By using multiple structures on the same mask, this technique may allow batch fabrication without any modification of the setup.

#### 3.4.4.4 Fabrication of Microfluidic Devices with Microstereo Lithography

Microstereo lithography method is currently considered a subsection of *additive manufacturing technologies* or *3-D printing methods*. Because of its flexibility in making any 3-D structure, microstereo lithography can be used for the fabrication of complex microfluidic devices. The technique allows the fabrication of microvalves [94] and integrated fluid systems [99], which have a microchannel network embedded in the transparent polymeric device. Microreactors with embedded photodiodes for biochemical analysis were reported in [95, 100]. Active devices such as SMA-actuated micropumps and microvalves can be integrated in the above systems for controlling

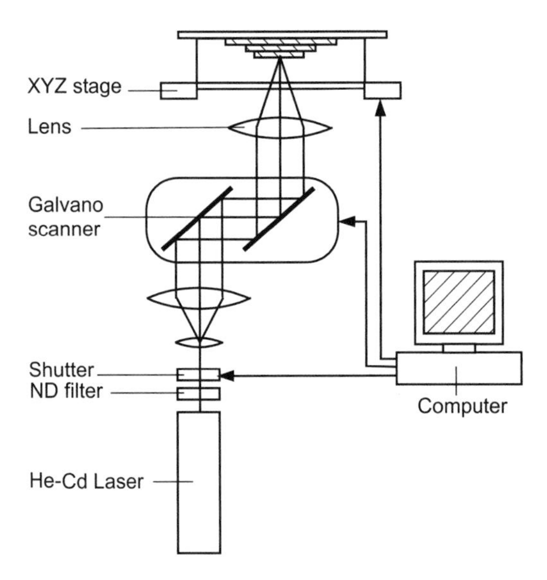


**Figure 3.25** System setup for microstereo lithography with free surface configuration and blue laser. (*After*: [95].)

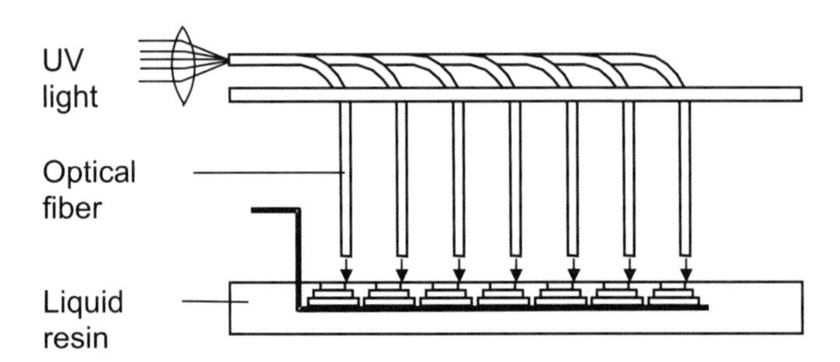


**Figure 3.26** Setup for mass production with microstereo lithography. (*After* [100].)

fluid flow [101]. The technique of layer-by-layer photolithography is used for fabricating static Y-mixers [102]. The flexibility of the technique allows scaling down of the complex geometry of a macroscopic mixer.

The advantage of microstereo lithography, and in general 3-D printing methods, is to fabricated the self-packaged devices. While fluidic interconnects are serious problems for microfluidic devices made with other microtechniques, interconnects are easily fabricated in the same process. Problems of leakage and expensive packaging can be solved with microstereo lithography. For 3-D printing of an object, initially a 3-D virtual model composed of overlaying slices of the object is produced. Then, the model is turned into a physical object through layer-by-layer printing of the slices [104, 105].

In contrary to photolithography, hot embossing and micro injection molding, 3-D printing methods enable the rapid fabrication of complex 3-D microfluidic structures in a self-packaged fashion without any need for having master molds or masks. However, more developments are required to enhance the resolution of 3-D printing methods as available for PDMS soft lithography. Importantly, it is expected that research teams from all around the world can design modular parts of microfluidic systems that can be shared through the Internet for 3-D printing and subsequent assembly.

**Table 3.13**

Typical Characteristics of Different Polymers for Micromolding (*After*: [109, 110].)

| *Polymers* | *PMMA* | *PC* | *PS* | *COC* | *PP* |
|---|---|---|---|---|---|
| Heat resistance (°C) | 105 | 140 | 100 | 130 | 110 |
| Density (kg/m$^3$) | 1,190 | 1,200 | 1,050 | 1,020 | 900 |
| Refractive index | 1.42 | 1.58 | 1.59 | 1.53 | opaque |
| *Resistant to:* | | | | | |
| Aqueous solutions | yes | limited | yes | yes | yes |
| Concentrated acids | no | no | yes | yes | yes |
| Polar hydrocarbons | no | limited | limited | yes | yes |
| Hydrocarbons | yes | yes | no | no | no |
| Suitable for micromolding | moderate | good | good | good | moderate |
| *Permeability coefficients* ($\times 10^{-17}$ m$^2$/s-Pa): | | | | | |
| He | 5.2 | 7.5 | — | — | — |
| $O_2$ | 0.12 | 1.1 | — | — | — |
| $H_2O$ | 480–1,900 | 720–1,050 | — | — | — |
| *Hot-embossing parameters:* | | | | | |
| Embossing temperature (°C) | 120–130 | 160–175 | — | — | — |
| Deembossing temperature (°C) | 95 | 135 | — | — | — |
| Embossing pressure (bars) | 25–37 | 25–37 | — | — | — |
| Hold time (s) | 30–60 | 30–60 | — | — | — |

As an example, Microstereo Lithography-based (MSL) 3-D printer could produce features with 0.35 μm roughness and height and width of 95μm and 150μm, respectively [103]. Such resolution and dimensions can meet the requirements for the fabrication of some microfluidic systems for bioanalytical or biological applications. On the other hand, lack of access to biocompatible resins is another challenge for 3-D printing while soft lithography of PDMS, thermoplastic materials and glass enables the fabrication of biocompatible microdevices.

### 3.4.5 Micromolding

Micromolding enables large-scale production of polymeric devices with high accuracy. Besides polymers, micromolding can also be used for ceramics. In the following, two common micromolding techniques are discussed: injection molding and compression molding.

#### 3.4.5.1 Injection Molding

Injection molding is the last technique used in a conventional LIGA process (see Section 3.4.1.1). The metal mold inserts are used as replica master for the molding process [106]. Injection molding is carried out at temperatures above the glass transition temperatures of amorphous thermoplastics such as polymethylmethacrylate (PMMA), polycarbonate (PC), and polysulfone (PSU). For semicrystalline thermoplastics such as polyoxymethylene (POM) and polyamide (PA), the molding temperature should be higher than the crystallite melting point. Table 3.13 compares the characteristics of different polymers commonly used for micromolding.

Mold inserts for injection molding can be fabricated with common microtechnologies such as bulk micromachining and LIGA. Alternatively, bulk-micromachined parts can be used as masters for electroplating [107]. After separation from the silicon part, the metal parts can be used as negative mold inserts to fabricate replicas of the silicon parts [107].

Most of the structures have high-aspect ratios and microscopic filling channels. Therefore, high pressures are required for the injection of melted plastics. An alternative for high-pressure injection molding is reaction injection molding (RIM), which involves the high-speed mixing of two

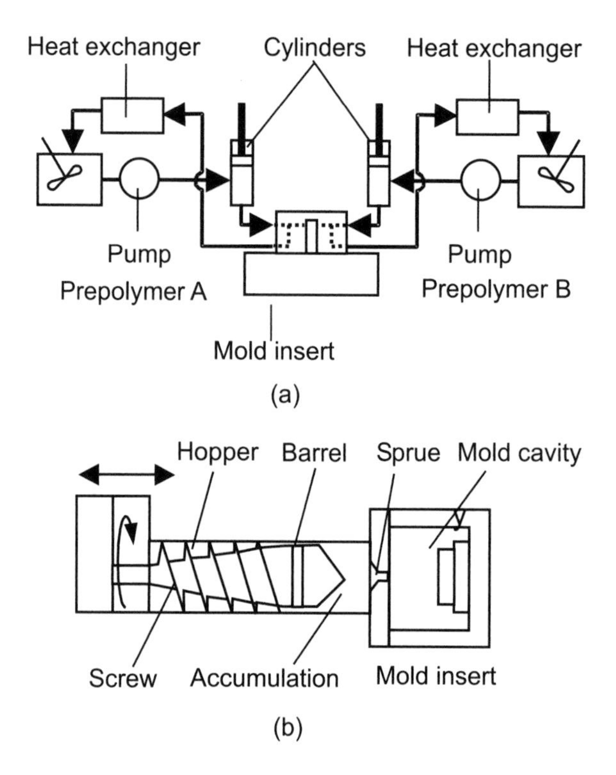


**Figure 3.27** Schematics of injection molding systems: (a) reaction injection molding (prepolymers are circulated in two loops); and (b) reciprocating screw unit.(*After*: [108].)

or more reactive chemicals, such as prepolymers, as an integral part of injecting them into a mold. In Figure 3.27(a), the two cylinders press the prepolymers into a mix head. The mixture flows into the mold at relatively low temperature, pressure, and viscosity. Curing occurs in the mold, again at relatively low temperature and pressure. The excess prepolymers are fed back in a recirculation loop by pumps. The entire process, from mixing to demolding, typically takes less than a minute. The low processing viscosity allows good filling and high molding accuracy.

Figure 3.27(b) shows the schematics of a typical injection molding system. The machine consists of a hopper, a screw, a sprue, and a mold insert [108]. To start with, polymer pellets are loaded into the hopper. Appendix C lists abbreviations of the most common plastics. The reciprocating screw shears, melts, and pumps the polymer into the accumulation zone. If the desired polymer amount is reached, the screw moves forward and pushes the polymeric melt into the mold cavity through the sprue. The required pressure is typically on the order of 500 to 2,000 bars [109, 110]. After cooling, the melt solidifies and can be taken out from the mold.

#### 3.4.5.2 Compression Molding

Compared to injection molding, compression molding or hot embossing is a simple technique. The structure is pressed under vacuum to a semifinished polymer material at elevated temperatures above its glass temperature, which is on the order of 50°C to 150°C [111, 112]. The vacuum is needed due to the formation of gas bubbles in the small structures. The vacuum also prevents corrosion of the master. The drawback of this technique is the relatively long cycle time (on the order of several minutes), while injection molding has fast cycle times (on the order of several seconds).

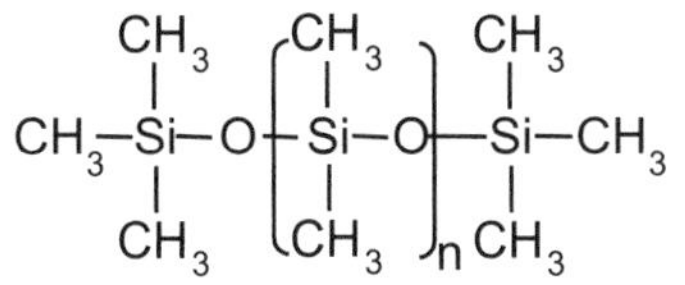

$$CH_3-Si(CH_3)_2-O-[Si(CH_3)_2-O]_n-Si(CH_3)_2-CH_3$$

**Figure 3.28** Chemical structures of PDMS.

The most important parameters of compression molding are embossing temperature, deembossing temperature, embossing pressures, and hold time. Table 3.13 lists the typical values of these parameters for hot embossing of PMMA and PC. Hot embossing can only make open channel structures. Fabricating covered channels and fluidic interconnects need additional packaging techniques, such as bonding to a sheet of the same material at temperatures above the glass temperature.

Injection compression molding combines the advantages of both injection and compression molding. The polymer melt is first injected into the mold. The mold melt is then compressed to shape the final part. The low viscosity of the melt results in good filling in the molded part.

#### 3.4.5.3 Microcasting

Microcasting is also called soft lithography, which is a direct transfer technique. The term "soft" refers to an elastomeric stamp with patterned relief structures on its surface. Polydimethylsiloxane (PDMS) has been used successfully as the elastomeric material. There are different techniques to transfer the pattern on this elastomeric stamp: microcontact printing and replica micromolding [114]. In many applications, the elastomeric PDMS part can be used directly as a microfluidic device with microchannels on it.

*PDMS.* Although all kinds of polymers can be used for the fabrication of an elastomeric stamp, PDMS exhibits unique properties suitable for this purpose. PDMS has an inorganic siloxane backbone with organic methyl groups attached to silicon (see Figure 3.28). Both prepolymers and curing agents are commercially available. PDMS has high optical transparency above a wavelength of 230 nm and low self-fluorescence. PDMS has a low interfacial free energy, which avoids molecules of most polymers sticking on or reacting with its surface. The interfacial free energy of PDMS can be manipulated with plasma treatment. The modified surface properties of PDMS are needed for certain applications. PDMS is stable against humidity and temperature. This material is optically transparent and can be cured by UV light. PDMS is an elastomer and can therefore attach on nonplanar surfaces. PDMS is mechanically durable. These characteristics make PDMS an ideal material for soft lithography [114].

However, PDMS also presents a number of drawbacks, such as volume change and elastic deformation. The design of a PDMS part should consider the shrinking effect upon curing. A number of organic solvents can swell PDMS as well. Furthermore, elastic deformation can limit the aspect ratio of the designed structure. A too-high aspect ratio leads to a pairing effect, in which two parallel structures attach to each other. An aspect ratio that is too low leads to a sagging of noncontact regions, which makes further steps of soft lithography impossible. The recommended aspect ratios for PDMS structures are between 0.2 and 2 [114].

*Fabrication of the PDMS Device.* The master for a PDMS device can be fabricated with conventional micromachining technologies, as described in Section 3.2. Figure 3.29 describes the fabrication process of a PDMS part. The silicon master is silanized by exposure to the vapor of $CF_3(CF_2)_6(CH_2)_2SiCl_3$ for about 30 minutes [113]. The prepolymer is coated on the silicon or

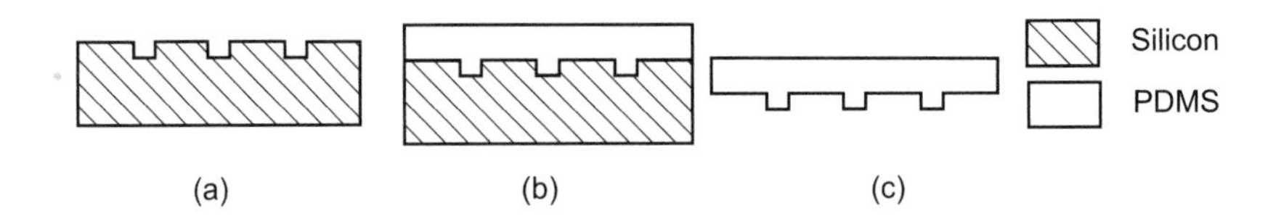


**Figure 3.29** Fabrication of PDMS stamps: (a) DRIE of silicon master; (b) coating; and (c) release.

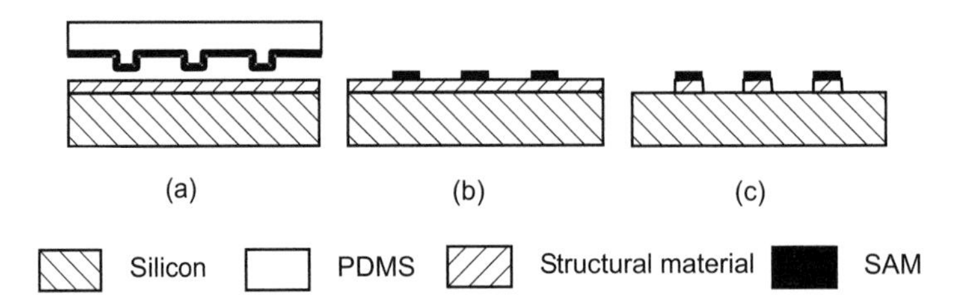


**Figure 3.30** Microcontact printing with PDMS stamps: (a) immersion; (b) stamping; and (c) etching.

glass master. After curing in an elevated temperature, the embossed PDMS layer can be peeled off and is ready for use in the subsequent steps.

The master can also be fabricated with SU-8 [115]. The PDMS part is used directly as structural material. The PDMS device is bonded to a glass plate after oxidizing their surfaces with oxygen plasma. In a similar approach, 3-D structures are fabricated by the lamination of different structured PDMS layers [116]. Fluidic interconnects are embedded directly in the PDMS device.

*Microcontact Printing with a PDMS Stamp.* Microcontact printing uses the relief structures on the surface of a PDMS part as a stamp to transfer a pattern of SAMs to the substrate surface by contact. SAMs can be created by immersion of the substrate in a solution containing a ligand $Y(CH_2)_nX$, where X is the head group and Y is the anchoring group. The head group determines the surface property of the monolayer. In microcontact printing, the stamp is wetted with the above solution and pressed on the substrate surface [Figures 3.30(a, b)].

The patterned SAM can be used as a resist layer to transfer its structure to an underlying structural layer. Because of its small thickness, SAM can be quickly destroyed by ion bombardment. Therefore, SAM resist is not suitable for reactive ion etching. However, SAM resist can be used for wet etching of the underlying structural layer [Figure 3.30(c)]. This thicker layer, in turn, can be used as a mask for the more aggressive RIE. The resolution of microcontact printing depends on the properties of stamp material, and can reach several tens of nanometers.

Surface properties of patterned SAM can be used as templates for selective deposition of other materials. For example, a patterned hydrophilic SAM traps liquid prepolymer on its surface. After curing, a polymer structure is formed on top of the patterned SAM. Furthermore, patterned SAM can be used for controlled deposition of metals and ceramics by selective CVD.

*Micromolding with a PDMS Replica Master.* Micromolding with a PDMS replica master can be categorized as: replica molding, microtransfer molding, micromolding in capillaries, and solvent-assisted micromolding [114].

In *replica molding*, the PDMS part is used as replica master for a prepolymer. The final structure is obtained by curing with a UV exposure or elevated temperature [Figure 3.31(a)]. This technique can achieve resolutions of less than 10 nm.

In *microtransfer molding*, liquid prepolymer is applied on the PDMS master [Figure 3.31(b), part 1]. The prepolymer layer is then planarized by removing excess prepolymer. Only prepolymer

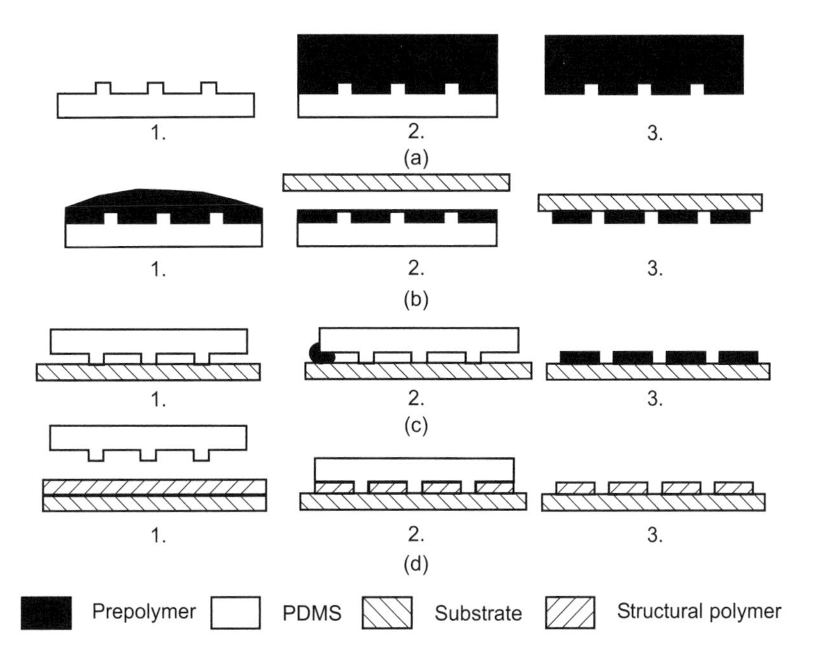


**Figure 3.31** Micromolding with PDMS replica master: (a) replica molding; (b) microtransfer molding; (c) micromolding in capillaries; and (d) solvent-assisted micromolding.

trapped between the relief structures remains on the surface of the PDMS master [Figure 3.31(b), part 2]. The master is then placed on a planar substrate. UV exposure or heating solidifies the prepolymer. Peeling off the elastic PDMS master results in polymer structures on the substrate surface [Figure 3.31(b), part 3]. This method does not remove completely the excess prepolymer on top of the PDMS stamp. A thin polymer layer on the order of 100 nm remains on the substrate surface. If the pattered polymer is to be used as a mask for subsequent etching, this thin polymer layer should be removed by oxygen plasma [114].

*Micromolding in capillaries* uses capillary forces to fill the gaps between the substrate and the PDMS master. First, the PDMS master is pressed tightly on a planar substrate [Figure 3.31(c), part 1]. Elastic PDMS seals off walls and creates capillary channels. A drop of liquid prepolymer is placed at the ends of these channels and fills them automatically due to capillary forces [Figure 3.31(c), part 2]. After curing and peeling of the PDMS master, polymer structures remain on the substrate surface [Figure 3.31(c), part 3]. This technique can be used to pattern silicon and glass with different materials [114].

*Solvent-assisted micromolding* is similar to the conventional embossing technique. Instead of using heat and compression, this technique uses a solvent to wet the PDMS stamp and soften the structural polymer. The solvent only dissolves the structural polymer and not PDMS. The stamp is pressed on a polymer film, which dissolves in the solvent and fills the gaps between relief structures of the stamp. After dissipation and evaporation of the solvent, solid polymer remains on the substrate [Figure 3.31(d)] [114].

*Fabrication of Microchannels with Soft Lithography.* Soft lithography can be used for rapid prototyping of microfluidic devices. Replica molding can be used for fabrication of microchannels [Figure 3.31(a)] [114]. PDMS is a good device material because it has a number of useful properties: low cost, low toxicity, transparency from the visible wavelengths into the near ultraviolet wavelengths, and chemical inertness.

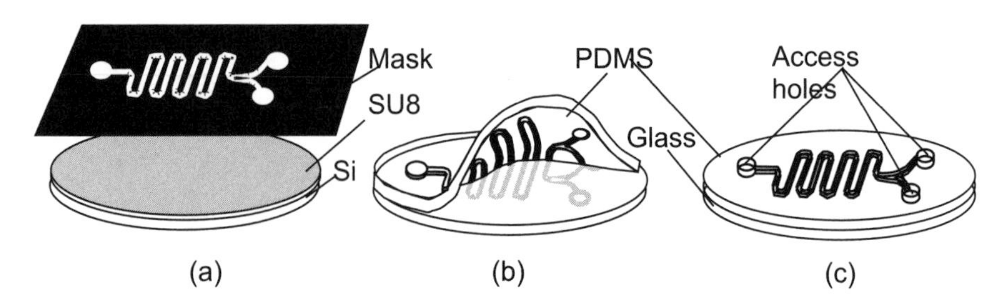


**Figure 3.32** Fabrication of microchannels with soft lithography: (a) spin coating a silicon wafer with SU-8, UV exposure with a clear field mask, development of SU-8 master; (b) pouring PDMS on the mold, peeling off the PDMS part; and (c) surface treatment of PDMS in oxygen plasma and bonding to glass.

First, PDMS is mixed from prepolymers. The weight ratio of the base and the curing agent could be 10:1 or 5:1. The solid master is fabricated from SU-8 [Figure 3.32(a)]. Glass posts are placed on the SU-8 master to define the inlets and reservoirs. The PDMS mixture is poured into the master and stands for a few minutes to self-level. The whole set is then cured at relatively low temperature (from 60°C to 80°C) for several hours. After peeling off and having surface treatment with low-temperature oxygen plasma [Figure 3.32(b)], the structured PDMS membrane is brought into contact with clean glass, silica, or another piece of surface-activated PDMS [Figure 3.32(c)]. The sealed channel can withstand pressures up to five bars. Without surface treatment, PDMS also forms a watertight seal when pressed against itself, glass, or most other smooth surfaces. These reversible seals are useful for detachable fluidic devices, which are often required in research and prototyping.

In a similar approach, the inlet and outlet tubes are embedded in the PDMS device [116]. Three-dimensional structures are formed by lamination of many PDMS sheets. Methanol is used as a surfactant for both bonding and self-alignment. The surface tension at superimposed holes in the PDMS sheets self-aligns them. Methanol prevents instant bonding between two PDMS sheets after plasma treatment. After evaporating methanol on a hot plate, the laminated stack is bonded. In [117], microchannels in PDMS are used for patterning surfaces. Micromolding in capillaries is utilized to fill the patterns. Colored UV-curable polymers with low viscosity (150 cps) are the filling liquids. This technique of soft lithography has much potential in patterning colored dyes for display applications and in patterning molecules for biochemical analysis [117].

## 3.5 FABRICATION TECHNIQUES FOR PAPER-BASED MICROFLUIDICS

Paper has gained attractions for implementation of various analytical and bioanalytical devices. Paper allows passive liquid transport through capillary wetting due to its hydrophilic porous microarchitecture. Various methods have been developed to create desired microfluidic patterns on paper for liquid handling and flow manipulation processes required for a (bio)analytical assay.

### 3.5.1 Wax Printing

The technique prints a layer of wax in desired patterns on the surface of a paper followed by heating on a hot plate. The wax is melted and penetrates the full thickness of the paper. Melted wax patterns create hydrophobic barriers allowing liquid to flow only through the hydrophilic zones of the paper, Figure 3.33. This method enables the fabrication of various paper-based microfluidic devices in a simple and cost-effective manner.

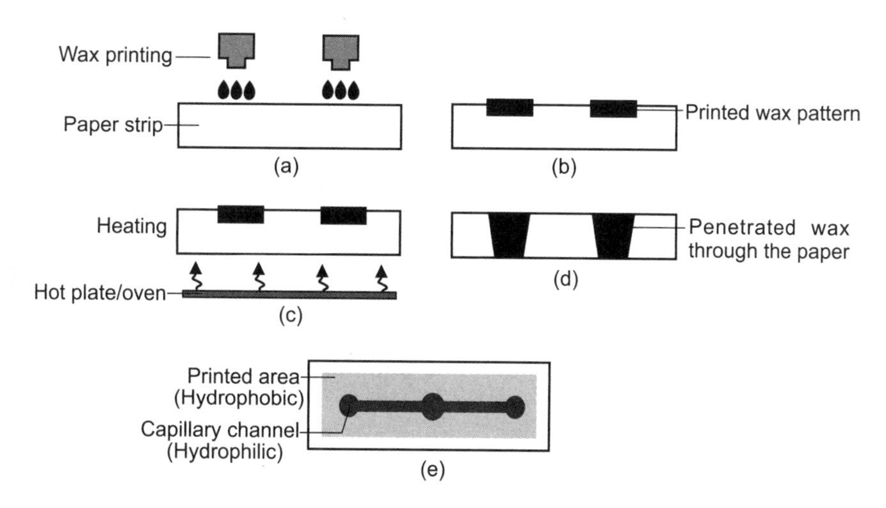


**Figure 3.33** Fabrication of the paper based device via wax printing: (a) and (b) wax printing and deposition; (c) heating; (d) wax penetration; and (e) top view of the printed paper with introduced liquid sample.

### 3.5.2 Inkjet Printing

Inkjet printing can digitally control both the position of a small nozzle and the ejection of fluid from the nozzle in the form of a droplet. To create a desired design on paper, drop-on-demand (DOD) technology is employed to form ink droplets in a dot by dot fashion. Inkjet printing provides higher resolution of patters with high-contrast hydrophilic channels as compared to wax printing methods. Importantly, the inkjet printer can precisely deposit pico liter volumes of various biochemicals within the printed patterns on paper strips to fabricate paper-based analytical devices [120].

### 3.5.3 Paper Photolithography

Photolithography has been employed to project microfluidic patterns onto the photoresist through a photomask on a chromatography paper. The chromatography paper is initially soaked in a photoresist solution that creates hydrophobic regions after UV exposure through the photomask. The paper is then developed and washed using propan-2-ol solution. This method provides precise paper patterning; however, the overall fabrication cost is higher than the printing methods due to the need to have access to cleanroom facilities such as mask aligner and the use of UV-curable photoresists. One approach to make this process cost-effective is to cure special photoresists without any need to use a UV source or cleanroom facilities [121, 122] .

### 3.5.4 Flexographic Printing

This printing method creates hydrophobic barriers through transferring inks onto the paper [Figure 3.34]. The hydrophobic polymeric ink partially or completely penetrates through the thickness of the paper to guide the liquid between the hydrophobic barriers. The process of flexography include plate making and printing on paper. The plate is needed to print the pattern on paper. A positive mirrored master of the desired pattern is formed (e.g., on rubber) using one of the following plate-making processes:

- Using a light-sensitive polymer: a film of negative mask is prepared for pattern transfer to a light-sensitive polymer using UV light;
- Using a computer-guided laser beam to create the desired pattern;

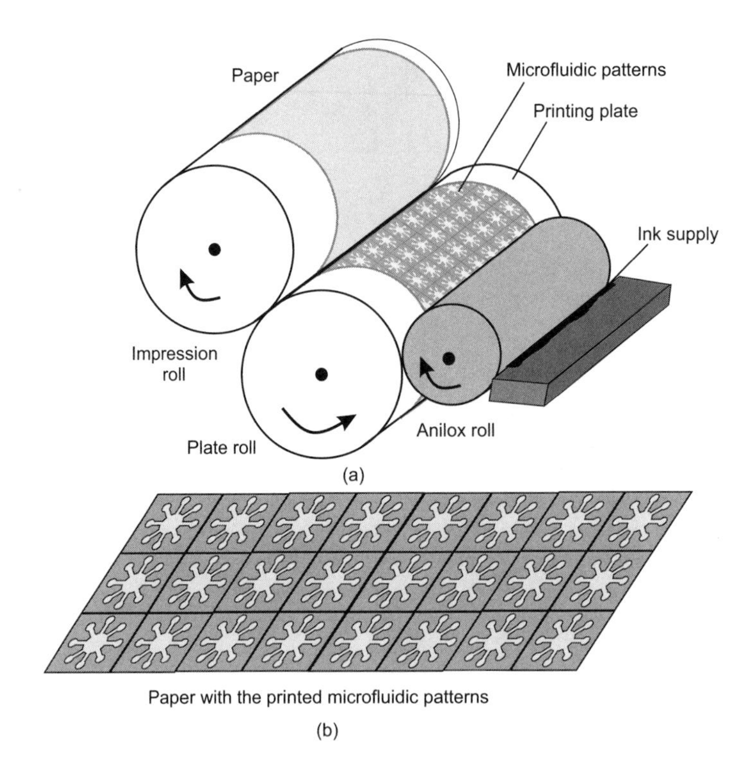


**Figure 3.34** Flexography process: (a) schematic illustration of the flexography unit for mass production of paper-based devices; and (b) printed paper with transferred microfluidic patterns. (*After:* [123].)

- Using molding process to fabricate a metal mold to compress the rubber or plastic compound to form the printing plate.

For the mass production of paper-based devices using flexography, the ink (hydrophobic polymer) is transferred from the anilox roll, which is partially immersed in the ink tank or fed by nozzles, to the printing plate. The anilox roll has thousands of small wells that hold and transfer the received ink to the printing plate. To transfer the pattern, the paper is compressed between the impression roll and the printing plate [123].

## 3.6 OTHER MICROMACHINING TECHNIQUES

Besides the above three major batch microtechniques, there are a number of serial microtechniques for fabrication of microfluidic devices. The advantage of these techniques is that they do not depend on substrate materials. Microstereo lithography is actually a serial technique, but it is placed in Section 3.4 to conform to the general term of polymeric micromachining. Here, serial microtechniques are categorized as subtractive (removal) techniques and additive (deposition) techniques.

### 3.6.1 Subtractive Techniques

#### 3.6.1.1 Micromilling

Micromilling employs rotating cutting tools to remove material from a workpiece [124] [Figures 3.35(a, b)]. The cutting tool is secured in the overhead spindle. Advanced milling systems are

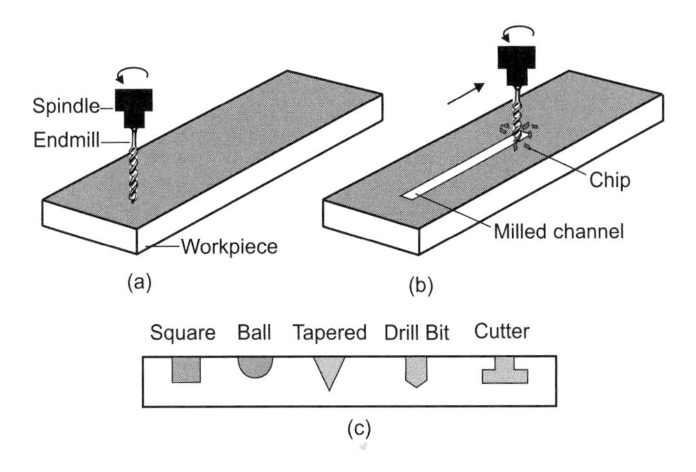


**Figure 3.35** Schematics of a micromilling process: (a) workpiece before micromilling; (b) material removal by the rotating endmill to create a microchannel; and (c) cross-sectional view of microchannels created by endmills in various profiles. (*After:* [124].)

equipped with computer numerical control (CNC) to automate the fabrication process. A CNC mill is able to fabricate an object directly from a 3-D computer-aided design (CAD) model with high precision and repeatability. As for microfluidic applications, micromilling can be employed for (1) machining the master molds required for other fabrication methods such as embossing and injection molding, and for (2) direct machining of microfluidic channels and features to prepare the final object [124]. Poly(methyl methacrylate) (PMMA), Polystyrene (PS), and cyclic olefin copolymer (COC) are transparent polymers suitable for micromilling of microfluidic devices.

The most common cutting tools for milling are endmills that can remove materials along any axis. They are commercially available in various materials, shapes, and dimensions enabling the fabrication of various microfluidic features with complex topography in a rapid and cost-effective fashion [Figure 3.35(c)]. Endmills with diameters of 0.1 mm to 0.3 mm are mainly employed for the micromilling of polymeric microfluidic devices. Endmills with diameters of 25 $\mu$m and smaller are also commercially available; however, to use such endmills requires running a high-precision milling system by an expert user.

#### 3.6.1.2 Laser Micromachining

Three types of lasers are commonly used for laser micromachining:

- Excimer lasers with UV wavelengths (351, 308, 248, 193 nm);
- Nd:YAG lasers with near infrared, visible, and UV wavelengths (1,067, 533, 355, 266 nm);
- $CO_2$ lasers with deep infrared wavelength (10.6 $\mu$m).

The two major parameters of laser micromachining are wavelength and laser power. The choice of wavelength depends on the minimum structure size and the optical properties of the substrate material, such as absorption and reflection characteristics. Theoretically, the minimum achievable focal spot diameter and consequently the smallest size are about twice the laser wavelength.

The choice of power depends on the desired structure size and the ablation rate. When excimer or Nd:YAG lasers with a pulse duration of a few tens of nanoseconds are utilized, a single laser pulse will typically vaporize the surface material to a depth of 0.1 to 1 $\mu$m (see Table 3.14). Since each pulse removes such a thin layer of material, the depth of the machined trench can be controlled accurately by the number of laser pulses. Furthermore, laser pulses of very short duration eliminate heat flow to surrounding materials. Consequently, clean and accurate structures can be achieved with

**Table 3.14**
Typical Ablation Depths Per Pulse of Different Material (Nanosecond Laser)

| *Material* | *Depth Per Pulse* ($\mu m$) |
|---|---|
| Polymers | 0.3–0.7 |
| Ceramics and glass | 0.1–0.2 |
| Diamond | 0.05–0.1 |
| Metals | 0.1–1.0 |

shortly pulsed lasers. There are two modes of laser micromachining: direct writing and using a mask [118]. In the direct writing mode, the laser beam is focused on the substrate surface. The pattern is scanned using a precision x-y stage or galvano scanning mirrors. The setup for this mode is similar to those of microstereo lithography depicted in Figures 3.24 and 3.25. In this mode, the smallest structure depends on the accuracy of the scanning system and is on the order of 25 to 50 μm. In the mask mode, a mask determines the shape of the structure. Therefore, the minimum structure size can be brought down to twice that of the laser wavelength.

Laser micromachining is suitable for fabrication of microchannels and fluidic access holes. A LIGA-like technique uses laser machining instead of X-ray lithography to machine PMMA [119]. Furthermore, the laser beam can be used for sealing polymeric devices fabricated with other techniques or making shadow masks.

Laser micromachining of PMMA using $CO_2$ laser has gained widespread popularity in research labs for rapid prototyping of microfluidic devices due to its low cost and ease of fabrication. Various thicknesses of commercially available PMMA films and sheets can be cut through or engraved though precise adjustment of power and speed of laser beam. Similar to micromilling, the laser beam can be controlled by computer for automated machining of objects without any need to have a mask or a master mold with high repeatability and precision, Figures 3.36(a), (b). $CO_2$ laser beam has a Gaussian-shape energy distribution that enables the fabrication of microchannel with a V-shaped cross section once the laser beam is focus on the surface of the workpiece [38] [Figures 3.36(c-1) and (d-1)]. Curved microchannels could be obtained employing unfocused laser beam [22], Figures 3.36(c-2) and (d-2). Curved microchannel is an essential component for implementation of some designs of microvalves in microfluidic devices. Further details of microvalve designs are discussed in Chapter 6. The surface roughness of the engraved channels depends on the laser and power speed and the material properties. Similar to micromilled devices, the roughness of the laser micromachinned surfaces can be improved using solvent vapor polishing.

#### 3.6.1.3 Focused Ion Beam Micromachining

In focused ion beam (FIB) micromachining, a highly focused ion beam scans and cuts the substrate surface. In a typical FIB system, the ion beam is ejected from a liquid metal ion source (usually gallium). With a spot size of less than 10 nm, the ion beam is scanned across the substrate material in a manner similar to scanning electron microscopy (SEM).

#### 3.6.1.4 Microelectro Discharge Machining

Microelectro discharge machining (EDM) or spark erosion is an electrothermal machining process that applies sparks created between a tool electrode and a workpiece electrode to remove substrate material. Each spark can create locally a temperature up to 10,000°C. For this purpose, both the tool and the substrate material should be electrically conductive. Both electrodes are immersed in

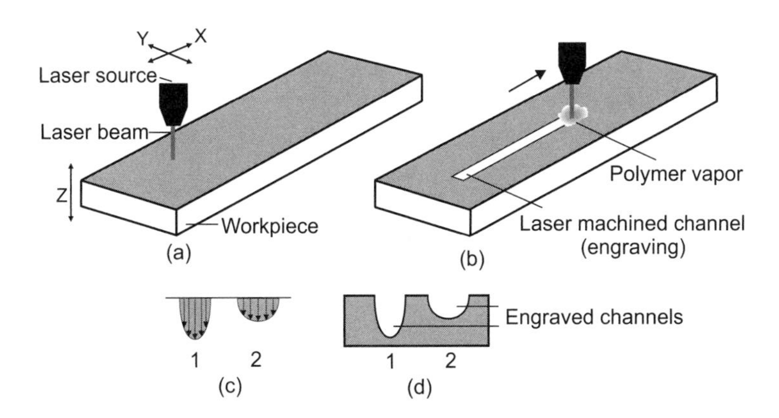


**Figure 3.36** Schematics of $CO_2$ laser micromachining: (a) workpiece before machining; (b) engraving process to create a microchannel; (c-1) energy distribution of focused $CO_2$ laser beam, (c-2) energy distribution of unfocused $CO_2$ laser beam; (d-1) cross-sectional view of engraved channel using focused beam; and (d-2) cross-sectional view of engraved channel using unfocused beam.

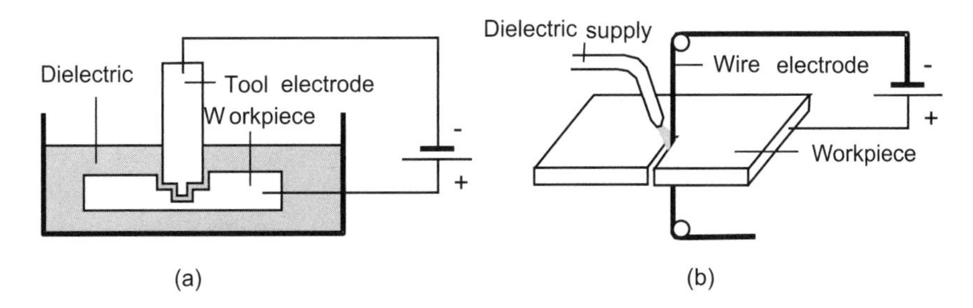


**Figure 3.37** Schematics of microelectro discharge machining: (a) die-sinking EDM; and (b) wire EDM.

a dielectric fluid such as oil, which cools the electrodes and removes the debris out of the sparking gap. The spark frequency can be up to 500 kHz.

The two major EDM techniques are die-sinking EDM and wire EDM. Die-sinking EDM is more relevant for fabrication of microfluidic devices. Microchannels and access holes can be fabricated with this technique [Figure 3.37(a)]. Wire EDM cuts through the workpiece and is not relevant for microfluidic applications [Figure 3.37(b)].

### 3.6.1.5 Powder Blasting

Powder blasting is another erosion technique that uses kinetic energy of powder particles to generate cracks on the substrate surface and consequently to remove material. The major process parameters of this technique are particle material, particle size, particle velocity, and incident angle [127].

The resolution of this technique depends on the particle size. As a rule of thumb, the smallest cut is about three times that of the particle size [128]. High resolutions are kept by the use of a hard metal mask, which is machined by another technique, such as laser micromachining. Other alternative masks are thick resist foils and polyimide resist [128]. With particle velocities on the order of 80 to 200 m/s, the erosion rate is on the order of 1 mm/min [127].

Using 30-μm alumina particles, microchannels of 100-μm width and 10-μm depth can be fabricated in glass [127]. Changing the incident angle of the powder beam can lead to channels with slanted walls. Microchannels for capillary electrophoresis can be fabricated with powder blasting. Microchannels with 85-μm width and 22-μm depth are fabricated in glass using 9-μm alumina

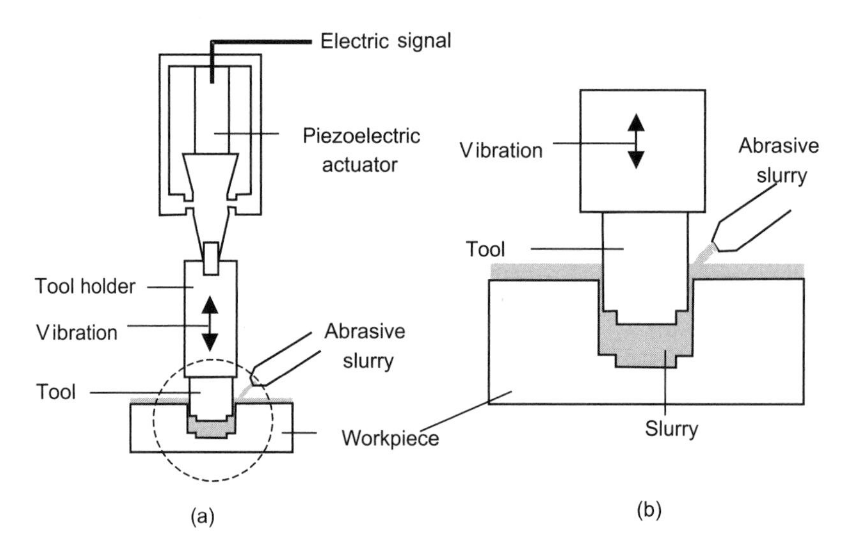


**Figure 3.38** Ultrasonic machining: (a) setup schematics; and (b) close-up of the tool and the workpiece.

particles [129]. In microfluidics, powder blasting can be used for drilling fluidic access through a substrate.

#### 3.6.1.6 Ultrasonic Micromachining

Another practical technique for drilling fluidic access holes is ultrasonic drilling, which uses ultrasonically induced vibrations delivered to a tool to machine the substrate. When combined with abrasive slurry, ultrasonic drilling can handle hard, brittle materials such as glass and silicon. Figure 3.38 illustrates a typical setup.

The drilling process starts with converting a high-frequency electrical signal into a mechanical oscillation, which is acoustically transmitted to the tool. A piezoelectric stack actuator can be used for this purpose. The linear oscillation is typically at a rate of 20 kHz. With the use of abrasive slurry flowing around the cutting tool, microscopic grinding occurs between the surfaces of the tool and the workpiece. The machined area becomes an exact counterpart of the tool.

In microfluidics, ultrasonic machining is often used for drilling via holes in glass and silicon. Hole diameters on the order of 100 to 200 μm can be drilled. The tool for ultrasonic machining can be fabricated with wire EDM, as described above.

### 3.6.2 Additive Techniques

#### 3.6.2.1 Laser-Assisted Chemical Vapor Deposition

Laser-assisted chemical vapor deposition (LCVD) uses a focused laser beam to control the chemical deposition from the vapor phase. The common laser types used for this purpose are the Nd:YAG laser and argon laser. Scanning the laser in space results in 3-D structures [Figure 3.39(a)].

#### 3.6.2.2 Localized Electrochemical Deposition

Localized electrochemical deposition (LECD) has the opposite effect of EDM. Localized growth of the metal structure is determined by a sharp microelectrode. Scanning the electrode tip in space allows fabricating 3-D structures [Figure 3.39(b)]. A microneedle with a 125-μm diameter and a 500-μm length was fabricated with LECD [130].

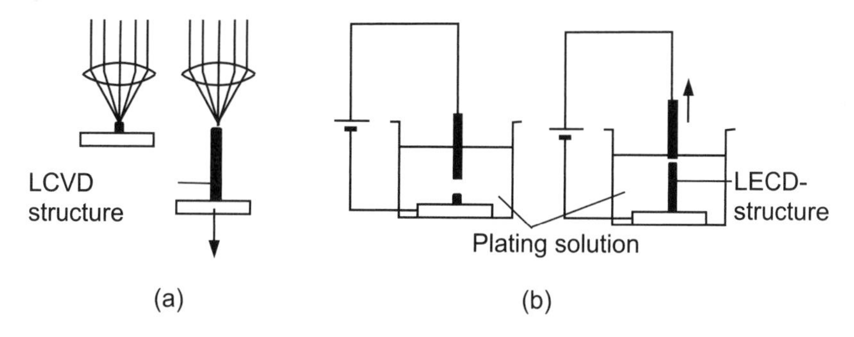


**Figure 3.39** Additive techniques: (a) LCVD; and (b) LECD.

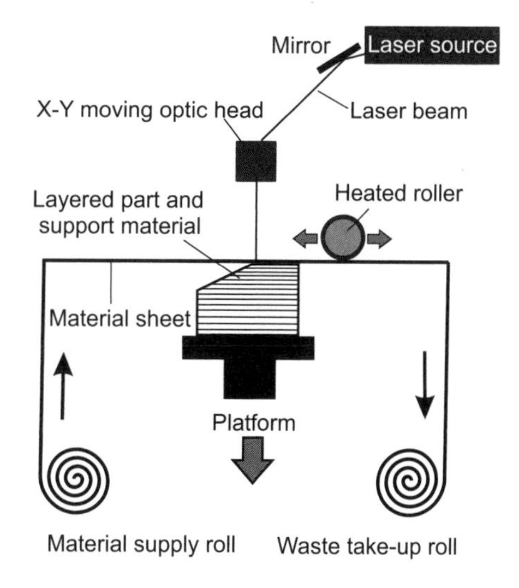


**Figure 3.40** Schematic illustration (front view) of LOM: continual bonding of cut thin layers to print an object.

#### 3.6.2.3 Laminated Object Manufacturing (LOM)

The operation of LOM is based on fabricating a 3-D object out of 2-D layers. The layers are usually made of thin polymeric sheets within the thickness range of 50 μm to 500 μm. The sheet is cut by the laser beam in desired geometry and bonded together under pressure and heat induced by a roller. The cut sheets are stacked on each other gradually to build the whole structure of a desired object in a layer-by-layer fashion. The unused area of the sheet is removed from the working area and taken up in the waste roll automatically [Figure 3.40]. Once the fabrication of one layer is completed, the platform is moved down at a step equal to the thickness of the sheet. Simultaneously, a new piece of sheet is fed to the top of the object for the next layer cutting. The remaining unused sections of the sheet are collected over the waste take-up roll.

## 3.7 ASSEMBLY AND PACKAGING OF MICROFLUIDIC DEVICES

Although batch fabrication and microtechnologies minimize the use of assembly processes, assembling and packaging techniques are still needed for the final microfluidic product. In fact, the

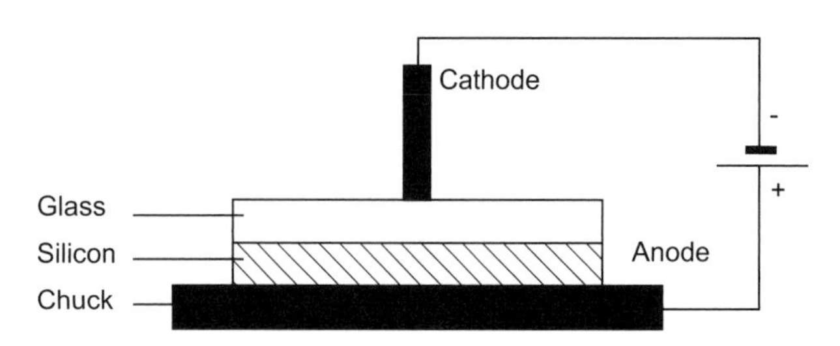


**Figure 3.41** Setup for anodic bonding.

commercialization of microfluidic devices is difficult due to the bottleneck effect between microfabrication and packaging. Because of additional fluidic interconnects, packaging may cause a much higher cost than the microdevice itself.

From the product perspective, the assembly and packaging hierarchy of a microfluidic device can be structured in the wafer level and the device level. In the wafer level, assembly and packaging still can be carried out in batch, concurrently with other device components. In the device level, each device should be packaged and readied to be shipped. The problem of fluidic interconnects and high cost arises in this level.

### 3.7.1 Wafer-Level Assembly and Packaging

#### 3.7.1.1 Anodic Bonding

Anodic bonding is the oldest bonding technique in MEMS. A glass wafer and a silicon wafer are bonded together at elevated temperatures on the order of 400°C and high electrical field with bonding voltage about 1 kV. Figure 3.41 describes the typical setup of anodic bonding. Silicon is connected to the positive electrode and works as an anode, which gives the name to this bonding technique.

Because the bonding process occurs at a relatively high temperature, the thermal expansion coefficients of glass and silicon should match, to avoid cracks after cooling the bonded stack to room temperature. Glasses suitable for this purpose are Corning 7740 (Pyrex), Corning 7750, Schott 8329, and Schott 8330. Thin metal layers on glass or silicon do not affect the bonding quality.

Anodic bonding can be used for sealing silicon to silicon. For this purpose, an intermediate glass layer between the two silicon wafers makes the seal possible. The glass layer can be deposited by different techniques, such as PVD and spin-on glass. Glass material with a matching thermal expansion coefficient is used as a source or target for the evaporation or sputtering process. The bonding processes work as usual with the glass-covered silicon wafer replacing the glass wafer in Figure 3.41. Because of the much thinner glass layer on the order of 0.5 to 4 μm, much lower bonding voltages are needed for the same field strength of bonding glass to silicon. The bonding voltage in this case is on the order of several tens of volts.

Anodic bonding is used in many of the microfluidic applications discussed in Chapters 5 to 13. Because of the optically transparent glass, this bonding technique is suitable for biochemical applications, where optical access for manipulation and evaluation of the fluid are required.

#### 3.7.1.2 Direct Bonding

Direct bonding refers to the bonding process between two substrates of the same material. In this section, the direct bonding of silicon, glasses, polymers, ceramics, and metals is discussed.

*Silicon Direct Bonding.* Silicon direct bonding, also called *silicon fusion bonding*, seals two silicon wafers directly. There is no need of an intermediate layer (glass in case of anodic bonding or polymers in case of adhesive bonding), which complicates the fabrication process. The advantage of this technique is the lack of thermal stress because of the same thermal expansion coefficient of the two silicon wafers.

Silicon direct bonding utilizes the reaction between hydroxyl (OH) groups at the surface of the oxide layers (native or deposited) of the two silicon wafers. First, the silicon wafers undergo hydration by being immersed in an $H_2O_2/H_2SO_4$ mixture, boiling nitric acid, or diluted $H_2SO_4$. The bonding process is accomplished at elevated temperatures between 300°C and 1,000°C. Annealing the bonded stack at high temperatures (800°C to 1,100°C) improves the bond quality.

*Glass Direct Bonding.* Because of the optical transparency, glasses are relevant for microfluidic applications in chemical analysis. Many applications use soda-lime glass, which consists of $SiO_2$, $Na_2O$, CaO, MgO, and a small amount of $Al_2O_3$ (see Table 3.6). The following bonding process is used for two soda-lime glass slides.

First, the glass wafers are cleaned in an ultrasonic bath and subsequently 10 minutes in a solution of [5 $H_2O$: 1 $NH_3$ (25%): 1 $H_2O_2$ (20%)] or [6$H_2O$: 1 HCl (37%): 1 $H_2O_2$ (20%)]. After removing moisture by annealing at 130°C, the two wafers are thermally bonded together at 600°C for 6 to 8 hours [131].

*Polymer Direct Bonding.* Many polymers are thermally bonded at temperatures above their glass transition temperatures. In cases of polymers with low surface energy, such as PDMS, a surface treatment with oxygen plasma seals the two polymer parts at room temperature. Another bonding method uses solvent to wet the bonding surfaces. Bonding is accomplished after evaporating the solvent.

*Direct Bonding of Ceramics and Metals.* Ceramic green tapes and metal sheets structured by serial techniques can be directly bonded together at high pressure and high temperatures. Ceramic green tapes are typically bonded at 138 bars, 70°C for 10 minutes [133]. Stainless steel sheets are typically bonded at 276 bars, 920°C for 4 hours [134].

PDMS direct bonding is achievable through oxygen plasma treatment. This method enables the bonding of PDMS to PDMS and PDMS to other materials such as glass. The bonding surface of PDMS-based components are treated with oxygen plasma to convert the monomer O-Si$(CH_3)_2$ of PDMS near the surface into silanol group (SiOH). The hydroxyl group (OH) is available for covalent bonding once exposed to another plasma-treated PDMS piece [132]. After plasma treatment, the components are stacked on top of each other with subsequent gentle compression to create the desired bonding.

Thermal fusion bonding present a common direct bonding for thermoplastics. While applying a pressure, the thermoplastic substrates are heated to reach near or above glass transition temperature ($T_g$) of one or both of the substrate materials [138]. The combination of pressure and temperature creates sufficient flow of polymer at the bonding interface. This results in the inter-diffusion of polymer chains between the bonding surfaces. The polymeric chains remains in place once the substrates are cooled to room temperature. Since the bonding surfaces are fused in each other, unoptimized temperature and pressure may result in undesired channel deformations [138]. Figure 3.42 shows the thermal fusion bonding process. Usually, vacuum is applied during heating process to avoid bubble entrapment between the bonding layers [22].

#### 3.7.1.3 Adhesive Bonding

Adhesive bonding uses an intermediate layer to "glue" the substrate. Depending on substrate materials and applications, the intermediate layer can be glass, epoxies, photoresists, or other polymers.

A thin intermediate glass layer can thermally bond silicon wafers. Glass frits with relatively low sealing temperatures ranging from 400°C to 650°C are commercially available. The glass layer can be sprayed, screen-printed, or sputtered on the substrate. Annealing the stack at sealing

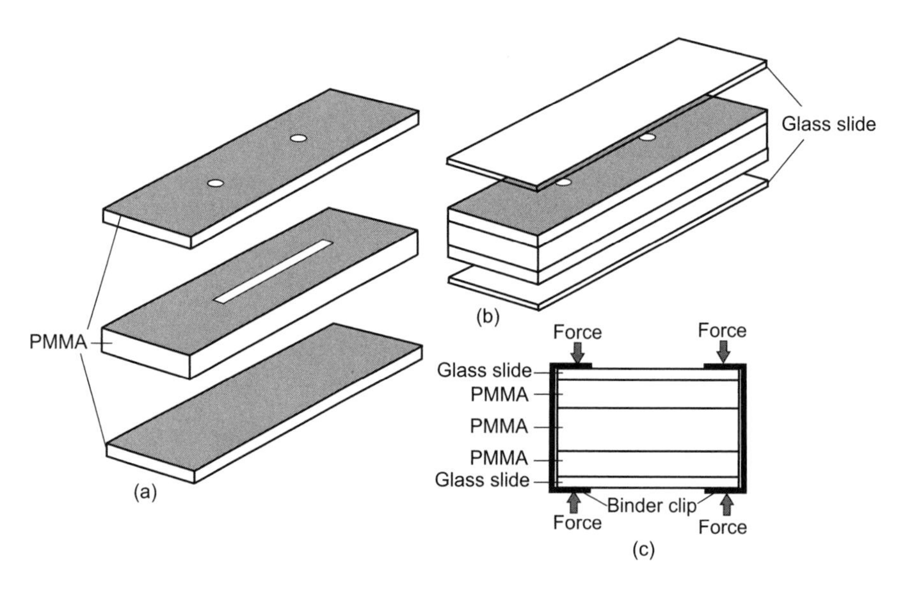


**Figure 3.42** Schematics of thermal fusion bonding process: (a) components are aligned and stacked; (b) the aligned layers are sandwiched between two flat smooth substrates such as glass slides to apply uniform force to the assembly; and (c) the assembly is placed in an oven for heating. Binder clips are used to compress the bonding layers.

temperatures makes the glass layer melt and flow. Cooling down to room temperature results in a strong bond between two substrates [135]. A number of epoxies [136], UV-curable epoxies [137], and photoresists can be used for adhesive bonding. SU-8 is used in many microfluidic applications as both spacer and adhesive layers. The advantage of using polymers as an intermediate layer is the low process temperature required. These low packaging temperatures are needed for many devices, which have, for instance, aluminum structures in them. The other advantage is that adhesive bonding is not limited to silicon and can be used for all kinds of substrate material. Adhesive bonding using lamination films is another indirect bonding method for polymeric chips [138]. The bonding process is rapid and of high throughput that makes the bonding method suitable for mass production of thermoplastic microfluidics. In particular, pressure sensitive adhesive films using dry lamination films provide an easy and simple-to-perform bonding method making it well suited for research laboratories. The dry films can be precisely cut in desired shapes using blade plotter or laser machining process.

#### 3.7.1.4 Eutectic Bonding

Eutectic bonding is a common packaging technique in electronics. Gold-silicon eutectic bonding is achieved at a relatively low temperature of 363 °C. A thin gold film can be sputtered on the silicon surface for this purpose. Furthermore, a gold-silicon preform with composition close to eutectic point can also be used as the intermediate layer.

### 3.7.2 Device-Level Packaging

#### 3.7.2.1 Fluidic Interconnects

Generally, a microsystem needs interconnects for power flow, information flow, and material flow to communicate with its environment. Interconnects for power flow supply the device with energy. Interconnects for information flow are actually signal ports, which allow information coming in and out of the device. Unlike other MEMS devices, microfluidic devices have interconnects for

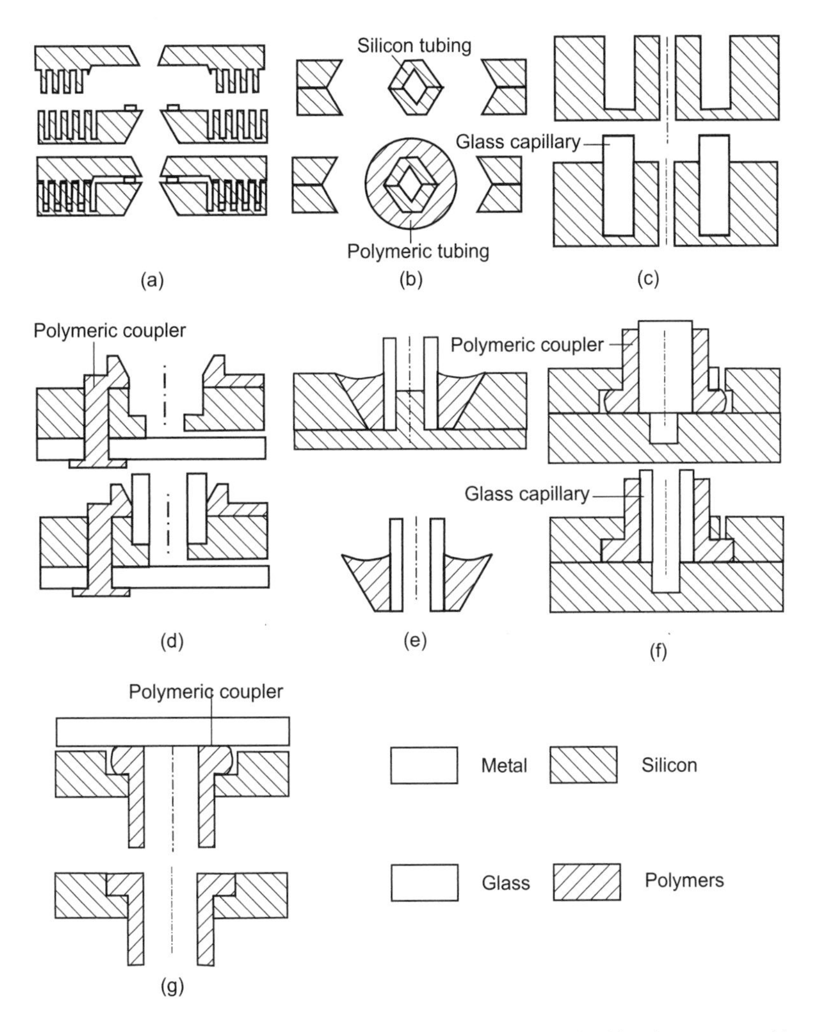


**Figure 3.43** Press-fit interconnects: (a) interlocking finger joins; (b) horizontal tubing interconnect; (c) vertical tubing interconnect; (d) plastic press fitting; (e) molded coupler; and (f) and (g) compression molded coupler.

material flow. The material flow is the fluid flow, which is processed in microfluidic devices. While many solutions for the first two types of interconnects exist in traditional microelectronics, fluidic interconnections pose a big challenge. In this section, fluidic interconnects are categorized by the coupling nature as press-fit interconnects and glued interconnects.

*Press-Fit Interconnects.* Press-fit interconnects utilize elastic forces of coupling parts to seal the fluidic access. Figure 3.43 shows several examples of press-fit interconnects. Due to the relatively small sealing forces, this type of interconnect is only suitable for low-pressure applications.

Figure 3.43(a) illustrates the concept of interlocking finger joints [139]. The fingers are fabricated in bulk silicon. A structured polysiloxane film works as an O-ring at the interconnect interface. This type of interconnect is suitable for coupling channel systems in different wafers.

For fluidic coupling out of the device, tubing interconnects are needed. Figure 3.43(b) shows a horizontal tubing interconnect fabricated by wet etching of silicon. The external polymeric tubing is press-fitted to the silicon tubing [139]. Vertical tubing interconnects can be etched in silicon using DRIE [Figure 3.43(c)]. These vertical tubes can hold fused silica capillaries, which are perpendicular to the device surface [140].

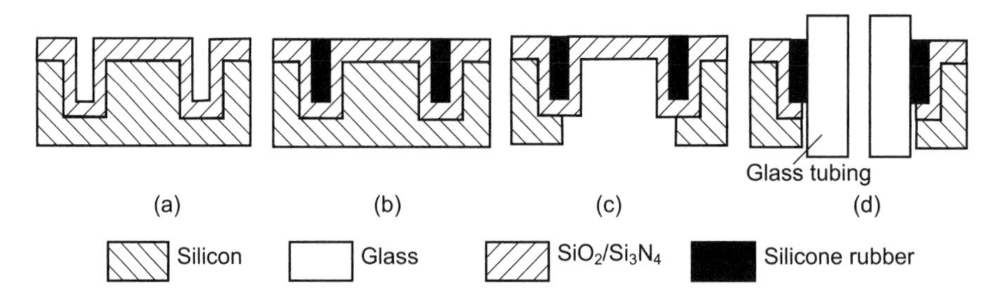


**Figure 3.44** Fabrication steps for an integrated O-ring: (a) DRIE and oxide/nitride deposition; (b) rubber coating; (c) DRIE from the back side; and (d) oxide/nitride etch. (*After*: [144].)

If external capillaries are to be inserted directly into etched openings, plastic couplers can be used to keep them in place [Figure 3.43(d)] [139]. The plastic couplers in [140] are injection-molded from POM. A PDMS gasket between the plastic coupler and the substrate surface can improve the seal [141].

The molded coupler shown in Figure 3.43(e) is fabricated from two bonded silicon wafers. The fused silica capillary is embedded in the plastic coupler. The capillary with the coupler is then inserted into the fluidic opening. Annealing the device at elevated temperatures allows the plastic to reflow. After cooling down to room temperature, the remolded plastic coupler hermetically seals the capillary and the opening [142]. Plastic couplers can also be compression-molded. In Figure 3.43(f), the thermoplastic tube is inserted into the opening. Under pressure and temperatures above the glass transition temperature, the plastic melts and fills the gaps between the coupler and silicon. Similar flanged tubes can be fabricated with the technique shown in Figure 3.43(g) [143].

In many cases, the elastic force of the couplers cannot withstand high pressures. One solution for high-pressure interconnects is the use of a mesoscale casing, which has conventional O-rings for sealing fluidic interconnects [144]. If a microfluidic system has multiple fluidic ports, many small O-rings are required. In this case, it is more convenient to have integrated O-rings in the device. Figure 3.44 presents the fabrication steps of such integrated O-rings [145]. First, the cavities for the O-rings are etched in silicon with DRIE. After depositing an oxide/nitride layer [Figure 3.44(a)], silicone rubber is squeezed into the cavities [Figure 3.44(b)]. The fluidic access is opened from the backside by DRIE [Figure 3.44(c)]. Subsequently, the oxide/nitride layer is etched in buffered hydrofluoric acid and $SF_6$ plasma. The silicone rubber O-ring remains on top of the opening. If a capillary is inserted into the opening, the rubber O-ring seals it tightly [Figure 3.44(d)].

*Glued Interconnects.* In many cases, press-fit interconnects are glued with adhesives. Besides their holding function, adhesives offer good sealing by filling the gap between the external tubes and the device opening. Figure 3.45 shows typical glued fluidic interconnects. The glued surface can be roughened to improve adhesion. A combination of surface roughening, compression molding, and adhesive bonding is used to make tight fluidic interconnects [Figure 3.45(c)] [146]. Kovar is an alloy consisting of 29% nickel, 17% copper, and the balance of iron. Because of its relatively low thermal expansion coefficient, Kovar is often used for glass-to-metal seals in electronics packaging. Figure 3.45(d) shows a solution with glass seal and Kovar tubes [147]. The Kovar tubes are fitted into the fluidic access. Glass beads are placed around them. A carbon fixture is used as the mold for the glass melt. Glass sealing is accomplished after annealing the assembly at 1,020°C [147].

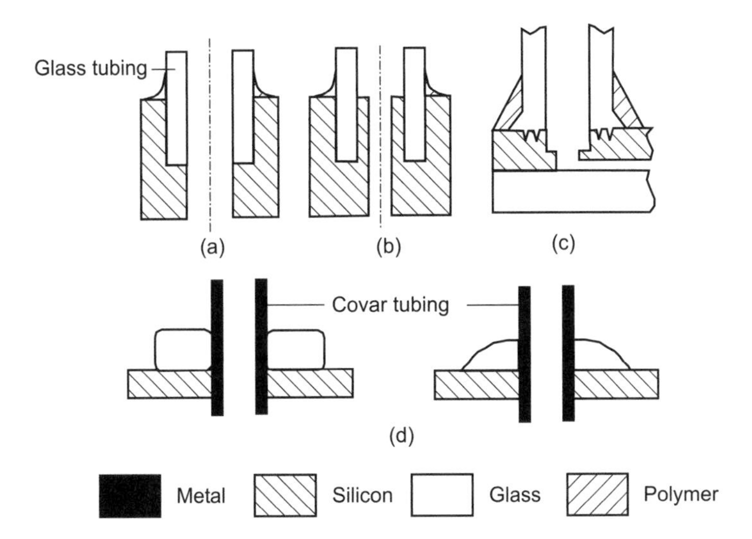


**Figure 3.45** Glued interconnects: (a) and (b) DRIE holes; (c) roughened surface; and (d) glass seal.

## 3.8 BIOCOMPATIBILITY

Microfluidic devices for biomedical and biochemical analysis are the major applications of microfluidics. Therefore, the biocompatibility of device and packaging is an important issue in the development and commercialization of microfluidic devices. Biocompatibility can be determined by a number of tests. However, there is no absolute biocompatibility. The biocompatibility tests are usually tailored to the applications. For commercialization, responsible agencies such as the U.S. Food and Drug Administration (FDA) only approve medical devices for specific purposes and not the devices themselves in isolation. The use of microfluidic devices in a biological environment triggers a material response in the devices and tissue as well as a cellular response in the host [148].

### 3.8.1 Material Response

The most common material responses to the biological environment are swelling and leaching. The simplest material response is a mass transfer across the tissue/material interface. Fluid diffuses from the host tissue into the device material, causing it to swell. The changes in dimension may cause microcracks on the material's surface, which in turn alter the mechanical properties of the device. Leaching is another reaction caused by fluid transfer. The fluid that had previously diffused into the device material can move back into the biological environment, and carries material particulates suspended within it. Removed particulates damage both the device and surrounding tissues.

### 3.8.2 Tissue and Cellular Response

Reactionary tissue response begins with inflammation at the device/tissue interface. The classic symptoms of inflammation are reddening, swelling, heating, and pain. Chemical signals released by the damaged tissue trigger the inflammation and attract white blood cells as body response. The device is covered with macrophages and foreign body giant cells leading to a fibrous encapsulation. The encapsulation can affect the functionality of the device. If the healing process occurs as described above, the device can be called biocompatible.

Alternatively, a long-lasting inflammation can be caused by chemical or physical properties of the device material or by motion of the device itself. Constant local cell damages make an inflammation reaction continue to be released. If the device material causes cells to die, it is called *cytotoxic*. If the device material is inert, its particulates cannot be digested by macrophages. The material response to the biological environment can cause chemical changes in the device material, which can be cytotoxic and damage cells.

### 3.8.3 Biocompatibility Tests

There are two general test methods for biocompatibility:

- In vitro (in laboratory glassware);
- In vivo (in a live animal or human).

The usual procedure starts with in vitro tests, which can weed out the clearly dangerous, but may not expose all problems that may occur in the complex living system. In vitro tests should also be carried out to evaluate the adhesion behavior of cells, DNA, and other polymers. Furthermore, cytotoxicity of the device material should be tested for applications such as tissue engineering. Some animal tests may be required before the device moves to a human test stage. There are standardized tests to follow for the in vitro assessment.

## Problems

**3.1** If the mask for KOH-etching of a $\{100\}$-wafer is a 200-nm-thick silicon dioxide layer, what is the maximum etched depth in silicon? Use averaged etch rates from Table 3.8.

**3.2** An optical measurement in microchannels needs an optical path between two surfaces that have a 45° angle to each other. Develop a process for the mirror surfaces with anisotropy etching in bulk silicon.

**3.3** Design the mask opening for fabricating a nozzle in silicon with KOH etching. The nozzle opening on the other side of the wafer has a dimension of 20 μm × 20 μm. The wafer is 500 μm thick.

**3.4** We consider the channel depicted in Figure 3.10. The access gap for isotropic etching in Figure 3.11(b) is 1 μm. After sealing with thermal oxide, we need to inspect the channel and to open the access gap. If the sealing thermal oxide layer is etched with a buffered HF solution, how large is the gap of the opening after etching away all silicon dioxide?

**3.5** A deep trench is etched in a silicon wafer using DRIE. The trench measures 20 μm × 100 μm. After removing all protecting layers on the trench wall, the trench is further etched in KOH. Describe the final shape of the trench after KOH etching if the wafer orientation is (a) $\{100\}$, (b) $\{110\}$, or (c) $\{111\}$.

**3.6** Explain why it is difficult to reflow negative resists such as SU-8, but easy for positive resists such as the AZ4000 series.

**3.7** In Figure 3.15(b), what is the final form of the channel if the channel in the second SU-8 layer is wider than that in the first SU-8 layer?

**3.8** In Figure 3.16(a), why can lift-off technique not be used for structuring the embedded metal mask?

## References

[1] Madou, M. J., *Fundamentals of Microfabrication: The Science of Miniaturization*, 2nd ed., Boca Raton, FL: CRC Press, 2002.

[2] Thompson, L. F., Willson, C. G., and Bowden, M. J., *Introduction to Microlithography*, Washington, D.C.: American Chemical Society, 1994.

[3] Friedrich, H., Widmann, D., and Mader, H., *Technologie hochintegrierter Schaltungen*, Berlin: Springer-Verlag, 1996.

[4] Pearce, C. W., "Epitaxy," in *VLSI Technology*, S. M. Sze (ed.), New York: McGraw-Hill, 1988, pp. 55–97.

[5] Adams, A. C., "Dielectric and Polysilicon Film Deposition," in *VLSI Technology*, S. M. Sze (ed.), New York: McGraw-Hill, 1988, pp. 233–271.

[6] Becker, H., and Gärtner, C., "Polymer Microfabrication Methods for Microfluidic Analytical Applications," *Electrophoresis*, Vol. 21, 2000, pp. 12–26.

[7] Monreal, G., and Mari, C. M., "The Use of Polymer Materials as Sensitive Elements in Physical and Chemical Sensors," *Journal of Micromechanics and Microengineering*, Vol. 7, No. 3, 1997, pp. 121–124.

[8] Soane, D., (ed.), *Polymer Applications for Biotechnology: Macromolecular Separation and Identification*, Upper Saddle River, NJ: Prentice-Hall, 1992.

[9] Kovacs, G. T. A., Maluf, N. I., and Petersen, K. E., "Bulk Micromachining of Silicon," *Proceedings of the IEEE*, Vol. 86, No. 8, 1998, pp. 1536–1551.

[10] Williams, K. R., and Muller, R. S., "Etch Rates for Micromachining Processing," *Journal of Microelectromechanical Systems*, Vol. 5, No. 4, 1996, pp. 256–269.

[11] Bean, K. E., "Anisotropic Etching of Silicon," *IEEE Transactions Electron Devices*, Vol. ED-25, 1978, pp. 1185–1193.

[12] Seidel, H., et al., "Anisotropic Etching of Crystalline Silicon in Alkaline Solutions II: Influence of Dopants," *Journal of Electrochemical Society*, Vol. 137, No. 11, 1990, pp. 3626–3632.

[13] Kaminsky, G., "Micromachining of Silicon Mechanical Structures," *Journal of Vacuum Science and Technology*, Vol. B3, No. 4, 1985, pp. 1015–1024.

[14] Kern, W., "Chemical Etching of Silicon, Germanium, Gallium Arsenide and Gallium Phosphide," *RCA Review*, Vol. 39, 1978, pp. 278–308.

[15] Schnakenberg, U., et al., "$NH_4OH$ Based Etchants for Silicon Micromachining: Influence of Additives and Stability of Passivation Layers," *Sensors and Actuators A*, Vol. 25, No. 1-3, 1990, pp. 1–7.

[16] Tabata, O., et al., "Anisotropic Etching of Silicon in TMAH Solutions," *Sensors and Actuators A*, Vol. 34, No. 1, 1992, pp. 51–57.

[17] Schnakenberg, U., Beneke, W., and Lange, P., "TMAHW Etchants for Silicon Micromachining," *Proceedings of Transducers '91, 6th International Conference on Solid-State Sensors and Actuators*, San Francisco, June 23–27, 1991, pp. 815–818.

[18] Petersen, K. E., "Silicon as a Mechanical Material," *Proceedings of IEEE*, Vol. 70, 1982, pp. 420–457.

[19] Mehregany M., and Senturia, S. D., "Anisotropic Etching of Silicon in Hydrazine," *Sensors and Actuators*, Vol. 13, No. 4, 1988, pp. 375–390.

[20] Linde, H., and Austin, L., "Wet Silicon Etching with Aqueous Amine Gallates," *Journal of Electrochemical Society*, Vol. 139, No. 4, 1992, pp. 1170–1174.

[21] Fleming, J. G., "Combining the Best of Bulk and Surface Micromachining Using Si {111} Substrates," *Proceedings SPIE Micromach. Microfab. IV Tech. Conf.*, Vol. 3511, 1998, pp. 162–168.

[22] Lehmann, V., "Porous Silicon — A New Material for MEMS," *Proceedings of MEMS'96, 9th IEEE International Workshop Micro Electromechanical System*, San Diego, CA, Feb. 11–15, 1996, pp. 1–6.

[23] Lärmer, P., *Method of Anisotropically Etching Silicon*, German Patent DE 4 241 045, 1994.

[24] Sobek, D., Senturia, S. D., and Gray, M. L., "Microfabricated Fused Silica Flow Chambers for Flow Cytometry," *Technical Digest of the IEEE Solid State Sensor and Actuator Workshop*, Hilton Head Island, SC, June 13–16, 1994, pp. 260–263.

[25] Harrison, D. J., et al., "Miniaturized Chemical Analysis Systems Based on Electrophoretic Separations and Electroosmotic Pumping," *Proceedings of Transducers '93, 7th International Conference on Solid-State Sensors and Actuators*, Yokohama, Japan, June 7–10, 1993, pp. 403–406.

[26] Jacobson, S. C., et al., "Electrically Driven Separations on a Microchip," *Technical Digest of the IEEE Solid State Sensor and Actuator Workshop*, Hilton Head Island, SC, June 13–16, 1994, pp. 65–68.

[27] Möbius, H., et al., "Sensors Controlled Processes in Chemical Reactors," *Proceedings of Transducers '95, 8th International Conference on Solid-State Sensors and Actuators*, Stockholm, Sweden, June 16–19, 1995, pp. 775–778.

[28] Jiang, L., Wong, M., and Zohar, Y., "A Micro-Channel Heat Sink with Integrated Temperature Sensors for Phase Transition Study," *Proceedings of MEMS'99, 12th IEEE International Workshop Micro Electromechanical System*, Orlando, FL, Jan. 17–21, 1999, pp. 159–164.

[29] Chen, J., and Wise, K. D., "A High Resolution Silicon Monolithic Nozzle Array for Inkjet Printing," *IEEE Transaction on Electron Devices*, Vol. 44, No. 9, 1997, pp. 1401–1409.

[30] Chen, J., et al., "A Multichannel Neural Probe for Selective Chemical Delivery at the Cellular Level," *IEEE Transaction on Biomedical Engineering*, Vol. 44., No. 8, 1997, pp. 760–769.

[31] Tjerkstra, R. W., et al., "Etching Technology for Microchannels," *Proceedings of MEMS'97, 10th IEEE International Workshop Micro Electromechanical System*, Nagoya, Japan, Jan. 26-30, 1997, pp. 147–151.

[32] De Boer, M. J., et al., "Micromachining of Buried Micro Channels in Silicon," *Journal of Microelectromechanical Systems*, Vol. 9, No. 1, 2000, pp. 94–103.

[33] Osterbroek, R. E., et al., "Etching Methodologies in $<111>$-Oriented Silicon Wafers," *Journal of Microelectromechnical Systems*, Vol. 9, No. 3, 2000, pp. 390–397.

[34] Cao, H., et al., "Fabrication of 10 nm Enclosed Nanofluidic Channels," *Applied Physics Letter*, Vol. 81, No. 1, 2002, pp. 174–176.

[35] Bustillo, J. M., Howe, R. T., and Muller, R. S., "Surface Micromachining for Microelectromechanical Systems," *Proceedings of the IEEE*, Vol. 86, No. 8, 1998, pp. 1552–1574.

[36] Tang, W. C., Nguyen, T. C. H., and Howe, R. T., "Laterally Driven Polysilicon Resonant Microstructures," *Sensors and Actuators*, Vol. 20, 1989, pp. 25–32.

[37] Keller, C., and Ferrari, M., "Milli-Scale Polysilicon Structures," *Technical Digest of the IEEE Solid State Sensor and Actuator Workshop*, Hilton Head Island, SC, June 13–16, 1994, pp. 132–137.

[38] Keller, C. G., and Howe, R. T., "Hexsil Bimorphs for Vertical Actuation," *Proceedings of Transducers '95, 8th International Conference on Solid-State Sensors and Actuators*, Stockholm, Sweden, June 16–19, 1995, pp. 99–102.

[39] Flannery, A. F., "PECVD Silicon Carbide for Micromachined Transducers," *Proceedings of Transducers '97, 9th International Conference on Solid-State Sensors and Actuators*, Chicago, IL, June 16–19, 1997, pp. 217–220.

[40] Rajan, N., et al., "Fabrication and Testing of Micromachined Silicon Carbide and Nickel Fuel Atomizers for Gas Turbine Engines," *Journal of Microelectromechanical Systems*, Vol. 8, No. 3, 1999, pp. 251–257.

[41] Shaw, K. A., Zhang, Z. L., and ld, N. C., "SCREAM-I: A Single Mask, Single-Crystal Silicon, Reactive Ion Etching Process for Microelectromechanical Structures," *Sensors and Actuators A*, Vol. 40, No. 1, 1994, pp. 63–70.

[42] Li, Y. X., et al., "Fabrication of a Single Crystalline Silicon Capacitive Lateral Accelerometer Using Micromachining Based on Single Step Plasma Etching," *Proceedings of MEMS'95, 8th IEEE International Workshop Micro Electromechanical System*, Amsterdam, the Netherlands, Jan. 29-Feb. 2, 1995, pp. 398–403.

[43] Jansen, H., et al., "The Black Silicon Method IV: The Fabrication of Three-Dimensional Structures in Silicon with High Aspect Ratios For Scanning Probe Microscopy and Other Applications," *Proceedings of MEMS'95, 8th IEEE International Workshop Micro Electromechanical System*, Amsterdam, the Netherlands, Jan. 29-Feb. 2, 1995, pp. 88–93.

[44] De Boer, M., Jansen, H., and Elwenspoek, M., "The Black Silicon Method V: A Study of the Fabricating of Moveable Structures for Micro Electromechanical Systems," *Proceedings of Transducers '95, 8th International Conference on Solid-State Sensors and Actuators*, Stockholm, Sweden, June 16–19, 1995, pp. 565–568.

[45] Jansen, H., de Boer, M., and Elwenspoek, M., "The Black Silicon Method VI: High Aspect Ratio Trench Etching for MEMS Applications," *Proceedings of MEMS'96, 9th IEEE International Workshop Micro Electromechanical System*, San Diego, CA, Feb. 11–15, 1996, pp. 250–257.

[46] Bell, T. E., et al., "Porous Silicon as a Sacrificial Material," *Journal of Micromechanics and Microengineering*, Vol. 6, 1996, pp. 361–369.

[47] Lee, S., Park, S., and Cho, D., "The Surface/Bulk Micromachining (SBM) Process: A New Method for Fabricating Released MEMS in Single Crystal Silicon," *Journal of Microelectromechanical Systems*, Vol. 8, No. 4, 1999, pp. 409–416.

[48] Lin, L., Pisano, A. P., and Muller, R. S., "Silicon Processed Microneedles," *Proceedings of Transducers '93, 7th International Conference on Solid-State Sensors and Actuators*, Yokohama, Japan, June 7–10, 1993, pp. 237–240.

[49] Wu, S., et al., "A Suspended Microchannel with Integrated Temperature for High-Pressure Flow Studies," *Proceedings of MEMS'98, 11th IEEE International Workshop Micro Electromechanical System*, Heidelberg, Germany, Jan. 25–29, 1998, pp. 87–92.

[50] Talbot, N. H., and Pisano, A. P., "Polymolding: Two Wafer Polysilicon Micromolding of Closed-Flow Passages for Microneedles and Microfluidic Devices," *Technical Digest of the IEEE Solid State Sensor and Actuator Workshop*, Hilton Head Island, SC, June 4–8, 1998, pp. 265–268.

[51] Spiering, V. L., et al., "Novel Microstructures and Technologies Applied in Chemical Analysis Techniques," *Proceedings of Transducers '97, 9th International Conference on Solid-State Sensors and Actuators*, Chicago, IL, June 16–19, 1997, pp. 511–514.

[52] Papautsky, I., et al., "Micromachined Pipette Arrays," *IEEE Transaction on Biomedical Engineering*, Vol. 47, No. 6, 2000, pp. 812–819.

[53] Carlen, E. T., and Mastrangelo, C. H., "Parafin Actuated Surface Micromachined Valves," *Proceedings of MEMS'00, 13th IEEE International Workshop Micro Electromechanical System*, Miyazaci, Japan, Jan. 23–27, 2000, pp. 381–385.

[54] Becker, E. W., et al., "Fabrication of Microstructures with High Aspect Ratios and Great Structural Heights by Synchrotron Radiation Lithography, Glavanoforming, and Plastic Moulding (LIGA Process)," *Microelectronic Engineering*, Vol. 4, pp. 35–56.

[55] Guckel, H., Christensen T. R., and Skrobis, K. J., *Formation of Microstructures Using a Preformed Photoresist Sheet*, U.S. Patent #5378583, January 1995.

[56] Chaudhuri, B., et al., "Photoresist Application for the LIGA Process," *Microsystem Technologies*, Vol. 4, 1998, pp. 159–162.

[57] Mohr, J., et al., "Requirements on Resist Layers in Deep-Etch Synchrotron Radiation Lithography," *Journal of Vacuum Science and Technology*, Vol. B6, 1988, pp. 2264–2267.

[58] Guckel, H., et al., "Plasma Polymerization of Methyl Methacrylate: A Photoresist for Applications," *Technical Digest of the IEEE Solid State Sensor and Actuator Workshop*, Hilton Head Island, SC, June 4–7, 1988, pp. 43–46.

[59] Ghica, V., and Glashauser, W., *Verfahren fr die Spannungsfreie Entwicklung von Bestrahlten Polymethylmethacrylate-Schichten*, German patent, #3039110, 1982.

[60] Lee, K., et al., "Micromachining Applications for a High Resolution Ultra-Thick Photoresist," *J. Vac. Scien. Technol. B*, Vol. 13, 1995, pp. 3012–3016.

[61] Shaw, J. M., et al., "Negative Photoresists for Optical Lithography," *IBM Journal of Research and Development*, Vol. 41, 1997, pp. 81–94.

[62] Lorenz, H., et al., "SU-8: A Low-Cost Negative Resist for MEMS," *Journal of Micromechanics and Microengineering*, Vol. 7, 1997, pp. 121–124.

[63] Lorenz, H., et al., "Fabrication of Photoplastic High-Aspect Ratio Microparts and Micromolds Using SU-8 UV Resist," *Microsystem Technologies*, Vol. 4, 1998, pp. 143–146.

[64] MicroChem Corp, NANO$^{TM}$ *SU-8 Negative Tone Photoresists Formulations 2-25*, Data sheets, 2001.

[65] MicroChem Corp, NANO$^{TM}$ *SU-8 Negative Tone Photoresists Formulations 50-100*, Data sheets, 2001.

[66] Chang, H. K., and Kim, Y. K., "UV-LIGA Process for High Aspect Ratio Structure Using Stress Barrier and C-Shaped Etch Hole," *Sensors and Actuators A*, Vol. 84, 2000, pp. 342–350.

[67] Ghantasala, M. K., et al., "Patterning, Electroplating and Removal of SU-8 Moulds by Excimer Laser Micromachining," *Journal of Micromechanics and Microengineering*, Vol. 11, 2001, pp. 133–139.

[68] Ayliffe, H. E., Frazier, A.B., and Rabbitt, R.D., "Electric Impedence Spectroscopy Using Microchannels with Integrated Metal Electrodes," *IEEE Journal of Microelectromechanical Systems*, Vol. 8, No. 1, 1999, pp. 50–57.

[69] Jackman, R. J., et al., "Microfluidic Systems with On-Line UV Detection Fabricated in Photodefinable Epoxy," *Journal of Micromechanics and Microengineering*, Vol. 11, 2001, pp. 263–269.

[70] Hostis, E. L., et al., "Microreactor and Electrochemical Detectors Fabricated Using Si and EPSON SU-8," *Sensors and Actuators B*, Vol. 64, 2000, pp. 156–162.

[71] Alderman, B. E. J., et al., "Microfabrication of Channels Using an Embedded Mask in Negative Resist," *Journal of Micromechanics and Microengineering*, Vol. 11, 2001, pp. 703–705.

[72] Guérin, L. J., et al., "Simple and Low Cost Fabrication of Embedded Micro Channels by Using a New Thick-Film Photoplastic," *Proceedings of Transducers '97, 9th International Conference on Solid-State Sensors and Actuators*, Chicago, IL, June 16–19, 1997, pp. 1419–1421.

[73] Chuang, Y. J., et al., "A Novel Fabrication Method of Embedded Micro-Channels by Using SU-8 Thick-Film Photoresists," *Sensors and Actuators A*, Vol. 103, 2003, pp. 64–69.

[74] Tay, F. E. H., et al., "A Novel Micro-Machining Method for the Fabrication of Thick-Film SU-8 Embedded Micro-Channels," *Journal of Micromechanics and Microengineering*, Vol. 11, 2001, pp. 27–32.

[75] Zhang, J., et al., "Polymerization Optimization of SU-8 Photoresist and Its Applications in Microfluidic Systems and MEMS," *Journal of Micromechanics and Microengineering*, Vol. 11, 2001, pp. 20–26.

[76] O'Brien, J., et al., "Advanced Photoresist Technologies for Microsystems," *Journal of Micromechanics and Microengineering*, Vol. 11, 2001, pp. 353–358.

[77] Qu, W., Wenzel, C., and Jahn, A., "One-Mask Procedure for The Fabrication of Movable High-Aspect-Ratio 3D Microstructures," *Journal of Micromechanics and Microengineering*, Vol. 8, 1998, pp. 279–283.

[78] Conédéra, V., Le Goff, B., and Fabre, N., "Potentialities of a New Positive Photoresist for the Realization of Thick Moulds," *Journal of Micromechanics and Microengineering*, Vol. 9, 1999, pp. 173–175.

[79] Loechel, B., "Thick-Layer Resists for Surface Micromachining," *Journal of Micromechanics and Microengineering*, Vol. 10, 2000, pp. 108–115.

[80] Weston, D. F., et al., "Fabrication of Microfluidic Devices in Silicon and Plastic Using Plasma Etching," *Journal of Vacuum Science and Technology B*, Vol. 19, 2001, pp. 2846–2851.

[81] Wang, X., Engel, J., and Liu, C., "Liquid Crystal Polymer (LCP) for MEMS: Processes and Applications," *Journal of Micromechanics and Microengineering*, Vol. 13, 2003, pp. 628–633.

[82] Zhao, Y., and Cui, "Fabrication of High-Aspect-Ratio Polymer-Based Electrostatic Comb Drives Using the Hot Embossing Technique," *Journal of Micromechanics and Microengineering*, Vol. 13, 2003, pp. 430–435.

[83] Truong, T. Q., and Nguyen, N. T., "A Polymeric Piezoelectric Micropump Based on Lamination Technology," *Journal of Micromechanics and Microengineering*, Vol. 14, 2004, pp. 632–638.

[84] Nguyen, N. T., Ho, S. S., and Low, L. N., "A Polymeric Microgripper with Integrated Thermal Actuators," *Journal of Micromechanics and Microengineering*, Vol. 14, 2004, pp. 969–974.

[85] Frazier, A. B., and Allen, M.G., "Metallic Microstructures Fabricated Using Photosensitive Polyimide Electroplating Molds," *Journal of Microelectromechanical Systems*, Vol. 2, No. 2, 1993, pp. 87–94.

[86] Ito, T., et al., "Fabrication of Microstructure Using Fluorinated Polyimide and Silicon-Based Positive Photoresist," *Microsystem Technologies*, Vol. 6, 2000, pp. 165–168.

[87] Stieglitz, T., "Flexible Biomedical Microdevices with Double-Sided Electrode Arrangements For Neural Applications," *Sensors and Actuators A*, Vol. 90, 2001, pp. 203–211.

[88] Metz, S., Holzer, R., and Renaud, P., "Polyimide-Based Microfluidic Devices," *Lab on a Chip*, Vol. 1, No. 1, 2001, pp. 29–34.

[89] Smela, E., "Microfabrication of PPy Microactuators and Other Conjugated Polymer Devices," *Journal of Micromechanics and Microengineering*, Vol. 9, 1999, pp. 1–18.

[90] Reed, H. A., et al., "Fabrication of Microchannels Using Polycarbonates as Sacrificial Materials," *Journal of Micromechanics and Microengineering*, Vol. 11, 2001, pp. 733–737.

[91] Bhusari, D., et al., "Fabrication of Air-Channel Structures for Microfluidic, Microelectromechanical, and Microelectronic Applications," *Journal of Micromechanics and Microengineering*, Vol. 10, 2001, pp. 400–408.

[92] Maruo, S., and Ikuta, K., "Three-Dimensional Microfabrication by Use of Single-Photon-Absorbed Polymerization," *Applied Physics Letters*, Vol. 76, No. 19, 2000, pp. 2656–2658.

[93] Maruo, S., and Kawata, S., "Two-Photon-Absorbed Near-Infrared Photopolymerization for Three-Dimensional Microfabrication," *Journal of Microelectromechanical Systems*, Vol. 7, No. 4, 1998, pp. 411–415.

[94] Ikuta, K., and Hirowatari, K., "Real Three Dimensional Micro Fabrication Using Stereo Lithography and Metal Molding," *Proceedings of MEMS'93, 6th IEEE International Workshop Micro Electromechanical System*, San Diego, CA, Jan. 25–28, 1993, pp. 42–47.

[95] Ikuta, K., Maruo, S., and Kojima, S., "New Micro Stereo Lithography for Freely Moveable 3D Micro Structure—Super IH Process with Submicron Resolution," *Proceedings of MEMS'98, 11th IEEE International Workshop Micro Electromechanical System, Heidelberg*, Germany, Jan. 25–29, 1998, pp. 290–295.

[96] Miwa, M., et al., "Femtosecond Two-Photon Stereo-Lithography," *Applied Physics A*, Vol. 73, 2001, pp. 561–566.

[97] Ikuta, K., et al., "Development of Mass Productive Micro Stereo Lithography (Mass-IH Process)," *Proceedings of MEMS'96, 9th IEEE International Workshop Micro Electromechanical System*, San Diego, CA, Feb. 11–15, 1996, pp. 301–306.

[98] Bertsch, A., Lorenz, H., and Renaud, H., "Combining Microstereolithography and Thick Resist UV Lithography for 3D Microfabrication," *Sensors and Actuators A*, Vol. 73, 1999, pp. 14–23.

[99] Ikuta, K., Hirowatari, K., and Ogata, T., "Three Dimensional Micro Integrated Fluid Systems (MIFS) Fabricated by Stereo Lithography," *Proceedings of MEMS'94, 7th IEEE International Workshop Micro Electromechanical System*, Oiso, Japan, Jan. 25–28, 1994, pp. 1–6.

[100] Ikuta, K., et al., "Micro Concentrator with Opto-Sense Micro Reactor for Biochemical IC Chip Family—3D Composite Structure and Experimental Verification," *Proceedings of MEMS'99, 12th IEEE International Workshop Micro Electromechanical System*, Orlando, FL, Jan. 17–21, 1999, pp. 376–381.

[101] Ikuta, K., et al., "Fluid Drive Chips Containing Multiple Pumps and Switching Valves for Biochemical IC Family—Development of SMA 3D Micro Pumps and Valves in Leak-Free Polymer Package," *Proceedings of MEMS'00, 13th IEEE International Workshop Micro Electromechanical System*, Miyazaci, Japan, Jan. 23–27, 2000, pp. 739–744.

[102] Bertsch, A., et al., "Static Micromixers Based on Large-Scale Industrial Mixer Geometry," *Lab on a Chip*, Vol. 1, 2001, pp. 56–20.

[103] Macdonald, N. P., et al., "Comparing Microfluidic Performance of Three-dimensional (3D) Printing Platforms." *Analytical Chemistry*, Vol. 89, 2017, pp. 3858–3866.

[104] Au, A. K., et al., "3D Printed Microfluidics," *Angewandte Chemie International Ed.*, Vol. 55, 2016, pp. 3862–3881.

[105] Bhattacharjee, N., et al., "The Upcoming 3D-printing Revolution in Microfluidics," *Lab Chip*, Vol. 16, 2016, pp. 1720–1742.

[106] Piotter, V., et al., "Injection Molding and Related Techniques for Fabrication of Microstructures," *Microsystem Technologies*, Vol. 4, 1997, pp. 129–133.

[107] Larsson, O., et al., "Silicon Based Replication Technology of 3D-Microstructures by Conventional CD-Injection Molding Techniques," *Proceedings of Transducers '97, 9th International Conference on Solid-State Sensors and Actuators*, Chicago, IL, June 16–19, 1997, pp. 1415–1418.

[108] Despa, M. S., Kelly, K. W., and Collier, J. R., "Injection Molding of Polymeric LIGA HARMS," *Microsystem Technologies*, Vol. 6, 1999, pp. 60–66.

[109] Niggemann, M., et al., "Miniaturized Plastic Micro Plates for Applications in HTS," *Microsystem Technologies*, Vol. 6, 1999, pp. 48–53.

[110] Gerlach, A., et al., "Gas Permeability of Adhesives and Their Application for Hermetic Packaging of Microcomponents," *Microsystem Technologies*, Vol. 7, 2001, pp. 17–22.

[111] Becker, H., and Heim, U., "Hot Embossing as a Method for the Fabrication of Polymer High Aspect Ratio Structures," *Sensors and Actuators A*, Vol. 83, 2000, pp. 130–135.

[112] Heckele, M., Bacher, W., and Müller, K. D., "Hot Embossing—The Molding Technique for Plastic Microstructures," *Microsystem Technologies*, Vol. 4, 1998, pp. 122–124.

[113] Man, P. F., Jones, D. K., and Mastrangelo, C. H., "Microfluidic Plastic Capillaries on Silicon Substrates: A New Inexpensive Technology for Bioanalysis Chips," *Proceedings of MEMS'97, 10th IEEE International Workshop Micro Electromechanical System*, Nagoya, Japan, Jan. 26–30, 1997, pp. 311–316.

[114] Xia, Y., and Whitesides, G. M., "Soft Lithography," *Annual Review of Material Sciences*, Vol. 28, 1998, pp. 153–194.

[115] Duffy, D. C., et al., "Rapid Prototyping of Microfluidic Switches in Poly (Dimethyl Siloxane) and Their Actuation by Electro-Osmotic Flow, *Journal of Micromechanics and Microengineering*, Vol. 9, 1999, pp. 211–217.

[116] Jo, B. H., et al., "Three-Dimensional Micro-Channel Fabrication in Polydimethylsiloxane (PDMS) Elastomer," *Journal of Microelectromechanical Systems*, Vol. 9, No. 1, 2000, pp. 76–81.

[117] Juncker, D., et al., "Soft and Rigid Two-Level Microfluidic Networks for Patterning Surfaces," *Journal of Micromechanics and Microengineering*, Vol. 11, 2001, pp. 532–541.

[118] Chichkov, B., et al., "Femtosecond, Picosecond and Nanosecond Laser Ablation of Solids," *Applied Physics A*, Vol. 63, 1996, pp. 109–115.

[119] Laws, R. A., Holmes, A. S., and Goodall, F. N., "The Formation of Moulds for 3D Microstructures Using Excimer Laser Ablation," *Microsystem Technologies*, Vol. 3, 1996, pp. 17–19.

[120] Yamad, K., et al., "Paper-Based Inkjet-Printed Microfluidic Analytical Devices," *Angewandte Chemie International Ed.*, Vol. 54, 2015, pp. 5294–5310.

[121] Martinez, A., et al., "FLASH: A Rapid Method for Prototyping Paper-Based Microfluidic Devices," *Lab Chip*, Vol. 8, 2008, pp. 2146–2150.

[122] Carriho, E., et al., "Understanding Wax Printing: A Simple Micropatterning Process for Paper-Based Microfluidics," *Analytical Chemistry*, Vol. 81, 2009, pp. 7091–7095.

[123] Olkkonen, J., et al., "Flexographically Printed Fluidic Structures in Paper," *Analytical Chemistry*, Vol. 82, 2010, pp. 10246–10250.

[124] Guckenberger, D. J., et al.,"Micromilling: A Method for Ultra-Rapid Prototyping of Plastic Microfluidic Devices," *Lab Chip*, Vol. 15, 2015, pp. 2343–2524.

[125] Chan, S. H., et al.,"Development of a Polymeric Micro Fuel Cell Containing Laser-Micromachined Flow Channels," *J. Micromech. Microeng.*, Vol. 15, 2005, pp. 231–236.

[126] Shaegh, S. A. M., et al., "Rapid Prototyping of Whole-Thermoplastic Microfluidics With Built-in Microvalves Using Laser Ablation and Thermal Fusion Bonding", *Sensors and Actuators B: Chemical*, Vol. 255, 2018, pp. 100–109.

[127] Belloy, E., et al., "The Introduction of Powder Blasting for Sensor and Microsystem Applications," *Sensors and Actuators A*, Vol. 84, 2000, pp. 330–337.

[128] Wensink, H., et al., "Mask Materials for Powder Blasting," *Journal of Micromechanics and Microengineering*, Vol. 10, 2000, pp. 175–180.

[129] Shlautmann, S., et al., "Powder-Blasting Technology as an Alternative Tool for Microfabrication of Capillary Electrophoresis Chips with Integrated Conductivity Sensors," *Journal of Micromechanics and Microengineering*, Vol. 11, 2001, pp. 386–389.

[130] Yeo, S. H., and Choo, J. H., "Effects of Rotor Electrode In the Fabrication of High Aspect Ratio Microstructures by Localized Electrochemical Deposition," *Journal of Micromechanics and Microengineering*, Vol. 11, 2001, pp. 435–442.

[131] Stjernström, M., and Roeraade J., "Method for Fabrication of Microfluidic System in Glass," *Journal of Micromechanics and Microengineering*, Vol. 8, 1998, pp. 33–38.

[132] Chen, C.-f, and Wharton, K., "Characterization and failure mode analysis of air plasma oxidized PDMS-PDMS bonding by peel testing," *RSC Adv.*, Vol. 7, 2017, pp. 1286–1289.

[133] Matson, D. W., et al., "Laminated Ceramic Components for Micro Fluidic Applications," *SPIE Conference Proceeding Vol 3877: Microfluidic Devices and Systems II*, Santa Clara, CA, Sept. 20-22, 1999, pp. 95–100.

[134] Martin, P. M., et al., "Laser Micromachined and Laminated Microfluidic Components for Miniaturized Thermal, Chemical and Biological Systems," *SPIE Conference Proceedings, Vol. 3680: Design, Test, and Microfabrication of MEMS and MOEMS*, Paris, France, Mar. 30-Apr. 1, 1999, pp. 826–833.

[135] Ko, W. H., et al., "Bonding Techniques for Microsensors," in *Micromachining and Micropackaging of Transducers*, Fung, C. D., et al. (eds.), Amsterdam: Elsevier, 1985, pp. 41–61.

[136] Weckwerth, M. V., et al., "Epoxy Bond and Stop-Etch (EBASE) Technique Enabling Backside Processing of (Al)GaAs Heterostructures," *Superlattices Microstructures*, Vol. 20, No. 4, 1996, pp. 561–567.

[137] Nguyen H., et al., "A Substrate-Independent Wafer Transfer Technique for Surface-Micromachined Devices," *Proceedings of MEMS'00, 13th IEEE International Workshop Micro Electromechanical System*, Miyazaci, Japan, Jan. 23–27, 2000, pp. 628–632.

[138] Tsao, C.-W, and DeVoe, D. L., "Bonding of Thermoplastic Polymer Microfluidics," *Microfluidic Nanofluidic*, Vol. 6, 2009, pp. 1–16.

[139] Gonzalez, C., Collins, S. D., and Smith, R. L., "Fluidic Interconnects for Modular Assembly of Chemical Microsystems," *Sensors and Actuators B*, Vol. 49, 1998, pp. 40–45.

[140] Gray, B. L., et. al., "Novel Interconnection Technologies for Integrated Microfluidic Systems," *Sensors and Actuators A*, Vol. 77, 1999, pp. 57–65.

[141] Mourlas, N. J., et al., "Reuseable Microfluidic Coupler with PDMS Gasket," *Proceedings of Transducers '99, 10th International Conference on Solid-State Sensors and Actuators*, Sendai, Japan, June 7–10, 1999, pp. 1888–1889.

[142] Meng, E., Wu, S., and Tai, Y.-C., "Micromachined Fluidic Couplers," in *Micro Total Analysis Systems 2000*, A. van den Berg et al. (eds.), Netherlands: Kluwer Academic Publishers, 2000, pp. 41–44.

[143] Puntambekar, A., and Ahn, C. H., "Self-Aligning Microfluidic Interconnects for Glass- and Plastic-Based Microfluidic Systems," *Journal of Micromechanics and Microengineering*, Vol. 12, No. 1, 2002, pp. 35–40.

[144] Trah, H. P., et al., "Micromachined Valve with Hydraulically Actuated Membrane Subsequent to a Thermoelectrically Controlled Bimorph Cantilever," *Sensors and Actuators A*, Vol. 39, 1993, pp. 169–176.

[145] Yao, T. J., et al., "Micromachined Rubber O-Ring Micro-Fluidic Couplers," *Proceedings of MEMS'00, 13th IEEE International Workshop Micro Electromechanical System*, Miyazaci, Japan, Jan. 23–27, 2000, pp. 745–750.

[146] Wijngarrt W., et al., "The First Self-Priming and Bi-Directional Valve-Less Diffuser Micropump for Both Liquid and Gas," *Proceedings of MEMS'00, 13th IEEE International Workshop Micro Electromechanical System*, Miyazaci, Japan, Jan. 23–27, 2000, pp. 674–679.

[147] London, A. P., "Development and Test of a Microfabricated Bipropellant Rocket Engine," Ph.D. thesis, Massachusetts Institute of Technology, 2000.

[148] Black, J., *Biological Performance of Materials*, New York: Marcel Dekker, 1992.

# Chapter 4

# Experimental Flow Characterization

## 4.1 INTRODUCTION

There are many areas in science and engineering where it is important to determine the flow field at the micron scale. These will be discussed in Chapters 5 through 13. Industrial applications of microfabricated fluidic devices are present in the aerospace, computer, automotive, and biomedical industries. In the aerospace industry, for instance, micron-scale supersonic nozzles measuring approximately 35 μm are being designed for JPL/NASA to be used as microthrusters on microsatellites, and for AFOSR/DARPA as flow control devices for palm-size microaircraft [1]. In the computer industry, inkjet printers, which consist of an array of nozzles with exit orifices on the order of tens of microns in diameter, account for 65% of the computer printer market [2]. The biomedical industry is currently developing and using microfabricated fluidic devices for patient diagnosis, patient monitoring, and drug delivery. The i-STAT device (i-STAT, Inc.) is the first microfabricated fluidic device that has seen routine use in the medical community for blood analysis. Other examples of microfluidic devices for biomedical research include microscale flow cytometers for cancer cell detection, micromachined electrophoretic channels for DNA fractionation, and polymerase chain reaction (PCR) chambers for DNA amplification [3]. The details of the fluid motion through these small channels, coupled with nonlinear interactions between macromolecules, cells, and the surface-dominated physics of the channels, create very complicated phenomena, which can be difficult to simulate numerically.

A wide range of diagnostic techniques have been developed for experimental microfluidic research. Some of these techniques have been designed to obtain the highest spatial resolution and velocity resolution possible, while other techniques have been designed for application in nonideal situations where optical access is limited [4], or in the presence of highly scattering media [5].

### 4.1.1 Pointwise Methods

*Laser Doppler velocimetry* (LDV) has been a standard optical measurement technique in fluid mechanics since the 1970s. In the case of a dual-beam LDV system, two coherent laser beams are aligned so that they intersect at some region. The volume of the intersection of the two laser beams defines the measurement volume. In the measurement volume, the two coherent laser beams interfere with each other, producing a pattern of light and dark fringes. When a seed particle passes through these fringes, a pulsing reflection is created that is collected by a photomultiplier, processed, and turned into a velocity measurement. Traditionally, the measurement volumes of standard LDV systems have characteristic dimensions on the order of a few millimeters. Compton and Eaton [6] used short focal length optics to obtain a measurement volume of 35 μm × 66 μm. Using very short focal length lenses, Tieu et al. [7] built a dual-beam solid-state LDA system with a measurement volume of approximately 5 μm × 10 μm. Their micro-LDV system was used to measure the flow through a 175-μm-thick channel, producing time-averaged measurements that compare well to the expected parabolic velocity profile, except within 18 μm of the wall. Advances in microfabrication technology are expected to facilitate the development of new generations of self-contained solid-state LDV systems with micron-scale probe volumes. These systems will likely serve an important role in the diagnosis and monitoring of microfluidic systems [8]. However, the size of the probe volume significantly limits the number of fringes that it can contain, which subsequently limits the accuracy of the velocity measurements.

*Optical Doppler tomography* (ODT) has been developed to measure micron-scale flows embedded in a highly scattering medium. In the medical community, the ability to measure in vivo blood flow under the skin allows clinicians to determine the location and depth of burns [5]. ODT combines single-beam Doppler velocimetry, with heterodyne mixing from a low-coherence Michelson interferometer. The lateral spatial resolution of the probe volume is determined by the diffraction spot size. The Michelson interferometer is used to limit the effective longitudinal length of the measurement volume to that of the coherence length of the laser. The ODT system developed by Chen et al. [5] has lateral and longitudinal spatial resolutions of 5 μm and 15 μm, respectively. The system was applied to measure flow through a 580-μm-diameter conduit.

### 4.1.2 Full-Field Methods

Full-field experimental velocity measurement techniques are those that generate velocities that are minimally two-component velocity measurements distributed within a 2-D plane. These types of velocity measurements are essential in the microfluidics field for several reasons. First, global measurements, such as the pressure drop along a length of channel, can reveal the dependence of flow physics upon length scale by showing that the pressure drop for flow through a small channel is smaller or larger than a flow through a large channel. However, global measurements are not very useful for pointing to the precise cause of why the physics might change, such as losing the no-slip boundary for high Knudsen number gas flows. A detailed view of the flow, such as that provided by full-field measurement techniques, is indispensable for establishing the reasons why flow behavior changes at small scales. Full-field velocity measurement techniques are also useful for optimizing complicated processes such as mixing, pumping, or filtering, on which some MEMS depend. Several of the common macroscopic full-field measurement techniques have been extended to microscopic length scales. These are scalar image velocimetry, molecular tagging velocimetry, and particle image velocimetry. These techniques will be introduced briefly in this section, and then discussed in detail in the following sections.

*Scalar image velocimetry* (SIV) refers to the determination of velocity fields by recording images of a passive scalar quantity and inverting the transport equation for a passive scalar. Dahm et al. [9] originally developed SIV for measuring turbulent jets at macroscopic length scales. Successful velocity measurements depend on having sufficient spatial variations in the passive-scalar field and relatively high Schmidt numbers. Since SIV uses molecular tracers to follow the flow, it has several advantages at the microscale over measurement techniques such as PIV or LDV, which use discrete flow-tracing particles. For instance, the molecular tracers will not become trapped in even the smallest passages within a MEMS or NEMS device. In addition, the discrete flow tracing particles used in PIV can acquire a charge and move in response to not only hydrodynamic forces but also electrical forces, in a process called *electrophoresis* (see Chapter 2). However, molecular tracers typically have much higher diffusion coefficients than do discrete particles, which can significantly lower the spatial resolution and velocity resolution of the measurements.

Paul et al. [10] analyzed fluid motion using a novel dye that, while not normally fluorescent, can be made fluorescent by exposure to the appropriate wavelength of light. Dyes of this class are typically called *caged dyes*, because the fluorescent nature of the dye is defeated by a photoreactive bond that can be easily broken. This caged dye was used in microscopic SIV procedures to estimate velocity fields for pressure-driven and electrokinetically driven flows in 75-μm-diameter capillary tubes. A 20-μm×500-μm sheet of light from a $\lambda = 355$-nm frequency-tripled Nd:YAG laser was used to uncage a 20-mm-thick cross-sectional plane of dye in the capillary tube. In this technique, only the uncaged dye is excited when the test section is illuminated with a shuttered beam from a continuous wave Nd:$YVO_4$ laser. The excited fluorescent dye is imaged using a 10×, $NA = 0.3$ objective lens onto a charge-coupled device (CCD) camera at two known time exposures. The velocity field is then inferred from the motion of the passive scalar. We approximate the spatial resolution of this experiment to be on the order of $100 \times 20 \times 20$ μm, based on the displacement of the fluorescent dye between exposures, and the thickness of the light sheet used to uncage the fluorescent dye.

*Molecular tagging velocimetry* (MTV) is another technique that has shown promise in microfluidics research. In this technique, flow-tracing molecules fluoresce or phosphoresce after being excited by a light source. The excitement is typically in the form of a pattern such as a line or grid written into the flow. The glowing gridlines are imaged twice, with a short time delay between the two images. Local velocity vectors are estimated by correlating the gridlines between the two images [11]. MTV has the same advantages and disadvantages, at least with respect to the flow tracing molecules, as SIV. In contrast to SIV, MTV infers velocity in much the same way as particle image velocimetry—a pattern is written into the flow and the evolution of that pattern allows for inferring the velocity field. MTV was demonstrated at microscopic length scales by Maynes and Webb [12] in their investigation of liquid flow through capillary tubes, as well as by Lempert et al. [13] in an investigation of supersonic micronozzles. Maynes and Webb [12] investigated aqueous glycerin solutions flowing at Reynolds numbers ranging from 600 to 5,000 through a 705-mm-diameter fused-silica tube of circular cross section. They stated that the spatial resolution of their technique is 10 mm across the diameter of the tube and 40 mm along the axis of the tube. The main conclusion of this work was that the velocity measured in their submillimeter tube agreed quite well with laminar flow theory, and that the flow showed a transition to turbulence beginning at a Reynolds number of 2,100. Lempert et al. [13] flowed a mixture of gaseous nitrogen and acetone through a 1-mm straight-walled "nozzle" at pressure ratios ranging from highly underexpanded to perfectly matched. Because the nozzle was not transparent, the measurement area was limited to positions outside the nozzle. A single line was written in the gas, normal to the axis of the nozzle, by a frequency quadrupled Nd:YAG (266 nm) laser. The time evolution of the line was observed by an intensified CCD camera. They reported measurements at greater than Mach 1 with an accuracy of ±8 m/s ($\pm 3\%$) and a spatial resolution of 10 μm perpendicular to the nozzle axis.

The machine vision community developed a class of velocimetry algorithms, called *optical-flow algorithms*, to determine the motion of rigid objects. The technique can be extended to fluid flows by assuming the effect of molecular diffusion is negligible, and by requiring that the velocity field is sufficiently smooth. Since the velocity field is computed from temporal and spatial derivatives of the image field, the accuracy and reliability of the velocity measurements are strongly influenced by noise in the image field. This technique imposes a smoothness criterion on the velocity field, which effectively lowpass filters the data, and can lower the spatial resolution of the velocity measurements [14]. Lanzillotto et al. [4] applied the optical-flow algorithms to infer velocity fields from 500 to 1,000-μm diameter microtubes by indirectly imaging 1 to 20-μm diameter X-ray-scattering emulsion droplets in a liquid flow. High-speed X-ray microimaging techniques were presented by Leu et al. [15]. A synchrotron is used to generate high-intensity X-rays that scatter off the emulsion droplets onto a phosphorous screen. A CCD camera imaging the phosphorous screen detects variations in the scattered X-ray field. The primary advantage of the X-ray imaging technique is that one can obtain structural information about the flow field, without having optical access. Hitt et al. [16] applied the optical flow algorithm to in vivo blood flow in microvascular networks, with diameters $\approx 100$ μm. The algorithm spectrally decomposes subimages into discrete

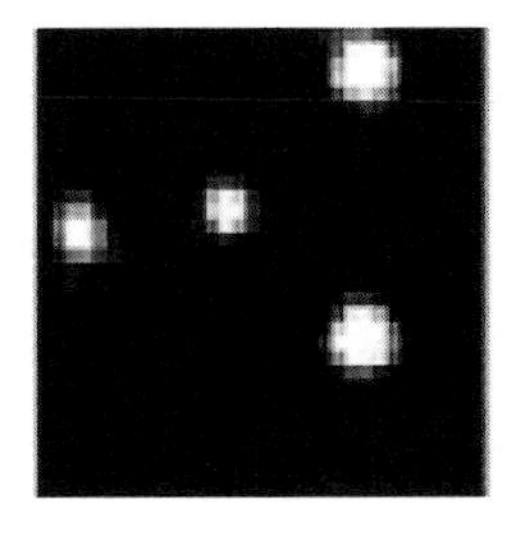

**Figure 4.1** The same interrogation region at two different times. Notice the displacement of the particle image pattern.

spatial frequencies by correlating the different spatial frequencies to obtain flow field information. The advantage of this technique is that it does not require discrete particle images to obtain reliable velocity information. Hitt et al. [17] obtained in vivo images of blood cells flowing through a microvascular network using a 20× water-immersion lens with a spatial resolution on the order of 20 μm in all directions.

*Particle image velocimetry* (PIV) has been used since the mid-1980s to obtain high spatial resolution 2-D velocity fields in macroscopic flows. The experimental procedure is, at its heart, conceptually simple to understand. A flow is made visible by seeding it with particles. The particles are photographed at two different times. The images are sectioned into many smaller regions called *interrogation regions*, as shown in Figure 4.1. The motion of the group of particles within each interrogation region is determined using a statistical technique called a *cross-correlation*. If the array of gray values comprising the first image is called $f(i, j)$ and the second image $g(i, j)$, the cross-correlation is given by

$$\Phi(m, n) = \sum_{j=1}^{q} \sum_{i=1}^{p} f(i, j) \cdot g(i + m, j + n) \tag{4.1}$$

The cross-correlation for a high-quality set of PIV measurements should look like Figure 4.2. The location of the peak indicates how far the particles have moved between the two images. Curve fitting with an appropriate model is used to obtain displacement results accurate to 0.1 pixels.

A PIV bibliography by Adrian [18] lists more than 1,200 references describing various PIV methods and the problems investigated by them. For a good reference describing many of the technical issues pertinent at macroscopic length scales, see [19]. This section will provide a brief explanation of how PIV works in principle and then concentrate on how PIV is different at small length scales.

In 1998, Santiago et al. [20] demonstrated the first μPIV system—a PIV system with a spatial resolution sufficiently small enough to be able to make measurements in microscopic systems. Since then, the technique has grown in importance at a tremendous rate. Figure 4.3 shows the number of journal papers using μPIV from 1997 to 2004. By 2005, there were well over 250 μPIV journal articles. Because of this large amount of activity, well over an order of magnitude more than the previously described techniques combined, the remainder of this chapter will concentrate on applications and extensions of μPIV.

The first μPIV system was capable of measuring slow flows—velocities on the order of hundreds of microns per second—with a spatial resolution of 6.9 × 6.9 × 1.5 μm [20]. The system used an epifluorescent microscope and an intensified CCD camera to record 300-nm-diameter polystyrene flow-tracing particles. The particles are illuminated using a continuous Hg-arc lamp. The continuous Hg-arc lamp is chosen for situations that require low levels of illumination light

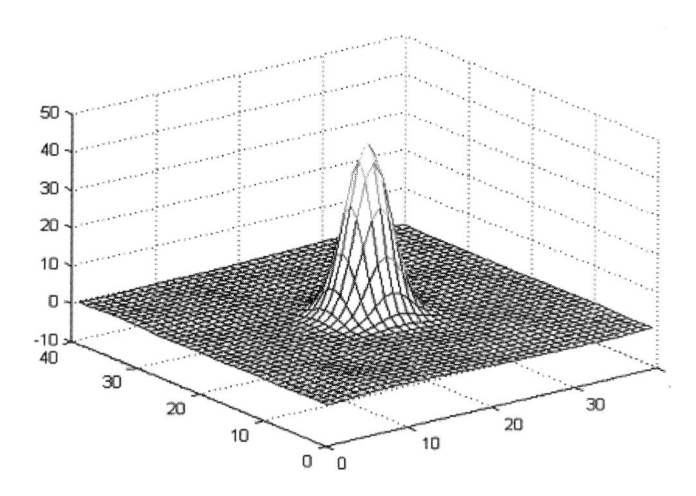


**Figure 4.2** PIV cross-correlation peak.

(e.g., flows containing living biological specimens), where the velocity is sufficiently small enough so that the particle motion can be frozen by the CCD camera's electronic shutter (Figure 4.4).

Koutsiaris et al. [21] demonstrated a system suitable for slow flows that used 10-μm glass spheres for tracer particles and a low spatial resolution, high-speed video system to record the particle images, yielding a spatial resolution of 26.2 μm. They measured the flow of water inside 236-μm round glass capillaries and found agreement between the measurements and the analytical solution within the measurement uncertainty.

Later applications of the μPIV technique moved steadily toward faster flows more typical of aerospace applications. The Hg-arc lamp was replaced with a New Wave two-headed Nd:YAG laser that allowed cross-correlation analysis of singly exposed image pairs acquired with submicrosecond time steps between images. At macroscopic length scales, this short time step would allow analysis of supersonic flows. However, because of the high magnification, the maximum velocity measurable with this time step is on the order of meters per second. Meinhart et al. [22] applied μPIV to measure the flow field in a 30 μm high × 300 μm wide rectangular channel, with a flow rate of 50 μl/hr, equivalent to a centerline velocity of 10 mm/s or three orders of magnitude greater than the initial effort a year before. The experimental apparatus, shown in Figure 4.4, images the flow with a 60×, NA = 1.4, oil-immersion lens. The 200-nm-diameter polystyrene flow-tracing particles were chosen small enough so that they faithfully followed the flow and were 150 times smaller than the smallest channel dimension. A subsequent investigation by Meinhart and Zhang [23] of the flow inside a microfabricated inkjet printer head yielded the highest speed measurements made with μPIV. Using a slightly lower magnification (40×) and consequently lower spatial resolution, measurements of velocities as high as 8 m/s were made. Considering the clear advantages that μPIV has over other techniques enumerated in Table 4.1, the rest of the chapter will concentrate on μPIV. However, some of the algorithms that will be presented are equally applicable to other microflow diagnostic techniques. In the following sections, we will give an overview of μPIV techniques and provide several application examples.

**Table 4.1** Comparison of High-Resolution Velocimetry Techniques [24].

| *Technique* | *Author* | *Flow Tracer* | *Spatial Resolution* (μm) | *Observation* |
|---|---|---|---|---|
| LDA | Tieu et al. (1995) | — | $5\times5\times10$ | 4–8 fringes limits velocity resolution |
| Optical Doppler tomography (ODT) | Chen et al. (1997) | 1.7-μm polystyrene beads | $5\times15$ | Can image through highly scattering media |
| Optical flow using video microscopy | Hitt et al. (1996) | 5-μm blood cells | $20\times20\times20$ | In vivo study of blood flow |
| Optical flow using X-ray imaging | Lanzillotto et al. (1996) | 1–20-μm emulsion droplets | $\sim$ 20–40 | Can image without optical access |
| Uncaged fluorescent dyes | Paul et al. (1997) | Molecular dye | $100\times20\times20$ | Resolution limited by molecular diffusion |
| Particle streak velocimetry | Brody et al. (1996) | 0.9-μm polystyrene beads | $\sim$ 10 | Particle streak velocimetry |
| PIV | Urushihara et al. (1993) | 1-μm oil droplets | $280\times280\times200$ | Turbulent flows |
| Super-resolution PIV | Keane et al. (1995) | 1-μm oil droplets | $50\times50\times200$ | Particle tracking velocimetry |
| μPIV | Santiago et al. (1998) | 300-nm polystyrene particles | $6.9\times6.9\times1.5$ | Hele-Shaw flow |
| μPIV | Meinhart et al. (1999) | 200-nm polystyrene particles | $5.0\times1.3\times2.8$ | Microchannel flow |
| μPIV | Westerweel et al. (2004) | 500-nm polystyrene particles | $0.5\times0.5\times2.0$ | Silicon microchannel flow |

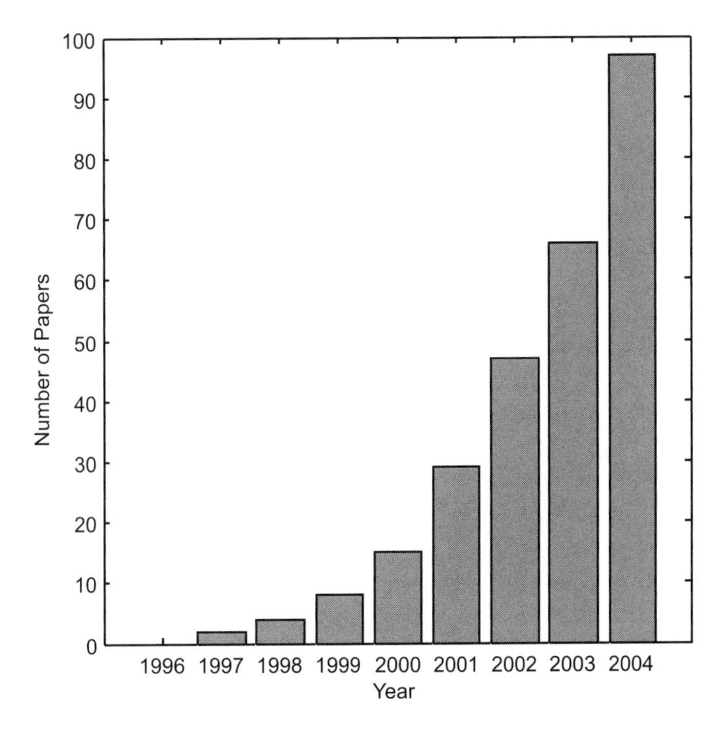


**Figure 4.3** Number of μPIV journal papers per year since its invention.

## 4.2 OVERVIEW OF μPIV

### 4.2.1 Fundamental Physics Considerations of μPIV

Three fundamental problems differentiate μPIV from conventional macroscopic PIV: the particles are small compared to the wavelength of the illuminating light; the illumination source is typically not a light sheet but rather an illuminated volume of the flow; and the particles are small enough that the effects of Brownian motion must be considered.

#### 4.2.1.1 Particles Small Compared to $\lambda$

Flow-tracing particles must also be large enough to scatter sufficient light so that their images can be recorded. In the Rayleigh scattering regime, where the particle diameter $d$ is much smaller than the wavelength of light, $d \ll \lambda$, the amount of light scattered by a particle varies as $d^{-6}$ [25]. Since the diameter of the flow-tracing particles must be small enough that the particles not disturb the flow being measured, they can frequently be on the order from 50 to 100 nm. Their diameters are then one-tenth to one-fifth the wavelength of green light, $\lambda_{\text{green}} = 532$ nm, and are therefore approaching the Rayleigh scattering criteria. This places significant constraints on the image recording optics, making it extremely difficult to record particle images.

One solution to this imaging problem is to use epifluorescence imaging to record light emitted from fluorescently labeled particles in which an optical wavelength-specific long-pass filter is used to remove the background light, leaving only the light fluoresced by the particles. This technique has been used successfully in liquid flows many times to record images of 200 to 300-nm-diameter fluorescent particles [18, 22]. While fluorescently labeled particles are well suited for μPIV studies in liquid flows, they are not readily applicable to gaseous flows for several reasons. First, commercially available fluorescently labeled particles are generally available as aqueous

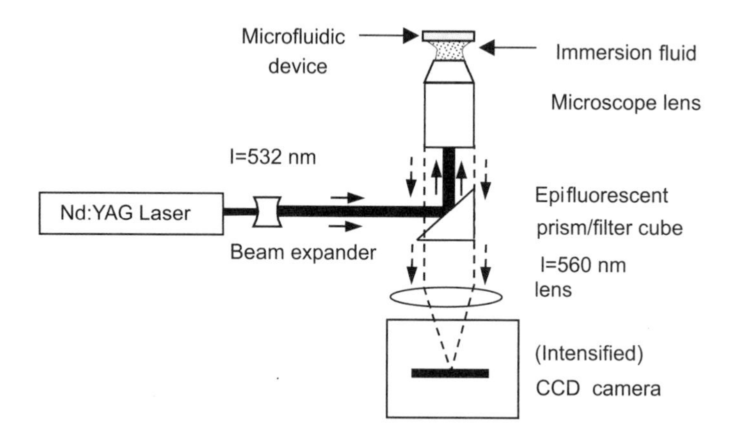


**Figure 4.4** Schematic of a μPIV system. A pulsed Nd:YAG laser is used to illuminate fluorescent 200-nm flow-tracing particles, and a cooled CCD camera is used to record the particle images.

suspensions. A few manufacturers (e.g., Duke Scientific) do have dry fluorescent particles available, but only in larger sizes, greater than 7 μm. In principle, the particle-laden aqueous suspensions can be dried, and the particles subsequently suspended in a gas flow, but this often proves problematic because the electrical surface charge that the particles easily acquire allows them to stick to the flow boundaries and to each other. Successful results have been reported using theatrical fog and smoke—both nonfluorescent [26]. Furthermore, the emission decay time of many fluorescent molecules is on the order of several nanoseconds, which may cause streaking of the particle images for high-speed flows.

*In-Plane Spatial Resolution.* The in-plane spatial resolution can be analyzed by considering how the tracer particles are imaged. The particles should be imaged with high numerical aperture diffraction-limited optics, and with sufficiently high magnification so that they are resolved with at least 3 to 4 pixels across their diameter. Following Adrian [27], the diffraction-limited spot size of a point source of light $d_s$ imaged through a circular aperture is given by

$$d_s = 2.44\,(M+1)\,f^{\#}\lambda \tag{4.2}$$

where *M* is the magnification, $f^{\#}$ is the f-number of the lens, and $\lambda$ is the wavelength of the light scattered or emitted by the particle. For infinity-corrected microscope objective lenses, Meinhart and Wereley [28] showed that

$$f^{\#} = \frac{1}{2}\left[\left(\frac{n}{\mathrm{NA}}\right)^2 - 1\right]^{\frac{1}{2}}. \tag{4.3}$$

The numerical aperture NA is defined as $\mathrm{NA} \equiv n\sin\theta$, where *n* is the index of refraction of the recording medium and $\theta$ is the half-angle subtended by the aperture of the recording lens. Numerical aperture is a more convenient expression to use in microscopy because of the different immersion media used. In photography, air is the only immersion medium generally used and hence $f^{\#}$ is sufficient. To avoid confusion when reading the μPIV literature, it must be noted here that (4.3) reduces to

$$f^{\#} \approx \frac{1}{2\mathrm{NA}} \tag{4.4}$$

for the immersion medium being air ($n_{air} \approx 1.0$) and small numerical apertures. This is a small angle approximation that is accurate to within 10% for $\mathrm{NA} \leq 0.25$ but approaches 100% error for $\mathrm{NA} \geq 1.2$ [28]. This approximation is used, for example, by [20, 29]. Combining (4.2) and (4.3)

**Table 4.2**
Effective Particle Image Diameters When Projected Back into the Flow, $d_e/M$ (μm) [30]

| M | 60 | 40 | 40 | 20 | 10 |
|---|---|---|---|---|---|
| NA | 1.40 | 0.75 | 0.60 | 0.50 | 0.25 |
| n | 1.515 | 1.00 | 1.00 | 1.00 | 1.00 |
| $d_p$ (μm) | *Effective particle image diameters* $d_e/M$ (μm) | | | | |
| 0.01 | 0.29 | 0.62 | 0.93 | 1.24 | 2.91 |
| 0.10 | 0.30 | 0.63 | 0.94 | 1.25 | 2.91 |
| 0.20 | 0.35 | 0.65 | 0.95 | 1.26 | 2.92 |
| 0.30 | 0.42 | 0.69 | 0.98 | 1.28 | 2.93 |
| 0.50 | 0.58 | 0.79 | 1.06 | 1.34 | 2.95 |
| 0.70 | 0.76 | 0.93 | 1.17 | 1.43 | 2.99 |
| 1.00 | 1.04 | 1.18 | 1.37 | 1.59 | 3.08 |
| 3.00 | 3.01 | 3.06 | 3.14 | 3.25 | 4.18 |

yields the expression

$$d_s = 1.22M\lambda\left[\left(\frac{n}{\mathrm{NA}}\right)^2 - 1\right]^{\frac{1}{2}} \tag{4.5}$$

for the diffraction-limited spot size in terms of the numerical aperture directly. The actual recorded image can be estimated as the convolution of the point-spread function with the geometric image. Approximating both these images as Gaussian functions, the effective image diameter $d_e$ can be written as [31]

$$d_e = \left[d_s^2 + M^2 d_p^2\right]^{\frac{1}{2}} \tag{4.6}$$

This expression is dominated by diffraction effects and reaches a constant value of $d_s$ when the size of the particle's geometric image $Md_p$ is considerably smaller than $d_s$. It is dominated by the geometric image size for geometric image sizes considerably larger than $d_s$ where $d_e \approx Md_p$.

The most common microscope objective lenses range from diffraction-limited oil-immersion lenses with $M = 60$, $\mathrm{NA} = 1.4$ to low magnification air-immersion lenses with $M = 10$, $\mathrm{NA} = 0.1$. Table 4.2 gives effective particle diameters recorded through a circular aperture and then projected back into the flow, $d_e/M$. Using conventional microscope optics, particle image resolutions of $d_e/M \approx 0.3$ $\mu$m can be obtained using oil-immersion lenses with numerical apertures of $\mathrm{NA} = 1.4$ and particle diameters $d_p < 0.2$ μm. For particle diameters $d_p > 0.3$ $\mu$m, the geometric component of the image decreases the resolution of the particle image. The low magnification air-immersion lens with $M = 10$, $\mathrm{NA} = 0.25$ is diffraction-limited for particle diameters $d_p < 1.0$ μm.

*Effective Numerical Aperture.* In experiments where the highest possible spatial resolution is desired, researchers often choose high numerical aperture oil-immersion lenses. These lenses are quite complicated and designed to conduct as much light as possible out of a sample (see Figure 4.5) by not allowing the light to pass into a medium with a refractive index as low as that of air ($n_{\mathrm{air}} \approx 1.0$). When the index of refraction of the working fluid is lower than that of the immersion medium, the effective numerical aperture that an objective lens can deliver is decreased from that specified by the manufacturer because of total internal reflection. Meinhart and Wereley [28] analyzed this numerical aperture reduction through a ray-tracing procedure.

As a specific example, assume that a 60× magnification, numerical aperture $\mathrm{NA}_D$=1.4 oil-immersion lens optimized for use with a 170-μm coverslip, and a maximum working distance $w_d$ of 200 μm is used with immersion oil having an index of refraction $n_o$ matching that of the coverslip and a working fluid (water) with a lower refractive index ($n_w$). This is a common situation in μPIV and was described in [22], among many others. The following analysis is not restricted to these

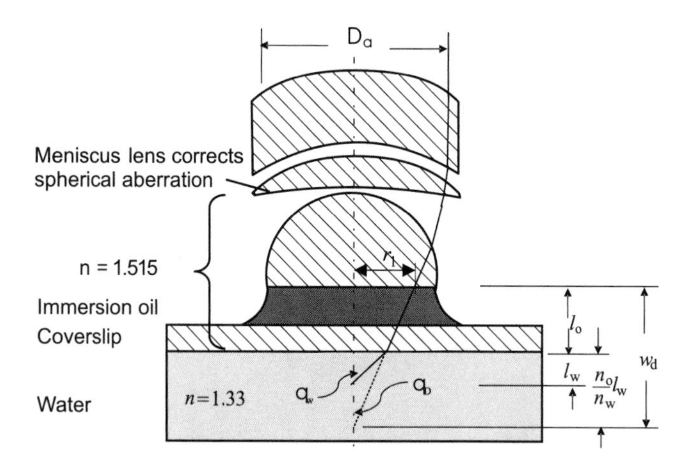


**Figure 4.5** Geometry of a high numerical aperture oil-immersion lens, immersion oil, coverslip, and water as the working fluid. A point source of light emanating from a depth in the water, $l_w$, appears to be at a distance $w_d$ from the lens entrance. (*After*: [28].)

**Table 4.3**

Estimates of the Diffraction-Limited Spot Size $d_s$ for Various Imaging Media as a Function of Imaging Depth $l_w$ [28]

| Imaging Depth $l_w$ | Imaging Medium | $\mathrm{NA_{eff}}$ | $d_s$ (μm) |
|---|---|---|---|
| All Depths | Oil | 1.40 | 18.8 |
| 0 μm (min) | Water | 1.33 | 24.8 |
| 200 μm (max) | Water | 1.21 | 34.2 |

specific parameters and is easily generalizable to any arbitrary immersion medium and working fluid as long as the immersion medium refractive index exceeds that of the working fluid.

Using a complicated ray-tracing procedure, Meinhart and Wereley [28] derived an implicit expression, relating the depth into the flow at which the focal plane is located, called the imaging depth $l_w$, to the effective numerical aperture $\mathrm{NA_{eff}}$. The expression is

$$\frac{w_d}{\left[\left(\frac{n_o}{NA_D}\right)-1\right]^{\frac{1}{2}}}=\frac{l_w}{\left[\left(\frac{n_w}{\mathrm{NA_{eff}}}\right)-1\right]^{\frac{1}{2}}}+\frac{w_d-\frac{n_o}{n_w}l_w}{\left[\left(\frac{n_o}{\mathrm{NA_{eff}}}\right)-1\right]^{\frac{1}{2}}} \tag{4.7}$$

where $n_o$ and $n_w$ are the refractive indices of the oil and water and $w_d$ is the lens's working distance. Since no closed-form analytical solution is possible for (4.7), it is solved numerically for the imaging depth $l_w$ in terms of $\mathrm{NA_{eff}}$. Figure 4.6 shows the numerical solution of (4.7) using $n_o$=1.515 and $n_w$=1.33. When imaging at the coverslip boundary (i.e., only slightly into the water), the effective numerical aperture is approximately equal to the refractive index of the water, 1.33. The effective numerical aperture decreases with increasing imaging depth. At the maximum imaging depth the effective numerical aperture is reduced to approximately 1.21. The effective numerical aperture and the diffraction-limited spot size are given as a function of the imaging medium and imaging depth in Table 4.3. The refractive index change at the water/glass interface significantly reduces the effective numerical aperture of lens, even when the focal plane is right at the surface of the glass. This in turn increases the diffraction-limited spot size, which reduces the spatial resolution of the μPIV technique.

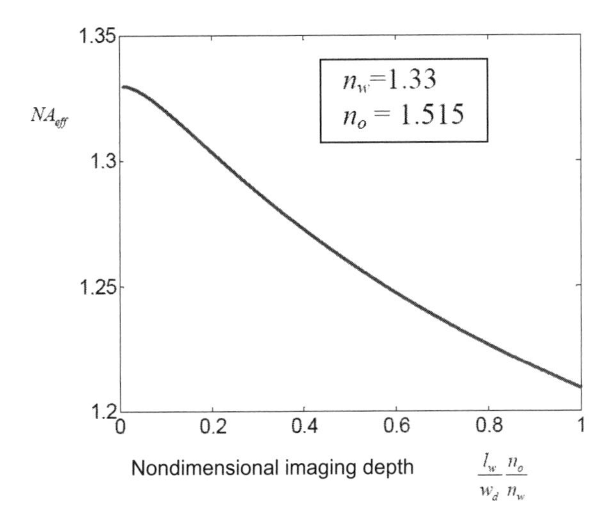


**Figure 4.6** Effective numerical aperture of an oil-immersion lens imaging into water as a function of the dimensionless imaging depth. (*After*: [28].)

In experiments where the working fluid is water, a similar diffraction-limited spot size can be achieved using a $\mathrm{NA} = 1.0$ water-immersion lens, $d_s = 39.8$ µm, compared to an oil-immersion lens, which may only achieve an effective numerical aperture, $\mathrm{NA}_{\mathrm{eff}} \approx 1.21$ where $d_s = 34.2$ µm. A water-immersion lens with $\mathrm{NA}_D = 1.2$ will exhibit better performance than the oil-immersion lens, having $d_s = 21.7$ µm. Further, water-immersion lenses are designed to image into water and will produce superior images to the oil-immersion lens when being used to image a water flow.

*Relationship Between Spot Size and Spatial Resolution.* The effective particle image diameter places a bound on the spatial resolution that can be obtained by PIV. Assuming that the particle images are sufficiently resolved by the CCD array, such that a particle diameter is resolved by 3 to 4 pixels, the location of the correlation peak can be sufficiently resolved to within one-tenth the particle image diameter [32]. Therefore, the uncertainty of the correlation peak location for a $d_p = 0.2$ µm diameter particle recorded with a $\mathrm{NA} = 1.4$ lens is $\delta x \approx d_e/10M = 35$ nm. The measurement error due to detectability $\varepsilon_d$ can be written as the ratio of the uncertainty in the correlation peak location $\delta x$ to the particle displacement $\Delta x$ according to

$$\varepsilon_d = \frac{\delta x}{\Delta x} \tag{4.8}$$

For an uncertainty of $\varepsilon_d = 2\%$, the required particle displacement is $\Delta x = 1.8$ µm for the highest resolved case. For a low resolution case with a $d_p = 3.0$ µm diameter particle, image with a $M = 10$, $\mathrm{NA} = 0.25$ objective lens, the required particle displacement would be $\Delta x = 21$ µm. From (4.8), the trade-off between spatial resolution and accuracy of the velocity measurements is clear. The spatial resolution of the measurements will be higher if the particles are allowed to displace only slightly (i.e., small $\Delta x$), but the accuracy of the measurements will be higher if the particles are allowed to displace significantly (i.e., large $\Delta x$) presuming $\delta x$ to remain constant.

#### 4.2.1.2 Volume Illumination of the Flow

Another significant difference between µPIV and macroscopic PIV is that, due to a lack of optical access along with significant diffraction in light sheet-forming optics, light sheets are typically not a practical source of illumination for microflows. Consequently, the flow must be volume illuminated,

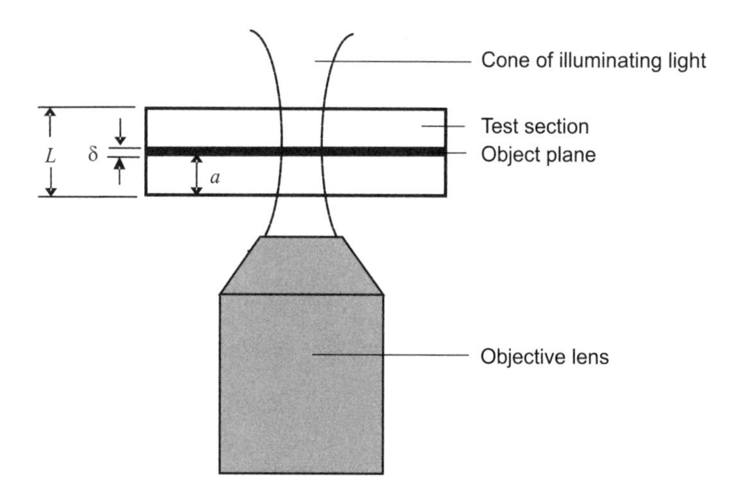


**Figure 4.7** Schematic showing the geometry for volume illumination particle image velocimetry. The particles carried by the flow are illuminated by light coming out of the objective lens (i.e., upward).

leaving two choices for how to visualize the seed particles—with an optical system whose depth of field exceeds the depth of the flow being measured, or with an optical system whose depth of field is small compared to that of the flow. Both of these techniques have been used in various implementations of μPIV. Cummings [33] used a large depth of field imaging system to explore electrokinetic and pressure-driven flows. The advantage of the large depth of field optical system is that all particles in the field of view of the optical system are well focused and contribute to the correlation function comparably. The disadvantage of this scheme is that all knowledge of the depth of each particle is lost, resulting in velocity fields that are completely depth-averaged. For example, in a pressure-driven flow where the velocity profile is expected to be parabolic with depth, the fast-moving particles near the center of the channel will be focused at the same time as the slow-moving particles near the wall. The measured velocity will be a weighted-average of the velocities of all the particles imaged. Cummings [33] addressed this problem with advanced processing techniques that will not be covered here.

The second choice of imaging systems is one whose depth of field is smaller than that of the flow domain, as shown in Figure 4.7. The optical system will then sharply focus those particles that are within the depth of field $\delta$ of the imaging system, while the remaining particles will be unfocused—to greater or lesser degrees—and contribute to the background noise level. Since the optical system is being used to define thickness of the measurement domain, it is important to characterize exactly how thick the depth of field, or more appropriately, the *depth of correlation* $z_{\mathrm{corr}}$, is. The distinction between depth of field and depth of correlation is an important although subtle one. The depth of field refers to the distance that a point source of light may be displaced from the focal plane and still produce an acceptably focused image, whereas the depth of correlation refers to how far from the focal plane a particle will contribute significantly to the correlation function. The depth of correlation can be calculated starting from the basic principles of how small particles are imaged [30, 34].

*Three-Dimensional Diffraction Pattern.* Following Born and Wolf [25], the intensity distribution of the 3-D diffraction pattern of a point source imaged through a circular aperture of radius a can be written in terms of the dimensionless diffraction variables $(u, v)$:

$$I(u, v) = \left(\frac{2}{u}\right)^2 \left[U_1^2(u, v) + U_2^2(u, v)\right] I_0 \tag{4.9a}$$

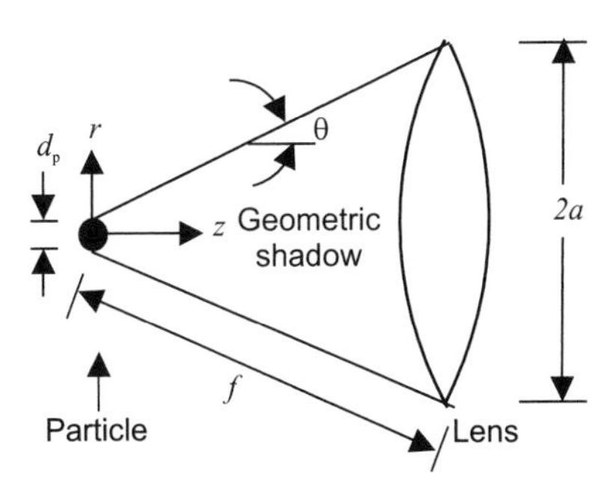


**Figure 4.8** Geometry of a particle with a diameter, $d_p$, being imaged through a circular aperture of radius, $a$, by a lens of focal length, $f$. (*After*: [34].)

$$I(u,v) = \left(\frac{2}{n}\right)^2 \left\{1 + V_0^2(u,v) + V_1^2(u,v) - 2V_0(u,v)\cos\left[\frac{1}{2}\left(u + \frac{v^2}{u}\right)\right] - 2V_1(u,v)\sin\left[\frac{1}{2}\left(u + \frac{v^2}{u}\right)\right]\right\} I_0 \tag{4.9b}$$

where $U_n(u,v)$ and $V_n(u,v)$ are called Lommel functions, which may be expressed as an infinite series of Bessel functions of the first kind:

$$U_n(u,v) = \sum_{s=0}^{\infty}(-1)^s \left(\tfrac{u}{v}\right)^{n+2s} J_{n+2s}(v)$$
$$V_n(u,v) = \sum_{s=0}^{\infty}(-1)^s \left(\tfrac{v}{u}\right)^{n+2s} J_{n+2s}(v) \tag{4.10}$$

The dimensionless diffraction variables are defined as:

$$u = 2\pi\tfrac{z}{\lambda}\left(\tfrac{a}{f}\right)^2$$
$$v = 2\pi\tfrac{r}{\lambda}\left(\tfrac{a}{f}\right)^2 \tag{4.11}$$

where $f$ is the radius of the spherical wave as it approaches the aperture (which can be approximated as the focal length of the lens), $\lambda$ is the wavelength of light, and $r$ and $z$ are the in-plane radius and the out-of-plane coordinate, respectively, with the origin located at the point source (Figure 4.8).

Although both (4.9a) and (4.9b) are valid in the region near the point of focus, it is computationally convenient to use (4.9a) within the geometric shadow, where $|u/v| < 1$, and to use (4.9b) outside the geometric shadow, where $|u/v| > 1$ [25].

Within the focal plane, the intensity distribution reduces to the expected result

$$I(0,v) = \left[\frac{2J_1(v)}{v}\right]^2 I_0 \tag{4.12}$$

which is the Airy function for Fraunhofer diffraction through a circular aperture. Along the optical axis, the intensity distribution reduces to

$$I(u,0) = \left[\frac{\sin u/4}{u/4}\right]^2 I_0 \tag{4.13}$$

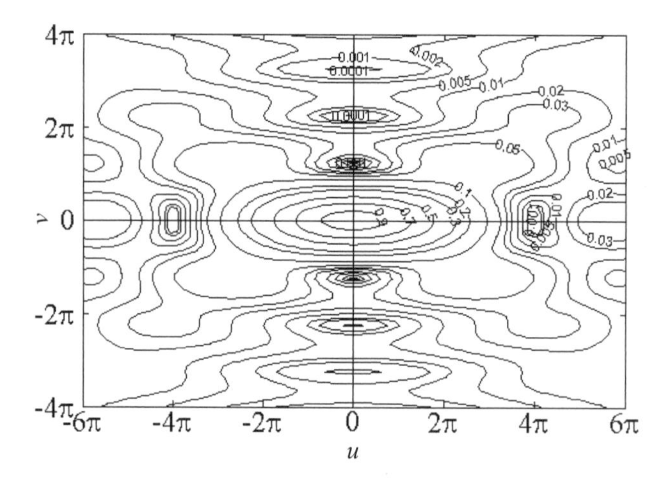


**Figure 4.9** Three-dimensional intensity distribution pattern expressed in diffraction units $(u, v)$, following Born and Wolf [25]. The focal point is located at the origin, the optical axis is located along $v = 0$, and the focal plane is located along $u = 0$. (*After*: [25].)

The 3-D intensity distribution calculated from (4.9a) and (4.9b) is shown in Figure 4.9. The focal point is located at the origin, the optical axis is located at $v = 0$, and the focal plane is located at $u = 0$. The maximum intensity, $I_0$, occurs at the focal point. Along the optical axis, the intensity distribution reduces to 0 at $u = \pm 4\pi$, $\pm 8\pi$, while a local maximum occurs at $u = \pm 6\pi$.

*Depth of Field.* The depth of field of a standard microscope objective lens was given by Inoué and Spring [35] as:

$$\delta z = \frac{n\lambda_0}{\mathrm{NA}^2} + \frac{ne}{\mathrm{NA} \cdot M} \tag{4.14}$$

where $n$ is the refractive index of the fluid between the microfluidic device and the objective lens, $\lambda_0$ is the wavelength of light in a vacuum being imaged by the optical system, NA is the numerical aperture of the objective lens, $M$ is the total magnification of the system, and $e$ is the smallest distance that can be resolved by a detector located in the image plane of the microscope (for the case of a CCD sensor, $e$ is the spacing between pixels). Equation (4.14) is the summation of the depths of field resulting from diffraction (first term on the right side), and geometric effects (second term on the right side).

The cutoff for the depth of field due to diffraction [first term on the right side of (4.14)] is chosen by convention to be one-quarter of the out-of-plane distance between the first two minima in the three-dimensional point spread function; that is, $u = \pm\pi$ in Figure 4.9 and (4.9a) and (4.9b). Substituting $\mathrm{NA} = n \sin\theta = n \cdot a/f$, and $\lambda_0 = n\lambda$ yields the first term on the right side of (4.14).

If a CCD sensor is used to record particle images, the geometric term in (4.14) can be derived by projecting the CCD array into the flow field, and then, considering the out-of-plane distance, the CCD sensor can be moved before the geometric shadow of the point source occupies more than a single pixel. This derivation is valid for small light collection angles, where $\tan\theta \approx \sin\theta = \mathrm{NA}/n$.

*Depth of Correlation.* The depth of correlation is defined as twice the distance that a particle can be positioned from the object plane so that the intensity along the optical axis is an arbitrarily specified fraction of its focused intensity, denoted by $\varepsilon$. Beyond this distance, the particle's intensity is sufficiently low that it will not influence the velocity measurement.

While the depth of correlation is related to the depth of field of the optical system, it is important to distinguish between them. The depth of field is defined as twice the distance from the object plane in which the object is considered unfocused in terms of image quality. In the case of volume-illuminated μPIV, the depth of field does not define precisely the thickness of the measurement plane. The theoretical contribution of an unfocused particle to the correlation function is estimated by considering: (1) the effect due to diffraction, (2) the effect due to geometric optics, and (3) the finite size of the particle. For the current discussion, we shall choose the cutoff for the on-axis image intensity, $\varepsilon$, to be arbitrarily one-tenth of the in-focus intensity. The reason for this choice is that the correlation function varies like the intensity squared, so a particle image with one-tenth the intensity of a focused image can be expected to contribute less than 1% to the correlation function.

The effect of diffraction can be evaluated by considering the intensity of the point spread function along the optical axis in (4.13). If $\phi = 0.1$, then the intensity cutoff will occur at $u \approx \pm 3\pi$. Using (4.11), substituting $\delta z = 2z$, and using the definition of numerical aperture, $\mathrm{NA} \equiv n \sin\theta = n \cdot a/f$, one can estimate the depth of correlation due to diffraction as:

$$\delta_{\mathrm{cg}} = \frac{3n\lambda_0}{\mathrm{NA}^2} \tag{4.15}$$

The effect of geometric optics upon the depth of correlation can be estimated by considering the distance from the object plane in which the intensity along the optical axis of a particle with a diameter, $d_{\mathrm{p}}$, decreases an amount, $\varepsilon = 0.1$, due to the spread in the geometric shadow, that is, the lens's collection cone. If the light flux within the geometric shadow remains constant, the intensity along the optical axis will vary as $\approx z^{-2}$. From Figure 4.9, if the geometric particle image is sufficiently resolved by the CCD array, the depth of correlation due to geometric optics can be written for an arbitrary value of $\phi$ as:

$$\delta z_{\mathrm{cd}} = \frac{(1-\sqrt{\phi})d_{\mathrm{p}}}{\sqrt{\phi}\tan\theta} \text{ for } d_{\mathrm{p}} > \frac{e}{M} \tag{4.16}$$

Following the analysis of Olsen and Adrian [29, 66] and using (4.3), the effective image diameter of a particle displaced a distance $z$ from the objective plane can be approximated by combining (4.6) with a geometric approximation to account for the particle image spreading due to displacement from the focal plane to yield

$$d_{\mathrm{e}} = \left\{ M^2 d_{\mathrm{p}}^2 + 1.49\,(M+1)^2\,\lambda^2 \left[ \left(\frac{n}{\mathrm{NA}}\right)^2 - 1 \right] + \left[ \frac{MD_a z}{s_{\mathrm{o}} + z} \right]^2 \right\}^{\frac{1}{2}} \tag{4.17}$$

where $s_{\mathrm{o}}$ is the object distance and $D_a$ is the diameter of the recording lens aperture.

The relative contribution $\varepsilon$ of a particle displaced a distance $z$ from the focal plane, compared to a similar particle located at the focal plane can be expressed in terms of the ratio of the effective particle image diameters raised to the fourth power

$$\varepsilon = \left[ \frac{d_{\mathrm{e}}(0)}{d_{\mathrm{e}}\left(z_{\mathrm{corr}}\right)} \right]^4 \tag{4.18}$$

Approximating $\frac{D_a^2}{(s_{\mathrm{o}}+z)^2} \approx \frac{D_a^2}{s_{\mathrm{o}}^2} = 4\left[\left(\frac{n}{\mathrm{NA}}\right)^2 - 1\right]^{-1}$, combining (4.17) and (4.18), and solving for $z_{\mathrm{corr}}$ yields an expression for the depth of correlation

$$z_{\mathrm{corr}} = \left\{ \left( \frac{1-\sqrt{\varepsilon}}{\sqrt{\varepsilon}} \right) \left[ \frac{d_{\mathrm{p}}^2 \left[ (n/\mathrm{NA})^2 - 1 \right]}{4} + \frac{1.49\,(M+1)^2\,\lambda^2 \left[ (n/\mathrm{NA})^2 - 1 \right]^2}{4M^2} \right] \right\}^{\frac{1}{2}} \tag{4.19}$$

**Table 4.4**

Thickness of the Measurement Plane for Typical Experimental Parameters, $2z_{\text{corr}}$ (μm) [30]

| $M$ | 60 | 40 | 40 | 20 | 10 |
|---|---|---|---|---|---|
| NA | 1.40 | 0.75 | 0.60 | 0.50 | 0.25 |
| $n$ | 1.515 | 1.00 | 1.00 | 1.00 | 1.00 |
| $d_{\text{p}}$ (μm) | *Measurement plane thickness* $2z_{\text{corr}}$ (μm) | | | | |
| 0.01 | 0.36 | 1.6 | 3.7 | 6.5 | 34 |
| 0.10 | 0.38 | 1.6 | 3.8 | 6.5 | 34 |
| 0.20 | 0.43 | 1.7 | 3.8 | 6.5 | 34 |
| 0.30 | 0.52 | 1.8 | 3.9 | 6.6 | 34 |
| 0.50 | 0.72 | 2.1 | 4.2 | 7.0 | 34 |
| 0.70 | 0.94 | 2.5 | 4.7 | 7.4 | 35 |
| 1.00 | 1.3 | 3.1 | 5.5 | 8.3 | 36 |
| 3.00 | 3.7 | 8.1 | 13 | 17 | 49 |

From (4.19), it is evident that the depth of correlation $z_{\text{corr}}$ is strongly dependent on numerical aperture NA and particle size $d_{\text{p}}$ and is weakly dependent upon magnification $M$. Table 4.4 gives the thickness of the measurement plane, $2z_{\text{corr}}$, for various microscope objective lenses and particle sizes. The highest out of plane resolution for these parameters is $2z_{\text{corr}}$=0.36 μm for a NA $= 1.4$, $M = 60$ oil-immersion lens and particle sizes $d_{\text{p}} < 0.1$ μm. For these calculations, it is important to note that the effective numerical aperture of an oil-immersion lens is reduced according to (4.7) when imaging particles suspended in fluids such as water, where the refractive index is less than that of the immersion oil.

One important implication of volume illumination that affects both large and small depth of focus imaging systems is that all particles in the illuminated volume will contribute to the recorded image. This implies that the particle concentrations will have to be minimized for deep flows, and lead to the use of low image density images as described below.

Olsen and Adrian [29] used a small angle approximation to derive the depth of correlation as

$$z_{\text{corr}} = \left\{ \left( \frac{1-\sqrt{\varepsilon}}{\sqrt{\varepsilon}} \right) \left[ f^{\#2} d_{\text{p}}^2 + \frac{5.95\,(M+1)^2\,\lambda^2 f^{\#4}}{M^2} \right] \right\}^{\frac{1}{2}} \tag{4.20}$$

where all the variables are as given above. Because it is given in terms of $f^{\#}$ instead of NA, it is only applicable for air-immersion lenses and not oil- or water-immersion lenses. This model for the depth of correlation has been indirectly experimentally confirmed for low magnification $(M \leq 20\times)$ and low numerical aperture (NA $\leq 0.4$) air-immersion lenses. The depth of correlation was not evaluated explicitly, but rather a weighting function used in the model for how the particle intensity varies with distance from the object plane was confirmed with experimental measurements [36].

*Particle Visibility*. The quality of μPIV velocity measurements strongly depends upon the quality of the recorded particle images from which those measurements are calculated. In macroscopic PIV experiments, it is customary to use a sheet of light to illuminate only those particles that are within the depth of field of the recording lens. The light sheet has two important effects—it minimizes the background noise from out of focus particles and ensures that every particle visible to the camera is well-focused. However, in μPIV, the microscopic length scales and limited optical access necessitate using volume illumination.

Experiments using the μPIV technique must be designed so that focused particle images can be observed even in the presence of background light from unfocused particles and test section surfaces. The background light scattered from test section surfaces can be removed by using fluorescence techniques to filter out elastically scattered light (at the same wavelength as the illumination) while leaving the fluoresced light (at a longer wavelength) virtually unattenuated [20].

Background light fluoresced from unfocused tracer particles is not so easily removed because it occurs at the same wavelength as the signal, the focused particle images, but it can be lowered to acceptable levels by choosing proper experimental parameters. Olsen and Adrian [29] presented a theory to estimate particle visibility, defined as the ratio of the intensity of a focused particle image to the average intensity of the background light produced by the unfocused particles. The analysis in this section refers to the dimensions labeled on Figures 4.5 and 4.7.

Assuming light is emitted uniformly from the particle, the light from a single particle reaching the image plane can be written as

$$J(z) = \frac{J_{\mathrm{p}} D_a^2}{16\left(s_{\mathrm{o}} + z\right)^2} \tag{4.21}$$

where $J_{\mathrm{p}}$ is total light flux emitted by a single particle. Approximating the intensity of a focused particle image as Gaussian,

$$I(r) = I_0 \exp\left(\frac{-4\beta^2 r^2}{d_{\mathrm{e}}^2}\right) \tag{4.22}$$

where the unspecified parameter $\beta$ is chosen to determine the cutoff level that defines the edge of the particle image. Approximating the Airy distribution by a Gaussian distribution, with the area of the two axisymmetric functions being equal, the first zero in the Airy distribution corresponds to [31]

$$\frac{I}{I_0} = \exp\left(-\beta^2 \approx -3.67\right) \tag{4.23}$$

Integrating (4.22) over an entire particle image and equating that result to (4.21) allows $I_0$ to be evaluated and (4.22) to be rewritten as

$$I(r, z) = \frac{J_{\mathrm{p}} D_a^2 \beta^2}{4\pi d_{\mathrm{e}}^2 \left(s_{\mathrm{o}} + z\right)^2} \exp \frac{-4\beta^2 r^2}{d_{\mathrm{e}}^2} \tag{4.24}$$

Making the simplifying assumption that particles located outside a distance $|z| > \delta/2$ from the object plane as being completely unfocused and contributing uniformly to background intensity, while particles located within a distance $|z| < \delta/2$ as being completely focused, the total flux of background light $J_{\mathrm{B}}$ can be approximated by

$$J_{\mathrm{B}} = A_v C \left\{ \int_{-a}^{-\frac{\delta}{2}} J(z) dz + \int_{\frac{\delta}{2}}^{L-a} J(z) dz \right\} \tag{4.25}$$

where $C$ is the number of particles per unit volume of fluid, $L$ is the depth of the device, and $A_{\mathrm{v}}$ is the average cross-sectional area contained within the field of view. Combining (4.21) and (4.25), correcting for the effect of magnification, and assuming $s_{\mathrm{o}} \gg \delta/2$, the intensity of the background glow can be expressed as [29]

$$I_B = \frac{C J_P L D_a^2}{16 M^2 \left(s_{\mathrm{o}} - a\right)\left(s_{\mathrm{o}} - a + L\right)}. \tag{4.26}$$

Following Olsen and Adrian [29], the visibility $V$ of a focused particle can be obtained by combining (4.17) and (4.24), dividing by (4.26), and setting $r = 0$ and $z = 0$,

$$V = \frac{I(0,0)}{I_B} = \frac{4 M^2 \beta^2 \left(s_{\mathrm{o}} - a\right)\left(s_{\mathrm{o}} - a + L\right)}{\pi C L s_{\mathrm{o}}^2 \left\{ M^2 d_{\mathrm{p}}^2 + 1.49\left(M+1\right)^2 \lambda^2 \left[\left(\frac{n}{\mathrm{NA}}\right)^2 - 1\right]\right\}} \tag{4.27}$$

**Table 4.5**

Maximum Volume Fraction of Particles $V_{\text{fr}}$, Expressed in Percent, Necessary to Maintain a Visibility $V$ Greater Than 1.5 When Imaging the Center of an $L = 100$ μm-Deep Device [30]

| $M$ | 60 | 40 | 40 | 20 | 10 |
|---|---|---|---|---|---|
| NA | 1.40 | 0.75 | 0.60 | 0.50 | 0.25 |
| $n$ | 1.515 | 1.00 | 1.00 | 1.00 | 1.00 |
| $s_o(mm)$ | 0.38 | 0.89 | 3 | 7 | 10.5 |
| $d_p$ (μm) | | | *Volume Fraction (%)* | | |
| 0.01 | 2.0E-5 | 4.3E−6 | 1.9E−6 | 1.1E−6 | 1.9E−7 |
| 0.10 | 1.7E−2 | 4.2E−3 | 1.9E−3 | 1.1E−3 | 1.9E−4 |
| 0.20 | 1.1E−1 | 3.1E−2 | 1.4E−2 | 8.2E−3 | 1.5E−3 |
| 0.30 | 2.5E−1 | 9.3E−2 | 4.6E−2 | 2.7E−2 | 5.1E−3 |
| 0.50 | 6.0E−1 | 3.2E−1 | 1.8E−1 | 1.1E−1 | 2.3E−2 |
| 0.70 | 9.6E−1 | 6.4E−1 | 4.1E−1 | 2.8E−1 | 6.2E−2 |
| 1.00 | 1.5E+0 | 1.2E+0 | 8.7E−1 | 6.4E−1 | 1.7E−1 |
| 3.00 | 4.8E+0 | 4.7E+0 | 4.5E+0 | 4.2E+0 | 2.5E+0 |

From this expression, it is clear that for a given recording optics configuration, particle visibility $V$ can be increased by decreasing particle concentration $C$ or by decreasing test section thickness $L$. For a fixed particle concentration, the visibility can be increased by decreasing the particle diameter $d_p$ or by increasing the numerical aperture NA of the recording lens. Visibility depends only weakly on magnification and object distance $s_o$.

An expression for the volume fraction $V_{\text{fr}}$ of particles in solution that produce a specific particle visibility can be obtained by rearranging (4.27) and multiplying by the volume occupied by a spherical particle to get

$$V_{\text{fr}} = \frac{2d_p^3 M^2 \beta^2 (s_o - a)(s_o - a + L)}{3VLs_o^2 \left\{ M^2 d_p^2 + 1.49 (M+1)^2 \lambda^2 \left[ \left( \frac{n}{\text{NA}} \right)^2 - 1 \right] \right\}} \tag{4.28}$$

Reasonably high quality velocity measurements require visibilities in excess of 1.5. Although this is an arbitrary threshold, it works well in practice. To see this formula in practice, assume that we are interested in measuring the flow at the centerline ($a = L/2$) of a microfluidic device with a characteristic depth of $L = 100$ μm. Table 4.5 shows for various experimental parameters the maximum volume fraction of particles that can be seeded into the fluid while maintaining a focused particle visibility greater than 1.5. Here, the object distance $s_o$ is estimated by adding the working distance of the lens to the designed coverslip thickness.

Meinhart et al. [34] verified these trends with a series of imaging experiments using known particle concentrations and flow depths. The particle visibility $V$ was estimated from a series of particle images taken of four different particle concentrations and four different device depths. A particle solution was prepared by diluting $d_p = 200$ nm diameter polystyrene particles in de-ionized water. Test sections were formed using two feeler gauges of known thickness sandwiched between a glass microscope slide and a coverslip. The images were recorded with an oil-immersion $M = 60\times$, $\text{NA} = 1.4$ objective lens. The remainder of the μPIV system was as described above.

The measured visibility is shown in Table 4.6. As expected, the results indicate that, for a given particle concentration, a higher visibility is obtained by imaging a flow in a thinner device. This occurs because decreasing the thickness of the test section decreases the number of unfocused particles, while the number of focused particles remains constant. Increasing the particle concentration also decreases the visibility, as expected. In general, thinner test sections allow higher

**Table 4.6**
Experimental Assessment of Particle Visibility [22]

| | *Particle Concentration (by Volume)* | | | |
|---|---|---|---|---|
| *Depth* ($\mu m$) | 0.01% | 0.02% | 0.04% | 0.08% |
| 25 | 2.2 | 2.1 | 2.0 | 1.9 |
| 50 | 1.9 | 1.7 | 1.4 | 1.2 |
| 125 | 1.5 | 1.4 | 1.2 | 1.1 |
| 170 | 1.3 | 1.2 | 1.1 | 1.0 |

particle concentrations to be used, which can be analyzed using smaller interrogation regions. Consequently, the seed particle concentration must be chosen judiciously so that the desired spatial resolution can be obtained, while maintaining adequate image quality (i.e., particle visibility).

#### 4.2.1.3 Effects of Brownian Motion

When the seed particle size becomes small, the collective effect of collisions between the particles and a moderate number of fluid molecules is unbalanced, preventing the particle from following the flow to some degree. This phenomenon, commonly called *Brownian motion*, has two potential implications for μPIV: one is to cause an error in the measurement of the flow velocity; the other is to cause an uncertainty in the location of the flow-tracing particles. In order to fully consider the effect of Brownian motion, it is first necessary to establish how particles suspended in flows behave. These three issues will be considered.

*Flow/Particle Dynamics.* In stark contrast to many macroscale fluid mechanics experiments, the hydrodynamic size of a particle (a measure of its ability to follow the flow based on the ratio of inertial to drag forces) is usually not a concern in microfluidic applications because of the large surface-to-volume ratios at small length scales. A simple model for the response time of a particle subjected to a step change in local fluid velocity can be used to gauge particle behavior. Based on a simple first-order inertial response to a constant flow acceleration (assuming Stokes flow for the particle drag), the response time $\tau_\mathrm{p}$ of a particle is:

$$\tau_\mathrm{p} = \frac{d_\mathrm{p}^2 \rho_\mathrm{p}}{18\eta} \tag{4.29}$$

where $d_\mathrm{p}$ and $\rho_\mathrm{p}$ are the diameter and density of the particle, respectively, and $\eta$ is the dynamic viscosity of the fluid. Considering typical μPIV experimental parameters of 300-nm-diameter polystyrene latex spheres immersed in water, the particle response time would be $10^{-9}$ seconds. This response time is much smaller than the time scales of any realistic liquid or low-speed gas flow field.

In the case of high-speed gas flows, the particle response time may be an important consideration when designing a system for microflow measurements. For example, a 400-nm particle seeded into an air micronozzle that expands from the sonic at the throat to Mach 2 over a 1-mm distance may experience a particle-to-gas relative flow velocity of more than 5% (assuming a constant acceleration and a stagnation temperature of 300K). Particle response to flow through a normal shock would be significantly worse. Another consideration in gas microchannels is the breakdown of the no-slip and continuum assumption as the particle Knudsen number $\mathrm{Kn_p}$ (see Chapter 2), defined as the ratio of the mean free path of the gas to the particle diameter, approaches (and exceeds) one. For the case of the slip-flow regime ($10^{-3} < \mathrm{Kn_p} < 0.1$), it is possible to use corrections to the Stokes drag relation to quantify particle dynamics [37]. For example, a correction offered by Melling [38] suggests the following relation for the particle response time:

$$\tau_{\mathrm{p}} = (1 + 2.76\mathrm{Kn_p})\frac{d_{\mathrm{p}}^2 \rho_{\mathrm{p}}}{18\eta} \tag{4.30}$$

*Velocity Errors.* Santiago et al. [18] briefly considered the effect of the Brownian motion on the accuracy of μPIV measurements. A more in-depth consideration of the phenomenon of Brownian motion is necessary to completely explain its effects in μPIV. Brownian motion is the random thermal motion of a particle suspended in a fluid [39]. The motion results from collisions between fluid molecules and suspended microparticles. The velocity spectrum of a particle due to the Brownian motion consists of frequencies too high to be resolved fully and is commonly modeled as Gaussian white noise [40]. A quantity more readily characterized is the particle's average displacement after many velocity fluctuations. For time intervals $\Delta t$ much larger than the particle inertial response time, the dynamics of Brownian displacement are independent of inertial parameters such as particle and fluid density, and the mean square distance of diffusion is proportional to $D\Delta t$, where $D$ is the diffusion coefficient of the particle. For a spherical particle subject to Stokes drag law, the diffusion coefficient $D$ was first given by Einstein [41] as:

$$D = \frac{KT}{3\pi\eta d_{\mathrm{p}}} \tag{4.31}$$

where $d_{\mathrm{p}}$ is the particle diameter, $K$ is Boltzmann's constant, $T$ is the absolute temperature of the fluid, and $\eta$ is the dynamic viscosity of the fluid. The random particle displacements measured with respect to the moving fluid can be described by the following 3D (Gaussian) probability density function:

$$p(x, y, z) = \frac{\exp\left[-(x^2 + y^2 + z^2)/4D\Delta t\right]}{(2\pi)^{3/2}(2D\Delta t)^{3/2}} \tag{4.32}$$

where the displacements in the $x$, $y$, and $z$ directions can be considered statistically independent random variables.

In all of the μPIV techniques developed to date and discussed here, particle displacements are determined by tracking the 2-D projections of particles located within some measurement depth. Therefore, the probability density function of the imaged particle displacements with respect to the moving fluid can be described by integrating (4.32) along the optical axis ($z$) to get:

$$p(x, y) = \int_{-\infty}^{\infty} p(x, y, z)\mathrm{d}z = \frac{\exp\left[-(x^2 + y^2)/4D\Delta t\right]}{4\pi D\Delta t} \tag{4.33}$$

where the coordinates $x$, $y$, and $z$ are taken to be displacements with respect to the moving fluid, assumed constant throughout the region over which the Brownian particle diffuses; that is, the diffusing particles are assumed to sample a region of low velocity gradient. In practice, (4.33) is an approximation to the imaged particle displacement distribution since particles may leave the measurement volume altogether during the interval $\Delta t$ due to diffusion or convection in the $z$-direction. Therefore, the 2-point probability density function described by (4.33) can be interpreted as being valid for cases where $\Delta z = w\Delta t < \delta z_{\mathrm{m}}$, where $w$ is the $z$-component of velocity. This limitation is not especially restrictive in microflows because of the absence of turbulence and the relatively large depth of correlation.

The random Brownian displacements cause particle trajectories to fluctuate about the deterministic pathlines of the fluid flow field. Assuming the flow field is steady over the time of measurement and the local velocity gradient is small, the imaged Brownian particle motion can be considered a fluctuation about a streamline that passes through the particle's initial location. An ideal, non-Brownian (i.e., deterministic) particle following a particular streamline for a time period $\Delta t$ has $x$- and $y$-displacements of:

$$\Delta x = u\Delta t$$
$$\Delta y = v\Delta t \tag{4.34}$$

where $u$ and $v$ are the $x$- and $y$- components of the time-averaged, local fluid velocity, respectively. The relative errors, $\varepsilon_x$ and $\varepsilon_y$, incurred as a result of imaging the Brownian particle displacements in a 2-D measurement of the $x$- and $y$-components of particle velocity, are given as:

$$l\varepsilon_x = \frac{\sigma_x}{\Delta x} = \frac{1}{u}\sqrt{\frac{2D}{\Delta t}} \tag{4.35a}$$

$$\varepsilon_y = \frac{\sigma_y}{\Delta y} = \frac{1}{u}\sqrt{\frac{2D}{\Delta t}} \tag{4.35b}$$

This Brownian error establishes a lower limit on the measurement time interval $\Delta t$ since, for shorter times, the measurements are dominated by uncorrelated Brownian motion. These quantities (ratios of the rms fluctuation-to-average velocity) describe the relative magnitudes of the Brownian motion and will be referred to here as Brownian intensities. The errors estimated by (4.35a) and (4.35b) show that the relative Brownian intensity error decreases as the time of measurement increases. Larger time intervals produce flow displacements proportional to $\Delta t$ while the rms of the Brownian particle displacements grows as $\Delta t^{1/2}$. In practice, Brownian motion is an important consideration when tracing 50 to 500,-nm particles in flow field experiments with flow velocities of less than about 1 mm/s. For a velocity on the order of 0.5 mm/s and a 500-nm seed particle, the lower limit for the time spacing is approximately 100 μs for a 20% error due to Brownian motion. This error can be reduced by both averaging over several particles in a single interrogation spot and by ensemble averaging over several realizations. The diffusive uncertainty decreases as $1/\sqrt{N}$, where $N$ is the total number of particles in the average [42].

Equation (4.35) demonstrates that the effect of the Brownian motion is relatively less important for faster flows. However, for a given measurement, when $u$ increases, $\Delta t$ will generally decrease. Equations (4.35a) and (4.35b) also demonstrate, that when all conditions but Dt are fixed, going to a larger $\Delta t$ will decrease the relative error introduced by the Brownian motion. Unfortunately, a longer $\Delta t$ will decrease the accuracy of the results because the PIV measurements are based on a first order accurate approximation to the velocity. Using a second-order accurate technique [called central difference interrogation (CDI) and presented in Section 4.2.3.1] allows for a longer $\Delta t$ to be used without increasing this error.

*Particle Position Error.* In addition to the flow velocity measurement error associated with particle displacement measurements, the Brownian motion incurred during the exposure time $t_{\text{exp}}$ may also be important in determining the particle location, especially for slow flows with long exposures and small tracer particles. For example, a 50-nm particle in water at room temperature will have an rms displacement of 300 nm if imaged with a 10-ms exposure time. For this particle image, the Brownian displacements projected into the image plane during the time of exposure are on the order of the image size estimated by (4.30) (given the best available far-field, visual optics with a numerical aperture of 1.4). This random displacement during image exposure can increase the uncertainty associated with estimating the particle location. For low velocity gradients, the centroid of this particle image trace is an estimate of the average location of the particle during the exposure. This particle location uncertainty is typically negligible for exposure times where the typical Brownian displacement in the image plane is small compared to the particle image diameter or a value of the diffusion time $d_{\text{e}}^2/(4DM^2)$ much less than the exposure time. For the experimental parameters mentioned above, $d_{\text{e}}^2/(4DM^2)$ is 300 ms and the exposure time is 5 ns for a typical Nd:YAG laser.

### 4.2.2 Special Processing Methods for μPIV Recordings

When evaluating digital PIV recordings with conventional correlation-based algorithms or image-pattern tracking algorithms, a sufficient number of particle images are required in the interrogation window or the tracked image pattern to ensure reliable and accurate measurement results. However, in many cases, especially in μPIV measurements, the particle image density in the PIV recordings is usually not high enough [Figure 4.10(a)]. These PIV recordings are called low-image density (LID) recordings and are usually evaluated with particle-tracking algorithms. When using particle-tracking algorithms, the velocity vector is determined with only one particle, and hence the reliability and accuracy of the technique are limited. In addition, interpolation procedures are usually necessary to obtain velocity vectors on the desired regular gridpoints from the random distributed particle-tracking results [Figure 4.11(a)], and therefore, additional uncertainties are added to the final results. Fortunately, some special processing methods can be used to evaluate the μPIV recordings, so that the errors resulting from the low-image density can be avoided [43]. In this section, two methods are introduced to improve measurement accuracy of μPIV: by using a digital image processing technique and by improving the evaluation algorithm.

#### 4.2.2.1 Overlapping of LID-PIV Recordings

In early days of PIV, multiple exposure imaging techniques were used to increase the particle image numbers in PIV recordings. Similar to multiply exposing a single frame, high-image density (HID) PIV recordings can be generated by computationally overlapping a number of LID-PIV recordings with:

$$g_0(x, y) = \max[g_k(x, y), k = 1, 2, 3, ..., N] \tag{4.36}$$

wherein $g_k(x, y)$ is the gray value distribution of the LID-PIV recordings with a total number $N$, and $g_0(x, y)$ is the overlapped recording. Note that in (4.36) the particle images are positive (i.e., with bright particles and dark background); otherwise, the images should be inverted or the minimum function used. An example of the image overlapping can be seen in Figure 4.7(b) for overlapping nine LID-PIV recordings. The size of the PIV recordings in Figure 4.11 is $256 \times 256$ pixels, and the corresponding measurement area is $2.5 \times 2.5$ mm$^2$. The effect of the image overlapping is shown in Figure 4.11. Figure 4.12(a) has the evaluation results for one of the LID-PIV recording pairs with a particle-tracking algorithm [44]. Figure 4.11(b) has the results for the overlapped PIV recording pair (out of nine LID-PIV recording pairs) with a correlation-based algorithm. The results in Figure 4.11(b) are more reliable, more dense, and more regularly spaced than those in Figure 4.11(a).

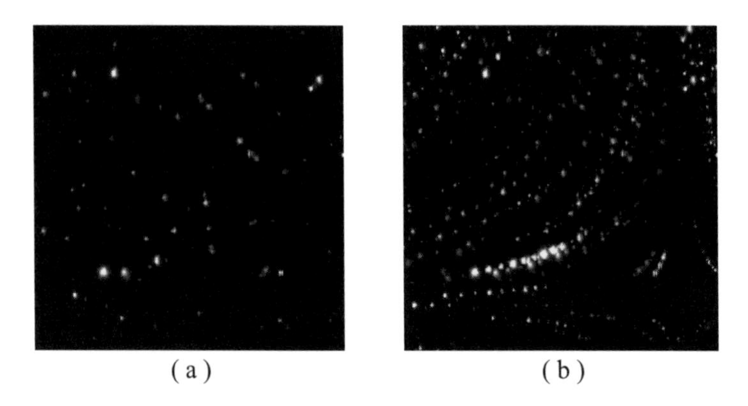


**Figure 4.10** Example of image overlapping: (a) one of the LID-PIV recordings; and (b) result of overlapping 9 LID-PIV recordings. Image size: $256 \times 256$ pixels. ([43]. ©2002, AIAA. Reprinted with permission.)

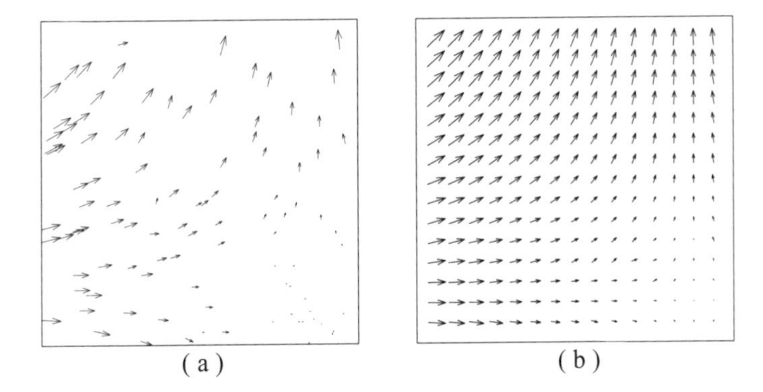


**Figure 4.11** Effect of image overlapping: (a) results for a single LID-PIV recording pair with a particle-tracking algorithm; and (b) results for the overlapped PIV recording pair with a correlation-based algorithm. ([43]. ©2002, AIAA. Reprinted with permission.)

The image overlapping method is based on the fact that flows in microdomains typically have very low Reynolds numbers, so that the flow can be considered as laminar and steady in the data acquisition period. Note that this method cannot be extended to measurements of turbulent or unsteady flows, and it may not work very well with overlapping HID-PIV recordings or too many LID-PIV recordings, because with large numbers of particle images, interference between particle images will occur [45]. Further study of this technique will be necessary to quantify these limitations but the promise of the technique is obvious.

#### 4.2.2.2 Ensemble Correlation Method

For correlation-based PIV evaluation algorithms, the correlation function at a certain interrogation spot is usually represented as:

$$\Phi_k(m,n) = \sum_{j=1}^{q}\sum_{i=1}^{p} f_k(i,j) \cdot g_k(i+m, j+n) \tag{4.37}$$

where $f_k(i,j)$ and $g_k(i,j)$ are the gray value distributions of the first and second exposures, respectively, in the $k$th PIV recording pair at a certain interrogation spot of a size of $p \times q$ pixels. The correlation function for a singly exposed PIV image pair has a peak at the position of the particle image displacement in the interrogation spot (or window), which should be the highest among all the peaks of $\Phi_k$. The subpeaks, which result from noise or mismatch of particle images, are obviously lower than the main peak (i.e., the peak of the particle image displacement). However, when the interrogation window does not contain enough particle images or the noise level is too high, the main peak will become weak and may be lower than some of the subpeaks, and as such, an erroneous velocity vector is generated. In the laminar and steady flows measured by the µPIV system, the velocity field is independent of the measurement time. That means the main peak of $\Phi_k(m,n)$ is always at the same position for PIV recording pairs taken at different times, while the subpeaks appear with random intensities and positions in different recording pairs. Therefore, when averaging $\Phi_k(m,n)$ over a large number of PIV recording pairs ($N$), the main peak will remain at the same position in each correlation function, but the noise peaks, which occur randomly, will average to zero. The averaged (or ensemble) correlation function is given as:

$$\Phi_{\text{ens}}(m,n) = \frac{1}{N}\sum_{k=1}^{N} \Phi_k(m,n) \tag{4.38}$$

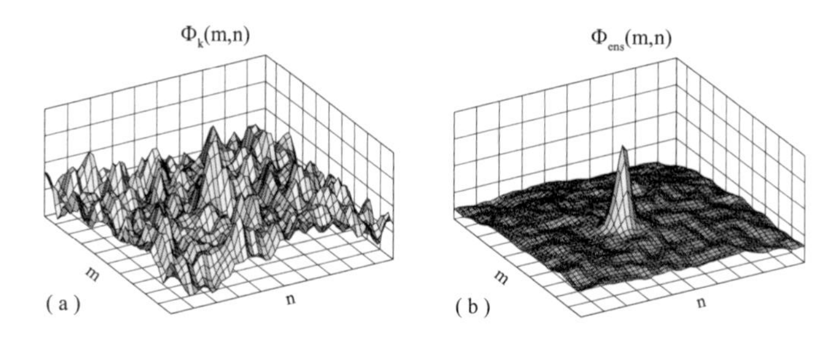


**Figure 4.12** Effect of ensemble correlation: (a) results with conventional correlation for one of the PIV recording pairs; and (b) results with ensemble correlation for 101 PIV recording pairs. ([43]. ©2002, AIAA. Reprinted with permission.)

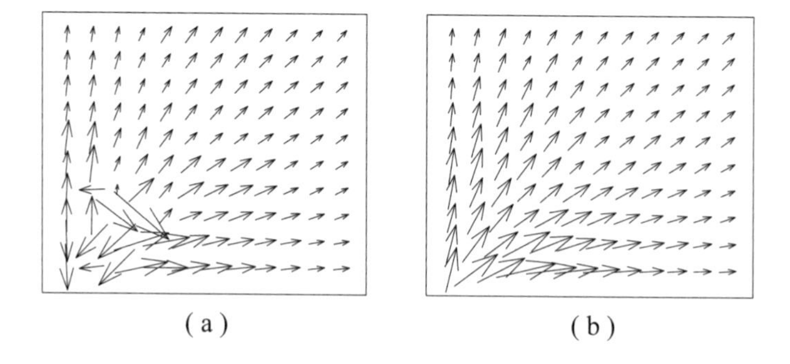


**Figure 4.13** Comparison of the evaluation function of (a) a single PIV recording pair with (b) the average of 101 evaluation functions. ([43]. ©2002, AIAA. Reprinted with permission.)

Just as with the image overlapping method detailed above, the ensemble correlation requires a steady flow. However, in contrast with the image overlapping method, the technique is not limited to LID recordings or to a small number of recordings. The concept of averaging correlation functions can also be applied to other evaluation algorithms such as correlation tracking and the MQD method. This method was first proposed and demonstrated by Meinhart et al. [45].

The ensemble correlation function technique is demonstrated for 101 LID-PIV recording pairs ($\Phi_{\text{ens}}$) in Figure 4.12 in comparison to the correlation function for one of the single recording pair ($\Phi_k$). These PIV recording pairs are chosen from the flow measurement in a microfluidic biochip for impedance spectroscopy of biological species [46]. With the conventional evaluation function in Figure 4.12(a), the main peak cannot easily be identified among the subpeaks, so the evaluation result is neither reliable nor accurate. However, the ensemble correlation function in Figure 4.12(b) shows a very clear peak at the particle image displacement, and the subpeaks can hardly be recognized.

The effect of the ensemble correlation technique on the resulting velocity field is demonstrated in Figure 4.9 with the PIV measurement of flow in the microfluidic biochip. All the obvious evaluation errors resulting from the low-image density and strong background noise [Figure 4.13(a)] are avoided by using the ensemble correlation method based on 101 PIV recording pairs [Figure 4.13(b)]. One important note here is that since the bad vectors in Figure 4.13(a) all occur at the lower left corner of the flow domain, the removal of these bad vectors and subsequent replacement by interpolated vectors will only coincidentally generate results that bear any resemblance to the true velocity field in the device. In addition, if the problem leading to low signal levels in the lower left of the images is systematic (i.e., larger background noise), even a large collection of images will not generate better results, because they will all have bad vectors at the same location.

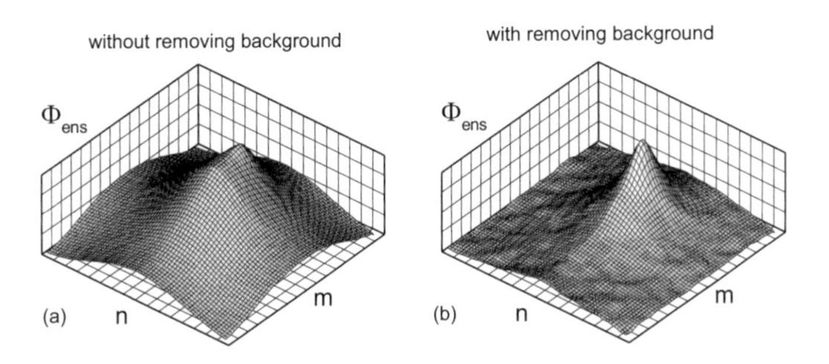


**Figure 4.14** Ensemble correlation function for 100 image sample pairs (a) without and (b) with background removal ([43]. ©2002, AIAA. Reprinted with permission.)

#### 4.2.2.3 Removing Background Noise

For using the recording overlapping or ensemble correlation techniques, a great number of μPIV recording pairs are usually obtained, enabling the removal of the background noise from the μPIV recording pairs. One of the possibilities for obtaining an image of the background from plenty of PIV recordings is averaging these recordings [44]. Because the particles are randomly distributed and quickly move through the camera view area, their images will disappear in the averaged recording. However, the image of the background (including the boundary, contaminants on the glass cover, and particles adhered to the wall) maintains the same brightness distribution in the averaged recording, because it does not move or change. Another method is building at each pixel location a minimum of the ensemble of PIV recordings, because the minimal gray value at each pixel may reflect the background brightness in the successively recorded images [47]. The background noise may be successfully removed by subtracting the background image from the PIV recordings.

A dataset from a flow in a microchannel is used to demonstrate this point. The size of the interrogation regions are $64 \times 64$ pixels and the total sample number is 100 pairs. The mean particle image displacement is about 12.5 pixels from left to right. In one particular interrogation region in the images, the particle images in a region at the left side of the interrogation region look darker than those outside this region. This may result from an asperity on the glass cover of the microchannel.

The ensemble correlation function for the 100 image sample pairs without background removal is given in Figure 4.14(a), which shows a dominant peak near zero displacement because the fleck does not move. When the background image is built with the minimum gray value method and subtracted from the image sample pairs, the influence of the asperity is reduced, so that the peak of the particle image displacement appears clearly in the evaluation function in Figure 4.14(b).

### 4.2.3 Advanced Processing Methods Suitable for Both Micro/Macro-PIV Recordings

For further improving the reliability and accuracy of μPIV measurements, a number of evaluation techniques, which also work well for standard PIV systems, are applied. It is known that the measurement uncertainty of PIV data includes both bias error and precision error. One of the most effective methods for reducing the bias error of PIV measurements in complex flows is the CDI method. For reducing the precision (or random) error, image correction methods are suggested. The CDI method and one of the image correction methods are introduced below.

#### 4.2.3.1 CDI

Currently, adaptive window offsetting is widely used with the FFT-based correlation algorithm for reducing the evaluation error and with the image pattern tracking algorithms for increasing the

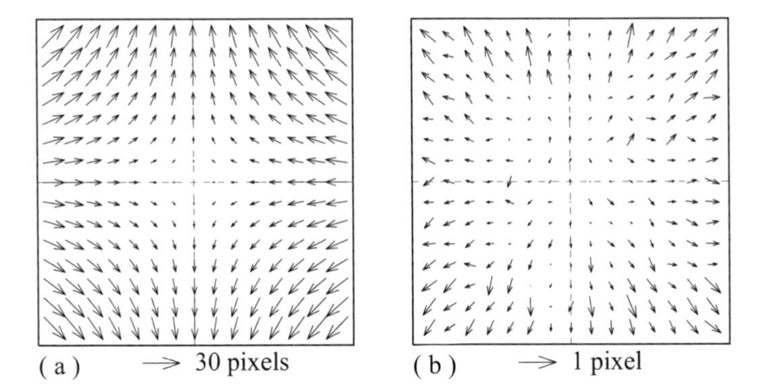


**Figure 4.15** Simulation of the four-roll-mill test: (a) desired flow pattern; and (b) evaluation errors with FDI. ([43]. ©2002, AIAA. Reprinted with permission.)

spatial resolution. The adaptive window offset method, as typically implemented, can be referred to as forward difference interrogation (FDI), because the second interrogation window is shifted in the forward direction of the flow an amount equal to the mean displacement of the particle images initially in the first window. Although the FDI method leads to significant improvements in the evaluation quality of PIV recordings in many cases, there are still some potentially detrimental bias errors that cannot be avoided when using an FDI method. The CDI method was initially introduced by Wereley et al. [48], and further developed and explored by Wereley and Meinhart [49], to avoid the shortcoming of FDI and increase the accuracy of the PIV measurement. The comparison between the CDI and FDI methods is analogous to the comparison between central difference and forward difference discretizations of derivatives, wherein the central difference method is accurate to the order of $\Delta t^2$ while the forward difference method is only accurate to the order of $\Delta t$. When using CDI, the first and second interrogation windows are shifted backwards and forwards, respectively, each by one-half of the expected particle image displacement (see Figure 3 in [49]). As with many adaptive window shifting techniques, this technique requires iteration to achieve optimum results.

In order to demonstrate the advantage of CDI over the FDI, a typical curvature flow (i.e., the flow in a four-roll-mill) is used here as an example. Based on actual experimental parameters, such as particle image size, concentration, and intensity, PIV recording pairs are simulated with the desired flow field shown in Figure 4.15(a). The maximal particle image displacement in the PIV recording pair of size of 1,024×1,024 pixels is about 30 pixels. The corresponding measurement area and the maximal velocity are $10 \times 10$ mm$^2$ and 0.04 mm/s, respectively. When combining the FFT-based correlation algorithm with FDI, the evaluation errors of a pair of the simulated recordings are determined by subtracting the desired flow field from the evaluation results and are given in Figure 4.15(b). The evaluation errors in Figure 4.15(b) are obviously dominated by bias errors that depend on the radial position (i.e., the distance between the vector location and the flow field center).

In this test, the bias errors are determined by averaging 500 individual error maps. As shown in Figure 4.16(b), distribution of rms values of the random errors is further computed. The dependence of the bias and random errors on the radial position are determined and shown in Figures 4.16(a and b) for the FDI and CDI, respectively. The total error is defined as the root-sum-square (RSS) of the bias and random error. It is shown in Figure 4.16 that the evaluation error of FDI is dominated by the bias error, at radial positions greater than 200 pixels. When CDI is used, the bias error is so small that it can be neglected in comparison to the random error that does not depend on the location.

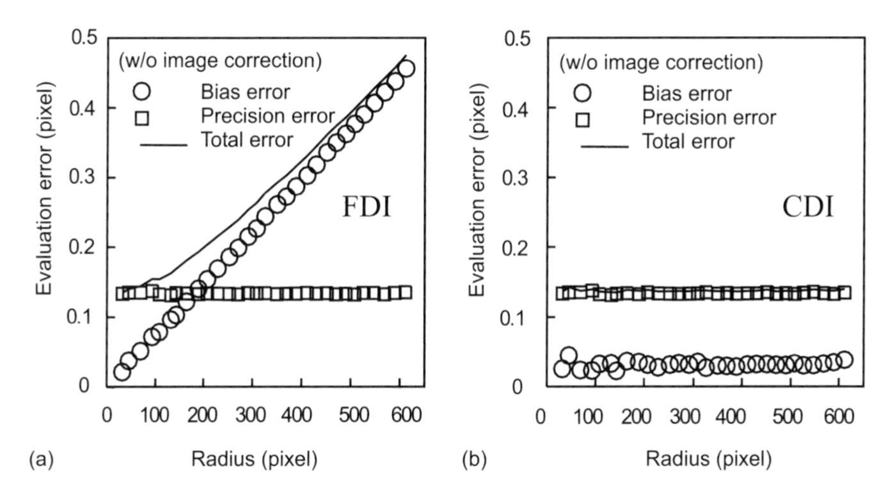


**Figure 4.16** Dependence of evaluation errors on the location (radius) of the evaluation with (a) FDI and (b) CDI for the four-roll-mill test. ([43]. ©2002, AIAA. Reprinted with permission.)

4.2.3.2 Image Correction Technique

In the above example, the bias error of the four-roll-mill test is minimized by using the CDI method. In order to further reduce the measurement uncertainty (i.e., the total error), the random errors must also be reduced. In the four-roll-mill test case, even when the flow is ideally seeded and the PIV recordings are made without any noise, evaluation errors may result from the deformation of the measured flow. To account for the deformation of the PIV image pattern, image correction techniques have been developed. The idea of image correction was presented by Wereley and Gui [50], and similar ideas were also applied by others. However, since the image correction was a complex and time-consuming procedure, it has not been widely used. In order to accelerate the evaluation, the authors modified the image correction method as follows. Based on previous iterations, the particle image displacements at the four corners of each interrogation window are calculated and used to deform the image patterns in the interrogation area for both exposures of the PIV recording pair using a simple bilinear interpolation, so that the image patterns have a good match despite spatial velocity gradients at the particle image displacement (see Figure 1 in [51]). Combining the modified image correction technique with the FFT-based correlation algorithm, the evaluation can be run at a very high speed. The effect of the image correction is presented in Figure 4.13. By comparing Figure 4.16 with Figure 4.17, the effect of the image correction can be seen to reduce the total error of the measurement scheme by about one-half.

## 4.3 μPIV EXAMPLES

### 4.3.1 Flow in a Microchannel

No flow is more fundamental than the pressure driven flow in a straight channel of constant cross section. Since analytical solutions are known for most such flows, they prove invaluable for gauging the accuracy of μPIV.

4.3.1.1 Analytical Solution to Channel Flow

Although the solution to flow through a round capillary is the well-known parabolic profile, the analytical solution to flow through a capillary of a rectangular cross section is less well known.

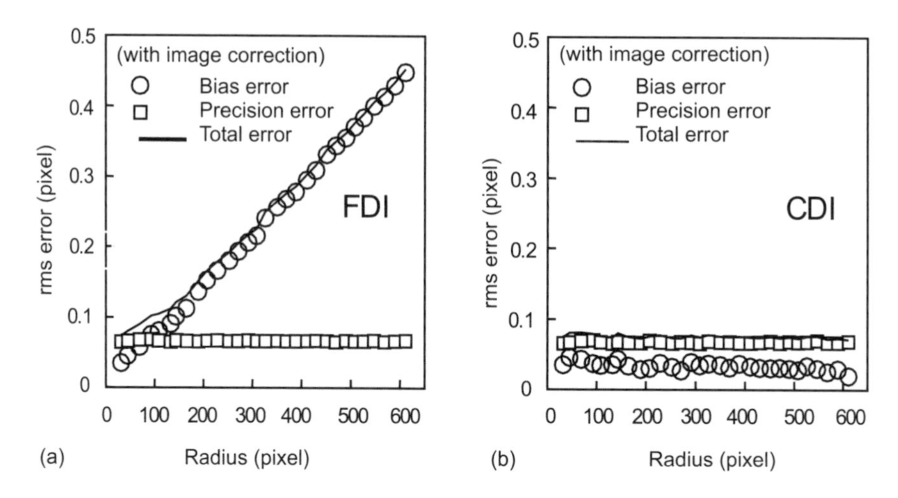


**Figure 4.17** Dependence of evaluation errors on the location of the evaluation with (a) FDI and (b) CDI by using image correction. ([43]. ©2002, AIAA. Reprinted with permission.)

Since one of the goals of this section is to illustrate the accuracy of μPIV by comparing to a known solution, it is useful here to briefly discuss the analytical solution. The velocity field of flow through a rectangular duct can be calculated by solving the Stokes equation (the low Reynolds number version of the Navier-Stokes equation discussed in Chapter 2), with no-slip velocity boundary conditions at the wall [52] using a Fourier series approach. Figure 4.18 shows a rectangular channel in which the width $W$ is much greater than the height $H$. Sufficiently far from the wall (i.e., $Z \gg H$), the analytical solution in the $Y$ direction (for constant $Z$) converges to the well-known parabolic profile for flow between infinite parallel plates. In the $Z$ direction (for constant $Y$), however, the flow profile is unusual in that it has a very steep velocity gradient near the wall ($Z < H$), which reaches a constant value away from the wall ($Z \gg H$).

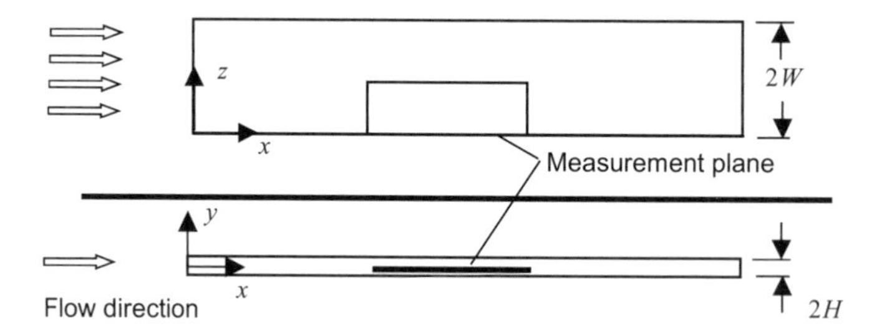


**Figure 4.18** Geometry of the microchannel. The microchannel is $2H$ high and $2W$ wide, and is assumed infinitely long in the axial direction. The measurement plane of interest is orientated in the $x - z$ plane and includes the microchannel wall at $z = 0$. The centerline of the channel is at $y = 0$. The microscope objective images the test section from below, in the lower part of the figure.

#### 4.3.1.2 Experimental Measurements

A 30 × 300 × 25 μm glass rectangular microchannel, fabricated by Wilmad Industries, was mounted flush to a 170-μm-thick glass coverslip and a microscope slide. By carefully rotating the glass coverslip and the CCD camera, the channel was oriented to the optical plane of the microscope within 0.2° in all three angles. The orientation was confirmed optically by focusing the CCD camera on the microchannel walls. The microchannel was horizontally positioned using a high-precision

$x - y$ stage, and verified optically to within $\approx$ 400 nm using epifluorescent imaging and image enhancement. The experimental arrangement is sketched in Figure 4.18.

The flow in the glass microchannel was imaged through an inverted epifluorescent microscope and a Nikon Plan Apochromat oil-immersion objective lens with a magnification $M = 60$ and a numerical aperture $\text{NA} = 1.4$. The object plane was placed at approximately $7.5 \pm 1$ mm from the bottom of the 30-μm-thick microchannel. The Plan Apochromat lens was chosen for the experiment because it is a high-quality microscope objective designed with low curvature of field and low distortion, and is corrected for spherical and chromatic aberrations.

Since deionized water (refractive index $n_{\text{w}} = 1.33$) was used as the working fluid but the lens immersion fluid was oil (refractive index $n_{\text{i}} = 1.515$), the effective numerical aperture of the objective lens was limited to $\text{NA} \approx n_{\text{w}}/n_{\text{i}} = 1.23$ [35].

A filtered continuous white light source was used to align the test section with the CCD camera and to test for proper particle concentration. During the experiment, the continuous light source was replaced by the pulsed Nd:YAG laser. A Harvard Apparatus syringe pump was used to produce a 200 ml hr$^{-1}$ flow through the microchannel.

The particle-image fields were analyzed using a custom-written PIV interrogation program developed specifically for microfluidic applications. The program uses an ensemble-averaging correlation technique to estimate velocity vectors at a single measurement point by: (1) cross-correlating particle-image fields from 20 instantaneous realizations, (2) ensemble averaging the cross-correlation functions, and (3) determining the peak of the ensemble-averaged correlation function. The signal-to-noise ratio is significantly increased by ensemble-averaging the correlation function before peak detection, as opposed to either ensemble-averaging the velocity vectors after peak detection, or ensemble-averaging the particle-image field before correlation. The ensemble-averaging correlation technique is strictly limited to steady, quasi-steady, or periodic flows, which is certainly the situation in these experiments. This process is described in detail in Section 4.2.2.2. For the current experiment, 20 realizations were chosen because that was more than a sufficient number of realizations to give an excellent signal, even with a first interrogation window of only $120 \times 8$ pixels.

The signal-to-noise ratio resulting from the ensemble-average correlation technique was high enough that there were no erroneous velocity measurements. Consequently, no vector validation postprocessing was performed on the data after interrogation. The velocity field was smoothed using a $3 \times 3$ Gaussian kernel with a standard deviation of one grid spacing in both directions.

Figure 4.19(a) shows an ensemble-averaged velocity-vector field of the microchannel. The images were analyzed using a low spatial resolution away from the wall, where the velocity gradient is low, and using a high spatial resolution near the wall, where the wall-normal velocity gradient is largest. The interrogation spots were chosen to be longer in the streamwise direction than in the wall-normal direction. This allowed for a sufficient number of particle images to be captured in an interrogation spot, while providing the maximum possible spatial resolution in the wall-normal direction. The spatial resolution, defined by the size of the first interrogation window, was $120 \times 40$ pixels in the region far from the wall, and $120 \times 8$ pixels near the wall. This corresponds to a spatial resolution of $13.6 \times 4.4$ μm and $13.6 \times 0.9$ μm, respectively. The interrogation spots were overlapped by 50% to extract the maximum possible amount of information for the chosen interrogation region size according to the Nyquist sampling criterion. Consequently, the velocity vector spacing in the wall-normal direction was 450 nm near the wall. The streamwise velocity profile was estimated by line-averaging the measured velocity data in the streamwise direction. Figure 4.20 compares the streamwise velocity profile estimated from the PIV measurements (shown as symbols) to the analytical solution for laminar flow of a Newtonian fluid in a rectangular channel (shown as a solid curve). The agreement is within 2% of full-scale resolution. Hence, the accuracy of μPIV is at worst 2% of full-scale for these experimental conditions. The bulk flow rate of the analytical curve was determined by matching the free-stream velocity data away from the wall. The wall position of the analytical curve was determined by extrapolating the velocity profile to zero near the wall.

Since the microchannel flow was fully developed, the wall-normal component of the velocity vectors is expected to be close to zero. The average angle of inclination of the velocity field was found to be small, 0.0046 radian, suggesting that the test section was slightly rotated in the plane of the CCD array relative to a row of pixels on the array. This rotation was corrected mathematically by rotating the coordinate system of the velocity field by 0.0046 radian. The position of the wall can be determined to within about 400 nm by direct observation of the image because of diffraction, as well as the blurring of the out-of-focus parts of the wall. The precise location of the wall was more accurately determined by applying the no-slip boundary condition, which is expected to hold at these length scales for the combination of water flowing through glass, and extrapolating the velocity profile to zero at 16 different streamwise positions [Figure 4.19(a)]. The location of the wall at every streamwise position agreed to within 8 nm of each other, suggesting that the wall is extremely flat, the optical system has little distortion, and the PIV measurements are very accurate. This technique is the precursor for the nanoscope technique, which will be explored in greater detail in Section 4.4.1.

Most PIV experiments have difficulty measuring velocity vectors very close to the wall. In many situations, hydrodynamic interactions between the particles and the wall prevent the particles from traveling close to the wall, or background reflections from the wall overshadow particle images. By using 200-nm-diameter particles and epifluorescence to remove background reflections, we have been able to make accurate velocity measurements to within about 450 nm of the wall; see Figure 4.19.

### 4.3.2 Flow in a Micronozzle

The utility of these new imaging and processing algorithms along with the μPIV technique itself can be demonstrated by measuring the flow through a micronozzle. The micronozzles were designed to be operated with supersonic gas flows. In the initial stages of this investigation, however, they were operated with a liquid in order to assess the spatial resolution capabilities of the μPIV technique without having to push the temporal envelope simultaneously. Consequently, the converging-diverging geometry of the micronozzle served as a very small venturi. The micronozzles were fabricated by Robert Bayt and Kenny Breuer (now at Brown University) at MIT in 1998. The 2-D nozzle contours were etched using DRIE in 300-μm-thick silicon wafers. The nozzles used in these experiments were etched 50 μm deep into the silicon wafer. A single 500-μm-thick glass wafer was anodically bonded to the top of the wafer to provide an end wall. The wafers were mounted to a macroscopic aluminum manifold, pressure sealed using #0 O-rings and vacuum grease, and connected with plastic tubing to a Harvard Apparatus syringe pump.

The liquid (de-ionized water) flow was seeded with relatively large 700-nm diameter fluorescently-labeled polystyrene particles (available from Duke Scientific). The particles were imaged using an air-immersion $\mathrm{NA} = 0.6$, $40\times$ objective lens, and the epifluorescent imaging system described in Section 4.2. A flow rate of 4 ml $\mathrm{hr}^{-1}$ was delivered to the nozzle by the syringe pump.

Figure 4.21 is the velocity field inside a nozzle with a 15° half-angle and a 28-μm throat. The velocity field was calculated using the CDI technique with image overlapping (10 image pairs) and image correction, as explained above. The interrogation windows measured $64 \times 32$ pixels in the $x$ and $y$ directions, respectively. When projecting into the fluid, the correlation windows were $10.9 \times 5.4$ μm in the $x$ and $y$ directions, respectively. The interrogation spots were overlapped by 50% in accordance with the Nyquist criterion, yielding a velocity-vector spacing of 5.4 μm in the streamwise direction and 2.7 μm in the spanwise direction. The Reynolds number, based upon bulk velocity and throat width, is Re = 22.

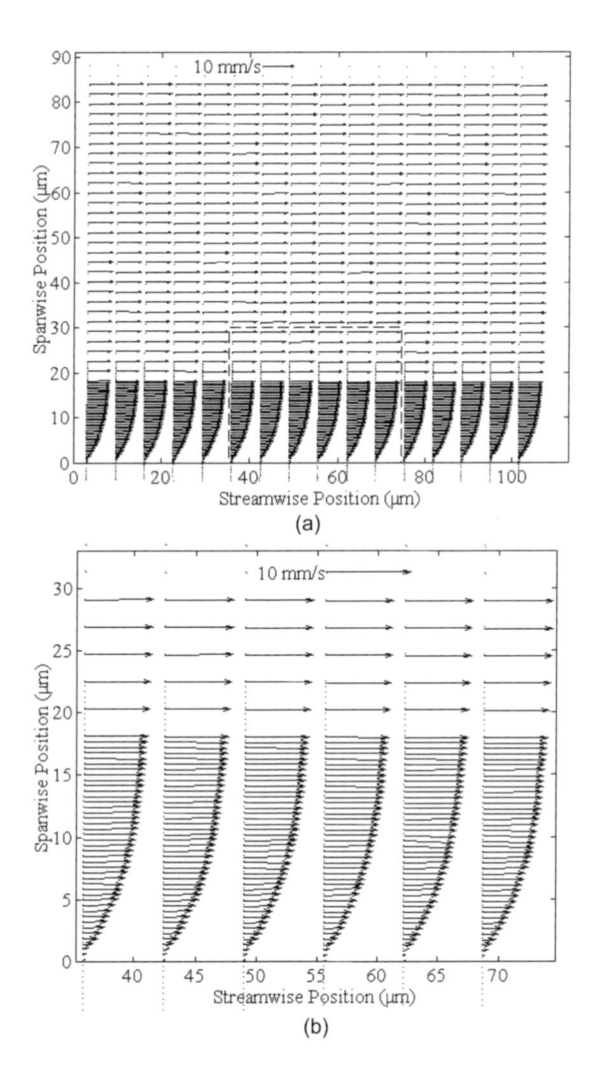


**Figure 4.19** (a) Large area view of ensemble-averaged velocity-vector field measured in a 30-μm deep × 300-μm wide × 25-μm channel. The spatial resolution, defined by the interrogation spot size of the first interrogation window, is 13.6 μm × 4.4 μm away from the wall, and 13.6 μm × 0.9 μm near the wall. (b) Near wall view of boxed region from (a). ([22] ©1999, Springer. Reprinted by permission.)

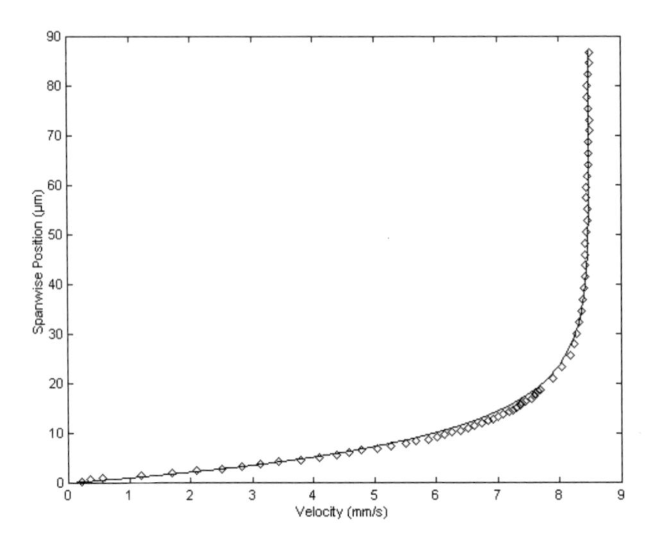


**Figure 4.20** Velocity profile measured in a nominally 30-μm × 300-μm channel. The symbols represent the experimental PIV data while the solid curve represents the analytical solution. ([22]. ©1999, Springer. Reprinted by permission.)

Turning now from a converging geometry to a diverging geometry, we can explore whether instabilities well predicted by the Reynolds number at macroscopic length scales are indeed as well-predicted by the Reynolds number at small length scales. The diffuser has a throat width of 28 μm and a thickness of 50 μm. The divergence half angle is quite large—40°. The expected behavior for this geometry would be that at low Reynolds numbers the flow would be entirely Stokes flow (i.e., no separation), but at larger Reynolds numbers, where inertial effects become important, separation should appear. Indeed, this is just what happens. At a Reynolds number of 22, the flow in the diverging section of the nozzle remains attached to the wall (not shown), while at a Reynolds number of 83, the flow separates as shown in Figure 4.22. This figure is based on a single pair of images, and as such represents an instantaneous snapshot of the flow. The interrogation region size measured $32 \times 32$ pixels$^2$ or $5.4 \times 5.4$ μm$^2$. A close inspection of Figure 4.22 reveals that the separation creates a stable, steady vortex standing at the point of separation. After the flow has dissipated some of its energy in the vortex, it no longer has sufficient momentum to exist as a jet, and it reattaches to the wall immediately downstream of the vortex. This is arguably the smallest vortex ever measured. Considering that the Kolmogorov length scale is frequently on the order of 0.1 to 1.0 mm, μPIV has more than enough spatial resolution to measure turbulent flows at, and even significantly below, the Kolmogorov length scale. The example shown has 25 vectors measured across the 60-μm extent of the vortex.

### 4.3.3 Flow Around a Blood Cell

A surface tension driven Hele-Shaw flow with a Reynolds number of $3 \times 10^{-4}$ was developed by placing deionized water seeded with 300-nm diameter polystyrene particles between a 500-μm-thick microscope slide and a 170-μm coverslip. Human red blood cells, obtained by autophlebotomy, were smeared onto a glass slide. The height of the liquid layer between the microscope slide and the coverslip was measured as approximately 4 μm by translating the microscope objective to focus on the glass surfaces immediately above and below the liquid layer. The translation stage of the microscope was adjusted until a single red blood cell was visible (using white light) in the center of the field of view. This type of flow was chosen because of its excellent optical access, ease of setup, and its 4-μm thickness, which minimized the contribution of out-of-focus seed particles to

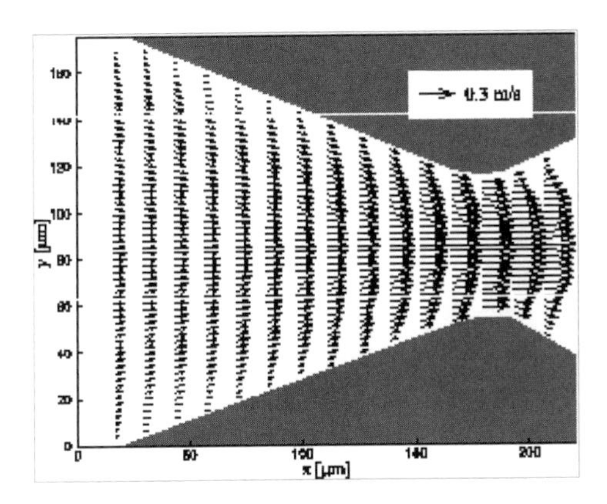

**Figure 4.21** Velocity field produced from 10 overlapped image pairs. The spatial resolution is 10.9 μm in the horizontal direction and 5.4 μm in the vertical direction. For clarity only every fifth column of measurement is shown. ([43]. ©2002, AIAA. Reprinted with permission.)

the background noise. Since red blood cells have a maximal tolerable shear stress, above which hemolysis occurs, this flow also is potentially interesting to the biomedical community.

The images were recorded in a serial manner by opening the shutter of the camera for 2 ms to image the flow, and then waiting 68.5 ms before acquiring the next image. Twenty-one images were collected in this manner. Since the camera is exposed to the particle reflections at the beginning of every video frame, each image can be correlated with the image following it. Consequently, the 21 images recorded can produce 20 pairs of images, each with the same time between exposures, $\Delta t$. Interrogation regions, sized 28 × 28 pixels, were spaced every 7 pixels in both the horizontal and vertical directions for a 75% overlap. Although technically, overlaps greater than 50% oversample the images, they effectively provide more velocity vectors to provide a better understanding of the velocity field.

Two velocity fields are shown in Figure 4.23. Figure 4.23(a) is the result of a forward difference interrogation and Figure 4.23(b) is the result of a CDI. The differences between these two figures will be discussed below while the commonalities will be considered now. The flow exhibits the features that we expect from a Hele-Shaw flow. Because of the disparate length scales in a Hele-Shaw flow, with the thickness much smaller than the characteristic length and width of the flow, an ideal Hele-Shaw flow will closely resemble a two-dimensional potential flow [53]. However, because a typical red blood cell is about 2 μm, while the total height of the liquid layer between the slides is 4 μm, there is a possibility that some of the flow will go over the top of the cell instead of around it in a Hele-Shaw configuration. Since the velocity field in Figure 4.23 closely resembles that of a potential flow around a right circular cylinder, we can conclude that the flow is primarily a Hele-Shaw flow. Far from the cylinder, the velocity field is uniformly directed upward and to the right at about a 75° angle from the horizontal. On either side of the red blood cell, there are stagnation points where the velocity goes to zero. The velocity field is symmetric with respect to reflection in a plane normal to the page and passing through the stagnation points. The velocity field differs from potential flow in that near the red blood cell there is evidence of the no-slip velocity condition. These observations agree well with the theory of Hele-Shaw.

For both algorithms, measurement regions that resulted in more than 20% of the combined area of the first and second interrogation windows inside of the blood cell are eliminated and replaced with a '×' symbol, because they will tend to produce velocity measurements with serious errors. The CDI scheme is able to accurately measure velocities closer to the surface of the cell than the FDI scheme. The CDI scheme has a total of 55 invalid measurement points, equivalent to 59.4

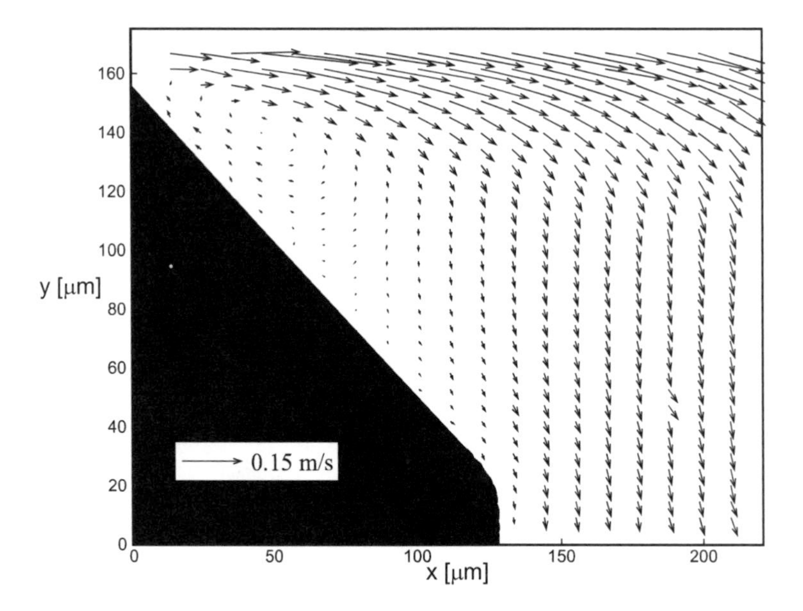


**Figure 4.22** Recirculation regions in a microdiffuser with spatial resolution of $5.4 \times 5.4$ mm$^2$: only every fourth column and every second row are shown for clarity. ([43]. ©2002, AIAA. Reprinted with permission.)

μm$^2$ of image area that cannot be interrogated, while the FDI has 57 invalid measurement points or 61.6 μm$^2$. Although this difference of two measurement points may not seem significant, it amounts to the area that the FDI algorithm cannot interrogate being 3.7% larger than the area the CDI algorithm cannot interrogate. Furthermore, the distribution of the invalid points is significant. By carefully comparing Figure 4.23(a) with Figure 4.23(b), it is apparent that the FDI has three more invalid measurement points upstream of the blood cell than does the CDI scheme, while the CDI scheme has one more invalid point downstream of the blood cell. This difference in distribution of invalid points translates into the centroid of the invalid area being nearly twice as far from the center of the blood cell in the FDI case (0.66 μm, or 7.85%, of the cell diameter) versus the CDI case (0.34 μm, or 4.05%, of the cell diameter). This difference means that the FDI measurements are less symmetrically distributed around the blood cell than the CDI measurements are. In fact, they are

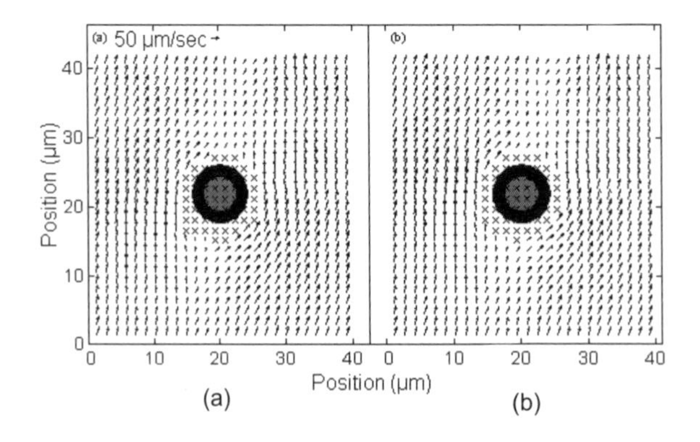


**Figure 4.23** Flow around a single human red blood cell: (a) forward difference adaptive window offset analysis; and (b) central difference adaptive window offset analysis. Cross symbols (×) indicate points that cannot be interrogated because they are too close to the cell. ([49]. ©2001, Springer. Reprinted with permission.)

biased toward the time the first image was recorded. Computing the average distance between the invalid measurement points and the surface of the blood cell indicates how closely to the blood cell surface each algorithm will allow the images to be accurately interrogated. On average, the invalid measurement points bordering the red blood cell generated using the FDI scheme are 12% farther from the cell than the invalid measurement points generated with the CDI scheme. Consequently, the adaptive CDI algorithm is more symmetric than the adaptive FDI algorithm and allows measurement of the velocity field nearer the cell surface.

### 4.3.4 Flow in Microfluidic Biochip

Microfluidic biochips are microfabricated devices that are used for delivery, processing, and analysis of biological species (molecules and cells). Gomez et al. [46] successfully used μPIV to measure the flow in a microfluidic biochip designed for impedance spectroscopy of biological species. This device was studied further in [54]. The biochip is fabricated in a silicon wafer with a thickness of 450 μm. It has a series of rectangular test cavities ranging from tens to hundreds of microns on a side, connected by narrow channels measuring 10 μm across. The whole pattern is defined by a single anisotropic etch of single crystal silicon to a depth of 12 μm. Each of the test cavities has a pair or series of electrodes patterned on its bottom. The array of test cavities is sealed with a piece of glass of about 0.2 mm thick, allowing optical access to the flow. During the experiment, water-based suspensions of fluorescein-labeled latex beads with a mean diameter of 1.88 μm were injected into the biochip and the flow was illuminated with a constant intensity mercury lamp. Images are captured with a CCD camera through an epifluorescence microscope and recorded at a video rate (30 Hz). One of the PIV images covering an area of 542 $\times$ 406 $\mu m^2$ on the chip with a digital resolution of 360 $\times$ 270 pixels is shown in Figure 4.24. The flow in a rectangular cavity of the biochip is determined by evaluating more than 100 μPIV recording pairs with the ensemble-correlation method, CDI, and the image correction technique, and the results are given in Figure 4.25. An interrogation window of 88 pixels is chosen for the PIV image evaluation, so that the corresponding spatial resolution is about 12$\times$12 $\mu m^2$. The measured velocities in the cavity range from about 100 mm/s to 1,600 μm/s.

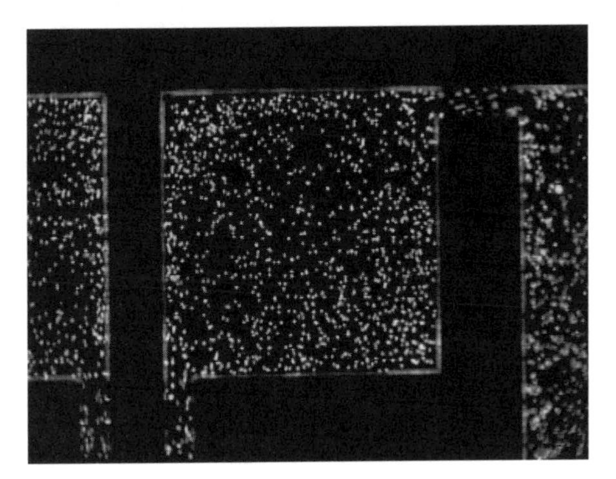

**Figure 4.24** Digital image of the seeded flow in the cavities and channels of the biochip (360 $\times$ 270 pixels, 542 $\times$ 406 $\mu m^2$). ([54]. ©2001, AIAA. Reprinted with permission.)

This biochip can function in several different modes, one of which is by immobilizing on the electrodes antibodies specific to an antigen being sought. The rate at which the antibodies capture the antigens will be a function of the concentration of the antigens in the solution as well as the solution

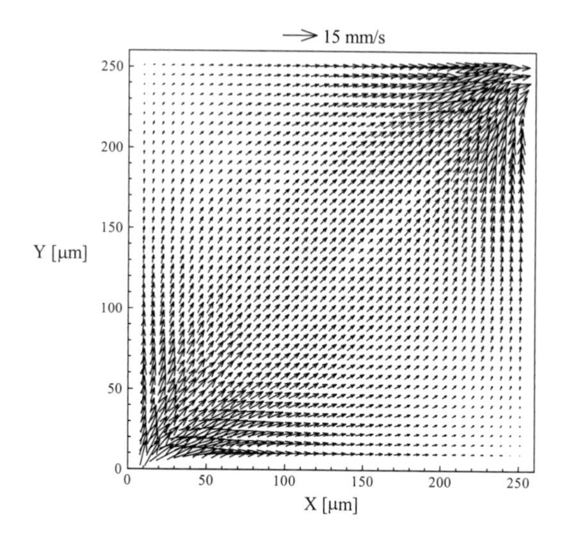


**Figure 4.25** PIV measurement results in a rectangular cavity of the biochip with a spatial resolution of 12×12 μm$^2$. ([54]. ©2001, AIAA. Reprinted with permission.)

flux past the electrodes. Consequently, knowledge of the velocity field is critical to characterizing the performance of the biochip. Since this is an example of the μPIV technique, the reader is directed to [46] for more details about the biochip performance characterization.

### 4.3.5 Conclusions

Currently, applying all the advanced techniques outlined here, the maximum spatial resolution of the μPIV technique stands at approximately 1 μm. By using smaller seed particles that fluoresce at shorter wavelengths, this limit could be reduced by a factor of 2 to 4. Still higher spatial resolutions could be obtained by adding a particle tracking step after the correlation-based PIV. Spatial resolutions of an order of magnitude smaller could then reasonably be reached.

The various μPIV apparatus and algorithms described above have been demonstrated to allow measurements at length scales on the order of 1 μm, significantly below the typical Kolmogorov length scale. These spatial resolutions are indispensable when analyzing flows in microdomains or the smallest scales of turbulence. The most significant problem standing in the way of extending μPIV to gas phase flows is seeding. With adequate seeding, the results presented here can be extended to gas flows. The significant issues associated with extending the results presented here to gas phase flows are further explored by Meinhart et al. [34].

## 4.4 EXTENSIONS OF THE μPIV TECHNIQUE

The μPIV technique is very versatile and can be extended in several meaningful ways to make different but related measurements of use in characterizing microflows. One extension involves turning the μPIV system into a high-resolution microscope, called a *microfluidic nanoscope*, for probing the location of flow boundaries. Another extension, called *microparticle image thermometry*, involves using the μPIV to measure fluid temperature in the same high spatial resolution sense as the velocity is measured. A third extension involves using a similar measurement technique but having wavelengths longer than visual, called the *near infrared*, to measure flows completely encased in silicon—a real benefit in the MEMS field. These extensions will be discussed next.

### 4.4.1 Microfluidic Nanoscope

Since the invention of the microscope, it has been possible to visualize objects with length scales on the order of microns. Modern optical microscope objectives have magnifications as high as 100× and can be used where there is sufficient optical access. With some care, optical microscopes can have diffraction-limited resolutions approaching 0.5 μm in the lateral directions [55], nearly the wavelength of the illuminating light. Unfortunately, this resolution is not sufficient when feature sizes are submicron, as is the case in many MEMS and biological systems. Traditional optical microscopes have even lower resolutions in the out-of-plane direction.

The scanning electron microscope (SEM) and the scanning probe microscope (SPM) provide much higher spatial resolution than traditional optical microscopes. An SEM uses a focused electron beam instead of light to displace electrons from the surface of an object to create an image. Features within the plane of view of the microscope are resolvable to length scales of about $\pm 5$ nm with the SEM [55]. The SEM has several disadvantages that make it undesirable for microfluidic applications. First, the imaging sample must be placed in a vacuum. In addition, the surface of the object must be highly conductive to prevent an accumulation of charge. Due to the nature of SEMs, it is difficult to make out-of-plane measurements with high resolution, and it is not possible to image an internal cavity within a device.

The SPM, of which there are many varieties, including the atomic force microscope (AFM), measures the deflection of an atomically sharp probe tip mounted on the end of a cantilever as the tip is scanned over the surface being imaged. It has the greatest resolution of the three microscopes described and the slowest speed. Its resolution is typically $\pm 0.1$ to 1.0 nm in the $x$, $y$ (in-plane) directions, and $\pm 0.01$ nm in the $z$ (out-of-plane) direction [55, 56]. The SPM is the only microscopy technique able to measure with nanometer accuracy in the $z$ direction. The SPM and its derivatives suffer from two primary drawbacks: they have a small depth of focus and the surface must be exposed and accessible to the probe in order to be imaged [56].

The microfluidic nanoscope is a recent development [57] for measuring the shape of internal cavities and flow channels with a spatial resolution of tens of nanometers. This new technique uses μPIV to measure the motion of fluid near a cavity wall. This information, combined with a model for the fluid's boundary condition, typically the no-slip boundary condition, can be used to determine the wall's location and topology. The microfluidic nanoscope has been demonstrated on flat walls, but can, in principle, detect other boundary shapes. The only limitations on possible shapes that can be detected are that the wall position should change slowly relative to the size of the interrogation regions used in the nanoscope technique.

The excellent spatial resolution of the microfluidic nanoscope technique can be demonstrated by considering again the microchannel flow discussed in Section 4.3.1.1. By overlapping interrogation regions by 50%, the vector spacing in the spanwise and streamwise directions was 450 nm and 6.8 μm, respectively. In this manner, more than 5,000 time-averaged velocity measurements were obtained simultaneously, each representing a volume of fluid with dimensions of 0.9 × 13.6 ×1.8 μm. Figure 4.26(b) shows the velocity data nearest the wall for a 35-mm streamwise section of the microchannel. Careful inspection of the velocity vectors within two to five particle diameters (i.e., from 400 to 1,000 nm) of the wall reveals that they seem to be inclined away from the wall. These points must be treated as suspect because of the potential hydrodynamic interactions between the flow-tracing particles and the wall. These points are ignored in the following analysis.

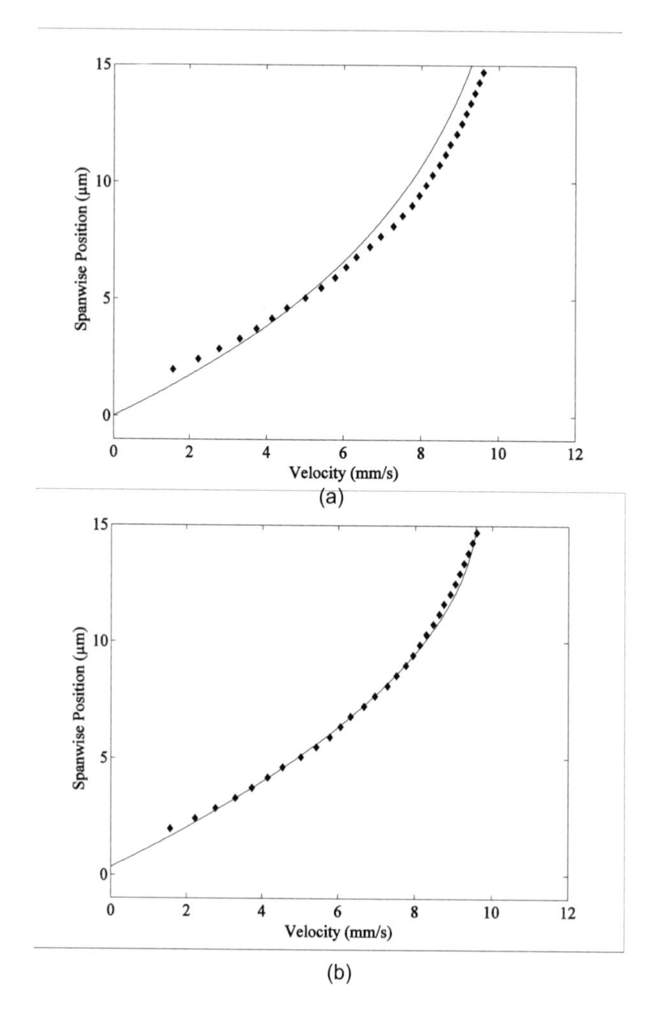


**Figure 4.26** Velocity measurements averaged in the streamwise direction (symbols) compared to (a) the analytical solution (solid curve); and (b) a second-order polynomial fit to the velocity measurements (solid curve). (*After*: [57]. Courtesy of Meinhart and Stone.)

#### 4.4.1.1 Estimation of the Wall Position

The goal of this technique is to estimate the wall location as accurately as possible using the fluid velocity field. Since at these length scales we expect the fluid to behave as a continuum, it is reasonable to assume that the velocity field will vary continuously. As a further refinement on that assumption, it is also reasonable to assume that the fluid will behave according to the Stokes equation because of the low Reynolds number of this flow. Consequently, we will try to estimate the wall location using two different curve fits, one derived from the analytical solution for laminar flow in a rectangular cross-section capillary (see Section 4.3.1.1), and one that is based on a simple polynomial.

*Analytical Fit.* A velocity profile from the analytical solution of the channel flow was overlaid with streamwise-averaged μPIV velocity measurements (see Figure 4.27). The measured velocity vectors agree with the analytical solution to within 2%. Initial wall positions were estimated as the edge of the particle-image field. By visual inspection of the recorded images, these positions were well inside the channel wall, but were still used as a starting point for the following calculations.

A discrete velocity profile was created from the analytical solution with velocity vectors spaced every nanometer in the spanwise direction. The experimentally measured velocity vectors were grouped into 16 spanwise profiles for the purpose of the fitting process. Each analytical profile was normalized by the average of the far-field velocity of the measured profile. Since the wall estimation algorithm essentially matches variations in the measured data versus the variations in the analytical solution, only the region where the velocity profile varies rapidly (close to the wall) is examined. As a result, the mean square error between the analytical and the measured velocity vectors was estimated by averaging over each measurement location only within the nearest 16 μm of the wall for each velocity profile. The zero velocity position of the analytical solution was incremented relative to the experimental data by the resolution of the analytical data (in this case, 1 nm), and a new mean square error was estimated. This process was repeated to determine the zero position that minimizes the mean square error, corresponding to the best agreement between the analytical solution and the measured velocity profile. Extrapolating to the location where the analytical solution goes to zero (assuming the no-slip condition) is an estimation of the location of the wall, averaged over the streamwise and out-of-plane resolution of the velocity measurements (13.6 $\times$ 1.8 μm, in our case). In essence, the wall imparts information into the fluid velocity field in varying locations and wavenumbers via the momentum deficit near the wall. The location and dynamics of the momentum deficit are directly related to the location of the no-slip boundary. Using this method, the location of the microchannel wall was determined with an rms error of 62 nm.

*Second-Order Polynomial Fit.* Even though an analytical solution matches the streamwise-averaged data profile within a 2% error, qualitatively the analytical solution fails to capture the trends of the measured profile very near the wall [Figure 4.26(a)]. In order to fit a more representative curve, a second-order polynomial was fit to the experimental data at each location in the streamwise direction. The polynomial had the following form:

$$x = C_1U^2 + C_2U + C_3 \tag{4.39}$$

where $x$ is the distance in the streamwise direction and $C_1$, $C_2$, and $C_3$ are constants. The velocity vectors ranging from 5 to 15 mm from the wall were used to fit the data at each streamwise location. The root of the second-order polynomial (where velocity is zero) was assumed to be the location of the wall. Qualitatively, the shape of the second-order polynomial corresponded more closely to the shape of the measured profile [Figure 4.26(b)] than did the shape of the analytical solution [Figure 4.26(a)]. Figure 4.27 shows the 16 velocity profiles of measured data (symbols) overlaying the second-order polynomial fit (solid line). The difference between the second-order polynomial and the fitted portion of the measured velocity profile was less than 0.1%. Assuming the wall to be flat, the wall position was measured to be flat with an rms uncertainty of 62 nm. Figure 4.28 shows the results of the wall estimation algorithm versus position in the streamwise direction. Nearly all of the measurements agree to within $\pm$100 nm. Higher-order polynomials, beyond second-order, were also explored. The higher-order polynomials required more datapoints to be included in the fit, to obtain the same rms uncertainty, and they often failed to agree qualitatively with the measured data. Near the wall, the higher-order polynomials matched the measured velocity profiles quite well. However, in the flatter region away from the wall, the two curves did not agree.

#### 4.4.1.2 Effect of Noise in Velocity Measurements

Numerical simulations were conducted to determine the effect that noise in the velocity measurements has upon the uncertainty of measuring wall position. A test dataset was created by adding unbiased white noise to the analytical channel flow. The magnitude of the random component was a percentage of the magnitude of each velocity vector. The vector spacing was every 450 nm, corresponding to the resolution of the velocity measurements. A randomly prescribed offset was also

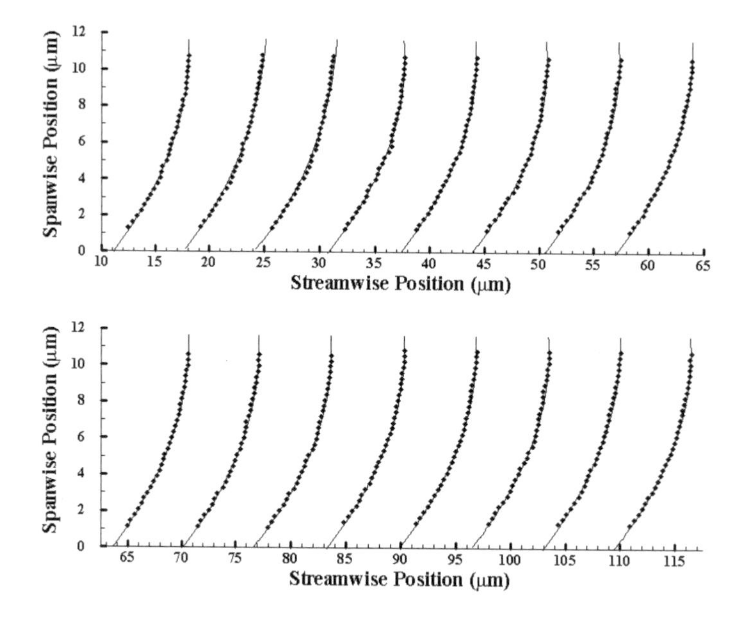


**Figure 4.27** Best-fit curve of the velocity profile (solid line) overlaying measured velocity data (black diamonds). The best-fit curve is a second order polynomial fit to the measured data. Extrapolating the best-fit curve to zero velocity gives an estimate of the location of the channel wall. Zero velocity occurs where the best-fit curve intersects the horizontal axis. (*After*: [57]. Courtesy of Meinhart and Stone.)

added to the spanwise position of the test profile. The uncertainty in the wall position was defined as the difference between the actual specified wall position and the wall position determined by the second-order polynomial fit. The uncertainty in the wall position scales approximately linearly with the noise in the velocity data. An rms uncertainty of 62 nm for the wall position of the microchannel corresponds to a noise level of 2.5% in the velocity measurements.

The surface roughness of the test section was assessed by scanning a line on the outside of one of the glass microchannels using a Sloan Dektak IIa profilometer. Since the glass microchannel was manufactured by an extrusion process, it was assumed that the inside wall roughness matches the outside wall roughness. The rms uncertainty of the outside of the glass microchannel was 2.5 nm, which is well below the noise of the nanoscope. Therefore, for all practical purposes, the walls are considered to be flat.

#### 4.4.1.3 Discussion

The nanoscope is capable of resolving lateral features of ±62-nm rms. The close velocity-vector spacing of 450 nm in the spanwise direction (which allows 62 nm resolution) comes at the cost of averaging over a large streamwise distance of 13.6 μm. Streamwise resolution can be improved by either choosing shorter interrogation regions, or by ensemble-averaging the correlation function of more image pairs. While the nanoscope presented in this section is 1-D, 2-D information can be obtained by varying the $y$-location of the measurement plane.

#### 4.4.1.4 Hydrogels

There are several scientific and engineering problems that may be examined using the microfluidic-based nanoscope. In principle, the combination of hydrogels and the nanoscope may lead to a new type of chemical sensor. Hydrogels are cross-linked polymers that can undergo volume changes in response to physical and chemical changes in their environment [58, 59]. The deformation of

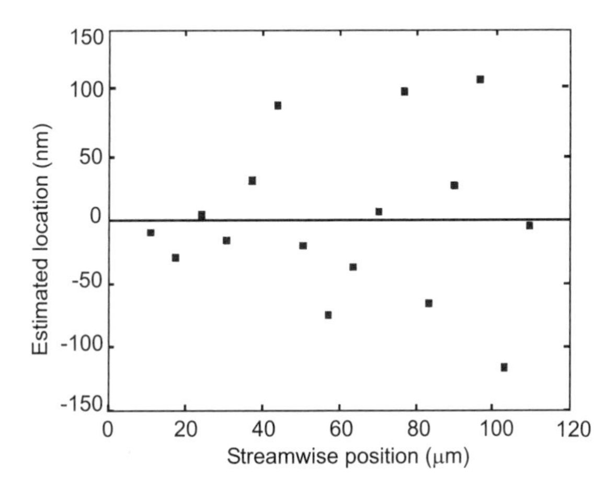


**Figure 4.28** Measurements of wall position by extrapolating the second order polynomial fit to the velocity field to zero. Assuming the wall is approximately flat, the rms uncertainty of the measurements is 62 nm. (*After*: [57]. Courtesy of Meinhart and Stone.)

the hydrogel can be detected to within tens of nanometers using the μPIV technique. Hydrogel performance is a diffusion-limited process that is rate-limited by surface area. Consequently, macroscale applications of hydrogels have response times of hours to days [60]. However, the high surface-to-volume ratios found in microscale applications promote a dramatic decrease in the response time. Liu et al. [60] reported a hydrogel-based microvalve with a closure time of 39 seconds, while Madou et al. [61] found that, by using micropores, the swelling time of poly (HEMA) hydrogel was approximately 3 minutes.

Olsen et al. [62] reported development of a technique for measuring hydrogel deformation rate related to the microfluidic nanoscope. Since hydrogels are typically made from a liquid prepolymer mixture, fluorescent seed particles can easily be incorporated into this mixture in volume fractions from 0.1% to 1.0%. The prepolymer mixture is then formed into functional structures inside microfluidic channels, as described by Beebe et al. [63]. Since the hydrogel is relatively transparent at visual wavelengths, all the principles of the μPIV technique developed thus far should be applicable to these seed particles, despite the fact that they are held fixed with respect to the hydrogel. Figure 4.29 demonstrates how the deformation rate can be measured using μPIV. The hydrogel structure is represented by the cylinder in the center of the figure. It is anchored to the top and bottom planes of the figure, which represent the top and bottom walls of a microchannel. The sides of the microchannel are not shown. The seed particles are represented as the small black circles distributed randomly throughout the hydrogel. As in the other μPIV apparatus described here, the entire flow is volume-illuminated by a microscope system (not shown) and imaged by the same system. The working distance of the microscope objective lens defines the focal plane of the measurement system, which is shown as the gray plane in the vertical center of the microchannel. Deformations of the hydrogel in the focal plane are measured the same way movement of fluid is measured—by acquiring two images and cross-correlating them.

In the experiments described in [62], 1-μm seed particles are used in a volume concentration of 1%. The seed particles are illuminated by a Nd:YAG laser and recorded by a CCD camera. The magnification is chosen such that the interrogation regions represent 30 $\times$ 30 $\times$ 38 μm. The hydrogel, initially at its smallest size of about 400 μm in a solution of pH $=$ 3.0, was immersed in a solution of pH $=$ 12.0 and allowed to expand over the course of 10 minutes. Using a relatively long time separation between the images of $\Delta t = 1$ second, maximum deformation rates of approximately 10 μm/s were measured. This technique has the potential to provide important

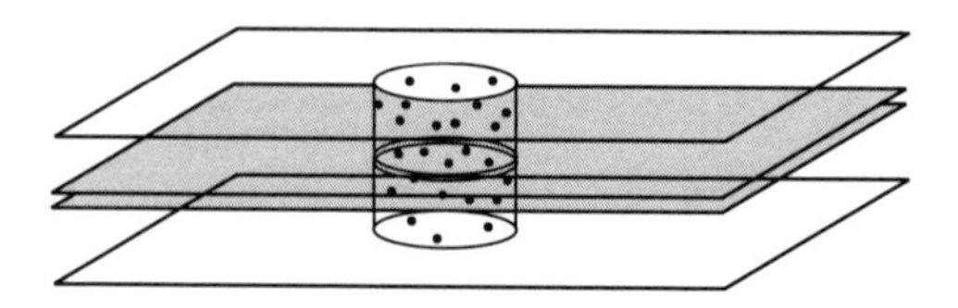

**Figure 4.29** Configuration for measuring hydrogel deformation with μPIV. (*After*: [62].)

quantitative information about the response of hydrogels to their environment which could be used to optimize hydrogel shape or other properties. The same technique could be used to visualize the deformation of other flexible materials commonly used in microfabrication, such as PDMS and PMMA.

#### 4.4.1.5 Conclusions

A novel microfluidic-based nanoscope is presented. With this instrument, surface shapes can be measured inside microscale fluidic devices, where the surface is immersed in a fluid and not exposed to a free surface. Such configurations cannot be measured with traditional scanning electron microscopes or scanning probe microscopes. The instrument has a demonstrated accuracy of ±62 nm rms over a measurement length of 120 μm. The surface position is determined by assuming the no-slip velocity boundary condition and then by extrapolating μPIV measured velocity profiles to zero.

### 4.4.2 Microparticle Image Thermometry

Regardless of the length scale of the experimental work, fluid dynamics and heat transfer intertwine to govern the physics of mechanical processes. Therefore, the ability of a μPIV system to produce high-resolution velocity data naturally begins the search for methods that provide temperature measurements with the same high spatial resolution. The small scales of such MEMS devices often make physically probing a flow impractical, as even the smallest probe would likely be on the order of the size of the device itself. This condition puts minimally invasive measurement techniques, such as PIV, at a premium. Several other noninvasive temperature measurement techniques have been developed but have serious restrictions. Liquid crystal thermometry [64] has been developed, but has a limited temperature range and the liquid crystal particles are too large to be practical in a microdevice. Ratiometric techniques [65], which depend on two dyes of different fluorescence temperature dependence, are not directly compatible with velocimetry and have relatively large uncertainties.

Early literature on μPIV noted that for small seed-particles ($<$ 1 μm) and low-speed flows ($<$ 10 μm/s), the Brownian particle motion was measurable and could be significant enough to introduce considerable error into velocity measurements [20]. Because it was viewed as an undesirable effect on velocity data, the effect of the Brownian motion on the PIV correlation function—namely, a width-wise spreading of the correlation peak that added to the uncertainty of locating the peak center—was often intentionally and substantially reduced through ensemble-averaging over multiple images for a given flow field. However, Olsen and Adrian [66] postulated that such spreading of the correlation peak could be used to deduce fluid temperature since Brownian motion has a direct, explicit temperature dependence. Furthermore, since peak location (which yields velocity information) and peak width are independent parameters, it is conceivable that PIV may be used to simultaneously measure both temperature and velocity [67].

The following section experimentally applies the latter approach to the creation and analysis of μPIV data. Instead of imaging a moving flow, measuring its velocity field, and ensemble-averaging away the effects of Brownian particle motions (which take the form of correlation peak broadening), this work—in implementing PIV-based thermometry—images stagnant flows and measures the variations in correlation peak broadening with variations in temperature. The end goal, which is theoretically possible but not experimentally treated in this work, is a μPIV algorithm that interprets both peak-location and peak-width data from the same image set to yield both velocity and temperature information for the flow field.

#### 4.4.2.1 Theory of PIV-Based Thermometry

The theoretical basis for the hypothesis of the current work—that temperature measurements can be deduced from the images of individual particles in a flow—rests on the theory of particle diffusion due to the Brownian motion. Constant random bombardment by fluid molecules results in a random displacement in the seed particles. This random particle displacement will be superimposed on any particle displacement due to the fluid velocity. This phenomenon, previously discussed in Section 4.2.1.2, is described by (4.31):

$$D = \frac{KT}{3\pi\eta d_{\mathrm{p}}}$$

where the diffusivity $D$ of particles of diameter $d_{\mathrm{p}}$ is immersed in a liquid of temperature $T$ and dynamic viscosity $\eta$. $K$ is Boltzmann's constant. Noting the dimension of $D$ is [length$^2$/time], the square of the expected distance traveled by a particle with diffusivity $D$ in some time window $\Delta t$ is given by:

$$\langle s^2 \rangle = 2D\Delta t \tag{4.40}$$

Combining (4.31) and (4.40), it can be observed that an increase in fluid temperature, with all other factors held constant, will result in a greater expected particle displacement, $\sqrt{\langle s^2 \rangle}$. However, the dynamic viscosity, $\eta_{\mathrm{f}}$, is a strong function of temperature and $\langle s^2 \rangle \propto T/\eta_{\mathrm{f}}$. The effect of this relationship depends on the phase of the fluid. For a liquid, increasing temperature decreases absolute viscosity—so the overall effect on the ratio $T/\eta_{\mathrm{f}}$ follows the change in $T$. For a gas, however, increasing temperature increases absolute viscosity, meaning that the effect on $T/\eta_{\mathrm{f}}$, and hence $\sqrt{\langle s^2 \rangle}$, would need to be determined from fluid-specific properties.

Relating changes in fluid temperature to changes in the magnitude of a random seed-particle motion due to the Brownian motion is the key to measuring temperature using PIV. To demonstrate the feasibility of using PIV for temperature measurement, an analytical model of how the peak width of the spatial cross-correlation varies with relative Brownian motion levels must be developed.

In a typical PIV experiment, two images are taken of the flow field at times $t_1$ and $t_2 = t_1 + \Delta t$. If these two images are denoted by $I_1(X)$ and $I_2(X)$, then the spatial cross-correlation can be estimated by the convolution integral (following Keane et al. [68]):

$$R(s) = \int I_1(X) I_2(X + s) \mathrm{d}X \tag{4.41}$$

$R(s)$ can be decomposed into three components, such that:

$$R(\mathbf{s}) = R_{\mathrm{C}}(\mathbf{s}) + R_{\mathrm{F}}(\mathbf{s}) + R_{\mathrm{D}}(\mathbf{s}) \tag{4.42}$$

where $R_{\mathrm{C}}(\mathbf{s})$ is the convolution of the mean intensities of the two images and is a broad function of $s$ with a diameter of the order of the interrogation spot diameter, $R_{\mathrm{F}}(\mathbf{s})$ is the fluctuating noise component, and $R_{\mathrm{D}}(\mathbf{s})$ is the displacement component of the correlation function, and as such gives the particle displacement from time $t_1$ to $t_2$. Thus, $R_{\mathrm{D}}(\mathbf{s})$ contains the information necessary to calculate the velocity vector; that is, it is the PIV signal peak.

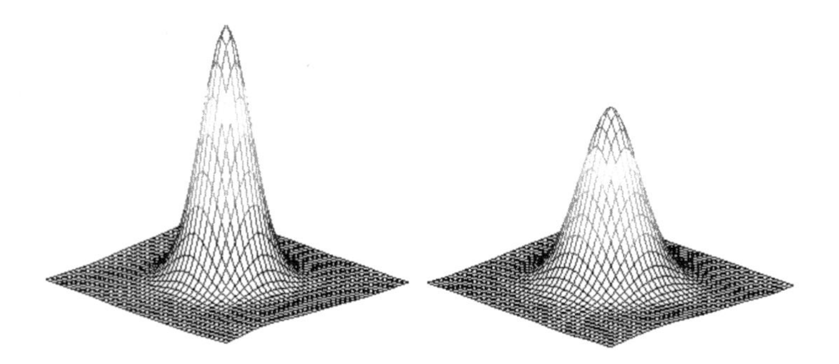

**Figure 4.30** A pair of correlation functions demonstrates possible variations in peak width due to Brownian motion. The autocorrelation (no Brownian motion) is on the left, while the cross-correlation (contains Brownian motion) is shown on the right. Both figures are shown in the same scale.

The displacement and width of the signal peak are dependent on the probability function $f(\mathbf{x}', t_2; \mathbf{x}, t_1 | \mathbf{u}(\mathbf{x}), T)$ where $f$ is the probability that a particle initially at $(\mathbf{x}, t_1)$ moves into the volume $\mathbf{x}', \mathbf{x}' + dx$ at $t_2$ for a given velocity field $\mathbf{u}(\mathbf{x})$ and temperature $T$. Letting $\Delta\mathbf{x} = \mathbf{u}(\mathbf{x}, t)\Delta t$ be the displacement undergone by a particle that was initially at $(\mathbf{x}, t)$, in the absence of Brownian motion $f$ is simply described by a delta function, that is:

$$f(\mathbf{x}', t_2; \mathbf{x}, t_1 | \mathbf{u}(\mathbf{x}), T) = \delta(\mathbf{x}' - \mathbf{x} - \Delta\mathbf{x}) \tag{4.43}$$

because for a given velocity field, there is only one possible end location for the particle.

When Brownian motion is present, however, any displacement $\Delta x$ is possible, and $f$ will no longer be a delta function. Instead, $f$ will have a probability distribution centered at $\mathbf{x}' = \mathbf{x} + \Delta\mathbf{x}$ with a shape defined by the space-time correlation for the Brownian motion derived by Chandrasekhar [69] (see also [70]):

$$f(\mathbf{x}', t_2; \mathbf{x}, t_1 | \mathbf{u}(\mathbf{x}), T) = (4\pi D\Delta t)^{-3/2} \times \exp\left(\frac{-(\mathbf{x}' - \mathbf{x} - \Delta\mathbf{x})^2}{4D\Delta t}\right) \tag{4.44}$$

This change in $f$ due to the Brownian motion has two effects on the signal peak. It broadens it and reduces its height. It is the broadening of the signal peak in the cross-correlation function that is the key to measuring temperature using PIV. This effect is demonstrated in Figure 4.30 for a pair of experimental PIV images.

Olsen and Adrian [66] derived analytical equations describing the shape and height of the cross-correlation function in the presence of the Brownian motion for both light-sheet illumination and volume illumination (as is used in μPIV). In both cases, the signal peak in the cross-correlation has a Gaussian shape with the peak located at the mean particle displacement. One of the key differences between light-sheet PIV and μPIV (volume illumination PIV) lies in the images formed by the seed particles. In light-sheet PIV, if the depth of focus of the camera is set to be greater than the thickness of the laser sheet, then all of the particle images will (theoretically) have the same diameter and intensity. The relationship between a particle's actual cross-sectional area and that of its image will be governed by the characteristics of the imaging optics [27]. The image of a particle will be the convolution of two quantities: the geometric image of the particle, and, given a diffraction-limited lens, the point response function of the imaging system. The point response function is an Airy function with diameter given by:

$$d_s = 2.44(1 + M)f^{\#}\lambda \tag{4.45}$$

where $M$ is the magnification of the lens, $f^{\#}$ is the f-number of the lens, and $\lambda$ is the wavelength of light reflected, scattered, or fluoresced by the particle. Adrian and Yao [31] found that the Airy function can accurately be approximated by a Gaussian function. If both the geometric image and the point response function are approximated by Gaussian functions, the following approximate formula for particle image diameter is obtained:

$$d_{\mathrm{e}} = (M^2 d_{\mathrm{p}}^2 + d_{\mathrm{s}}^2)^{1/2} \tag{4.46}$$

where $d_{\mathrm{p}}$ is the actual particle diameter and $M$ is the magnification of the lens.

In μPIV, the particle images are a bit more complex. Because the entire volume of fluid is illuminated in μPIV, all of the particles throughout the depth of the flow field will contribute to the resulting PIV image. Those particles close to the focal plane will form sharp, bright images, while particles farther from the focal plane will form blurry, dim images. Olsen and Adrian [29] found that the image formed by a particle in μPIV can be approximated by:

$$d_{\mathrm{e}} = (M^2 d_{\mathrm{p}}^2 + d + \mathrm{s}^2)^{1/2} \tag{4.47}$$

where

$$d_{\mathrm{z}} = \frac{M_{\mathrm{z}} D_{\mathrm{a}}}{x_0 + z} \tag{4.48}$$

and $z$ is the distance of the particle from the object plane, $D_{\mathrm{a}}$ is the aperture diameter of the microscope objective, and $x_0$ is the object distance (typically, $x_0 \gg z$, and the image diameter as a function of distance from the focal plane is a hyperbola).

From their analysis of cross-correlation PIV, Olsen and Adrian [29] found that one effect of the Brownian motion on cross-correlation PIV is to increase the correlation peak width $\Delta s_{\mathrm{o}}$—taken as the $e^{-1}$ diameter of the Gaussian peak. For the case of light-sheet PIV, they found that:

$$\Delta s_{\mathrm{o,a}} = \sqrt{2}\frac{d_{\mathrm{e}}}{\beta} \tag{4.49}$$

when the Brownian motion is negligible, to:

$$\Delta s_{\mathrm{o,c}} = \sqrt{2}\frac{(d_{\mathrm{e}}^2 + 8M^2\beta^2 D\Delta t)^{1/2}}{\beta} \tag{4.50}$$

when the Brownian motion is significant. Note that in any experiment, even one with a significant Brownian motion, $\Delta s_{\mathrm{o,a}}$, can be determined by computing the autocorrelation of one of the PIV image pairs. It can be seen that (4.49) reduces to (4.50) in cases where the Brownian motion is a negligible contributor to the measurement (i.e., when $D \cdot \Delta t \to 0$). The constant $\beta$ is a parameter arising from the approximation of the Airy point-response function as a Gaussian function [see (4.32)]. Adrian and Yao [31] found a best fit to occur for $\beta^2 = 3.67$.

For the case of volume illumination PIV, the equations for $\Delta s_0$ are a bit more complex. The equations have the same form, but because of the variation of $d_{\mathrm{e}}$ with distance from the focal plane, instead of being constants, the $d_{\mathrm{e}}$ terms in (4.49) and (4.50) are replaced with integrals over the depth of the device.

The difficulty in calculating the integral term for $d_{\mathrm{e}}$ for volume illumination PIV can be avoided by strategic manipulation of (4.49) and (4.50). Squaring both equations respectively, taking their difference, and multiplying by the quantity $\pi/4$, converts the individual peak width (peak diameter for a three-dimensional peak) expressions to the difference of two correlation peak areas—namely, the difference in area between the auto- and cross-correlation peaks. Performing this operation and substituting (4.31) for $D$ yields:

$$\Delta A = \frac{\pi}{4}\left(\Delta s_{\mathrm{o,c}}^2 - \delta s_{\mathrm{o,a}}^2\right) = C_0 \frac{T}{\eta} \Delta t \quad (4.51)$$

where $C_0$ is the parameter $2M^2K/3d_{\mathrm{p}}$. The expected particle displacement, $\sqrt{\langle s^2 \rangle}$, of the classical diffusion theory can now be tied to the peak width of the correlation function, $\Delta s_0$, or the change in peak area, $\Delta A$, through the diffusivity, $D$. These relationships are established in (4.50) and (4.51) and qualify the current experimental work.

#### 4.4.2.2 Modeling

Before any measurements were made using the μPIV system, an experimental model and feasibility analysis were designed using simulated particle images and the cross-correlation PIV algorithm. The initial analysis allowed for the characterization of the algorithm's sensitivity to particle displacement and anticipated the likelihood that good experimental images would result in measurable trends.

The particle image simulation code starts with an $n \times n$ pixel black image ($n$ indicating a user-prescribed image size) and creates a number of white particles analogous in size and number density (particles/pixel area) to an expected experimental particle image. Particle placement within the image frame is set by uniformly distributed random numbers assigned to $x$ and $y$ coordinates. This is done to simulate the actual random distribution of particle locations within an experimental image. The light intensity of the simulated particles is given a Gaussian distribution because a spherical particle image is a convolution of an Airy point-spread function and a geometric particle image—both of which are well-modeled as Gaussian functions.

After an initial particle image was generated, a random number generator (mean-zero, normal distribution) was used to impart a random shift to the $x$ and $y$ position of each particle. The particle shift routine was repeated several times, each time imparting a random shift from the previous particle positions, in order to produce an image set. These successive shifts approximated the spatial shifts that the experimental particles undergo between successive frames of video.

The simulated particle images were then cross-correlated and the trends in peak broadening and peak location observed. Assuming truly random shifts, it was expected that the average correlation would indicate a net-zero velocity. Furthermore, because the shifts were done in an additive fashion, a simulated sequence of images, 1-2-3, should yield an increase in peak-width from a correlation of 1-2 to a correlation of 1-3, and so on.

The major difference between the simulated images and experimental μPIV images is that the effects of out-of-focus particles are not present—out-of-focus particles negatively contribute to the image by effectively varying the particle image size and luminous intensity, raising the signal-to-noise ratio, and ultimately affecting the correlation [29]. However, (4.51) demonstrates that these effects, which are present in both the autocorrelation and cross-correlation peak morphology, are removed from consideration by using the difference of the cross-correlation and autocorrelation peak widths to infer temperature. Therefore, even though the simulated images more closely approximate PIV images under light-sheet illumination than volume illumination, they should provide considerable insight into the volume-illumination case.

#### 4.4.2.3 Experimental Technique

An Olympus BX50 system microscope with BX-FLA fluorescence light attachment (housing the dichroic mirror/prism and optical filter cube) was used to image the particle-laden solution. All experiments were carried out with a 50× objective (NA=0.8). A Cohu (model 4915-3000) 8-bit CCD video camera was used—the CCD array consisted of 768 (horizontal) × 494 (vertical) pixels and a total image area of 6.4×4.8 mm. The particles-700-nm-diameter polystyrene latex microspheres

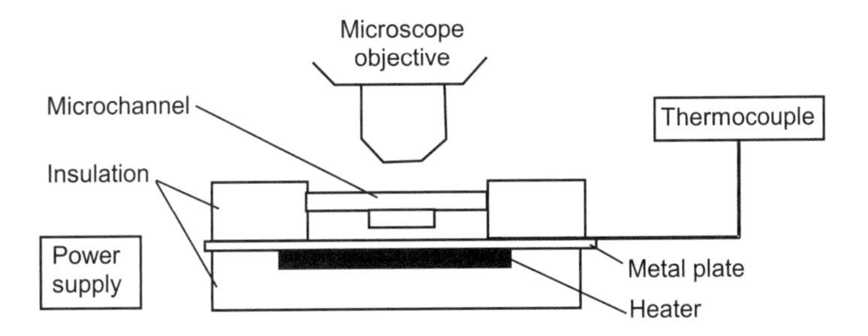


**Figure 4.31** Microchannel, heater, and thermocouple arrangement used for experimentation. (*After*: [67].)

(Duke Scientific, Palo Alto, California)—had a peak excitation wavelength at 542 nm and peak emission at 612 nm. A variable intensity halogen lamp was used for illumination-optical filters were used to isolate the wavelength bands most applicable to the particles—from 520 to 550 nm for the incident illumination, and greater than 580 nm for the particle fluorescence.

In the experiments, the power supplied to a patch heater (Figure 4.31) was adjusted incrementally, and several successive images of the random particle motion were captured at each of several temperature steps. At each temperature, the system was allowed 5 minutes to come to a steady temperature as recorded by a thermocouple.

The gating between frames was fixed by the camera frame rate of 30 frames per second. In order to get numerical data, the individual frames were cross-correlated using a custom-written PIV interrogation code capable of outputting the correlation function. The first frame in each sequence was correlated with itself (i.e., autocorrelated) to provide a reference value of correlation peak width. Then the first frame was cross-correlated with the second frame, and then with the third, and the fourth, and so on, and time between cross-correlated frames linearly increased. Each individual cross-correlation of such successive pairs of images produced one correlation peak (the average over each interrogation region peak). The width of each peak was recorded, and trends, both temporal and thermal dependence, in the measurements of peak widths were observed. Temperatures from 20°C to 50°C were measured in this way.

#### 4.4.2.4 Results

The results of the simulated particle images are shown in Figure 4.32(a), while the experimental results are shown in Figure 4.32(b) for a certain fixed temperature. In both figures, the PIV-determined peak area data is plotted as the rhomboid symbols, while the line is a best-fit linear profile. The salient feature of these two plots is that the area increase is very nearly proportional to time, as predicted by (4.51). The quality of the linear fit gives an estimate of the uncertainty of the technique. The rms difference between experimental and theoretical values of peak area increase (at each datapoint) was taken for many such curves. This rms difference in square-pixels was converted to an equivalent temperature difference, which is the nominal value of temperature resolution given above. The average error over all the test cases is about ±3°C. By combining many such datasets, the temperature at many different values can be assessed and compared to the thermocouple measurements, as in Figure 4.33.

#### 4.4.2.5 Experimental Uncertainty

To examine the uncertainty of the temperature measurement technique, rms difference between experimental and theoretical values of peak area increase (at each datapoint) was computed for a wide range of operating conditions. This rms difference in square-pixels was converted to an

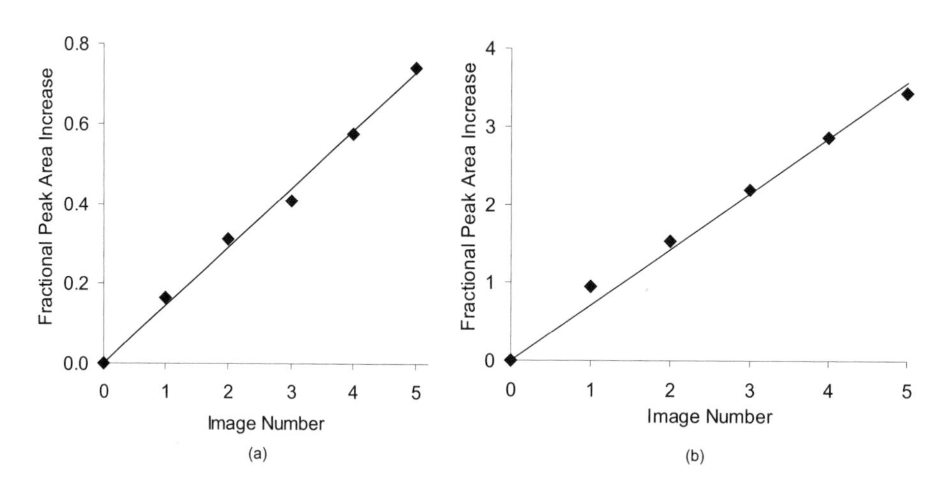


**Figure 4.32** Measured versus theoretical peak broadening trend for particle diffusion in (a) simulated images and (b) actual experimental images. (*After* [67].)

equivalent temperature difference, which is the nominal value of temperature resolution given above. The average error over all the test cases is about $\pm 3°C$.

To further explore the uncertainty of the measurements, (4.51) can be used in a standard uncertainty analysis to find:

$$\delta T = \pm T \left(1 - \frac{T}{\eta}\frac{\mathrm{d}\eta}{\mathrm{d}T}\right)^{-1} \cdot \sqrt{\left(\frac{\delta(\Delta A)}{\Delta A}\right)^2 + \left(2\frac{\delta M}{M}\right)^2 + \left(\frac{\delta(\Delta t)}{\Delta t}\right)^2 \left(\frac{\delta d_{\mathrm{p}}}{d\mathrm{p}}\right)^2} \tag{4.52}$$

The one unusual result in this uncertainty calculation is the term raised to the negative one power that comes from the temperature dependence of the viscosity, combined with the assumption that the viscosity is a known property of the fluid dependent only on the temperature. Upon substituting the values of the parameters present in (4.52) used in the current work, the expected accuracy of the technique can be assessed. At this stage, we will assume that the particles are monodisperse ($\delta d_{\mathrm{p}} = 0$), that the laser pulse separation is exactly known [$\delta(\Delta t) = 0$], and that the magnification of the microscope is exactly known ($\delta M = 0$). These assumptions are reasonable because the uncertainty in the area difference $\delta(\Delta A)$ dominates the right-hand side of (4.53). As an estimate of the uncertainty in area difference $\delta(\Delta A)$, we can use the standard error between the datapoints and the curve fits in Figure 4.32. For the simulations, $\delta(\Delta A) = 0.95$ pixels$^2$ and $\delta(\Delta A) = 42$ pixels$^2$, while for the experiments, $\delta(\Delta A) = 5.0$ pixels$^2$ and $\delta(\delta A) = 126$ pixels$^2$. Using these values, we can plot in Figure 4.34 the expected uncertainty of the measurements and simulations over the range for which water is normally a liquid. The measurement uncertainty of the simulations is about one-half as large as that of the experiments. There are several possible explanations for this. One is that the experiments were performed using volume-illuminated μPIV while the simulations were done more in keeping with light sheet PIV, wherein the particles have uniform brightness and size within the light sheet and no particles are visible in front of or behind the light sheet. Both of these factors tend to lower the signal-to-noise ratio and raise the uncertainty of the measurements. The second reason is that the statistics may not have been stationary for the experimental results. Further experiments and simulation using more particle images and better simulation approximations will shed further light on the best possible results achievable.

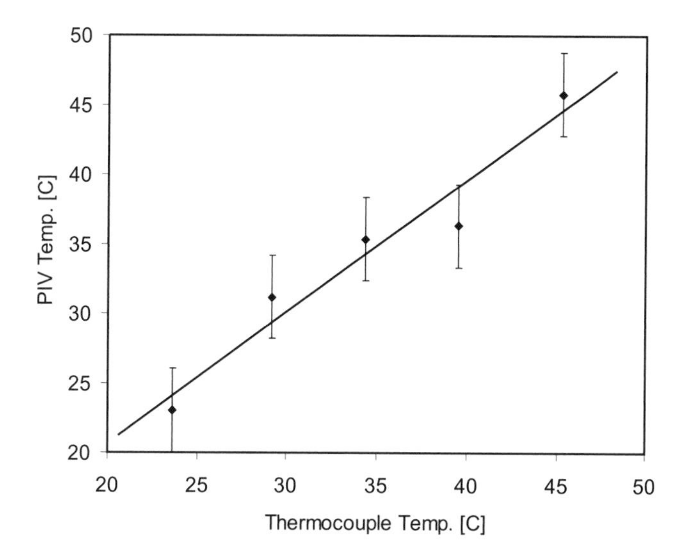


**Figure 4.33** Temperature inferred from PIV measurements plotted versus thermocouple-measured temperature. Error bars indicate the range of average experimental uncertainty, $\pm 3°C$. (*After*: [67].)

### 4.4.2.6 Simultaneous Velocity and Temperature Measurement

In Section 4.1, the potential for the simultaneous measurement of both velocity and temperature was noted. While it is true that all experimental measurements reported in the current work were made using fluids with a net-zero velocity, it is also true that the correlation peak location (from which the velocity measurement is derived) and the correlation peak width (which yields the temperature measurement) are independent measurements whose convolution results in the correlation function. Therefore, it is not mere speculation to say that the two may be simultaneously measured. The key is finding the range of flow parameters over which both may be measured. The following analysis, based on the results of this work, establishes a range over which both velocity and temperature may be measured using PIV.

For a control surface, take a standard $n \times n$ pixel interrogation region (IR) in a larger image of a particle-seeded fluid at some temperature $T$ and some steady, bulk velocity $V$. A set of images is recorded—with some time delay between each image—for the purposes of measuring $T$ and $V$. The time delay, $\Delta t$, must satisfy two criteria: (1) it needs to be sufficiently long to allow measurable particle motion to take place, yet (2) short enough that a significant fraction of the particles in the first image also be present in the second. The former criterion is imposed by Brownian particle motion, the latter by bulk fluid motion.

The first criterion establishes a minimum $\Delta t$ and requires that particles move beyond a measurable threshold—a property of the optical system—in order to be accurately resolved. With a precision of microns and stationary fluorescing particles (dried onto a microscope slide) under the experimental ($50\times$) microscope objective, the threshold of the current system was found to be about 100 nm.

The second criterion establishes a maximum $\Delta t$ and requires that particles not leave the interrogation region between frames. While there are two possible ways a particle can leave an IR—either by convection, following a streamline of the flow or by diffusion, following a Brownian path—for bulk flows with velocities greater than 10 mm/s (using particles on the order of 100 nm), bulk motion will dominate diffusion in violating the second criterion [20]. Furthermore, judicious

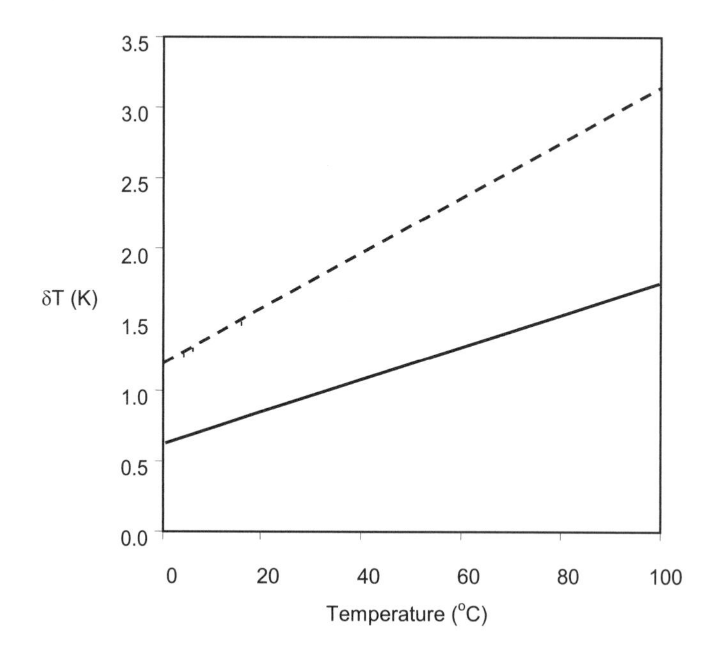


**Figure 4.34** Measurement uncertainty over the range of temperatures for which water is a liquid. (*After* [67].)

window shifting can allow particles to move a distance bounded only by the size of the CCD array—much farther than the size of an interrogation region, which is the bound that exists without a window-shifting scheme. Combining (4.30) and (4.40) yields:

$$\sqrt{\langle s^2 \rangle} = \left( \frac{KT\Delta t}{3\pi\eta d} \right)^{\frac{1}{2}} \tag{4.53}$$

which gives the average particle displacement due to the Brownian motion in terms of experimental parameters. This value is bounded below by the measurement threshold of the system, about 100 nm, and above by the size of the CCD. Constraining the analysis to the current experimental parameters, the minimum $\Delta t$ required to get a $\sqrt{\langle s^2 \rangle}100$ nm, for 700-nm-diameter particles in liquid water, ranges from 0.016 second for 20°C to 0.008 second for 50°C. Furthermore, using the size of the CCD array (6.4 mm) as the limiting factor for bulk flow, a $\Delta t$ of 0.016 second under 50× magnification limits velocities to 8 mm/s. A $\Delta t$ of 0.008 second allows for velocities up to 16 mm/sec. Extending the calculation to the limit of liquid water, a 100°C pool would allow velocity measurements to approach 3.5 cm/s.

#### 4.4.2.7 Conclusions

The random component of the Brownian particle motion is directly related to the temperature of a fluid through the self-diffusion mechanism. Changes in fluid temperature result in changes in the magnitude of Brownian particle motions. Such fluctuations can be detected by analyzing a spatial cross-correlation of successive particle images.

A cross-correlation-based PIV algorithm provides a reliable spatial cross-correlation. Whereas standard PIV looks at correlation peak location to measure velocity, PIV thermometry focuses on changes in correlation peak area—the cross-sectional peak area being a measure of the average magnitude of Brownian motion. Experimentally, temperatures were measured for zero-velocity pools of particle-seeded water over the range from 20°C to 50°C using a standard PIV system.

Results for experimentally measured temperatures, when compared with thermocouple-measured temperatures, agreed within a range of $\pm 3°C$.

The experimental error for PIV thermometry is tied closely to the optical resolution of the PIV system being used—the higher the optical resolution, the more accurately fluid temperatures can be measured using this technique. The effects of thermal gradients on the accuracy of the temperature measurement have not yet been assessed. Increasing optical resolution also has effects on the potential for PIV to simultaneously measure both velocity and temperature. The results of the current work indicate that, using water as the fluid, measurements of fluid velocity less than or equal to 8 mm/s can be made while simultaneously measuring temperatures greater than 20°C.

### 4.4.3 Infrared μPIV

Another recent extension of μPIV that has potential to benefit microfluidics research in general and silicon MEMS research in particular is that of infrared PIV (IR-PIV) [71]. The main difference between the established technique of μPIV and IR-PIV is the wavelength of the illumination, which is increased from visible to infrared wavelengths to take advantage of silicon's relative transparency at IR wavelengths. While this difference may seem trivial, it requires several important changes to the technique while enabling several important new types of measurements to be made.

#### 4.4.3.1 Differences Between μPIV and IR-PIV

The fluorescent particles that allow the use of epifluorescent microscopes for μPIV are not available with both absorption and emission bands at IR wavelengths [71]. Consequently, elastic scattering must be used in which the illuminating light is scattered directly by the seed particles with no change in wavelength. Using this mode of imaging, it is not possible to separate the images of the particles from that of the background using colored barrier filters as in the μPIV case. The intensity of elastic scattering intensity $I$ of a small particle of diameter $d$ varies according to:

$$I \propto \frac{d^6}{\lambda^4} \tag{4.54}$$

where $\lambda$ is the wavelength of the illuminating light [25]. Thus, a great price is exacted for imaging small particles with long wavelengths. The main implication of (4.54) is that there is a trade-off between using longer wavelengths where silicon is more transparent and using shorter wavelengths where the elastic scattering is more efficient. Typically, infrared cameras are also more efficient at longer wavelengths. Han and Breuer [71] found a good compromise among these competing factors by using 1-μm polystyrene particles and an illumination wavelength of $\lambda = 1,200$ nm. An experimental apparatus suitable for making IR-PIV measurements is described by Han and Breuer [71] and is shown in Figure 4.35. As with μPIV, a dual-headed Nd:YAG laser is used to illuminate the particles. However in this case, the 532-nm laser light is used to drive an optoparametric oscillator (OPO)—a nonlinear crystal system that transforms the 532-nm light into any wavelength between 300 and 2,000 nm. The laser light retains its short pulse duration when passing through the OPO. The output of the OPO is delivered via fiber optics to the microfluidic system being investigated. Han and Breuer [71] used an off-axis beam delivery, as shown in Figure 4.35, with an angle of 65° between the normal surface of the device and the axis of the beam. Alternatively, dark field illumination could be used. The light scattered by the particles is collected by a Mititoyo near infrared (NIR) microscope objective (50×, NA $=$ 0.42) mounted on a 200-mm microscope tube and delivered to an Indigo Systems indium gallium arsenide (InGaAs) NIR camera.

The camera has a $320 \times 256$ pixel array with 30-μm pixels—a relatively small number of relatively large pixels compared to the high resolution cameras typically used for μPIV applications. The NIR is a video rate camera that cannot be triggered, meaning that the PIV technique needs to be

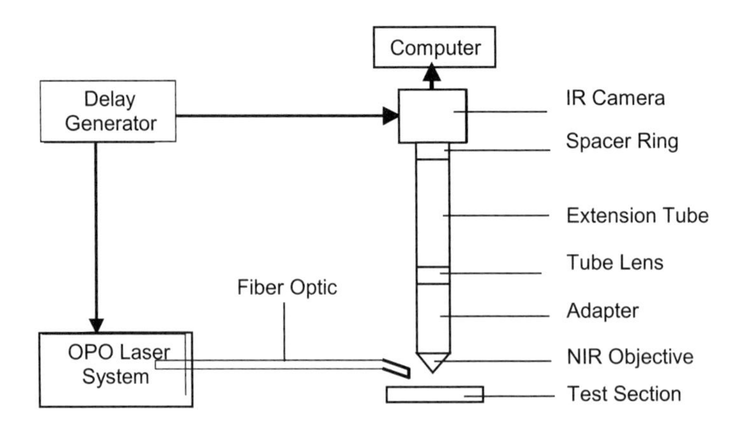


**Figure 4.35** Schematic of the experimental apparatus for IR-PIV. (*After*: [67].)

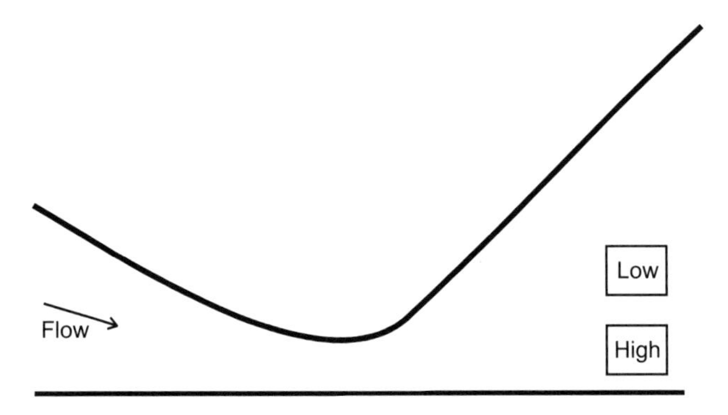


**Figure 4.36** Schematic view of Han and Breuer [71] nozzle with the two interrogation regions used—one for high-speed measurements and one for low-speed measurements.

modified slightly. Instead of using the computer as the master for the PIV system, the camera is the master, running at its fixed frequency of 60 Hz. The laser pulses are synchronized to the video sync pulses generated by the camera and can be programmed to occur at any point within a video frame. For high-speed measurements, a process called *frame straddling* is used, in which the first laser pulse is timed to occur at the very end of one frame and the second laser pulse is timed to occur at the very beginning of the next video frame. Using the frame-straddling technique, the time between images can be reduced to as little as 0.12 ms—suitable for measuring velocities on the order of centimeters per second. Higher-speed flows can be measured by recording the images from both laser pulses on a single video frame. The signal-to-noise ratio of the double-exposed images is decreased somewhat when compared to the single exposed images, but flows on the order of hundreds of meters per second can be measured with the double-exposed technique.

Han and Breuer [71] demonstrated their system on a flow through a silicon micronozzle. The nozzle was constructed using DRIE. The nozzle was 300 μm deep into a 500-μm-thick silicon wafer. The nozzle was sealed by fusion-bonding a second, planar silicon wafer to the wafer with the nozzle geometry. The nozzle, shown in Figure 4.36, had a 40-μm throat width. The measurements were made downstream of the throat in an area where the nozzle had expanded to 1,000 μm wide. For the initial experiments, Han and Breuer [71] measured a water flow through the nozzle. The flow-tracing particles were 1-mm polystyrene particles with a refractive index of 1.56 at 589 nm. These particles, combined with the imaging system, produce images on the CCD array of 181 μm in diameter. This value would be prohibitively large for a visual wavelength CCD imager, which typically has pixels of 6 to 8 μm. Fortunately, pixels in IR imagers tend to be larger—on the order of 30 mm—and so the particle images are only 6 pixels across instead of 30 pixels across. Han and Breuer [71] successfully made measurements in two regions of the nozzle—a high-speed region with flow on the order of 3 cm/s and a low-speed region with flow on the order of 50 μm/s. These regions are labeled "High" and "Low" in Figure 4.36. Although this technique is still being developed, it shows great promise for making measurements inside devices that would otherwise be inaccessible.

### 4.4.4 Particle-Tracking Velocimetry

A typical method for increasing the resolution of PIV measurements at macroscopic length scales is called *super-resolution PIV* [68]. In this technique, a typical PIV analysis is first used to determine the average velocity of the particles within each interrogation region. This spatially averaged velocity information is then used by a particle tracking velocimetry (PTV) algorithm as an initial estimate of the velocity of each particle in the interrogation region. This estimate is progressively refined by one of several methods [72, 73] to eventually determine the displacement of as many individual seed particles as possible. Recently this super-resolution technique has been applied at microscopic length scales by Devasenathipathy et al. [74] to produce very high spatial resolution velocity measurements of flow through microfabricated devices. The super-resolution PIV algorithm presented in their work is based on Kalman filtering for prediction of the particle trajectory and testing for validation of particle identification [75]. An estimate of the velocity field is obtained from μPIV, and individual particle images are identified and located in both images by a particle mask correlation method. The approximate velocity of each particle in the first image is then found by interpolating the μPIV data. This velocity is used as a predictor for the particle pairing search in the second image by Kalman filtering. An iterative procedure is implemented until the number of particle pairs matched is maximized.

Devasenathipathy et al. [74] measured the flow in prototypical microfluidic devices made of both silicon and acrylic in several shapes. The acrylic microfluidic devices used were fabricated by ACLARA BioSciences. These microchannels were embossed on an acrylic substrate and have the D-shaped cross section typical of an isotropic etch (depth at centerline: 50 μm; top width: 120 μm). A 115-μm-thick acrylic sheet thermally laminated to the top of the channels acted as an optical coverslip and the top wall of the channels. The flow in these microchannels was driven using the hydrostatic pressure head of liquid columns established in reservoirs mounted at the ends of the microchannels.

Figure 4.37(a) shows the raw experimental measurements obtained using a combined μPIV-PTV analysis. Because of the high concentration of measurements, it is not possible to see individual vectors. Figure 4.37(b) zooms in on a smaller region of the flow. In this figure, it is possible to get a good idea of the spatial resolution of the technique. The velocity field is clearly unstructured with the vectors nonuniformly spaced due to the random distribution of seed particles in the flow. The main advantages with this combined analysis are the higher sampling density and the higher spatial resolution achieved. The number of vectors obtained by μPIV is 480, while in the same area, the number of vectors produced by the super-resolution Kalman filtering $\chi^2$ method is 5,200, more than a tenfold increase. Note that the vectors from the PTV analysis are individual particle displacements, randomly positioned in the flow field, and include the full Brownian motion component. The spatial resolution of these PTV measurements along the flow direction is equal to the local particle displacement, which ranges from negligible values in the side channels to 10 μm at the centerline. Normal to the local flow direction, the spatial resolution is well approximated by the seed-particle size, which is 500 nm. The number density of particle displacements in the test section is 20/100

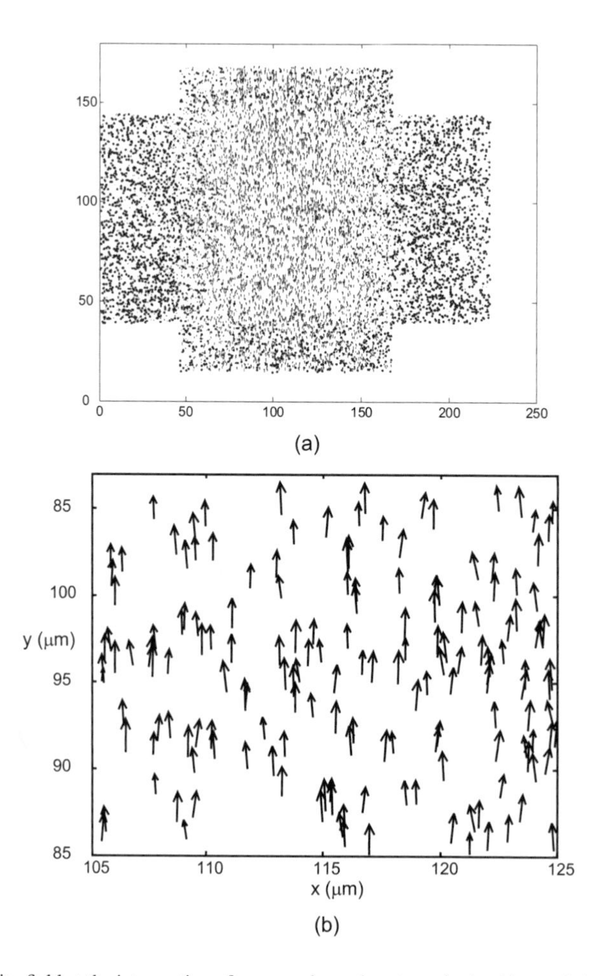


**Figure 4.37** (a) Velocity field at the intersection of a cross-channel system obtained by applying a combined PIV and PTV analysis. More than 5,200 velocity measurements were recorded using 10 image pairs, 10 times the number produced by μPIV analysis; and (b) close-up view of the velocity field at the intersection of the cross-channel system from part (a). The spatial coordinates correspond to those in the previous figure. (*After*: [74]. Courtesy of Devasenathipathy et al.)

$\mu m^2$ (for the combined dataset consisting of all 10 image pairs) so that we can expect a mean vector spacing on the order of 2 μm.

Figure 4.38 compares the experimental data (both μPIV and PTV) at the intersection of the acrylic microchannel system with results from a simulation of the velocity field. The randomly located PTV data was interpolated onto the same gridpoints as the PIV data, with a Gaussian convolution scheme that allowed the redistribution of the unstructured PTV data onto a regular grid structure [68]. The simulation was calculated using CFD Research Corporation's (Huntsville, Alabama) CFD-ACE+ code. The model geometry for the numerical predictions replicates the intersection region of the D-shaped cross section of the channels. The boundary conditions were that the inlet and the outlet were held at fixed pressures and zero net flow was allowed to pass through the side channels. Both the μPIV and PTV measurements compare very well with the predicted velocity distributions as shown in Figure 4.38. This agreement shows that a volume illumination may be used in conjunction with the depth of field of the objective lens to resolve 2-D velocity measurements in at least weakly 3-D velocity fields using μPIV. The agreement of the flow measurements with theory

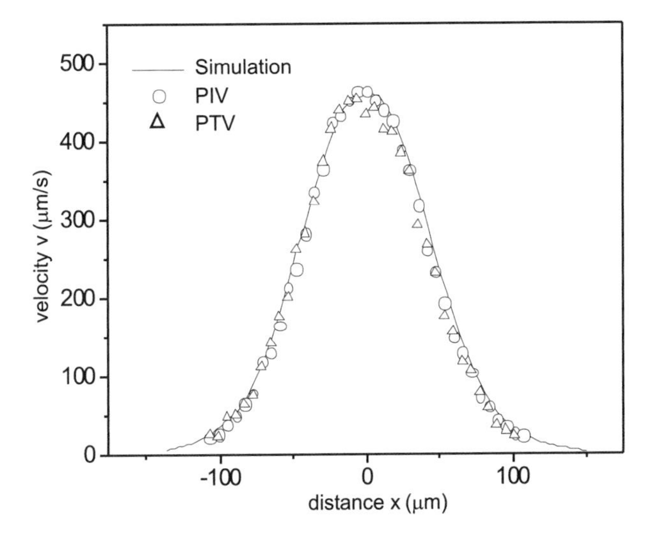


**Figure 4.38** Comparison of experimental data with numerical predictions for flow at the intersection—streamwise velocity, $v$, against streamwise distance, $y$. (*After*: [74]. Courtesy of Devasenathipathy et al.)

support the expected result that the liquid microflow at these length scales is well described by the macroscale flow theory.

## Problems

**4.1** Compute the particle response times for polystyrene latex seed particles ($\rho_p$ = 1,050 kg/m$^3$) suspended in water ($\eta = 10^{-3}$ kg/ms). Consider particle diameters of 1 μm, 300 nm, and 100 nm.

**4.2** Compute the diffusion coefficient $D$ for a 300-nm polystyrene latex seed particle in water at room temperature. Given a flow speed of 10 mm/s and 10 particles in an interrogation region, what $\Delta t$ is required to achieve 10% relative error? 1% relative error?

**4.3** Given a μPIV imaging system consisting of a 40× magnification, air-immersion lens, 0.7 NA microscope objective, and 600-nm illumination, what is the depth of correlation for particles of diameters 1 μm, 300 nm, and 100 nm?

**4.4** What particle concentration will give a signal-to-noise ratio of 1.5 in a 50-μm-deep microchannel (Table 4.6)? What are the characteristics of the imaging system necessary to achieve this signal-to-noise ratio? What happens to the signal-to-noise ratio if this particle concentration is used in a channel that is 150 μm deep?

**4.5** What are several methods that can be used to compensate for having too few particles in a μPIV velocity measurement? What are their respective advantages and disadvantages?

**4.6** A flow of water carries 500-nm seed particles that are imaged using a 50× magnification lens and a $\Delta t$ of 100 ms. The μPIV autocorrelation peak is found to measure 5 pixels in diameter and the cross-correlation peak is found to measure 7 pixels in diameter. What is temperature-to-viscosity ratio ($T/\eta$) for this flow? How would you determine the temperature from this information?

**4.7** Given a 500-nm particle being imaged with 500-nm light, by what fraction is the elastic scattering intensity $I$ increased or decreased when the wavelength of the illumination is doubled? How is $I$ changed when the particle diameter is reduced to 100 nm?

## References

[1] Bayt, R. L., and Breuer, K. S., "Fabrication and Testing of Micron-Sized Cold-Gas Thrusters in Micropropulsion of Small Spacecraft," *Advances in Aeronautics and Astronautics*, Vol. 187, pp. 381–398, Micci, M. and Ketsdever, A., Eds., Washington, D.C., AIAA Press, 2001.

[2] Kamisuki S., et al., "A Low Power, Small, Electrostatically-Driven Commercial Inkjet Head," *Proceedings of MEMS'98, 11th IEEE International Workshop Micro Electromechanical System*, Heidelberg, Germany, Jan. 25–29, 1998, pp. 63–68.

[3] Northrup, M. A., et al., "A MEMS-Based DNA Analysis System," *Proceedings of Transducers '95, 8th International Conference on Solid-State Sensors and Actuators*, Stockholm, Sweden, June 16-19, 1995, pp. 764–767.

[4] Lanzillotto, A. M., et al., "Applications of X-Ray Micro-Imaging, Visualization and Motion Analysis Techniques to Fluidic Microsystems," *Technical Digest of the IEEE Solid State Sensor and Actuator Workshop*, Hilton Head Island, SC, June 3–6, 1996, pp. 123–126.

[5] Chen, Z., et al., "Optical Doppler Tomographic Imaging of Fluid Flow Velocity in Highly Scattering Media," *Optics Letters*, Vol. 22, 1997, pp. 64–66.

[6] Compton, D. A., and Eaton, J. K., "A High-Resolution Laser Doppler Anemometer for Three-Dimensional Turbulent Boundary Layers," *Exp. Fluids*, Vol. 22, 1996, pp. 111–117.

[7] Tieu, A. K., Mackenzie, M. R., and Li, E. B., "Measurements in Microscopic Flow with a Solid-State LDA," *Exp. Fluids*, Vol. 19, 1995, pp. 293–294.

[8] Gharib M, et al., "Optical Microsensors for Fluid Flow Diagnostics," *40th AIAA Aerospace Sciences Meeting and Exhibit*, Reno, NV, AIAA Paper 2002–0252, Jan. 14–17, 2002.

[9] Dahm, W. J. A., Su, L. K., and Southerland, K. B., "A Scalar Imaging Velocimetry Technique for Fully Resolved Four-Dimensional Vector Velocity Field Measurements in Turbulent Flows," *Physics of Fluids A (Fluid Dynamics)*, Vol. 4, No. 10, 1992, pp. 2191–2206.

[10] Paul, P. H., Garguilo, M. G., and Rakestraw, D. J., "Imaging of Pressure- and Electrokinetically Driven Flows Through Open Capillaries," *Anal. Chem.*, Vol. 70, 1998, pp. 2459–2467.

[11] Koochesfahani, M. M., et al., "Molecular Tagging Diagnostics for the Study of Kinematics and Mixing in Liquid Phase Flows," *Developments in Laser Techniques in Fluid Mechanics*, R. J. Adrian et al. (eds.), New York: Springer-Verlag, 1997, pp. 125–134.

[12] Maynes, D., and Webb, A. R., "Velocity Profile Characterization in Sub-Millimeter Diameter Tubes Using Molecular Tagging Velocimetry," *Exp. Fluids*, Vol. 32, 2002, pp. 3–15.

[13] Lempert, W. R., et al., "Molecular Tagging Velocimetry Measurements in Supersonic Micro Nozzles," *Proceedings of 39th AIAA Aerospace Sciences Meeting and Exhibit*, Reno, NV, Jan. 8–11, 2001, pp. 2001–2044.

[14] Wildes, R. P., et al., "Physically Based Fluid Flow Recovery from Image Sequences," *Proceedings of IEEE Computer Society Conference on Computer Vision and Pattern Recognition*, Los Alamitos, CA, June 17–19, 1997, pp. 969–975.

[15] Leu, T. S., et al., "Analysis of Fluidic and Mechanical Motions in MEMS by Using High Speed X-Ray Micro-Imaging Techniques," *Proceedings of Transducers '97, 9th International Conference on Solid-State Sensors and Actuators*, Chicago, IL, June 16–19, 1997, pp. 149–150.

[16] Hitt, D. L., et al., "A New Method for Blood Velocimetry in the Microcirculation," *Microcirculation*, Vol. 3, No. 3, 1996, pp. 259–263.

[17] Hitt, D. L., Lowe M. L., and Newcomer, R., "Application of Optical Flow Techniques to Flow Velocimetry," *Phys. Fluids*, Vol. 7, No. 1, 1995, pp. 6–8.

[18] Adrian, R. J., *Bibliography of Particle Image Velocimetry Using Imaging Methods: 1917-1995*, Urbana, IL: University of Illinois at Urbana-Champaign, 1996.

[19] Raffel, M., Willert, C., and Kompenhans, J., *Particle Image Velocimetry: A Practical Guide*, New York: Springer, 1998.

[20] Santiago, J. G., et al., "A Particle Image Velocimetry System for Microfluidics," *Exp. Fluids*, Vol. 25, 1998, pp. 316–319.

[21] Koutsiaris, A. G., Mathioulakis, D. S., and Tsangaris, S., "Microscope PIV for Velocity-Field Measurement of Particle Suspensions Flowing Inside Glass Capillaries," *Meas. Sci. Technol.*, Vol. 10, 1999, pp. 1037–1046.

[22] Meinhart, C. D., Wereley, S. T., and Santiago, J. G., "PIV Measurements of a Microchannel Flow," *Exp. Fluids*, Vol. 27, 1999, pp. 414–419.

[23] Meinhart, C. D., and Zhang, H., "The Flow Structure Inside a Microfabricated Inkjet Printer Head," *J. Microelectromechanical Systems*, Vol. 9, 2000, pp. 67–75.

[24] Meinhart, C. D., Wereley, S. T., and Santiago, J. G., "Micron-Resolution Velocimetry Techniques," *Laser Techniques Applied to Fluid Mechanics*, R. J. Adrian et al. (eds.), New York: Springer-Verlag, 2000, pp. 57–70.

[25] Born, M., and Wolf, E., *Principles of Optics*, New York: Pergamon Press, 1997.

[26] Kim, Y. H., Wereley, S. T., and Chun, C.H., "Phase-Resolved Flow Field Produced by a Vibrating Cantilever Plate Between Two Endplates," *Phys. Fluids*, Vol. 16, 2004, pp. 145–162.

[27] Adrian, R. J., "Particle-Imaging Techniques for Experimental Fluid Mechanics," *Annual Review of Fluid Mechanics*, Vol. 23, 1991, pp. 261–304.

[28] Meinhart, C. D., and Wereley, S. T., "Theory of Diffraction-Limited Resolution in Micro Particle Image Velocimetry," *Meas. Sci. Technol.*, Vol. 14, 2003, pp. 1047–1053.

[29] Olsen, M. G., and Adrian, R. J., "Out-of-Focus Effects on Particle Image Visibility and Correlation in Particle Image Velocimetry," *Exp. Fluids [Suppl.]*, 2000, pp. 166–174.

[30] Wereley, S. T. and Meinhart, C. D., "Micron-Resolution Particle Image Velocimetry" in *Micro- and Nano-Scale Diagnostic Techniques*, K. S. Breuer (ed.), Springer-Verlag, New York, 2005.

[31] Adrian, R. J., and Yao, C. S., "Pulsed Laser Technique Application to Liquid and Gaseous Flows and the Scattering Power of Seed Materials," *Applied Optics*, Vol. 24, 1985, pp. 44–52.

[32] Prasad, A. K., Adrian, R. J., Landreth, C. C., and Offutt, P. W., "Effect of Resolution on the Speed and Accuracy of Particle Image Velocimetry Interrogation," *Exp. Fluids*, Vol. 13, 1992, pp. 105–116.

[33] Cummings, E. B., "An Image Processing and Optimal Nonlinear Filtering Technique for PIV of Microflows," *Exp. Fluids*, Vol. 29 [Suppl.], 2001, pp. 42–50.

[34] Meinhart, C. D., Wereley S. T., and Gray, M. H. B., "Depth Effects in Volume Illuminated Particle Image Velocimetry," *Meas. Sci. Technol.*, Vol. 11, 2000, pp. 809–814.

[35] Inoué, S., and Spring, K. R., *Video Microscopy*, 2nd ed., New York: Plenum Press, 1997.

[36] Bourdon, C. J., Olsen, M. G. and Gorby, A. D., "Validation of an Analytical Solution for Depth of Correlation in Microscopic Particle Image Velocimetry," *Meas. Sci. Technol.*, Vol. 15, 2004, pp. 318-327.

[37] Beskok, A., Karniadakis, G., E., and Trimmer, W., "Rarefaction and Compressibility," *Journal of Fluids Engineering*, Vol. 118, 1996, pp. 448–456.

[38] Melling, A., *Seeding Gas Flows for Laser Anemometry*, Ft. Belvoir Defense Technical Information Center, 1986.

[39] Probstein, R. F., *Physicochemical Hydrodynamics: An Introduction*, New York: Wiley, 1994.

[40] Van Kampen, N. G., *Stochastic Processes in Physics and Chemistry*, Amsterdam: Elsevier, 1997.

[41] Einstein, A., "On the Movement of Small Particles Suspended in a Stationary Liquid Demanded by the Molecular-Kinetic Theory of Heat," *Theory of Brownian Movement*, New York: Dover, 1905, pp. 1–18.

[42] Bendat, J. S., and Piersol, J. G., *Random Data: Analysis and Measurement Procedures*, New York: Wiley, 1986.

[43] Wereley, S. T., Gui, L., and Meinhart, C. D., "Advanced Algorithms for Microscale Particle Image Velocimetry," *AIAA J.*, Vol. 40, 2002, pp. 1047–1055.

[44] Gui L., Merzkirch, W., and Shu, J. Z., "Evaluation of Low Image Density PIV Recordings with the MQD Method and Application to the Flow in a Liquid Bridge," *J. Flow Vis. And Image Proc.*, Vol. 4, No. 4, 1997, pp. 333–343.

[45] Meinhart, C. D., Wereley, S. T., and Santiago, J. G., "A PIV Algorithm for Estimating Time-Averaged Velocity Fields," *Journal of Fluids Engineering*, Vol. 122, 2000, pp. 285–289.

[46] Gomez, R., et al., "Microfluidic Biochip for Impedance Spectroscopy of Biological Species," *Biomedical Microdevices*, Vol. 3, No. 3, 2001, pp. 201–209.

[47] Cowen, E. A., and Monismith, S. G., "A Hybrid Digital Particle Tracking Velocimetry Technique," *Exp. Fluids*, Vol. 22, 1997, pp. 199–211.

[48] Wereley, S. T., et al., "Velocimetry for MEMS Applications," *Proc. of ASME/DSC Vol. 66, Micro-Fluidics Symposium*, Anaheim, CA, Nov. 1998, pp. 453–459.

[49] Wereley S. T., and Meinhart, C. D., "Adaptive Second-Order Accurate Particle Image Velocimetry," *Exp. Fluids*, Vol. 31, 2001, pp. 258–268.

[50] Wereley, S. T., and Gui, L. C., "PIV Measurement in a Four-Roll-Mill Flow with a Central Difference Image Correction (CDIC) Method," *Proceedings of 4th International Symposium on Particle Image Velocimetry*, Gttingen, Germany, paper number 1027, Sept. 2001.

[51] Meinhart, C. D., Gray, M. H. B., and Wereley, S. T., "PIV Measurements of High-Speed Flows in Silicon-Micromachined Nozzles," *Proceedings of 35th AIAA/ASME/SAE/ASEE Joint Propulsion Conference and Exhibit*, Los Angeles, CA, AIAA-99-3756, June 20–24, 1999.

[52] Deen, W., *Analysis of Transport Phenomena*, New York: Oxford University Press, 1998.

[53] Batchelor, G. K., *An Introduction to Fluid Dynamics*, Cambridge, England: Cambridge University Press, 1987.

[54] Wereley S. T., Gui, L., and Meinhart, C. D., "Flow Measurement Techniques for the Microfrontier," 39th AIAA, *Aerospace Sciences Meeting and Exhibit*, Reno, NV, Jan. 8–11, 2001, AIAA Paper 2001-0243.

[55] Wiesendanger, R., *Scanning Probe Microscopy and Spectroscopy: Methods and Applications*, Cambridge, England: Cambridge University Press, 1994.

[56] Revenco, I., and Proksch, R., "Magnetic and Acoustic Tapping Mode Microscopy of Liquid Phase Phospholipid Bilayers and DNA Molecules," *J. App. Phys.*, Vol. 87, 2000, pp. 526–533.

[57] Stone, S. W., Meinhart, C. D., and Wereley, S. T., "A Microfluidic-Based Nanoscope," *Exp. Fluids*, Vol. 33, No. 5, 2002, 613–619.

[58] Kataoka, K., et al., "Totally Synthetic Polymer Gels Responding to External Glucose Concentration: Their Preparation and Application to On-Off Regulation of Insulin Release," *J. Am. Chem. Soc.*, Vol. 120, 1998, pp. 12694–12695.

[59] Osada, Y., and Ross-Murphy, S., "Intelligent Gels," *Scientific American*, Vol. 268, 1993, pp. 82–87.

[60] Liu, R., et al., "In-Channel Processing to Create Autonomous Hydrogel Microvalves," *Micro Total Analysis Systems 2000*, A. van den Berg et al. (eds.), Boston, MA: Kluwer Academic Publishers, 2000, pp. 45–48.

[61] Madou, M., He, K., and Shenderova, A. "Fabrication of Artificial Muscle Based Valves for Controlled Drug Delivery," *Micro Total Analysis Systems 2000*, A. van den Berg et al. (eds.), Boston, MA: Kluwer Academic Publishers, 2000, pp. 147–150.

[62] Olsen, M. G., Bauer, J. M., and Beebe, D. J., "Particle Imaging Technique for Measuring the Deformation Rate of Hydrogel Microstructures," *Applied Physics Letters*, Vol. 76, 2000, pp. 3310–3312.

[63] Beebe, D. J., et al., "Functional Hydrogel Structures for Autonomous Flow Control Inside Microfluidic Channels," *Nature*, Vol. 404, 2000, pp. 588–590.

[64] Fujisawa, N., and Adrian, R. J., "Three-Dimensional Temperature Measurement in Turbulent Thermal Convection by Extended Range Scanning Liquid Crystal Thermometry," *Journal of Visualization*, Vol. 1, No. 4, 1999, pp. 355–364.

[65] Sakakibara, J., and Adrian, R. J., "Whole Field Measurement of Temperature in Water Using Two-Color Laser Induced Fluorescence," *Experiments in Fluids*, Vol. 26, No. 1–2, 1999, pp. 7–15.

[66] Olsen, M. G., and Adrian, R. J., "Brownian Motion and Correlation in Particle Image Velocimetry," *Optics and Laser Tech.*, Vol. 32, 2000, pp. 621–627.

[67] Keane, R. D., and Adrian, R. J., "Theory of Cross-Correlation Analysis of PIV Images," *Applied Scientific Research*, Vol. 49, 1992, pp. 1–27.

[68] Keane, R. D., Adrian, R. J., and Zhang. Y., "Super-Resolution Particle Imaging Velocimetry," *Measurement Science Technology*, Vol. 6, 1995, pp. 754–768.

[69] Chandrasekhar, S., "Stochastic Problems in Physics and Astronomy," *Review of Modern Physics*, Vol. 15, 1941, pp. 1–89.

[70] Edwards, R., et al., "Spectral Analysis from the Laser Doppler Flowmeter: Time-Independent Systems," *Journal of Applied Physics*, Vol. 42, 1971, pp. 837–850.

[71] Han, G. and Breuer, K. S., "Infrared PIV for Measurement of Fluid and Solid Motion Inside Opaque Silicon Microdevices," *Proceedings of 4th International Symposium on Particle Image Velocimetry*, Gttingen, Germany, paper number 1146, Sept. 2001.

[72] Guezennec, Y. G., et al., "Algorithms for Fully Automated 3-Dimensional Particle Tracking Velocimetry," *Exp. Fluids*, Vol. 17, 1994, pp. 209–219.

[73] Ohmi, K., and Li, H. Y., "Particle-Tracking Velocimetry with New Algorithm," *Measurement Science Technology*, Vol. 11, 2000, pp. 603–616.

[74] Devasenathipathy, S., et al., "Particle Tracking Techniques for Microfabricated Fluidic Systems," *Exp. Fluids*, Vol. 34, 2003, pp. 504–514.

[75] Takehara, K., et al., "A Kalman Tracker for Super-Resolution PIV," *Exp. Fluids*, Vol. 29, 2000, pp. 34–41.

# Chapter 5

# Microfluidics for External Flow Control

## 5.1 VELOCITY AND TURBULENCE MEASUREMENT

In a turbulent regime, tiny flow structures are responsible for viscous drag increase of aerodynamic surfaces, such as wings and engine inlet ducts in an aircraft. These structures are called eddies. Eddies are typically several hundred microns in width and several millimeters in length. The length scale of eddies is broad, ranging from the same size as the width of the flow field to the microscale, where viscous effects become dominant and energy is transferred from kinetic into internal [1].

Besides the small sizes, the short lifetime of eddies also pose a challenge for measurement and control. The lifetime of eddies is on the order of milliseconds [2]. Conventional instruments with low response time are not able to detect such a fast dynamic.

Because of their small size, micromachined devices are suitable for measuring and manipulating eddies in a turbulent flow. In addition to the spatial and temporal advantages, micromachined sensors and actuators can be integrated with microelectronics, which make them "smart." The small size and the batch machining process allow fabricating arrays of such devices, which can give more detailed information about the flow field.

The following sections discuss the design of microsensors for velocity and turbulence measurement. In contrast to the flow sensors in Chapter 8, velocity measurement in this section is referred to as point velocity measurement. The other type of turbulence measurement is shear stress sensing on walls. Although flow sensors, velocity sensors, and shear stress sensors serve different purposes, they can share the same operation concept and design. For example, hot-film and hot-wire sensors can be used for measuring both flow velocity and shear stress. In order to avoid redundancy, Section 5.1.1 focuses on design considerations with illustrating calculation examples, while Section 5.1.2 and Chapter 8 deal with design examples of thermal shear stress sensors and thermal flow sensors, respectively.

### 5.1.1 Velocity Sensors

#### 5.1.1.1 Design Considerations

The two common sensor types of classical point velocity measurement are Pitot-probe anemometer and thermal anemometer. The Pitot probe utilizes the dependence of the static pressure on the velocity:

$$\Delta p = \frac{\rho u^2}{2} \tag{5.1}$$

where $\rho$ is the fluid density, $u$ is the flow velocity, and $\Delta p$ is the pressure drop across the Pitot tube as shown in Figure 5.1(a). In microscale, the design of a Pitot probe becomes the design of micropressure sensors. Examples for designing pressure sensors are discussed in Section 8.1.

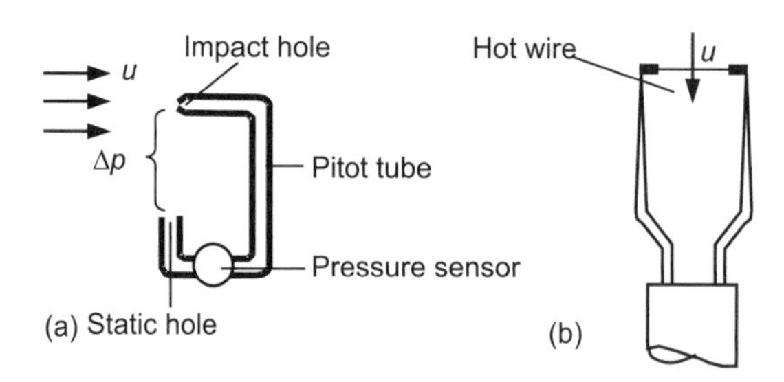


**Figure 5.1** Velocity sensors: (a) Pitot-probe anemometer; and (b) thermal anemometer (hot wire).

The thermal anemometer measures flow velocities based on the correlation between velocity and heat transfer. Figure 5.1(b) depicts a conventional hot-wire anemometer. The sensor consists of a heater, which is a thin wire made of platinum or tungsten. The typical diameter of the wire ranges from 5 to 25 μm. The wire is 1 to 3 mm long. The resistance of the wire is typically on the order of several tens of ohms. Using a temperature sensor as the reference, the wire can be controlled at a constant wire temperature or at a constant heating power. The corresponding electric powers (in the case of constant temperature) or the wire temperatures (in the case of constant heating power) are related directly to the flow velocities. In terms of spatial and temporal resolutions, thermal anemometry has more advantages. A micro hot wire fabricated in silicon can resolve few microns in space and few microseconds in time. Electrothermal behavior of a hot wire is analyzed in the following examples.

**Example 5.1: Characteristics of a Hot-Wire Sensor**

A polycrystalline silicon wire with a length of $L = 50$ μm and a cross section of $A = 2\times\ 5$ μm is used as a hot-wire sensor. The temperature coefficient and thermal conductivity of polysilicon are $8 \times 10^{-4}$ $\mathrm{K}^{-1}$ and 32 $\mathrm{Wm}^{-1}\mathrm{K}^{-1}$, respectively. The bulk substrate temperature is 25°C. The resistivity of polysilicon at 25°C is $4 \times 10^{-5}$ Wm. Determine the maximum temperature of the wire and its resistance at a current of 10 mA.

**Solution.** With the temperature $T$ as a function of $x$, $\kappa$ the thermal conductivity of the wire, and $\rho$ its resistivity, the energy balance between the flux and heat generated by the current is described as:

$$\begin{aligned}
-\kappa A[T'(x+\Delta x) - T'(x)] &= I^2R = I^2\tfrac{\rho\Delta x}{A} \\
-\kappa A\tfrac{[T'(x+\Delta x)-T'(x)]}{\Delta x} &= I^2\tfrac{\rho}{A} \\
-\kappa AT'' &= I^2\tfrac{\rho}{A} \\
-\kappa AT'' &= \tfrac{I^2}{A^2}\rho \\
-\kappa AT'' &= J^2\rho
\end{aligned}$$

where $J = I/A$ is the current density in the wire. We assume that resistivity is a linear function of temperature with a temperature coefficient of $\alpha$:

$$\rho = \rho_0[1 + \alpha(T - T_0)]$$

Thus, the final differential equation describing the heat balance has the form:

$$\kappa T'' = J^2\rho_0[1 + \alpha(T - T_0)]$$

Solving the above equation with boundary condition of $T(0) = T(L) = T_\mathrm{b}$, one gets [3]:

$$T(x) = \left(T_0 - \frac{1}{\alpha}\right) + \left(T_\mathrm{b} - T_0 + \frac{1}{\alpha}\right)\frac{\cos\left[\beta\left(\frac{2x}{L} - 1\right)\right]}{\cos\beta}$$

with:

$$\beta = \frac{JL}{2}\sqrt{\frac{\alpha\rho_0}{\kappa}} = \frac{10^{-2}\times 5\times 10^{-5}}{2\times 2\times 5\times 10^{-12}}\sqrt{\frac{8\times 10^{-4}\times 4\times 10^{-5}}{31}} = 0.79\ \mathrm{rad}$$

The maximum temperature at the wire center is:

$$T_\mathrm{max} = T(L/2) = \left(T_0 - \frac{1}{\alpha}\right) + \frac{\left(T_\mathrm{b} - T_0 + \frac{1}{\alpha}\right)}{\cos\beta}$$

$$T_\mathrm{max} = \left(25 - \frac{1}{8\times 10^{-4}}\right) + \frac{\left(25 - 25 + \frac{1}{8\times 10^{-4}}\right)}{\cos 0.79}$$

The resistance of the wire is:

$$R = \int_0^L \frac{\rho(x)}{A}\mathrm{d}x = \frac{\rho_0 L}{A}[1 + \alpha(T_\mathrm{b} - T_0)]\frac{\tan\beta}{\beta} + R_0$$

Assuming that $R_0$ is zero, the resistance of the wire at 10 mA is:

$$R = \frac{4\times 10^{-5}\times 5\times 10^{-5}}{2\times 5\times 10^{-12}}[1 + 8\times 10^{-4}\times(552 - 25)]\frac{\tan 0.79}{0.79}$$

**Example 5.2: Characteristics of a Hot-Wire Sensor, Simplified Approach**

For the hot wire of Example 5.1, determine the maximum temperature, and its resistance, by assuming constant resistivity. Compare the results with Example 5.1 and estimate the thermal time constant of the wire. The thermal diffusivity of polysilicon is $1.85\times 10^{-5}$ m$^2$/s.

**Solution.** We start with the governing differential equation:

$$-\kappa T'' = J^2\rho$$

Solving the above equation by assuming a quadratic trial function and constant resistivity, we get [4]:

$$T(x) = T_\mathrm{b} + T_\mathrm{max}\left[1 - \left(\frac{2x}{L} - 1\right)^2\right]$$

where the maximum temperature at the center of the wire is:

$$T_\mathrm{max} = \frac{I^2\rho_0 L^2}{8A^2\kappa} = \frac{(10^{-2})^2\times 4\times 10^{-5}\times(5\times 10^{-5})^2}{8\times(2\times 5\times 10^{-12})^2\times 32} = 391^\circ\mathrm{C}$$

Using the quadratic temperature profile above, the resistance at the given condition ($T_\mathrm{b} = T_0 = 25^\circ\mathrm{C}$) can be estimated as:

$$R = \frac{1}{A}\int_0^L \rho_0[1 + \alpha(T(x) - T_0)] = \frac{\rho_0 L}{A}\left[1 + \frac{2}{3}\alpha(T_\mathrm{max} - T_\mathrm{b})\right]$$

$$R = \frac{4\times 10^{-5}\times 5\times 10^{-5}}{2\times 5\times 10^{-12}}\left[1 + \frac{2}{3}\times 8\times 10^{-4}(391 - 25)\right] = 239\ \Omega$$

Compared to the exact solution in Example 5.1, the simplified model results in an error of 29% for the maximum temperature and an error of 34% for the resistance. For further simplification, the above parabolic profile can be developed to a Fourier series. Using the first eigenmode and an exponential cooling function, we have a transient trial function for the temperature profile:

$$T(x,t) = T_{\mathrm{b}} + (T_{\mathrm{max}} - T_{\mathrm{b}}) \cos\left(\frac{\pi(x - L/2)}{L}\right) \exp\left(\frac{-t}{\tau_{\mathrm{th}}}\right)$$

where $\tau_{\mathrm{th}}$ is the thermal time constant of the wire. The partial differential equation describing the cooling (heating) process is [4]:

$$\frac{\partial^2 T}{\partial x^2} = \frac{c\rho}{\kappa}\frac{\partial T}{\partial t}$$

where $c$ is the specific heat of polysilicon. Using the above two equations, we get the estimated time constant of the wire as a function of the length $L$ and thermal diffusivity $a = \kappa/c\rho$ :

$$\tau_{\mathrm{th}} = \frac{c\rho L^2}{\pi^2 \kappa} = \frac{L^2}{\pi^2 a} = \frac{(5 \times 10^{-5})^2}{\pi^2 \times 1.85 \times 10^{-5}}\mathrm{s} = 13.7\ \mu\mathrm{s}$$

**Example 5.3: Characteristics of a Hot-Film Sensor**

A hot-film sensor is made of a platinum heater on a thin silicon nitride membrane. Determine the time function of the temperature distribution.

**Solution.** The partial differential equation describing the cooling (heating) process is (see Section 2.2.2.2):

$$\frac{\partial^2 T}{\partial x^2} + \frac{\partial^2 T}{\partial y^2} = \frac{c\rho}{\kappa}\frac{\partial T}{\partial t}$$

Solving the above equation, we get the simplified form of a temperature distribution of the hot-film plate:

$$T(x,t) = T_{\mathrm{max}} \cos\left(\frac{\pi x}{w}\right) \cos\left(\frac{\pi y}{L}\right) \exp\left(\frac{-t}{\tau}\right)$$

where $T_{\mathrm{max}}$ is the maximum temperature at the center of the membrane and $-w/2 < x < w/2$; $-L/2 < y < L/2$.

#### 5.1.1.2 Design Examples

The micromachined hot-wire anemometer reported in [5] has a free-standing polysilicon heater. The heater wire is 0.5 μm thick, 1 μm wide, and 10 to 160 μm long. Heating time and cooling time of the sensors are a few microseconds. With a 10-μm wire, the wire reaches its fastest time constant of 0.5 μs, which corresponds to a temporal bandwidth of 1.4 MHz. More examples are discussed in Section 5.1.2.2 and in Chapter 8.

### 5.1.2 Shear Stress Sensors

One of the most important requirements for external flow control is the measurement of wall shear stress. Phenomena such as viscous drag, transition from laminar to turbulence, flow separation, and turbulent eddies can be detected with a shear stress sensor. Micromachined shear stress sensors offer

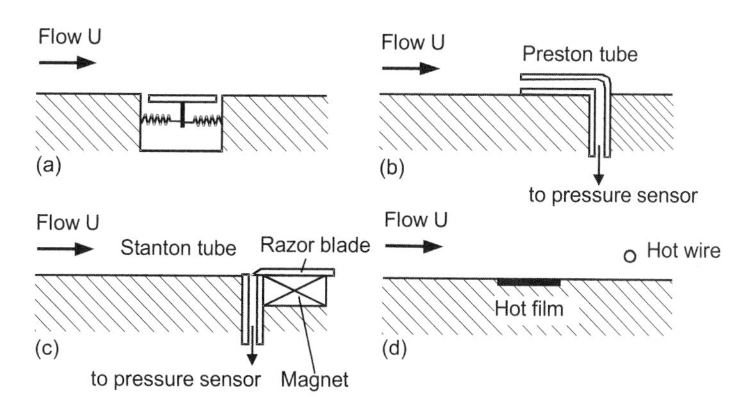


**Figure 5.2** Operation principles of shear stress sensor: (a) direct type with flush-mounted floating plate; (b) indirect type with Preston tube for measuring the stagnation pressure; (c) indirect type with Stanton tube; and (d) indirect type based on heat and mass transfer.

the advantages of miniaturization for higher resolution and improved dynamics. Shear stress sensors can be divided into two categories: direct and indirect types [6]. Figure 5.2 illustrates basic principles of shear stress sensors.

The direct types measure local shear stress using a floating plate, which is flush-mounted on the wall bounding the flow [Figure 5.2(a)]. The shear force displaces the floating plate. The displacement can then be measured capacitively, piezoresistively, or optically. The indirect types measure shear stress by means of its subsequent effects, such as stagnation pressure, heat transfer, or near-wall velocity. The stagnation pressure type utilizes an obstacle embedded in the boundary layer. The stagnation pressure caused by the obstacle increases with higher shear stress and can be measured with a pressure sensor. Figures 5.2(b, c) show two typical setups for this type: the Preston tube and the Stanton tube.

The heat-transfer type uses heat and mass transfer analogies to measure the correlation between measured properties and the actual shear stress [7]. The velocity type measures the near-wall velocity profile. The measurement requires multiple velocity sensors or optical techniques such as PIV (see Chapter 4). Because of the external instruments, the third indirect type is not suitable for an integrated solution.

#### 5.1.2.1 Design Considerations

Hot-film and hot-wire sensors are the simplest indirect shear-stress sensors, which can be implemented easily with micromachining technology. However, there are two drawbacks related to this concept.

The first drawback of thermal shear stress sensors is the dependence on empirical correlation, which requires extensive calibration measurements. The correlation function for the heating power $P$ of a hot-film sensor and the wall shear stress $\tau_\mathrm{w}$ is given as [8]:

$$P = \Delta T (C_0 + C_1 \tau_\mathrm{w}^{1/3}) \tag{5.2}$$

where $\Delta T$ is the temperature difference between the heater and the fluid, and $C_0$ and $C_1$ are calibration constants.

Since sensor signals depend strongly on the thermal behavior of the sensor, the second drawback of hot-film sensors is the error caused by heat conduction into the sensor substrate. Careful thermal insulation should be considered in the design of this sensor type.

**Example 5.4: Influence of Thermal Insulation on Thermal Shear Stress Sensor**

A thin-film heater with an overall dimension of $100 \times 100$ μm is deposited on a silicon substrate over an oxide layer. The sensor characteristics follow (5.2) with $C_0 = 10$ mW/K and $C_1 = 1$ mW/K-Pa$^{1/3}$. Estimate the thickness of silicon oxide. If the oxide layer is used as a sacrificial layer and is etched away, how much would the heat ratio improve? The thermal conductivity of silicon oxide and air is 1.1 W/m$^{-1}$K$^{-1}$ and 0.0257 W/m$^{-1}$K$^{-1}$, respectively.

**Solution.** Following (5.2), we estimate that the offset $\Delta T \times C_0$ is caused by heat conduction. In this analysis, we only consider the heat conduction to silicon substrate and assume that the substrate has the same temperature of the fluid. $C_0$ represents the thermal conductance of the silicon oxide layer:

$$C_0 = \frac{\kappa A}{d} \rightarrow d = \frac{\kappa A}{C_0} = \frac{1.1 \times (10^{-4})^2}{10^{-2}} = 1.1 \times 10^{-6}\text{m} = 1.1\ \mu\text{m}$$

If the gap is replaced by air, this factor would be:

$$C_{0,\text{air}} = \frac{\kappa_{\text{air}} A}{d} = \frac{0.0257 \times (10^{-4})^2}{1.1 \times 10^{-6}} = 0.234 \times 10^{-3}\ \text{W} = 0.234\ \text{mW}$$

Thus, the ratio $C_1/C_0$ improves:

$$\frac{C_1/C_{0,\text{air}}}{C_1/C_0} = \frac{C_0}{C_{0,\text{air}}} \approx 43 \text{ times}$$

Designing other indirect sensors, which are based on the measurement of stagnation pressure, leads to the development of pressure sensors. However, the realization of an obstacle and the pressure sensors make the design complex and impractical.

The most suitable technique for micromachining is direct sensing with a floating element flush-mounted in the wall. Refer to [6] for detailed considerations of floating element sensors. The design problems associated with this sensor type are discussed here.

The presence of a gap around the floating element can disturb the flow field and cause errors in the measurement. The error can be neglected if the gap is smaller than the characteristic length $l$:

$$l = \frac{\nu}{u_\tau} \tag{5.3}$$

where $\nu$ is kinematic viscosity of the fluid, and $u_\tau$ is the friction velocity:

$$u_\tau = \sqrt{\frac{\tau_\text{w}}{\rho}} \tag{5.4}$$

where $\rho$ is fluid density and the wall shear stress $\tau_\text{w}$:

$$\tau_\text{w} \approx 0.03 \frac{\rho u^2}{\text{Re}(x)^{0.2}} \tag{5.5}$$

The Reynolds number $\text{Re}(x)$ based on streamwise distance $x$ is defined as:

$$\text{Re}(x) = \frac{ux}{\nu} \tag{5.6}$$

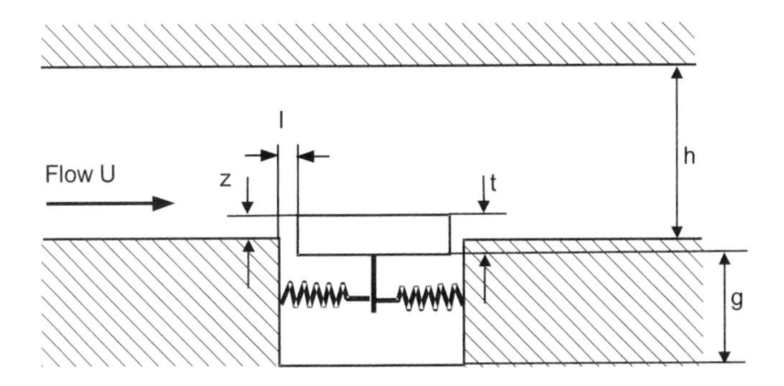


**Figure 5.3** Geometry parameters of a floating-element shear stress sensor.

where $u$ is the free-stream velocity. If the floating element is not perfectly flush-mounted on the wall, the resulting drag force can lead to the error $\Delta\tau_{\mathrm{w}}$ [6]:

$$\frac{\Delta_{\mathrm{w}}}{\tau_{\mathrm{w}}} = \frac{\rho z^3}{3(\eta^2 L_{\mathrm{e}}^2)} \tag{5.7}$$

where $z$ is the misalignment of the floating element, $L_{\mathrm{e}}$ is the streamwise length of the floating element, and $\eta$ is the dynamic viscosity.

The pressure gradient along the floating element can cause further error $\Delta\tau_{\mathrm{w}}$, which is determined by the geometry of the sensor [6]:

$$\frac{\Delta_{\mathrm{w}}}{\tau_{\mathrm{w}}} = \frac{g}{h} + \frac{2t}{h} \tag{5.8}$$

The geometry parameters are defined in Figure 5.3. Further error sources are the acceleration of the floating element, fluctuation of the normal pressure, and temperature dependence [6].

**Example 5.5: Typical Parameters of a Floating-Element Shear Stress Sensor**

A floating-element shear stress sensor is flush-mounted on a flat plate at a distance of 1m from the plate front. The free stream velocity is 10 m/s. Determine the Reynolds number at the measurement position and the shear stress expected at this position. What is the largest gap allowable around the floating element? Kinematic viscosity and density of air are assumed to be $1.6\times10^{-5}$ m$^2$/s and 1.164 kg/m$^3$, respectively.

**Solution.** According to (5.6), the Reynolds number at the measurement position is:

$$\mathrm{Re}(x) = \frac{ux}{\nu} = \frac{10 \times 1}{1.6 \times 10^{-5}} = 6.25 \times 10^5$$

According to (5.5), the shear stress expected for this position is:

$$\tau_{\mathrm{w}} \approx 0.03\frac{\rho u^2}{\mathrm{Re}(x)^{0.2}} = 0.03 \times \frac{1.164 \times 10^2}{(6.25 \times 10^5)^{0.2}} \approx 0.24 \text{ Pa}$$

For this shear stress and Reynolds number, the largest gap allowed for the floating element is:

$$l = \frac{\nu}{u_\tau} = \frac{\nu}{\sqrt{\tau_{\mathrm{w}}/\rho}} = 35 \times 10^{-6} \text{ m} = 35 \text{ µm}$$

**Example 5.6: Errors Related to Geometrical Parameters**

If the floating element of the above sensor is 5 mm long and the plate is mounted 20 μm too high into the flow, determine the measurement error caused by the drag force.

**Solution.** The dynamic viscosity of the air is:

$$\eta = \nu\rho = 1.6 \times 10^{-5} \times 1.164 = 1.8624 \times 10^{-5} \text{ Ns/m}^2$$

According to (5.7), the error due to drag force is:

$$\Delta\tau_{\mathrm{w}} = \tau_{\mathrm{w}}\frac{\rho z^3}{3\eta^2 L_{\mathrm{e}}^2} = 0.24\frac{1.164 \times (20 \times 10^{-6})^3}{3 \times (1.8624 \times 10^{-5})^2 \times (5 \times 10^{-3})^2} = 8.66 \times 10^{-2} \text{ Pa}$$

The drag error is about 36%, and it cannot be neglected.

**Example 5.7: Designing an Optical Floating-Element Shear Stress Sensor**

The above sensor is made of a silicon plate with a dimension of 1 × 1 mm. The plate is suspended on four flexures as shown in the figure below. The exposed width of each of the photodiodes is 10 μm. Design the flexures so that the velocity in Example 5.5 results in a change of 100% in the photocurrent.

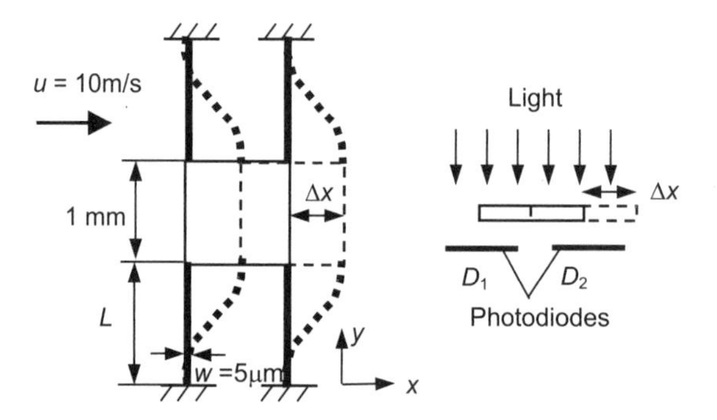


**Solution.** The shear force acted on the floating plate at 10 m/s is:

$$F_{\mathrm{p}} = \tau_{\mathrm{w}} A = 0.24 \times (10^{-3})^2 = 0.24 \times 10^{-6} \text{ N}$$

Assuming that the force is distributed equally among four flexures, each of them has a load of:

$$F_{\mathrm{f}} = \frac{F_{\mathrm{p}}}{4} = \frac{0.24 \times 10^{-6}}{4} = 0.06 \times 10^{-6} \text{ N}$$

In order to use the small deflection theory of beams, we choose a flexure width of $t = 5$ μm and a thickness of $w = 10$ μm. The displacement of a beam with a tip guide is (see Appendix D):

$$\Delta x = \frac{F_{\mathrm{f}} L^3}{Ewt^3}$$

The change of light-induced current is proportional to the displaced area:

$$\frac{\Delta I}{I} = \frac{2\Delta x}{w_{\mathrm{diode}}} = 100\% \rightarrow \Delta x = \frac{w_{\mathrm{diode}}}{2} = 5 \text{ μm}$$

Thus, the required length of the flexure is:

$$L = \sqrt[3]{\frac{E\Delta x w t^3}{F_{\mathrm{f}}}} = \sqrt[3]{\frac{170 \times 10^9 \times 5 \times 10 \times 5^3 \times (10^{-6})^5}{0.06 \times 10^{-6}}} = 26 \times 10^{-4}\ \mathrm{m} = 2.6\ \mathrm{mm}$$

**Example 5.8: Designing a Piezoresistive Floating-Element Shear Stress Sensor**

If piezoresistors are integrated on the sides of the four flexures in the above sensor, is the resistance change measurable?

**Solution.** The deflection of a beam with a guided tip can be described as (see Appendix D):

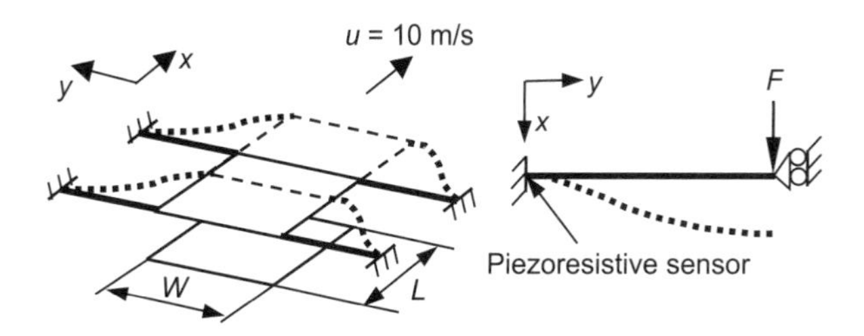


$$x(y) = \frac{F}{2EI}\left(\frac{-y^3}{3} + \frac{Ly^2}{2}\right)$$

The function of the radius of curvature can be described as:

$$\frac{1}{r(y)} = \frac{\mathrm{d}^2 x}{\mathrm{d}y^2} = \frac{F}{2EI}(-2y + L)$$

Thus, the strain along the beam is:

$$\varphi(y) = \frac{t}{2r(y)} = \frac{Ft}{4EI}(-2y + L)$$

The maximum strain at the beam end $x = 0$ is:

$$\varphi_{\max} = \varepsilon(0) = \frac{FtL}{4EI} = \frac{3FtL}{Ewt^3} = \frac{3FL}{Ewt^2}$$

Assuming a piezoresistive coefficient of 100, we get the change of a resistor integrated on the beam end:

$$\frac{\Delta R}{R} = 100\varphi_{\max} = 100 \times \frac{3FL}{Ewt^2} = 100 \times \frac{3 \times 0.06 \times 10^{-6} \times 2.6 \times 10^{-3}}{170 \times 10^9 \times 10^{-5} \times (5 \times 10^{-6})^2} = 0.0011 = 0.11\%$$

The change of 0.11% is small but measurable.

**Example 5.9: Designing a Capacitive Floating-Element Shear Stress Sensor**

If the floating plate is used as an electrode of a capacitor, what is the change of the capacitance?

**Solution.** The capacitance of a parallel plate capacitor of a dimension $W \times L$ and a gap $d$ is:

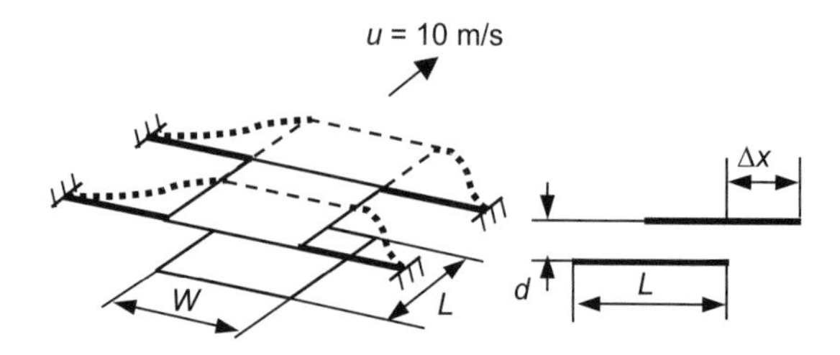


Thus, the capacitance is proportional to the plate length. The capacitance change is proportional to the displacement:

$$\frac{\Delta C}{C} = \frac{\Delta x}{L} = \frac{5 \times 10^{-6}}{10^{-3}} = 5 \times 10^{-3} = 0.5\%$$

#### 5.1.2.2 Design Examples

*Thermal Shear Stress Sensors.* A hot-film flow sensor can be used as a thermal shear-stress sensor [8]. The sensor has thermopiles as temperature sensors. Due to the large heat capacity of the substrate, the sensor has a very low bandwidth (see Example 5.2) [Figure 5.4(a)]. Using a hot wire with smaller geometry and better thermal isolation, higher bandwidth up to 500 Hz can be achieved [9] [Figure 5.4(b)]. Smaller polysilicon wires with widths on the order of 1 μm can increase bandwidth up to 500 kHz [3].

An array of micromachined hot wires can measure the distribution of wall stress over the sensor surface. Such a sensor array can resolve a spatial distribution of 300 μm and a temporal bandwidth of 30 kHz [10]. Several design variations are shown in Figures 5.4(c–f).

The design illustrated in Figure 5.4(c) has a 2-μm deep vacuum cavity and a 0.25-μm-thick polysilicon wire embedded in the nitride membrane. The design of Figure 5.4(d) is similar to that shown in Figure 5.4(c)—the polysilicon wire stands up 4 μm above the nitride membrane. Consequently, better thermal isolation can be achieved. The design shown in Figure 5.4(e) is similar to that of Figure 5.4(d), but there is no vacuum cavity. The design shown in Figure 5.4(f) is a conventional hot-wire design with a freestanding polysilicon bridge.

The sensor presented in [11] uses diodes as temperature sensors and is able to measure up to 40 kHz. The sensor is fabricated with SOI technology. The bottom wafer is etched to the silicon oxide layer to form an isolation membrane. An isolation trench is etched into the top wafer, and the trench is then filled with polyimide. The heater is a polysilicon resistor located inside the isolation trench. An integrated diode detects the temperature [Figure 5.4(g)].

Since most of the airfoil objects have a curved form, there is a need of a flexible sensor or sensor arrays, which can attach smoothly to curved surfaces. Developments at the University of California at Los Angeles and the California Institute of Technology led to a new technology called *flexible skin*. These flexible sensor arrays are silicon islands, which are connected by a flexible polyimide membrane [Figure 5.4(h)]. The array consists of 32 thermal shear stress sensors, which are able to wrap around a 0.5-inch edge radius. This array can detect the separation line on the rounded leading edge of a delta wing. The information is crucial for employing microactuators to control a delta-wing aircraft [12].

*Direct Shear Stress Sensor.* The direct shear stress sensor in [13] has a plate suspended by four beams as the floating element. The displacement of the floating element is detected by means of differential capacitive principle. The sensor can measure shear stress in low-speed turbulent boundary layers with a bandwidth of 10 kHz. The sensor is fabricated with surface micromachining technology,

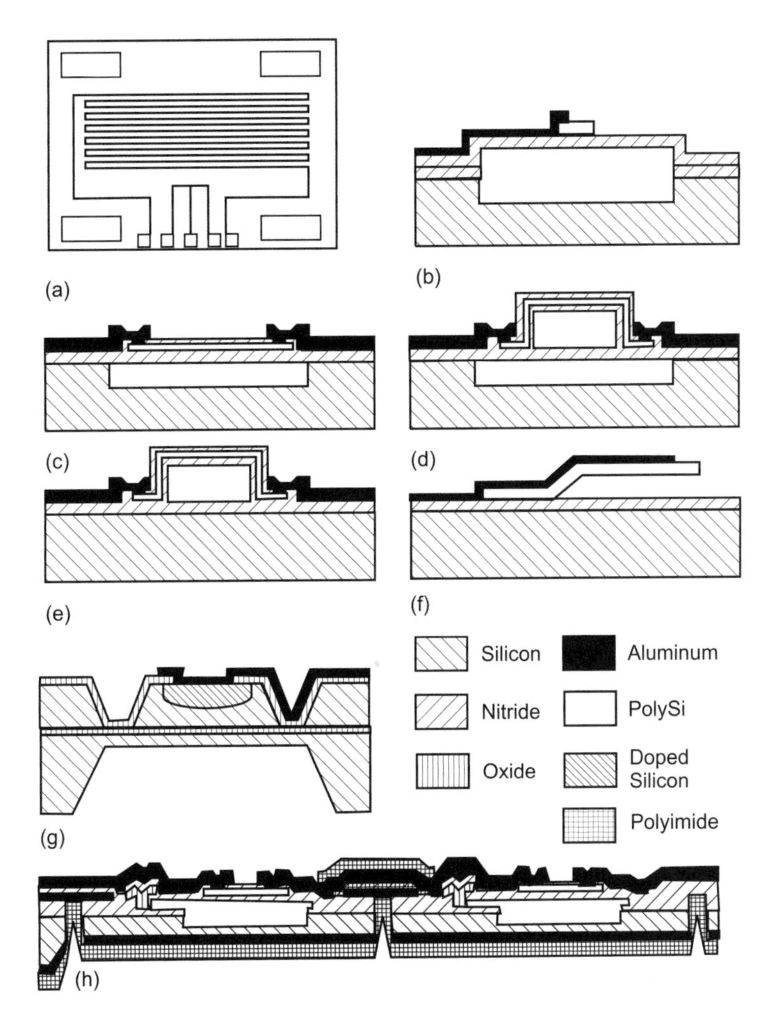


**Figure 5.4** Typical thermal shear stress sensors: (a) hot film with thermopiles; (b)-(f) hot wires; (g) hot film with diodes; and (h) hot wire with flexible skin.

which utilizes aluminum as a sacrificial layer. The floating element is made of polyimide. A 30-nm-thick evaporated chrome layer is used as the embedded electrode in the floating element. The counter electrode is made of polysilicon in a standard CMOS process. The low spring constant allows the sensor to measure small shear stress on the order of 1 Pa.

The sensors presented in [14, 15] have piezoresistors in the suspending flexures. The sensor in [14] is designed for a polymer extrusion flow, which has very high shear stress. The sensor is fabricated with the SOI techniques. Since the suspending flexures are parallel to the flow direction or parallel to the shear force, high shear stress on the order of 10 kPa can be measured. The arrangement causes tensile stress on the two upstream beams and compressive stress on the two downstream beams [Figure 5.5(b)]. The change in strain of each beam pair leads to change in resistance, which can be detected using a Wheatstone bridge.

The sensor presented in [16] is fabricated with surface micromachining. The floating plate is made of polysilicon. The arrangement in the measured flow is similar to the design of [14]. The displacement is detected with the comb-finger structure depicted in Figure 5.5(c). This design allows

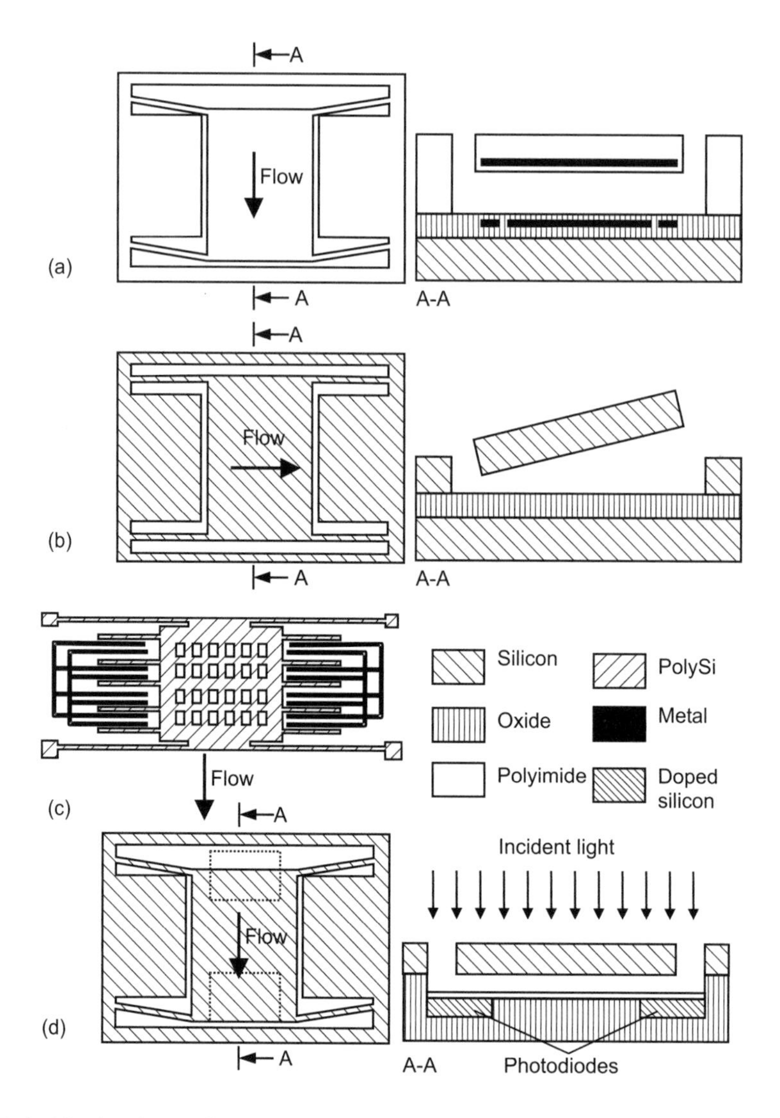


**Figure 5.5** Typical floating-element shear stress sensors: (a) capacitive; (b) piezoresistive; (c) capacitive; and (d) optical.

the sensor to work in the force-balanced mode, where the electrostatic force keeps the plate at its initial position by a closed loop control circuitry.

The sensor reported in [17] has the same floating element structure as the design of [14], and the displacement is detected by photodiodes, which are connected in the differential mode. The floating element is made of single-crystalline silicon, which is bonded to a handle wafer. The handle wafer carries the detecting photodiodes [Figure 5.5(d)]. In the absence of the shear stress, the exposed area of the two photodiodes are equal and their differential photoelectric current is zero. In the presence of the shear stress, the displacement of the floating element causes an increase in the exposed area on the leading-edge diode and a decrease in area in the trailing-edge diode. The resulting differential photoelectric current is proportional to the wall shear stress [17]. Table 5.1 summarizes the most important parameters of the above shear stress sensors.

**Table 5.1**
Typical Parameters of Micromachined Shear Stress Sensors

| *Refs.* | *Type* | *Size (mm×mm)* | $\tau_{\max}$ *(Pa)* | $f_{\max}$ *(kHz)* | *Technology* |
|---|---|---|---|---|---|
| [5] | Hot film (thermoelectric) | $4 \times 3$ | 2 | 0.001 | Bulk |
| [6] | Hot wire (thermoresistive) | $0.2 \times 0.2$ | 1.4 | 0.5 | Surface |
| [3] | Hot wire(thermoresistive) | $0.01 \times 0.001$ | — | 500 | Surface |
| [7] | Hot wire (thermoresistive) | $0.15 \times 0.003$ | — | 30 | Surface |
| [8] | Hot film (thermoelectronic) | $0.3 \times 0.06$ | 4 | 40 | Bulk |
| [10] | Floating element (capacitive) | $0.5 \times 0.5$ | 1 | 10 | Surface |
| [11] | Floating element (piezoresistive) | $0.12 \times 0.14$ | $10^5$ | — | Bulk |
| [11] | Floating element (piezoresistive) | $0.5 \times 0.5$ | $10^5$ | — | Bulk |
| [13] | Floating element (capacitive) | $0.1 \times 0.1$ | 10 | — | Bulk |
| [14] | Floating element (optical) | $0.12 \times 0.12$ | 140 | 52 | Bulk |

## 5.2 TURBULENCE CONTROL

Conventional air vehicles use large flaps at their wings to generate enough torque for aerodynamic control. This control mechanism requires large and powerful actuators. Taking advantage of the sensitivity of flow separation to perturbation at its origin, microactuators can control vortices at the leading edge separation line. Small forces and displacements at the separation line can lead to large global vortex pair asymmetry. Using shear stress sensors described in the previous part, the separation line on a wing can be detected. The global field can then be controlled by an array of microactuators, which deliver disturbance deflections on the order of the boundary layer thickness [18].

Since the boundary layer thickness of common air vehicles is on the order of one to several millimeters, the microactuators must be able to deliver the same out-of-plane displacement, and withstand a load on the order of several hundred micronewtons. Although the displacements for this purpose are in the range of mesoscopic actuators, micromachined actuators are still attractive because of the possibility of integration of both shear stress sensor and actuator in a single device.

Figure 5.6 explains the use of microactuators for controlling a macroscopic flying object. The object in this example is a delta-wing airfoil. In a laminar flow and at a certain angle of attack, two counterrotating vortices separate from the laminar flow. The vortices start at the leading edge and propagate over the wing's top. The vortices cause underpressure over the wing, and consequently, lifting forces. The sum of these forces can contribute up to 40% of the total lifting force of the delta wing. By disturbing the boundary layer at the leading edges, the microactuators can make the vortices asymmetrical. This causes an imbalance of lifting forces on the two sides of the wing and a rolling moment, which can be used to maneuver the flying object [19].

In the following section, different actuating schemes for turbulence control are discussed. All microactuators can be fabricated with microtechnology, but have a typical size on the order of 1 mm to match the required boundary layer thickness. Typical devices are microflaps, microballoons, and microsynthetic jets.

### 5.2.1 Microflaps

Microflaps are used to induce a disturbance on the order of 1 mm. For microdevices, this length scale is relatively large. The mechanical load on the flap is several hundred micronewtons, which is a large force for microactuators. Therefore, microflaps require an actuator with a large force and a

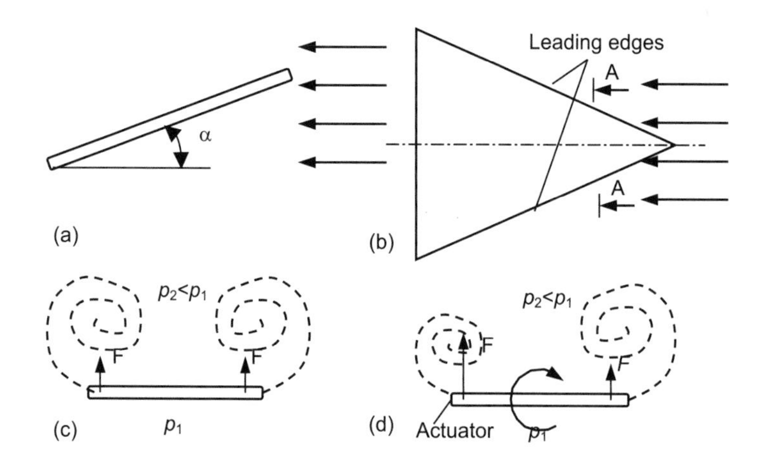


**Figure 5.6** Controlling a delta-wing with microactuators: (a) side view of the delta-wing at an angle of attack $\alpha$; (b) top view of the delta-wing; (c) Section A-A with symmetrical vortices (actuator off); and (d) Section A-A with asymmetrical vortices (actuator on).

large displacement. The most suitable actuating scheme is the electromagnetic principle. Using an external magnetic field, the magnetic microactuator can be realized in two ways:

- Active actuator with microcoil;
- Passive actuator with magnetic material.

#### 5.2.1.1 Design Considerations

The required large displacement should compromise the high dynamics in designing the spring of the microflap. Large displacement requires a soft design or a low spring constant. High dynamics in turn needs stiff design with a large spring constant. For the force balance, the intrinsic stress in the flap should be taken in account. Because high currents are expected for actuators with microcoils, the self-heating effect and, consequently, the thermomechanical bending of the structure can strongly influence the actuator's operation.

For actuators with microcoil, the magnetic force can be calculated as [20]:

$$F = (\vec{m} \cdot \vec{\nabla})\vec{B} \tag{5.9}$$

where $\vec{B}$ (in tesla) is magnetic flux density of the external field and $\vec{m}$ (in Am$^2$) is magnetic moment of the coil [Figure 5.7(a)]: as [20]:

$$\vec{m} = NI\pi r^2 \vec{n} \tag{5.10}$$

where $N$ is the number of coil turns, $I$ is the current, $r$ is the average coil radius, and $\vec{n}$ is the normal vector of the current loop. In a permanent magnet, the magnetization $M_{\mathrm{m}}$ represents the density of microscopic magnetic moments in the magnet and has the unit of A/m [Figure 5.7(b)]. The force in the $z$ direction of a magnetic flux density $B$ on a permanent magnet with a magnetization $M_{\mathrm{m}}$ is [21] (see Section 6.7.1):

$$F = M_{\mathrm{m}} \int (\mathrm{d}B_z/\mathrm{d}z)\mathrm{d}V \tag{5.11}$$

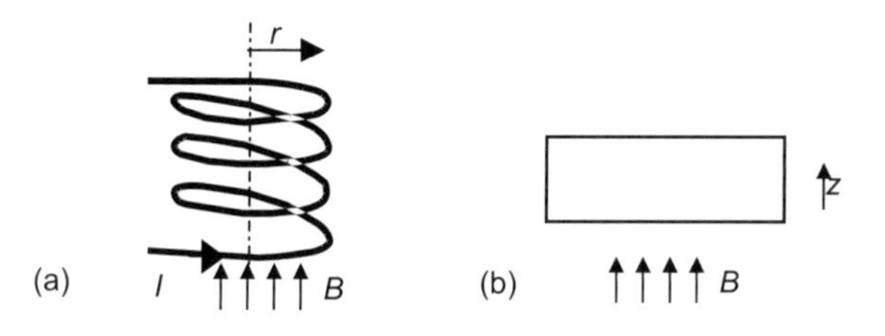


**Figure 5.7** Magnetic actuation principle: (a) microcoil; and (b) permanent magnet.

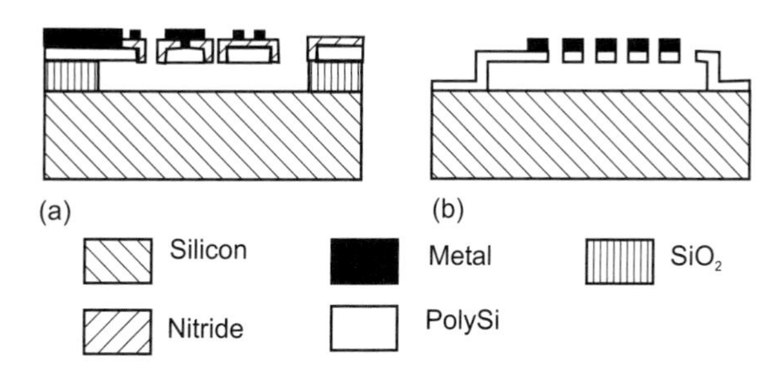


**Figure 5.8** Magnetic microactuators with (a) microcoil and (b) micromagnet.

#### 5.2.1.2 Design Examples

Figure 5.8(a) shows a magnetic microactuator fabricated with surface micromachining [22]. The flap is made of polysilicon and silicon nitride. A metal layer (Al or Cr/Au) forms the coil. The flap sizes vary from 250 to 900 μm. A 420 × 420-μm flap can achieve 100-μm vertical deflection. Because of the undesirable thermomechanical deflection, this type of magnetic actuator is not robust and breaks easily.

The improved version shown in Figure 5.8(b) [19] has a more robust design. The actuator is based on the concept of a permanent magnet. The magnet is made of Permalloy (80% nickel and 20% iron), which is electroplated on the polysilicon flap. At a magnetic field intensity of 60,000 A/m, a 1 mm × 1 mm × 5 μm can reach 65° angular displacement and exert 87 μN force in the direction perpendicular to the substrate. The actuator is able to control a delta wing [19].

### 5.2.2 Microballoon

The microflaps described in Section 5.2.1 are not robust enough for real aircraft with speeds higher than 50 m/s. With magnetic actuators, the deflection could not exceed 1 mm, which is required for efficient control of the boundary layer [23]. A simple design of silicone rubber microballoons would solve the above-mentioned problem [23]. However, microballoons are pneumatic actuators, which require external pressure supply. The relatively low response of a thermopneumatic actuator may not meet the requirement on dynamics of turbulence control.

The actuator consists of a silicon wafer, which is etched from the backside. Silicone rubber is spin-coated (Figure 5.9). Each balloon membrane is 23 × 8.6 mm large and 120 μm thick. With 48 kPa supply pressure, a deflection of 1.8 mm can be achieved. The device survived real flights with velocities close to the speed of sound and temperatures ranging from −23°C to 43°C.

### 5.2.3 Microsynthetic Jet

A microsynthetic jet is a further control concept that does not expose moving parts to the mean flow. A synthetic jet device has a cavity, which has a flexible actuating membrane on one side and an

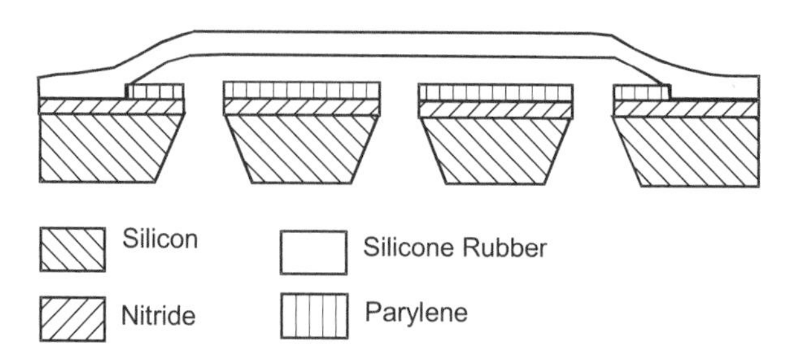


**Figure 5.9** Microballoon for turbulence control.

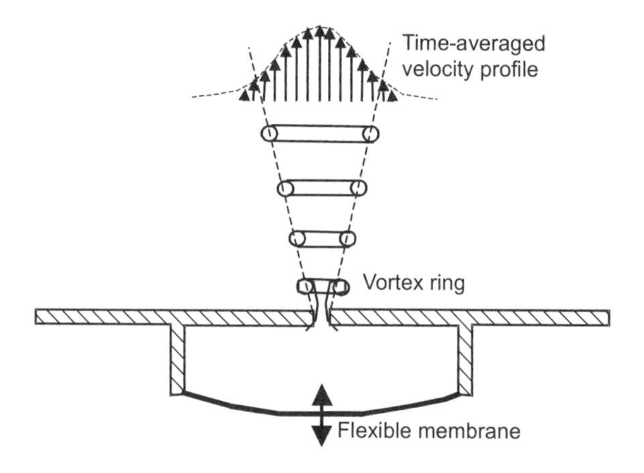


**Figure 5.10** Principle of synthetic jets.

orifice on the other side (Figure 5.10). If the membrane is actuated, the fluid is pumped in and out of the cavity through the same orifice. This fluid movement forms a series of vortices, which propagate away from the orifice. This principle allows the actuator to transfer momentum into the surrounding fluid without a net mass injection. Therefore, this device is called zero mass-flux actuator.

The addressable microjet array described in [24] consists of an array of small orifices positioned on top of an array of actuator cavities as shown in Figure 5.11.

Both the orifices and the cavities are batch fabricated with silicon bulk micromachining. The length of an orifice is defined by the wafer thickness, typically 250 μm. The orifice width ranges from 50 to 800 μm. The actuator cavity is approximately 15 μm in depth and has typical lateral dimensions ranging from 1 to 4 mm. Individual jet control is achieved by using a metallized flexible polyimide membrane. The metal electrodes on the diaphragm are patterned so that voltage can be

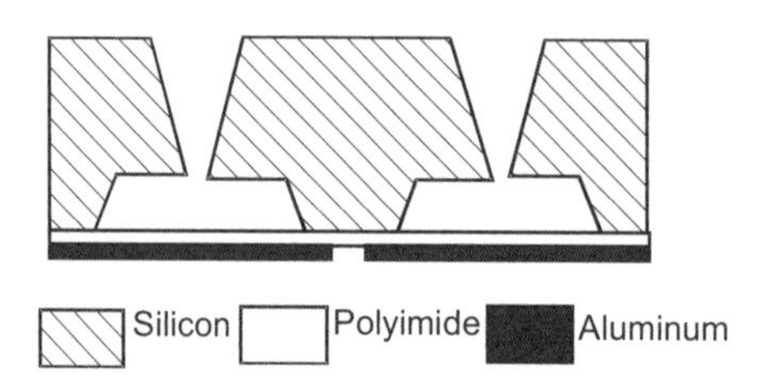


**Figure 5.11** Addressable microjet array.

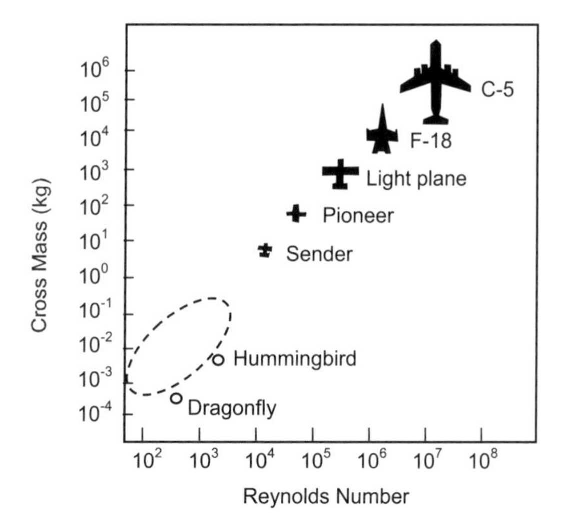


**Figure 5.12** MAV flight regime compared to existing flight vehicles [25].

**Table 5.2**
Principal Characteristics of MAVs (*After*: [25].)

| *Wing span (cm)* | *Payload (g)* | *Flight Speed (m/s)* | *Distance (km)* | *Weight (g)* | *Endurance (min)* |
|---|---|---|---|---|---|
| 15–20 | 20 | 10–20 | 10 | $< 100$ | 20–60 |

individually applied to the region over each actuator cavity. The membranes are actuated either electrostatically or piezoelectrically.

## 5.3 MICROAIR VEHICLES

The development of microair vehicles (MAVs) is one of the most exciting fields, where microfluidic components for external flow control could find their application. Possible applications of MAVs are surveillance in military missions, biological-chemical agent detection, and space exploration. The Defense Advanced Research Projects Agency (DARPA) program, which started the initiative, defines the principal characteristics of MAVs, listed in Table 5.2 [25]. Two of the biggest challenges in designing an MAV are the stable flight at low Reynolds numbers and the design of energy sources to meet the required endurance.

The term "micro" indicates that MAVs are not a mere miniaturized version of conventional macroscale aircraft. The MAV itself has a size in the centimeter range, but its components can be small and fabricated with microtechnology.

Figure 5.12 compares the cross mass versus the typical Reynolds number of typical aircrafts, natural flyers, and MAVs. This section discusses the following concepts of the MAV: fixed-wing MAV, flapping-wing MAV, microrotorcraft, and microrockets.

### Example 5.10: Characteristics of an MAV

An MAV flies with a speed of 5 m/s. The device is 15 cm long. Determine the corresponding Reynolds number. The kinematic viscosity of air is $1.6 \times 10^{-5}$ m$^2$/s.

**Solution.** The typical Reynolds number is:

$$\mathrm{Re} = \frac{uL}{\nu} = \frac{5 \times 15 \times 10^{-2}}{1.6 \times 19^{-5}} = 46,875$$

The Reynolds number is on the order of those of small birds.

### 5.3.1 Fixed-Wing MAV

#### 5.3.1.1 Design Considerations

The first approach of designing MAVs is the miniaturization of conventional fixed-wing vehicles. The major concern of this approach is the power source. Assuming a constant energy density in power sources (i.e., batteries, charged capacitors, and fuels), the available power decreases with the third power of miniaturization. The power requirement $P$ for a propeller driven aircraft can be estimated as [25]:

$$P = \frac{W \frac{C_{\mathrm{D}}}{C_{\mathrm{L}}^{3/2}} \sqrt{\frac{W}{A_{\mathrm{wing}}}} \sqrt{\frac{2}{\rho_{\mathrm{air}}}}}{\eta} \tag{5.12}$$

where $W$ is the aircraft weight, $A_{\mathrm{wing}}$ is the wing surface area, $\rho_{\mathrm{air}}$ is the air density, and $\eta$ is the propeller efficiency. The lift coefficient $C_{\mathrm{L}}$ is defined as:

$$C_{\mathrm{L}} = \frac{2F_{\mathrm{L}}}{\rho_{\mathrm{air}} A_{\mathrm{wing}} u^2} \tag{5.13}$$

where $u$ is the flight speed and $F_{\mathrm{L}}$ is the lift force. The drag (or thrust) coefficient can be expressed as:

$$C_{\mathrm{D}} = \frac{2F_{\mathrm{D}}}{\rho_{\mathrm{air}} A_{\mathrm{wing}} u^2} \tag{5.14}$$

where $F_{\mathrm{D}}$ is the drag (or thrust) force.

Equation (5.12) reveals that for a long endurance one should increase the total stored energy, minimize the weight, and minimize the weight-to-wing surface ratio. For the maximum energy required, high-density power sources are needed. Beside batteries, fuel cells are possible candidates. The small weight-to-wing surface ratio leads to a low aspect ratio design, which looks more like a flying wing [25].

**Example 5.11: Power Requirement and Endurance of a Fixed-Wing MAV**

An MAV is designed as a delta wing, which is driven by a propeller. The delta wing is an equilateral triangle with side dimension of 15 cm. The design velocity is 10 m/s. The propeller is driven by an electric motor and has an efficiency of 80%. The lift and drag coefficients of the wing are 0.5 and 0.1, respectively. The device uses energy from a NiCd battery, which supplies 1.5 Ah at 1.5V. The total mass of the device is 100g. Determine the flight endurance and the maximum mission distance. Air density is assumed to be 1.118 kg/m$^3$.

**Solution.** The total weight of the MAV is:

$$W = mg = 100 \times 10^{-3} \times 9.8 = 0.98 \text{ N}$$

The total wing area of the device is:

$$A_{\mathrm{wing}} = \sqrt{3}a^2/4 = \sqrt{3} \times (15 \times 10^{-2})^2/4 = 97.4 \times 10^{-4} \text{ m}^2$$

The expected power consumption for the given characteristics and speed is (5.12):

$$P = \frac{W \frac{C_{\mathrm{D}}}{C_{\mathrm{L}}^{3/2}} \sqrt{\frac{W}{A_{\mathrm{wing}}}} \sqrt{\frac{2}{\rho_{\mathrm{air}}}}}{\eta} = \frac{0.98 \times \frac{0.1}{0.5^{3/2}} \sqrt{\frac{0.98}{97.4 \times 10^{-4}}} \sqrt{\frac{2}{1.118}}}{0.8} = 4.65\mathrm{W}$$

The energy stored in the battery is:

$$E = 1.5 \times 1.5 \times 3,600 = 8,100 \text{ J}$$

Thus, the flight endurance is:

$$t = \frac{E}{P} = \frac{8,100}{4.65} = 1,742 \text{ sec} \approx 29 \text{ min}$$

The MAV is able to maintain autonomy for 29 minutes. The maximum mission distance is the distance that the device can reach and then return to its initial location:

$$D = \frac{1}{2}ut = \frac{1}{2} \times 10 \times 1,742 = 8,710 \text{ m}$$

5.3.1.2 Design Examples

The fixed-wing MAV developed by AeroVironment Inc. (Simi Valley, California) has a total mass of 80g [26]. The device, called Black Widow, has a 15-cm wing span and has the form of a disk. A propeller drives the MAV. A lithium battery delivers energy for the entire device. The propeller measures 9.5 cm and has a best efficiency of 83%. The device is able to carry a video camera system with downlink transmitter, which has a maximum communication range of 2 km. The entire electronic subsystem includes a 2g video camera, 2g video transmitter, 5g radio control systems, and 0.5g actuators. The device is able to fly with a speed of 13.4 m/s in 30 minutes. Parts of the device are fabricated with conventional precision technology as well as manually. The propeller is molded from unidirectional and woven carbon-fiber composites. The motor and the electronic subsystem are commercially available.

### 5.3.2 Flapping-Wing MAV

5.3.2.1 Design Considerations

The second approach of flapping wings mimics natural flyers such as birds and insects. Observing natural flyers leads to a statistical relation between mass and speed [27]:

$$u = 4.77m^{\frac{1}{6}} \tag{5.15}$$

where $u$ is the flight speed in m/s and $m$ is the mass in grams. For a flapping-wing MAV, the design speed and mass can be estimated from the above relation.

The other useful statistical relations taken from nature are wingtip speed and mass [28, 29]:

$$11.7m^{-0.065} < u_{\mathrm{wingtip}} < 9.6m^{-0.0434} \tag{5.16}$$

where $u_{\mathrm{wingtip}}$ is the wingtip speed in meters per second and $m$ is the mass in grams.

The relation reveals that larger flyers have lower wingtip speeds and lower flapping frequencies. Figure 5.13 shows the typical flying regimes of natural flyers and MAVs. Flapping-wing MAVs

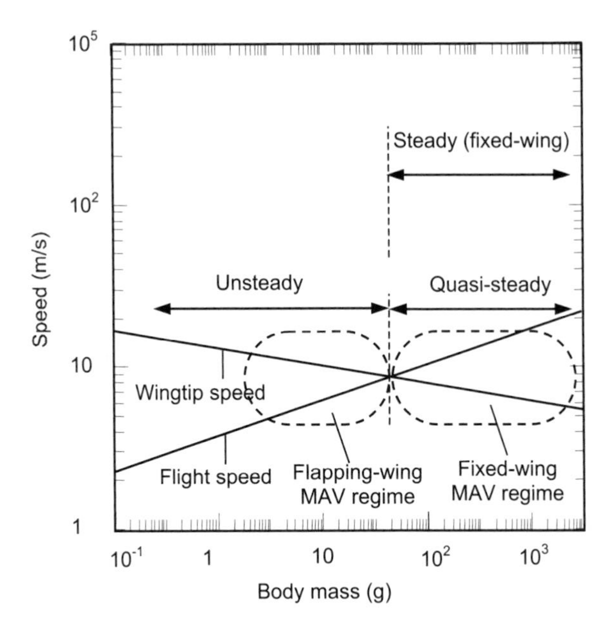


**Figure 5.13** The flying regimes of natural flyers and MAVs. (*After*: [25].)

work in an unsteady regime, where the airflow over the wing is not constant in time. Most insects and small birds fly in this regime.

The characteristic Reynolds number for flapping-wing flyers is defined as [30]:

$$\mathrm{Re} = \frac{Lu_{\mathrm{wingtip}}}{\nu} = \frac{4\Phi f R^2}{\nu\Omega} \tag{5.17}$$

where $L$ is the mean chord length, $R$ is the length of one wing, $\Omega = 2R/L4$ is the aspect ratio, $f$ is the wing beat frequency, and $\Phi$ is the peak-to-peak wing beat amplitude in radians. The maximum mass that can be supported by flapping wings during hovering with a horizontal stroke plane is estimated as [30]:

$$m = 0.387\frac{\Phi^2 f^2 R^4 C_{\mathrm{L}}}{\Omega} \tag{5.18}$$

where $C_{\mathrm{L}}$ is the lift coefficient defined in (5.13). The total power in watts, which is needed for overcoming lift and drag, can then be estimated with the mass $m$ as:

$$P = P_{\mathrm{lift}} + P_{\mathrm{drag}} = mfR\left[14.0\left(\frac{\Phi C_{\mathrm{L}}}{\Omega}\right)^{1/2} + 18.2\Phi\frac{C_{\mathrm{D}}}{CL}\right] \tag{5.19}$$

The drag coefficient $C_{\mathrm{D}}$ of typical insect wings can be estimated with [27]:

$$C_{\mathrm{D}} = 7\sqrt{\mathrm{Re}} \tag{5.20}$$

where the Reynolds number is defined in (5.6).

**Example 5.12: Power Requirement and Endurance of a Flapping-Wing MAV**

A flapping-wing MAV is designed with 15-cm mean chord length and a wing length of 5 cm. The flapping frequency is optimized for 50 Hz. The stroke amplitude is 120°. The lift coefficient of the wing is 1.5. The drag coefficient is 0.15. The device uses energy from a NiCd battery, which supplies 1.5 Ah at 1.5V. Determine the maximum mass of the MAV and the flight endurance in this case.

**Solution.** The aspect ratio of the device is:

$$\Omega = \frac{2R}{L} = \frac{2 \times 5}{15} = 0.67$$

With a stroke amplitude of 120° or $0.67\pi$, the maximum mass of the MAV is:

$$m = 0.387\tfrac{\Phi^2 f^2 R^4 C_{\rm L}}{\Omega}$$

$$m = 0.387\tfrac{0.67^2\pi^2 50^2 (5\times10^{-2})^4 \times 1.5}{0.67} = 0.0597 \text{ kg} = 59.7 \text{ g}$$

The power requirement for the device during flight is:

$$P = mfR\left[14.0\left(\tfrac{\Phi C_{\rm L}}{\Omega}\right)^{1/2} + 18.2\Phi\tfrac{C_{\rm D}}{C_{\rm L}}\right]$$

$$P = 0.155 \times 50 \times 5 \times 10^{-2}\left[14 \times \left(\tfrac{0.67\pi\times1.5}{0.67}\right)^{1/2} + 18.2 \times 0.67\pi\tfrac{1.5}{0.15}\right] = 5.1\text{W}$$

The energy stored in the battery is:

$$E_{\rm energy} = 1.5 \times 1.5 \times 3,600 = 8,100 \text{ J}$$

Thus, the flight endurance is:

$$t = \frac{E_{\rm energy}}{P} = \frac{8,100}{5.1} = 1,588 \text{ sec} = 26.5 \text{ min}$$

The results of the flapping-wing MAV are comparable to those of the fixed-wing MAV in Example 5.11.

5.3.2.2 Design Examples

Figure 5.14 shows the process for fabricating wing structures [28]. The structures made of silicon are too fragile for the relatively large size required for the wing, which is on the order of several centimeters. A similar process utilizing titanium alloy as the substrate and carbon fiber rod is shown in Figure 5.15. The titanium alloy structure is formed in an enchant solution (HF, $HNO_3$, $H_2O$) [29]. The most critical points in the development of such a flapping-wing MAV are the power source and energy-saving actuating schemes. The device reported in [28, 29] can only fly for 5 to 18 seconds, while the required minimum flight duration is about 20 minutes.

### 5.3.3 Microrotorcraft

Considering the definitions of lift and drag coefficients in (5.13) and (5.14), the ratio between lift and thrust or the ratio between weight and drag is derived as:

$$\frac{W}{F_{\rm D}} = \frac{C_{\rm L}}{C_{\rm D}} \tag{5.21}$$

The ratio for typical aircrafts and mesoscale MAVs ranges from 10 to 20. Since the weight is proportional to volume and the drag force is proportional to surface area, the ratio in [21] obeys the

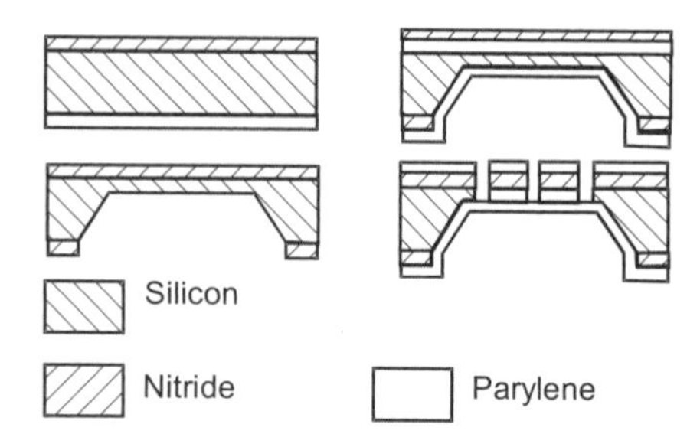


**Figure 5.14** Fabrication process of silicon MEMS wings.

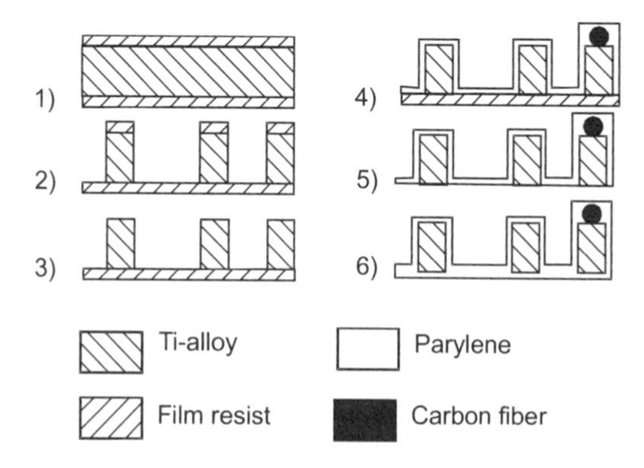


**Figure 5.15** Fabrication process of Ti-alloy MEMS wings.

cubic square relation of scaling law. In other words, the ratio $W/F_D$ decreases with miniaturization. In microscale, the drag force can be many orders larger than the weight of the device. The above flight concepts with wings, which utilize smaller thrust to balance drag force, will not work well in the microscale.

If drag force is much larger than weight, the thrust alone can be used to balance drag force and weight. The advantage of scaling law can be used in designing a microrotorcraft, which is more efficient than helicopters in the macroscale.

The work at Stanford University dealt with a centimeter-size electric helicopter [31]. This size scale fills the gap between the true microscale (millimeters or less) and the conventional scale (tens of centimeters or more). The air vehicle is therefore called Mesicopter. There is no report indicating that this device can fly.

Table 5.3 compares the most important parameters of the design examples of the fixed-wing MAV, flapping-wing MAV, and the microrotorcraft discussed above and illustrated in Figure 5.16.

**Table 5.3**
Typical Parameters of MAVs

| *Refs.* | *Type* | *Size (mm×mm)* | *Speed (m/s)* | *Endurance (min)* | *Mass (g)* | *Maximum Altitude (m)* |
|---|---|---|---|---|---|---|
| [26] | Fixed-wing | 150×150 | 13.4 | 30 | 80 | 234 |
| [29] | Flapping-wing | 150×150 | 4.4 | $\approx 0.3$ | 10.5 | $\approx 10$ |

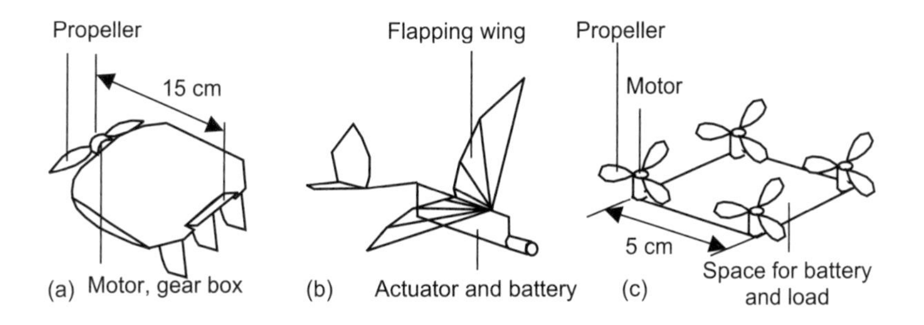


**Figure 5.16** Design examples of MAV: (a) fixed-wing MAV Black Widow (*After*: [26]); (b) flapping-wing MAV Ornithopter (*After:* [29]); and (c) microrotorcraft Mesicopter (*After*: [31]).

### 5.3.4 Microrockets

#### 5.3.4.1 Design Considerations

The mean application of microrockets will be providing precise thrust for controlling satellite position. The basic idea behind microrockets is the gain of the thrust-to-weight ratio $F/W$ with miniaturization. The thrust $F$ is proportional to square of length scale $l$:

$$F \propto p_{\mathrm{c}} A_{\mathrm{t}} \propto l^2 \tag{5.22}$$

where $p_{\mathrm{c}}$ is the combustion chamber pressure and $A_{\mathrm{t}}$ is the throat area. Since the weight is proportional to the volume $l^3$, the thrust-to-weight ratio is inversely proportional to the length scale:

$$\frac{F}{W} \propto \frac{l^2}{l^3} = \frac{1}{l} \tag{5.23}$$

Conventional rockets with a typical dimension of 1m have thrust-to-weight ratios on the order of 50. Microrockets with a typical dimension of 1 mm will result in a thrust-to-weight ratio of 50,000. The better performance promises the use of many microrockets in parallel to achieve the desired thrust in macroscale applications.

In a rocket engine, thermal energy is converted into the kinetic energy of a working gas. Kinetic energy of the exhaust gas through a nozzle provides thrust needed for the application. Based on the source of thermal energy, rockets are categorized as [32]:

- Cold-gas rockets;
- Chemical rockets;
- Electric rockets;
- Nuclear rockets;
- Solar thermal rockets.

The first three types are relevant for miniaturization. Cold-gas rockets only need a high-pressure gas supply. Chemical rockets use a combustion reaction to provide heat. Electric rockets utilize the high temperature of an electric arc to generate heat for the engine. This section focuses on the design of cold-gas microrockets and chemical microrockets.

Based on the type of fuels, chemical rockets can be further categorized as cold-gas rockets, liquid propellant rockets, and solid propellant rockets. Liquid propellants have two forms: bipropellant

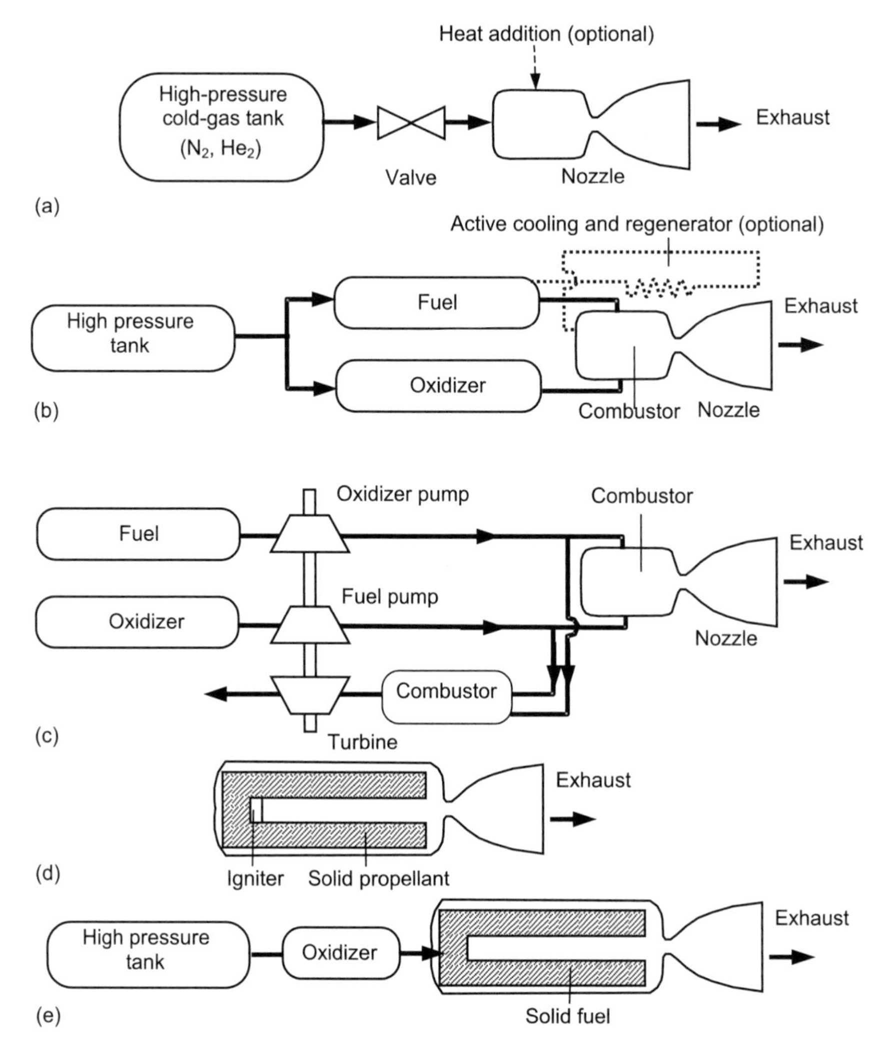


**Figure 5.17** Schematics of common rocket configurations: (a) cold-gas rocket; (b) bipropellant rocket; (c) bipropellant rocket with turbo pumps; (d) solid propellant rockets; and (e) hybrid rockets.

and monopropellant. A bipropellant consists of a liquid fuel and a liquid oxidizer. A monopropellant is a single liquid containing both fuel and oxidizer.

Figure 5.17 shows the basic rocket configurations. The cold-gas thruster can be implemented easily in the microscale with an external gas supply [Figure 5.17(a)]. The bipropellant rocket [Figure 5.17(b)] is another good candidate. Special consideration should be taken in designing the combustor because of the high operation temperature there. An active cooling system with fuel can be used as a regenerator as shown in Figure 5.17(b). The high-pressure tank in Figure 5.17(b) would increase the device weight, and is not practical in case of a microrocket. The system depicted in Figure 5.17(c) can provide a compact solution. A microgas turbine system (see Section 13.2.1.1) can provide work to two pumps, which in turn deliver the fuel and the oxidizer to the combustor.

The solid propellant rocket has both fuel and oxidizer in solid forms in the combustor [Figure 5.17(d)]. The hybrid solution described in Figure 5.17(e) uses liquid oxidizer with solid fuel.

Assuming an isentropic expansion process of the propellant gas from the combustor, to the nozzle throat, to the exit area, the characteristics of an ideal rocket can be calculated with basic thermodynamic relations.

*Combustion Chamber Characteristics.* The most important properties of a combustor are pressure $p_\mathrm{c}$ and temperature $T_\mathrm{c}$. The chamber temperature can be estimated with the maximum adiabatic temperature of a combustion reaction. However, because of massive cooling, the actual temperatures in a combustor are much lower than the estimated adiabatic combustion temperature. The chamber pressure can be estimated with the mass flow rate of propellant and chamber temperature using the ideal gas equation.

*Throat Characteristics.* If the throat velocity reaches the speed of sound (Mach number Ma = 1), the following relations at the nozzle throat can be derived for pressure $p_\mathrm{t}$, specific volume $v_\mathrm{t}$, temperature $T_\mathrm{t}$, and velocity $u_\mathrm{t}$ [32]:

$$\frac{p_\mathrm{t}}{p_\mathrm{c}} = \left[\frac{2}{k+1}\right]^{\frac{k}{k-1}} \tag{5.24}$$

$$\frac{v_\mathrm{t}}{v_\mathrm{c}} = \left[\frac{k+1}{2}\right]^{\frac{1}{k-1}} \tag{5.25}$$

$$\frac{T_\mathrm{t}}{T_\mathrm{c}} = \frac{2}{k+1} \tag{5.26}$$

$$u_\mathrm{t} = \sqrt{\frac{2k}{k+1}RT_\mathrm{c}} \tag{5.27}$$

where $k$ is the specific heat ratio, and subscripts t and c represent the throat and combustor, respectively.

*Exit Area Characteristics.* Using the energy balance equation for the process across the rocket throat and assuming that the velocity in the combustion chamber is zero, the exit velocity can be estimated as [32]:

$$u_\mathrm{exit} = \sqrt{\frac{2k}{k-1}RT_\mathrm{c}\left[1-\left(\frac{p_\mathrm{e}}{p_\mathrm{c}}\right)^{\frac{k-1}{k}}\right]} \tag{5.28}$$

where $R$ is the gas constant of the propellant. The gas temperature $T_\mathrm{exit}$ and the specific volume at the exit $v_\mathrm{exit}$ can be estimated from the isentropic relations:

$$\frac{T_\mathrm{exit}}{T_\mathrm{c}} = \left[\frac{p_\mathrm{exit}}{p_\mathrm{c}}\right]^{\frac{k-1}{k}} \tag{5.29}$$

$$\frac{v_\mathrm{exit}}{v_\mathrm{c}} = \left[\frac{p_\mathrm{c}}{p_\mathrm{exit}}\right]^{\frac{1}{k}} \tag{5.30}$$

*Performance of a Microrocket.* The first important parameter for a microrocket is its specific impulse:

$$Y_\mathrm{sp} = \frac{F}{\dot{m}g} \tag{5.31}$$

where $F$ is the thrust, $\dot{m}$ is the mass flow rate, and $g$ is the acceleration of gravity. The ratio $F/\dot{m}$ is effective exhaust velocity $u_{\mathrm{exit}}$. Thus, the specific impulse can be defined as:

$$Y_{\mathrm{sp}} = \frac{u_{\mathrm{exit}}}{g} \tag{5.32}$$

The specific impulse has the unit of time, and represents the time during which a certain amount of fuel can deliver the same thrust as its own weight. Thus, the specific impulse is the primary performance indicator for microrockets. For evaluating the effectiveness of the nozzle, the thrust coefficient $C_{\mathrm{F}}$ can be used:

$$C_{\mathrm{F}} = \frac{F}{p_{\mathrm{c}} A_{\mathrm{t}}} \tag{5.33}$$

where $A_{\mathrm{t}}$ is the throat area. The thrust coefficient is given for an ideal isentropic rocket flow by [29]:

$$C_{\mathrm{F}} = \sqrt{\frac{2k^2}{k-1}\left[1-\left(\frac{p_{\mathrm{e}}}{p_{\mathrm{c}}}\right)^{\frac{k-1}{k}}\right]\left(\frac{2}{k+1}\right)^{\frac{k+1}{2(k-1)}}} + \frac{p_{\mathrm{e}}-p_{\mathrm{atm}}}{p_{\mathrm{c}}}\frac{A_{\mathrm{e}}}{\mathrm{At}} \tag{5.34}$$

where $p_{\mathrm{e}}$ and $A_{\mathrm{e}}$ are the exit pressure and exit area of the nozzle, respectively, and $k$ is the specific heat ratio of the gas. If the characteristic exhaust velocity $c^*$ is defined as:

$$c^* = \frac{p_{\mathrm{c}} A_{\mathrm{t}}}{\dot{m}} \tag{5.35}$$

the characteristic exhaust velocity for an ideal rocket is [32]:

$$c^* = \sqrt{RT_{\mathrm{c}}}\left(\frac{k+1}{2k}\right)^{\frac{k+1}{2(k-1)}} \tag{5.36}$$

Combining the above relations, the exit velocity can be described as:

$$u_{\mathrm{exit}} = I_{\mathrm{sp}} g = C_{\mathrm{F}} c^* \tag{5.37}$$

**Example 5.13: Characteristics of a Microrocket**

A microrocket is made of silicon. The height of the combustion chamber, throat, and exit area is 1 mm. The exit width is 5 mm. The rocket is designed to work with 10 bars chamber pressure and 1,000K chamber temperature. Determine the theoretical thrust coefficient, the exit velocity, and the specific impulse. For the propellant gas, use $k = 1.3$ and $R = 355.4$ J/kg-K.

**Solution.** Assuming that the rocket operates at atmospheric conditions ($p_{\mathrm{exit}} = p_{\mathrm{atm}} = 1$ bar), the theoretical thrust coefficient is (5.34):

$$C_{\mathrm{F}} = \sqrt{\frac{2k^2}{k-1}\left[1-\left(\frac{p_{\mathrm{e}}}{p_{\mathrm{c}}}\right)^{\frac{k-1}{k}}\right]\left(\frac{2}{k+1}\right)^{\frac{k+1}{2(k-1)}}} + \frac{p_{\mathrm{e}}-p_{\mathrm{atm}}}{p_{\mathrm{c}}}\frac{A_{\mathrm{e}}}{\mathrm{At}}$$

$$C_{\mathrm{F}} = \sqrt{\frac{2\times 1.3^2}{1.3-1}\left[1-\left(\frac{1}{10}\right)^{\frac{1.3-1}{1.3}}\right]\left(\frac{2}{1.3+1}\right)^{\frac{1.3+1}{2(1.3-1)}}} = 1.26$$

The exit velocity of the microrocket is (5.28):

$$u_{\text{exit}} = \sqrt{\frac{2k}{k-1} R T_{\text{c}} \left[1 - \left(\frac{p_{\text{e}}}{p_{\text{c}}}\right)^{\frac{k-1}{k}}\right]}$$

$$u_{\text{exit}} = \sqrt{\frac{2 \times 1.3}{1.3-1} 355.4 \times 1000 \left[1 - \left(\frac{1}{10}\right)^{\frac{1.3-1}{1.3}}\right]} = 1,127 \text{ m/sec} \tag{5.38}$$

Assuming an acceleration of gravity of 9.8 $\text{m/s}^2$, the specific impulse is:

$$I_{\text{sp}} = \frac{u_{\text{exit}}}{g} = \frac{1,127}{9.8} = 115 \text{ sec}$$

**Example 5.14: Thrust and Thrust-to-Weight Ratio**

We consider the microrocket described above. If the total device weight is 1.5g, determine the mass flow rate, the thrust, and the thrust-to-weight ratio.

**Solution.** The cross-sectional exit area of the microrocket is:

$$A_{\text{exit}} = 5 \times 1 \times 10^{-6} = 5 \times 10^{-6} \text{ m}^2$$

Thus, the volumetric flow rate at exit is:

$$(Au)_{\text{exit}} = 5 \times 10^{-6} \times 1,127 = 5,635 \times 10^{-6} \text{ m}^2/\text{s}$$

We assume an isentropic expansion across the rocket nozzle. Using the relation of (5.29), the exit temperature can be estimated as:

$$T_{\text{exit}} = T_{\text{c}} \left[\frac{p_{\text{exit}}}{p_{\text{c}}}\right]^{\frac{k-1}{k}} = 1,000 \left[\frac{1}{10}\right]^{\frac{1.3-1}{1.3}} = 588 \text{ K}$$

For this temperature, the specific volume of propellant gas is:

$$u_{\text{exit}} = \frac{RT_{\text{exit}}}{p_{\text{exit}}} = 2.09 \text{ m}^3/\text{kg}$$

Thus, the mass flow rate of exhaust gas is:

$$\dot{m}_{\text{exit}} = \frac{(Au)_{\text{exit}}}{u_{\text{exit}}} = \frac{5,635 \times 10^{-6}}{2.09} = 2.7 \times 10^{-3} \text{ kg/s}$$

With the exit velocity of the previous example, the thrust of the microrocket is:

$$F = \dot{m}_{\text{exit}} u_{\text{exit}} = 2.7 \times 10^{-3} \times 1,127 \approx 3.04 \text{ N}$$

Considering only the weight of the combustion chamber and nozzle, the thrust-to-weight ratio of the microrocket is:

$$\frac{F}{W} = \frac{2.48}{1.5 \times 10^{-3} \times 9.8} = 207$$

**Example 5.15: Designing an Ideal Microrocket**

An ideal microrocket, which operates at a pressure of 1 bar, should have a thrust of 1N at a pressure of 10 bars and a temperature of 1,000K in the combustion chamber. The chamber and nozzle heights are 1 mm. Assuming that $k = 1.3$, and $R = 355.4$ J/kg-K, determine the exit velocity, throat width, and exit width.

**Solution.** The microrocket is designed for the critical case of the speed of sound at the nozzle throat. The throat velocity is determined by (5.27):

$$u_\mathrm{t} = \sqrt{\frac{2k}{k+1}RT_\mathrm{c}} = \sqrt{\frac{2\times 1.3}{1.3+1}355.4\times 1,000} = 634 \text{ m/s}$$

The exit velocity is estimated with (5.28):

$$u_\mathrm{exit} = \sqrt{\frac{2k}{k-1}RT_\mathrm{c}\left[1-\left(\frac{p_\mathrm{e}}{p_\mathrm{c}}\right)^{\frac{k-1}{k}}\right]} =$$

$$\sqrt{\frac{2\times 1.3}{1.3-1}\times 355.4\times 1,000\times\left[1-\left(\frac{1}{10}\right)^{\frac{1.3-1}{1.3}}\right]} = 407 \text{ m/s}$$

The mass flow rate of propellant gas can be estimated from the relations in (5.31) and (5.32) as:

$$\dot{m} = \dot{m}_\mathrm{exit} = \frac{F}{u_\mathrm{exit}} = \frac{1}{407} = 2.457\times 10^{-3} \text{ kg/s}$$

The specific volume of propellant gas in combustion chamber is:

$$v_\mathrm{c} = \frac{RT_\mathrm{c}}{p_\mathrm{c}} = \frac{355.4\times 1,000}{10\times 10^5} = 355.4\times 10^{-3}\mathrm{m}^3/\mathrm{kg}$$

The specific volume at throat can be estimated with (5.25):

$$v_\mathrm{t} = v_\mathrm{c}\left[\frac{k+1}{2}\right]^{\frac{1}{k-1}} = 355.4\times 10^{-3}\left[\frac{1.3+1}{2}\right]^{\frac{1}{1.3-1}} = 566.3\times 10^3 \text{ m}^3\text{/kg}$$

The specific volume at exit can be estimated with (5.30):

$$v_\mathrm{exit} = v_\mathrm{c}\left(\frac{p_\mathrm{c}}{p_\mathrm{exit}}\right)^{\frac{1}{k}} = 355.4\times 10^{-3}(10)^{\frac{1}{1.3}} \text{ m}^3\text{/kg}$$

The width of throat is:

$$w_\mathrm{t} = \frac{A_\mathrm{t}}{h} = \frac{\dot{m}v_\mathrm{t}}{u_\mathrm{t}h} = \frac{0.8878\times 10^{-3}\times 566.3\times 10^{-3}}{634\times 10^{-3}} = 2.195\times 10^{-3} \text{ m} = 2.195 \text{ mm}$$

The width of exit is:

$$w_\mathrm{exit} = \frac{A_\mathrm{exit}}{h} = \frac{\dot{m}v_\mathrm{exit}}{u_\mathrm{exit}h} = \frac{0.8878\times 10^{-3}\times 2089\times 10^{-3}}{1,127\times 10^{-3}} = 12.61\times 10^{-3} \text{ m} = 12.61 \text{ mm}$$

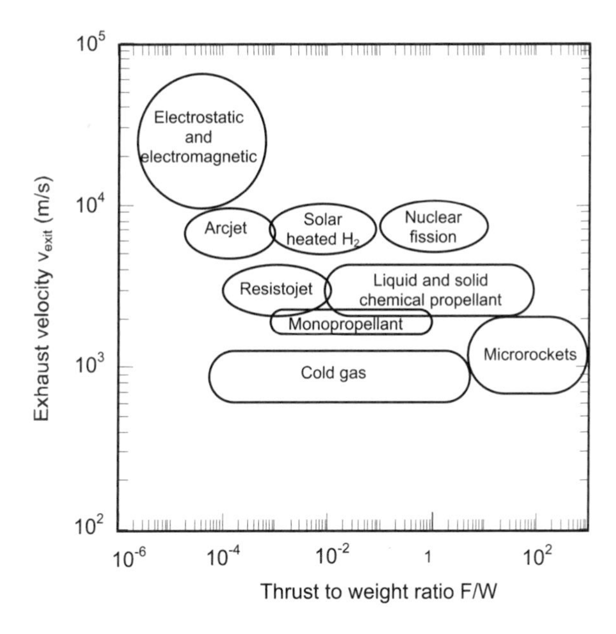


**Figure 5.18** Exhaust velocity and thrust-to-weight ratio of different rockets compared to microrockets. (Data of conventional rockets is from [29].)

The major drawbacks of silicon technology in fabricating microrockets are its high thermal conductivity and its relatively low melting point. Consequently, the combustion chamber needs to be cooled actively. Furthermore, the high thermal conductivity causes difficulties for ignition of the combustion process.

The other drawback is the limit of planar structures in silicon technology. For places with high pressure such as the combustion chamber of a microrocket, the planar silicon/glass structure cannot resist the pressure emerging after ignition. Silicon/silicon structures fabricated with fusion bonding are better for microrockets.

However, the special characteristics of high thrust-to-weight ratio and relatively low exhaust velocity are interesting and give microrockets functions complementary to conventional rockets (Figure 5.18).

### 5.3.4.2 Design Examples

The simplest microrocket is a cold-gas microthruster. The thruster consists of a micronozzle and high-pressure gas supply. The microcold-gas thruster described in [33] has a nozzle throat width of 20 μm, and an exit width of 500 μm. At a chamber pressure of 7 bars, a thrust of 7 mN and a specific impulse of 75 seconds can be achieved. The micronozzle is etched in silicon using DRIE. Two glass plates are anodically bonded on two sides of the silicon wafer. Heated gas can improve the performance of the microrocket (5.28). The heater is made of bulk silicon fins etched with the nozzle [Figure 5.19(a)]. The heater fins also work as a heat exchanger. With a chamber temperature ranging from 650K to 700K, the specific impulse is improved by 50%.

The microthruster system presented in [34] has four micronozzles, four microvalves, and two microfilters integrated in a hybrid system. Similar to the example of [30], the nozzle has narrow curved fins as a heat exchanger. With a throat width ranging from 10 to 40 μm and an exit angle of 13°, thrusts ranging from 0.1 to 10 mN are achieved.

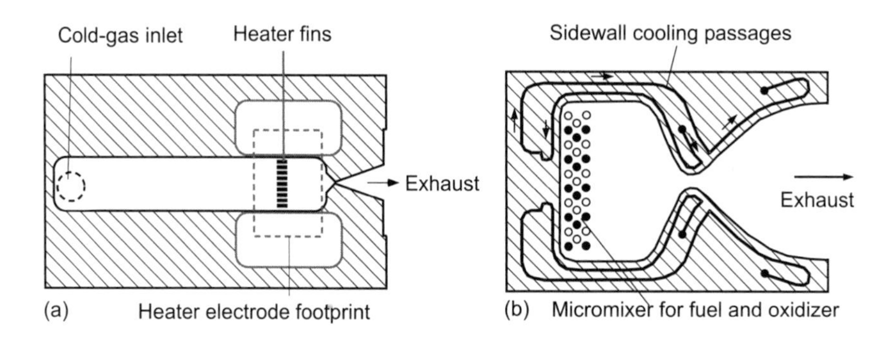


**Figure 5.19** Microrockets (not to scale): (a) microthruster with electrothermal augmentation (the dimension of the silicon chip is 1.5 × 6 × 10 mm); and (b) bipropellant microrocket with cooling passages and a micromixer (the dimension of the silicon chip is 3 mm × 13.5 mm × 18 mm).

**Table 5.4**

Typical Parameters of Microrockets

| *Refs.* | *Type* | *Size (mm×mm)* | *F (mN)* | $I_s$*(sec)* | *Technology/Material* |
|---|---|---|---|---|---|
| [33] | Gas microthruster | 1.5 × 6 × 10 | 7 | 75 DRIE/Silicon | |
| [34] | Gas microthruster | — | 0.1–10 | ≈ 100 | DRIE/Silicon |
| [35] | Bipropellant microrocket | 3 × 13.5 × 18 | 1,000 | 30 | DRIE/Silicon |
| [36] | Solid fuel microrocket | 3.2 × 12.7(25.4) | 15 | 10–20 | Hybrid/ceramic, silicon |
| [37] | Vaporizing microthruster | — | — | — | Bulk/Silicon |
| [38] | Vaporizing microthruster | — | — | — | Bulk/Silicon |

The micronozzle developed for cold-gas microthrusters can be used in a bipropellant microrocket [35]. The microrocket uses liquid oxygen and methane as a propellant. Ethanol is fed through a channel system, which actively cools the combustion chamber. The active cooling system keeps the chamber wall temperature at 900K. In further development, ethanol could serve as fuel. Fuel and an oxidizer are mixed with a parallel micromixer (see Chapter 10). The mixer has 484 injectors, 242 each for fuel and oxidizer. At a chamber pressure of 12.16 bars, a thrust of 1N can be achieved. The corresponding thrust power is 750W.

A ceramic microrocket with solid fuel was reported in [36]. The microrocket is made of alumina ceramic, the thermal conductivity of which is five times lower than that of silicon. The combustion chambers have inner diameters of 3.2 mm and lengths of either 12.7 or 25.4 mm. The propellant is a solid composite fuel containing 74% oxidizer ammonium perchlorate (AP) and 14% binder hydroxyl-terminated polybutadiene (HTPB). The nozzles have throat diameters of 1.57 mm and lengths of either 6.35 or 12.7 mm. This rocket is able to deliver a maximum thrust of from 4 to 10 mN and specific impulses ranging from 10 to 20 seconds.

Microthrusters can be propelled with vaporizing liquid as well [37, 38]. Thermal bubbles generate the thrust. The concept is similar to that of an inkjet printer head. The energy is supplied by resistive heating. Thus, the efficiency of such a microthruster is relatively low. Table 5.4 compares the most important parameters of the above microrockets.

An array of microrockets can improve the performance with their redundancy. Furthermore, digital propulsion is possible with an array of microrockets [39]. The corresponding number of fired "microrocket bits" can deliver precisely the desired thrust.

## Problems

**5.1** The pressure drop across a Pitot probe is 30 mbar. Determine the flow velocity if the fluid is (a) water and (b) air.

**5.2** For a polysilicon wire with a length of $L = 60$ μm, a cross section of $A = 2 \times 4$ μm is used as a hot-wire sensor. The temperature coefficient and thermal conductivity of polysilicon are $8 \times 10^{-4}\text{K}^{-1}$ and $32\text{W m}^{-1}\text{K}^{-1}$, respectively. The bulk substrate temperature is 25°C. The resistivity of polysilicon at 25°C is $4 \times 10^{-5}$ Ωm. Determine the current required for a maximum temperature of 200°C at the center of the wire.

**5.3** The polysilicon wire described in Example 5.1 is used as a hot wire for measuring turbulent velocities. Determine the highest turbulence frequency that the sensor can detect.

**5.4** Determine the Reynolds number and the shear stress expected at 2m from the front of a flat plate. The free stream velocity is 5 m/s. If a shear stress sensor is flush-mounted at this position and the gap around the floating plate is 50 μm, can the error caused by flow field disturbance be ignored? Kinematic viscosity and density of air are assumed to be $1.6 \times 10^{-5}$ m²/s and 1.164 kg/m³, respectively.

**5.5** The sensor mentioned in Problem 5.4 is made of a silicon plate with a dimension of 2 × 2 mm. The plate is suspended on four flexures. Each flexure has a dimension of 1 mm × 10 μm × 10 μm. Displacement of the floating element is detected with an interdigitated comb electrode structure. Each side of the floating plate has 20 electrodes. The gap between two counterelectrodes is 2 mm. Design the capacitive sensor so that the capacitance change at 10 m/s is 10%.

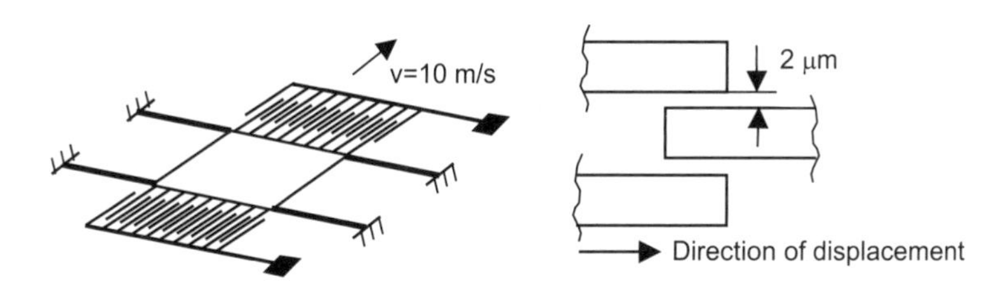


**5.6** Repeat the design process if sensing comb structures are perpendicular to flow direction.

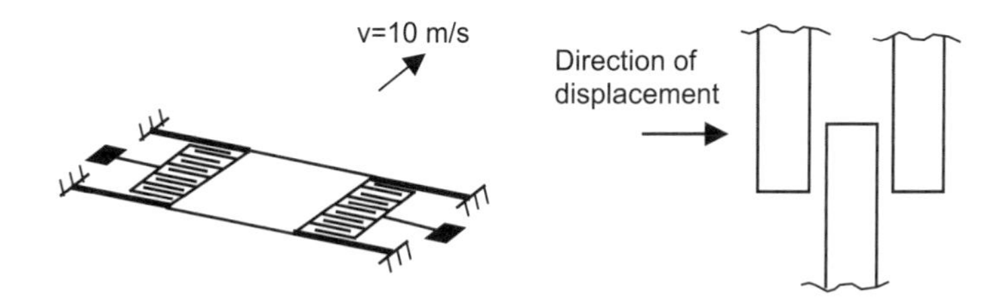


**5.7** An MAV is designed as a disk wing, which is driven by a propeller. The circular disk wing has a diameter of 10 cm. The design velocity is 15 m/s. The propeller is driven by an electric motor and has an efficiency of 60%. The lift and drag coefficients of the wing are 1.5 and 0.1, respectively. The device is designed for a total mass of 100g. Determine the capacity in mAh of a battery for a flight endurance of 20 minutes. The battery supplies a voltage of 3.3V. Air density is assumed to be 1.118 kg/m³.

**5.8** A flapping-wing MAV is designed with 15-cm mean chord length and a wing length of 10 cm. The flapping frequency is optimized for 30 Hz. The stroke amplitude is 80°. The lift coefficient of the wing is 1.5. The drag coefficient is 0.15. Determine the maximum mass of the device, the flight speed, and the energy capacity needed for a flight endurance of 20 minutes.

**5.9** Design an ideal microrocket, which operates at a pressure of 0.01 bar and has a thrust of 2N. The pressure and temperature in combustion chamber are 10 bar and 1,000K, respectively. The chamber and nozzle heights are 1 mm. Assuming that $k = 1.3$ and $R = 355.4$ J/kg-K, determine the exit velocity, throat width, and exit width of the microrocket.

## References


[1] Gad-el-Hak, M., "The Fluid Mechanics of Microdevices - The Freeman Scholar Lecture," *Journal of Fluids Engineering*, Vol. 121, 1999, pp. 5–32.

[2] Ho, C. M., and Tai, Y. C, "Review: MEMS and Its Applications for Flow Control," *Journal of Fluids Engineering*, Vol. 118, 1996, pp. 437–447.

[3] Fedder, G. K., "Simulation of Microelectromechanical Systems," Ph.D. thesis, University of California at Berkeley, 1994.

[4] Senturia, S. D., *Microsystem Design*, Boston: Kluwer Academic Publishers, 2001, pp. 632–634.

[5] Jiang, F., et al., "A Micromachined Polysilicon Hot-Wire Anemometer," *Technical Digest of the IEEE Solid State Sensor and Actuator Workshop*, Hilton Head Island, SC, June 13–16, 1994, pp. 264–267.

[6] Padmanabhan, A., "Silicon Micromachined Sensors and Sensor Arrays for Shear Stress Measurement in Aerodynamic Flow," Ph.D. thesis, Massachusetts Institute of Technology, 1997.

[7] Haritonidis, J. H., "The Measurement of Wall Shear Stress," *Advances in Fluid Mechanics Measurements*, M. Gad-el-Hak (ed.), New York: Springer Verlag, 1989, pp. 229–261.

[8] Oudheusden B., and J.Huijsing, "Integrated Flow Friction Sensor," *Sensors and Actuators A*, Vol. 15, 1988, pp. 135–144.

[9] Liu, C., et al., "Surface-Micromachined Thermal Shear Stress Sensor," *Journal of Microelectromechanical Systems*, Vol. 8, No. 1, 1999, pp. 90–99.

[10] Jang, F., et al., "A Surface Micromachined Shear Stress Imager," *Proceedings of MEMS'96, the 9th IEEE International Workshop Micro Electromechanical System*, San Diego, CA, Feb. 11–15, 1996, pp. 110–115.

[11] Kalvesten, E., "Pressure and Wall Shear Stress Sensors for Turbulence Measurements," Ph.D. thesis, Royal Institute of Technology, Stockholm, Sweden, 1996.

[12] Jang, F., et al., " A Flexible Micromachine-Based Shear-Stress Sensor Array and Its Application to Separation-Point Detection," *Sensors and Actuators A*, Vol. 79, No. 3, 2000, pp. 194–203.

[13] Schmidt, M. A., et al., "Design and Calibration of a Microfabricated Floating-Element Shear-Stress Sensor," *IEEE Transactions on Electron Devices*, Vol. ED-35, 1988, pp. 750–757.

[14] Ng, K.-Y., "A Liquid Shear-Stress Sensor Using Wafer-Bonding Technology," M.S. thesis, Massachusetts Institute of Technology, 1990.

[15] Goldberg, H. D., Breuer, K. S., and Schmidt, M. A., "A Silicon Wafer-Bonding Technology for Microfabricated Shear-Stress Sensors with Backside Contacts," *Technical Digest of the IEEE Solid State Sensor and Actuator Workshop*, Hilton Head Island, SC, June 13–16, 1994, pp. 111–115.

[16] Pan, T., et al., "Calibration of Microfabricated Shear Stress Sensors," *Proceedings of Transducers '95, 8th International Conference on Solid-State Sensors and Actuators*, Stockholm, Sweden, June 16–19, 1995, pp. 443–446.

[17] Padmanabhan, A., et al., "A Wafer-Bonded Floating-Element Shear Stress Microsensor with Optical Position Sensing by Photodiodes," *Journal of Microelectromechanical Systems*, Vol. 5, No. 4, 1996, pp. 307–315.

[18] Lee, G. B., et al., "Sensing and Control of Aerodynamic Separation by MEMS," *The Chinese Journal of Mechanics*, Vol. 16, No. 1, 2000, pp. 45–52.

[19] Liu, C., et al., "Out-of-Plane Magnetic Actuators with Electroplated Permalloy for Fluid Dynamics Control," *Sensors and Actuators A*, Vol. 78, No. 2–3, 1999, pp. 190–197.

[20] Dwight, H. B., *Electrical Coils and Conductors*, New York: McGraw-Hill, 1945.

[21] Benecke, W., "Silicon Microactuators: Activation Mechanisms and Scaling Problems," *Proceedings of Transducers '91, 6th International Conference on Solid-State Sensors and Actuators*, San Francisco, CA, June 23–27, 1991, pp. 46–50.

[22] Liu, C., et al., "Surface Micro-Machined Magnetic Actuators," *Proceedings of MEMS'94, 7th IEEE International Workshop Micro Electromechanical System*, Oiso, Japan, Jan. 25–28, 1994, pp. 57–62.

[23] Grosjean, C., et al., "Micro Balloon Actuators for Aerodynamic Control," *Proceedings of MEMS'98, 11th IEEE International Workshop Micro Electromechanical System*, Heidelberg, Germany, Jan. 25–29, 1998, pp. 166–171.

[24] Coe, D. J., et al., "Addressable Micromachined Jet Arrays," *Proceedings of Transducers '95, 8th International Conference on Solid-State Sensors and Actuators*, Stockholm, Sweden, June 16–19, 1995, pp. 329–332.

[25] Marischal, J. M., and Francis, M. S., "Micro Air Vehicles—Toward a New Dimension in Flight," http://www.darpa.com,1997.

[26] Grasmeyer, J. M., and Keennon, M. T., "Development of the Black Widow Micro Air Vehicle," http://www.aerovironment.com/area-aircraft/prod-serv/bwidpap.pdf, 2001.

[27] Dickinson, M. H., Lehmann, F. O., and Sane, S. P., "Wing Rotation and the Aerodynamic Basis of Insect Flight," *Science*, Vol. 284, 1999, pp. 1954–1960.

[28] Pornsin-Sirirak, T. N., et al., "Titanium-Alloy MEMS Wing Technology for a Micro Aerial Vehicle Application," *Sensors and Actuators A*, Vol. 89, No 1–2, 2001, pp. 95–103.

[29] Pornsin-Sirirak, T. N., et al., "MEMS Wing Technology for a Battery-Powered Ornithopter," *Proceedings of MEMS'00, 13th IEEE International Workshop Micro Electromechanical System*, Miyazaci, Japan, Jan. 23–27, 2000, pp. 799–804.

[30] Ellington, C. P., "The Novel Aerodynamics of Insect Flight: Applications to Micro-Air Vehicles," *The Journal of Experimental Biology*, Vol. 202, 1999, pp. 3439–3448.

[31] Kroo, I., Mesicopter project, http://aero.stanford.edu/mesicopter/, Stanford University.

[32] Sutton, G. P., and Biblarz, O., *Rocket Propulsion Elements*, 7th ed., New York: Wiley, 2001.

[33] Bayt, R., "Analysis, Fabrication, and Testing of a MEMS-Based Micropropulsion System," Ph.D. thesis, Massachusetts Institute of Technology, 1999.

[34] Köhler, J., et al., "A Hybrid Cold Gas Microthruster System for Spacecraft," *Proceedings of Transducers '01, 11th International Conference on Solid-State Sensors and Actuators*, Munich, Germany, June 6–7, 2001, pp. 886–889.

[35] London, A. P., "Development and Test of a Microfabricated Bipropellant Rocket Engine," Ph.D. thesis, Massachusetts Institute of Technology, 2000.

[36] Teasdale, D., "Solid Propellant Microrockets," M.Sc. thesis, University of California at Berkeley, 2000.

[37] Wallace, A. P., et al., "Design, Fabrication, and Demonstration of a Vaporizing Liquid Attitude Control Microthruster," *Proceedings of Transducers '99, 10th International Conference on Solid-State Sensors and Actuators*, Sendai, Japan, June 7–10, 1999, pp. 1800–1803.

[38] Kim, S., Kang, T., Cho, Y., "High-Impulse Low-Power Microthruster Using Liquid Propellant with High-Viscous Fluid-Plug," *Proceedings of Transducers '01,11th International Conference on Solid-State Sensors and Actuators*, Munich, Germany, June 6–7, 2001, pp. 898–901.

[39] Lewis, D. H., "Digital Micropropulsion," *Proceedings of MEMS'99, 12th IEEE International Workshop Micro Electromechanical System*, Orlando, FL, Jan.17–21, 1999, pp. 517–522.

# Chapter 6

# Microfluidics for Internal Flow Control: Microvalves

Microvalves are one of the most important microfluidic components. Besides pumps and flow sensors, active valves are critical components for controlling fluid flows in microfluidic systems. In particular, microvalves are considered as vital components for on-chip flow manipulation that is required for various chemical, analytical and biological assays.

Today, the industry requirements for microfluidic systems continue to force an evolution and a revolution in valve design because of the new effects in microscale. Smaller device size needed for high-density integration of microvalves on a chip, higher pressures, reliability, cost of fabrication, biocompatibility, response, and, most importantly, the microtechnology, are all contributing to the valve design in microscale. Since *passive valves* or check-valves are a part of micropumps, this chapter only deals with active microvalves for flow control.

With the conventional design, an *active valve* is a pressure-containing mechanical device used to shut off or otherwise modify the flow of a fluid that passes through it. The working state of the valve is determined by a closure element—the valve seat, which is driven by an actuator. This definition reveals that a valve is a very simple device. It has a body to contain the fluid and its pressure, a valve seat to manipulate the fluid, and an actuator to control the position of the valve seat.

There are many ways to categorize active microvalves. Based on their initial working state, there are three microvalve types: normally open, normally closed, and bistable (Figure 6.1) . Bistable microvalves can actively open and close the valve seat; an off-state is not defined.

Similar to an electronic transistor, valves can control flow in two ways: analog and digital. In the analog, or proportional, mode, at a constant inlet pressure, the valve actuator varies the gap between the valve seat and valve opening to change the fluidic resistance and, consequently, the flow rate. In the digital mode, there are only two valve states: fully open and fully closed. However, a digital active valve can be driven in pulse-width-modulation (PWM) mode or as a digitally weighted valve array for proportional flow control.

In the PWM mode, the open time is controlled, and as a result, the net flow rate can be varied proportionally with the opening time. In an array, many digital valves are used to control the flow. If the flow rates of each valve are equal, the net flow rate is proportional to the number of opened valves. A more elegant solution is to weight the flow rate of each valve with the binary system. Thus, the array can represent a fluidic digital/analog converter. With, for example, 8 binary valves, 256 different flow rate values can be controlled (Figure 6.2).

In this chapter, active microvalves are categorized by their actuation principles as:

- Pneumatic microvalves;

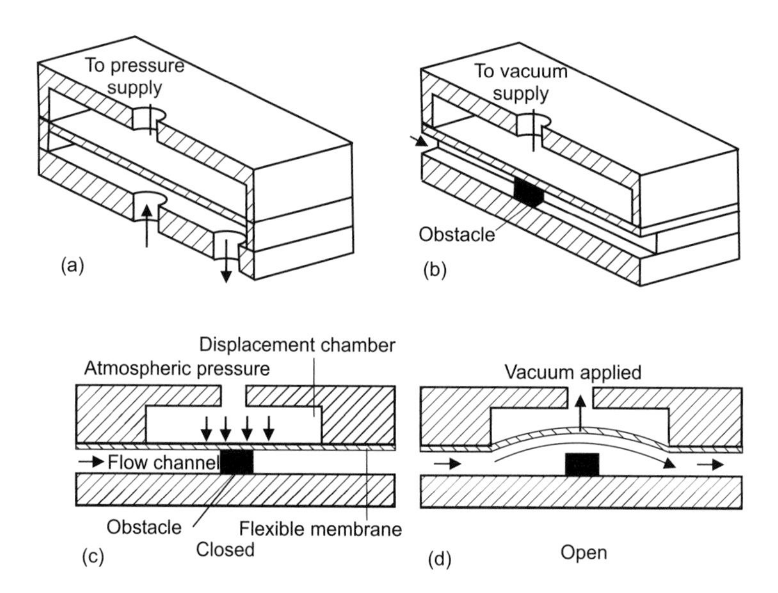


**Figure 6.1** Two design examples of microvalves: (a) normally open valve (NO); (b) normally closed valve (NC); (c) side view of the normally closed valve once it is closed; and (d) side view of the normally closed valve once it is open.

- Thermopneumatic microvalves;
- Thermomechanical microvalves;
- Piezoelectric microvalves;
- Electrostatic microvalves;
- Electromagnetic microvalves;
- Electrochemical and chemical microvalves;
- Capillary force microvalves.

The major specification bases for microvalves are leakage, valve capacity, power consumption, closing force (pressure range), temperature range, response time, reliability, biocompatibility, and chemical compatibility.

The ideal active valve should have zero leakage in the closed position. The *leakage ratio* $L_{\text{valve}}$ is defined as the ratio between the flow rate of the closed state $\dot{Q}_{\text{closed}}$ and of the fully open state $\dot{Q}_{\text{open}}$ at a constant inlet pressure:

$$L_{\text{valve}} = \frac{\dot{Q}_{\text{closed}}}{\dot{Q}_{\text{open}}} \tag{6.1}$$

Sometimes, the leakage is defined as the on/off ratio. In order to avoid the confusion between normally closed valves and normally open valves, the above definition of leakage (6.1) is kept throughout this chapter.

The *valve capacity* characterizes the maximum flow rate the valve can handle. The valve capacity $C_{\text{valve}}$ is defined as:

$$C_{\text{valve}} = \frac{\dot{Q}_{\text{max}}}{\sqrt{\Delta p_{\text{max}}/(L\rho g)}} \tag{6.2}$$

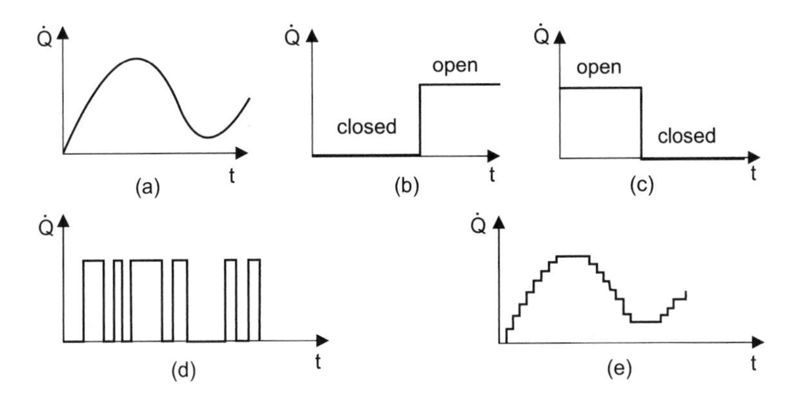


**Figure 6.2** Functional classification of microvalves: (a) analog (proportional) valve; (b) digital normally closed (NC) valve; (c) digital normally open (NO) valve; (d) PWM proportional valve; and (e) fluidic digital-analog converter.

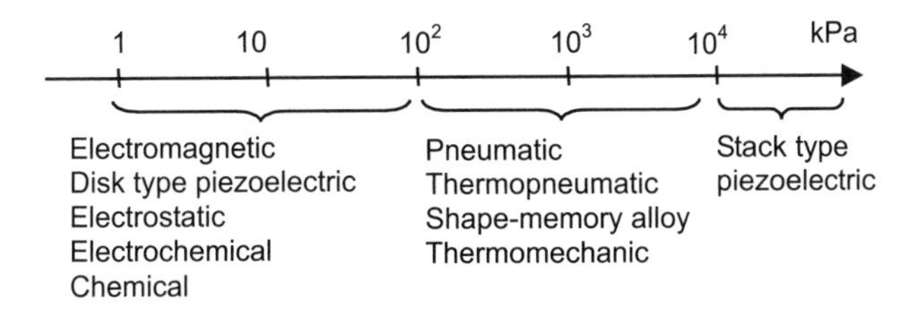


**Figure 6.3** Pressure range of different actuators used in microvalves.

where $\dot{Q}_{\mathrm{max}}$ and $\Delta p_{\mathrm{max}}$ are the flow rate and the pressure drop across the valve at the fully open position, respectively; $L$ is the characteristic length of the valve; $\rho$ is the fluid density; and $g$ is the acceleration of gravity. The power consumption is the total input power of the valve in its active, power-consuming state. Depending on the actuating principle, the power consumption may vary several orders of magnitude from very small (electrochemical) to very large (thermopneumatic).

The *closing force* depends on the pressure generated by the actuator. Figure 6.3 gives an overview of the pressure range of different actuators used in microvalves [1].

The *temperature range* of the valve depends strongly on the material and its actuation concept. Pneumatic valves are often used for high-temperature applications because the temperature range depends only on their material. The valve response time is actually the response time of the actuator in use. Figure 6.4 compares the response time of different actuators.

The actuators and the operating conditions determine the reliability of a microvalve. In the microscale, the operation failure is often caused by particulate contamination and not by the reliability of its actuator. Therefore, it is important to use filters to keep particulates out of microvalves.

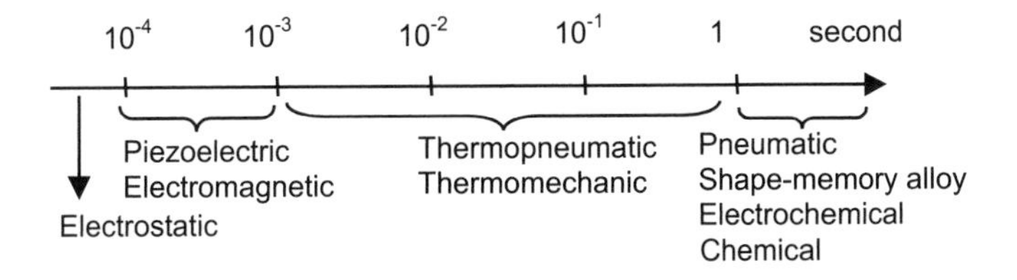


**Figure 6.4** Time response range of different actuators used in microvalves.

## 6.1 DESIGN CONSIDERATIONS

At the start of a microvalve design project, some important questions should be answered.

- What specifications should the valve meet for the intended application?
- What material should be used for valve fabrication?
- How much will it cost?
- What type of actuator best suits the specifications and constraints?
- What is the optimal form of valve spring and valve seat for the maximum closure pressure and minimum leakage?

All the specifications and constraints of the application should be analyzed carefully before starting to design the valve. Since a large number of microvalves are used for biomedical and chemical analysis, the choice of valve material and its biocompatibility as well as chemical compatibility are crucial for the final product. The overall cost of the valve is also a very important factor. Constrained by high safety requirements, biomedical analysis often needs disposable devices, which should be mass fabricated at a low cost. In addition, the size of a microvalve is a critical design constraint, since for many applications a large number of valves are employed on a single microfluidic circuit for different on-chip flow manipulation schemes. In this section, design considerations of functional elements, such as actuators, valve springs, and valve seats, are discussed in detail.

### 6.1.1 Actuators

Figure 6.5 illustrates the basic actuation concepts for an active microvalve. The actuators have several purposes in valve operation.

- *Moving function:* The actuator should be able to move the valve seat to the desired position. For this function, the actuator should provide enough force, displacement, and controllability.
- *Holding function:* The actuator should be able to keep the valve seat in the desired position, especially in the case of proportional valves. The actuator and the valve spring should be able to overcome the inlet pressure.
- *Dynamic function:* Since the response time of the valve is determined mainly by the actuator, the actuator should meet the dynamic requirement of the application.

The parameter that characterizes actuator performance is the work produced:

$$W_{\mathrm{a}} = F_{\mathrm{a}} s_{\mathrm{a}} \tag{6.3}$$

where $F_{\mathrm{a}}$ is the actuator force and $s_{\mathrm{a}}$ is the maximum displacement. Because the size of the actuator is crucial for miniaturization, the energy density $E'_{\mathrm{a}}$ stored in each actuator can be used for rating their performance:

$$E'_{\mathrm{a}} = \frac{W_{\mathrm{a}}}{V_{\mathrm{a}}} = \frac{F_{\mathrm{a}} s_{\mathrm{a}}}{V_{\mathrm{a}}} \tag{6.4}$$

where $V_{\mathrm{a}}$ is the total actuator volume. Figure 6.6 represents the range of energy density for different types of microactuator [1].

The energy density of thermomechanical actuators can be estimated from Young's modulus $E$, the thermal expansion coefficient $\gamma$, and the temperature difference $\Delta T$:

$$E'_{\mathrm{a}} = \frac{1}{2} E (\gamma \Delta T)^2 \tag{6.5}$$

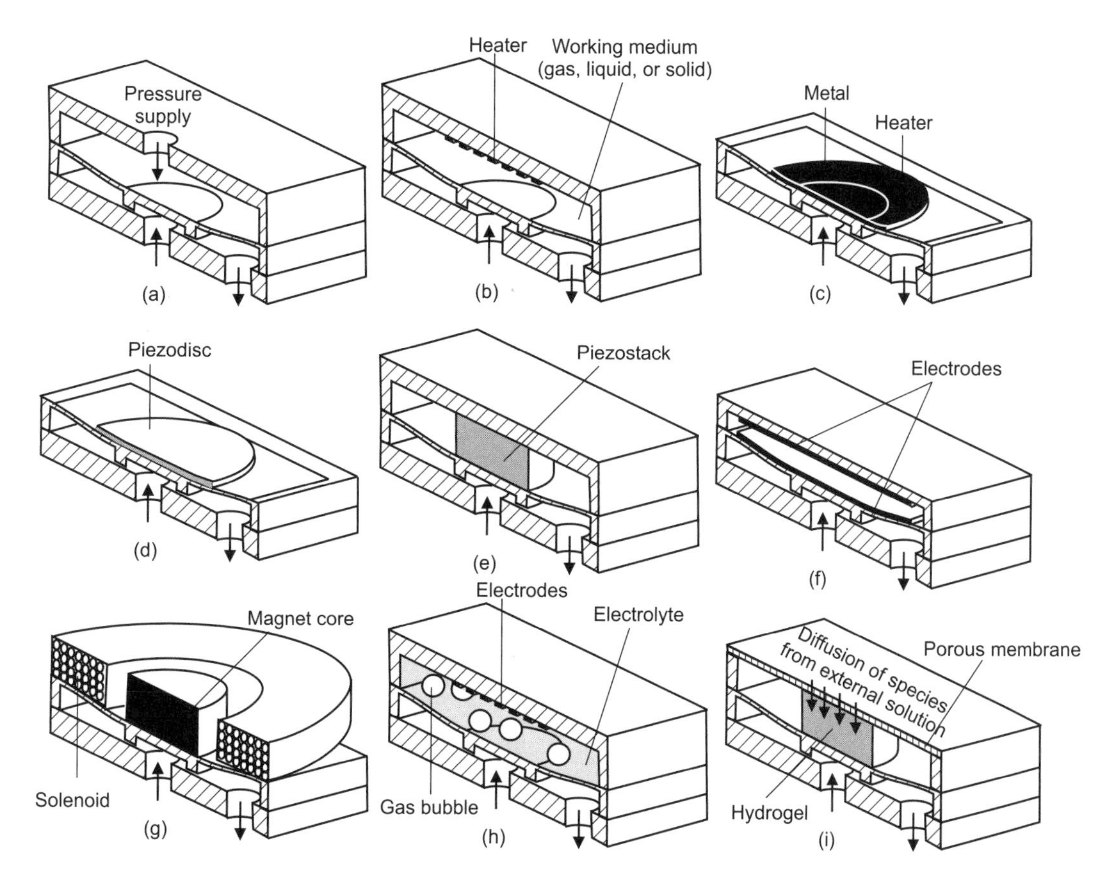


**Figure 6.5** Basic actuation concepts for an active microvalve: (a) pneumatic; (b) thermopneumatic; (c) thermomechanic; (d) piezoelectric; (e) piezoelectric; (f) electrostatic; (g) electromagnetic; (h) electrochemical; and (i) chemical.

The energy density of piezoelectric actuators can be calculated with the electric field strength $E_{\mathrm{el}}$, from Young's modulus $E$, and the piezoelectric coefficient $d_{33}$:

$$E'_{\mathrm{a}} = \frac{1}{2}\frac{E_{\mathrm{el}}}{(d_{33}E)^2} \tag{6.6}$$

The energy density stored in electrostatic actuators can be calculated based on the energy stored in the capacitance between their two electrodes:

$$E'_{\mathrm{a}} = \frac{1}{2}\varepsilon E_{\mathrm{el}}^2 \tag{6.7}$$

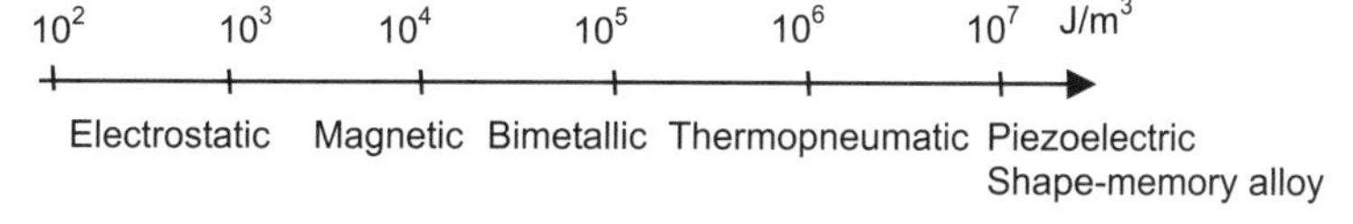


**Figure 6.6** Energy density range of different actuators used in microvalves.

where $\varepsilon = \varepsilon_r \varepsilon_0$ is the dielectric constant of the material between two electrodes (i.e., air for electrostatic actuators and piezoelectric material for piezoelectric actuators), $\varepsilon_0 = 8.85418 \times 10^{-12}$ F/m is the permittivity of vacuum, and $\varepsilon_r$ is the relative dielectric constant of the material.

The energy density stored in an electromagnetic actuator is:

$$E'_a = \frac{1}{2} \frac{B^2}{\mu} \tag{6.8}$$

where $B$ is the magnetic flux density (in tesla), $\mu = \mu_r \mu_0$ is the permeability of the operating medium, $\mu_0 = 4\pi \times 10^{-7} = 1.26 \times 10^{-6}$ H/m is the permeability of vacuum, and $\mu_r$ is the relative permeability of the material.

#### 6.1.1.1 Pneumatic Actuators

Pneumatic actuation is the simplest actuation concept. However, this actuation concept is the most popular method for microfluidic applications. A pneumatic valve needs an external pressure source for actuation. Thus, this actuation concept may limit its use for some applications such as hand-held or pocket-sized microfluidic systems where compactness is a critical design constraint. Pneumatic actuation has been adopted for both normally open and normally closed microvalves. For the actuation of a normally open pneumatic microvalve, a gas source with pressure higher than the atmospheric value is required. The high-pressure gas is usually introduced above a flexible membrane to create the required deformation to impede fluid movement in a flow channel. The level of required gas pressure is mainly determined by the overall size of the flow channel, the membrane diameter, its thickness and the mechanical properties of the membrane material. As for the actuation of normally closed pneumatic microvalves, a vacuum supply is needed to create sufficient deformation in a flexible membrane allowing for a fluid passing through the flow channel. Membrane deformation happens in the displacement chamber where the vacuum supply is applied, Figure 6.1(c, d).

In terms of temperature ranges, pneumatic valves offer the widest range of operation, since an applied external pneumatic pressure always deflects the valve membrane, regardless of the thermal state of the system that may influence other actuation principles. Pneumatic valves have a membrane structure as the valve seat. The valve should be optimized for the following specifications:

- Low actuating pressure, high differential pressure across the valve;
- Low leakage or high on/off ratio of the flow rate;
- Low spring constant of the membrane.

The response time of pneumatic microvalves depends on external switching devices, which are much slower than the microvalve itself. Due to the relatively large external pneumatic supply system, pneumatic microvalves expect a response time on the order from several hundred milliseconds to several seconds.

**Example 6.1: Designing a Pneumatic Microvalve**

A pneumatic microvalve has a circular silicon membrane as the valve seat. The membrane is 20 μm thick and has a diameter of 4 mm. The valve is normally open with a gap of 20 μm between the membrane and the valve inlet. Determine the pressure required for closing the valve at an inlet pressure of $p_{in} = 1$ bar. The opening diameter is 200 μm.

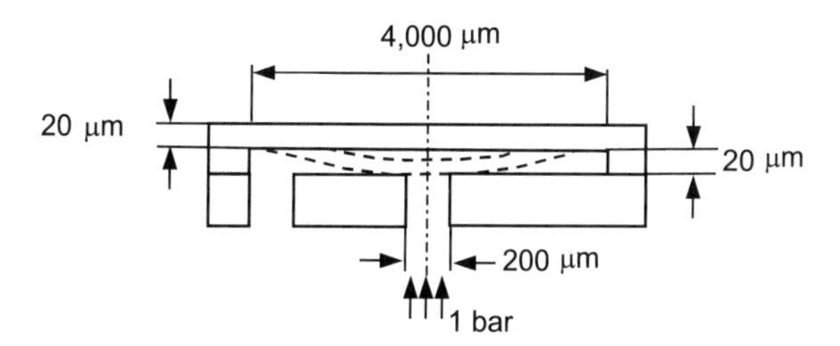


**Solution.** Assume a distributed load on the valve membrane, a Poisson's ratio of 0.25, and a bulk Young's modulus of silicon of 170 GPa. For a small deflection, the spring constant of the valve membrane is estimated as (see Appendix D):

$$k = \frac{16\pi E t^3}{3r^2(1-\nu^2)} = \frac{16\pi \times 170 \times 10^9 \times (2 - \times 10^{-6})^3}{3 \times (2 \times 10^{-3})^2(1 - 0.25^2)} = 6.08 \times 10^3 \text{ N/m}$$

If the microvalve is closed at $p_{\text{act}}$, the force balance on the membrane is:

$$p_{\text{act}} A_{\text{m}} = p_{\text{in}} A_{\text{m}} + F_{\text{spring}}$$

Thus:

$$p_{\text{act}} = p_{\text{in}} + \frac{F_{\text{spring}}}{A_{\text{m}}} = p_{\text{in}} + \frac{kg}{\pi r_{\text{m}}^2}$$

$$p_{\text{act}} = 10^5 + \frac{6.08 \times 10^3 \times 20 \times 10^{-6}}{\pi \times (2 \times 10^{-3})^2} = 100,000 + 9,677 = 109,677 \text{ Pa}$$

The reader can calculate the pressure required for keeping the valve closed and compare it with the above result.

#### 6.1.1.2 Thermopneumatic Actuators

Thermopneumatic actuation relies on the change in volume of sealed liquid or solid under thermal loading. The principle also uses the phase change from liquid to gas or from solid to liquid to gain a larger volume expansion. The mechanical work generated in such a thermodynamic process exceeds the energy stored in an electrostatic or electromagnetic field.

Although the nucleation and boiling process in the microscale behaves differently than in the macroscale, an explosion-like phase transition was widely used for droplet generation (see Chapter 11). This section considers the simple phase transition model because most of the thermopneumatic actuators are on the scale of millimeters.

The actuation material of the thermopneumatic valve can be gas, liquid, or solid. Due to the large change in specific volume of the phase transition, thermopneumatic actuators can utilize the solid/liquid and liquid/gas phase change to obtain the maximum performance. Figure 6.7 represents the typical thermodynamic states of a thermopneumatic valve with a liquid as the initial actuating material. The volume $V$ and mass $m$ determine the specific volume $v$ and the density $\rho$ of the working material in the actuation chamber:

$$v = \frac{V}{m} = \frac{1}{\rho} \tag{6.9}$$

Because the actuating material is sealed hermetically, the mass remains constant. The specific volume is proportional to the volume. Assuming a linear spring load, the states and closing processes are depicted in Figure 6.7.

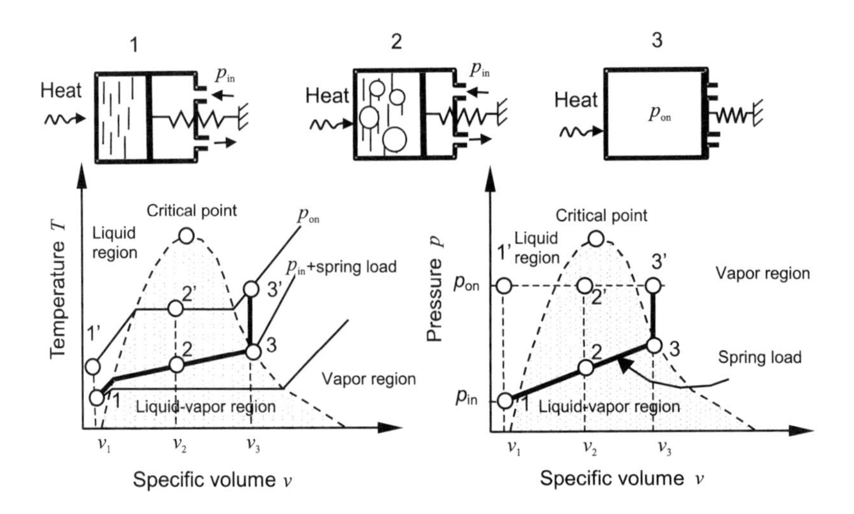


**Figure 6.7** Working principle of thermopneumatic valves.

*Expansion in the Liquid Region (Process 1-1').* In the initial state, the fluid in the actuator chamber is in the liquid phase. With the increasing temperature caused by the heater, the volume expands with an amount of $\Delta V$:

$$\Delta V = V_0 \gamma_f (T_2 - T_1) \tag{6.10}$$

where $V_0$ is the initial fill volume at temperature $T_1$, $\gamma_f$ is the thermal expansion coefficient of the liquid, and $T_2$ is the operating temperature of the actuator.

If in this state the valve membrane already closes the valve inlet, and the specific volume remains constant, the pressure and the temperature increase as a result. Based on the specific volume $v$ defined in (6.9) and the known operating temperature, the pressure inside the actuator chamber can be determined from the properties of compressed liquids.

*Expansion in the Liquid-Vapor Mixture Region (Process 1-2-2').* If there is enough space inside the chamber, the fluid transforms into the liquid-vapor phase. If the valve is closed in this state, temperature and pressure increase as a result of the heat transfer. Because the pressure increases with temperature in state 2', keeping the temperature constant can close the valve with a constant pressure. In this state, the dryness fraction $x$ or the mass percentage of the vapor phase can be determined from its specific volume $v$ and the specific volumes of saturated liquid $v_f$ and saturated vapor $v_g$:

$$x = \frac{v - v_f}{v_g - v_f} \tag{6.11}$$

*Expansion in the Superheated Vapor Region (Process 1-2-3-3').* If the heat transfer continues, the fluid can jump from the liquid-vapor mixture region into the superheated vapor region. In this state, the generated pressure $p$ in the actuator chamber is given by:

$$p = p_0 \exp\left(\frac{-\bar{L}_0}{\bar{R}T}\right) \tag{6.12}$$

where $\bar{R} = 8.314$ kJ/kmol-K is the universal gas constant, $\bar{L}_0$ is the molar latent heat of evaporation of the actuating fluid, and $T$ is the absolute temperature in Kelvin. Equation (6.12) assumes that the vapor behaves as an ideal gas. For a practical design, we can use the thermodynamic approach illustrated in the following examples.

**Example 6.2: Designing a Thermopneumatic Microvalve with Air as Working Fluid**

The valve described in Example 6.1 is designed with a thermopneumatic actuator on top of the membrane. The actuator chamber is a cylinder with a height of 500 μm. If the chamber is filled with air and hermetically sealed, determine the temperature required for closing the valve at an inlet pressure of 1 bar. The initial pressure and temperature in the chamber are 1 bar and 27°C.

**Solution.** Assuming that the volume of the actuating chamber is constant, the relation between temperature and pressure is:

$$\frac{T_1}{T_2} = \frac{p_1}{p_2} \rightarrow T_2 = T_1\frac{p_2}{p_1} = 300 \times \frac{109,677}{100,000} = 329 \text{ K} = 56°\text{C}$$

In order to close the valve at an inlet pressure of 1 bar, the temperature in the actuating chamber should be at least 56°C.

**Example 6.3: Thermopneumatic Valve with a Liquid-Vapor Mixture as Working Fluid**

The valve described in Example 6.2 is now redesigned with a small actuation chamber and a large gap for minimum pressure drop in the opening state. The membrane is made of silicone rubber and the spring force is negligible. The initial actuating chamber height is 20 μm. The gap at the opening position is 500 μm. The water is filled and hermetically sealed at 1 bar, 25°C. In operation, the chamber is heated up to 120°C. Determine the actuating pressure. (Thermodynamic data of water at 120°C: specific volume of saturated liquid $v_f = 1.985 \times 10^{-3}$ m$^3$/kg, specific volume of saturated vapor: $v_g = 0.8919$ m$^3$/kg, saturation pressure $p = 1.986$ bar.)

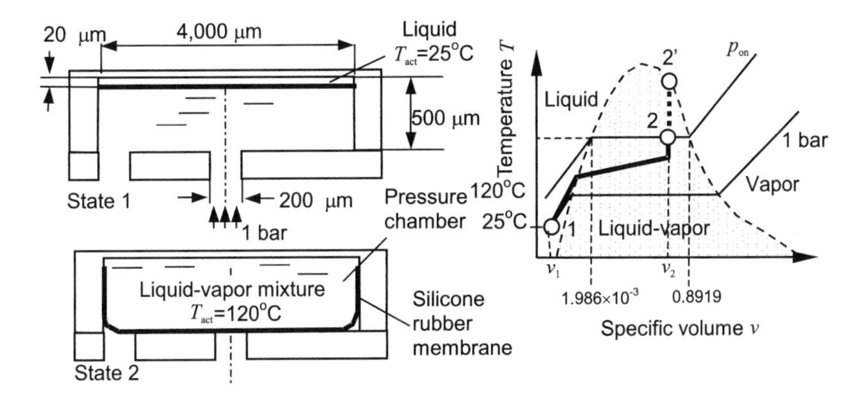


**Solution.** The mass of water in the chamber is:

$$m = \frac{V}{v_1} = \frac{(20 \times 10^{-6}) \times \pi \times (2 \times 10^{-3})^2}{1.0029 \times 10^{-3}} = \frac{251.33 \times 10^{-12}}{1.0029 \times 10^{-3}} = 250 \times 10^{-9} \text{ kg}$$

At the actuating position, we assume that the entire pressure chamber is filled with water; thus, the specific volume at this final state is:

$$v_2 = \frac{V_2}{m} = \frac{[520 \times 10^{-6} \times \pi \times (2 \times 10^{-3})]^2}{250.6 \times 10^{-9}} = 0.0261 \text{ m}^3/\text{kg}$$

Since the specific volume of saturated liquid and saturated vapor at 120°C are $1.985 \times 10^{-3}$ m$^3$/kg and 0.8919 m$^3$/kg, respectively, the final state is in the mixture region and the actuating pressure is equal to the saturation pressure of 1.986 bar.

**Table 6.1**
Thermal Properties of Some Materials at 300K (*After*: [2, 3].)

| *Material* | *Density (kg/m$^3$)* | *Heat Capacity (J/kgK)* | *Conductivity (W/mK)* | *Thermal Expansion Coefficient (10$^{-6}$K$^{-1}$)* |
|---|---|---|---|---|
| Silicon | 2,330 | 710 | 156 | 2.3 |
| Silicon oxide | 2,660 | 750 | 1.2 | 0.3 |
| Silicon nitride | 3,100 | 750 | 19 | 2.8 |
| Aluminum | 2,700 | 920 | 230 | 23 |
| Copper | 8,900 | 390 | 390 | 17 |
| Gold | 19,300 | 125 | 314 | 15 |
| Nickel | 8,900 | 450 | 70 | 14 |
| Chrome | 6,900 | 440 | 95 | 6.6 |
| Platinum | 21,500 | 133 | 70 | 9 |
| Parylene-N | 1,110 | 837.4 | 0.12 | 69 |
| Parylene-C | 1,290 | 711.8 | 0.082 | 35 |
| Parylene-D | 1,418 | — | — | 30–80 |

**Example 6.4: Thermopneumatic Valve with Water Vapor as Working Fluid**

What are the temperature and pressure at which the working fluid of the valve described in Example 6.3 totally evaporates? Thermodynamic properties are given in the following table.

| *Boiling Point (°C)* | *Specific Volume v (m$^3$/kg)* | *Pressure (bars)* |
|---|---|---|
| 280 | 0.03017 | 64.12 |
| 290 | 0.02557 | 74.36 |

**Solution.** The operation state will be point 2' in the $T - v$ diagram of Example 6.3. Interpolating in the table for the specific volume, we get the operation temperature and actuating pressure:

$$T_2 \approx 289°\text{C}, p_2 = 73.18 \text{ bar}$$

With this high temperature and pressure, it is unlikely that microvalves can be actuated using the vapor phase of a working liquid.

#### 6.1.1.3 Solid-Expansion Actuators

Thermal-expansion actuators use the volume change of a solid body to induce stress in it. The generated force is proportional to the temperature difference $\Delta T$ between the heater and the ambient temperature:

$$F \propto \gamma_s \Delta T \tag{6.13}$$

where $\gamma_s$ is the thermal expansion coefficient of the solid material (Table 6.1).

Careful design of heaters, their location, and thermal isolation are needed for the optimal operation of microvalves based on solid-expansion actuation. The valve depicted in Figure 6.8 uses a polysilicon heater on a membrane, which is suspended on four flexures [4]. The hot membrane expands while the colder frame remains fixed. The high compressive stress within the membrane is accumulated up to a critical value and buckles instantaneously. The buckling stress opens the valve. Due to the relatively high heat conductivity of silicon, the valve consumes several watts of heating power.

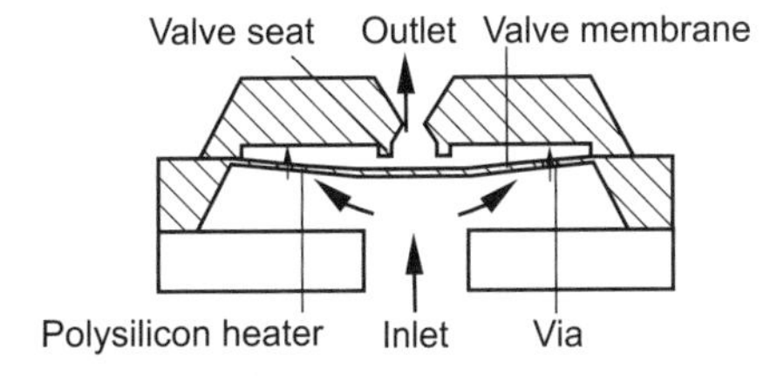


**Figure 6.8** Typical solid-expansion microvalve.

#### 6.1.1.4 Bimetallic Actuators

Bimetallic actuation uses the difference in thermal coefficient of expansion of two bonded solids. This principle is often called thermal bimorph actuation. The heater is usually integrated between the two solid materials or on one side of the bimorph. Because most bimorph structures are thin membranes, the temperature gradient along their thickness can be assumed to be zero. Thus, it is not important where the heater is integrated. Bimetallic actuators offer an almost linear deflection dependence on heating power. Their disadvantages are high power consumption and slow response. Integrated temperature sensors are often integrated with the heater to keep the actuator temperature constant using closed control loop. The resulting force is proportional to the difference between the thermal expansion coefficients of the two materials $\gamma_2 - \gamma_1$ and the temperature difference $\Delta T$:

$$F \propto (\gamma_2 - \gamma_1)\Delta T \tag{6.14}$$

Figure 6.9 depicts a typical bimetallic beam actuator with a length $L$ and layer thickness of $t_1$ and $t_2$. The displacement of the beam tip is calculated as:

$$y(L) \approx \frac{L^2}{2R} \tag{6.15}$$

$R$ is the radius of curvature [5]:

$$R = \frac{(b_1E_1t_1)^2 + (b_2E_2t_2)^2 + 2b_1b_2E_1E_2t_1t_2(2t_1^2 + 3t_1t_2 + 2t_2^2)}{6(\gamma_2 - \gamma_1)\Delta T b_1b_2E_1E_2t_1t_2(t_1 + t_2)} \tag{6.16}$$

where $b_1$ and $b_2$ are the widths of the two material layers. $E_1$ and $E_2$ are Young's moduli. The equivalent force on the tip of the beam is:

$$F = \frac{3(EI)_{\text{beam}}y(L)}{L^3} \tag{6.17}$$

The flexural rigidity of the beam $(EI)_{\text{beam}}$ is:

$$(EI)_{\text{beam}} = \frac{(b_1E_1t_1^2)^2 + (b_2E_2t_2)^2 + 2b_1b_2E_1E_2t_1t_2(2t_1^2 + 3t_1t_2 + 2t_2^2)}{12(t_1b_1E_1 + t_2b_2E_2)} \tag{6.18}$$

Assuming that the two layers have the same width $b$, the radius of curvature and the equivalent tip force can be simplified to:

$$r = \frac{t_1 + t_2}{6} \frac{3\left(1 + \frac{t_1}{t_2}\right)^2 + \left(1 + \frac{t_1E_1}{t_2E_2}\right)\left[\left(\frac{t_1}{t_2}\right)^2 + \frac{t_2E_2}{t_1E_1}\right]}{(\gamma_2 - \gamma_1)\Delta T\left(1 + \frac{t_1}{t_2}\right)^2} \tag{6.19}$$

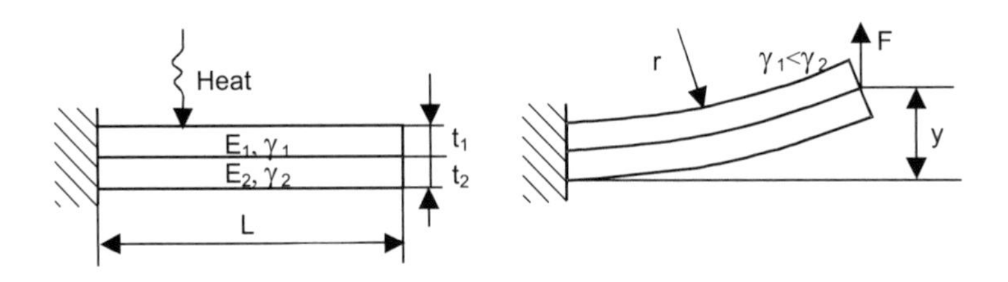


**Figure 6.9** Bimetallic beam actuator.

$$F = \frac{3b}{4L} \frac{t_1 + t_2}{\frac{1}{t_1 E_1} + \frac{1}{t_2 E_2}} (\gamma_2 - \gamma_1) \Delta T \tag{6.20}$$

### Example 6.5: Design of a Thermomechanical Valve with Bimetallic Actuator

A thermomechanical microvalve has a rigid square seat of 500 × 500 μm. The valve seat is suspended on four flexures. Each flexure is 500 μm long and 200 μm wide. The flexure is made of 10-μm silicon and 2-μm aluminum. The silicon heater is integrated in the flexure. The aluminum layer is evaporated at 400°C. If a normally closed microvalve is to be designed at 25°C, what is the maximum gap between valve opening and the surface of the valve seat wafer? Material properties are given in the figure below.

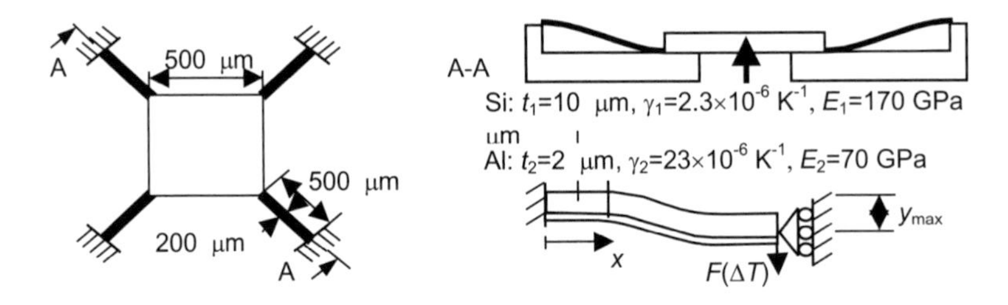


**Solution.** From (6.20), we get the equivalent force at the tip of each actuating flexure as a function of the temperature difference:

$$F = \frac{3b}{4L} \frac{t_1 + t_2}{\frac{1}{t_1 E_1} + \frac{1}{t_2 E_2}} (\gamma_2 - \gamma_1) \Delta T = 9.639 \times 10^{-6} \Delta T \text{ N}$$

Assuming that the two layers are stress-free under evaporation conditions, the force acting at each flexure tip at room temperature is:

$$F = 9.639 \times 10^{-6} \Delta T = 9.639 \times 10^{-6} (400 - 25) = 3.615 \times 10^{-3} \text{ N}$$

Using (6.18), the flexural rigidity of the composite beam is:

$$\begin{aligned}(EI)_{\text{beam}} &= \frac{(b_1 E_1 t_1^2)^2 + (b_2 E_2 t_2)^2 + 2 b_1 b_2 E_1 E_2 t_1 t_2 (2t_1^2 + 3t_1 t_2 + 2t_2^2)}{12(t_1 b_1 E_1 + t_2 b_2 E_2)} \\ &= 3.774 \times 10^{-9} \text{ Pa.m}^4\end{aligned}$$

Since the rectangular valve seat is rigid, we need to assume that the slope at the tip is zero. If $y$ is the function of the deflection along the beam, the differential equation for $y$ is:

$$(EI)_{\text{beam}} y''' = -F$$

Solving the above equation with boundary conditions at ($x = 0$) and ($x = L$):

$$y'(0) = 0; y'(L) = 0; y(0) = 0$$

we get the beam deflection as a function of $x$ (see Appendix D):

$$y(x) = \frac{F}{2(EI)_{\text{beam}}}\left(\frac{-x^3}{3} + \frac{Lx^2}{2}\right)$$

Thus, the maximum deflection at the beam tip is:

$$y_{\text{max}} = \frac{FL^3}{12(EI)_{\text{beam}}} = \frac{3.615 \times 10^{-3} \times (5 \times 10^{-}4)^{-3}}{12 \times 3.774 \times 10^{-9}} = 10 \times 10^{-6}\ \text{m} = 10\ \mu\text{m}$$

The gap between the valve opening and the surface of the valve seat wafer should be less than 10 µm to have a normally closed valve.

**Example 6.6: Minimum Opening Pressure and Temperature**

If the gap between valve opening and surface of the valve seat wafer is 5 µm, how large should the inlet pressure be to open the valve? What is the minimum temperature difference for opening the valve at zero inlet pressure? The valve opening is 200 × 200 µm.

**Solution.** The force for keeping the flexure deflection at 5 µm is:

$$F = y\frac{12(EI)_{\text{beam}}}{L^3} = 5 \times 10^{-6}\frac{12 \times 3.774 \times 10^{-9}}{(5 \times 10^{-4})^3} = 1.812 \times 10^{-3}\ \text{N} = 1.812\ \text{mN}$$

Thus, the closing force of the valve seat is:

$$F_{\text{closing}} = 4 \times (3.615 - 1.812) \times 10^{-3} = 7.212 \times 10^{-3}\ \text{N}$$

The corresponding minimum opening pressure required is:

$$\Delta p_{\text{opening}} = \frac{F_{\text{closing}}}{A} = \frac{7.212 \times 10^{-3}}{(200 \times 10^{-6})^2} = 1.803 \times 10^5\ \text{Pa} = 1.803\ \text{bar}$$

From the solution of Example 6.5, the thermomechanical force as a function of temperature difference is given as:

$$F = 9.639 \times 10^{-6}\Delta T\ \text{N}$$

The temperature difference needed for opening the valve is:

$$\Delta T = 400 - \frac{1.812 \times 10^{-3}}{9.639 \times 10^{-6}} = 188^{\circ}\text{C}$$

#### 6.1.1.5 Shape-Memory Alloy Actuators

Shape-memory alloys (SMA) are materials such as titanium/nickel alloy, which, once mechanically deformed, return to their original undeformed shape upon a change of temperature. SMAs undergo phase transformations from a “soft” state (martensite) at low temperatures to a “hard” state (austenite) at higher temperatures. Since the alloy structure represents an electrical resistor, passing a current

**Table 6.2**
Typical SMAs (*After*: [6].)

| *Alloy* | *Composition* | *Transformation Temperature Range (°C)* | *Transformation Hysteresis (°C)* |
|---|---|---|---|
| Ag-Cd | 44/49 at.% Cd | −190 to −50 | 15 |
| Au-Cd | 46.5/50 at.% Cd | 30 to 100 | 15 |
| Cu-Al-Ni | 14/14.5 wt.% Al 3/4.5 wt.% Ni | −140 to 100 | 35 |
| Cu-Sn | approx. 15 at.% Sn | −120 to 30 | — |
| Cu-Zn | 38.5/41.5 wt.% Zn | −180 to −10 | 10 |
| In-Ti | 18/23 at.% Ti | 60 to 100 | 4 |
| Ni-Al | 36/38 at.% Al | −180 to 100 | 10 |
| Ni-Ti | 49/51 at.% Ni | −50 to 110 | 30 |
| Fe-Pt | approx. 25 at.% Pt | approx. −130 | 4 |
| Mn-Cu | 5/35 at.% Cu | −250 to 180 | 25 |
| Fe-Mn-Si | 32 wt.% Mn, 6 wt.% Si | −200 to 150 | 100 |

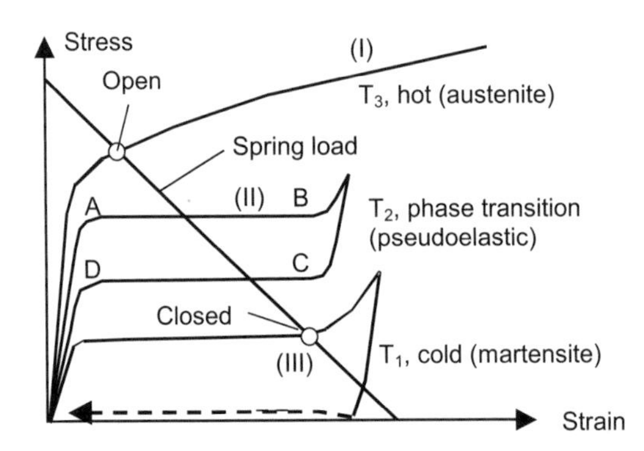


**Figure 6.10** Typical stress-strain curves at different temperatures relative to the transformation, showing (I) austenite, (II) pseudoelastic, and (III) martensite behavior.

through it can generate heat. Table 6.2 lists the most important parameters of some common SMAs [6].

Of all these alloys, the Ni/Ti alloys have received the most development effort and commercial exploitation. Figure 6.10 shows the typical stress-strain curves of an SMA at different phases. At a low temperature $T_1$, the martensite is soft with a low Young's modulus. At a high temperature $T_3$, the austenite has a much higher Young's modulus. The dashed line on the martensite curve indicates that upon heating and after removing the stress, the alloy remembers and takes its unstrained shape as the SMA transforms back to austenite. The shape recovery is not found in the austenite phase.

#### 6.1.1.6 Piezoelectric Actuators

Piezoelectricity is a reversible effect in which a mechanical stress on a material generates electrical charge, and an applied electric field generates a mechanical strain in the material. Piezoelectric actuators generally generate small strain (usually less than 0.1%) and high stresses (several megapascals). Therefore, they are suitable for applications which require large forces but small displacements. The relation between the strain and electric field strength is described with the piezoelectric coefficients $d_{31}$ and $d_{32}$. The indices indicate the directions of polarization. In the case of an electric field $E_{\mathrm{el}}$ as

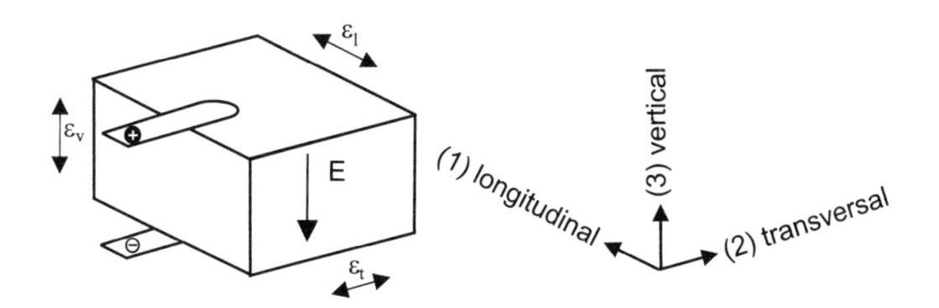


**Figure 6.11** Piezoelectric actuator.

**Table 6.3**
Properties of Common Piezoelectric Materials (*After*: [3, 7])

| *Material* | $d_{31}$ *(10$^{-12}$ C/N)* | $d_{33}$ *(10$^{-12}$ C/N)* | *Relative Permittivity* $\varepsilon_r$ |
|---|---|---|---|
| PZT | $-60$–$-270$ | 380 to 590 | 1,700 |
| ZnO | $-5$ | 12.4 | 1,400 |
| PVDF | 6–10 | 13–22 | 12 |
| $BaTiO_3$ | 78 | 190 | 1,700 |
| $LiNbO_3$ | $-0.85$ | 6 | — |

depicted in Figure 6.11, the strains in three directions are given by:

$$\begin{aligned} \varphi_\mathrm{l} &= \varphi_\mathrm{t} = d_{31}E_\mathrm{el} \\ \varphi_\mathrm{v} &= d_{33}E_\mathrm{el} \end{aligned} \tag{6.21}$$

where $\varphi_\mathrm{l}$, $\varphi_\mathrm{t}$, and $\varphi_\mathrm{v}$ are the longitudinal, transversal, and vertical strains, respectively. Table 6.3 lists the properties of some common piezoelectric materials. Common piezoelectric materials, which can be integrated with microtechnology, are polyvinylidene fluoride (PVDF), lead zirconate titanate (PZT), and zinc oxide (ZnO). PZT offers high piezoelectric coefficients but is very difficult to deposit as a thin film. PVDF and ZnO are often used in microfabrication.

Because of the large force and commercial availability, piezoelectric actuators were among the first to be used in microvalves. Thin-film piezoelectric actuators do not deliver enough force for valve applications. All reported piezoelectric microvalves used external actuators, such as piezostacks, bimorph piezocantilevers, or bimorph piezodiscs. Compared to a piezostack, bimorph piezocantilevers have the advantage of large displacements. The relatively small forces are not critical due to the small opening in microvalves. Example 6.7 illustrates the use of a bimorph piezocantilever.

### Example 6.7: Designing a Piezoelectric Valve with a Piezobimorph Actuator

The specification of a piezobimorph is given in the following table:

| *Dimension (mm)* | *Voltage (V)* | *C (nF)* | $Y_\mathrm{max}$ *(μm)* | $F_\mathrm{max}$ *(N)* | *Frequency (Hz)* |
|---|---|---|---|---|---|
| $25 \times 7.5 \times 0.4$ | $\pm 70$ | 20 | $\pm 200$ | 0.15 | 300 |

The actuator is used for closing a silicon microvalve, which has a 300 × 300 μm valve seat suspended on four flexures as described in the following figure. The initial gap between the valve seat and the valve inlet is 20 μm. The dimensions of the flexures are 500 × 100 × 20 μm. The valve inlet is drilled in a glass substrate with a diameter of 200 μm. Determine the maximum inlet pressure.

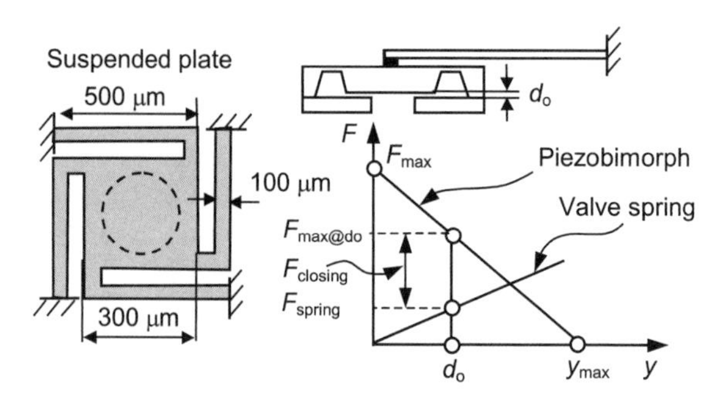


**Solution.** The spring constant of the valve seat can be modeled as that of four beams:

$$k = \frac{4Ewt^3}{l^3} = \frac{4 \times 170 \times 10^9 \times 10^{-4} \times (2 \times 10^{-5})^3}{(5 \times 10^{-4})^3} = 4,352 \text{ N/m}$$

The characteristics of the actuator and the valve spring are described in the figure above. The maximum achievable force at displacement $d_0$ is:

$$F_{\text{max@}d_0} = F_{\text{max}} \left(1 - \frac{d_0}{y_{\text{max}}}\right) = 0.15 \times (1 - 20/200) = 0.135 \text{ N}$$

The maximum closing force is:

$$F_{\text{max@}d_0} = F_{\text{max@}d_0} - kd_0 = 0.135 - 4,352 \times 20 \times 10^{-6} = 0.048 \text{ N}$$

The maximum inlet pressure the valve can withstand is:

$$\Delta p_{\text{max}} = F_{\text{closing,max}}/A = 0.048/\pi(10^{-4})^2 = 15.28 \times 10^5 \text{ Pa} = 15.28 \text{ bar}$$

#### 6.1.1.7 Electrostatic Actuators

Electrostatic actuation is based on the attractive force between two oppositely charged plates. The simplest approximation for electrostatic forces is the force between two plates with the overlapping plate area $A$, distance $d$, applied voltage $V$, relative dielectric coefficient $\varepsilon_r$ (Table 6.4), and the permittivity of vacuum $\varepsilon_0 = 8.85418 \times 10^{-12}$ F/m :

$$F = \frac{1}{2}\varepsilon_r \varepsilon_0 A \left(\frac{V}{d}\right)^2 \tag{6.22}$$

If an insulator layer of thickness $d_i$ and a relative dielectric coefficient $\varepsilon_i$ separate the two plates, the electrostatic force becomes:

$$F = \frac{1}{2}\varepsilon_r \varepsilon_0 A \left(\frac{V}{d}\right)^2 \left(\frac{\varepsilon_i d}{\varepsilon_r d_i + \varepsilon_i d}\right)^2 \tag{6.23}$$

The biggest advantage of an electrostatic actuator is its fast response. However, the high drive voltage and small displacement make it difficult to integrate electrostatic microvalves into a microfluidic system. One solution for this problem is the use of curved electrodes. Figure 6.12(a) shows an example of an electrostatic actuator with a curved electrode. The gap between the two electrodes is small at the clamped edge of the moving electrode and increases along its length. When a voltage is applied, the moving electrode deforms and attaches to the curved electrode [Figure 6.12(b)]. An insulator layer on the fixed electrode is required to prevent a short circuit between the two electrodes.

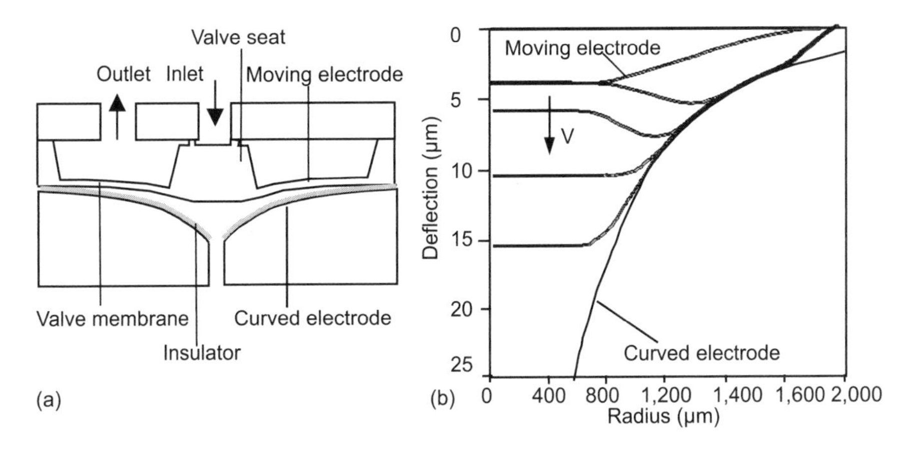


**Figure 6.12** Concept of a curved electrode for a large electrostatic displacement: (a) valve concept, and (b) simulated membrane deflection of the membrane; the arrow indicates the increase of drive voltage.

**Table 6.4**

Permittivity of Common Materials (*After*: [3, 7].)

| *Material* | *Relative Permittivity* |
|---|---|
| Silicon oxide | 3.8 |
| Silicon nitride | 4 |
| Polyimide | 3.4 (at 1 MHz) |
| Parylene-N (C, D) | 2.65 (3.15, 2.84) |
| Silicone rubber | 2.8 (at 1 MHz) |
| $Al_2O_3$ | 10 (at 1 MHz) |
| Quartz | 3.8 (at 1 MHz) |
| Silicon | 11.8 |
| Silicon carbide | 9.7 |

### Example 6.8: Designing an Electrostatic Valve

A normally open electrostatic valve for air flow is designed with a silicon nitride/aluminum valve seat. The aluminum layer and the silicon substrate represent the actuator electrodes. The valve opening is 100 × 100 μm. The valve seat is a 500- × 500-μm square plate, which is suspended on four flexures. Each flexure is 100 μm long and 20 μm wide. The aluminum layer is 5 μm thick, and the silicon nitride layer is 2 μm thick. The gap between the valve seat and the valve opening is 5 μm. Assume that there is no residual stress at room temperature. Determine the pull-in voltage of the actuator and the maximum inlet pressure that the valve can withstand at this drive voltage ($E_{\text{nitride}} = 270$ GPa, $\varepsilon_{\text{nitride}} = 4$, $E_{\text{Al}} = 704$ GPa). If the voltage decreases, what is the pull-up voltage?

**Solution.** Neglecting the areas of the flexures, the active area of the electrostatic actuator is:

$$\Delta A = A_{\text{seat}} - A_{\text{orifice}} = (500 \times 10^{-6})^2 - (100 \times 10^{-6})^2 = 24 \times 10^{-8}\ \text{m}^2$$

Based on the model of a composite beam described in Example 6.5, the spring constant of each flexure is:

$$\begin{aligned} k_0 &= \frac{12(EI)_{\text{beam}}}{L^3} \\ &= \frac{(b_1E_1t_1^2)^2+(b_2E_2t_2)^2+2b_1b_2E_1E_2t_1t_2(2t_1^2+3t_1t_2+2t_2^2)}{L^3(t_1b_1E_1+t_2b_2E_2)} \\ &= 468.78 \text{ N/m} \end{aligned}$$

The spring constant of the valve seat is:

$$k = 4k_0 = 4 \times 468.78 = 1,875 \text{ N/m}$$

Assuming that the relative dielectric coefficient of air is 1, and rearranging (6.23), the electrostatic force acting on the valve seat at a gap $d$ is:

$$F = \frac{\epsilon_0 A V^2}{2\left(d + \frac{d_i}{\varepsilon_i}\right)^2}$$

With the initial gap $d_0 = 5$ mm, the net force acting on the valve seat is the sum of electrostatic force and spring force:

$$F = \frac{-\epsilon_0 A V^2}{2\left(d + \frac{d_i}{\varepsilon_i}\right)^2} + k(d_0 - d)$$

Since the net force is a function of the gap $d$, the change of this force is calculated as:

$$\Delta F_{\text{net}}(d) = F'_{\text{net}}(d)\Delta d = \left[\frac{\varepsilon_0 A V^2}{4} \times \frac{1}{(d + d_i/\varepsilon_i)^3} - k\right]\Delta d$$

At the pull-in voltage, the actuator reaches an unstable equilibrium, where the following conditions are fulfilled:

$$\begin{cases} F_{\text{net}} = 0 \\ \Delta F_{\text{net}} = 0 \end{cases}$$

or

$$\begin{cases} k = \frac{\varepsilon_0 A V^2}{4} \times \frac{1}{(d+d_i/\varepsilon_i)^3} & \text{(a)} \\ \frac{\varepsilon_0 A V^2}{2(d+d_i/\varepsilon_i)^2} = k(d_0 - d) & \text{(b)} \end{cases}$$

Substituting (a) in (b), we get a quadratic equation in $d$. Solving this equation and taking the positive solution, we get the critical gap $d_{\text{cr}}$:

$$d_{\text{cr}} = \frac{d_0}{3} - \frac{2d_i}{3\varepsilon_i}$$

$$d_{\text{cr}} = \frac{5\times10^{-6}}{3} - \frac{2\times2\times2\times10^{-6}}{3\times4} = 1.33 \times 10^{-6} \text{ m}$$

Substituting the critical gap $d_{\mathrm{cr}}$ into (a) and rearranging it for the critical pull-in voltage, which is the voltage required to close the valve, we get:

$$V_{\mathrm{cr}} = 2\sqrt{\frac{k\left(d_{\mathrm{cr}} + d_{\mathrm{i}}/\varepsilon_{\mathrm{i}}\right)^3}{\varepsilon_0 A}}$$

With the given data, we get the critical pull-in voltage of the valve seat:

$$V_{\mathrm{cr}} = 2\sqrt{\frac{1,875 \times \left(1.33 \times 10^{-6} + 2 \times 10^{-6}/4\right)^3}{8.85418 \times 10^{-12} \times 24 \times 10^{-8}}}$$

At this voltage, the closing force of the valve is:

$$\begin{aligned} F_{\mathrm{closing}} &= \tfrac{\varepsilon_{\mathrm{i}}\varepsilon_0 A V^2}{2d_{\mathrm{i}}^2} - kd_0 \\ &= \tfrac{4\times 8.85418\times 10^{-12}\times 24\times 10^{-8}\times 147^2}{2(2\times 10^{-6})^2} - 1,875 \times 5 \times 10^{-6} = 13.6 \times 10^{-3}\ \mathrm{N} \end{aligned}$$

Thus, the maximum inlet pressure in this case is:

$$\Delta p = \frac{F_{\mathrm{closing}}}{A_{\mathrm{opening}}} = \frac{13.6 \times 10^{-3}}{(10^{-4})^2} = 13.6 \times 10^5\ \mathrm{Pa} = 13.6\ \mathrm{bars}$$

The pull-up voltage, which is required to open the valve, is reached if the electrostatic force is equal to the spring force:

$$\frac{\varepsilon_{\mathrm{i}}\varepsilon_0 A V^2}{2d_{\mathrm{i}}^2} - kd_0 \rightarrow V = \sqrt{\frac{2kd_0 d_{\mathrm{i}}^2}{\varepsilon_{\mathrm{i}}\varepsilon_0 A}} = \sqrt{\frac{2 \times 1875 \times 5 \times 10^{-6} \times (2 \times 10^{-6})^2}{4 \times 8.8541 \times 10^{-12} \times 24 \times 10^{-8}}}$$

This example describes the general behavior of a normally open, electrostatic valve. Increasing the drive voltage up to pull-in voltage causes the valve seat to snap down. The closing force with this voltage is large enough to keep the valve closed against an inlet pressure. If this voltage decreases, the valve remains closed until the pull-up voltage is reached. The spring force pushes the valve seat back to its opening position.

#### 6.1.1.8 Electromagnetic Actuators

Electromagnetic actuators offer a large deflection. The vertical force of a magnetic flux $B$ in direction $z$ acting on a magnet with its magnetization $M_{\mathrm{m}}$ (Table 6.5) and volume $V$ is given by:

$$F = M_{\mathrm{m}} \int \frac{\mathrm{d}B}{\mathrm{d}z}\mathrm{d}V \tag{6.24}$$

The relation between the magnetic flux $B$ and the magnetic field strength is:

$$B = \mu H = \mu_0 \mu_{\mathrm{r}} H = \mu(1 + \chi_{\mathrm{m}})H \tag{6.25}$$

where $\mu$, $\mu_0$, $\mu_{\mathrm{r}}$, and $\chi_{\mathrm{m}}$ are the permeability, permeability of free space ($4\pi \times 10^{-7}$ H/m), relative permeability, and the magnetic susceptibility of the medium, respectively.

Electromagnetic microvalves often use a solenoid actuator with a magnetic core and a coil for generating the magnetic field. One or both components can be integrated in silicon for a compact

**Table 6.5**
Magnetization of Common Materials at $B = 0$ T (*After*: [7])

| *Material* | *Magnetization* $M_m$ *(A/m)* | *Note* |
|---|---|---|
| Nickel | 3,000 | Electroplated, annealed |
| Iron | 320 | — |
| Fe-Ni78 | $< 80$ | Electroplated, annealed |
| Fe-Ta-Ni | 46 | Sputtered |
| Fe-Al-Si | 40 | Sputtered |

design. The drawback of electromagnetic actuators is the low efficiency due to heat loss in the coil. The magnetic field strength inside a solenoid with $N$ turns, a length $L$ and a driving current $I$ can be estimated as:

$$H = \frac{NI}{L} \tag{6.26}$$

6.1.1.9 Electrochemical Actuators

The pressure drop $\Delta p$ across a liquid-gas interface in a capillary can be used as the active valve pressure:

$$\Delta p = \frac{2\sigma_{\mathrm{lg}} \cos\theta}{r_0} \tag{6.27}$$

where $\sigma_{\mathrm{lg}}$ is the gas-liquid surface tension, $\theta$ is the contact angle, and $r_0$ is the channel radius.

Electrochemical valves are actuated using gas microbubbles generated by the electrolysis of water:

$$2\mathrm{H_2O} \rightarrow 2\mathrm{H_2} \uparrow + \mathrm{O_2} \uparrow \tag{6.28}$$

The generated gases have a volume that is about 600 times that of the water volume needed for it. This ratio exceeds that of thermopneumatic actuators. The pressure inside the bubble is proportional to the surface tension $\sigma$ and the radius of curvature $R$ of the meniscus. Equation (6.27) becomes:

$$\Delta p = \frac{2\sigma}{R} \tag{6.29}$$

The reversed reaction converts the two gases back to water. The reaction needs a catalyst such as platinum. Platinum is able to absorb hydrogen. The hydrogen-platinum bond is weaker than the hydrogen-hydrogen bond. Therefore, the energy barrier required for freeing hydrogen atoms from $H_2$ and bonding with oxygen is lower than in the gas phase:

$$2\mathrm{H_2} + \mathrm{O_2} \overset{\mathrm{Pt,heat}}{\longrightarrow} 2\mathrm{H_2O} \tag{6.30}$$

**Example 6.9: Designing an Electrochemical Valve**

Determine the energy required for generating an electrolysis bubble with an approximated dimension of $200 \times 100 \times 28$ μm. Compare it to a thermal bubble of the same size. The specific density of hydrogen and oxygen at 1 bar and 25°C are 0.08988 kg/m$^3$ and 1.429 kg/m$^3$, respectively. The surface tension of water is assumed to be constant at $72 \times 10^{-3}$ N/m. Enthalpy of formation of water is 285.83 kJ/kmol. The thermodynamic properties of liquid water at 1 bar are: v(25°C) = $1.0029 \times 10^{-3}$ m$^3$/kg, u(25°C) = 104.88 kJ/kg; and of vapor: v(100°C) = 1.673 m$^3$/kg, u(25°C) = 2506.5 kJ/kg.

**Solution.** Assuming a contact angle of zero, the pressure inside the bubble can be estimated as:

$$\Delta p = \frac{2\sigma}{r} = \frac{4\sigma}{h} = \frac{4 \times 72 \times 10^{-3}}{28 \times 10^{-6}} \approx 10^5 \text{ Pa} = 1 \text{ bar}$$

Thus, all properties at 1 bar can be taken for the analysis. The specific density of the electrolysis gas mixture is:

$$\rho_{\text{mixture}} = \frac{\rho_{\text{oxygen}} + 2\rho_{\text{hydrogen}}}{3} = \frac{1.429 + 2 \times 0.08988}{3} = 0.5363 \text{ kg/m}^3$$

Thus, the amount of water required for generating the electrolysis bubble is:

$$m = V\rho_{\text{mixture}} = 2 \times 10^{-4} \times 10^{-4} \times 28 \times 10^{-6} \times 0.5363 = 30.03 \times 10^{-14} \text{ kg}$$

The mole number is:

$$n = \frac{m}{M_{\text{H}_2\text{O}}} = \frac{30.03 \times 10^{-14}}{18} = 1.67 \times 10^{-14} \text{ kmol}$$

The energy required for making an electrolysis bubble is:

$$\Delta U_{\text{electrolysis}} = nh_{\text{f}}^0 = 1.67 \times 10^{-14}(2506.5 - 104.88) \times 10^3 = 0.804 \times 10^{-6} \text{ J}$$

The specific volume of water vapor at 100°C is 1.673 $\text{m}^3$/kg, and the mass of water is:

$$m = \frac{V}{v} = \frac{2 \times 10^{-4} \times 10^{-4} \times 28 \times 10^{-6}}{1.673} = 33.47 \times 10^{-14} \text{ kg}$$

Ignoring the heat losses, the energy required for making a thermal bubble is:

$$\Delta U_{\text{thermal}} = m(u_2 - u_1) = 33.47 \times 10^{-14}(2506.5 - 104.88) \times 10^3 = 0.804 \times 10^{-6}$$

In both cases, the expansion work done by the bubble is:

$$W = p\Delta V = 10^5 \times (2 \times 10^{-4} \times 10^{-4} \times 28 \times 10^{-6}) = 56 \times 10^{-9} \text{ J}$$

The efficiency of an electrolysis bubble is:

$$\eta_{\text{electrolysis}} = \frac{W}{W + \Delta U_{\text{electrolysis}}} = \frac{56 \times 10^{-9}}{56 \times 10^{-9} + 477 \times 10^{-11}} \approx 100\%$$

The maximum efficiency of a thermal bubble is:

$$\eta_{\text{thermal}} = \frac{W}{W + \Delta U_{\text{thermal}}} = \frac{56 \times 10^{-9}}{56 \times 10^{-9} + 0.804 \times 10^{-6}} = 6.51\%$$

#### 6.1.1.10 Chemical Actuators

In all the above actuating concepts, electrical energy is converted into mechanical energy using a number of physical effects. In contrast, chemical actuators convert chemical energy into mechanical energy. Furthermore, the increasing use of polymers in the fabrication of microfluidic systems (see Chapter 3) is compatible with the integration of polymer-based chemical actuators.

One of the major drawbacks of polymeric material in microfluidic systems is their swelling behavior. However, this behavior is attractive for actuator applications. Hydrogels are polymers with high water content. Hydrogels can change their volume if changes in temperature, solvent concentration, and ionic strength are induced. The swelling behavior is controlled by diffusion and thus is very slow in macroscale. In microscale, the response time to stimuli is drastically reduced due to the shorter diffusion path. Another advantage of this actuation scheme is that these polymers can be designed to suit conventional microtechnologies such as photolithography. For instance, a hydrogel called PNIPAAm [poly-(N-isopropylacrylamide)] [8] can be designed to have a compromise between film coating, UV polymerization, and working conditions at room temperatures. Hoffmann et al. reported a hydrogel system, which can be coated with a thickness of several tens of microns and cross-linked with conventional UV exposure units [8]. This polymer has a phase transition temperature of approximately 33°C. At room temperature below this value, the polymer is swollen. The polymer shrinks if the temperature increases above the transition temperature. Thus, together with a microheater, the hydrogel can be used as a thermomechanical actuator [9].

The swelling effect can also be achieved with the diffusion of a solvent such as water. Since the temperature diffusivity is on the order of $10^{-3}$ cm$^2$/s, the diffusion coefficient of a solvent is on the order of $10^{-5}$ cm$^2$/s, the cooperative diffusion coefficient of polymer chains is on the order of $10^{-7}$ cm$^2$/s, and the swelling dynamics are determined by the later coefficient. The characteristic time constant of swelling response can be estimated as [10]:

$$\tau = \frac{d^2}{D_{\text{coop}}} \tag{6.31}$$

where $d$ is the characteristic dimension of the hydrogel, and $D_{\text{coop}}$ is the cooperative diffusion coefficient of the polymer chains. From (6.31), the dynamics of chemical actuators depends only on the size and the bulk polymer. Using packed small particles instead of a bulk material would improve the response of the actuator.

Other types of hydrogels such as polymethacrylic acid tri-ethyleneglycol (pMAA-g-EG) [11], polyacrylic acid-2-hydroxymethylmethacrylate (pAA-2-HEMA) [12], and polyacrylamide-3-methaacrylamidophenylboronic acid (pAAm-3-MPBA) [13] swell and shrink in response to changes in concentrations of pH (potenz of hydrogen). The pH of distilled water is 7, which is neutral. Any solution with a pH below 7 is an acid and any solution with a pH above 7 is an alkali. The hydrogel swells in alkali solution and shrinks in acidic solution. The transition region of pAA-2-HEMA and pMAA-g-EG is approximately $4 < \text{pH} < 7$ [11, 12], while AAm-3-MPBA [13] switches its volume at approximately $7 < \text{pH} < 9$. The swelling effect results from the tendency of the polymer chains to mix with the solution. Swelling stops when the elastic restoring force is in equilibrium with the osmotic force.

#### 6.1.1.11 Capillary-Force Actuators

When considering miniaturization, surface tension becomes an important force in microscale. While capillary forces and surface tension cause sticking problems in other MEMS applications, they are useful for microfluidic applications. Surface tension and capillary forces can be controlled actively or passively using different effects: electrocapillary, thermocapillary, and passive capillary. Applications of surface tension driven flow for pumping are discussed in Chapter 7.

*Electrocapillary Effect.* The *electrocapillary effect* [14], also known as *electrowetting*, changes the surface tension between two immiscible, conductive liquids, or between a solid surface and a liquid, by varying their potential difference. The effect is based on the adsorption characteristics of ions in the wetting interface. This interface is an electric double layer. This layer, typically 10Å to 100Å thick, works as an electrical insulator between the two conductive liquids. By changing the electrical potential across this double layer, the surface tension $\sigma$ between the two liquids becomes [14]:

$$\sigma = \sigma_0 - \frac{C}{2}(V - V_0)^2 \tag{6.32}$$

where $\sigma_0$ is the maximum value of surface tension at $V = V_0$, $C$ is the capacitance per unit area of the double layer, and $V$ is the voltage applied across the liquid interface. Readers can refer to Chapter 7 for more configurations and applications of electrocapillary effect.

**Example 6.10: Characteristics of an Electrocapillary Actuator**

The contact angle of the nonwetted side of a capillary is 140° and the angle on the wetting electrode is 30°. The gap between the bottom electrode and the nonwetted side is 10 μm. Determine the pressure gain from the electrocapillary effect if the liquid is a water-based electrolyte.

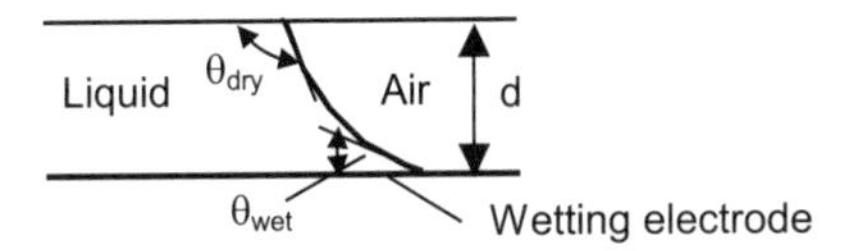


**Solution.** The pressure gain can be calculated from the contact angles and the surface tension of water-air $\sigma = 72 \times 10^{-3}$ N/m [15]:

$$\Delta p = \sigma \left[2 \cos\left(\frac{\theta_{\mathrm{dry}} + \theta_{\mathrm{wet}}}{2}\right) \cos\left(\frac{\theta_{\mathrm{dry}} - \theta_{\mathrm{wet}}}{2}\right) / d\right]$$

$$\Delta p = 72 \times 10^{-3} \left[2 \cos(85°) \cos(55°) / 10 \times 10^{-6}\right] = 144 \times 10^2 \text{ Pa} = 144 \text{ mbar}$$

*Thermocapillary Effect.* The thermocapillary effect is caused by the temperature dependence of the surface tension. The surface tension reflects the surface energy. At a higher temperature, the molecules of the liquid move faster and their attractive force becomes smaller. The smaller attractive force causes lower viscosity and lower surface tension. Figure 6.13 demonstrates the principle of the thermocapillary effect, or the so-called Marangoni effect. Imagine a vapor bubble in a capillary [Figure 6.13(a)]. Across the bubble exists a temperature gradient. At the left-hand side, the temperature is higher, so the surface tension is lower. That causes a pressure gradient across the bubble, which leads to bubble movement from right to left. Since the rate of change of surface tension with temperature is not large, the effect requires considerable heating power to get the desired force. The same effect causes a liquid droplet to move away from the heat source. This effect can be utilized for designing droplet micropumps described in Chapter 7.

*Passive Capillary Effect.* The passive capillary effect utilizes dependence on the geometry. The surface tension increases with the smaller radius of curvature, which can be controlled by channel width or the solid/liquid contact angle [16]. Figure 6.14 explains the principle of the passive capillary effect. If a bubble is confined in a channel with a hydrophilic wall, the bubble is forced to adopt the shape of the channel [Figure 6.14(a)]. The surface tension or surface energy usually tends to have its minimum values. That explains the spherical surface of a bubble. In the case of Figure 6.14(b),

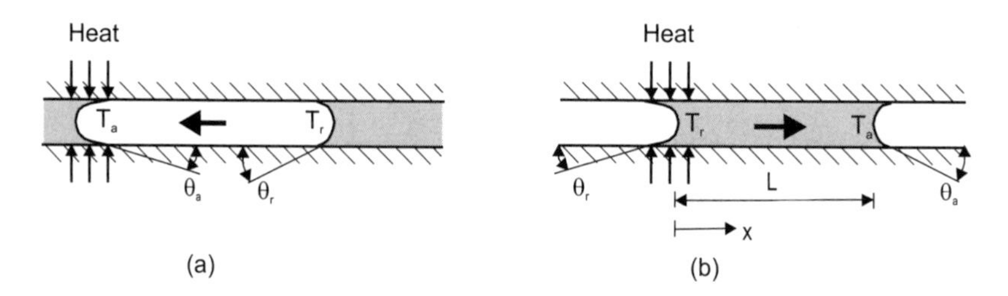


**Figure 6.13** Thermocapillary effect (Marangoni effect): (a) a gas bubble is attracted to the heat source; and (b) a liquid plug is driven away from a heat source.

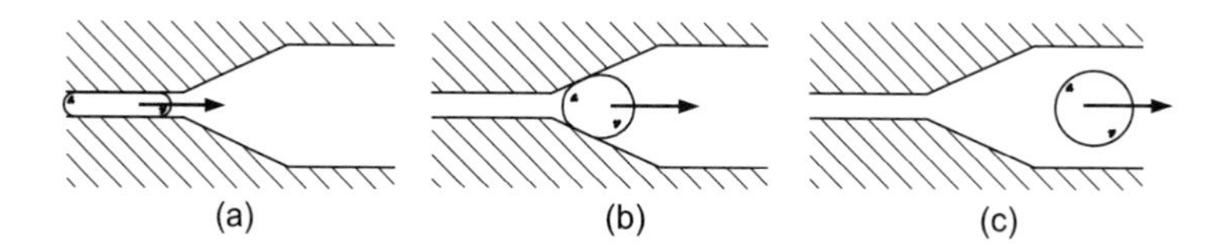


**Figure 6.14** Passive capillary effect: (a) bubble in small channel; (b) bubble in the intersection; and (c) bubble in the large channel.

the bubble is forced to have different surface tensions on the left-hand and right-hand sides because of the different radii of curvature. The surface tension gradient moves the bubble toward the larger section of the channel. In the large section, the bubble forms a sphere to minimize its surface energy [Figure 6.14(c)]. Refer to Chapter 7 for more applications of this effect.

**Example 6.11: Characteristics of a Passive Capillary Valve**

A bubble valve is designed with a vapor bubble between channel sections with different widths of 50 and 200 μm (Figure 6.13). Determine the pressure difference the valve can withstand.

**Solution.** Assuming that the radii of curvature of the bubble are equal to channel widths, and the surface tension of the water-air interfaces $\sigma = 72 \times 10^{-3}$ N/m, and using (6.29) for each side of the bubble, we get the pressure difference across it:

$$\begin{aligned} \Delta p &= 2\sigma \left( \frac{1}{r_1} - \frac{1}{r_2} \right) \\ &= 2 \times 72 \times 10^{-3} \left( \frac{1}{50 \times 10^{-6}} - \frac{1}{200 \times 10^{-6}} \right) = 2,160 \text{ Pa} = 21.6 \text{ mbar} \end{aligned}$$

More than one-fiftieth of atmospheric pressure can be achieved with this passive capillary valve.

### 6.1.2 Valve Spring

In most microvalve designs, the valve seat structure can be thought of as a spring. The spring force closes the valve in normally closed valves. In the case of normally open valves, the spring works against the actuator and decreases the closing force. Figure 6.15 illustrates different arrangements of forces acting on the valve seat.

Figure 6.16 depicts common forms of the valve spring and their corresponding spring constants. In normally closed valves, the spring constant should be large to resist the inlet pressure. In normally open valves, the spring constant is to be optimized for a minimum value, which allows a larger closing force of the actuator. A small spring constant can be realized with a soft material such as rubber. The solution with soft materials offers a further advantage of excellent fitting characteristics. The leakage ratio can be improved from three to four orders of magnitude compared to those made of hard material such as silicon, glass, or silicon nitride.

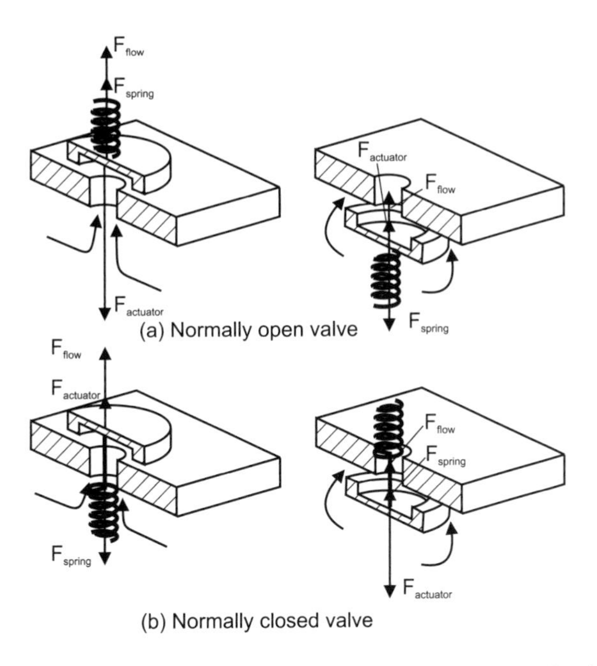


**Figure 6.15** Functions of valve springs and force balance in (a) NO valves and (b) NC valves; $F_{\text{inlet}}$, $F_{\text{spring}}$, and $F_{\text{actuator}}$ are forces of fluid, spring, and actuator, respectively.

If the valve is designed for bistable operation, there is no need for a valve spring because the two valve states are controlled actively. Since the nonpowered state is undefined, a valve spring can still be considered for the initial, nonpowered state to assure safe operation.

A bistable valve spring allows the valve seat to snap into its working position. In this case, the actuator just needs to be powered in a short period to have enough force to trigger the position change. The force generated by the spring is then high enough to seal the valve inlet. Thus, the actuator does not need to be powered over the entire operation. A spring in the form of a prestressed beam [Figure 6.16(a)] or prestressed circular membrane [Figure 6.16(d, e)] has for instance, this bistable characteristic. For the buckling behavior of a beam, readers may refer to Timoshenko's classic *Theory of Elastic Stability* [17].

For a prestressed circular valve membrane with a radius of $R$ and a thickness of $t$, the two stable deflection positions of the membrane center are [18]:

$$y_{\text{snap}} = \pm\frac{1}{24}\sqrt{-105\left(3R^2\sigma_0\frac{1-\nu^2}{E}+4t^2\right)} \tag{6.33}$$

where $\sigma_0$, $E$, and $\nu$ are the prestress of the membrane at nondeflected position $[y(r) = 0]$, the Young's modulus, and the Poisson ratio, respectively. The deflection of the membrane under a distributed load can be estimated as:

$$y(r) = y_0\left(1-\frac{r^2}{R^2}\right)^2 \tag{6.34}$$

At a given center deflection $y_0$, the pressure and the force exerted by the valve membrane are:

$$p = \frac{4ty_0}{R^2}\left(\frac{4}{3}\frac{t^2}{R^2}\frac{E}{1-\nu^2}+\sigma_0+\frac{64}{105}\frac{y_0^2}{R^2}\frac{E}{1-\nu^2}\right) \tag{6.35}$$

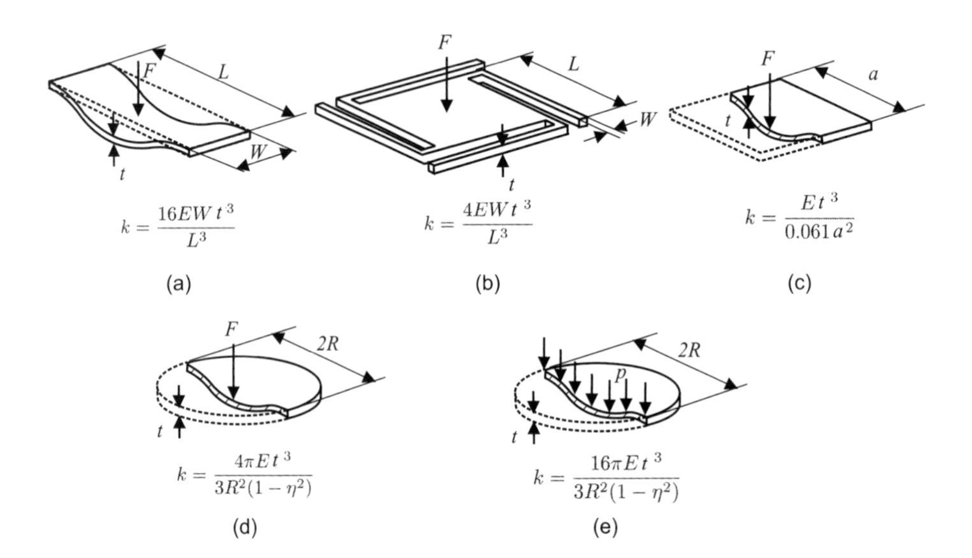


**Figure 6.16** Typical valve springs and their spring constants (linear theory for small deflection, which is smaller than the structure thickness; $k$: spring constant, $E$: Young's modulus, $t$: structure thickness, $W$: structure width, $L$: structure length, $\nu$: Poisson's ratio, $R$: radius, and $a$: membrane width; see Appendix D): (a) cantilever; (b) suspended plate; (c) rectangular membrane, $\nu = 0.3$; (d) circular membrane, point load; and (e) circular membrane, distributed load.

$$F = -\frac{2\pi}{3}\left(\frac{8}{3}t^3\frac{E}{1-\nu^2}\frac{y_0}{R^2} + 2y_0 t\sigma_0 + \frac{128}{105}\frac{E}{1-\nu^2}\frac{ty_0^3}{R^2} - \frac{1}{2}pR^2\right) \tag{6.36}$$

where $p$ is the pressure acting on the membrane. Thus, with a gap $d$ between valve membrane, and an inlet and a inlet radius of $R_{\text{inlet}}$, the inlet pressure can be estimated as:

$$p_{\text{inlet}} = \frac{4}{3}\frac{dt}{R_{\text{inlet}}^2}\left(\frac{4}{3}\frac{t^2}{R^2}\frac{E}{1-\nu^2} + \sigma_0 + \frac{64}{105}\frac{E}{1-\nu^2}\frac{d^2}{R^2}\right) \tag{6.37}$$

### Example 6.12: Operation Conditions of a Bistable Microvalve

A microvalve is designed with a circular membrane [Figure 6.16(e)] working as both valve spring and valve seat. The membrane is made of hardbacked SU-8, and has a radius of $R = 2$ mm and a thickness of $t = 20$ μm. The gap between the membrane and the valve inlet is 100 μm, while the radius of the valve inlet is 100 μm. The center deflection of the membrane at nonloading condition is $y_{0,\text{ini}} = 150$ μm. Determine critical switching pressure and the maximum inlet pressure that the valve can withstand in its closing state.

**Solution.** The Young's modulus and Poisson ratio of hardbacked SU-8 are assumed to be $E = 4$ GPa and $\nu = 0.22$, respectively. The membrane is deflected at nonloading condition. Thus, it is prestressed with an initial stress of $\sigma_0$. Setting the loading pressure $p$ in (6.35) to zero and rearranging for the initial stress results in:

$$\begin{aligned}\sigma_0 &= -\frac{4}{105}\frac{E}{1-\nu^2}\frac{16y_{0,\text{ini}}^2+35t^2}{R^2}\\ &= -\frac{4}{105}\frac{4\times10^9}{1-0.22^2}\frac{16\times(100\times10^{-6})^2+35(20\times10^{-6})^2}{(2\times10^{-3})^2} = -1.50\times10^7\text{Pa} = -15.0\text{ MPa}\end{aligned}$$

From (6.33), the critical switching positions are:

$$y_{\text{snap}} = \pm\tfrac{1}{24}\sqrt{-105\left(3R^2\sigma_0\tfrac{1-\nu^2}{E} + 4t^2\right)}$$

$$= \pm\tfrac{1}{24}\sqrt{-105\left[3(2\times10^{-3})^2(-1.5\times10^7)\tfrac{1-0.22^2}{4\times10^9} + 4(20\times10^{-6})^2\right]}$$

$$= \pm 8.66\times10^{-5}\ \text{m} = \pm 86.6\ \mu\text{m}$$

Inserting the critical position $y_{\text{snap}}$ in (6.35) results in the critical switching pressure for the valve membrane:

$$p_{\text{snap}} = \tfrac{4ty_0}{R^2}\left(\tfrac{4}{3}\tfrac{t^2}{R^2}\tfrac{E}{1-\nu^2} + \sigma_0 + \tfrac{64}{105}\tfrac{y_0^2}{R^2}\tfrac{E}{1-\nu^2}\right)$$

$$= \tfrac{4\times20\times10^{-6}\times86.6\times10^{-6}}{(2\times10^{-3})^2}\left[\tfrac{4}{3}\tfrac{(20\times10^{-6})^2}{(2\times10^{-3})^2}\tfrac{4\times10^9}{1-0.22^2} - 15\times10^6 + \tfrac{64}{105}\tfrac{(86.6\times10^{-6})^2}{(2\times10^{-3})^2}\tfrac{4\times10^9}{1-0.22^2}\right]$$

$$= -1.66\times10^4\ \text{Pa} = -16.6\ \text{kPa}$$

From (6.37), the maximum inlet pressure the valve can withstand is:

$$p_{\text{inlet}} = \tfrac{4}{3}\tfrac{dt}{R_{\text{inlet}}^2}\left[\tfrac{4}{3}\tfrac{t^2}{R^2}\tfrac{E}{1-\nu^2} + \sigma_0 + \tfrac{64}{105}\tfrac{E}{1-\nu^2}\tfrac{d^2}{R^2}\right]$$

$$= \tfrac{4}{3}\tfrac{-100\times10^{-6}\times20\times10^{-6}}{(100\times10^{-6})^2}\left[\tfrac{4}{3}\tfrac{(20\times10^{-6})^2}{(2\times10^{-3})^2}\tfrac{4\times10^9}{1-0.22^2} - 15\times10^6 + \tfrac{64}{105}\tfrac{4\times10^9}{1-0.22^2}\tfrac{(100\times10^{-6})^2}{(2\times10^{-3})^2}\right]$$

$$= 2.13\times10^6\ \text{Pa} = 21.3\ \text{bars}$$

**Example 6.13: Optimization of a Bistable Microvalve [18]**

Determine for the bistable valve of Example 6.12 the optimal membrane thickness $t$ and the optimal gap $d$ between the valve membrane and the inlet. If both parameters are to be optimal, what is the maximum achievable inlet pressure?

**Solution.** If the gap is fixed, the optimal membrane thickness results from setting the differentiation of (6.37) for $t$ to be zero:

$$t_{\text{optimal}} = \sqrt{-\tfrac{16}{105}d^2 - \tfrac{1}{4}\sigma_0\tfrac{1-\nu^2}{E}R^2}$$

$$= \sqrt{-\tfrac{16}{105}(-100\times10^{-6})^2 - \tfrac{1}{4}\times(-15\times10^6)\tfrac{1-0.22^2}{4\times10^6}(2\times10^{-3})^2}$$

$$= 4.51\times10^{-5}\ \text{m} = 45.1\ \mu\text{m}.$$

If the membrane thickness is fixed, the optimal gap results from setting the differentiation of (6.37) for $d$ to be zero:

$$d_{\text{optimal}} = \tfrac{\sqrt{105}}{24}\sqrt{-3\sigma_0\tfrac{1-\nu^2}{E}R^2 - 4t^2}$$

$$= \tfrac{\sqrt{105}}{24}\sqrt{-3\times-15\times10^6\tfrac{1-0.22^2}{4\times10^9}(2\times10^{-3})^2 - 4\times(20\times10^{-6})^2}$$

$$= 8.66\times10^{-5}\ \text{m} = 86.6\ \mu\text{m}$$

If both thickness $t$ and the gap $d$ are to be optimized, the above two equations result in:

$$\begin{aligned} t_{\text{ideal}} &= \tfrac{1}{4}R\sqrt{-3\sigma_0\tfrac{1-\nu^2}{E}} \\ &= \tfrac{1}{4}(2\times10^{-2})\sqrt{-3(-15\times10^6)\tfrac{1-0.22^2}{4\times10^9}} \\ &= 5.17\times10^{-5}\ \text{m} = 51.7\ \mu\text{m} \end{aligned}$$

and

$$\begin{aligned} d_{\text{ideal}} &= \tfrac{R}{16}\sqrt{-105\sigma_0\tfrac{1-\nu^2}{E}} \\ &= \tfrac{2\times10^{-3}}{16}\sqrt{-105\times(-15\times10^6)\tfrac{1-0.22^2}{4\times10^9}} \\ &= 7.64\times10^{-5}\ \text{m} = 76.4\ \mu\text{m} \end{aligned}$$

The maximum achievable inlet pressure is then:

$$\begin{aligned} p_{\max} &= \tfrac{\sqrt{35}}{32}\sigma_0^2\tfrac{1-\nu^2}{E}\tfrac{R^2}{R_{\text{inlet}}^2} \\ &= \tfrac{\sqrt{35}}{32}(-5\times10^6)^2\tfrac{1-0.22^2}{4\times10^9}\tfrac{(2\times10^{-3})^2}{(100\times10^{-6})^2} \\ &= 3.94\times10^6\ \text{Pa} = 39.4\ \text{bars} \end{aligned}$$

### 6.1.3 Valve Seat

Valve seats represent a large challenge to microvalve design and fabrication. The valve seat should satisfy two requirements: zero leakage and resistance against particles trapped in the sealing area. For a minimum leakage rate, the valve should be designed with a large sealing area, which has to be extremely flat. Softer materials such as rubber or other elastomers are recommended for the valve seat.

Resistance against particles can be realized in many ways. First, a hard valve seat can simply crush the particles. For this purpose, the valve needs actuators with large force such as piezostacks and hard coating layer for the valve seat. Second, the particles can be surrounded and sealed by a soft coating on the valve seat. Third, small particle traps such as holes or trenches can be fabricated on the valve seat or on the opposite valve base [Figure 6.17(c)]. A combination of the third measure with the first and the second measures is recommended, so that tiny particles can be trapped and buried after being crushed by the large actuation force [Figures 6.17(d, e)]. For the first and third measures, the valve seat needs to be coated with hard, wear-resistant material such as silicon nitride, silicon carbide, or diamond-like carbon.

### 6.1.4 Pressure Compensation Design

The closing pressure generated by the actuator should be greater than the inlet pressure to keep the valve functioning. If the inlet pressure varies, there is a need for closed loop control for the actuator to maintain the closing force [Figure 6.18(a)]. If the inlet pressure exceeds the maximum closing pressure of the actuator, the compensation approach described in Figure 6.18(b) can be used. With this method, the closing pressure is the sum of the inlet pressure $p_{\text{in}}$ and the actuator pressure $p_{\text{act}}$.

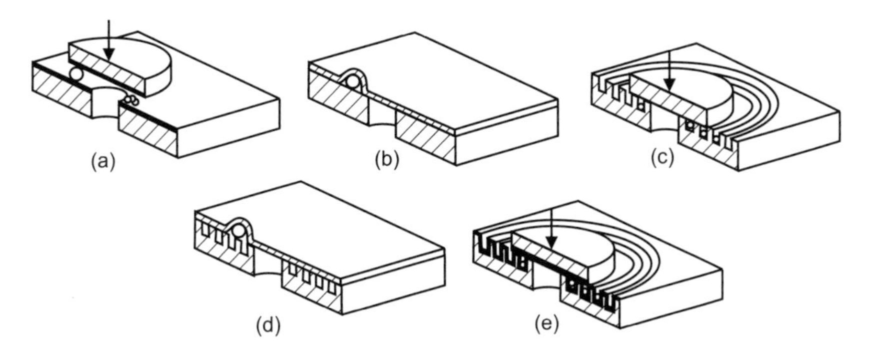


**Figure 6.17** Methods for leakage-free valve seat: (a) large force, hard surface; (b) particle traps; (c) soft valve seat; (d) combination of (a) and (b); and (e) combination of (c) and (b).

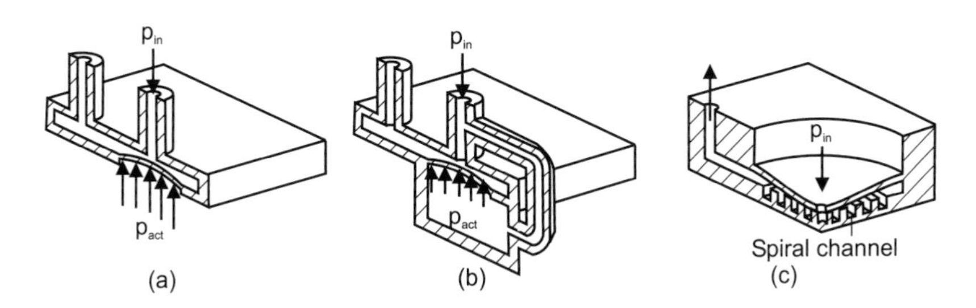


**Figure 6.18** Pressure compensation in microvalves: (a) behavior at the valve seat; and (b) passive flow controller.

This concept assures that the closing pressure is always larger than the inlet pressure. The closing pressure only depends on the actuator, which can be controlled by a simple open loop. The differential pressure concept is very useful for normally closed valves, where the closing force relies entirely on the relatively small spring force.

Another pressure compensation approach is the passive regulation of flow rates. This passive component can also be categorized as a pneumatic proportional valve. Figure 6.18(c) shows a design example [19]. If the pressure difference across the device increases, the deformation of the silicon membrane increases as well. The deformed membrane covers more of the channel beneath the membrane. Thus, the channel length increases, and the higher fluidic impedance compensates the increase of the pressure. The principle results in a constant flow rate. The device is relevant for disposable drug delivery systems.

## 6.2 DESIGN EXAMPLES

### 6.2.1 Methods of Valve Integration

Valves are generally classified as stand-alone and built-in ones [20, 21, 22]. Stand-alone valves are initially fabricated and then integrated with a prefabricated microfluidic chip. The main motivation for the fabrication of stand-alone valves is their ease of integration with a microfluidic system. Stand-alone valves are more suitable for plug-and-play applications. They usually have minimum overall dimensions of few millimeters. Depending on its design, a stand-alone valve can be actuated manually using a screw, by an electrical solenoid or pneumatics. A stand-alone microvalve with pneumatic actuation for plug-and-play application is shown in Figure 6.19.

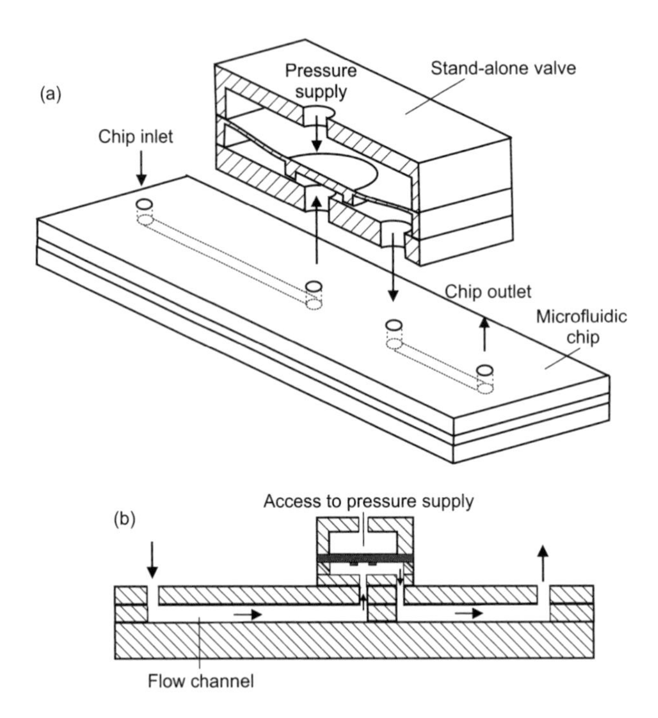


**Figure 6.19** A stand-alone microvalve: (a) a pneumatic normally open stand-alone microvalve for integration with a prefabricated microfluidic chip; and (b) side-view of the microvalve integrated with the chip.

In contrast to the stand-alone microvalves, built-in valves are fabricated during the course of chip fabrication. In this way, valves and other microfluidic components are implemented in a microfluidic chip at a same time (Figure 6.20). Such valves have a smaller footprint compared to stand-alone valves which make them suitable for applications where the integration of a large number of microvalves are required.

### 6.2.2 Pneumatic Valves

A low leakage rate can be achieved by designing the valve seat carefully. Soft materials such as rubber improve the leakage ratio significantly, as shown in Table 6.6, which compares the leakage ratio of rubber/silicon and glass/silicon valves.

The valve shown in Figure 6.21(a) has a bossed membrane [23]. The circular membrane has a diameter of 3.6 mm and a thickness of 25 μm. The stack of three silicon wafers is directly bonded at 1,000°C (see Section 3.6.1). The inlets and the outlet are drilled in a glass plate, which is mounted on the silicon stack using anodic bonding (see see Section 3.6.1). A chromium layer on the glass substrate avoids bonding the valve seat. A similar design, reported in [24], shows that the behavior of a pneumatically driven valve is analogous to that of field effect transistor (FET) in electronics.

The 225 × 125-μm valve membrane shown in Figure 6.21(b) is made of a 25-μm silicone rubber layer, which is spin-coated on the silicon substrate [25]. The pneumatic access in the top wafer is drilled with laser. The valve can also be fabricated entirely from silicone rubber for improving sealing properties as reported in [26].

The two membranes in Figure 6.21(c) are fabricated with polymeric surface micromachining (see Section 3.4). A 30-μm-thick silicone rubber layer works as the structural material, while thick positive photoresists work as sacrificial layers. The heights of the pressure chamber in the lower

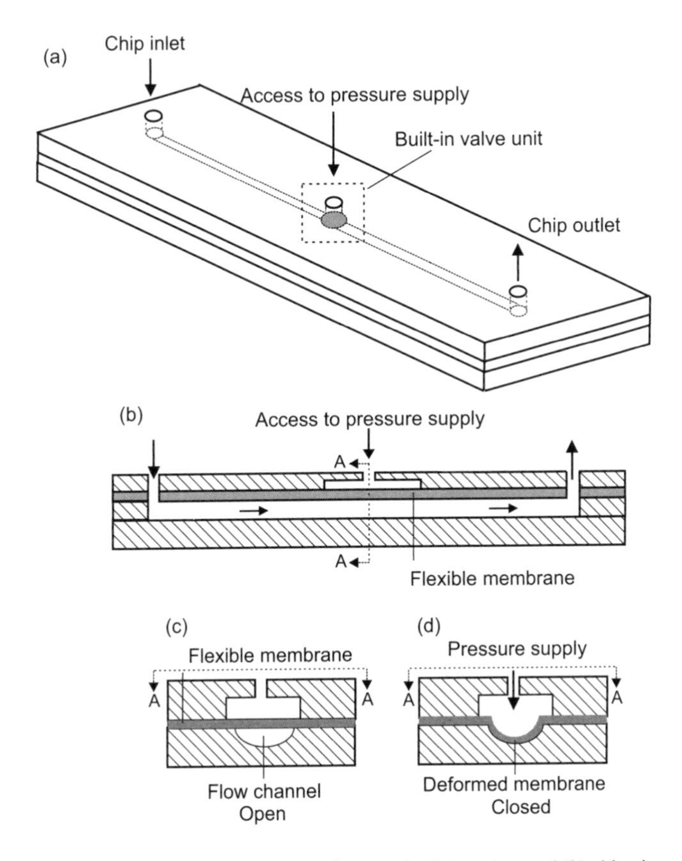


**Figure 6.20** A built-in microvalve: (a) a pneumatic normally-open built-in valve; and (b) side view of the microvalve; (c) valve operation once open; (d) valve operation once closed.

part, and of the channel in the upper part, are defined by a photoresist thickness of 5 and 30 μm, respectively.

Besides soft materials such as silicon rubber [25, 27], thinner membranes or corrugated membranes [25, 28] also provide low spring constants. In the valve of Figure 6.21(d), the membrane thickness is defined by deep-boron diffusion in silicon (see Section 3.3.1).

For parallel displacement of the valve seat, the membrane can be designed with a bossed region at its center. In the case of a corrugated membrane, the corrugation is structured in silicon by dry etching or isotropic etching (HF, $HNO_3$) prior to the deep-boron diffusion. The wet-etching process in KOH from the backside forms the corrugated membrane (Figure 6.22). Additional coating with parylene improves the leakage ratio [28]. In general, pneumatic microvalves need connection of the microfluidic device to an external pressure or vacuum source. This may impede the use of pneumatic microvalves for applications where compactness and size constraint is very critical.

Currently, the majority of microvalves is fabricated based on the Quake valve design, Figure 6.21(e, f). Quake valve is considered as a built-in valve that takes advantage of a simple bilayer structure constituting a liquid layer and a control layer and a thin elastic membrane [31]. The working principle is based on pneumatic deformation of the membrane, sandwiched between the control layer and the liquid layer, into a curved liquid microchannel to block the flow. Quake valves in push-up and push-down architectures can be fabricated monolithically using replica molding of PDMS. By combining push-up and push-down valves in a single microfluidic chip, a large number of integrated valves can be obtained. Such PDMS-based Quake valves can benefit

**Table 6.6**
Typical Parameters of Pneumatic Valves ($L_{valve}$: Leakage ratio)

| *Refs.* | *Type* | *Size (mm×mm)* | $\dot{Q}_{max}$ *(mL/min)* | $p_{max}/p_{actuator}$ *(kPa)* | $L_{valve}$ | *Material* | *Technology* |
|---|---|---|---|---|---|---|---|
| [23] | NC | $15 \times 15$ | 120 air | 241/69 | $> 300$ | Glass, silicon | Bulk |
| [24] | NC | $2 \times 5$ | 4 air | 50/80 | — | Glass, silicon | Bulk |
| [25] | NO | $0.225 \times 0.225$ | 0.26 water | 100/50 | 10,000 | Rubber, silicon | Bulk |
| [26] | NC | $1 \times 1$ | 0.01 water | 70/20 | — | PDMS | Polymeric |
| [29] | NC | $20 \times 20$ | 35 $N_2$ | 65/12 | 35 | Glass, silicon | Bulk |
| [27] | NO | $8.5 \times 4.2$ | 0.5 water | 60/10 | 10,000 | Rubber, silicon | Bulk |
| [28] | NO | $10 \times 10$ | 5 $N_2$ | 107/275 | 100,000 | Glass, silicon | Bulk |

from a small footprint (100 micron by 100 microns) that enables the high-density integration of hundreds to thousands of microvalves and control components in a single system using multilayer soft lithography (MLS). Such dense integration of microfluidic components is also known as *microfluidic large-scale integration (mLSI)*. This technology can be employed for development of miniaturized automation paradigms where hundreds of fluid handling processes are required [31]. As an example, a microfluidic device with 1,000 independent chambers and 3,574 microvalves was achieved as an addressable $25 \times 40$ chamber microarrays using the mLSI approach [32]. Precise operation of the microvalves are achieved using a programmable software system synchronized with external solenoid on/off pneumatic valves in connection with a pressure source. Other materials such as thermoplastics have also been employed to make microvalves based on Quake design. In this approach, the structural layers containing control chamber and round flow channel can be fabricated in plastic using hot embossing, micro milling or laser machining. The PDMS thin membrane can be sandwiched between the structural layers using adhesives or chemical treatment (e.g., PMMA/PDMS/PMMA) [Figure 6.23(c)]. Thermoplastic elastomers such as thermoplastic polyurethane (TPU) can be employed to fabricate the flexible membrane to be sandwiched between the control channel and flow channel using thermal fusion bonding (e.g. PMMA/TPU/PMMA) [Figure 6.23(d)]. Once all microvalve components are made using thermoplastic materials, a whole-thermoplastic microvalve is achieved.

### 6.2.3 Pinch Valves

Pinch valves are suitable for applications, where the supply of an external pressure or vacuum supply is not possible [33]. In comparison with Quake designs that employ remote control of membrane deformation to rectify flow, pinch valves are mainly used to produce local deformation of the flow channel to stop liquid movement [34] [Figure 6.24(a, b)]. Similar to Quake valves, pinch valves constitutes of a flow channel and control chamber with flexible walls. The control chamber can be partially deformed using a solenoid, screw, a moving rod, or finger to create sufficient pressure to deform the ceiling of the flow channel. In some designs, the moving rod or pin can be inserted directly on top of the flow channel for direct deformation of the flow channel ceiling to block the flow [Figures 6.24(c), (e)]. The microvalves can be designed to be operated either manually or automatically. Manual operation of the valves may provide some advantages for disposable, low-cost and simple microfluidic devices.

### 6.2.4 Thermopneumatic Valves

Figure 6.25 shows design examples of thermopneumatic valves. The metal membrane and the silicon membrane of the valves depicted in Figures 6.25(a, b) have large spring constants. Their deflection

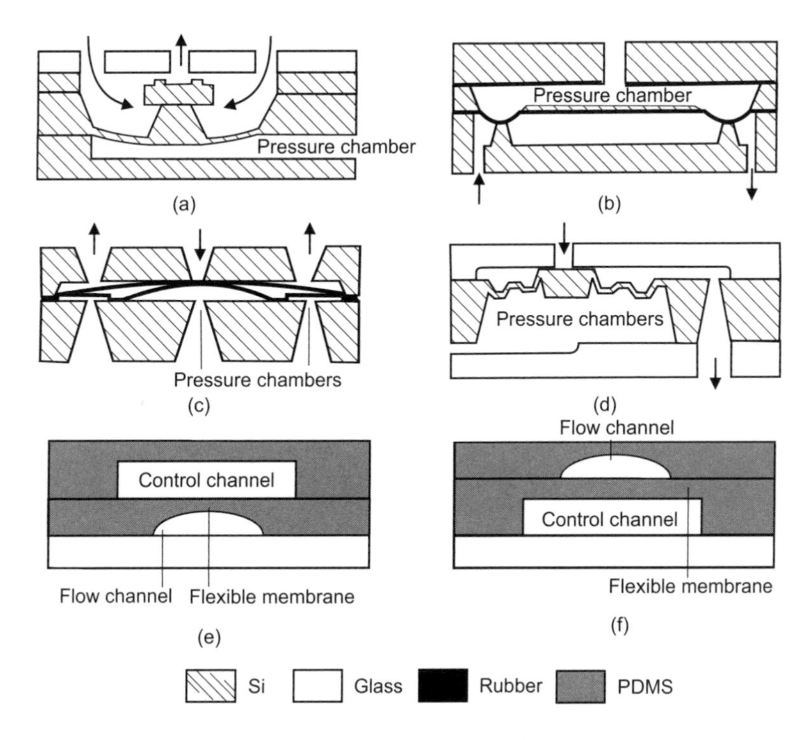


**Figure 6.21** Typical pneumatic valve: (a) bossed silicon membrane; (b) rubber membrane; (c) rubber membrane, two ways; (d) corrugated silicon membrane; (e) push-down Quake valve; and (f) push-up Quake valve.

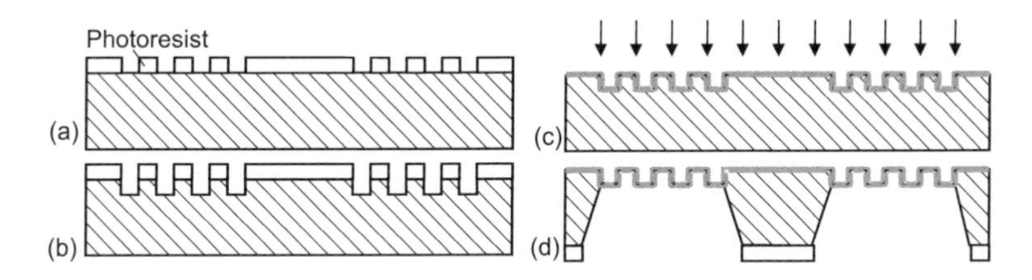


**Figure 6.22** Fabrication of corrugated membrane: (a) photoresist patterning; (b) dry etching of silicon substrate; (c) ion implantation of boron; and (d) anisotropic etching from backside.

and the closing force are small. Soft materials such as rubber offer a higher displacement, lower spring force, and, consequently, higher closing force [Figure 6.25(c)]. Since silicon is a thermal conductive material, the heat loss and consequently the required operating power of a valve with a silicon membrane are high. Table 6.7 shows that power consumption is one or two orders of magnitude larger for a silicon membrane compared to those with rubber membrane.

The heater in the valve depicted in Figure 6.25(b) is made of platinum, which is sputtered on the Pyrex glass wafer. The filling ports are drilled in the glass wafer. Fluorocarbon (FC) works as actuation fluid. The valve membrane and valve seat are fabricated with bulk silicon micromachining (see Section 3.3.1). The valve is assembled with silicon direct bonding and anodic bonding (see Section 3.6.1) [35, 36].

The valve membrane in Figure 6.25(c) is made of a 50-μm-thick silicone rubber and measures 1.5 × 2.5 μm [37]. For better thermal insulation, the Cr/Au heater is placed on a low-stress silicon nitride membrane [38]. FC is also used as actuation fluid. Kim et al. reported a microvalve made entirely in PDMS [39]. The indium-tin-oxide (ITO) heater is deposited on a glass substrate. Despite

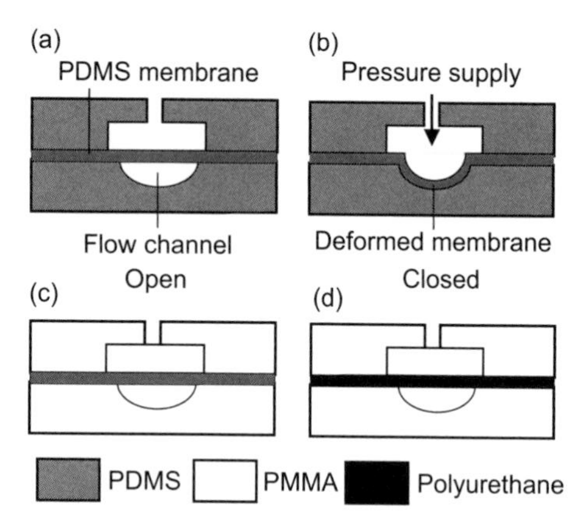


**Figure 6.23** Use of various materials for microvalve fabrication based on Quake valve concept: (a, b) whole-PDMS microvalve (PDMS/PDMS/PDMS) once it is open and closed; (c) PMMA/PDMS/PMMA microvalve where PDMS is used as a flexible membrane; and (d) PMMA/TPU/PMMA once TPU is used as a flexible membrane.

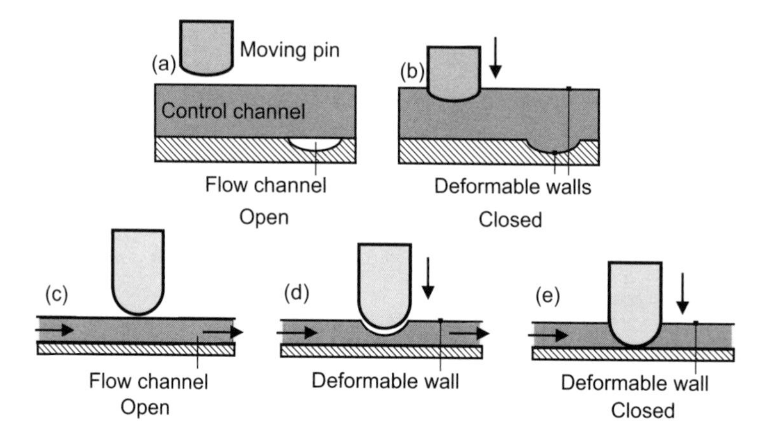


**Figure 6.24** Typical pinch valves: (a, b) pinchvalve with moving pin and control chamber to impede flow; (c) and (e) pinch valve with moving pin for direct deformation of the flow channel to impede flow.

the heat loss in the bulk glass substrate, the operation power can be kept at a reasonable level due to the soft valve material.

The valve shown in Figure 6.25(d) utilizes as actuation fluid the same fluid, which is to be controlled [40]. The pressure $p$ generated by the microbubble can be described as a function of surface tension $\sigma$ and the radius of curvature of the bubble $R$:

$$p = \frac{2\sigma}{R} \tag{6.38}$$

The bubble works as a piston to move the gate valve. The force generated by this type of actuator is on the order of 100 μN [40]. The valve is fabricated with DRIE of a SOI wafer (see Section 3.3.1). The piston head measures 100 × 75 μm. The whole structure is covered with a glass plate using adhesive bonding with a two-component epoxy (see Section 3.6.1). Subsequent treatment in oxygen plasma removes the epoxy from open areas such as microchannels and releases the moving structures. The lateral valve arrangement allows the integration of the valve in a planar microfluidic system. However, the poor leakage ratio of 1.15 makes this type of valve unsuitable for most applications.

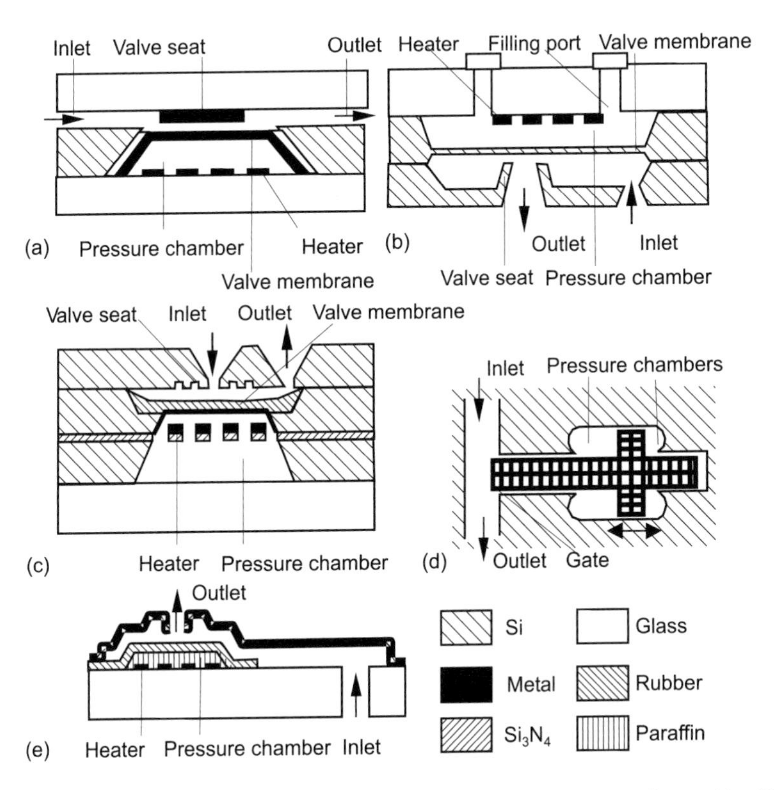


**Figure 6.25** Typical thermopneumatic valves: (a) aluminum membrane; (b) silicon membrane; (c) rubber membrane; (d) gate valve; and (e) parylene-C membrane, paraffin as actuation material.

One disadvantage of a thermopneumatic valve using liquid as actuation fluid is the complexity of the technology. Each valve must be primed, filled, and sealed individually. This process increases the cost and lowers the repeatability of the valve. The valve shown in Figure 6.25(e) has solid paraffin as an actuating medium [41]. The thin film solid paraffin can be integrated in a batch fabrication process. The valve membrane is made of a 3-μm-thick parylene-C layer (see Chapter 3). The circular membrane has a diameter of 400 μm. A 9-μm-thick paraffin layer works as actuation material. The heater is made of sputtered Cr/Au. This actuator allows 2-μm deflection at a 50-mW heating power. The valve seat and channel structures are fabricated with surface micromachining. Thick-film resist AZ9260 (see Chapter 3) works as sacrificial material. The 20-μm-thick electroplated titanium layer works as structural material.

### 6.2.5 Thermomechanical Valves

Thermomechanical microvalves utilize actuators, which convert thermal energy directly into mechanical stress. There are three types of thermomechanical actuators: solid-expansion, bimetallic, and shape-memory alloy.

The valve depicted in Figure 6.26(a) has an actuator in the form of a bossed membrane. Doped silicon resistors integrated in the membrane act as heaters. An aluminum layer is the second part of the bimetallic actuator [43, 44]. The deflection of the membrane in Figure 6.26(b) at zero force is given by [44]:

$$y = \frac{K_y}{1+\nu}\Theta a^2 \tag{6.39}$$

**Table 6.7**
Typical Parameters of Thermopneumatic Valves

| *Ref.* | *Type* | *Size (mm×mm)* | $\dot{Q}$*max (mL/min)* | $P_{max}$ *(kPa)* | $L_{valve}$ | *P (mW)* | *Membrane Material* | *Actuation Fluid* |
|---|---|---|---|---|---|---|---|---|
| [30] | NO | 5×5 | 1,500 air | 700 | – | 200 | Aluminum | Methyl chloride |
| [35, 36] | NO | 8×6 | 10 $N_2$ | 1.3 | 33,000 | 3,500 | Silicon | FC |
| [37, 38] | NO | 8×8 | 1,800 $N_2$ | 227 | – | 100 | Rubber | FC |
| [39] | NO | 2.4×2.4 | 0.3 water | 3 | – | 100 | PDMS | Air |
| [40] | NO | 0.1×0.8 | 0.24 water | 1.4 | 1.15 | 100 | – | Water |
| [41, 42] | NO | 8.5×4.2 | 2 $N_2$ | 100 | 14 | 50 | Rubber | Paraffin |

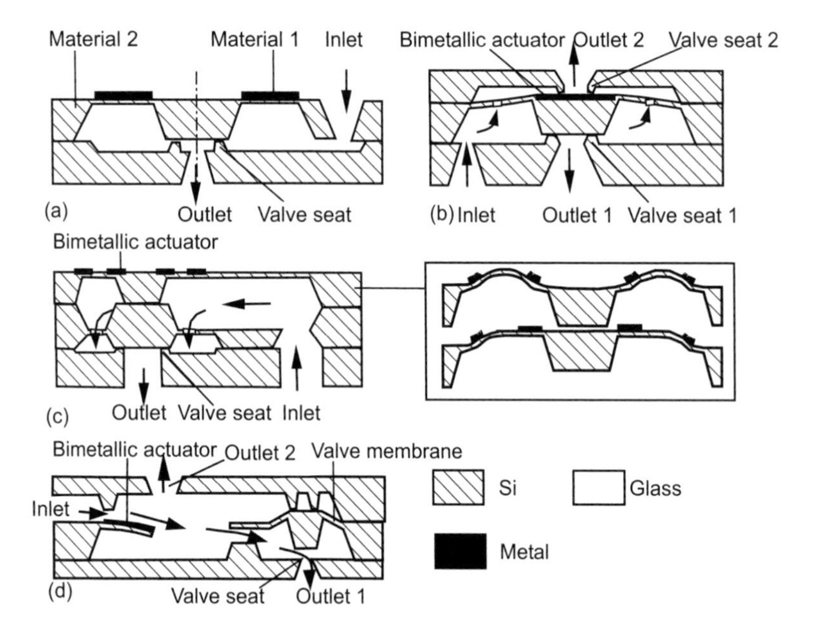


**Figure 6.26** Typical bimetallic microvalves: (a) and (b) bossed membrane; (c) bossed membrane, temperature compensation; and (d) beam actuator.

where $a$ is the radius of the membrane, $K_y$ is a constant that is a function of the ratio $(b/a)$, and $\Theta$ is the temperature term:

$$\Theta = \frac{6(\gamma_2 - \gamma_1)\Delta T(t_1 + t_2)(1 + \nu_e)}{t_2^2 K_l} \tag{6.40}$$

where $\gamma_1$ and $\gamma_2$ are the thermal expansion coefficients of aluminum and silicon, $t_1$ and $t_2$ are the corresponding thickness values, $\nu_e$ is the effective Poisson ratio of the composite membrane, and $K_l$ is a factor that depends on the relative stiffness of the two components.

The valve shown in Figure 6.26(b) [45] has a bimetallic actuator membrane. The heaters are integrated in the silicon substrate by boron diffusion. The 4 × 4-mm silicon membrane is 20 μm thick. The bossed area measures 500 × 500 μm for closing outlets 1 and 2, which measure 200 × 200 μm and 400 × 400 μm, respectively. The 10-μm aluminum layer is sputtered and wet-etched over the bossed area of the membrane. The four 400 × 400-μm orifices are etched into the membrane, allowing a three-way operation. The flow coming in can be switched between outlet 1 and outlet 2, alternatively. This valve also consumes several watts heating power for its operation.

The above bimetallic actuators have one drawback, in that the membrane also deflects at a higher ambient temperature. The valve described in [46] has an improved bimetallic actuator, which allows the buckling effect to be controlled only by a heater. With the symmetric arrangement of bimetallic layers, the membrane only buckles at thin membrane areas. The bossed region remains unaffected by higher ambient temperature. Only if the bimetallic layer close to the boss is heated can the valve seat be moved upward, as shown in Figure 6.26(c). The actuator membrane is defined by an n-type epitaxial layer grown on a p-type silicon substrate. The heaters are integrated in the membrane by diffusion.

An unconventional approach of designing a three-way valve is the fluidic switch [Figure 6.26(d)] [47]. The inlet fluid flow is switched between two outlets by a bimetallic cantilever. The fluid jet is passing along the bimetallic cantilever. By actuating the cantilever, the jet can be aimed above or under the valve membrane. In the first case, the jet drags on the valve membrane and closes outlet 1. The fluid flow can only exit the valve through outlet 2. In the latter case, the kinetic energy of the jet is partially transformed into static pressure energy, which opens the valve.

The advantages of SMA actuators are high force and large stroke. Disadvantages include the low efficiency and the low operation bandwidth (1 to 5 Hz). Figure 6.27(a) describes the working principle of a normally closed SMA microvalve. The valve is kept closed with an external beryllium-copper spring [48]. Heating the SMA lifts off the bossed valve membrane against the spring and opens the valve. Because the temperature profile across the membrane and the SMA structure is nonuniform, some portions of the SMA reach the transition temperature while other portions are below this temperature. Thus, the exerted force by the SMA can be proportionally controlled by the heating power. These characteristics make the proportional operation possible. The response time of this valve ranges from 0.5 to 5 Hz. The example depicted in Figure 6.27(b) also uses Ni/Ti as the SMA system [49], but the spring is made of silicon and bonded to the valve body.

The normally open valve shown in Figure 6.27(b) [2] uses a polyimide film as the valve membrane. The valve membrane has a diameter of 2 mm and a thickness of 2 μm. The polyimide membrane is transferred from a silicon wafer to the valve body, which is fabricated with conventional cutting techniques. The SMA is the material system Ti/Ni/Pd. Varying the composition of this system can adjust the transition temperatures in a range below 222°C. This relatively low temperature is needed for this valve because PMMA works as structural material. After patterning, the SMA film is transferred from a ceramic substrate onto the valve membrane. A spacer prestrains the SMA film and is used for adjusting the maximum back pressure at the inlet. Via holes are opened in the polyimide membrane to compensate the pressures above and below the SMA film.

The device is then sealed with adhesive bonding. In the cooled state, due to the low spring constants of both valve membrane and SMA film, the inlet pressure deflects the membrane and opens the valve. Thus, the valve is normally open. In the heated state, the force exerted by the SMA film closes the valve seat. Table 6.8 compares typical parameters of different thermomechanical valves.

### 6.2.6 Piezoelectric Valves

The valve depicted in Figure 6.28(a) has two piezostacks as its actuator [50]. The 50-μm valve membrane and the valve seat are fabricated in silicon with bulk micromachining (see Section 3.2.1). The external actuators are mounted on the valve membrane with epoxy. The improved versions with rubber valve seats are shown in Figures 6.28(b, c) [51]. The valves are designed as normally open [Figure 6.28(b)] or normally closed [Figure 6.28(c)]. In the upper silicon part, two 50-μm-thick beams work as valve springs, which suspend a mesa structure. Piezostacks are glued on the upper part. In the lower part, polymeric surface micromachining (see Chapter 3) is used to fabricate the flexible valve seat. A 5-μm-thick positive photoresist (AZ4562, see Chapter 3) works as a sacrificial layer. A 7-μm-thick photoresist works as the structural material. After coating and patterning the

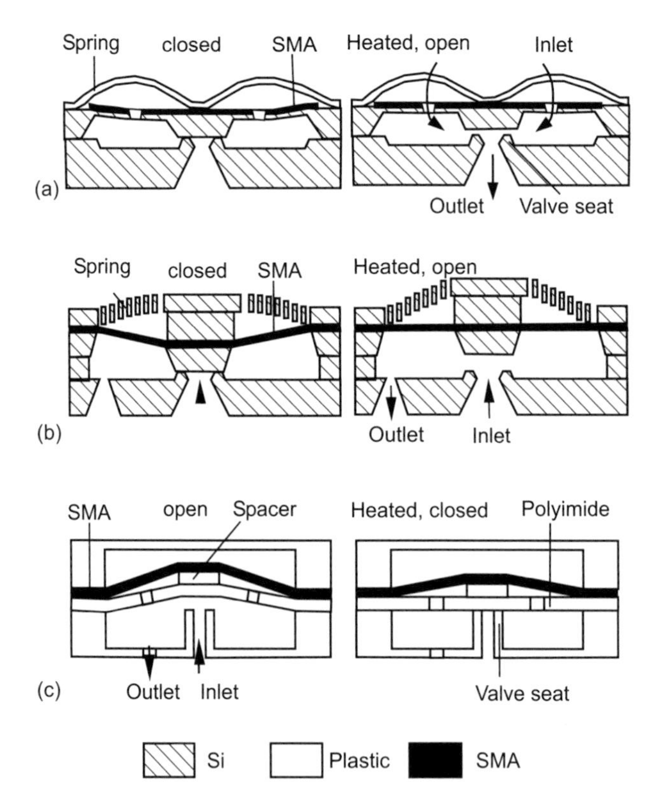


**Figure 6.27** Typical SMA microvalves: (a) external metal spring; (b) silicon spring; and (c) polymer spring.

two polymeric layers, the inlet and outlet are anisotropically etched from the back of the silicon wafer. The sacrificial positive resist is removed with acetone through the etched openings.

The two parts of the valve are press-fitted readily for operation. The normally closed version of this valve needs three piezostacks [Figure 6.28(c)]. The two piezostacks on both sides are used to pull up the mesa and to open the valve. The piezostack in the middle exerts a bias force to close the valve if it is leaky in the closed state. The large external piezostack's manual assembly and high drive voltage make this type of valve unattractive for integrated microfluidic systems. However, external actuators are attractive for disposable devices because they can be integrated in the reusable platform. This concept would make the design of the disposable part simpler and lower the total cost.

The valve shown in Figure 6.28(d) has a corrugated membrane as a valve spring. The valve seat is designed with particle traps in the form of trenches to improve the resistance to contamination [52]. The three silicon parts of the valve are fabricated with DRIE (see Section 3.3.1). The parts are assembled using eutectic bonding of a 0.26-mm Ti/Pt/Au system (see Section 3.6.2) at high temperature (350°C) and high pressure (10 to 200 bars).

In Figures 6.28(d, e), piezobimorph actuators are used. This type of actuator consists of two bonded piezoelectric layers with antiparallel polarization [53] or of a piezoelectric layer bonded to a nonpiezoelectric layer [54]. The valve membranes and valve seats are fabricated with bulk micromachining of silicon and glass. In the design of Figure 6.28(d), the glass part is used directly as the valve membrane. The actuator is a piezobimorph cantilever. The valve membrane in Figure 6.28(e) is etched in silicon. The actuator is a piezobimorph disc. Table 6.9 compares the typical parameters of piezoelectric microvalves.

**Table 6.8**

Typical Parameters of Thermomechanical Valves

($W$: Width, $l$: Length, $\dot{Q}_{max}$: Maximum flow rate, $p_{max}$: Maximum pressure, $L$: Leakage ratio, $P$: Input power)

| *Ref.* | *Type* | $W \times l$ *(mm²)* | $\dot{Q}_{max}$ *(mL/min)* | $p_{max}$ *(kPa)* | $L$ | $P$ (mW) | *Thermomechanical Material* | *Material* | *Technology* |
|---|---|---|---|---|---|---|---|---|---|
| [4] | NC | 5×5 | 700 $N_2$ | 100 | 70 | 4,000 | 12 μm Si | Silicon | Bulk |
| [43, 44] | NC | — | 80 air | 207 | 3,000 | 400 | 5 μm Al/8 μm Si | Silicon | Bulk |
| [45] | NC | 6×6 | 800 air | 1,000 | 35 $N_2$ | 1,500 | 10 μm Al/20 μm Si | Silicon | Bulk |
| [46] | NC | 10×6 | 5 water | 100 | — | 1,000 | Al, Si | Silicon | Bulk |
| [47] | NC | 14.5×8.5 | 75 water | 6000 | — | 10,000 | Al, Si | Silicon | Bulk |
| [48] | NC | — | 6,000 air | 550 | — | 290 | Ti/Ni | Silicon | Bulk |
| [49] | NC | — | 5 water | 1.3 | 1,000 | — | Ti/Ni | Silicon | Bulk |
| [2] | NO | 3×3 | 360 gas | 250 | — | 40; 250 | Ti/Ni, Ti/Ni/Pd | Plastic | Precision |

**Table 6.9**

Typical Parameters of Piezoelectric Valves

($W$: Width, $l$: Length, $\dot{Q}_{max}$: Maximum flow rate, $p_{max}$: Maximum pressure, $L$: Leakage ratio, $V$: Actuating voltage)

| *Ref.* | *Type* | $W \times l$ *(mm²)* | $\dot{Q}_{max}$ *(mL/min)* | $p_{max}$ *(kPa)* | $L$ | $V$ (V) | *Actuator* | *Material* | *Technology* |
|---|---|---|---|---|---|---|---|---|---|
| [50] | NC | 20×20 | 85 $N_2$ | 74 | > 850 | 100 | Ext. piezostack | Silicon, glass | Bulk |
| [51] | NO NC | 10×10 | 0.03 water | 25 | 100 | 100 | Ext. piezostack | Silicon, polymer | Bulk |
| [52] | NC | 10×10 | 0.025 water | 10 | 10 | 100 | Ext. piezostack | Silicon, polymer | Bulk |
| [53] | NC | 5.3×5.3 | 1.5 water | 10 | > 200 | 150 | Ext. piezo-bimorph cantilever | Silicon, Perspex | Bulk |
| [54] | NC | 5×5 | — | 50 | — | 100 | Ext. piezobimorph disc | Silicon, glass | Bulk |

### 6.2.7 Electrostatic Valves

The microvalve shown in Figure 6.29(a) has a valve seat in the form of a cantilever. The cantilever structure and the embedded metal electrodes are fabricated by surface micromachining. The valve inlet is opened with the common KOH wet etching process. The small gap between the electrode and the valve inlet allows the valve to close at a relatively low voltage of 30 V [55]. The same principle with a flexible electrode is used in the valve described in [56, 57]. The valve seat has the form of a flexible metal bridge, which can be bent in an S-shape [Figure 6.29(b)]. Because of the relatively large gap of 2 mm caused by the silicone resin spacer, the valve requires large actuating voltages, on the order of several hundreds of volts.

The valve depicted in Figure 6.29(c) has a silicon valve seat in the form of a bossed plate. The upper silicon plate acts as one fixed electrode. A Ti/Pt/Au layer works as the other fixed electrode on the lower glass substrate. The two electrodes above and under the valve seat allow the active control of both open and closed states. Thus, the design presents a bistable microvalve. With a gap of 2 μm, the pull-in and pull-up voltages (see Example 6.8) are 82V and 73V, respectively.

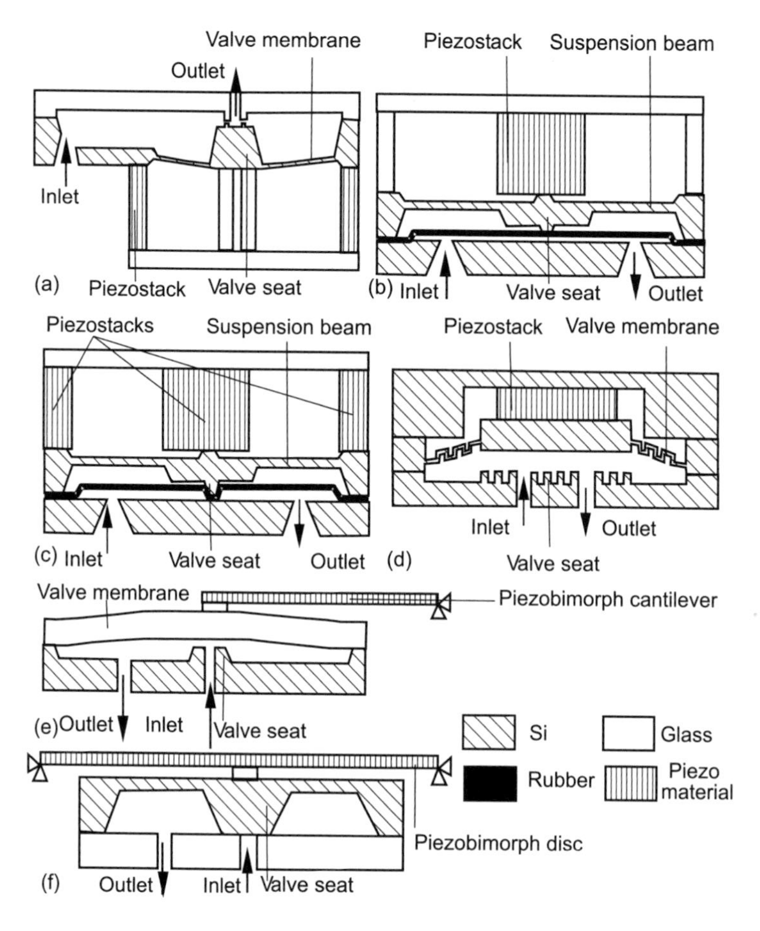


**Figure 6.28** Typical piezoelectric microvalves: (a) bossed membrane; (b) rubber membrane, NO; (c) rubber membrane, NC; (d) corrugated membrane, particle trap; (e) piezobimorph cantilever, glass membrane; and (f) piezobimorph disc.

Combining the electrostatic force, the buckling effect, and pneumatic force, large deflection can be achieved with relatively small drive voltages [59]. The two air chambers with underlying electrodes shown in Figure 6.29(d) are hermetically sealed but pneumatically connected. Thus, if one of the chambers is closed by attaching the membrane to the curved counterelectrode, the air is compressed in the other chamber and causes large deflections. The moving electrode is a membrane, which has a diameter of 2 or 3 mm. The membrane is made of an SOI wafer with a 7-μm silicon layer and a 1-μm-thick silicon dioxide layer. After etching back, the 7-μm silicon layer remains as the moving electrode. The fixed electrode is curved. The curvature is made with grayscale lithography of a 13-μm-thick photoresist. Subsequent dry etching of photoresist and silicon transfers the curvature to the silicon substrate. The pull-in voltages for the 2 and 3-mm membranes are 50V and 15V, respectively.

The valve shown in Figure 6.29(e) utilizes the same concept of curved electrodes for large valve seat deflection. The pressure is compensated at both sides of the actuating membrane. Thus, there is no need for a large force to keep the valve closed [54]. The curvature in the silicon substrate is fabricated with a silicon shadow mask and dry etching. The curved electrode is formed by deposition of polysilicon on the curved surface. An oxide/nitride insulating layer covers the polysilicon layer. The fixed electrode is made of n-type silicon. At a back pressure of 100 mbar, the pull-in and pull-up voltages are 162V and 145V, respectively.

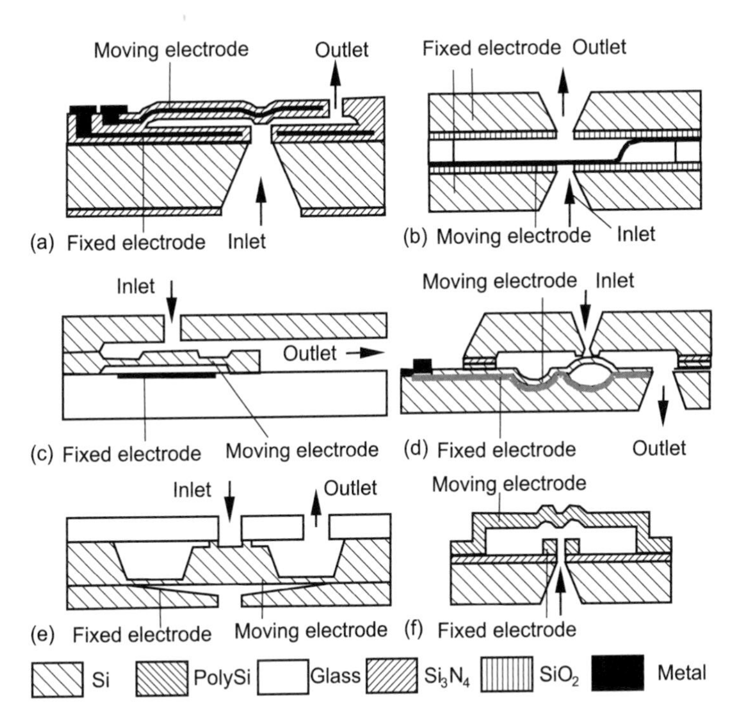


**Figure 6.29** Typical electrostatic valves: (a) surface micromachined; (b) flexible electrode; (c) bistable; (d) curved electrode, pneumatic; (e) curved electrode; and (f) surface micromachined.

**Table 6.10**

Typical Parameters of Electrostatic Valves

| *Refs.* | *Type* | *Size (mm×mm)* | $\dot{Q}_{max}$ *(mL/min)* | $P_{max}$ *(kPa)* | *L* | *V (V)* | *Valve Form* | *Material* | *Technology* |
|---|---|---|---|---|---|---|---|---|---|
| [55] | NO | 3.6×3.6 | 130 air | 12 | 24 | 30 | Cantilever | Silicon | Surface, bulk |
| [56, 57] | 3-way | 76×45 | 1 air | 100 | — | 250 | Bridge | Silicon | Bulk |
| [58] | Bistable | 10×10 | 1 air | 13 | 167 | 60 | Beam | Silicon, glass | Bulk |
| [60] | NO | 10×10 | 150 air | 10 | 15 | 250 | Membrane | Silicon | Surface |

The valve depicted in Figure 6.29(f) is made with surface micromachining. The polysilicon layer represents the valve membrane and the moving electrode. The fixed electrode is the bulk silicon. A 0.5-μm silicon nitride layer works as the insulator between the two electrodes. The valve is designed with rectangular membranes whose edge ranges from 300 to 500 μm. The air gap between the moving electrode and the nitride surface is 5 μm. For a 300 × 300-μm valve, the pull-in voltages are about 150V and 250V for back pressures of 0 and 135 mbar, respectively. These valves are arranged in a 5×5 array, which can control the flow rate digitally, as discussed at the beginning of this chapter. Table 6.10 compares the above valves.

### 6.2.8 Electromagnetic Valves

Figure 6.30(a) shows a typical microvalve with an external solenoid actuator. The valve nozzle and valve seat are fabricated in silicon using bulk micromachining [61]. The valve seat has the form of a plate suspended on four folded beams [Figure 6.30(b)]. The valve works proportionally by

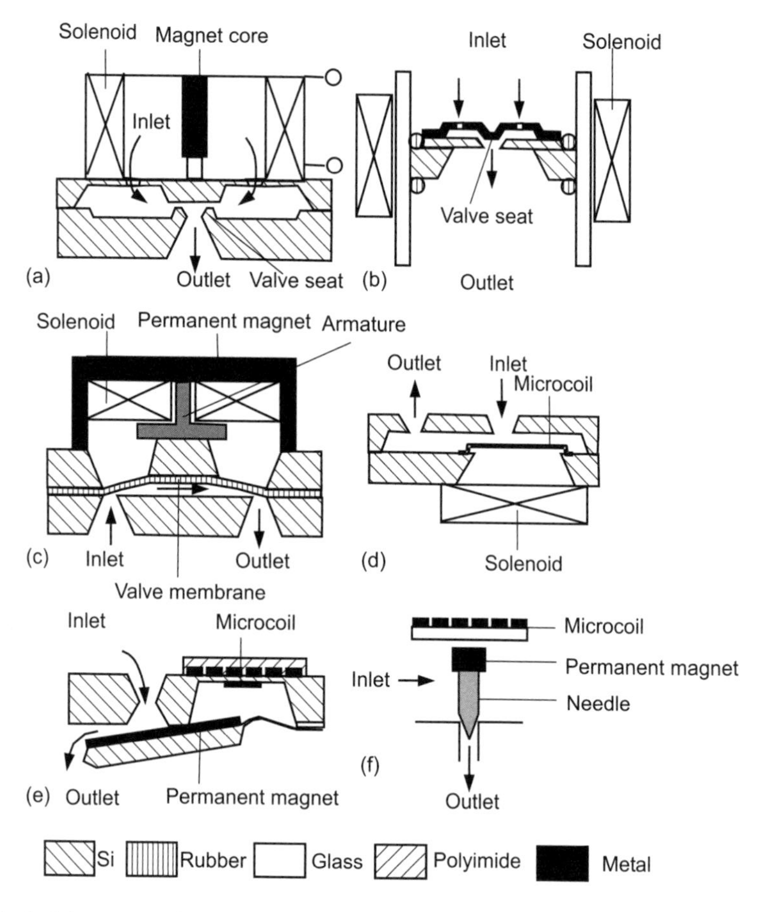


**Figure 6.30** Typical electromagnetic valves: (a) external solenoid; (b) external coil, integrated core; (c) external solenoid, rubber membrane; (d) external solenoid, integrated coil; (e) integrated coil and magnet; and (f) microcoil, permanent magnet.

changing the gap between the seat and the nozzle [42]. In the example of Figure 6.30(b), the valve seat represents the magnetic core itself. The valve seat is made of sputtered NiFe. The magnetic coil is placed outside a pipe, which carries the fluid flow and the valve inside [62]. This design allows electrically isolated operation. The valve shown in Figure 6.30(c) also has an external actuator. The actuator allows bistable operation, which keeps the valve open or closed. The rubber membrane improves the tightness and resistance against particles [Figure 6.30(c)] [63].

The design shown in Figure 6.30(d) uses an integrated microcoil and an external magnetic core. The coil is made of electroplated gold and is placed on the valve seat. The valve seat is a plate suspended on four folded beams. With a field gradient of 310 T/m and a current of 25 mA, a closing force of 800 μN can be achieved [64].

The valve shown in Figure 6.30(e) has a flapper valve seat, which rotates under magnetic force. The NiFe alloy coil is integrated in the valve body. Magnetic foil is used for fabricating the magnetic core. The complete package is then magnetized [65].

The needle valve depicted in Figure 6.30(f) has only the microcoil as a micromachined part. With a PWM drive signal, the valve is able to generate proportional flow rates at a constant inlet pressure [66]. Table 6.11 compares the above electromagnetic valves.

**Table 6.11**
Typical Parameters of Electromagnetic Valves

| *Refs.* | *Type* | *Size (mm×mm)* | $\dot{Q}_{max}$ *(mL/min)* | $p_{max}$ *(kPa)* | *L* | *P (mW)* | *Valve Type* | *Material* | *Technology* |
|---|---|---|---|---|---|---|---|---|---|
| [61] | NO | 4 × 4 | 500 air | — | — | 1,000 | Flapper | Silicon | Bulk |
| [63] | NO | 7 × 7 | 0.7 | 10 | 100 | — | Membrane | Silicon/rubber | Bulk |
| [65] | NC | 8 × 8 | 0.03 | 2 | — | 0.4 | Flapper | Silicon/Fe-Ni alloy | Bulk |
| [66] | NC | 5 × 5 | 0.9 water | 980 | 90/2 | — | Needle | Silicon/Fe-Ni alloy | Bulk |

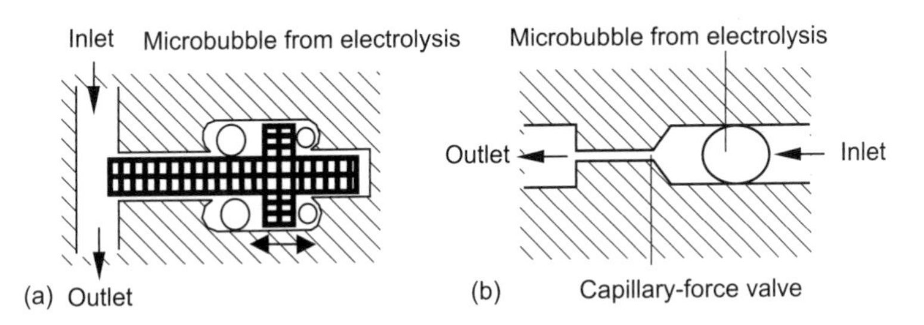


**Figure 6.31** Typical electrochemical valves: (a) gate form; and (b) capillary form.

### 6.2.9 Electrochemical and Chemical Valves

The valve shown in Figure 6.31(a) [67] has the same form as its thermopneumatic predecessor shown in Figure 6.25(d) [40]. The valve structure lies in the horizontal plane. The gas bubble generated in each chamber pushes the valve into open and closed positions. Therefore, the valve is bistable and no valve spring is needed for this design. The valve consumes 4.3 μW electrical power, which is 10,000 times smaller than the same type with its thermopneumatic actuator. The low operation voltage of 2.5V is a further advantage. Electrochemical actuators are suitable for power-saving applications, which do not require a fast response time.

In the example shown in Figure 6.31(b), the electrochemically generated gas bubble is used for overcoming the pressure difference caused by surface tension differences between wide and small channels. The structure can be used as a normally closed valve (without gas bubble). The gas bubble triggers the valve to open [16].

A number of microvalves with chemical actuators have been reported. All designs are based on the volume change of a hydrogel due to a stimulus such as temperature and pH value. Richter et al. [9] reported a microvalve with PNIPAAm as hydrogel. The hydrogel shrinks at higher temperature. The switching region is between 25°C and 35°C. Temperature change is induced by integrated microheaters. The valve consumes several hundreds of milliwatts power and has a response time of several tens of seconds. The microvalves reported by Liu et al. use purely chemical stimuli to operate. The solution with controlling pH is fed directly into the actuator chamber filled with pAA-HEMA copolymer to control the PDMS valve membrane. The response time of this valve is several minutes. The hydrogel can also be placed directly in a microchannel to block the flow if the pH value in the flow changes. A similar design was reported by Baldi et al. [13] with AAm-3-MPBA. The long response time of this valve is caused by the relatively large hydrogel amount in the actuator chamber.

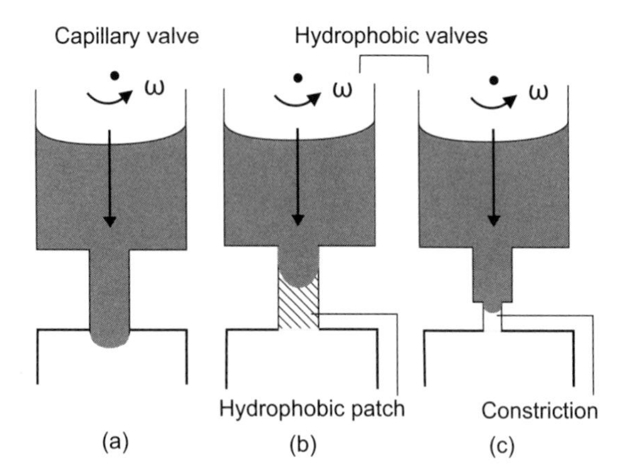


**Figure 6.32** Examples of typical valving methods for centrifugal microfluidics: (a) capillary-based microvalve; and (b, c) hydrophobic valves.

### 6.2.10 Capillary-Force Valves

Using the pressure generated by the surface tension gradient, bubbles in microchannels can be used as active valves. Table 6.12 lists the surface tensions of selected liquids at 20°C. The implementation of this concept is simpler than other conventional principles because there are no moving parts involved. The microbubble can be generated by means of electrolysis of water or by heating liquid water to the vapor phase. The transition from a small channel section to a larger section can act as a valve. The principle of this burst valve was explained previously.

In centrifugal microfluidic systems, capillary microvalves are widely used to control fluid movement. Figure 6.32(a) shows a capillary valve allows for fluid movement once the induced pressure exceeds the capillary pressure. An abrupt increase of cross section at a hydrophobic microchannel increases the capillary pressure at the liquid-air interface. The function of hydrophobic valves is based on either creating functionalized hydrophobic zones or abrupt reduction in a hydrophobic microchannel to inhibit fluid movement [69], Figure 6.32(b, c). In such designs, fluid can pass through the microvalve once the rotational speed is above a critical value.

The electrocapillary effect uses potential differences between liquid-metal (mercury) and electrolyte solution to move a mercury drop [14, 15]. The mercury drop works as a piston, which compresses the electrolyte and deflects the valve membrane pneumatically. This actuator type is not used directly in microvalves, but its application in a check-valve micropump [68] proves the feasibility of this valve principle; see Chapter 7.

## 6.3 SUMMARY

This chapter discusses the design of active microvalves. One of the key functional elements in a microvalve is the actuator. The choice of a suitable actuator depends on the application requirements, such as closing pressure or response time. Design and dimensioning an actuator require understanding of the physical effect behind the actuation concept. While most actuators use electricity as input energy, other types of energy, such as chemical energy, potential energy, or manual actuation, are also of interest. Designing of the valve spring and valve seat allows the valve to have a leakage-free operation. With the emergence of polymeric technologies, making active microvales from soft material such as PDMS could improve miniaturization, sealing properties and reduce actuator power consumption. While analog microvalves require precise position control of

**Table 6.12**

Surface Tensions of Selected Liquids (at 20°C if not otherwise indicated)

| *Liquid* | *σ* ($10^{-3}$ *N/m*) | *Liquid* | *σ* ($10^{-3}$ *N/m*) |
|---|---|---|---|
| Trifluoroacetic | Acid 13.63 (24°C) | Methyl Ethyl Ketone | 24.0 (25°C) |
| Pentane | 15.48 (25°C) | n-Butyl Alcohol | 24.57 |
| Ethyl Ether | 17.06 | Cyclohexane | 24.98 |
| Hexane | 17.91 (25°C) | n-Butyl Acetate | 25.09 |
| Iso-Octane | 18.77 | Methyl n-Propyl Ketone | 25.09 |
| Acetonitrile | 19.10 | Tetrahydrofuran | 26.4 (25°C) |
| Methyl t-Butyl Ether | 19.4 (24°C) | o-Dichlorobenzene | 26.84 |
| Heptane | 20.30 | Chloroform | 27.16 |
| Triethylamine | 20.66 | Dichloromethane | 28.12 |
| Isopropyl Alcohol | 21.79 (15°C) | Toluene | 28.53 |
| Ethyl Alcohol | 22.32 | o-Xylene | 30.03 |
| Cyclopentane | 22.42 | 2-Methoxyethanol | 31.8 (15°C) |
| Methanol | 22.55 | Ethylene Dichloride | 32.23 |
| Isobutyl Alcohol | 22.98 | Dimethyl Acetamide | 32.43 (30°C) |
| Acetone | 23.32 | Chlorobenzene | 33.28 |
| Methyl Isobutyl Ketone | 23.64 | 1,4-Dioxane | 34.45 (15°C) |
| n-Propyl Alcohol | 23.70 | N,N-Dimethylformamide | 36.76 |
| n-Butyl Chloride | 23.75 | Pyridine | 36.88 |
| Ethyl Acetate | 23.75 | Propylene Carbonate | 41.93 |
| | | Water | 72.8 |

the valve seat, digital valves in the pulse-width-modulation mode can achieve the same performance. Digital valves designed with a bistable valve spring can minimize power consumption, because the actuator is only needed for switching valve position. The lower Young's modulus of polymers allows large deflection of a bistable spring and a small triggering force, and thus makes polymers even more attractive for making active microvalves.

## Problems

**6.1** For the microvalve depicted in Figure 6.25(d), determine the actuating force on the gate. The area of the piston-like structure on each side is 100 × 73 μm. The gap, which determines the minimum radius of curvature, is 8 μm wide.

**6.2** The thermodynamic properties of liquid water are: $v(25°\text{C}) = 1.0029 \times 10^{-3}$ m$^3$/kg, $u(25°\text{C}) =$ 104.88 kJ/kg, and of vapor are $v(100°\text{C}) = 1.673$ m$^3$/kg, $u(25°\text{C}) = 2,506.5$ kJ/kg. Determine the maximum efficiency of a thermal bubble with an actuating pressure of 1 bar.

**6.3** Regarding the situation described in Figure 6.14, calculate the work done by an air bubble escaping a narrow channel of 10 μm in diameter into a larger section of 40 μm. The diffuser section is 10 μm long.

## References

[1] Shoji, S., and Esashi, M., "Microflow Devices and Systems," *Journal of Micromechanics and Microengineering*, Vol. 4, No. 4, 1994, pp. 157–171.

[2] Kohl, M., et al., "Thin Film Shape Memory Microvalves with Adjustable Operation Temperature," *Sensors and Actuators A*, Vol. 83, 2000, pp. 214–219.

[3] Maluf, N., *An Introduction to Microelectromechanical Systems Engineering*, Norwood, MA: Artech House 2000.

[4] Lisec, T., et al., "Thermally Driven Microvalve with Buckling Behaviour for Pneumatic Applications," *Proceedings of MEMS'94, 7th IEEE International Workshop Micro Electromechanical System*, Oiso, Japan, Jan. 25–28, 1994, pp. 13–17.

[5] Chu, W. H., Mehregany, M., and Mullen, R. L., "Analysis of Tip Deflection and Force of a Bimetallic Cantilever Microactuator," *Journal of Micromechanics and Microengineering*, Vol. 3, 1993, pp. 4–7.

[6] Shimizu, K., and Tadaki, T., *Shape Memory Alloys*, H. Funakubo, (ed.), Gordon and Breach Science Publishers, 1987.

[7] Gerlach, G., and Dötzel, W., *Grundlagen der Mikrosystemtechnik*, München: Hanser Verlag, 1997.

[8] Hoffmann, J., et al., "Photopatterning of Thermally Senstive Hydrogels Useful for Microactuators," *Sensors and Actuators A*, Vol. 77, 1999, pp. 139–144.

[9] Richter, A., et al., "Electronically Controllable Microvalves Based on Smart Hydrogels: Magnitudes and Potential Applications," *Journal of Microelectromechanical Systems*, Vol. 12, 2003, pp. 748–753.

[10] Richter, A., et al., "Influence of Volume Phase Transition Phenomena on the Behavior of Hydrogel-Based Valves," *Sensors and Actuators B*, Vol. 99, 2004, pp. 451–458.

[11] Cao, X., Lai, S., Lee, L. J., "Design of a Self-Regulated Drug Delivery Device," *Biomedical Microdevices*, Vol. 3, 2001, pp. 109–118.

[12] Liu, R. H., Yu, Q., Beebe, D. J., "Fabrication and Characterization of Hydrogel-Based Microvalves," *Journal of Microelectromechanical Systems*, Vol. 11, 2002, pp. 45–53.

[13] Baldi, A., et al., "A Hydrogel-Actuated Environmentally Sensitive Microvalve for Active Flow Control," *Journal of Microelectromechanical Systems*, Vol. 12, 2003, pp. 613–621.

[14] Lee, J., and Kim, C. J., "Surface-Tension-Driven Microactuation Based on Continuous Electrowetting," *Journal of Microelectromechanical Systems*, Vol. 9, No. 2, 2000, pp. 171–180.

[15] Lee, J., et al., "Addressable Micro Liquid Handling by Electric Control of Surface Tension," *Proceedings of MEMS'01, 14th IEEE International Workshop Micro Electromechanical System*, Interlaken, Switzerland, Jan. 21–25, 2001, pp. 499–502.

[16] Man, P. F., et al., "Microfabricated Capillary-Driven Stop Valve and Sample Injector," *Proceedings of MEMS'98, 11th IEEE International Workshop Micro Electromechanical System*, Heidelberg, Germany, Jan. 25–29, 1998, pp. 45–50.

[17] Timoshenko, S. P., and Gere, J. M., *Theory of Elastic Stability*, 2nd ed., New York: McGraw-Hill, 1963.

[18] Schomburg, W. K., and Goll, C., "Design Optimization of Bistable Microdiaphragm Valves," *Sensors and Actuators A*, Vol. 64, 1998, pp. 259–264.

[19] Cousseau, P., et al., "Improved MicroFlow Regulator for Drug Delivery Systems," *Proceedings of MEMS'01, 14th IEEE International Workshop Micro Electromechanical System*, Interlaken, Switzerland, Jan. 21–25, 2001, pp. 527–530.

[20] Hulme, S. E., et al., "Incorporation of Prefabricated Screw, Pneumatic, and Solenoid Valves into Microfluidic Devices," *Lab Chip*, Vol. 9, 2009, pp. 79–86.

[21] Shaegh, S. A. M., et al., "Plug-and-Play Microvalve and Micropump for Rapid Integration with Microfluidic Chips," *Microfluidics and Nanofluidics*, Vol. 19, No. 3, 2015, pp. 557–564.

[22] Shaegh, S. A. M., et al., "Rapid Prototyping of Whole-Thermoplastic Microfluidics with Built-in Microvalves Using Laser Ablation and Thermal Fusion Bonding," *Sensors and Actuators B: Chemical*, Vol. 255, 2018, pp. 100–109.

[23] Huff, M. A., and Schmidt, M. A., "Fabrication, Packaging, and Testing of a Wafer-Bonded Microvalve," *Technical Digest of the IEEE Solid State Sensor and Actuator Workshop*, Hilton Head Island, SC, June 22–25, 1992, pp. 194–197.

[24] Takao, H., and Ishida, M., "Microfluidic Integrated Circuits for Signal Processing Using Analogous Relationship Between Pneumatic Microvalve and MOSFET", *Journal of Microelectromechanical Systems*, Vol. 12, 2003, pp. 497–505.

[25] Vieider, C., Oehman, O., and Elderstig, H., "A Pneumatic Actuated Micro Valve with a Silicon Rubber Membrane for Integration with Fluid-Handling Systems," *Proceedings of Transducers '95, 8th International Conference on Solid-State Sensors and Actuators*, Stockholm, Sweden, June 16–19, 1995, pp. 284–286.

[26] Hosokawa, K., and Maeda, R., "A Pneumatically-Actuated Three-Way Microvalve Fabricated with Polydimethylsiloxane Using the Membrane Transfer Technique," *Journal of Micromechanics and Microengineering*, Vol. 10, 2000, pp. 415–420.

[27] Ohori, T., et al., "Three-Way Microvalve for Blood Flow Control in Medical Micro Total Analysis Systems (mTAS)," *Proceedings of MEMS'97, 10th IEEE International Workshop Micro Electromechanical System*, Nagoya, Japan, Jan. 26–30, 1997, pp. 333–337.

[28] Rich, C. A., and Wise, K. D., "A High-Flow Thermopneumatic Microvalve with Improved Efficiency and Integrated State Sensing," *Journal of Microelectromechanical Systems*, Vol. 12, 2003, pp. 201–208.

[29] Sim, D. Y., Kurabayashi, T., and Esashi, M., "Bakable Silicon Pneumatic Microvalve," *Proceedings of Transducers '95, 8th International Conference on Solid-State Sensors and Actuators*, Stockholm, Sweden, June 16–19, 1995, pp. 280–283.

[30] Zdeblic, M. J., et. al, "Thermopneumatically Actuated Microvalves and Integrated Electro-Fluidic Circuits," *Technical Digest of the IEEE Solid State Sensor and Actuator Workshop*, Hilton Head Island, SC, June 13–16, 1994, pp. 251–255.

[31] Melin, J., and Quake, S. R., "Microfluidic Large-Scale Integration: the Evolution of Design Rules for Biological Automation," *Annu Rev Biophys Biomol Struct.*, Vol. 36, 2007, pp. 213–31.

[32] Thorsen, T., Maerkl, S. J., and Quake, S. R., "Microfluidic Large-Scale Integration", *Science*, Vol. 298, 2002, No. 5593, pp. 580–584.

[33] Au, A. K., et al., "Microvalves and Micropumps for BioMEMS," *Micromachines*, Vol. 2, No. 2, 2011, pp. 179–220.

[34] Gu, W., et al., "Multiplexed Hydraulic Valve Actuation Using Ionic Liquid Filled Soft Channels and Braille Displays", *Appl. Phys. Lett.*, Vol. 90, 2007, pp. 1–3.

[35] Henning, A. K., "Microfluidic MEMS," *Proceedings of the IEEE Aerospace Conference*, Snowmass at Aspen, CO, March 1998, pp. 471–486.

[36] Henning, A. K., et al., "Microfluidic MEMS for Semiconductor Processing," *IEEE Transactions on Components, Hybrids, and Manufacturing Technology, Part B: Advanced Packaging*, Vol. 21, No. 4, 1998, pp. 329–337.

[37] Yang, X., et al., "A MEMS Thermopneumatic Silicone Membrane Valve," *Proceedings of MEMS'97, 10th IEEE International Workshop Micro Electromechanical System*, Nagoya, Japan, Jan. 26–30, 1997, pp. 114–118.

[38] Grosjean, C., Yang, X., Tai, Y. C., "A Practical Thermopneumatic Valve," *Proceedings of MEMS'99, 12th IEEE International Workshop Micro Electromechanical System*, Orlando, FL, Jan. 17–21, 1999, pp. 147–152.

[39] Kim, J. H., et al., "A Disposable Thermopneumatic-Actuated Microvalve Stacked with PDMS Layers and ITO-Coated Glass," *Microelectronic Engineering*, Vol. 73–74, 2004, pp. 864–869.

[40] Papavasiliou, A. P., Liepmann, D., and Pisano, A. P., "Fabrication of a Free Floating Gate Valve," *Proceedings of IMECE, International Mechanical Engineering Congress and Exposition*, Vol. 1, Nashville, TN, Nov. 14–19, 1999, pp. 435–440.

[41] Carlen, E. T., and Mastrangelo, C. H., "Surface Micromachined Paraffin-Actuated Microvalve," *Journal of Microelectromechanical Systems*, Vol. 11, 2002, pp. 408–420.

[42] Carlen, E. T., and Mastrangelo, C. H., "Simple, High Actuation Power, Thermally Activated Parafin Microactuator," *Proceedings of Transducers '99, 10th International Conference on Solid-State Sensors and Actuators*, Sendai, Japan, June 7–10, 1999, pp. 1364–1367.

[43] Jerman, H., "Electrically-Activated Micromachined Diaphragm Valves," *Technical Digest of the IEEE Solid State Sensor and Actuator Workshop*, Hilton Head Island, SC, June 4–7, 1990, pp. 65–69.

[44] Jerman, H., "Electrically-Activated Normally-Closed Diagphragm Valves," *Proceedings of Transducers '91, 6th International Conference on Solid-State Sensors and Actuators*, San Francisco, CA, June 23–27, 1991, pp. 1045–1048.

[45] Messner, S., et al., "A Normally-Closed, Bimetallically Actuated 3-Way Microvalve for Pneumatic Application," *Proceedings of MEMS'98, 11th IEEE International Workshop Micro Electromechanical System*, Heidelberg, Germany, Jan. 25–29, 1998, pp. 40–44.

[46] Franz, J., Baumann, H., and Trah, H.-P., "A Silicon Microvalve with Integrated Flow Sensor," *Proceedings of Transducers '95, 8th International Conference on Solid-State Sensors and Actuators*, Stockholm, Sweden, June 16–19, 1995, pp. 313–316.

[47] Trah, H. P., et al., "Micromachined Valve with Hydraulically Actuated Membrane Subsequent to a Thermoelectrically Controlled Bimorph Cantilever," *Sensors and Actuators A*, Vol. 39, 1993, pp. 169–176.

[48] Barth, P. W., "Silicon Microvalves for Gas Flow Control," *Proceedings of Transducers '95, 8th International Conference on Solid-State Sensors and Actuators*, Stockholm, Sweden, June 16–19, 1995, pp. 276–279.

[49] Kahn, H., Huff, M. A., and Heuer, A. H., "The Ti Ni Shape-Memory Alloy and Its Applications for MEMS," *Journal of Micromechanics and Microengineering*, Vol. 8, 1998, pp. 213–221.

[50] Esashi, M., Shoji, S., and Nakano, A., "Normally Closed Microvalve and Micropump Fabricated on a Silicon Wafer," *Proceedings of MEMS'89, 1st IEEE International Workshop Micro Electromechanical System*, Salt Lake City, UT, Feb. 1989, pp. 29–34.

[51] Shoji, S., et al., "Smallest Dead Volume Microvalves for Integrated Chemical Analyzing Systems," *Proceedings of Transducers '91, 6th International Conference on Solid-State Sensors and Actuators*, San Francisco, CA, June 23–27, 1991, pp. 1052–1055.

[52] Chakraborty, I., et al., "MEMS Micro-Valve for Space Applications," *Sensors and Actuators A*, Vol. 83, 2000, pp. 188–193.

[53] Stehr, M., et al., "The VAMP—A New Device for Handling Liquids or Gases," *Sensors and Actuators A*, Vol. 57, 1996, pp. 153–157.

[54] Nguyen, N. T., et al., "Hybrid-Assembled Micro Dosing System Using Silicon-Based Micropump/Valve and Mass Flow Sensor," *Sensors and Actuators A*, Vol. 69, 1998, pp. 85–91.

[55] Ohnstein, T. R. et al., "Micromachined Silicon Microvalve," *Proceedings of MEMS'90, 2th IEEE International Workshop Micro Electromechanical System*, Napa Valley, CA, Feb. 11–14, 1990, pp. 95–98.

[56] Shikida, M., et al., "Electrostatically Driven Gas Valve with High Conductance," *Journal of Micro Electromechanical Systems*, Vol. 3, No. 2, June 1994, pp. 76–80.

[57] Shikida, M., and Sato, K., "Characteristics of an Electrostatically-Driven Gas Valve Under High-Pressure Conditions," *Proceedings of MEMS'94, 7th IEEE International Workshop Micro Electromechanical System*, Oiso, Japan, Jan. 25–28, 1994, pp. 235–240.

[58] Robertson, J. K., and Wise, K. D., "A Nested Electrostatically-Actuated Microvalve for an Integrated Microflow Controller," *Proceedings of MEMS'94, 7th IEEE International Workshop Micro Electromechanical System*, Oiso, Japan, Jan. 25–28, 1994, pp. 7–12.

[59] Wagner, B., et al., "Bistable Microvalve with Pneumatically Coupled Membranes," *Proceedings of MEMS'96, 9th IEEE International Workshop Micro Electromechanical System*, San Diego, CA, Feb. 11–15, 1996, pp. 384–388.

[60] Vandelli, N., et al., "Development of a MEMS Microvalve Array for Fluid Flow Control," *Journal of Micro Electromechanical Systems*, Vol. 7, No. 4, Dec. 1998, pp. 395–403.

[61] Pourahmadi, F., et al., "Variable-Flow Micro-Valve Structure Fabricated with Silicon Fusion Bonding," *Technical Digest of the IEEE Solid State Sensor and Actuator Workshop*, Hilton Head Island, SC, June 4–7, 1990, pp. 78–81.

[62] Yanagisawa, K., Kuwano, H., and Tago, A., "An Electromagnetically Driven Microvalve," *Proceedings of Transducers '93, 7th International Conference on Solid-State Sensors and Actuators*, Yokohama, Japan, June 7–10, 1993, pp. 102–105.

[63] Böhm, S., et al., "A Micromachined Silicon Valve Driven by a Miniature Bistable Electromagnetic Actuator," *Sensors and Actuators A*, Vol. 80, 2000, pp. 77–83.

[64] Meckes, A., et al., "Microfluidic System for the Integration on Cyclic Operation of Gas Sensors," *Sensors and Actuators A*, Vol. 76, 1999, pp. 478–483.

[65] Capanu, M., et al., "Design, Fabrication, and Testing of a Bistable Electromagnetically Actuated Microvalve," *Journal of Microelectromechanical Systems*, Vol. 9, No. 2, 2000, pp. 181–189.

[66] Shinozawa, Y., Abe, T., and Kondo, T., "A Proportional Microvalve Using a Bi-Stable Magnetic Actuator," *Proceedings of MEMS'97, 10th IEEE International Workshop Micro Electromechanical System*, Nagoya, Japan, Jan. 26–30, 1997, pp. 233–237.

[67] Papavasiliou, A. P., Liepmann, D., and Pisano, A. P., "Electrolysis-Bubble Actuated Gate Valve," *Technical Digest of the IEEE Solid State Sensor and Actuator Workshop*, Hilton Head Island, SC, June 4–8, 2000, pp. 48–51.

[68] Yun, K. S., et al., "A Micropump Driven by Continuous Electrowetting Actuation for Low Voltage and Low Power Operations," *Proceedings of MEMS'01, 14th IEEE International Workshop Micro Electromechanical System*, Interlaken, Switzerland, Jan. 21–25, 2001, pp. 487–490.

[69] Gorkin, R., et al., "Centrifugal Microfluidics for Biomedical Applications", *Lab Chip*, Vol. 10, No. 14, 2010 pp. 1758–1773.

# Chapter 7

## Microfluidics for Internal Flow Control: Micropumps

The next active components of a microfluidic system are micropumps. With the growing importance of genomics, proteomics, discovery of new drugs, and on-chip diagnostics, controlled transport of fluids in microscale becomes an important and crucial task. New transport effects, such as electrokinetic effects, interfacial effects, acoustic streaming, magnetohydrodynamic effects, and electrochemical effects, which previously were neglected in macroscopic applications, now gain in importance on the microscale. Overviews on micropumps were reported in several excellent review papers. Gravesen et al. gave a general overview on fluidic problems in the microscale [1]. Shoji and Esashi discussed microfluidics from the device point of view, and considered micropumps, microvalves, and flow sensors [2]. This chapter only deals with micropumps, and discusses their design considerations as well as the published design examples.

In contrast to other MEMS devices, micropumps are one of the components with the largest variety of operating principles. The most important actuation principles are discussed in Chapter 6. Thus, in this chapter, micropumps are categorized according to their pumping principles.

Similar to other MEMS applications, the first approach in designing a micropump is the miniaturization of well-known mechanical principles from the macroscale. The next approaches were to apply new pumping effects, which are more effective on the microscale than the macroscale. Most of the micropumps developed with the latter approach are nonmechanical pumps. Therefore, micropumps are categorized in this chapter as either mechanical pumps or nonmechanical pumps.

*Mechanical pumps* can be further categorized according to the principles by which mechanical energy is applied to the fluid. Under this system, mechanical pumps are divided into two major categories: displacement pumps and dynamic pumps [4]. In *displacement pumps*, energy is periodically added by the application of force to one or more moveable boundaries of any desired number of enclosed, fluid-containing volumes, resulting in a direct increase in pressure, up to the value required to move the fluid through check valves or ports into the discharge line. Check-valve pumps, peristaltic pumps, valveless rectification pumps, and rotary pumps belong to the displacement category (Table 7.1). In *dynamic pumps*, mechanical energy is continuously added to increase the fluid velocities within the machine. The higher velocity at the pump outlet increases the pressure. Centrifugal pumps and ultrasonic pumps belong to the dynamic category.

*Nonmechanical pumps* add momentum to the fluid by converting another nonmechanical energy form into kinetic energy. While mechanical pumping is mostly used in macroscale pumps and micropumps with a relatively large size and high flow rates, this second category discovers its advantages in the microscale. Since the viscous force in microchannels increases in the second order with miniaturization, the first pump category cannot deliver enough power to overcome the high

**Table 7.1**
Mechanical Pumping Principles

| *Displacement Pumps* | *Dynamic Pumps* |
|---|---|
| Check-valve pumps | Ultrasonic pumps |
| Peristaltic pumps | Centrifugal pumps |
| Valveless rectification pumps | |
| Rotary pumps | |

**Table 7.2**
Nonmechanical Pumping Principles

| | *Pressure Gradient* | *Concentration Gradient* | *Electrical Potential Gradient* | *Magnetic Potential* |
|---|---|---|---|---|
| Fluid flow | Surface tension driven flow (electrowetting, capillarity, Marangoni effect, surface modification) | Osmosis (semipermeable membrane, surfactants) | Electro-osmosis (electrolyte), electrohydrodynamic (dielectric fluid) | Ferrofluidic |
| Solute flux | Ultrafiltration | Diffusion | Electrophoresis, dielectrophoresis | Magneto-hydrodynamic flow |

fluidic impedance in the microscale. Table 7.2 lists the common nonmechanical pumping principles. Figure 7.1 illustrates the typical flow rate range of micropumps.

For flow rates more than 10 mL/min, miniature pumps or macroscale pumps are the most common solutions. The typical operation range of displacement micropumps lies between hundreds of nanoliters per minute to tens of milliliters per minute. For flow rates less than 10 μL/min, alternative dynamic pumps or nonmechanical pumps can also been employed for an accurate control of these small fluid amounts.

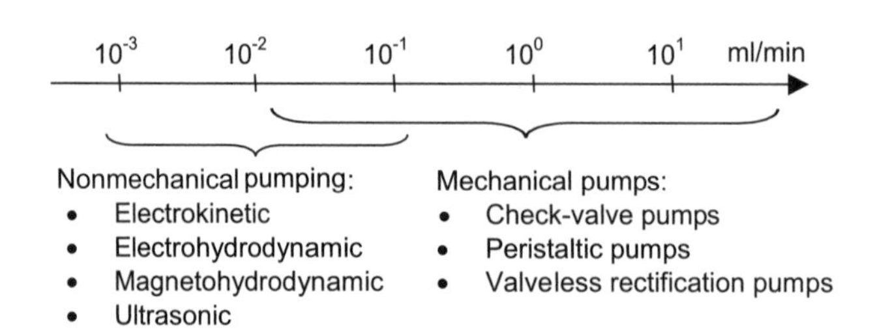


**Figure 7.1** Flow rate range of different pump principles.

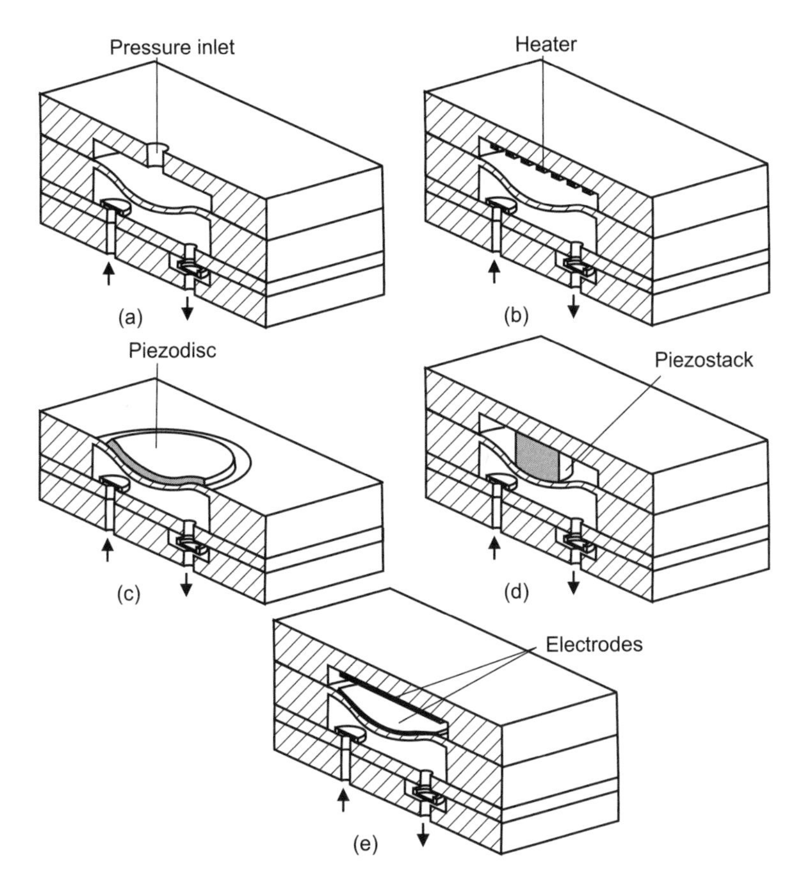


**Figure 7.2** Actuation schemes for check-valve micropumps: (a) pneumatic; (b) thermopneumatic; (c) piezoelectric disc; (d) piezoelectric stack; and (e) electrostatic.

## 7.1 DESIGN CONSIDERATIONS

### 7.1.1 Mechanical Pumps

#### 7.1.1.1 Actuators

All mechanical pumps require an electromechanical actuator, which generally converts electrical energy into mechanical work. Actuators can be categorized by their physical principles as seen in Chapter 6. Alternatively, regarding their integration ability, actuators can be categorized as external actuators and integrated actuators [2]. Figure 7.2 depicts a check-valve micropump equipped with the different actuators.

*External actuators* include electromagnetic actuators with solenoid plungers and external magnetic field, disc-type or cantilever-type piezoelectric actuators, stack-type piezoelectric actuators, pneumatic actuators, and SMA actuators. The biggest drawback of external actuators is their large size, which restricts the size of the whole micropump. Their advantages are a relatively large force and displacement.

*Integrated actuators* are micromachined with the pumps. The most common integrated actuators are electrostatic actuators, thermopneumatic actuators, electromagnetic actuators, and thermomechanical (bimetallic) actuators. Despite their fast response time and reliability, electrostatic actuators generate small forces and very small strokes. However, with special curved electrodes, large strokes on the order of several tens of microns can be achieved. Electrostatic actuators are suitable for designing micropumps with a low power consumption. Thermopneumatic actuators generate large pressure and relatively large strokes. This actuator type was therefore often used for mechanical pumps. Thermopneumatic and bimetallic actuators require a large amount of thermal energy for their operation, and, consequently, consume a lot of electrical power. High temperature and complicated thermal management are further drawbacks of thermopneumatic actuators. Electromagnetic actuators require an external magnetic field, which also restricts the pump size. The large electric current in the actuator's coil causes thermal problems and high power consumption.

#### 7.1.1.2 Pump Membrane

Many mechanical micropumps are based on the reciprocating concept. This pumping concept requires a flexible reciprocating membrane to displace a defined volume called the *Stroke volume*. The displaced volume is transported forward using rectifying elements such as check valves. The stroke volume can be estimated for a circular membrane, clamped at its perimeter and subjected to a pressure load from the deflection curve (6.34):

$$y(r) = y_0\left(1 - \frac{r^2}{R^2}\right)^2 \tag{7.1}$$

where $y_0$ is the center deflection of the membrane, and $R$ is the membrane radius. Also from the model discussed in Section 6.1.2 (6.35), the relation between the applied pressure $p$ and the center deflection $y_0$ of a circular membrane without prestress also can be estimated as:

$$p = \frac{16ty_0}{3R^4}\frac{E}{1-\nu^2}\left(t^2 + \frac{16}{35}y_0^2\right) \tag{7.2}$$

where $E$, $\nu$, and $t$ are the Young's modulus, the Poisson ratio, and the thickness of the membrane, respectively. With $p^* = p/E$, $y_0^* = y_0/R$, and $t^* = t/R$, the relation between the dimensionless pressure $p^*$, the dimensionless thickness $t^*$, and the dimensionless center deflection $y_0^*$ can be formulated as (Figure 7.3):

$$p^* = \frac{16t^*y_0^*}{3(1-\nu^2)}\left(t^{*2} + \frac{16}{35}y_0^{*2}\right) \tag{7.3}$$

For the dynamic behavior, the resonant frequency of the membrane without prestress can be estimated as [3]:

$$f_{kn} = 2\pi\beta_{kn}^2\sqrt{D/m} \tag{7.4}$$

where $D$ is the flexural rigidity:

$$D = \frac{Et^3}{12(1-\nu^2)} \tag{7.5}$$

and $m$ is the mass of the membrane. The integers $k = 0, 1, ...$ and $n = 1, 2, ...$ denote the vibration modes of the pump membrane. The eigenvalues $\beta_{kn}$ for the first three vibration modes are:

$$\begin{aligned} \beta_{01} &= 1.015\pi/R \\ \beta_{11} &= 1.468\pi/R \\ \beta_{02} &= 2.007\pi/R \end{aligned} \tag{7.6}$$

#### 7.1.1.3 Parameters of Micropumps

The most important specification bases for micropumps are the maximum flow rate, the maximum backpressure, the pump power, and the efficiency. The maximum flow rate, or pump capacity $\dot{Q}_{max}$,

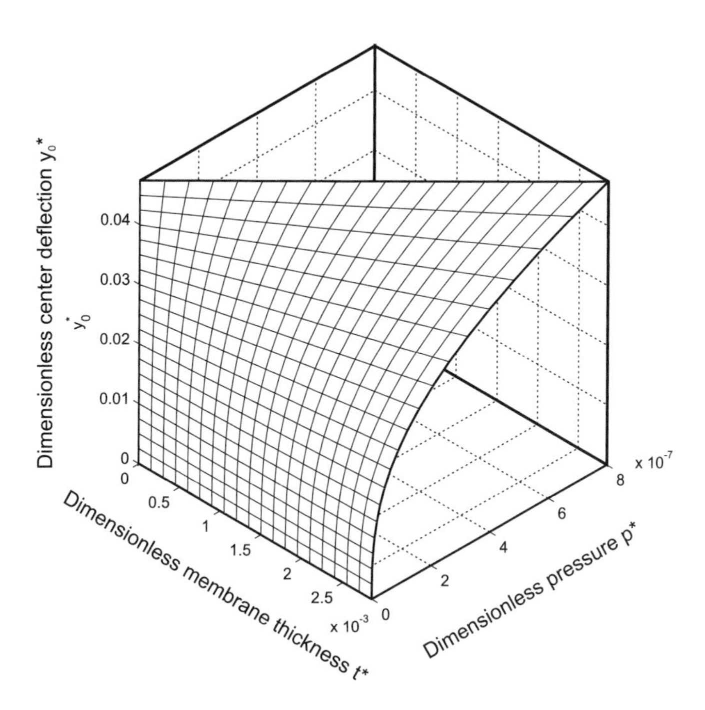


**Figure 7.3** Dimensionless center deflection of a circular membrane as a function of dimensionless thickness and dimensionless pressure.

is the volume of liquid per unit time delivered by the pump at zero backpressure. The maximum backpressure is the maximum pressure the pump can work against. At this pressure, the flow rate of the pump becomes zero. The term "pump head" is also often used as a parameter for pump performance. The pump head $h$ represents the net work done on a unit weight of liquid in passing from the inlet or suction tube to the discharge tube. It is given as [4]:

$$h = \left(\frac{p}{\rho} + \frac{u^2}{2g} + z\right)_{\text{out}} - \left(\frac{p}{\rho} + \frac{u^2}{2g} + z\right)_{\text{in}} \tag{7.7}$$

where $\rho$ is the fluid density. The term $p/\rho$ is called the *pressure head* or *flow work*, and represents the work required to move a unit weight of fluid across an arbitrary plane perpendicular to the velocity vector $u$ against a pressure $p$. The term $u^2/2g$ is the velocity head, which represents the kinetic energy of the fluid. The term $z$ is the elevation head or potential head. The first parenthetical term in (7.7) represents the outlet or discharge head, and the second, the inlet or suction head. Assuming that the static pressure at the inlet and outlet are equal, $u$ becomes zero, and the maximum pump head can be derived from (7.7) as:

$$h_{\max} = z_{\text{out}} - z_{\text{in}} \tag{7.8}$$

Based on the maximum flow rate $\dot{Q}_{\max}$, the maximum back-pressure $p_{\max}$, or the maximum pump head $h_{\max}$, the power of the pump $P_{\text{pump}}$ can be calculated as:

$$P_{\text{pump}} = \frac{p_{\max}\dot{Q}_{\max}}{2} = \frac{\rho g \dot{Q}_{\max} h_{\max}}{2} \tag{7.9}$$

The pump efficiency $\eta$ is defined as:

$$\eta = \frac{P_{\text{pump}}}{P_{\text{actuator}}} \tag{7.10}$$

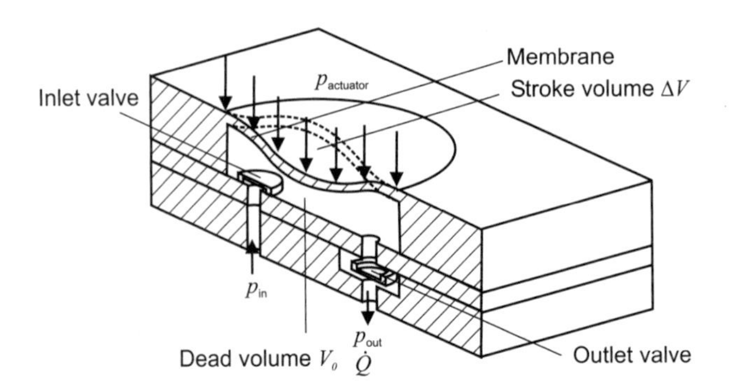


**Figure 7.4** General structure of a check-valve pump.

where $P_{\text{actuator}}$ is the input power of the actuator.

**Example 7.1: Pump Efficiency**

A thermopneumatic check-valve pump delivers a maximum flow rate of 34 mL/min and a maximum backpressure of 5 kPa. The heater resistance is 15Ω. The pump works with a symmetric square signal with a maximum voltage of 6V at 0.5 Hz. Determine the pump efficiency (data from [5]).

**Solution.** The power of the pump is:

$$P_{\text{pump}} = \frac{p_{\max}\dot{Q}_{\max}}{2} = \frac{5 \times 10^3 \times 34 \times 10^{-9}/60}{2} = 1.42 \times 10^{-6}\ \text{W}$$

Considering only the active half of the drive signal, the electrical power input is:

$$P_{\text{actuator}} = \frac{1}{2}\frac{V^2}{R} = \frac{6^2}{2 \times 15} = 1.2\text{W}$$

Thus, the pump efficiency is:

$$\eta = \frac{P_{\text{pump}}}{P_{\text{actuator}}} = \frac{1.42 \times 10^{-6}}{1.2} = 1.18 \times 10^{-6} = 1.18 \times 10^{-4}\ \%$$

7.1.1.4 Check-Valve Pumps

Check-valve pumps are the most common pump type in the macroscale. Therefore, the first attempts in designing a mechanical micropump were the realization of check-valve pumps. Figure 7.4 illustrates the general concept of a check-valve pump. A check-valve pump consists of an actuator unit, a pump membrane that creates the stroke volume $\Delta V$, a pump chamber with the dead volume $V_0$, and two checkvalves, which start to be opened by the critical pressure difference $\Delta p_{\text{crit}}$.

Check-valve pumps function under conditions of a small compression ratio and of a high pump pressure [6]. The compression ratio $\psi$ is the ratio between the stroke volume and the dead volume:

$$\psi = \frac{\Delta V}{V_0} \tag{7.11}$$

A high pump pressure $p$ fulfills the conditions:

$$\begin{cases} |p - p_{\text{out}}| > \Delta p_{\text{crit}} \\ |p - p_{\text{in}}| > \Delta p_{\text{crit}} \end{cases} \tag{7.12}$$

where $\Delta p_{\text{crit}}$ is the critical pressure required for opening the check valves. Reference [7] lists a number of methods for measuring the critical pressure. The critical pressures of a cantilever valve (1,700 × 1,000 × 15 μm) and an orifice (400 × 400 μm) are in the range from 10 to 100 mbar [6].

The criterion for the minimum compression ratio $\psi_{\text{gas}}$ of a gas pump is given as [6]:

$$\psi_{\text{gas}} > \left(\frac{p_0}{p_0 - |\Delta p_{\text{crit}}|}\right)^{\frac{1}{k}} - 1 \tag{7.13}$$

where $p_0$ is the atmospheric pressure, and $k$ is the specific heat ratio of the gas ($k = 1.4$ for air). At low pump frequency and small critical pressure $\Delta p_{\text{crit}}$, the above relation can be simplified as [6]:

$$\psi_{\text{gas}} > \frac{1}{k}\frac{|\Delta p_{\text{crit}}|}{p_0} \tag{7.14}$$

For liquids, the criterion for the minimum compression ratio $\psi_{\text{liquid}}$ is:

$$\psi_{\text{liquid}} > \Theta|\Delta p_{\text{crit}}| \tag{7.15}$$

where $\Theta$ is the compressibility of the liquid (for water, $\Theta = 0.5 \times 10^{-8}$ m$^2$/N). Since the value of $\Theta$ is very small compared to the pressure generated by different actuators, this criterion is easily met. However, this situation is only true if the pump chamber is filled entirely with water. In practice, there are always gas bubbles in the pump chamber due to outgassing or to bubbles that are transported by the liquid. Therefore, the condition in (7.14) should be kept for designing a self-priming micropump.

The above conditions lead to the following design rules for checkvalve micropumps.

- Minimize the critical pressure $\Delta p_{\text{crit}}$ by using more flexural valve design or valve material with a small Young's modulus.
- Maximize the stroke volume $\Delta V$ by using actuators with a large stroke or more flexible pump membrane.
- Minimize the dead volume $V_0$ by using a thinner spacer or wafer.
- Maximize the pump pressure $p$ by using actuators with large forces.

**Example 7.2: Designing a Micropump for Gases**

A check-valve pump has circular orifices as inlet and outlet. The orifices have diameters of 400 μm. Determine the minimum compression ratio for self-priming.

**Solution.** The critical opening pressure is important for the compression ratio. There are many sources for the initial closing force of a checkvalve, such as preloaded spring force, van der Wal forces, and capillary forces. For pumps working in moist conditions, the capillary force can have the biggest impact. Richter et al. [6] gave the following theory.

The surface energy of a wetted inlet orifice is:

$$U_{\text{surface}} = \sigma\pi dz$$

where $d$ is the orifice diameter and $z$ is the small gap between the checkvalve and the orifice, which is filled with a liquid. The external force applied on the check valve causes a mechanical work over the gap length $z$:

$$W = \Delta p_{\text{crit}}\frac{\pi d^2}{4}z$$

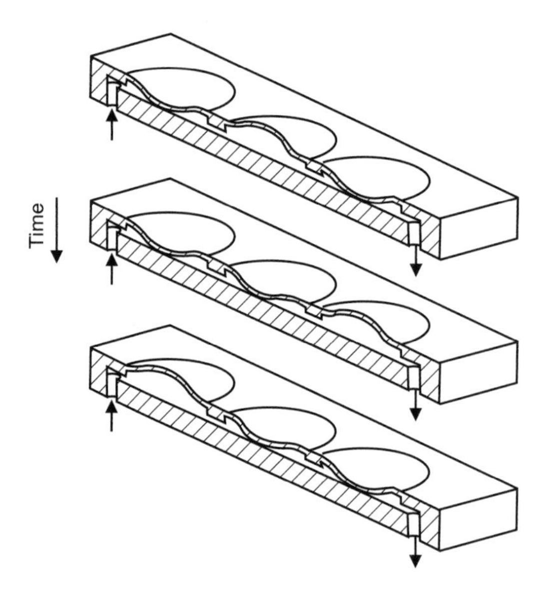


**Figure 7.5** Working principle of a peristaltic pump with three pump chambers.

This work should overcome the surface energy; thus:

$$W = U_{\text{surface}} \to \Delta p_{\text{crit}} \frac{\pi d^2}{4} z = \sigma \pi d z \to \Delta p_{\text{crit}} = \frac{4\sigma}{d}$$

Using the surface tension of water ($72 \times 10^{-3}$ N/m), the critical pressure for the pump is:

$$\Delta p_{\text{crit}} = \frac{4\sigma}{d} = \frac{4 \times 72 \times 10^{-3}}{400 \times 10^{-6}} = 720 \text{ Pa}$$

Using (7.14) and the specific heat ratio of air ($k = 1.4$) and ambient pressure of 1 bar, we get the minimum compression ratio:

$$\psi_{\min} = \frac{1}{k} \frac{|\Delta p_{\text{crit}}|}{p_0} = \frac{1}{1.4} \frac{720}{10^5} = 5.14 \times 10^{-3}$$

That means that the total stroke volume should be at least 0.5% of the dead volume.

#### 7.1.1.5 Peristaltic Pumps

In contrast to checkvalve pumps, peristaltic pumps do not require passive valves for flow rectification. The pumping concept is based on the peristaltic motion of the pump chambers, which squeezes the fluid in the desired direction (Figure 7.5). Theoretically, peristaltic pumps need three or more pump chambers with actuating membranes. Most of the realized pumps have three chambers.

Peristaltic pumps are easily implemented because of the lack of the more complicated check valves. From an operational point of view, peristaltic pumps are active valves connected in series with consecutive operation, as described in Chapter 6. The design rules and different actuation schemes can therefore be adopted from the design of microvalves described in Chapter 6. The most serious problem of peristaltic micropumps is leakage. A small pressure difference between the outlet and the inlet will cause a backflow in the nonactuated state. A one-way check valve should be connected in series with a peristaltic pump for applications such as drug delivery systems and chemical analysis, which do not allow a backflow. The one-way check valve can be implemented in the pump design.

The general optimization strategies for peristaltic pumps are maximizing the compression ratio and increasing the number of pump chambers. Since a peristaltic pump does not require a high chamber pressure, the most important optimization factors are the large stroke volume and the large compression ratio.

**Example 7.3: Designing a Peristaltic Micropump**

A peristaltic pump has three pump chambers and three circular unimorph piezodiscs as actuators. The pump membrane has a diameter of 4 mm. The pump works with a frequency of 100 Hz. Determine the volume flow rate at zero backpressure if the maximum membrane deflection is 40 μm.

**Solution.** We assume that the membrane deflection follows the deflection function of a thin circular plate (7.1):

$$y(r) = y_0 \left[1 - \left(\frac{r}{R}\right)^2\right]^2$$

According to the definition in Figure 7.4, the maximum liquid volume that a pump chamber can take in is:

$$\begin{aligned}\Delta V &= 2 \times \int_0^{2\pi} \int_0^R y_0 \left[1 - \left(\tfrac{r}{R}\right)^2\right]^2 r\mathrm{d}r\mathrm{d}r\phi = \tfrac{2\pi}{3} y_0 R^2 \\ &= \tfrac{2\pi}{3} 4 \times 10^{-5} \times (2 \times 10^{-3})^2 = 3.35 \times 10^{-10}\ \mathrm{m}^3\end{aligned}$$

At the relatively low frequency of 100 Hz, we assume a linear relation between flow rate and pump frequency. The estimated volume flow rate is:

$$Q = \Delta V f = 3.35 \times 10^{-10} \times 100 = 3.35 \times 10^{-8}\ \mathrm{m^3/sec} = 2\ \mathrm{mL/min}$$

#### 7.1.1.6 Valveless Rectification Pumps

The structure of valveless rectification pumps is similar to those of check-valve pumps. The only difference is the use of diffusers/nozzles or valvular conduits instead of check valves for flow rectification. The pressure loss $\Delta p$ at a rectification structure is given by:

$$\Delta p = \xi \frac{\rho u^2}{2} \tag{7.16}$$

where $\xi$ is the pressure loss coefficient, $\rho$ is the fluid density, and $u$ is the average velocity at the rectification structure. The ratio between the pressure loss coefficients between the inlet and outlet of the pump is the fluidic diodicity $\eta_{\mathrm{F}}$, which characterizes the performance of the rectification structure:

$$\eta_{\mathrm{F}} = \frac{\xi_-}{\xi_+} \tag{7.17}$$

where $\xi_-$ and $\xi_+$ are the pressure loss coefficients of the rectification structure in negative and positive flow directions, respectively. For common diffuser/nozzle structures, $\eta$ ranges from 1 to 5. With the known fluidic diodicity $\eta_{\mathrm{F}}$, the pump frequency $f$, and the stroke volume $\Delta V$, the flow

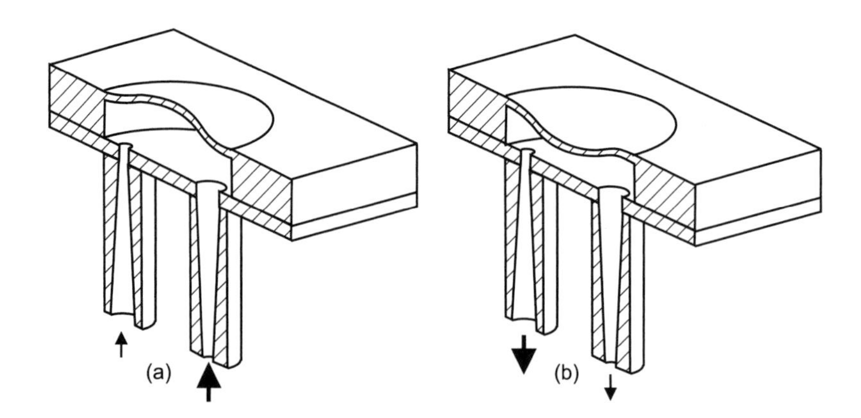


**Figure 7.6** Working principle of a diffuser/nozzle pump: (a) supply phase; and (b) pump phase. The larger arrow is indicative of higher flow rate. (*After:* [8, 9].)

rate of the pump can be estimated as [8]:

$$\dot{Q} = 2\Delta V f \frac{\sqrt{\eta_F} - 1}{\sqrt{\eta_F} + 1} \tag{7.18}$$

The last term in (7.18) can be defined as the static rectification efficiency $\chi$ [10]:

$$\chi = \frac{\sqrt{\eta_F} - 1}{\sqrt{\eta_F} + 1} \tag{7.19}$$

Figure 7.6 shows the net flow in a diffuser/nozzle pump in the supply and pump mode. For small diffuser angles, the nozzle works as pump inlet (in supply phase), while the diffuser works as pump outlet (in pumping phase). Figure 7.7 depicts the theoretical function of static rectification efficiency versus the diffuser angle $\theta$. The datapoints are collected from the design examples presented later in this section. Depending on the fabrication constraints, the two operation ranges for a diffuser/nozzle are $5° < \theta < 10°$ and $\theta \approx 70°$.

Valveless rectification pumps are more sensitive to backflow than peristaltic pumps because of the low fluidic impedance in both directions. Similar to peristaltic pumps, a one-way valve is recommended for this type of micropump. Maximizing the stroke volume and minimizing the dead volume are the major optimization measures for valveless rectification micropumps.

**Example 7.4: Designing a Diffuser/Nozzle Micropump**

A diffuser/nozzle pump has a pump membrane diameter of 10 mm and a maximum deflection of 10 μm. Determine the volume flow rate if the pump frequency is 100 Hz and the fluidic diodicity of the diffuser/nozzle structure is 1.58.

**Solution.** With the assumption made in Example 7.3, the volume per stroke of the pump is calculated as:

$$\Delta V = \frac{2\pi}{3} d_{\mathrm{max}} R^2 = \frac{2\pi}{3} 10^{-5} \times (5 \times 10^{-3})^2 = 5.24 \times 10^{-10}\ \mathrm{m}^3$$

Thus, after (7.18), the volume flow rate can be estimated as:

$$\begin{aligned} \dot{Q} &= 2\Delta V f \tfrac{\sqrt{\eta_F} - 1}{\sqrt{\eta_F} + 1} = 5.25 \times 10^{-10} \times 100 \times \tfrac{\sqrt{1.58} - 1}{\sqrt{1.58} + 1} \\ &= 1.19 \times 10^{-8}\ \mathrm{m}^3/\mathrm{s} = 715\ \mu\mathrm{L/min} \end{aligned}$$

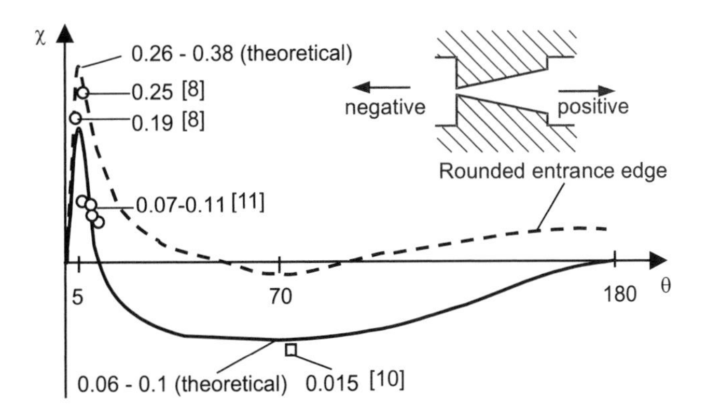


**Figure 7.7** Theoretical function of rectification efficiency and experimental data. (*From*: [8, 10, 11].) (Schematic, adapted from [11].)

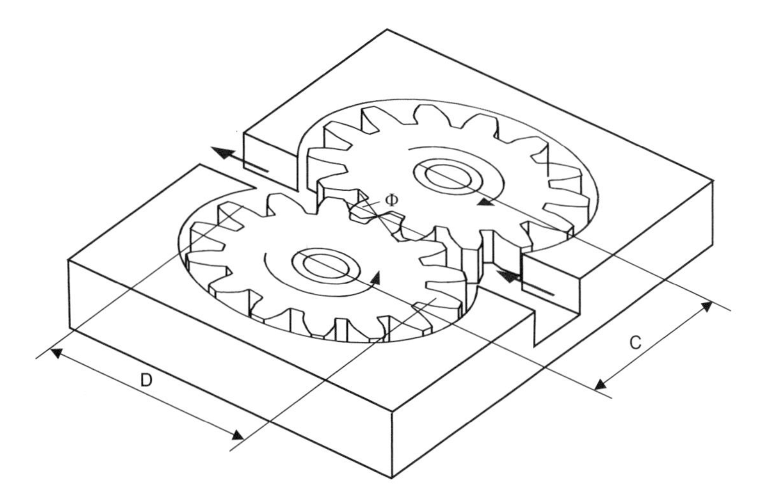


**Figure 7.8** Rotary pump with spur gears.

### 7.1.1.7 Rotary Pumps

In macroscale, rotary pumps are suitable for pumping highly viscous fluids. However, the high torque caused by the viscous forces requires strong external actuators. A fully integrated actuator is impractical in this case because of this high load. A rotary pump is designed as a pair of spur gears (Figure 7.8). The flow rate of gear pumps is given by [12]:

$$\dot{Q} = h\pi n\left(\frac{D^2}{2} - \frac{C^2}{2} - \frac{m_{\mathrm{gear}}^2\pi^2}{6}\cos\phi^2\right) \tag{7.20}$$

where $h$ is the channel height, $n$ is the number of revolutions per minute, $D$ is the gear diameter (pitch diameter), $C$ is the center distance, $\phi$ is the pressure angle, and $m_{\mathrm{gear}}$ is the module, which is determined by the gear diameter and the teeth number $N$:

$$m_{\mathrm{gear}} = \frac{D}{N} \tag{7.21}$$

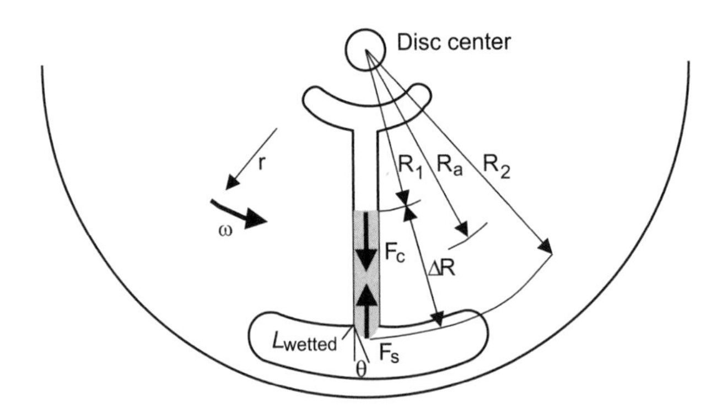


**Figure 7.9** Concept of a centrifugal pump on CD platform.

**Example 7.5: Designing a Rotary Gear Pump**

A rotary gear pump has the following parameters: 596-μm gear diameter, 12 teeth, 20° pressure angle, 515-μm center distance, and 500-μm thickness (data from [12]). Determine the volume flow rate at a speed of 300 rpm.

**Solution.** The volume flow rate is calculated from (7.20) and (7.21):

$$\begin{aligned}\dot{Q} &= h\pi n\left(\frac{D^2}{2} - \frac{C^2}{2} - \frac{m_{\text{gear}}^2\pi^2}{6}\cos\phi^2\right)\\ &= 500\pi 300\left(\frac{596^2}{2} - \frac{515^2}{2} - \frac{(596/12)^2\pi^2}{6}\cos\phi^2\right)\\ &= 19.5\times 10^9\ \mu\text{m}^3/\text{min}\end{aligned}$$

Because of the high viscosity and the small radius of the rotor, centrifugal pumps with internal actuators are not powerful enough to drive liquids in microchannel. A simple but more practical solution is the use of an external spinning motor. The pump only has channel structures in a compact disc (CD) platform. The flow velocity is controlled by the angular velocity of the disc. The change of the pumping force $\mathrm{d}F_{\mathrm{c}}$, which is the change of the centrifugal force, is calculated as [13]:

$$\frac{\mathrm{d}F_{\mathrm{c}}}{\mathrm{d}r} = \rho\omega^2 r \tag{7.22}$$

where $r$ is the radial coordinate and $\omega$ is the angular velocity of the CD platform. The capillary force $F_{\mathrm{s}}$ at a channel outlet due to the surface tension is given by:

$$F_{\mathrm{s}} = \frac{\sigma\cos\theta\cdot L_{\text{wetted}}}{A} \tag{7.23}$$

where $\sigma$ is the surface tension, $\theta$ is the contact angle, $A$ is the cross-section area of the channel, and $L_{\text{wetted}}$ is the wetted perimeter of the channel (Figure 7.9).

At the channel exit, the liquid will burst out if the centrifugal force $F_{\mathrm{c}}$ is larger than the capillary force $F_{\mathrm{s}}$. The required burst frequency of the rotating disc is [13]:

$$f_{\mathrm{b}} = \sqrt{\frac{\sigma\cos\theta}{\pi^2\rho R_{\mathrm{a}}\Delta R D_{\mathrm{h}}}} \tag{7.24}$$

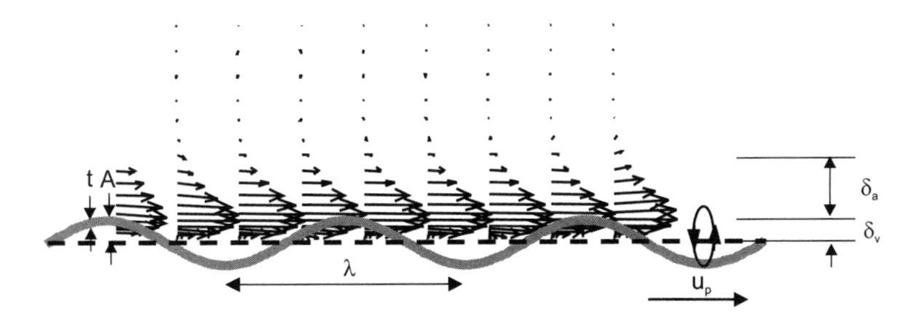


**Figure 7.10** Acoustic streaming and its key parameters.

where $D_{\rm h}$ is the hydraulic diameter of the channel, and $R_{\rm a} = (R_1 + R_2)/2$ is the average radial distance of the liquid drop. Other geometrical parameters are given in Figure 7.9. Based on the burst frequency or the angular velocity, liquid flows in the channel systems can be controlled precisely.

**Example 7.6: Designing a Centrifugal Pump**

Determine the burst frequency of a water column, which is 2 mm long and an average distance to disc center of 4 mm. The channel cross section is 100 ×50 μm. The contact angle of water surface is 30°. Surface tension and density of water are 72 × $10^{-3}$ N/m and 1,000 kg/m$^3$, respectively. What is the frequency to burst of the same water column at an average distance of 16 mm?

**Solution.** The hydraulic diameter of the channel is:

$$d_{\rm h} = 4A/U = 4 \times 100 \times 50/(2 \times 150) = 66.67\ \mu\text{m}$$

From (7.24), the burst frequency is:

$$f_{\rm b} = \sqrt{\frac{\sigma \cos\theta}{\pi^2 \rho R_{\rm a} \Delta R D_{\rm h}}}$$

$$= \sqrt{\frac{72\times10^{-3}\times\cos 30^\circ}{\pi^2\times1{,}000\times4\times10^{-3}\times2\times10^{-3}\times66.67\times10^{-6}}} = 108.8\ \text{Hz}$$

According to (7.24), the ratio of burst frequencies is:

$$f_2/f_1 = \sqrt{R_1/R_2}$$

Thus, the burst frequency at $R_2 = 16\ \mu$m is:

$$f_2 = f_1\sqrt{R_1/R_2} = 108.8 \times \sqrt{4/16} = 54.4\ \text{Hz}$$

7.1.1.8 Ultrasonic Pumps

The ultrasonic concept is a gentle pumping principle with no moving parts, heat, or strong electric field involved. The pumping effect is caused by acoustic streaming, which is induced by a mechanical traveling wave (Figure 7.10). The mechanical wave can be a flexural plate wave (FPW) [14, 15] or a surface acoustic wave [16]. This mechanical wave is excited by interdigitated transducers (IDT) placed on a thin membrane coated with a piezoelectric film [14], or on a piezoelectric bulk material [16].

The effect of acoustic streaming is based on the time-independent second-order component of the velocity field. Figure 7.10 shows the typical velocity field of acoustic streaming. The fast-moving

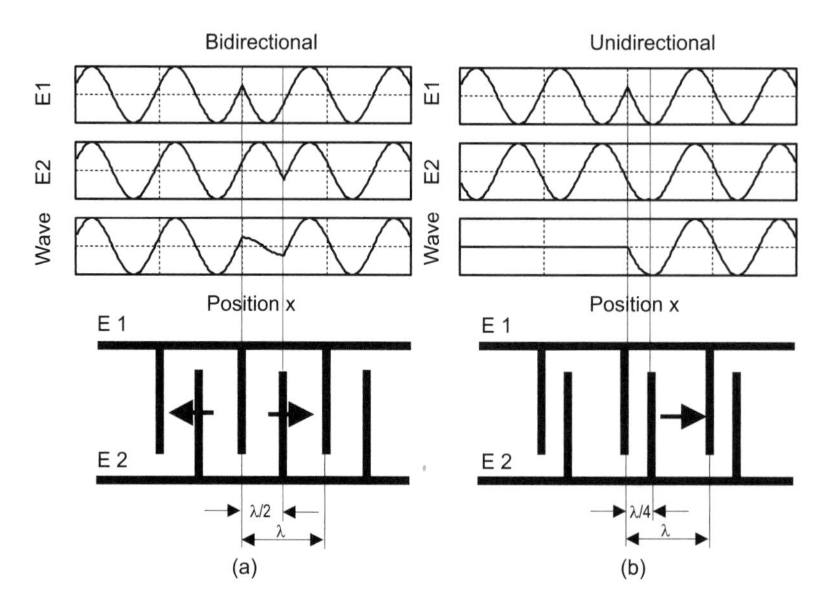


**Figure 7.11** Pumping modes with acoustic streaming: (a) bidirectional; and (b) unidirectional.

layer close to the actuating membrane is typical for flexural plate wave (FPW) acoustic streaming. The velocity profile of acoustic streaming is characterized by the acoustic evanescent decay length and the viscous evanescent decay length. The acoustic evanescent decay length can be estimated as:

$$\delta_{\mathrm{a}} = \frac{\lambda}{2\pi\sqrt{1 - u_{\mathrm{p}}/u_{\mathrm{s}}}} \tag{7.25}$$

where $\lambda$, $u_{\mathrm{p}}$, and $u_{\mathrm{s}}$ are the wavelength, the phase velocity of the wave, and the sound speed in the fluid, respectively. The viscous evanescent decay length depends on the properties of the fluid and the excitation frequency

$$\delta_{\mathrm{v}} = \sqrt{\frac{\nu}{\pi f}} \tag{7.26}$$

where $\nu$ is the kinematic viscosity of the fluid. The acoustic streaming velocity can then be estimated as [14]:

$$u_{\mathrm{a}} = \frac{5}{4}\frac{(2\pi f A)^2}{u_{\mathrm{p}}}(k\delta_{\mathrm{a}})^3\left(\frac{1}{k\delta_{\mathrm{a}}} + k\delta_{\mathrm{v}}\right)\left(1 - \frac{t}{2\delta_{\mathrm{a}}}\right)^2 \tag{7.27}$$

where $A$ is the wave amplitude, $k = 2\pi/\lambda$ is the wave number, and $t$ is the thickness of the flexural membrane. It is apparent that acoustic streaming velocity is a square function of the wave amplitude.

Conventional FPW systems have interdigitated electrodes with finger distance equal to the wavelength $\lambda$. The fingers of the counter electrode are placed at a distance $\lambda/2$ apart from the other electrode's fingers [Figure 7.11(a)]. The phase shift of the wave generated by the two electrodes is $180^\circ$. The resulting wave can be described as:

$$A(x,t) = \frac{A_0}{2}\left\{\sin(2\pi f t - k|x|) + \sin\left[2\pi f\left(t - \frac{1}{2f}\right) - k\left|x - \frac{\lambda}{2}\right|\right]\right\} \tag{7.28}$$

As depicted in Figure 7.11(a), the wave propagates in both directions. Thus, acoustic streaming is bidirectional.

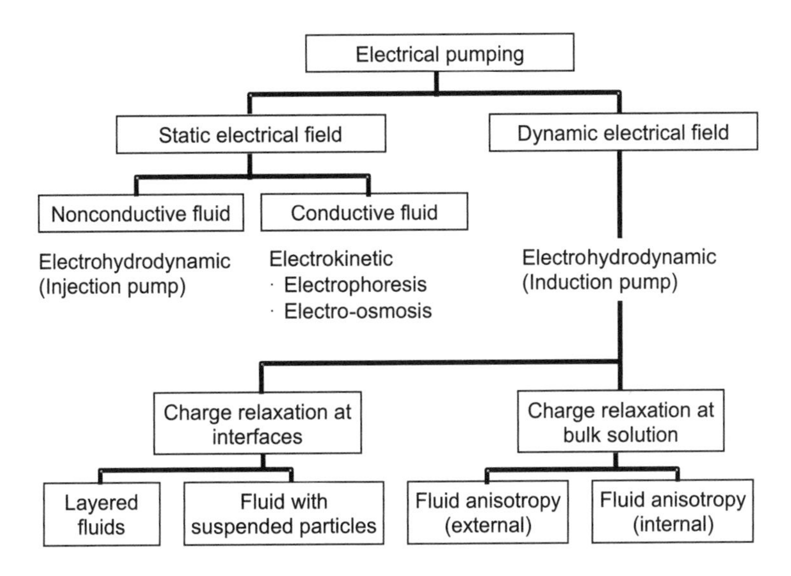


**Figure 7.12** Electrical pumping principles. (*After*: [17].)

If the finger of the counter electrode is placed at a distance $\lambda/4$ apart from the other electrode's fingers, the phase shift of the waves is $90°$. The superposition of the two waves results in wave propagation in one direction only:

$$A(x,t) = \frac{A_0}{2}\left\{\sin(2\pi ft - k|x|) + \sin\left[2\pi f\left(t - \frac{1}{4f}\right) - k\left|x - \frac{\lambda}{4}\right|\right]\right\} \tag{7.29}$$

Acoustic streaming in this mode is unidirectional.

### 7.1.2 Nonmechanical Pumps

#### 7.1.2.1 Electrohydrodynamic Pumps

Electrical micropumps utilize both static and dynamic electrical fields for pumping. Both types of electrical fields can be used for generating a pumping force. Based on the actual implementation and the type of fluid, different electrical pumping schemes can be achieved (Figure 7.12). Electrohydrodynamic (EHD) pumps are based on electrostatic forces acting on nonconductive fluids. The force density $F$ acting on a dielectric fluid with free space-charge density $q_\mathrm{f}$ in an inhomogeneous electric field $E_\mathrm{el}$ is given as [17, 18]:

$$F = \underbrace{q_\mathrm{f}E_\mathrm{el}}_{\text{Coulomb force}} + \underbrace{\vec{\mathbf{P}}\nabla E_\mathrm{el}}_{\text{Kelvin polarization force}} - \underbrace{\frac{1}{2}E_\mathrm{el}^2\nabla\varepsilon}_{\text{Dielectric force}} + \underbrace{\nabla\left(\frac{1}{2}\rho\frac{\partial\varepsilon}{\partial\rho}E_\mathrm{el}^2\right)}_{\text{Electrostrictive force}} \tag{7.30}$$

where $\varepsilon = \varepsilon_\mathrm{r}\varepsilon_0$ is the fluid permittivity, $\vec{\mathbf{P}}$ is the polarization vector, and $\rho$ is the fluid density. For pumping applications, Coulomb force and dielectric force play the most significant roles. EHD pumps can be categorized as EHD induction pumps and EHD injection pumps.

*EHD Induction Pumps.* The EHD induction pump is based on the induced charge at the material interface. A traveling wave of electric field drags and pulls the induced charges along the wave direction [Figure 7.13(a)]. EHD induction pumps can be categorized as:

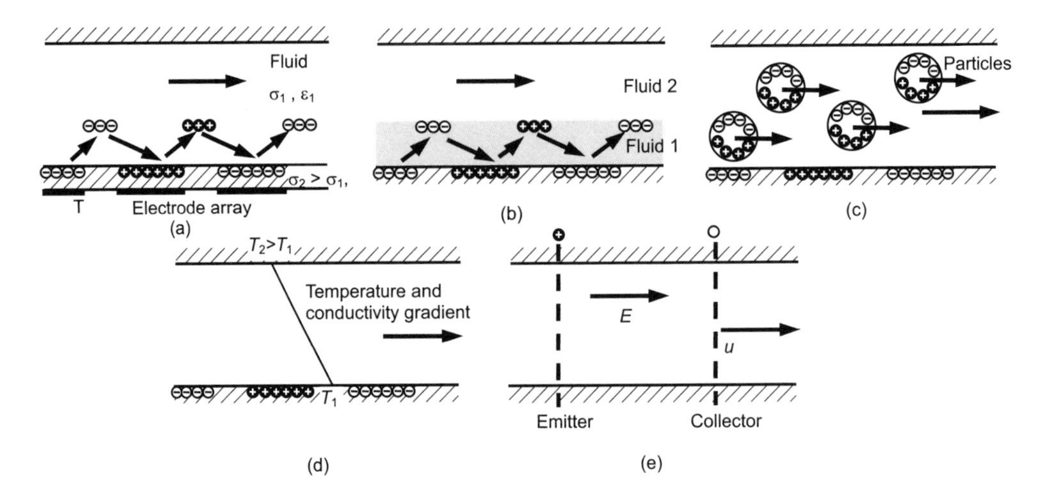


**Figure 7.13** Principles of electrohydrodynamic pumps: (a) general concept of EHD induction pump; (b) EHD induction pump with layered fluids; (c) EHD induction pump with suspended particles; (d) EHD induction pump with anisotropy of fluid properties; and (e) EHD injection pump.

- Pumps with layered fluids;
- Pumps with suspended particles;
- Pumps with anisotropy of fluid properties.

*Pumps with Layered Fluids.* This pump type requires a fluid-fluid interface [Figure 7.13(b)]. If there is a second liquid phase with low permittivity and low electrical conductance close to the isolating channel wall, a double charge layer exists at the interface between the two fluids. The traveling charge wave on the interface moves the fluid along the channel.

*Pumps with Suspended Particles.* This pumping concept utilizes the dielectrophoretic effect of small dielectric particles suspended in the fluid [Figure 7.13(c)]. If a traveling charge wave exists on the channel wall, the particles are polarized. Depending on the delay of the particle reaction to the induced field and the wave period, the particle moves in the same direction or in the opposite direction of the wave propagation. Due to viscosity forces, the movement of the particles causes the pumping effect in the bulk fluid.

*Pumps with Anisotropy of Fluid Properties.* The pump utilizes the temperature dependence of electric properties of a fluid [Figure 7.13(d)]. A temperature gradient from the top to the bottom of the channel causes a gradient in electric conductivity and permittivity [Figure 7.13(d)].

A higher temperature leads to higher electric conductivity $k_{\mathrm{el}}$, lower permittivity $\varepsilon$, and, consequently, to a lower charge relaxation time $\tau$ [17]:

$$\tau = \frac{\varepsilon}{k_{\mathrm{el}}} \tag{7.31}$$

As a result, a temperature gradient and a traveling charge wave on the channel wall lead to a gradient of charge relaxation time in the fluid, which in turn causes the pumping effect.

The pumping direction depends on the propagation velocity of the traveling charge wave. A temperature rise can be reached at the wall-fluid interface with an intense electric field in the electrically conducting fluid. The local anisotropy of the fluid properties can also lead to the pumping effect described above. In contrast to the EHD induction pumps, electrokinetic pumps utilize a static electrical field for pumping conductive fluid. The electrokinetic phenomenon can be divided into electrophoresis and electro-osmosis [19].

#### 7.1.2.2 Electrokinetic Pumps

*Electrical Double Layer.* An electric double layer (EDL) develops between an electrolyte liquid and a charged solid surface. Counter-ions from the liquid are attracted to the solid surface and form a thin charge layer called the Stern layer. The Stern layer is immobile, and holds tightly to the solid surface due to electrostatic force. The Stern layer attracts a thicker charge layer in the liquid called the Gouy-Chapman layer. The Gouy-Chapman layer is mobile and can be moved by an electrical field. Both charge layers form the EDL. The potential at the shear surface between these two layers is called the Zeta potential $\zeta$. The potential distribution in the liquid is determined by the Poisson-Boltzmann equation:

$$\frac{\mathrm{d}^2\Phi}{\mathrm{d}y^2} = \frac{2zen_\infty}{\varepsilon}\sinh\left(\frac{ze\Phi}{KT}\right) \tag{7.32}$$

where $z$ and $n_\infty$ are the valence number and the number density of the ionic species, $e$ is the elementary charge, $K$ is the Boltzmann constant, and $T$ is the absolute temperature. Assuming a small EDL compared to the channel dimension, the hyperbolic function can be approximated by a linear function $x = \sinh(x)$. Thus, the Poisson-Boltzmann equation reduces to:

$$\frac{\mathrm{d}^2\Phi}{\mathrm{d}y^2} = \frac{\Phi}{\lambda_\mathrm{D}} \tag{7.33}$$

where the Debye length $\lambda_\mathrm{D}$ is the characteristic thickness of the EDL:

$$\lambda_\mathrm{D} = \sqrt{\frac{\varepsilon KT}{2z^2e^2n_\infty}} \tag{7.34}$$

Solving (7.33) results in the potential distribution:

$$\Phi = \Phi_\mathrm{wall}\exp\left(-y/\lambda_\mathrm{D}\right) \tag{7.35}$$

*Electrophoresis.* Electrophoresis is the effect by which charged species in a fluid are moved by an electrical field relative to the fluid molecules. The electric field accelerates charged species until the electric force is equal to the frictional force. The acceleration process after applying the electrical field is on the order of a few picoseconds [19]. Thus, the velocity of the charged species is proportional to the field strength $E_\mathrm{el}$:

$$u = \mu_\mathrm{ep}E_\mathrm{el} \tag{7.36}$$

where $\mu_\mathrm{ep}$ is the electrophoretic mobility of the species, which is determined by the charge of the species $q_\mathrm{s}$, the radius of the species $r_0$, and the fluid viscosity $\eta$ [19]:

$$\mu_\mathrm{ep} = \frac{q_\mathrm{s}}{6\pi r_0\eta} \tag{7.37}$$

The most common application of electrophoresis is the separation of large molecules like DNA or proteins in gel matrices, in which these biomolecules are separated by their sizes. The separation in microchannels is called capillary electrophoresis. Webster et al. [20] used gel electrophoresis for

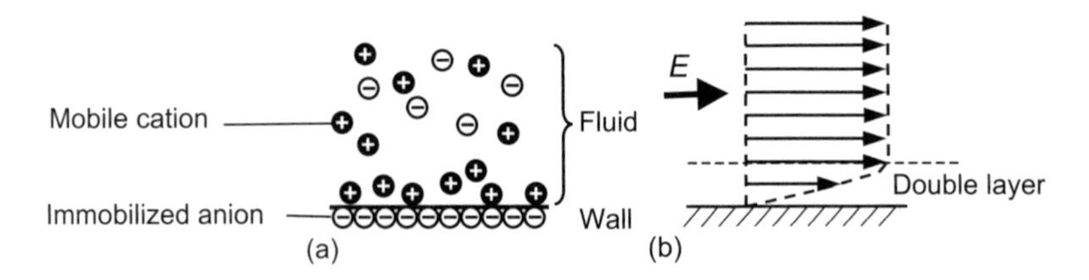


**Figure 7.14** Electro-osmotic flow: (a) surface charge on the channel wall; and (b) electro-osmotic flow field.

separating DNA molecules in microchannels, with relatively low field strength (5 to 10 V/cm). Since electrophoresis and electro-osmosis occur at the same time, electro-osmosis usually determines the overall direction of the fluid.

*Electro-osmosis.* Electro-osmosis is the pumping effect generated in a fluid in a channel under the application of an electrical field. If pH$>$ 2, a negative surface charge characterized by the zeta potential $\zeta$ exists at the plane of shear between the stationary and mobile layers of the electric double layer (see Section 2.4). The zeta potential is typically on the order of $-20$ to $-150$ mV. The surface charge comes either from the wall property or the absorption of charged species in the fluid. In the presence of an electrolyte solution, the surface charge induces the formation of a double layer on the wall by attracting oppositely charged ions from the solution [Figure 7.14(a)]. This layer has a typical thickness on the order of nanometers. An external electrical field forces the double layer to move. Due to the viscous force of the fluid, the whole fluid in the channel moves until the velocity gradient approaches zero across the microchannel. This effect results in a flat velocity profile (plug flow) [Figure 7.14(b)]. The momentum transfer process after applying the electrical field is on a time scale between 100 μm and 1 ms. The electro-osmotic flow velocity is calculated as (see Chapter 2):

$$u_{\text{eof}} = \mu_{\text{eo}} E_{\text{el}} \tag{7.38}$$

where $\mu_{\text{eo}}$ is the electro-osmotic mobility of the fluid, and $\mu_{\text{eo}}$ is a function of the dielectric constant of the solvent $\varepsilon$, its viscosity $\eta$, and the zeta potential $\zeta$ [19]:

$$\mu_{\text{eo}} = \frac{\varepsilon\zeta}{\eta} \tag{7.39}$$

Due to its nature, the electro-osmosis effect is used for pumping fluid into small channels without a high external pressure. In microanalysis systems, electro-osmosis is used for delivering a buffer solution, and, in combination with the electrophoretic effect, for separating molecules.

Because the linear flow velocity is independent of channel geometry, a large channel cross section will result in high flow rate and low backpressure. A low flow rate and a high backpressure require a small channel cross section. The high backpressure can be achieved by making small and long channels.

**Example 7.7: Designing an Electrokinetic Pumping Network**

The figure below describes an electrokinetic pumping network for capillary electrophoresis. The circular microchannels are etched in glass and have a diameter of 100 mm. The channel lengths are depicted in the figure. Measuring the electrical impedance across ports 1 and 2 results in 400 MΩ. The potentials at ports 1, 2, 3, and 4 are 1,000V, 1,000V, 1,500V, and 0V, respectively. The zeta potential is assumed to be $-100$ mV. Determine the direction and flow rate of the electrokinetic flow in the longest channel. The viscosity and the relative dielectric constant of the fluid are $10^{-3}$ kg/m-sec and 50, respectively. The dielectric constant of vacuum is $\varepsilon_0 = 8.85418 \times 10^{-14}$ F/cm.

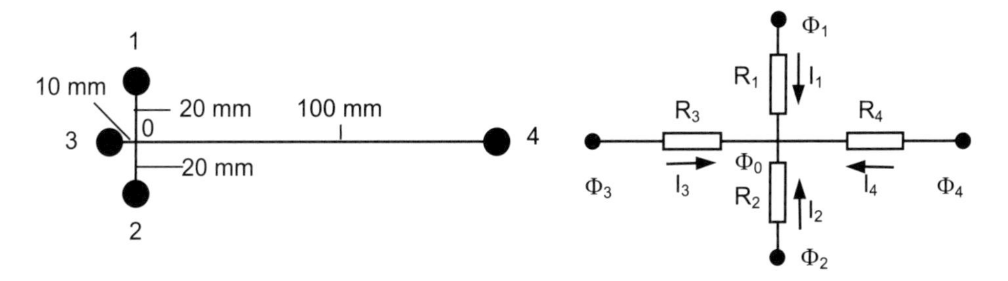


**Solution.** We assume that the fluid in the microchannel system is homogeneous; thus, the electrical impedance of each filled channel is proportional to its length. The system can be analyzed as a resistor network shown in the above picture, with:

$$R_1 = 200\ \text{M}\Omega,\ R_2 = 200\ \text{M}\Omega,\ R_3 = 100\ \text{M}\Omega,\ R_4 = 1,000\ \text{M}\Omega \tag{7.40}$$

Using the potentials and current directions assigned in the above figure, the relationship between the four currents can be described using Kirchhoff's law:

$$\begin{aligned} I_1 + I_2 + I_3 + I_4 &= 0 \\ \Phi_1 - \Phi_2 &= I_1 R_1 - I_2 R_2 \\ \Phi_3 - \Phi_4 &= I_3 R_3 - I_4 R_4 \\ \Phi_1 - \Phi_3 &= I_1 R_1 - I_3 R_3 \end{aligned}$$

Solving the above linear equation system, we get the current in the longest channel:

$$\begin{aligned} I_4 &= \tfrac{(\Phi_4-\Phi_1)R_2R_3+(\Phi_4-\Phi_2)R_1R_2+(\Phi_4-\Phi_3)R_1R_2}{R_1R_2R_3+R_2R_3R_4+R_3R_4R_1+R_4R_1R_2} \\ &= -1.2 \times 10^{-6}\ \text{A} = -1.2\ \mu\text{A} \end{aligned}$$

The current is negative. Thus, the current and consequently the fluid flows toward port 4. The electric field strength in the longest channel ($R_4$) can be calculated as:

$$E_{\text{el},4} = \frac{\Phi_0 - \Phi_4}{L} = \frac{I_4 R_4}{L} = \frac{1.2 \times 10^{-6} \times 1,000 \times 10^6}{100 \times 10^{-3}} = 11.9 \times 10^3\ \text{V/m}$$

According to (7.39), the electro-osmotic mobility of the system is:

$$\mu_{\text{eo}} = \tfrac{\varepsilon\varepsilon_0\zeta}{\eta}$$

$$\mu_{\text{eo}} = \tfrac{50\times 8.85418\times 10^{-12}\times 100\times 10^{-3}}{10^{-3}} = 4.43 \times 10^{-8}\ \text{m}^2/\text{V-s}$$

According to (7.38), the velocity of the electro-osmotic flow in the longest channel is:

$$u_4 = \mu_{\text{eo}} E_{\text{el},4} = 4.43 \times 10^{-8} \times 11.9 \times 10^3 = 5.27 \times 10^{-4}\ \text{m/sec} = 527\ \mu\text{m/s}$$

Thus, the flow rate in the longest channel is:

$$\begin{aligned} \dot{Q}_4 &= u_4 A = u_4 \pi R^2 \\ &= 5.27 \times 10^{-4} \times 3.14 \times (50 \times 10^{-6})^2 = 4.14 \times 10^{-12}\ \text{m}^3/\text{s} \approx 248\ \text{nL/min} \end{aligned}$$

**Example 7.8: Application of an Electrokinetic Pumping Network for Capillary Electrophoresis**

The system described above is used for capillary electrophoresis (see Section 12.2.2). Ports 1 and 2 are the inlet and the outlet for the sample, respectively. Port 3 is the inlet of the carrier buffer. The longest channel is the separation line. Port 4 is the waste port. In the situation described in Example 7.7, the sample is supposed to be separated in the longest channel. Do the fluids from ports 1 and 2 contaminate the separation process? Assume that the sample solution and buffer solution have identical electrokinetic properties.

**Solution.** The condition for a clean separation is that the sample solutions should flow away from the junction "0." Since flow direction corresponds to the current direction, we just need to check the direction of currents $I_1$ and $I_2$. Because of the symmetry of the potentials $\Phi_1$ and $\Phi_2$, only $I_1$ needs to be investigated. Solving the linear equation system in Example 7.7 for $I_1$, we get:

$$\begin{aligned} I_4 &= \frac{(\Phi_1-\Phi_4)R_2R_3+(\Phi_1-\Phi_2)R_4R_3+(\Phi_1-\Phi_3)R_4R_2}{R_1R_2R_3+R_2R_3R_4+R_3R_4R_1+R_4R_1R_2} \\ &= -9.5\times10^{-7}\ \text{A} = -0.95\ \mu\text{A} \end{aligned}$$

The negative sign indicates the current and the fluid flow away from the junction to ports 1 and 2. Thus, the separation is not contaminated.

For calculating the volumetric flow rates of an electro-osmotic pump, there is a need to get the velocity distribution across the channel cross section. Assuming the simplified Poisson-Boltzmann equation (7.33), the velocity distribution between two parallel plates is:

$$u(y) = u_{\text{eof}} \frac{\cosh\left(\frac{H}{2\lambda_{\text{D}}}y\right)}{\cosh\left(H/\lambda_{\text{D}}\right)} - 1 \tag{7.41}$$

where $H$ is the gap between the two parallel plates. With the same assumption for the potential distribution, the velocity distribution in a cylindrical capillary with a radius $R$ is:

$$u(r) = 2u_{\text{eof}} \left(\frac{R}{\lambda_{\text{D}}}\right)^2 \frac{I_1(R/\lambda_{\text{D}})}{I_0(R/\lambda_{\text{D}})} \sum_{n=1}^{\infty} \frac{1}{\lambda_n} \frac{J_0(\lambda_n r/R)}{J_1(\lambda_n)} \left[\frac{1}{\lambda_n^2 + (R/\lambda_{\text{D}})^2}\right] \tag{7.42}$$

where $I_n$ is the $n$-order modified Bessel functions of the first kind, and $J_n$ is the $n$-order Bessel function of the first kind. $\lambda_n$ is the $n$th positive root of the zero-order Bessel function of the first kind [$J_0(\lambda_n) = 0$].

#### 7.1.2.3 Surface Tension-Driven Pumps

According to the scaling laws, forces that are functions of the surface decrease more slowly than forces that depend on the volume. Thus, surface tension effects are dominant over inertial effects in microscale. Surface tension can be used as a driving force at these scales. Surface tension-driven pumping is based on capillary effects. Liquids are pumped passively by capillary effect or by transpiration concept borrowed from nature. Active control of surface tension also allows pumping liquid. There are a number of effects for controlling surface tension actively. Thermocapillary effect utilizes the temperature dependence of surface tension, while electrocapillary effects or electrowetting controls the surface energy by electrostatic forces. Electrochemical or photochemical capillary effects use the dependence of surface tension on the concentration of surfactants. Electrochemical or photochemical reactions can be used to control the surfactant

**Table 7.3**
Surface Tensions of Common Liquids at 20°C

| *Liquid* | $\sigma(10^{-3})$ N/m | *Liquid* | $\sigma(10^{-3})$ N/m |
|---|---|---|---|
| Acetone | 23.32 | Acetonitrile | 19.10 |
| Chloroform | 27.16 | Cyclohexane | 24.98 |
| Cyclopentane | 22.42 | Dichloromethane | 28.12 |
| Ethyl ether | 17.06 | Ethyl alcohol | 22.32 |
| Ethylene dichloride | 32.23 | Ethyl acetate | 23.75 |
| Glycerin | 63 | Isobutyl alcohol | 22.98 |
| Mercury | 484 | Water | 72.8 |

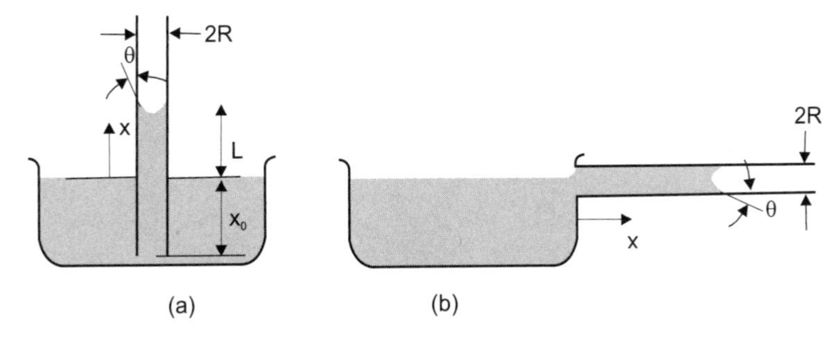


**Figure 7.15** Passive capillary flow (a) in a vertical microchannel and (b) in a horizontal microchannel.

concentration. Photomechanical capillary effect controls the hydrophobicity of the solid surface by means of photochemical reaction. Following, the different capillary effects and their applications for pumping liquids are discussed in detail.

*Passive Capillary Effect.* Passive capillary effect occurs if the movement of a fluid system is caused by the difference in surface tension. The fluid system should have a liquid, which is in contact with another liquid, gas, or solid. The driven pressure in passive capillary effect depends on three factors:

- Surface tensions;
- Geometry of the interface between the different phases;
- Geometry of the solid phase at the border line between the three phases (liquid, gas, solid).

The force balance at the interface is represented by the contact angle $\theta$. This balance equation is called the Young's equation:

$$\sigma_{\text{sg}} - \sigma_{\text{sl}} = \sigma_{\text{lg}} \cos\theta \tag{7.43}$$

where $\sigma_{\text{sg}}$, $\sigma_{\text{sl}}$, and $\sigma_{\text{lg}}$ are the tensions at the solid/gas, solid/liquid, and liquid/gas interface. The solid surface is hydrophobic if $90^\circ < \theta < 180^\circ$. If the solid surface is hydrophilic ($0^\circ < \theta < 90^\circ$), $\cos\theta$ is positive and the solid-liquid interface has a lower surface energy than the solid-gas interface. Table 7.3 lists the tension values at the air/liquid interface of some common liquids.

The surface tension can be determined by the capillary rise experiment shown in Figure 7.15(a). The force balance of the liquid column in a vertical circular microchannel is:

$$2\pi RL(\sigma_{\text{sg}} - \sigma_{\text{sl}}) = \Delta p\pi R^2 L = \rho gh\pi R^2 L \tag{7.44}$$

Substituting (7.43) in (7.44) results in the liquid/gas surface tension:

$$\sigma_{\mathrm{lg}} = \frac{\rho g L R}{2\cos\theta} \tag{7.45}$$

where $\rho$ is the density of the liquid, $g$ is the acceleration of gravity, $R$ is the radius of the microchannel, and $L$ is the column length [Figure 7.15(a)].

In the following two examples, the dynamics of passive capillary flow in a circular microchannel are considered. The two basic configurations of vertical and horizontal microchannels are depicted in Figure 7.15(a). Vertical configuration can be used for sample collection, while horizontal configuration can be used for self-filling of sample liquids in lab-on-chip applications.

**Example 7.9: Dynamics of Passive Capillary Flow in Vertical Configuration**

Determine the time function of the position and velocity of the meniscus shown in Figure 7.15(a).

**Solution.** The dynamic behavior of the meniscus can be described by the 1-D Navier-Stokes equation, which is a balance equation between the gravitational force, the viscous force, the interfacial force, and the inertial force:

$$\begin{aligned}\tfrac{\mathrm{d}}{\mathrm{d}t}\left(m\tfrac{\mathrm{d}x}{\mathrm{d}t}\right) &= -m'g - 2\tau\pi R(x+x_0) + 2\pi R\sigma_{\mathrm{lg}}\cos\theta \\ \tfrac{\mathrm{d}}{\mathrm{d}t}\left\{\left[\rho\pi R^2(x+x_0)\right]\tfrac{\mathrm{d}x}{\mathrm{d}t}\right\} &= -\left(\rho\pi R^2 x\right)g - \pi R^2\tfrac{\mathrm{d}p}{\mathrm{d}x}(x+x_0) + 2\pi R\sigma_{\mathrm{lg}}\cos\theta\end{aligned}$$

where $x$ is the position variable, $x_0$ is the immersed length, $m$ is the mass of the liquid column, $m'$ is the mass of the liquid column part above the reservoir surface, $\tau$ is the shear force, and $\mathrm{d}p/\mathrm{d}x$ is the pressure gradient in the flow direction. The viscous force can assume the Hagen-Poiseuille-model:

$$\frac{\mathrm{d}p}{\mathrm{d}x} = \frac{8\mu}{R^2}\frac{\mathrm{d}x}{\mathrm{d}t}$$

The above equations result in the inhomogeneous second-order differential equation:

$$\frac{\mathrm{d}^2x}{\mathrm{d}t^2} + \frac{1}{x+x_0}\left(\frac{\mathrm{d}x}{\mathrm{d}t}\right)^2 + \frac{8\mu}{R^2\rho}\frac{\mathrm{d}x}{\mathrm{d}t} + g\frac{x}{x+x_0} - \frac{2\sigma_{\mathrm{lg}}\cos\theta}{\rho R}\frac{1}{x+x_0} = 0$$

The initial conditions of the above equation are:

$$x\Big|_{t=0} = 0,\quad \frac{\mathrm{d}x}{\mathrm{d}t}\Big|_{t=0} = 0$$

The above inhomogeneous second-order differential equation can be solved numerically with the Runge-Kutta method. Figure 7.16 shows the typical results of the position and the velocity of the meniscus.

As mentioned above, the governing equation of the meniscus position is inhomogeneous and cannot be solved analytically. However, further simplification of this equation can deliver an analytical solution. First, a small initial position $x_0$ is assumed ($x_0 \ll x_\infty$). The final column length $x_\infty = L$ is depicted in Figure 7.15(a). Analytical solutions can be found for the two asymptotic cases ($t \to 0$) and ($t \to \infty$).

At the beginning of the filling process ($t \to 0$), the inertial term and the gravitational term can be neglected with a small initial position $x_0$. Thus, the governing equation can be simplified as:

$$\frac{8\mu}{R^2\rho}\frac{\mathrm{d}x}{\mathrm{d}t} - \frac{2\sigma_{\mathrm{lg}}\cos\theta}{\rho R}\frac{1}{x+x_0} = 0$$

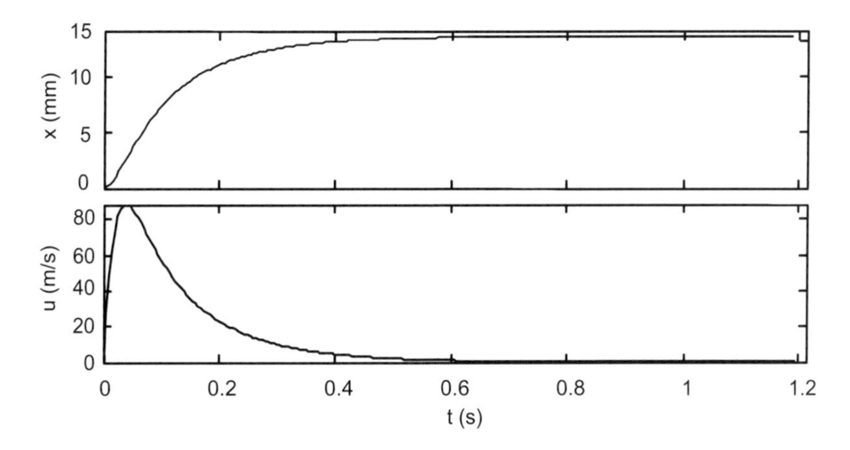


**Figure 7.16** Dynamics of passive capillary filling in a vertical microchannel (water, 400-μm channel radius).

replacing $x + x_0$ by a variable $X$ results in:

$$X\frac{\mathrm{d}X}{\mathrm{d}t} = \frac{\sigma_{\mathrm{lg}}\cos\theta}{4\mu}R$$

The solution of this asymptotic case is:

$$X = \sqrt{\frac{\sigma_{\mathrm{lg}}\cos\theta R}{2\mu}t} \rightarrow x(t) = \sqrt{\frac{\sigma_{\mathrm{lg}}\cos\theta R}{2\mu}t} + x_0$$

At the end of the filling process ($t \rightarrow \infty$), the solution of $X_\infty = x_0 + x_\infty = x_0 + L$ is known. With $X = x + x_0$ and $\nu = \mu/\rho$, the governing equation is reduced to:

$$\frac{4\nu}{gR^2}\frac{\mathrm{d}(X^2)}{\mathrm{d}t} + X - X_\infty = 0$$

Taking the distance between the final position and the transient position $\chi = X_\infty - X$ as the variable, the differential term of the above equation can be simplified as:

$$\left.\frac{\mathrm{d}(X^2)}{\mathrm{d}t}\right|_{\chi\rightarrow 0} = 2X\mathrm{d}\dot{X}\Big|_{\chi\rightarrow 0} = -2X_\infty\dot{\chi}$$

Substituting the above term back into the differential equation results in the equation:

$$\frac{8\nu}{gR^2}\dot{\chi} + \chi = 0$$

and its solution:

$$\chi(t) = \chi_\infty \exp\left(-\frac{R^2 g}{8\nu\chi_\infty}t\right)$$

Bringing this solution back to the original variable $x$ leads to the solution at the second asymptote ($t \rightarrow \infty$):

$$X(t) = X_\infty\left[1 - \exp\left(-\tfrac{R^2 g}{8\nu X_\infty}t\right)\right]$$

$$x(t) = L - (L + x_0)\exp\left[-\tfrac{R^2 g}{8\nu(L+x_0)}t\right]$$

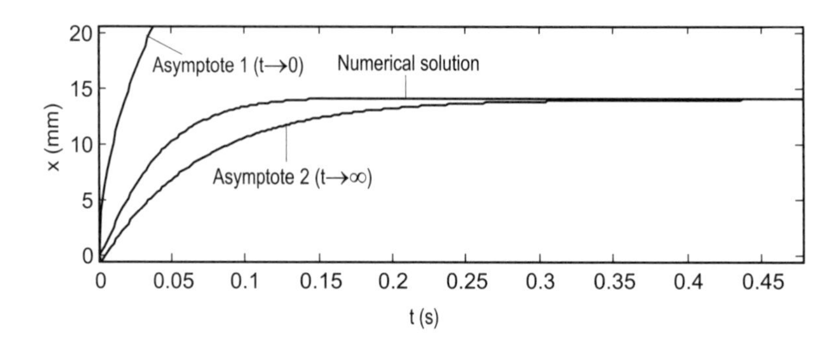


**Figure 7.17** Asymptotic solutions of the meniscus position in a vertical microchannel (water, 400 μm channel radius).

Figure 7.17 compares the numerical solution obtained by the Runge-Kutta method and the above two asymptotic solutions. The solutions show that the time constant:

$$\tau_{\text{rising}} = \frac{8(L + x_0)\nu}{R^2 g} \tag{7.46}$$

is characteristic for the filling process based on passive capillary in a vertical microchannel.

**Example 7.10: Dynamics of Passive Capillary Flow in Horizontal Configuration**

Determine the time function of the position and velocity of the meniscus shown in Figure 7.15(b).

**Solution.** Using the kinematic viscosity $\nu = \mu/\rho$, setting $x_0 = 0$, and neglecting the gravitational term in the governing equation of Example 7.9 leads to the governing equation for the meniscus position $x$ in a horizontal microchannel:

$$\frac{\mathrm{d}^2 x}{\mathrm{d}t^2} = -\frac{1}{x}\left(\frac{\mathrm{d}x}{\mathrm{d}t}\right)^2 - \frac{8\nu}{R^2}\frac{\mathrm{d}x}{\mathrm{d}t} + \frac{1}{x}\frac{2\sigma_{\text{lg}}\cos\theta}{\rho R}$$

In the above equation, the term $\frac{1}{x}\left(\frac{\mathrm{d}x}{\mathrm{d}t}\right)^2$ is small and can be neglected. The initial conditions are:

$$x\Big|_{t=0} = x_0, \quad \frac{\mathrm{d}x}{\mathrm{d}t}\Big|_{t=0} = 0$$

Substituting $X = x^2$ in the above equation leads to the ordinary differential equation:

$$\ddot{X} + \alpha\dot{X} = \beta$$

with

$$A = \frac{8\nu}{R^2}, \quad B = \frac{4\sigma_{\text{lg}}\cos\theta}{\rho R}$$

The solutions for the meniscus position and meniscus velocity are:

$$x = \sqrt{X} = \sqrt{\tfrac{B}{A^2}\exp\left(-At\right) + \tfrac{Bt}{A} + x_0^2 - \tfrac{B}{A^2}}$$
$$u = B\left[1 - \exp(-At)\right]/(2Ax)$$

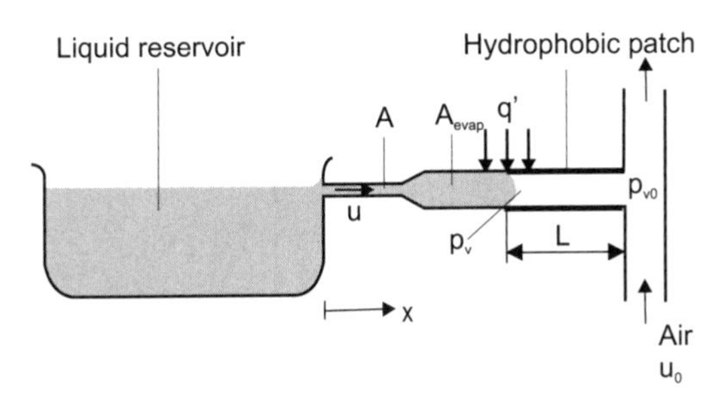


**Figure 7.18** Concept of transpiration pump.

*Transpiration Effect.* Passive capillary effect, as described above, needs a liquid/gas surface to function. If the microchannel is filled, the liquid stops flowing. In nature, water supply to tree leaves still continues, due to the phenomenon called *transpiration*. Water lost by vaporizing through pores on the leaf surface is compensated by a fresh supply. Figure 7.18 shows the concept of pumping based on transpiration reported by Namasivayam [21]. Liquid from a reservoir fills a horizontal microchannel due to passive capillary effect (see Example 7.10). The liquid meniscus is stopped by a hydrophobic patch at one end of the channel. If the liquid is heated at the meniscus, the vapor pressure at the interface increases. Air supplied from an external source carries away the liquid vapor and generates a gradient in vapor concentration as well as vapor pressure from the meniscus to the air flow channel. The gradient allows vapor to diffuse out and fresh liquid supply to flow into the channel. The velocity of the liquid can be controlled by the evaporation rate, which in turn is controlled by the vapor pressure or temperature at the meniscus. The temperature at the meniscus can be controlled by an integrated heater.

With a flow channel cross section $A$, and the cross section at the evaporation side $A_{\mathrm{evap}}$, the balance between supply liquid and evaporation is:

$$\rho u A = M A_{\mathrm{evap}} \dot{n}_{\mathrm{evap}} \tag{7.47}$$

where $\rho$, $M$ are the density and molecular weight of the liquid, $A$ is the channel cross section, and $\dot{n}$ is molar evaporation flux at the meniscus. Rearranging (7.47) results in the average supply velocity:

$$u = M A_{\mathrm{evap}} \dot{n}_{\mathrm{evap}} / (\rho A) \tag{7.48}$$

The relation between the vapor concentration and the vapor pressure is determined by the ideal gas equation:

$$p_{\mathrm{v}} V = n \bar{R} T \rightarrow c = \frac{n}{V} = \frac{p_{\mathrm{v}}}{\bar{R} T} \tag{7.49}$$

where $\bar{R} = 8.314$ J/mol-K is the universal gas constant. Solving the mass transport equation (see Chapter 10) results in the relation between the vapor pressure gradient and the vapor pressure at the meniscus $p_{\mathrm{v}}$:

$$\frac{\mathrm{d}p}{\mathrm{d}x} = \left(\frac{p_{\mathrm{v}} - p_{\mathrm{v0}}}{L}\right)\left(1 + \sqrt{\frac{u_0 D_{\mathrm{h}}}{D}}\right) \tag{7.50}$$

where $p_{\mathrm{v0}}$ is the vapor pressure at the air flow side, $u_0$ is the air velocity, and $D$ is the diffusion coefficient of the vapor. Thus, the molar evaporation flux can be calculated as:

$$\dot{n}_{\mathrm{evap}} = D \frac{\mathrm{d}c}{\mathrm{d}x} = \frac{D}{\bar{R} T} \frac{\mathrm{d}p_{\mathrm{v}}}{\mathrm{d}x} \tag{7.51}$$

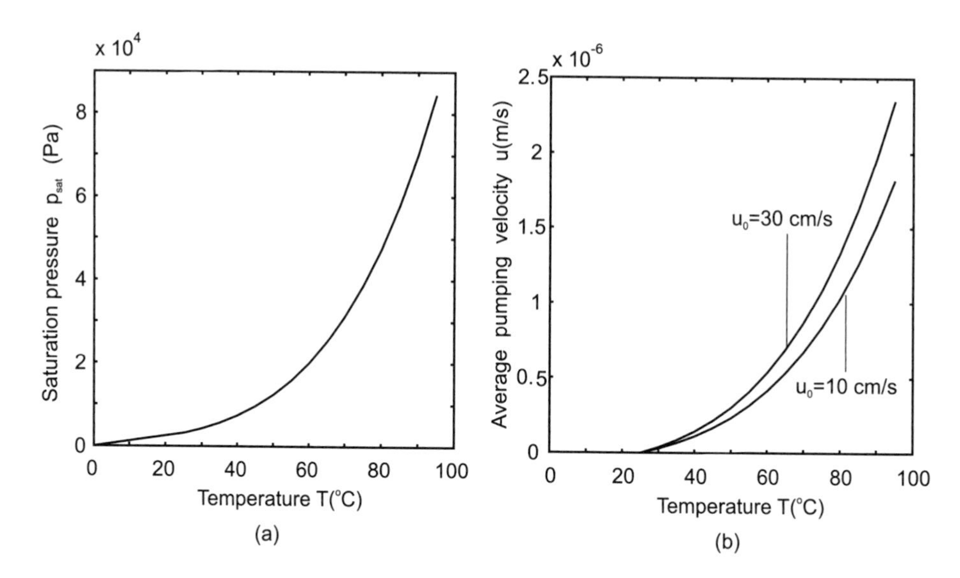


**Figure 7.19** Performance of transpiration pump: (a) saturation pressure of water as a function of temperature; and (b) average velocity as a function of temperature at the mensicus.

Substituting (7.50) and (7.51) in (7.48) results in the average velocity of transpiration pumping [21]:

$$u = \frac{MD}{\rho \bar{R}T} \frac{A_{\text{evap}}}{A} \left( \frac{P_{\text{v}} - P_{\text{v0}}}{L} \right) \left( 1 + \sqrt{\frac{u_0 D_{\text{h}}}{D}} \right) \tag{7.52}$$

where $D_{\text{h}}$ is the hydraulic diameter of the channel.

**Example 7.11: Design of a Transpiration Pump**

A transpiration pump for water is designed based on the concept shown in Figure 7.18. The microchannel is a circular capillary with a diameter of 100 μm and cross-section ratio of $A_{\text{evap}}/A = 1$. The length of the hydrophobic patch is 1 cm. The diffusion coefficient of water vapor in air is $0.229 \times 10^{-4}$ m$^2$/s. Assume that the relative humidities at the meniscus and at the air flow side are both 100%. The air temperature is 25°. Plot the average water velocity in the microchannel as a function of temperature at the meniscus and of the air flow velocity.

**Solution.** At a relative humidity of $\phi = 100\%$, the vapor pressure at the meniscus is equal to the saturation pressure at the given temperature $T$:

$$p_{\text{v@}T} = \phi p_{\text{sat@}T} = 100\% p_{\text{sat@}T} = p_{\text{sat@}T}$$

The relation between saturation pressure and temperature is depicted in Figure 7.19(a) [22]. For instance, the vapor pressure at the air flow side is:

$$p_{\text{v@25}^\circ} = \phi p_{\text{sat@25}^\circ} = 100\% p_{\text{sat@25}^\circ} = 3.169 \text{ kPa}$$

Substituting the values for vapor pressure at different temperatures and different air flow velocities in (7.52) results in the flow velocity depicted in Figure 7.19(b). The results show clearly that the temperature is the key parameter for controlling the flow velocity as well as the flow rate. The low velocity on the order of 1 μm/s can be improved by using a higher cross-section ratio $A_{\text{evap}}/A$ and a high air velocity $u$.

**Table 7.4**
Surface Tension of Water as Function of Temperature

| T(°C) | 0 | 10 | 20 | 30 | 40 | 50 | 60 | 70 | 80 | 100 |
|---|---|---|---|---|---|---|---|---|---|---|
| $\sigma_{\mathrm{lg}}(10^{-3})$ | 75.6 | 74.22 | 72.75 | 71.18 | 69.56 | 67.91 | 66.18 | 64.4 | 62.6 | 58.9 |

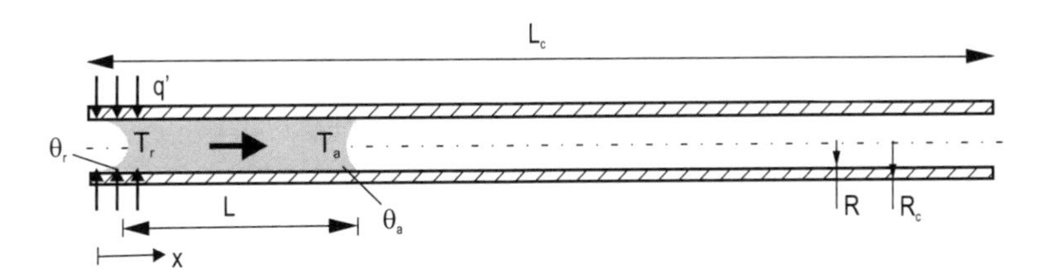


**Figure 7.20** Model of a liquid plug in a transient temperature field.

*Thermocapillary Effect.* Thermal pumping principles are based on the temperature dependence of fluid properties such as density, viscosity, or surface tension. Temperature gradients in a fluid cause density gradients, which in turn generate buoyancy force on the hotter and lighter fluid. However, in the microscale, the inertial effect such as a buoyancy force becomes insignificant. Thus, free convection is not suitable as a pumping effect. The temperature dependence of viscosity can be used for designing valveless rectification pumps. In the previous section, the temperature dependence of vapor pressure was used as the pumping concept. Temperature dependence of surface tension can be directly used for driving a liquid droplet. The temperature dependence of surface energy or surface tension (Table 7.4) causes the thermal-capillary effect (Marangoni effect), as mentioned in Section 6.1.1.11.

**Example 7.12: Dynamics of a Liquid Droplet in a Microchannel Under Thermocapillary Effect [23]**

A horizontal cylindrical capillary with a length $L_{\mathrm{c}}$, an inner radius $R$, and an outer diameter of $R_{\mathrm{o}}$, is heated on one end with a heat flux of $q'$ as shown in Figure 7.20. A liquid droplet of the length $L$ is placed at the same end of the capillary. Applying the heat flux drives the droplet away from the heater due to thermocapillary effect. Determine the dynamic behavior of the droplet.

**Solution.** Thermocapillary effect is caused by the temperature gradient along the capillary. First, the transient temperature field needs to be determined. For relatively low heater temperature, heat radiation can be neglected. Thus, the energy equation for heat transport in the capillary is formulated as:

$$\frac{\partial T}{\partial t} = a\frac{\partial^2 T}{\partial x^2} - \frac{2hR_{\mathrm{o}}}{\rho c(R_{\mathrm{o}}^2 - R^2)}T$$

where $T$ is the temperature difference compared to the ambient temperature, and $a$, $\rho$, and $c$ are the thermal diffusivity, density, and specific heat capacity of the capillary material, respectively. The outer and inner radii of the capillary are $R_{\mathrm{o}}$ and $R_{\mathrm{i}}$, respectively. The first term of the right side of the above equation describes heat conduction, while the second term describes free heat convection at the capillary outer wall. The initial and boundary conditions of temperature field are:

$$t = 0 : T(x) = 0$$
$$t > 0 : \begin{cases} x = 0 & \mathrm{d}T/\mathrm{d}x = -q'/k \\ x = L_\mathrm{c} & T = 0 \end{cases}$$

where $k$ is heat conductivity of capillary material. By introducing the dimensionless variables $T^* = T/(q' L_\mathrm{c}/k)$, $x^* = X/L_\mathrm{c}\, t^* = t/(L_\mathrm{c}^2 a)$, and

$$\beta = \sqrt{\frac{2hL_\mathrm{c}^2 R_\mathrm{o}^2}{k(R_\mathrm{o}^2 - R^2)}}$$

where $h = 0.631 k_\mathrm{a}/(2R_\mathrm{o})$ is the heat transfer coefficient on the outer capillary surface [24], $k_\mathrm{a}$, the heat conductivity of air, the dimensionless equation for the temperature field is:

$$\frac{\partial T^*}{\partial t^*} = \frac{\partial^2 T^*}{\partial x^{*2}} - \beta^2 T^*$$

The dimensionless boundary conditions are:

$$t^* = 0 : T^*(x^*) = 0$$
$$t^* > 0 : \begin{cases} x^* = 0 \ \mathrm{d}T^*/\mathrm{d}x^* = -1 \\ x^* = 1 \ T^* = 0 \end{cases}$$

Using separation of variables results in the dimensionless temperature and temperature gradient:

$$T^*(x^*, t^*) = \tfrac{1}{\beta}\left[\sinh(\beta)\cosh(\beta x^*)/\cosh(\beta) - \sinh(\beta x^*)\right] + $$
$$+ \textstyle\sum_{n=1}^{\infty} D_n \exp\left\{-\left[(n-1/2)^2 \pi^2 + \beta^2\right] t^*\right\} \cos\left[(n-1/2)\pi x^*\right]$$

$$\tfrac{\mathrm{d}T^*(x^*,t^*)}{\mathrm{d}x^*} = \left[\sinh(\beta)\sinh(\beta x^*)/\cosh(\beta) - \cosh(\beta x^*)\right] -$$
$$- \textstyle\sum_{n=1}^{\infty} (n-1/2) D_n \exp\left\{-\left[(n-1/2)^2 \pi^2 + \beta^2\right] t^*\right\} \sin\left[(n-1/2)\pi x^*\right]$$

with

$$D_n = 2\int_0^1 \left\{-\frac{1}{\beta}\left[\sinh(\beta)\cosh(\beta x^*)/\cosh(\beta) - \sinh(\beta x^*)\right]\right\} \cos\left[(n-1/2)\pi x^*\right] \mathrm{d}x^*$$

If a weak thermal interaction between the plug and the capillary can be assumed, surface tensions at the two droplet ends are functions of temperature, which in turn is a function of position $x$ as determined above:

$$\sigma_\mathrm{lg}(T) = f[T(x)] = g(x) = \sigma_\mathrm{lg}(x)$$

The simplest model for the friction between the liquid plug and the capillary wall is the Hagen-Poiseuille model:

$$\frac{\mathrm{d}p}{\mathrm{d}x} = \frac{8\mu}{R^2}\frac{\mathrm{d}x}{\mathrm{d}t},$$

where $R$ is the radius of the plug and $\mu$ is the dynamic viscosity of the liquid. The force balance equation for the liquid droplet is:

$$\rho\pi R^2 L\frac{\mathrm{d}^2x}{\mathrm{d}t^2} = -8\pi\mu L\frac{\mathrm{d}x}{\mathrm{d}t} - 2\pi R[\sigma_{\mathrm{lg}}(x+L)\cos\theta_{\mathrm{a}} - \sigma_{\mathrm{lg}}(x)\cos\theta_{\mathrm{r}}]$$

where $\theta_{\mathrm{r}}$ und $\theta_{\mathrm{a}}$ are the contact angles at the receding and advancing ends of the liquid plug, respectively. Introducing the velocity $u = \mathrm{d}x/\mathrm{d}t$, the kinetic viscosity $\nu = \mu/\rho$, and rearranging the governing equation results in:

$$\frac{\mathrm{d}u}{\mathrm{d}t} + \left(\frac{8\nu}{R^2}\right)u + \frac{2}{\rho RL}[\sigma_{\mathrm{lg}}(x+L)\cos\theta_{\mathrm{a}} - \sigma_{\mathrm{lg}}(x)\cos\theta_{\mathrm{r}}] = 0$$

The three terms on the left-hand side of the above equation represent the acceleration, the friction and the surface tension, respectively. For a small temperature range, the surface tension can be assumed as a linear function of temperature:

$$\sigma_{\mathrm{lg}}(T) = \sigma_{\mathrm{lg0}} - \gamma(T - T_0)$$

where $\sigma_{\mathrm{lg0}}$ is the surface tension at the reference temperature $T_0$. The temperature coefficient $\gamma$ can be determined from the temperature function of the surface tension values. Solving the above ordinary differential equation results in the velocity $u$:

$$u = \frac{B}{A}[1 - \exp(-At)]$$

with

$$A = \frac{8\nu}{R^2},\quad B = \frac{2}{\rho RL}[\sigma_{\mathrm{lg}}(x+L)\cos\theta_{\mathrm{a}} - \sigma_{\mathrm{lg}}(x)\cos\theta_{\mathrm{r}}]$$

The time, position, and velocity can be nondimensionalized as $t^* = t/(L_{\mathrm{c}}^2/a)$, $x^* = x/L_{\mathrm{c}}$ and $u^* = u/(L_{\mathrm{c}}/a)$. The reference velocity $u_0 = L_{\mathrm{c}}/a$ can be considered as the diffusion speed of the temperature. Assuming the same contact angle at the receding and advancing ends $\theta$ and a short droplet $L \ll L_{\mathrm{c}}$, the dimensionless form of the governing equation is:

$$\frac{\mathrm{d}u*}{\mathrm{d}t^{*2}} + \frac{8}{R^{*2}}\frac{\nu}{a}u^* + \frac{2}{R^*}\frac{L^3 q'}{\rho k a^2}\gamma\cos(\theta)\frac{\mathrm{d}T^*}{\mathrm{d}x^*} = 0$$

The temperature gradient $\mathrm{d}T^*/\mathrm{d}x^*$ was obtained in the first part of this analysis. The solution for the dimensionless velocity $u^*$ is then:

$$u^* = \frac{B^*}{A^*}[1 - \exp(-A^* t^*)]$$

with

$$A^* = \frac{8}{R^{*2}}\frac{\nu}{a},\quad B^* = \frac{2}{R^*}\frac{L^3 q'}{\rho k a^2}\gamma\cos(\theta)\frac{\mathrm{d}T^*}{\mathrm{d}x^*}$$

where $R^* = R/L_{\mathrm{c}}$ is the dimensionless capillary radius. Figure 7.21 shows the typical results of position and velocity for liquid plugs with different viscosities. The dynamic behavior of the

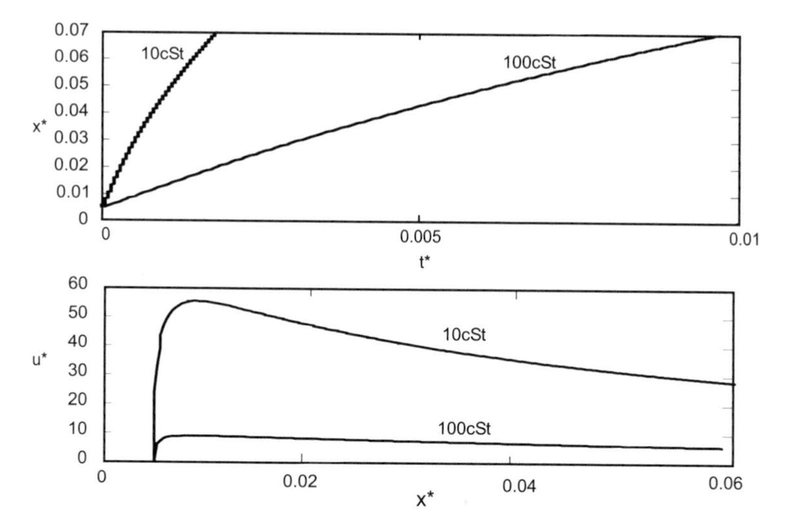


**Figure 7.21** Thermocapillary effect of a liquid plug: (a) position as a function of time; and (b) velocity as a function of position.

liquid plug can be divided into two periods: the acceleration period and the stabilizing period. The acceleration period is determined by $A^*$, while the stabilizing period is determined by $B^*$. A less viscous plug will accelerate faster and reach a higher velocity initially. Since the liquid plug initially moves faster than the thermal diffusion, which is represented by $Lc/a$, the velocity then decreases due to the lower temperature gradient.

*Electrocapillary Effect.* Electrocapillary effect, also called electrowetting, is based on the electrostatic force at the interface of different phases. For microfluidics, the three relevant electrocapillary effects are:

- Continuous electrowetting;
- Direct electrowetting;
- Electrowetting on dielectric.

*Continuous electrowetting* is the electrocapillary effect at the interface between two electrically conducting liquids. If a liquid metal droplet such as a mercury droplet comes in contact with an electrolyte, an EDL exists at the interface between these two liquids. With this EDL, the droplet can be considered as a charged particle. Similar to the electrophoresis effect, an electric field causes

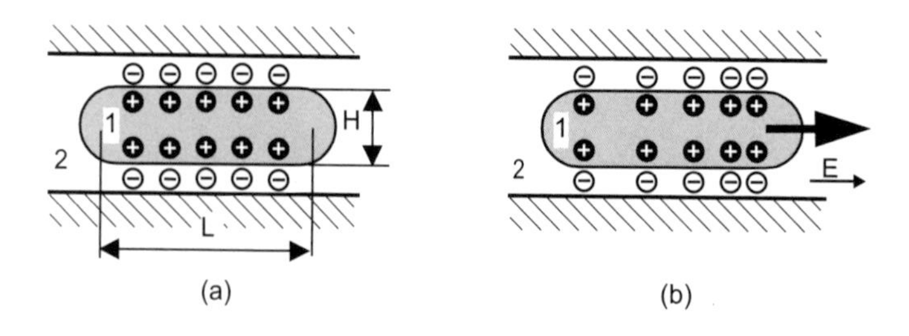


**Figure 7.22** Continous electrowetting: (a) formation of an electric double layer at the interface; and (b) the droplet moves in an electric field.

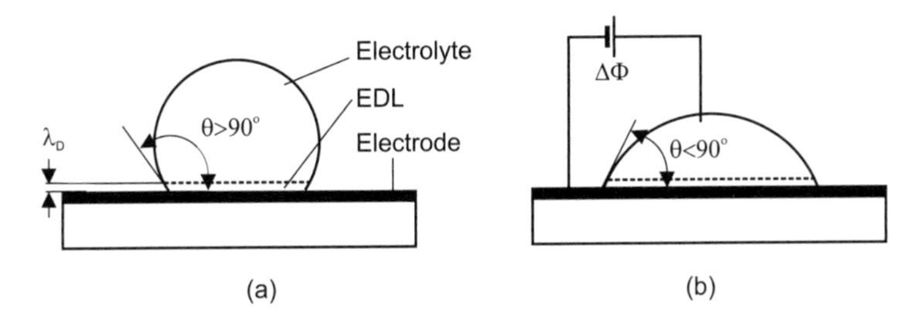


**Figure 7.23** Direct electrowetting: (a) formation of an electric double layer at the interface; and (b) an applied voltage changes the contact angle.

the droplet to move. For a droplet with a length $L$ in a flat channel with a height $H$ (Figure 7.22), the force balance of the droplet can be described as:

$$q_{\mathrm{s}} E_{\mathrm{el}} = 6\mu u/H \tag{7.53}$$

where $q_{\mathrm{s}}$ is the surface charge of the droplet, $E_{\mathrm{el}}$ is the electric field strength, $\mu$ is the droplet viscosity, and $u$ is the velocity. The left side of (7.53) describes the friction force and assumes the simple Hagen-Poiseuille model. If the voltage difference across the droplet is $\Delta\Phi$, the droplet velocity can be estimated based on the electric field $E_{\mathrm{el}} = \Delta\Phi/L$ as:

$$u = \frac{q_{\mathrm{s}} H}{6\mu L} \Delta\Phi \tag{7.54}$$

*Direct electrowetting* is the electrocapillary effect at the interface between an electrically conducting liquid and a solid electrode. Similar to the case of electro-osmosis, a thin EDL exists between the electrolyte and the electrode [Figure 7.23(a)]. The EDL of a thickness $\lambda_{\mathrm{D}}$ acts as a capacitor with the capacitance per unit surface:

$$c_{\mathrm{EDL}} = \frac{\varepsilon_0 \varepsilon_{\mathrm{r}}}{\lambda_{\mathrm{D}}} \tag{7.55}$$

where $\varepsilon_0$ and $\varepsilon_{\mathrm{r}}$ are the dielectric constant of vacuum and the relative dielectric constant of the electrolyte, respectively. The surface energy is reduced by the amount of energy stored in the capacitor. The relation between surface tension and the applied voltage across the interface is described by the Lippmann equation [25]:

$$\sigma_{\mathrm{sl}} = \sigma_{\mathrm{sl0}} - \frac{c_{\mathrm{EDL}} \Delta\Phi^2}{2} \tag{7.56}$$

where $\Delta\Phi$ is the voltage across the interface, and $\sigma_{\mathrm{sl0}}$ is the initial surface tension at $\Delta\Phi = 0$. Combining the Lippmann equation (7.56) with the Young equation (7.43) results in the relation between the contact angle $\theta$ and the applied voltage $\Delta\Phi$ [Figure 7.23(b)]:

$$\theta = \arccos\left(\cos\theta_0 + \frac{1}{\sigma_{\mathrm{lg}}} \frac{c\Delta\Phi^2}{2}\right) \tag{7.57}$$

From (7.57), the condition for the applied voltage to make an originally hydrophobic surface $\theta_0 > 90°$ hydrophilic ($\theta < 90°$) is:

$$\Delta\Phi > \sqrt{-2\sigma_{\mathrm{lg}} \cos\theta_0 / c} \tag{7.58}$$

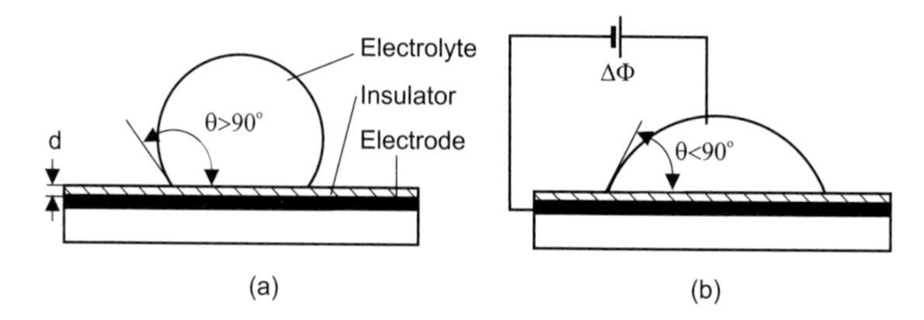


**Figure 7.24** Electrowetting on dielectric: (a) the hydrophobic dielectric layer acts as a capacitor; and (b) an applied voltage changes the contact angle.

In practice, increasing the voltage decreases the contact angle until a critical voltage is reached. At voltages higher than the critical value, the contact angle does not follow the Lippmann relation (7.56).

### Example 7.13: Direct Electrowetting on an Electrode

An electrolyte droplet is placed on a hydrophobic electrode. The initial contact angle is $\theta_0 = 120°$. The electrolyte is a NaCl solution with a concentration of $10^{-6}$ M and a relative dielectric constant of 80. The surface tension of the electrolyte is 72 mN/m. Determine the minimum voltage needed to make the surface hydrophilic.

**Solution.** The Debye length of the EDL is:

$$\lambda_{\mathrm{D}} = \sqrt{\frac{\varepsilon KT}{2z^2e^2n_\infty}}$$

$$= \sqrt{\frac{80\times8.854\times10^{-12}\times1.38\times10^{-23}\times298}{2\times1^2\times(1.602\times10^{-19})^2(N_{\mathrm{A}}\times10^{-6}\times10^{-3})}} = 3.07\times10^{-7}\ \mathrm{m}.$$

The capacitance per surface unit is:

$$c = \frac{\varepsilon_0\varepsilon_{\mathrm{r}}}{\lambda_{\mathrm{D}}} = \frac{8.854\times10^{-12}\times80}{0.307^{-6}} = 2.3\times10^{-3}\ \mathrm{F/m^2}$$

The minimum voltage needed to make the surface hydrophilic is:

$$\Delta\Phi_{\min} = \sqrt{-\frac{2\sigma_{\mathrm{lg}}}{c}\cos\theta_0} = \sqrt{-\frac{2\times0.072}{2.3\times10^{-3}\cos120°}} = 5.6\mathrm{V}$$

*Electrowetting on dielectric.* The major drawback of direct electrowetting is the fixed capacitance per unit surface and electrolysis at the electrodes. If the electrode is coated with a hydrophobic dielectric material such as Teflon ($\varepsilon_{\mathrm{r}} = 2$), the capacitance per unit surface can be controlled by the thickness $d$ of the dielectric coating (Figure 7.24):

$$c = \frac{\varepsilon_{\mathrm{r}}\varepsilon_0}{d} \tag{7.59}$$

The relation between the contact angle and the applied voltage in the case of electrowetting on dielectric can then be described with the Lippmann equation (7.56) and (7.57).

*Electrochemical and Photochemical Capillary Effect.* Besides temperature and electrostatic force, surface tension can be controlled by surface-active molecules, also called surfactants. A surfactant is a linear molecule with a hydrophilic head and a hydrophobic end. Surfactant molecules tend to form a surface between the liquid and the gas phase. The hydrophobic tails are in the gas phase, while hydrophilic heads are in the liquid phase. Increasing the surfactant concentration decreases the surface tension, until the critical concentration called Critical Micelle Concentration (CMC) is reached. Beyond CMC, the surface tension does not change significantly.

The surfactant concentration can be controlled by electrochemical reactions. Gallardo et al. [26] demonstrated that oxidation and reduction reactions at electrodes can transform a surfactant such as $Fc(CH_2)_11\text{-}N^+(CH_3)_3Br^-$ between the surface-active state and the surface-inactive state. This concept only needs electrical potentials less than 1V. Compared to thermocapillary effect, electrochemical capillary effect is more efficient, due to the minimum energy lost.

The contact angle to a solid surface can also be controlled by the surface property. The solid surface is coated with an insoluble monolayer, which is light-sensitive. Irradiating the monolayer with light transforms the layer between states that are rich in either *cis* or *trans* isomer. Correspondingly, the monolayer expands or contracts. The nanoscale change of morphology affects the contact angle at the liquid/gas/solid interface [27].

Photochemical reactions can control the water-soluble azobenzene-based surfactants. Under light irradiation, the change in the aggregation state of the surfactant in the bulk solution leads to a decrease in surface tension [27].

#### 7.1.2.4 Electrochemical and Chemical Pumps

Electrochemical actuating concepts were explained in Chapter 6. The generation of gas bubbles is used for displacing liquids. Chemical pumping is referred to as osmosis or osmotic pumping. Osmosis is the movement of a solvent through a membrane, which is permeable only to the solvent, from an area of low solute concentration to an area of high solute concentration. Osmosis is a fundamental effect in biological systems. This effect is applied to water purification and desalination, waste material treatment, and many other chemical and biochemical processes. The pressure that stops the osmotic flow is called the osmotic pressure. The osmotic pressure $p_{\text{osmotic}}$ is given by the van't Hoff formula [28]:

$$p_{\text{osmotic}} = c\bar{R}T \tag{7.60}$$

where $c$ is the molar solute concentration, $\bar{R}$ is the universal gas constant, and $T$ is the absolute temperature. This formula is similar to that of the pressure of an ideal gas derived previously for transpiration effect (7.49).

#### 7.1.2.5 Magnetohydrodynamic Pumps

The pumping effect of a magnetohydrodynamic (MHD) pump is based on the Lorentz force acted on a conducting solution:

$$\vec{\mathbf{F}} = \vec{\mathbf{I}} \times \vec{\mathbf{B}}w \tag{7.61}$$

where $\vec{\mathbf{I}}$ is the electric current across the pump channel, $\vec{\mathbf{B}}$ is the magnetic flux density, and $w$ is the distance between the electrodes. The flow direction depends on the direction of the magnetic field and the direction of the applied electrical field. Assuming that the flow direction is equal to the direction of the Lorentz force acted on a positively charged particle, the flow direction can be determined with the right-hand rule. First, the thumb of the right hand points in the direction of the current $\vec{\mathbf{I}}$. Second, the index finger points in the direction of the magnetic field $\vec{\mathbf{B}}$. Then the palm or the middle finger points in the direction of the Lorentz force [Figure 7.25(a)]. In the rectangular channel shown in Figure 7.25(b), the maximum pressure and maximum flow rate can be estimated as:

$$p_{\max} = JBL \tag{7.62}$$

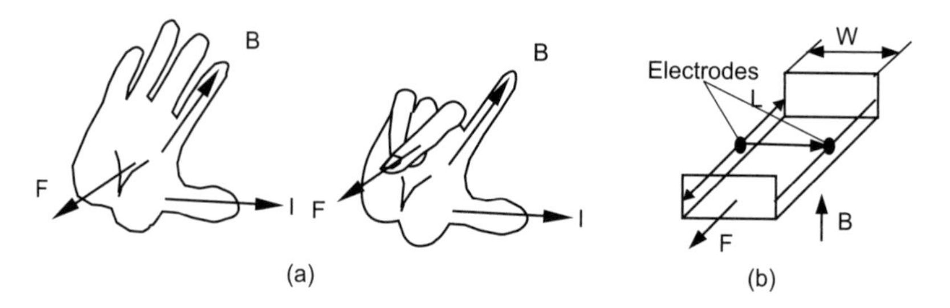


**Figure 7.25** Magnetohydrodynamic pump: (a) right-hand rule; and (b) concept.

$$\dot{Q}_{\mathrm{max}} = JB\frac{\pi D_{\mathrm{h}}^{4}}{128\mu} \tag{7.63}$$

where $L$ is the channel length, $J$ is the current density, and $B$ is the magnetic flux density.

Because the Lorentz force acts on the bulk fluid and creates a pressure gradient, MHD pumps generate a parabolic velocity profile, similar to pressure-driven flows.

## 7.2 DESIGN EXAMPLES

### 7.2.1 Mechanical Pumps

#### 7.2.1.1 Check-Valve Pumps

The terms for passive microvalves used here are adopted from Shoji and Esashi in [2]. Figures 7.26 and 7.27 depict some typical microcheck valves realized in silicon and polymer suitable for employment in pump design. One of the first micropumps made in silicon is shown in Figure 7.28(a) [5]. The pump has check valves in the form of a ring diaphragm, which is relatively stiff and consumes a large lateral area almost equivalent to that of a pump chamber. The same valves were also used in the pumps reported in [29], which have thermopneumatic actuators instead of piezodiscs. Figure 7.27 shows a polymeric microcheck valve using elastomeric membrane for embedding in polymeric microfluidic systems [30, 31]. Similar to stand-alone polymeric microvalves, polymeric microcheck valves have gained popularity to be used in various microfluidic systems to avoid back flow or as an inherent element of microcheck valve pumps.

*Check Valves with Low Spring Constant.* The next design improvement is shown in Figure 7.28(e) [9]. The pump has check valves made of polysilicon by using surface micromachining. The valve

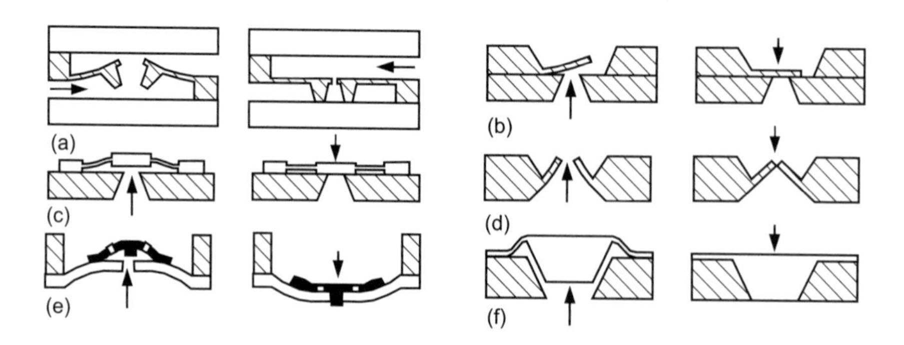


**Figure 7.26** Microcheck valves: (a) ring mesa; (b) cantilever; (c) disc; (d) V-shape; (e) membrane; and (f) float. (*After*: [2].)

**Table 7.5**
Young's Modulus of Different Materials

| *Material* | *Young's Modulus (GPa)* | *Material* | *Young's Modulus (GPa)* |
|---|---|---|---|
| Stainless steel | ≈240 | Silicon oxide | ≈70 |
| Aluminum | ≈70 | Polyimide | ≈10 |
| Silicon | ≈200 | Parylene | ≈3 |
| Silicon nitride | ≈300 | Silicone rubber | ≈0.0005 |

is a disc supported by four thin polysilicon beams. This design allows small valves to be integrated under the pump chamber. The pump uses an external piezoelectric actuator, which can deliver a large pump pressure.

References [32, 33] presented other small and more flexible designs. The valve has a form of a cantilever [Figure 7.28(f)]. The pump utilizes an integrated electrostatic actuator, which has a 5-μm gap between the electrodes and requires a peak actuation voltage of 200V. If the pump frequency is higher than the resonance frequency of the flap valves, the pump is able to pump backward. The same valve type was used in micropumps reported in [34, 35], which have bimorph piezoelectric discs as actuators.

*Check Valves Made of Soft Material.* Another way to make check valves flexible is to use a material with a lower Young's modulus. Table 7.5 compares the common materials used for check valves in micropumps. Polymers such as polyimide, polyester, and parylene are one order of magnitude more flexible than silicon. Pumps presented in [36, 37] used polyimide as material for the disc valve [Figure 7.4(b)]. While the pumps reported in [36, 37] used gold and titanium as device material, the pump reported in [38, 39] was made of plastic (PSU; see Appendix C) using LIGA and injection molding. The latter has polyimide ring diaphragm valves [Figure 7.4(c)] able to deliver a maximum flow rate of 44 mL/min and a maximum pressure head of 38 mbar with an average power consumption of 0.45W. The check valve shown in Figure 7.27 was completely made of polymer. The flexible membrane was from PDMS [30] or thermoplastic polyurethane (TPU) [31] while the liquid and displacement chambers were, respectively, fabricated in PDMS or PMMA, forming the checkvalve in PDMS/PDMS/PDMS or PMMA/TPU/PMMA assemblies. The PDMS check valve was fabricated using multilayer softlithography while the whole-thermoplastic microcheck valve employing PMMA and TPU was fabricated using laser micromachining.

A similar design with polyester valves was reported in [40]. The pump was fabricated by plastic injection molding. Conventional cutting techniques were used for fabricating the aluminum mold. The pump utilizes external actuators such as an electromagnetic actuator and piezoelectric

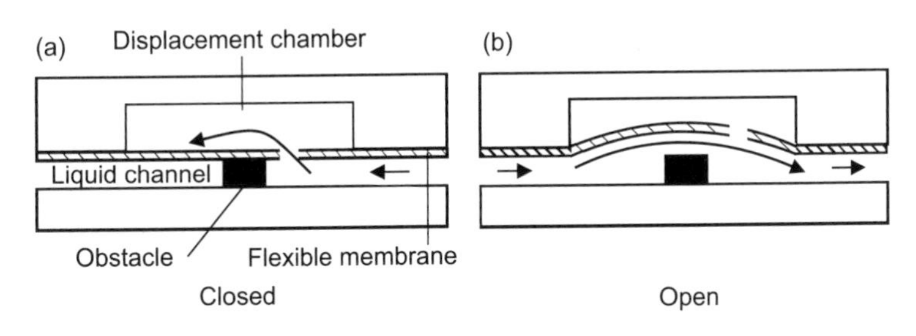


**Figure 7.27** Polymeric microcheck valve using elastomeric membrane: (a) closed valve in backward mode; and (b) open valve in forward mode. (*After*: [31].)

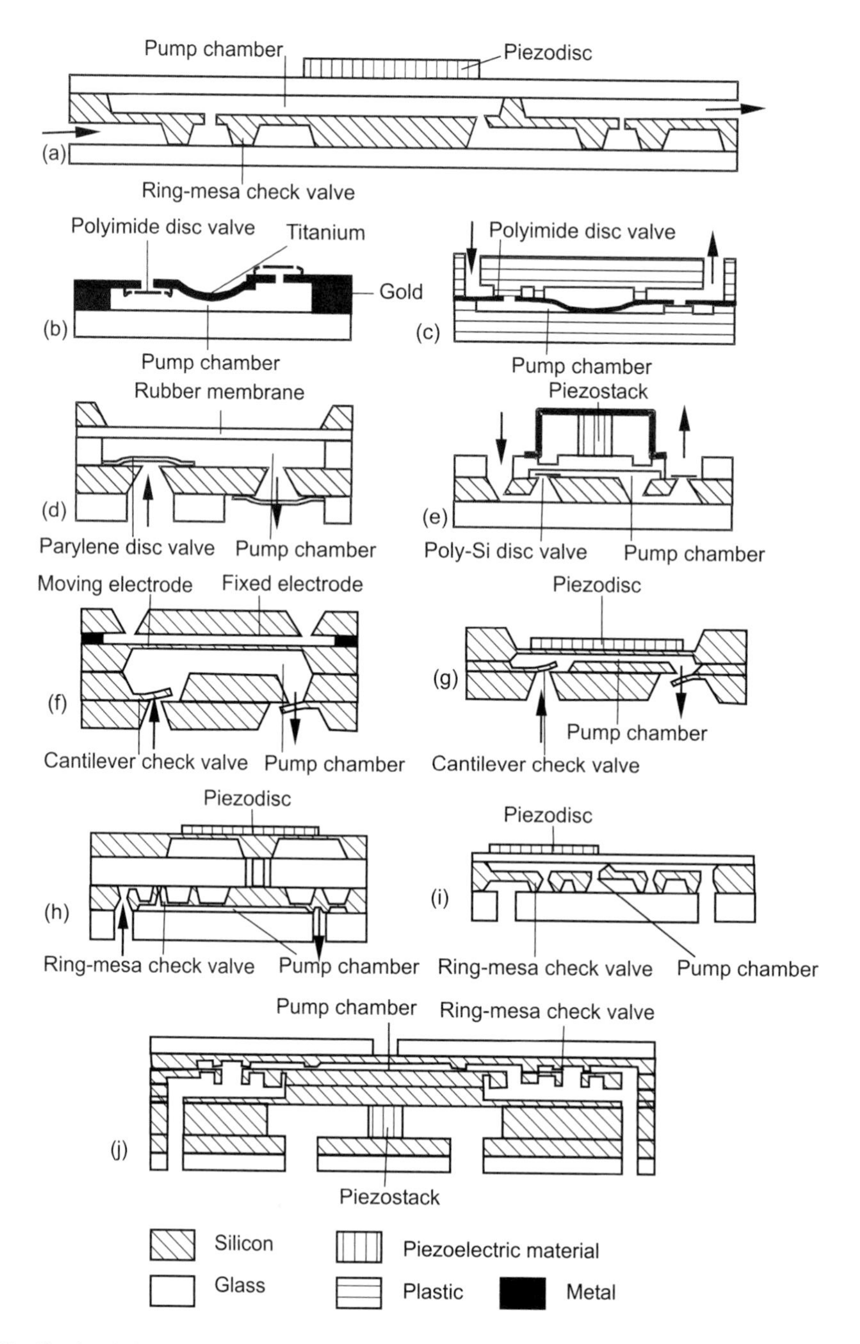


**Figure 7.28** Check-valve micropumps: (a) ring-mesa check valve; (b) and (c) polyimide disc valve; (d) parylene disc valve; (e) polysilicon disc valve; (f) and (g) silicon cantilever check valve; and (h), (i), and (j) ring-mesa check valve.

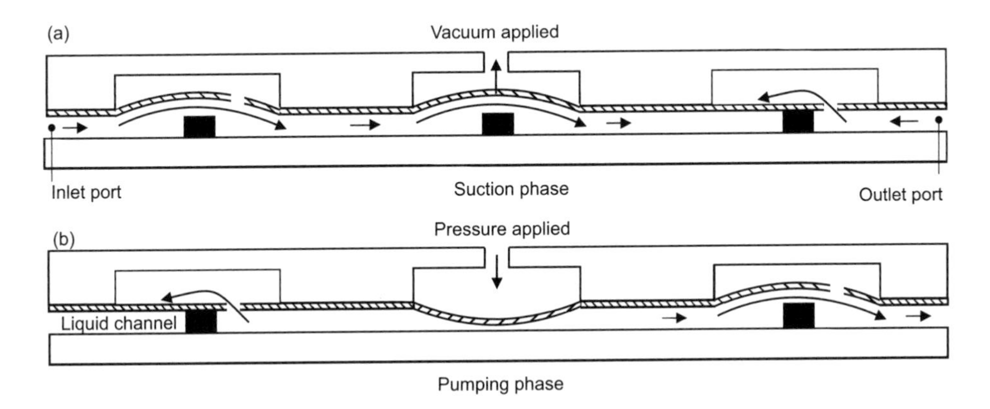


**Figure 7.29** Polymeric check-valve micropump: (a) suction phase; and (b) pumping phase. (*After*: [31].)

bimorph disc. In the pump presented in [41], the disc valve was realized in parylene, which can be deposited with a CVD process [Figure 7.28(d)]. This pump takes the advantage of a soft pump membrane made of silicone rubber.

*Pump Membranes Made of Soft Material.* The next optimization measure is the fabrication of the pump membrane with flexible material such as polyimide [38] [Figure 7.28(c)] or silicone rubber [41] [Figure 7.28(d)]. These membranes require small actuating pressures and have large deflection as well as large stroke volume. This type of membrane is suitable for pneumatic and thermopneumatic actuators. Pumps based on fully polymeric microcheck valves have been realized as well. As an example shown in Figure 7.29, two microcheck valves are integrated on a single substrate for consecutive pumping operation [31]. The pump has two phases of operation. In the suction phase, vacuum is created in the displacement chamber. Thus, the liquid channel is filled through the inlet, while the outlet microcheck valve is closed. During the pumping phase, the flexible membrane moves down to push the liquid towards the outlet while the inlet microcheck valve is closed. Liquid is continuously pumped through periodic actuation of the membrane.

*Pumps with Smaller Dead Volume.* Using thinner spacer or thinner wafers for the pump chamber, one can minimize the dead volume and consequently maximize the compression ratio. The pump shown in Figure 7.28(f) [42] is improved in the newer version depicted in Figure 7.28(g) [43]. The middle wafer is polished and thinned to 70 μm. As a result, the compression ratio increases from 0.002 to 0.085 [7]. The improved pump design is able to pump gas and is self-priming. The design shown in Figure 7.28(a) [5] was improved in the later version [44] [Figure 7.28(h)] and has a compression ratio of 1.16. This pump is self-priming and insensitive to ambient pressure because of the implementation of a special pump membrane limiter, which protects the membrane against overloading. Other good designs combine the valves with the pump chamber [45] [Figure 7.28(i)], or make a small pump chamber with DRIE [45] [Figure 7.28(j)].

Table 7.6 lists the most important parameters of the above micropumps. The pump designs depicted in Figure 7.4 also illustrate the "evolution" in designing check-valve micropumps. The development shows clearly how the dead volume of the pump chamber has become smaller over time and how the check valves and the pump membrane have become more flexible. Most of the developed micropumps tend to have a piezoelectric disc as an actuator, which is reasonable for the performance and size needed for this pump type.

**Table 7.6**
Typical Parameters of Check-Valve Micropumps
(Values for water, except [37] for air) ($S$: Chamber size; $h$: Chamber height; $D$: Maximum deflection; $\dot{Q}_{max}$: Maximum flow rate; $p_{max}$: Maximum backpressure; $f$: Typical frequency; ICPF: Ionic conducting polymer gel film)

| *Refs.* | $S$ (μm) | $h$ (μm) | $D$ (μm) | $\dot{Q}_{max}$ (μL/min) | $p_{max}$ (Pa) | $f$ (Hz) | *Technology* | *Actuator* |
|---|---|---|---|---|---|---|---|---|
| [5] | 12,500 | 130 | | 8 | 9,800 | 1 | Bulk, anodic bonding | Piezoelectric |
| [29] | 7,500 | 273 | 23 | 30 | 4,000 | 0.5 | Bulk, anodic bonding | Thermopneumatic |
| [46] | 7,500 | 261 | 23 | 58 | 2,940 | 5 | Bulk, anodic bonding | Thermopneumatic |
| [42] | 5,000 | | — | 20 | 14,700 | 40 | Bulk, surface | Piezoelectric |
| [32] | 4,000 | 425 | 4 | 70 | 2,500 | 25 | Bulk, adhesive bonding | Electrostatic |
| [33] | 4,000 | 425 | 5 | 850 | 31,000 | 1,000 | Bulk, adhesive bonding | Electrostatic |
| [34] | 4,000 | 380 | 1.7 | 150 | 2,000 | 200 | Bulk, adhesive bonding | Piezoelectric |
| [35] | 5,000 | 370 | — | 365 | 2,380 | 25 | Bulk, adhesive bonding | Piezoelectric |
| [37] | 5,000 | 100 | 100 | 44 | 3,800 | 5 | LIGA | Pneumatic |
| [39] | 10,000 | — | 30 | 400 | 210,000 | 50 | LIGA | Piezoelectric |
| [40] | 10,000 | 200 | — | 2,100 | 11,000 | 100 | Plastic molding | Piezoelectric, Electromagnetic |
| [41] | 7,000 | 400 | — | 13,000 | 5,900 | 12 | Bulk, silicone | Pneumatic, Electromagnetic |
| [43] | 5,700 | 15 | 15 | 1,300 | 90,000 | 200 | Bulk, direct bonding | Piezoelectric |
| [44] | 4,000 | — | — | 4 | 35,000 | 2 | Bulk, anodic bonding | Piezoelectric |
| [45] | 14,000 | 200 | | 100 | 9,000 | 40 | Bulk, anodic bonding | Piezoelectric |
| [47] | 3,600 | 18 | 18 | 1,400 | 300,000 | 3,500 | Bulk, direct bonding | Piezoelectric |
| [48] | 10,000 | 1,500 | — | 780 | 5,500 | 264 | Plastic molding | Electromagnetic |
| [49, 50] | 6,000 | — | 100 | 40 | | 2 | Precision machining | ICPF |
| [51] | 10,000 | — | — | 600 | 9,800 | 2 | Precision machining | Piezoelectric |
| [52] | 8,400 | 450 | — | 50 | 519 | 1 | Bulk, adhesive bonding | SMA |
| [31] | 3,100 | 120 | 120 | 88 | 6,860 | 1 | Laser micromachining | Pneumatic |

#### 7.2.1.2 Peristaltic Pumps

The peristaltic micropump shown in Figure 7.30(a) [53] has piezoelectric actuators and pump chambers that are etched in silicon. The pump represents three connected piezoelectric microvalves. The piezoelectric discs are glued to the pump membrane and work like piezobimorph discs. In batch fabrication, piezoelectric material can be deposited on the membrane. However, the required thickness of the piezoelectric layer makes this process impractical. An alternative is screen-printing a thick piezoelectric film [34].

The peristaltic pump shown in Figure 7.30(b) utilizes surface micromachining and electrostatic actuators [54]. The polysilicon electrodes are embedded in silicon nitride. The pump chamber is defined by the thickness of the sacrificial layer. Consequently, the dead volume can be kept very small. No results for maximum flow rate and backpressure were reported for this pump. The pump shown in Figure 7.30(c) has thermopneumatic actuators. Air was used as the actuation fluid [55]. The achieved deflection and the pump chamber height are about 10 mm. However, the heat loss caused by the good thermal conductivity of silicon minimizes the thermopneumatic effect and increases the power consumption. The pump depicted in Figure 7.30(c) also uses thermopneumatic actuators to drive four chambers [56]. The only difference to the conventional thermopneumatic actuators is that this pump has external laser light as its heat source. The peristaltic pump shown in Figure 7.30(d) employed thermal fusion bonding and micromilling of TPU and PMMA [57]. The flexible

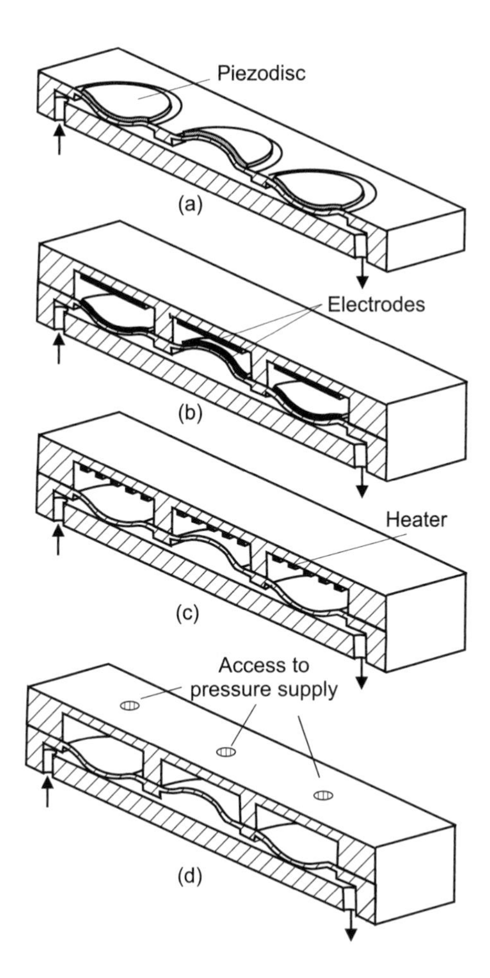


**Figure 7.30** Design examples of peristaltic micropumps: (a) piezoelectric; (b) electrostatic; (c) thermopneumatic; and (d) pneumatic.

membrane is sandwiched and then bonded between the PMMA layers using a low-temperature thermal fusion bonding method. Micromilling is used to engrave necessary patterns in PMMA. Air is introduced to the control chambers to actuate the flexible membranes.

Various peristaltic pneumatic micropumps have been developed based on the integration of multiple interconnected Quake valves on a single substrate using multi-layer softlithography, Figure 7.31. Design and operation of Quake valve were explained in detail in Chapter 6. A control channel is located on top of each liquid channel to regulate the flow [Figure 7.31(a)]. Air or nitrogen are introduced into the control channel for actuation purposes. The integration and operation of three valves in series allow for the implementation of a peristaltic pump. This gas-actuated micropump is able to create very precise flow rates on the order of few nanoliters per second [58]. In another design, Shaegh et al, designed a micropump based on the Quake valve design [59]. The curved liquid channels were fabricated using unfocused laser micromachining method and all polymeric layers were bonded together using thermal fusion bonding. The pump was able to deliver flow rates of up to 7.15 μL/min at an actuation frequency of 5 Hz.

Similar to the methods discussed in the previous section on checkvalve micropumps, silicone rubber can be used to form a more flexible pump membrane [60]. With external pneumatic sources, the pump can generate a flow rate up to 120 μL/min. In thermopneumatic mode, it only delivers a

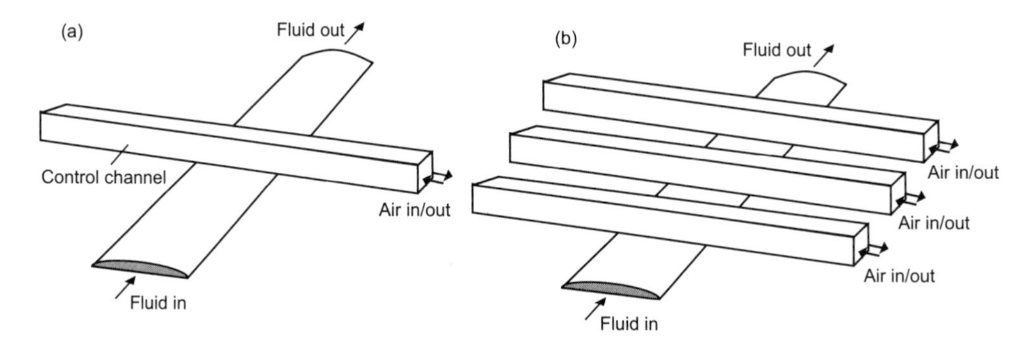


**Figure 7.31** Peristaltic micropump based on Quake valve: (a) a single Quake valve; and (b) a peristaltic micropump using three interconnected Quake valves. (*After*: [58].)

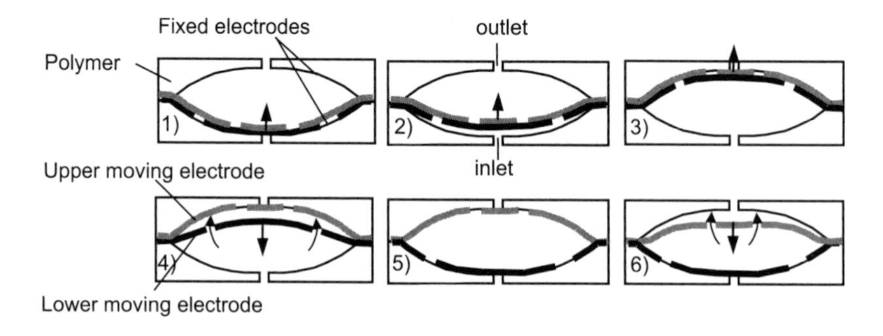


**Figure 7.32** Concept of dual-diaphragm pump: 1) to 3) pump phase; and 4) to 6): supply phase. (*After*: [62].)

few microliters per minute like similar designs in [56], because of the insufficient pressure generated by heated air in the pressure chamber.

The pump reported by Cabuz et al. increases the compression ratio to 10 by using curved pump chambers and flexible plastic pump membrane for electrostatic actuation [61]. The numerous pump chambers are designed by using a 3-D array structure. With these optimization measures, the pump is able to deliver 8 mL/min with only 75 V drive voltage and 4-mW electrical power. Electrostatic actuators allow a fast operation frequency of 100 Hz, which is almost two orders of magnitude faster than its thermopneumatic counterparts. The improved version of this pump has two actuating membranes in the pump chamber. These membranes have orifices and act alternatively as check valves (Figure 7.32). This new pumping principle was called the dual-diaphragm pump (DDP) [62].

The pump body is made of a polymer by injection molding (see Section 3.4.5). The fixed electrodes are deposited by evaporation. A thin insulator layer is deposited on top of the metal electrodes by ion beam sputtering. The moving electrodes are made of metalized Kapton films, which are coated with thin insulator layers as well. The fixed electrodes are connected to the ground, and the two moving electrodes are connected to two different drive signals. The operation phases of the pump are depicted in Figure 7.32. This pump is able to deliver 28-mL/min air flow at a drive voltage of 200V and drive frequency of 95 Hz. The pump reported in [63] utilized active valves in its design, similar to that of [53]. Table 7.7 lists the most important parameters of the above examples.

#### 7.2.1.3 Valveless Rectification Pumps

The first pump with diffuser/nozzle structures is reported in [8]. The pump is fabricated in brass using precision machining [Figure 7.33(a)]. The same design can be realized by printed circuit board

**Table 7.7**
Typical Parameters of Peristaltic Micropumps
(Values for water, $S$: Chamber size; $h$: Chamber height; $D$: Maximum deflection; $\dot{Q}_{\max}$: Maximum flow rate)

| *Ref.* | $S$ (μm) | $h$ (μm) | $D$ (μm) | $\dot{Q}_{\max}$ (μL/min) | $p_{\max}$(Pa) | $f$ (Hz) | *Technology* | *Actuator* |
|---|---|---|---|---|---|---|---|---|
| [53] | 5,000 | — | — | 100 | 5,880 | 15 | Bulk micromach. | Piezoelectric |
| [54] | 400 | 4 | 4 | | — | — | Surface micromach. | Electrostatic |
| [55] | 1,000 | 10 | 10 | 7 | — | 20 | | Thermopneumatic |
| [56] | 800 | 18 | 35 | 3 | 4,822 | 3 | Bulk micromach. | Thermopneumatic |
| [60] | 500 | 60 | 60 | 4.2 | 3,447 | 2 | Bulk, precision mach. | Thermopneumatic |
| [61] | 800 | 500 | 500 | 8,000 | — | 100 | Plastic molding | Electrostatic |
| [62] | 800 | 500 | 500 | 30,000 air | — | 100 | Plastic molding | Electrostatic |
| [63] | 10,000 | 400 | — | 40 | 2,450 | 1 | Bulk, anodic bonding | Piezoelectric |
| [58] | 100 | — | 10 | 0.14 | — | 100 | Softlithography | Pneumatic |
| [59] | 500 | 200 | 200 | 7.15 | 10,700 | 5 | Laser micromachining | Pneumatic |

technology [9]. Further development of this pump led to the flat design in silicon [Figure 7.33(b)] [11]. Using small opening angles (7° to 15°), the fluid is pumped out of the diffuser side [Figure 7.33(a)]. Deep reactive ion etching can achieve small chamber height, and, consequently, small dead volume and large compression ratio. Two such pumps connected in parallel, operating in opposite modes (supply mode and pump mode), allow relatively smooth flow rates. Piezobimorph discs are pump actuators, which allow the pumps to operate at resonant frequencies on the order of kilohertz. This pump design can also be realized with a thermopneumatic actuator. The actuator has the form of an expanding and collapsing bubble. The heater is integrated in the pump chamber, and there is no need for a pump membrane or other moving parts [64].

The pump effect occurs in the opposite direction if the opening angle is large. The pump presented in [65] has an opening of 70.5°, which is determined by the $\{111\}$ surface freed with anisotropic wet etching of a $\{100\}$-wafer [Figure 7.33(c)]. The optimized version of this design has a thermopneumatic actuator and a corrugated pump membrane, which increase the stroke volume [66]. However, the thermopneumatic actuator results in pump frequencies that are two orders of magnitude lower than piezoelectric actuators. The next feasible actuation scheme for this pump type is magnetostrictive actuation [67]. Magnetostrictive effect is the process by which a ferromagnetic material transforms from one shape to another in the presence of a magnetic field. This effect is a result of the rotation of small magnetic domains, causing internal strains in the material. These strains cause the material to expand in the direction of the magnetic field. As the field is increased, more domains rotate and become aligned until magnetic saturation is achieved. Since the magnetostrictive forces are molecular in origin, this type of actuator has a fast response, on the order of microseconds. Thick TbFe/FeCo film is sputtered on the membrane, which is used as an actuating layer. The rectification efficiency of the above-mentioned pumps is depicted in Figure 7.7.

Valveless rectification micropumps can also be fabricated in a planar manner where inlet and outlet ports and the displacement chamber are located on a horizontal plane. To this end, Zhao et al. developed a micropump for droplet generation using polyvinylidene fluoride (PVDF) organic piezoelectric film [Figure 7.34]. Flow rates within the range of 0–300 μL/min could be precisely achieved through tuning the voltage and its frequency applied on the diaphragm [68].

The pump shown in Figure 7.35 [69, 70] applies the valvular conduit structure in microscale. This principle was first invented by Tesla [71]. Similar to diffuser/nozzle structures, the valvular conduits shown in Figure 7.35 cause rectification effects without check valves. This pump type

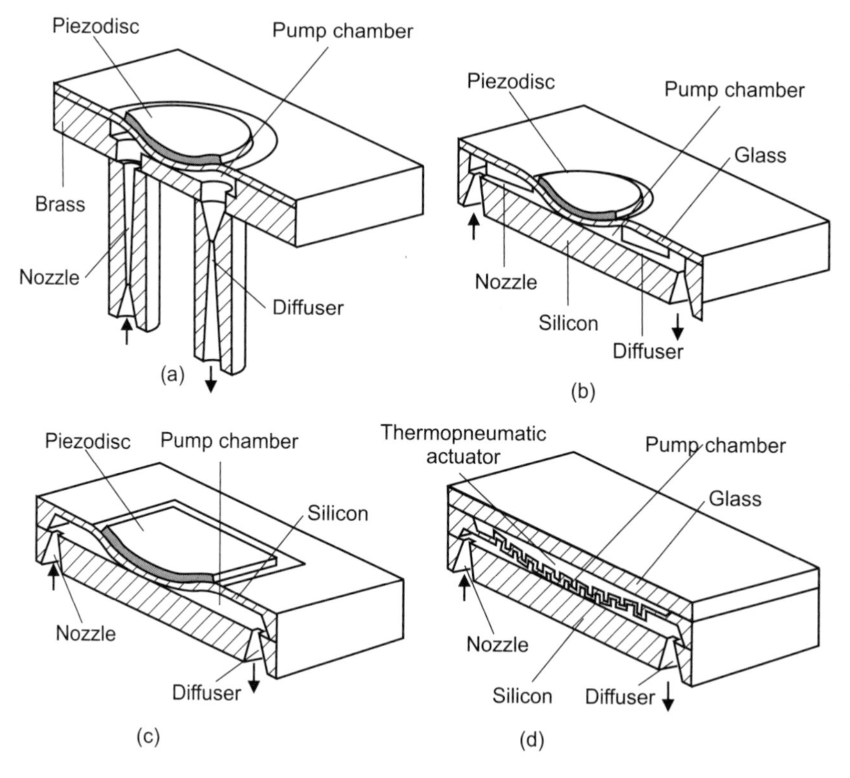


**Figure 7.33** Design examples of valveless rectification micropumps: (a) and (b) small diffuser angle; and (c) and (d) large diffuser angle.

can be realized easily in silicon with DRIE technology. A maximum fluidic diodicity $\eta = 1.2$ and a maximum rectification efficiency $\chi = 0.045$ can be reached with the Tesla structure [Figure 7.35(b)]. The next valveless concept takes advantage of the temperature dependency of viscosity [72]. The fluidic impedance at the outlet and inlet are modulated by means of heat. The heating cycles are synchronized with the pump cycle [Figure 7.36(a)]. This pump is optimized by thin-walled structures for lower heat capacity at the inlet and outlet. A similar design is used in [73]. The inlet and outlet are heated until they generate microbubbles. The bubbles work as active valves. The pressure difference between the gas phase and liquid phase generated by surface tension can restrict the flow. The pump uses the same bubble actuator as described in [64]. No pump membrane is required. By synchronizing the bubbles of the actuator and of the valves, pump effects can be achieved. This design is suitable for the general concept of planar microfluidics.

Another approach of valveless pumping is shown in Figures 7.36(b–d) [74]. Their authors called the pump concepts the elastic buffer mechanism [Figure 7.36(b)] and the variable gap mechanism [Figure 7.36(c)]. The pumps are actual microvalves, as described in Chapter 6. The dynamic interaction between the valve membrane and the fluid system causes the pumping effect. This pump type is able to pump liquids in two directions depending on its drive frequency. The pump effect is also achieved with a piezobimorph disc in a similar structure [Figure 7.36(d)] [75]. Table 7.8 gives an overview of typical parameters of the above-mentioned valveless rectification micropumps.

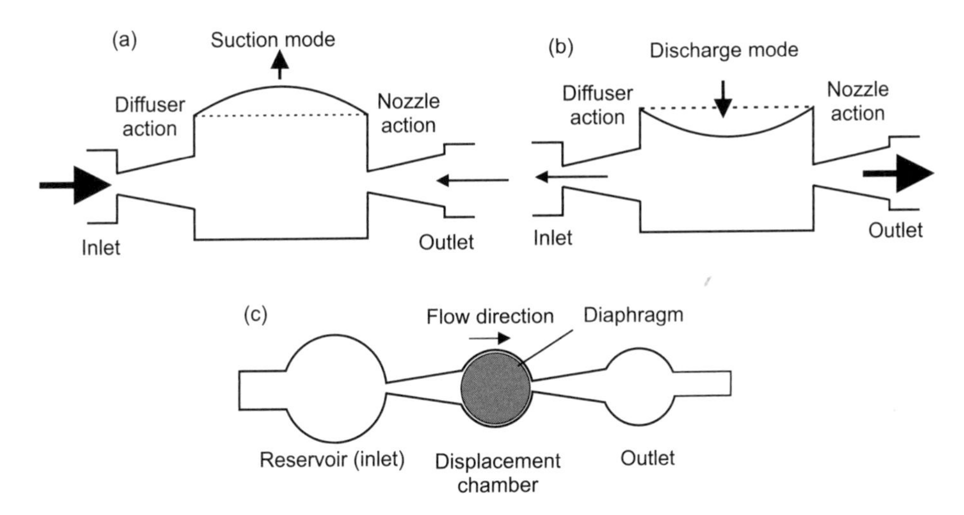


**Figure 7.34** An example of valveless rectification micropump with horizontal inlet and outlet ports: (a) suction mode of operation; (b) pumping (discharge) mode of operation; and (c) top view of the pump design. The larger arrow is indicative of higher flow rate. (*After*: [68].)

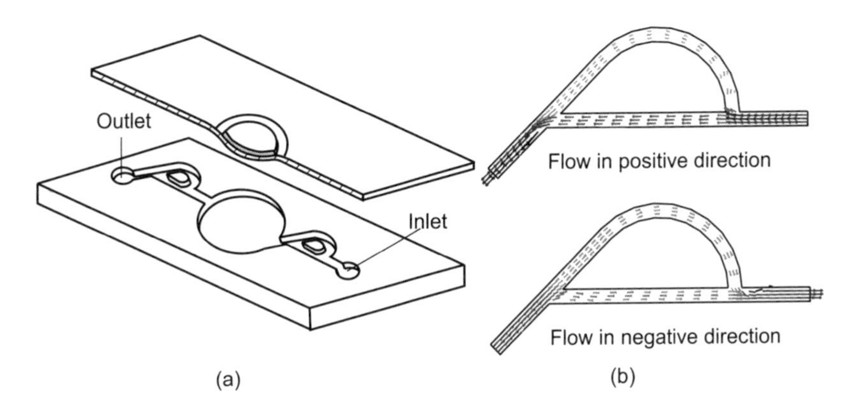


**Figure 7.35** Tesla pump: (a) design example; and (b) rectification effect.

### 7.2.1.4 Rotary Pumps and Centrifugal Pumps

The micropump based on the concept shown in Figure 7.8 has two gears made of iron-nickel alloy using LIGA technique [12]. An external motor drives the gears. The gears forced the fluid along by squeezing it into an outlet. The gears have a diameter of 600 μ and run at a speed of 2,250 rpm. For a glycerol-water solution, the pump can deliver a maximum flow rate of 180 μL/min and a maximum backpressure of 100 kPa. The pump works best for moderately highviscosity liquids.

The gears reported by Terray et al. [76] are made of several 3-μm silica microspheres arranged in a pump chamber. The gears are controlled by a laser beam. This pumping system was able to generate a flow rate on the order of 1 nL/h.

Actuating by means of an external magnetic field is possible. Magnetic force can be applied directly on a ferrofluid, which in turn is a plug for moving fluid in a microchannel. Figure 7.37 explains the concept of ferrofluidic pumping. A hydrocarbon-based ferrofluid is suspended in the channel. An external permanent magnet fixes the position of the ferrofluid plug. Moving the permanent magnet pushes the ferrofluid plug in the desired direction. The ferrofluid plug pushes the fluid on the one side, and pulls it on the other side. The effect causes a pumping effect with a pump

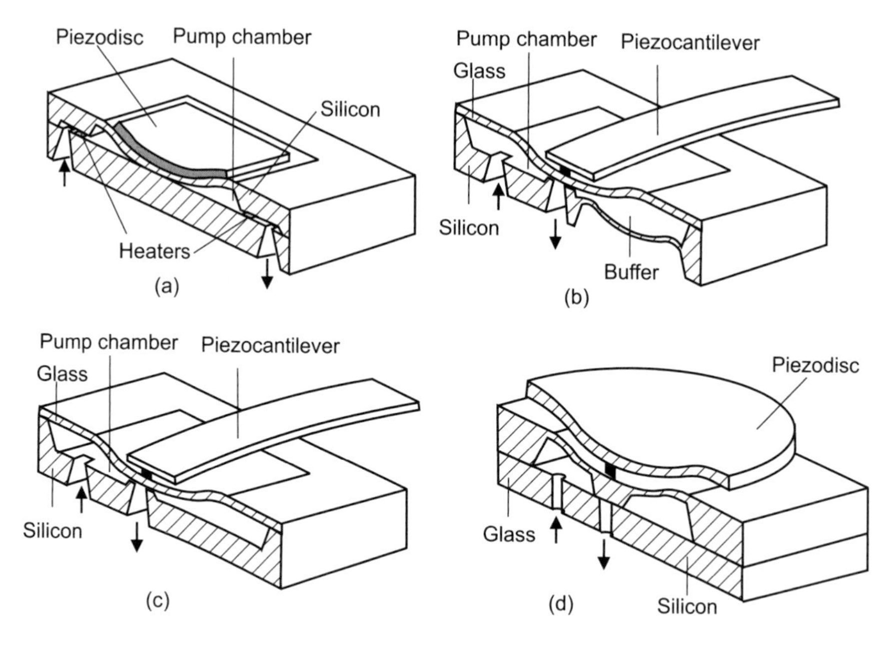


**Figure 7.36** Other valveless pumps: (a) based on temperature dependency of viscosity; (b) elastic buffer mechanism; and (c) and (d) variable gap mechanism.

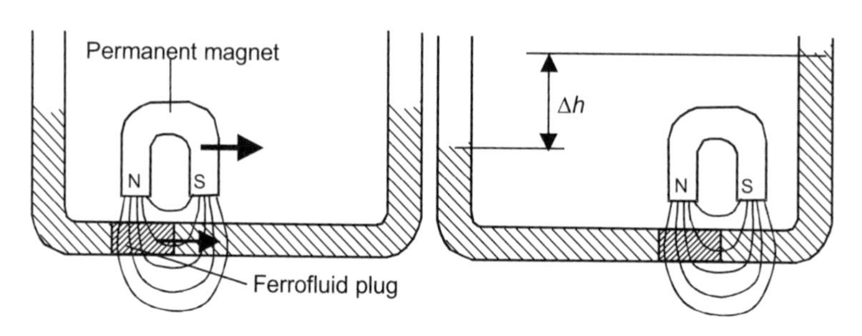


**Figure 7.37** Ferrofluidic magnetic pumping. (*After*: [77].)

head $\Delta h$. If the magnet stops, the magnetic field holds the plug at its current location and prevents the pumped fluid from going back. In this way, the ferrofluid plug also works as an active valve. The pumping power depends on the gradient of the external magnetic field and the magnetization of the ferrofluid. The field strength can be adjusted by the external permanent magnet, so the pump design becomes independent of the type of pumped fluid.

Magnetic pumping using a ferrofluid can be implemented in microscale [77]. The pump is realized in a planar concept. The channel structure is etched in silicon and covered by a borosilicate glass plate. The glass wafer and the silicon wafer are bonded anodically. Two permanent magnets are used as valve actuator and plug actuator.

The working principle of this rotary concept is explained in Figure 7.38. A fixed permanent magnet creates a stationary ferrofluid plug between the inlet and the outlet (1). First, a moving permanent magnet merges a second plug with the stationary plug (2). After merging completely, the large plug blocks both inlet and outlet (3). By separating the large plug into one stationary and one moving plug, the fluid is primed into the inlet and pushed out of the outlet (4).

The pump design is similar to the concept of a rotary pump and a check-valve pump. However, this design is a pure magnetic micropump because no mechanical moving part is involved. The

**Table 7.8**

Typical Parameters of Valveless Rectification Micropumps

($S$: Chamber size; $h$: Chamber height; $D$: Maximum deflection; $\dot{Q}_{max}$: Maximum flow rate)

| *Ref.* | $S$ *(µm)* | $h$ *(µm)* | $D$ *(µm)* | $\dot{Q}_{max}$ *(µL/min)* | $p_{max}$*(Pa)* | *f (Hz)* | *Technology* | *Actuator* |
|---|---|---|---|---|---|---|---|---|
| [8] | 19,000 | — | 13 | 16,000 | 7,840 | 110 | Precision eng. | Piezoelectric |
| [9] | 14,000 | 40 | 40 | 200 | 5,000 | 3,900 | PCB | Piezoelectric |
| [11] | 5,000 | 80 | 13 | 1,200 | 16,000 | 1,400 | Plastic molding | Piezoelectric |
| [64] | 1,000 | 50 | — | 5 | 350 | 400 | Bulk, deep RIE | Thermopneumatic |
| [65] | 5,000 | 430 | 45 | 400 | 3,200 | 3,500 | Bulk | Piezoelectric |
| [66] | 4,000 | 115 | 100 | 14 | — | 4 | Bulk | Thermopneumatic |
| [67] | 10,000 | 400 | 16 | 290 | 4,900 | 10,000 | Bulk | Magnetostrictive |
| [70] | 6,000 | 156 | — | 900 | 14,700 | 3,700 | Bulk, deep RIE | Piezoelectric |
| [72] | 5,000 | 250 | — | 5.5 | — | 5 | Bulk | Piezoelectric |
| [74] | 5,300 | 425 | 40 | 900 | 6,000 | 50 | Bulk | Piezoelectric |

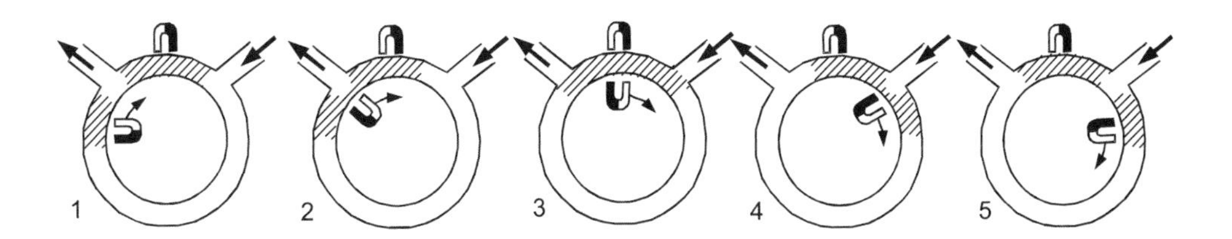


**Figure 7.38** Ferrofluidic magnetic pump. (*After*: [77].)

micropump is able to deliver a maximum flow rate of 70 mL/min and a maximum pump head of 1.3 kPa.

Ahn and Allen reported a centrifugal pump design with microturbine as a rotor in an integrated electromagnetic motor [78]. The pump simply adds momentum to the fluid by means of fast-moving blades. The rotor, stator, and coils are fabricated by electroplating of iron-nickel alloy. The high aspect ratio structures were fabricated at a low cost by using conventional photolithography of polyimide. Table 7.9 compares the performance of the above-mentioned rotary pumps and centrifugal pumps.

**Table 7.9**

Typical Parameters of Rotary Micropumps

(Typical size is the size of the turbine or the gear; $S$: Typical size; $h$: Chamber height; $\dot{Q}_{max}$: Maximum flow rate; $p_{max}$: Maximum backpressure; $n$: Rotation speed)

| *Ref.* | $S$ *(µm)* | $h$ *(µm)* | $\dot{Q}_{max}$ *(µL/min)* | $p_{max}$*(Pa)* | $n$ *(rpm)* | *Technology* | *Actuator* |
|---|---|---|---|---|---|---|---|
| [12] | 596 | 500 | 55 | 12,500 | 2,250 | LIGA | External motor |
| [78] | 500 | 160 | 24 | 10,000 | 5,000 | Bulk | Integrated motor |
| [13] | 50 | 200 | — | — | 524–1,126 | Plastic | External motor |

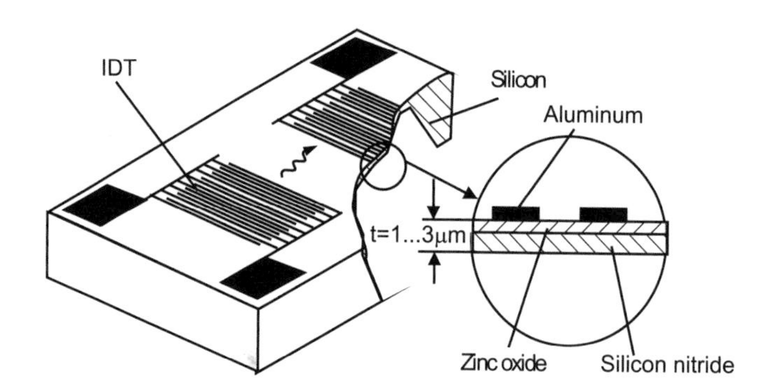


**Figure 7.39** Ultrasonic micropump based on acoustic streaming.

#### 7.2.1.5 Ultrasonic Pumps

A micropump based on acoustic streaming needs a thin flexible actuating membrane. Figure 7.39 depicts the conventional design of a flexural actuating membrane. The thin membrane consists of silicon nitride, zinc oxide, and aluminum. While aluminum is the electrode material, zinc oxide works as the piezoelectric material. The membrane is typically 1 to 3 μm thick. With a wave frequency of around 3 MHz, the typical wavelength is 100 μm. The interdigitated electrodes are placed one wavelength apart. Depending on the type of the pump (bidirectional or unidirectional), the fingers of the counter electrode are placed one-half of the wavelength or one-fourth of the wavelength apart from the first electrode fingers. The electrodes are sputtered directly on piezoelectric zinc oxide and allowing electric excitation and wave generation. Nguyen et al. reported a FPW pump with an integrated flow sensor for feedback control [15]. The membrane has an additional polysilicon layer acting as the ground electrode. Aluminum and polysilicon form a thermocouple for detecting temperature change in the integrated thermal flow sensor (see Chapter 8).

Acoustic streaming can be used in an active micromixer [79] or for handling cells and large molecules [80]. Chapters 10 and 12 give more examples of acoustic streaming in designing microfilters and micromixers.

### 7.2.2 Nonmechanical Pumps

#### 7.2.2.1 Electrohydrodynamic Pumps

The first micromachined EHD induction pump was reported by Bart et al. [18]. Fuhr et al. [17, 81, 82] and Ahn et al. [83] reported similar designs. Fluid velocities of several hundred microns per second can be achieved with this pump type. For a better pumping effect, external heat sources and heat sinks generate a temperature gradient and consequently a conductivity gradient across the channel height. The heat sink is actively cooled by a Peltier element [17].

In EHD injection pumps, the Coulomb's force is responsible for moving ions injected from one or both electrodes by means of electrochemical reaction. Richter et al. demonstrated this pumping concept with micromachined silicon electrodes [84]. The electrodes are grids made of silicon. Electrode structures allow a high electric field between them and concurrently let the fluid pass through the grids. The pressure gradient built up in the electric field causes the pumping effect.

Furuya et al. used electrode grids standing perpendicular to a device surface to increase the pressure gradient [85]. The pump can deliver 0.12 mL/min with a drive voltage of 200V. Table 7.10 lists the most important parameters of the EHD pumps discussed above. All micropumps discussed above are realized with silicon bulk micromachining.

**Table 7.10**

Typical Parameters of EDH Micropumps

($S$: Chamber size; $h$: Chamber height; $\dot{Q}_{max}$: Maximum flow rate; $p_{max}$: Maximum backpressure)

| *Ref.* | $S$ *(μm)* | $h$ *(μm)* | $\dot{Q}_{max}$ *(μL/min)* | $p_{max}$*(Pa)* | *Principle* | *Technology* |
|---|---|---|---|---|---|---|
| [18] | 500 | — | — | — | EHD induction | Bulk |
| [17, 86] | 600 | 50 | 0.45 | — | EHD induction | Bulk |
| [83] | 3,000 | 200 | 50 | 220 | EHD injection | Bulk |
| [84] | 3,000 | — | 15,000 | 500 | EHD injection | Bulk |
| [85] | 400 | 100 | 0.12 | — | EHD injection | Bulk |

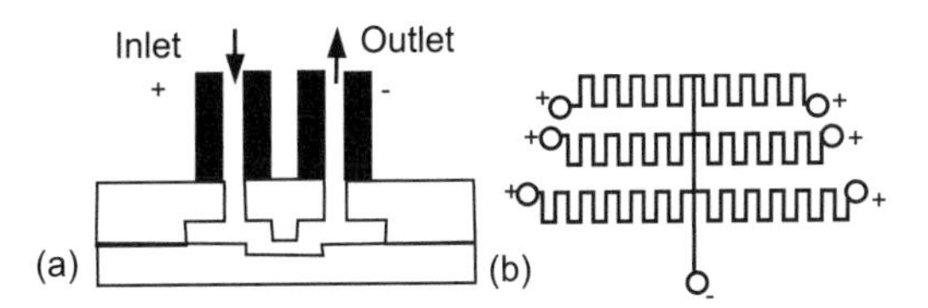


**Figure 7.40** Electrokinetic micropumps: (a) short and wide channel; and (b) parallel pumping.

### 7.2.2.2 Electrophoresis Pumps

Gunji et al. [87] presented an electrophoresis pump with parallel finger electrodes, which are able to generate and transport liquid droplets on the order of 60 nL at a frequency of 100 kHz. Smaller droplet volumes are possible with higher frequencies. The drawback of this pumping concept is the high field strength, caused by a drive voltage on the order of 500V to 700V over a relatively small gap of 100 μm. Dielectrophoresis can be extended to manipulation of small bioparticles such as cells, DNA strands, or viruses. Direct manipulation can be realized by combining microfluidic pumping and dielectrophoresis trapping. Chapter 12 will discuss more about these applications of electrophoresis.

### 7.2.2.3 Electrokinetic Pumps

Because of the high voltages required for electrokinetic pumping, silicon is not a suitable device material. Silicon itself is electrically conductive. Even in the case of coating an insulator layer such as silicon dioxide, the high electrical field strength needed for pumping and separation may reach the breakdown limit. Furthermore, the condition for an electro-osmotic flow is the presence of immobilized surface charges at the interface between the channel wall and an electrolyte solution. Glasses offer the best conditions for these surface charges. Because of these reasons, most of the devices for electrokinetic pumping are made of glass. With emerging polymeric technology, polymeric devices are promising.

The device with a long capillary proposed by Harrison et al. [88] could generate a fluid velocity of 100 μm/s in a field strength of 150 V/cm. Figure 7.40(a) illustrates an electrokinetic micropump with a short, wide, and extremely shallow channel [89]. The micropump with a 40-mm length, 1-mm width, and 1-μm depth is fabricated on soda-lime glass substrates using standard micromachining techniques. Deionized water ($10^{-3}$ S/m) is used as the working fluid. The pump is able to deliver a maximum flow rate of 2.5 μL/min and a maximum backpressure of 150 kPa. The pump requires high operating voltages on the order of several kilovolts.

**Table 7.11**
Typical Parameters of Electrokinetic Pumps
($W$: Channel width; $H$: Channel height; $L$: Channel length; $N$: Number of channels; $\dot{Q}_{\max}$: Maximum flow rate; $p_{\max}$: Maximum backpressure; $V_{\max}$: Maximum voltage)

| *Refs.* | $W$ *(μm)* | $H$ *(μm)* | $L$ *(μm)* | $N$ | $\dot{Q}_{\max}$ *(μL/min)* | $p_{\max}$*(Pa)* | $V_{\max}$ *(kV)* | *Technology* |
|---|---|---|---|---|---|---|---|---|
| [89] | 40,000 | 1 | 1 | 1 | 2.5 | 150 | 3 | Bulk, glass |
| [90, 91] | 300 | 50 | 83 | 6 | 1.8 | 0.441 | 1.5 | Bulk, glass |

Figure 7.40(b) illustrates the other pump with six small and long channels. The pump is micromachined in Pyrex glass [90, 91]. Borax solution ($Na_2B_4O_7$) is the working liquid. Because of the weaker field strength, the backpressure of this pump is two orders of magnitude smaller than the pump shown in Figure 7.40(a).

Electro-osmotic micropumps can reach higher flow rates with porous structures or packed beads in the microchannel. The pores work as tortuous capillaries for electro-osmotic flows. Table 7.11 compares the most important parameters of the above-mentioned electrokinetic micropumps.

#### 7.2.2.4 Capillary Pumps

*Passive Capillary Micropumps.* Passive capillary is the simplest solution for liquid filling in microfluidic systems. Tseng et al. used microchannels made in PDMS and SU-8 to transport blood samples and enzyme solution for glucose detection [92]. The SU-8 surface was treated with oxygen plasma to make it more hydrophilic.

Capillary pumping on paper has gained significant popularity within the last decade. Paper-based microfluidics is a branch of microfluidics attempting to develop analytical devices through the integration of valving, pumping and sensing schemes on paper. Various pumping methods taking advantage of capillary wetting of paper have been developed. In particular, two main methods of pumping are more favorable. As shown in Figure 7.41(a), hydrophilic capillary channels are created by depositing of a hydrophobic material on the paper substrate. Wax printing is a common method that is suitable for creating hydrophobic zones with desired patterns. A small volume of liquid sample can only flow through the hydrophilic zones by capillarity and reaches the detection zones for evaluation [Figures 7.41(b, c)]. Figure 7.42 shows another method of capillary pumping [93]. In this method, a paper disk acts like a reservoir to induce passive liquid pumping in a microchannel. Initially, the microchannel is partially primed by adding a priming droplet to its outlet, Figure 7.42(a). The sample solution is introduced into the reservoir followed by placing a paper disk over the priming droplet [Figures 7.42(b, c)]. The liquid starts to flow through the paper disk [Figure 7.42(d)].

*Thermocapillary Micropumps.* Takagi et al. [94] presented the first phase transfer pump. The alternate phase change was generated by an array of 10 integrated heaters [Figure 7.43(a)]. The same pump principle was realized with stainless steel and three heaters in [95]. Jun and Kim [96] fabricated a much smaller pump based on surface micromachining. The pump has six integrated polysilicon heaters in a channel that is 2 μm in height and 30 μm in width [Figure 7.43(b)]. The pump is capable of delivering a flow velocity of 160 μm/s or flow rates less than 1 nL/min. Song and Zhao reported an analytical model for pumping with a vapor bubble [97]. The concept was tested with a mesoscale capillary system (1 mm in diameter) and external heating wires. This pumping system can deliver a maximum flowrate of 5 μL/s and a maximum backpressure of 60-mm water.

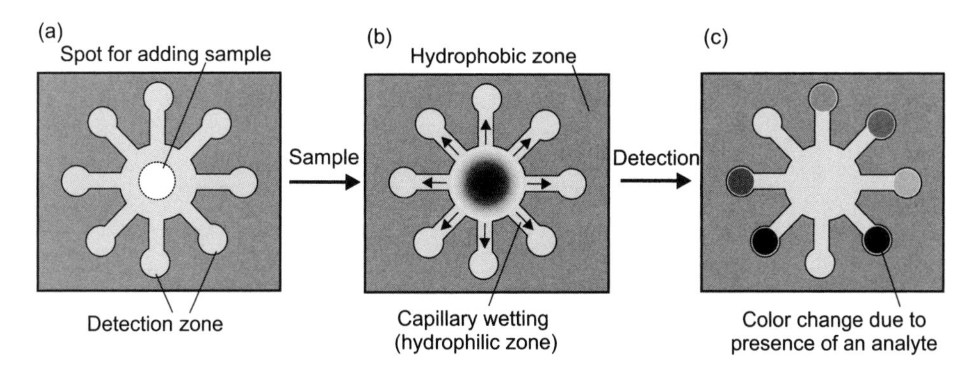


**Figure 7.41** An example of sample spreading and detection on a paper chip: (a) paper chip with hydrophobic pattern and detection zones; (b) capillary wetting on paper chip upon adding sample solution; and (c) detection of desired analytes in detection zones based on calorimetric method.

**Table 7.12**

Typical Parameters of Phase Transfer Micropumps

($S$: Chamber size; $H$: Chamber height; $\dot{Q}_{\max}$: Maximum flow rate; $\dot{p}_{\max}$: Maximum backpressure; $f$: Typical frequency)

| *Refs.* | $S$ *(μm)* | $h$ *(μm)* | $\dot{Q}_{\max}$ *(μL/min)* | $p_{\max}$*(Pa)* | *f (Hz)* | *Technology* |
|---|---|---|---|---|---|---|
| [94, 95] | 180 | 180 | 210 | — | 4 | Bulk micromachining |
| [96] | 30 | 2 | $5.05 \times 10^{-4}$ | 800 | 1–4 | Surface micromachining |
| [98] | 1,600 | | 150 | 265 | 4–10 | Precision machining |

Geng et al. [98] reported a self-heating pump. Ohmic heating of electrically conducting liquids generates the actuating bubble. The growth and collapse of a bubble at a conical section between a large channel and a smaller channel induces the pumping effect. With channel diameters of 1.6 and 0.8 mm, the pump is able to deliver a maximum flow rate of 150 μL/min and a maximum backpressure of 265 Pa. The pumped liquid is a saturated sodium chloride (NaCl) solution. Yokoyama et al. reported a thermocapillary pump made of copper [99]. The microchannels are fabricated by electroplating using a SU-8 mold. The microheaters were made of indium tin oxide (ITO) and sputtered on a glass substrate.

Darhuber et al. [100] manipulated liquid droplets using microheater arrays. The metal heaters were sputtered on a glass substrate and protected by a silicon oxide layer. The droplets are guided by hydrophilic stripes, which are defined by hydrophobic coating of PFOTS (1H,1H,2H,2H-perfluorooctyltrichlorosilane). Tseng et al. [101] reported a similar system with titanium/platinum heaters and silicon nitride protecting layer. A maximum velocity of about 1 mm/s was reported for silicone oil droplets. Table 7.12 compares the parameters of the above phase transfer micropumps.

*Electrocapillary Micropumps.* An electrocapillary pump, also called the electrowetting pump, was proposed by Matsumoto and Colgate [102]. The pump concept exploits the dependence of the surface tension at the solid/liquid interface on the charge of the surface. Lee and Kim [103] reported a microactuator based on the electrowetting of a mercury drop, which can be used for driving a mechanical pump with check valves, as proposed in [102] (Figure 7.44). The pump can deliver a maximum flow rate of 63 μL/min and a maximum backpressure of 600 Pa. The actuator only requires a power supply of 2.3V at 10 Hz.

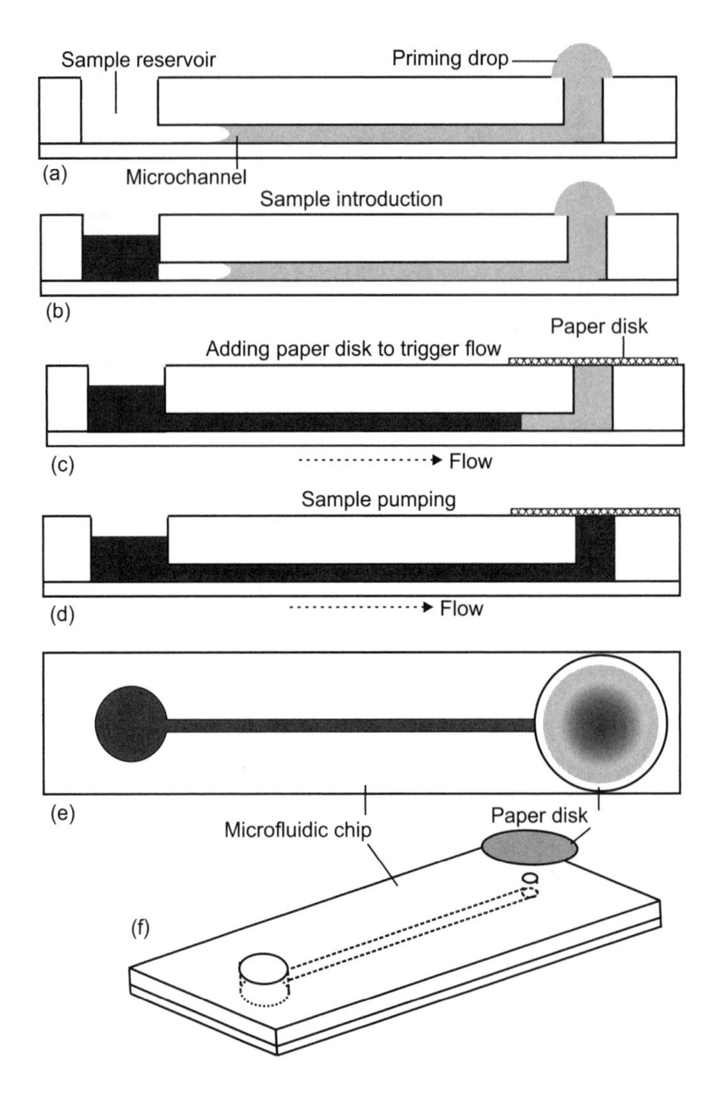


**Figure 7.42** An example of pumping fluid through a micochannel using a paper disk pump: (a) channel priming; (b) sample introduction; (c) initial wetting of paper disk; (d) sample pumping; (e) top view; and (f) isometric view of the microfluidic channel with paper disk. (*After*: [93].)

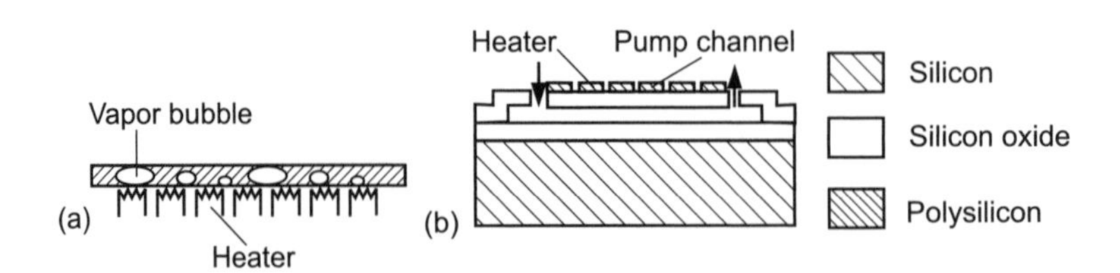


**Figure 7.43** Phase transfer pumps: (a) external heater; and (b) integrated heater.

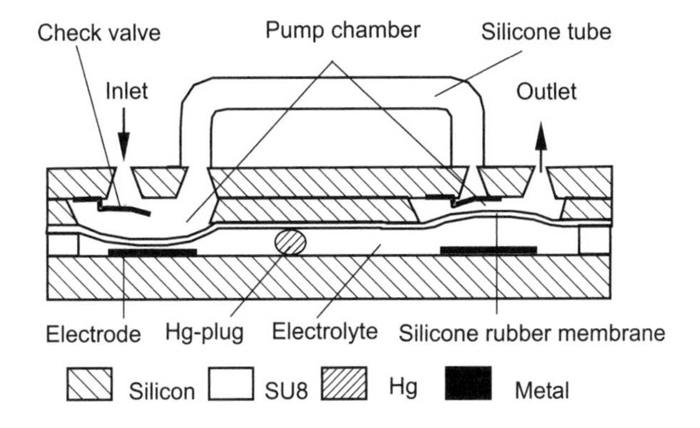


**Figure 7.44** Electrowetting pumps. (*After*: [103].)

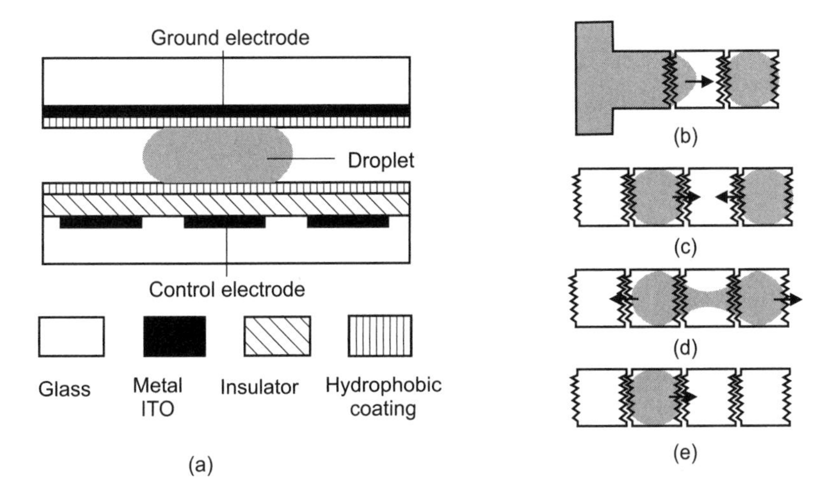


**Figure 7.45** Droplet-based microfluidics using electrowetting on dielectric: (a) device concept; (b) droplet dispensing; (c) droplet merging; (d) droplet cutting; and (e) droplet transport.

Direct electrowetting and electrowetting on dielectric are well suited for droplet-based microfluidics. Electrowetting can be used for dispensing and transporting a liquid droplet. Figure 7.45(a) depicts the device concept reported by Ren et al. [104]. The aqueous droplet is surrounded by immiscible oil. The droplet is aligned with the control electrode at the bottom. The $1 \times 1$ mm control electrode changes the hydrophobicity of the solid/liquid interface. A 800-nm Parylene C layer works as the insulator. The ground electrode is made of transparent ITO for optical investigation. A 60-nm Teflon layer was coated over the surface to make it hydrophobic. Electrowetting allows different droplet handling operations, such as droplet dispensing [Figure 7.45(b)], droplet merging [Figure 7.45(c)], droplet cutting [Figure 7.45(d)], and droplet transport [Figure 7.45(e)]. These basic operations allow merging and fast mixing of liquid droplets. Chapter 10 will discuss more about mixing with droplet-based microfluidics. Lee et al. [105] reported a similar device utilizing both direct electrowetting and electrowetting on dielectric. The device is able to transport liquid droplets surrounded by air. The liquid/air system may have a disadvantage of evaporation. However, the evaporation rate is slow due to the encapsulated small space around the droplet.

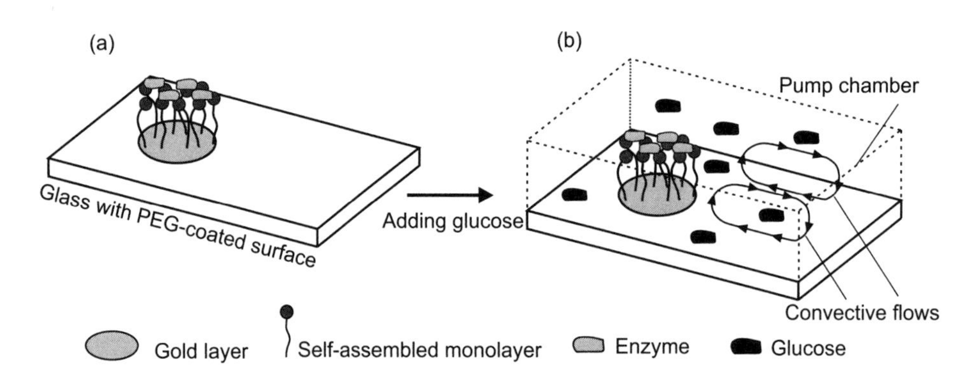


**Figure 7.46** An example of enzymatic micropump design and operation: (a) fabrications steps of gold patterning by e-beam evaporation, self-assembly monolayer formation, and enzyme immpbolization; and (b) fluid motion induced by the presence of both GOx and glucose that produces hydrogen peroxide inside the chamber. (*After*: [107].)

### 7.2.2.5 Chemical and Magnetohydrodynamic Pumps

Various chemical mechanism including self-electrophoresis and self-electro-osmosis, electrolyte self-diffusionophoresis, nonelectrolyte self-diffusionophoresis, bubble propulsion, and density-driven have been realized to drive particles and induce fluid motion [106]. Density-driven and bubble propulsion are more favorite for microfluidic applications where self-pumping is required. Density-driven convective flow is caused by inhomogenous distribution of local fluid density [106]. Particularly, enzyme-functionalized pumps have better biocompatibility than other micropumps in living organism. The pumps can sense and react to specific bioanalytes such as glucose. Figure 7.46 shows the assembly and function of an enzymatic chemical pump [107]. The immobilized enzymes can induce fluid motion upon contacting substrate. Chemical gradient of solutes and the resulting density difference, generated by the enzymatic reactions, generate convective flows within the pump chamber. Such pumping scheme can enable the implementation of stimuli responsive devices for drug delivery.

Böhm et al. reported on electrochemical pumps that use the pressure of gas bubbles generated by electrolysis water. Bidirectional pumping can be achieved by reversing the actuating current, which makes the hydrogen and oxygen bubbles react to water. The pumped fluid volume can be measured by estimating the gas volume from the measurement of the conductivity between electrodes 2 and 3 (Figure 7.47) [108]. Furdui et al. reported a similar system for isolating rare cells in blood [109]. Ateya et al. [110] reported a micropump based on the phase transfer concept described in the previous section. Instead of using heat, the bubbles were generated electrochemically. The pump is a straight microchannel with $25 \times 25\mu m$ cross section and five electrode pairs. The generated bubbles are kept in five corresponding pockets. Applying synchronized voltages ranging from 3.3V to 4.5V can deliver liquids with a maximum flow rate of 24 nL/min and a maximum backpressure of 110 kPa.

Bubble removal from a microfluidic system also serve as a pumping scheme. This method is suitable for devices with in situ spontaneous bubble generation due to chemical or electrochemical reactions. Self-delivery of aqueous fuel was realized for a microfluidic fuel cell through removal of generated bubbles from the systems [111]. In the microfluidic fuel cell running on formic acid and oxygen, bubbles of carbon dioxide are formed as byproducts in reaction chamber. The resulting bubbles could only move toward the outlet in one direction. Once the bubbles are removed through a hydrophobic nanoporous membrane embedded in the reaction chamber, fuel is pumped into the microchannel from the reservoir, Figure 7.48. The nanoporous membrane works as a vent that only allows the removal of gas through its pores. Thus, no liquid leakage through the membrane occurs.

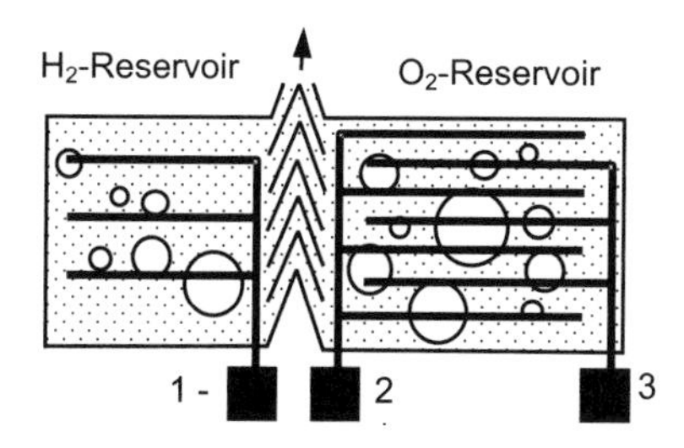


**Figure 7.47** Electrochemical pump for pushing liquid by the pressure of bubbles produced by water electrolysis. (*After*: [108].)

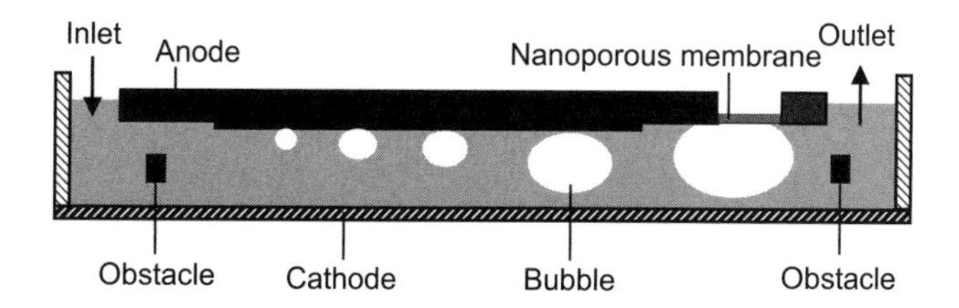


**Figure 7.48** Pumping by removal of generated bubbles from reaction chamber in a microfluidic fuel cell. (*After*: [111].)

Lemoff and Lee [112] realized the magnetohydrodynamic concept in silicon. The pump is able to generate a nonpulsatile flow like that of EHD pumps and electrokinetic pumps. The pump can achieve a maximum flow velocity of 1.5 mm/s (1M NaCl solution, 6.6V). Jang and Lee [113] reported a MHD pump with a channel width on the order of 1 mm. The pump operates with a magnetic flux density of 0.44T provided by a permanent magnet. The pump delivered a maximum flow rate of 63 μL/min and a maximum backpressure of 170 Pa. Eijkel [114] reported a MHD micropump made of a glass/gold/glass system. The channel has a cross section of $200 \times 30$ μm. At a magnetic flux density of 0.1T, the pump can deliver a maximum flow velocity of 40 μm/s.

#### 7.2.2.6 Finger-Powered Pumps

Finger-powered pumping schemes employ the manual power of the user hand to drive fluids through microfluidic systems (Figure 7.49). Fluid pumping is obtained either through direct squeezing a fluid reservoir or compressing an intermediate chamber to push liquid into the microchannel. In some designs, one finger is used for each inlet as *one-finger-per-inlet-port*, which limits this pumping scheme for applications with multiple flows [115]. For some microfluidic applications that required multiple infusion of fluids with more precise control, multiport pumping system with one finger has been implemented. In this method, a combination of microfluidic diodes are embedded within the microfluidic chip enabling pumping in a push-and-release manner [115]. In general, finger-powered pumping is more suitable for a microfluidic system that only requires a few reagents and corresponding pumping sequences.

## 7.3 SUMMARY

Mechanical pumps are the most widely reported micropump types. Most mechanical pumps have a size ranging from 5 mm and 1 cm, due to the relatively large piezoelectric actuator unit. With the use of alternatives such as thermopneumatic actuators [64, 73], electrowetting actuators [103], electrochemical actuators [108], or pneumatic actuators [58] the pump size can be reduced to less

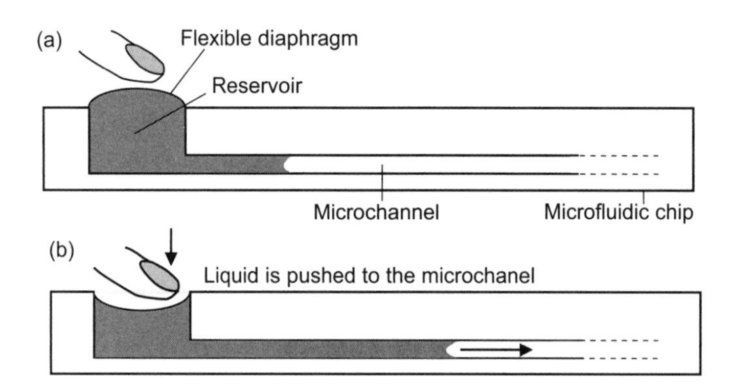


**Figure 7.49** Finger-powered actuation for direct flow pumping: (a) before; (b) after squeezing of the reservoir.

than 100 μm. The availability of fabrication methods such as soft lithography, micromilling and hot embossing has enabled the fabrication of high precision polymeric micropumps with dimensions less than 500 μm. Pneumatic valves taking advantage of the Quake valve [58] is able to generate flow rates less than 1 μL/min accurately.

In addition, low flow rates can be produced by nonmechanical pumps, such as electrohydrodynamic, magnetohydrodynamic, electrokinetic, and electrochemical pumps. The other challenge for micropumps is the relation between pump power and the pump size. Power of mechanical pumps also decreases with miniaturization. Reported data show clearly that the pumped energy varies between 1 nW and 0.1 mW, while the pump size is restricted between 0.6 mm and 10 mm. Smaller thermopneumatic actuators (almost 10 times smaller than piezoelectric actuators) can be integrated in the micromachining process, but the pump power is also 10 times less than that of piezoelectric actuators.

Gas-actuated microvalves have gained popularity for some microfluidic applications due to their robustness and simple operation, ease of fabrication and ability to produce a wide range of flow rates. Nonmechanical pumps have a good potential for lab-on-a-chip applications due to their simplicity and robustness, which is achieved by the lack of moving parts. Electro-osmotic micropumps and mechanical micropumps can deliver comparable performance for the same size. Surface tension effects are promising but still not fully explored. The concept of droplet-based microfluidics promises interesting applications. Despite electrowetting, droplet-based systems can also be realized by effects such as thermocapillary, electrochemical capillary, or photochemical capillary, as discussed in the first section of this chapter.

## Problems

**7.1** Two diffuser/nozzle pumps A and B have the same piezoelectric actuator and pump chamber size. The rectification efficiencies of the pumps are 0.19 and 0.25, respectively. Pump A delivers a maximum flow rate of 11.5 mL/min and a maximum pressure head of 0.5 $mH_2O$. Assuming that both pumps are working with the same frequency, what is the maximum flow rate and maximum pressure head of pump B?

**7.2** Design an ultrasonic micropump using FPW to achieve a pumped fluid layer of 300 microns.

**7.3** The stroke volume of a piezoelectric check-valve pump is estimated by:

$$V_0 = C\frac{Vd^4}{t^2}$$

where $d$ is the diameter of the pump chamber, $V$ is the drive voltage, $t$ is the thickness of the pump membrane, and $C$ is a dimensionless constant depending on the geometry and material of the piezoelectric bimorph. The pump has a pick drive voltage of 200V, the membrane thickness is 200 mm, and the pump chamber diameter is 15 mm. Assuming that $C = 3 \times 10^{-6}$ and the critical pressure for opening the check valve is 10 mbar, determine the largest dead volume allowable for pumping air.

**7.4** For the pump described in Problem 7.3, design a cantilever check valve as depicted in Figure 7.6(b). Assume that the valve is made of parylene and has a thickness of 10 mm. The inlet and outlet orifices are $100 \times 100$ μm in size.

**7.5** In a CD-lab platform, the pressure across the liquid/gas interface in a channel (50 μm wide, 100 μm deep) is 25 mbar. Assuming a surface tension of the liquid/air interface of $71.97 \times 10^{-7}$ J/cm$^2$, determine the number of revolutions per minute needed for bursting a liquid drop of a length of 1 mm at an average radial distance of 2 cm.

**7.6** Determine the pump efficiency of a check-valve pump with an electrostatic actuator. The actuator has a maximum stroke of 10 μm at a peak voltage of 100V, and the gap between the electrodes is filled with air. The pump is able to deliver a maximum flow rate of 850 μL/min and a maximum backpressure of 300 mbar at a pump frequency of 1 kHz. Assume a parallel plate actuator for the solution.

**7.7** For the situation in Example 7.9, determine the flow velocity at a height of 5 mm. How much time does the water column need to reach this height?

## References

[1] Gravesen, P., Brandebjerg, J., and Jensen, O. S., "Microfluidics—A Review," *Journal of Micromechanics and Microengineering*, Vol. 3, 1993, pp. 168–182.

[2] Shoji, S., and Esashi, M., "Microflow Devices and Systems," *Journal of Micromechanics and Microengineering*, Vol. 4, 1994, pp. 157–171.

[3] Geradin, M., and Rixen, D., *Mechanical Vibrations: Theory and Applications to Structural Dynamics*, 2nd ed., New York: Wiley, 1997.

[4] Karassik, I. J., et al., *Pump Handbook*, 2nd ed., New York: McGraw-Hill, 1986.

[5] van Lintel, H. T. G., van den Pol, F. C. M., and Bouwstra, S., "A Piezoelectric Micropump Based on Micromachining in Silicon," *Sensors and Actuators A*, Vol. 15, 1988, pp. 153–167.

[6] Richter, M., Linnemann, R., and Woias, P., "Robust Design of Gas and Liquid Micropumps," *Sensors and Actuators A*, Vol. 68, 1998, pp. 480–486.

[7] Linneman, R., et al., "A Self-Priming and Bubble Tolerant Piezoelectric Silicon Micropump for Liquids and Gases," *Proceedings of MEMS'98, 11th IEEE International Workshop Micro Electromechanical System*, Heidelberg, Germany, Jan. 25–29, 1998, pp. 532–537.

[8] Stemme, E., and Stemme, G., "A Valveless Diffuser/Nozzle-Based Fluid Pump," *Sensors and Actuators A*, Vol. 39, 1993, pp. 159–167.

[9] Nguyen, N. T., and Huang, X. Y., "Miniature Valveless Pumps Based on Printed Circuit Board Technique," *Sensors and Actuators A*, Vol. 88, No. 2, 2001, pp. 104–111.

[10] Gerlach, T., "Microdiffusers as Dynamic Passive Valves for Micropump Applications," *Sensors and Actuators A*, Vol. 69, 1998, pp. 181–191.

[11] Olsson, A., et al., "Micromachined Flat-Walled Valve-Less Diffuser Pumps," *Journal of Micro Electromechanical Systems*, Vol. 6, No. 2, 1997, pp. 161–166.

[12] Doepper, J., et al., "Micro Gear Pumps for Dosing of Viscous Fluids," *Journal of Micromechanics and Microengineering*, Vol. 7, 1997, pp. 230–232.

[13] Madou, M. J., et al., "Design and Fabrication of CD-Like Microfluidic Platforms for Diagnostics: Microfluidic Functions," *Journal of Biomedical Microdevices*, Vol. 3, No. 3, 2001, pp. 245–254.

[14] Moroney, R. M., "Ultrasonic Microtransport," Ph.D. Thesis, University of California Berkeley, 1995.

[15] Nguyen, N. T., et al., "Integrated Flow Sensor for In-Situ Measurement and Control of Acoustic Streaming in Flexural Plate Wave Micro Pumps," *Sensors and Actuators A*, Vol. 79, 2000, pp. 115–121.

[16] Miyazaki, S., Kawai, T., and Araragi, M., "A Piezoelectric Pump Driven by a Flexural Progressive Wave," *Proceedings of MEMS'91, 4th IEEE International Workshop Micro Electromechanical System*, Nara, Japan, Jan. 30–Feb. 4, 1991, pp. 283–288.

[17] Fuhr, G., Schnelle, T., and Wagner, B., "Travelling Wave-Driven Microfabricated Electrohydrodynamic Pumps for Liquids," *Journal Micromechanics Microengineering*, Vol. 4, 1994, pp. 217–226.

[18] Bart, S. F., et al., "Microfabricated Electrohydrodynamic Pumps," *Sensors and Actuators A*, Vol. 21–23, 1990, pp. 193–198.

[19] Manz, A., et al., "Electroosmotic Pumping and Electrophoretic Separations for Miniaturized Chemical Analysis Systems," *Journal of Micromachanics Microengineering*, Vol. 4, 1994, pp. 257–265.

[20] Webster, J. R., et al., "Electrophoresis System with Integrated On-Chip Fluorescence Detection," *Proceedings of MEMS'00, 13rd IEEE International Workshop Micro Electromechanical System*, Miyazaci, Japan, Jan. 23–27, 2000, pp. 306–310.

[21] Namasivayam, V., "Transpiration-Based Micropump for Delivering Continous Ultra-Low Flow Rates," *Journal of Micromechanics and Microengineering*, Vol. 13, 2003, pp. 261–271.

[22] Çengel, Y. A., and Boles, M. A., *Thermodynamics*, 4th ed., New York: McGraw-Hill, 2002.

[23] Nguyen, N. T., and Huang, X. Y., "Thermocapillary Effect of a Liquid Plug in Transient Temperature Fields," *Japanese Journal of Applied Physics*, Vol. 44, 2005, pp. 1139–1142.

[24] Yarin, A. L., Liu, W., and Reneker, D. H., "Motion of Droplets Along Thin Fibers with Temperature Gradient," *Journal of Applied Physics*, Vol. 91, 2002, pp. 4751–4760.

[25] Lippmann, M. G., "Relations Entre Lesénomèn Électriques et Capillares," *Ann. Chim. Phys.*, Vol. 5, 1875, pp. 494–549.

[26] Gallardo, B. S., et al., "Electrochemical Principles for Active Control of Liquids on Submillimeter Scale," *Science*, Vol. 283, 1999, pp. 57–60.

[27] Rosslee, C., and Abbott, N. L. "Active Control of Interfacial Properties," *Current Opinion in Colloid & Interface Science*, Vol. 5, 2000, pp. 81–87.

[28] van't Hoff, J. H., "The Role of Osmotic Pressure in the Analogy Between Solutions and Gases," *Zeitschrift fr physikalische Chemie*, Vol. 1, 1887, pp. 481–508.

[29] van den Pol, F. C. M., et al., "A Thermopneumatic Micropump Based on Micro-Engineering Techniques," *Sensors and Actuators A*, Vol. 21-23, 1990, pp. 198–202.

[30] Mosadegh, B., et al., "Integrated Elastomeric Components for Autonomous Regulation of Sequential and Oscillatory Flow Switching in Microfluidic Devices," *Nature Physics*, Vol. 6, 2010, pp. 433-437.

[31] Pourmand, A., et al., "Fabrication of Whole-Thermoplastic Normally Closed Microvalve, Micro check valve, and Micropump," *Sensors and Actuators B*, Vol. 262, 2018, pp. 625–636.

[32] Zengerle, R., Richter, A., and Sandmaier, H., "A Micromembrane Pump with Electrostatic Actuation," *Proceedings of MEMS'92, 5th IEEE International Workshop Micro Electromechanical System*, Travemuende, Germany, Jan. 25–28,1992, pp. 31–36.

[33] Zengerle, R., et al., "A Bi-Directional Silicon Micropump," *Proceedings of MEMS'95, 8th IEEE International Workshop Micro Electromechanical System*, Amsterdam, The Netherlands, Jan. 29–Feb. 2, 1995, pp. 19–24.

[34] Koch, M., et al., "A Novel Micromachined Pump Based on Thick-Film Piezoelectric Actuation," *Sensors and Actuators A*, Vol. 70, 1998, pp. 98–103.

[35] Wang, X., et al., "A PZT-Driven Micropump," *Proc. of Micro Mechatronics and Human Science 98*, 1998, pp. 269–272.

[36] Rapp, R., et al., "LIGA Micropump for Gases and Liquids," *Sensors and Actuators A*, Vol. 40, 1994, pp. 57–61.

[37] Schomburg, W. K., et al., "Active Valves and Pumps for Microfluidics," *Journal of Micromechanics and Microengineering*, Vol. 3, 1993, pp. 216–218.

[38] Schomburg, W. K., et al., "Microfluidic Components in LIGA Technique," *Journal of Micromechanics and Microengineering*, Vol. 4, 1994, pp. 186–191.

[39] Kämper, K. P., et al., "A Self Filling Low Cost Membrane Micropump," *Proceedings of MEMS'98, 11th IEEE International Workshop Micro Electromechanical System*, Heidelberg, Germany, Jan. 25-29, 1998, pp. 432–437.

[40] Böhm, S., Olthuis, W., and Bergveld, P., "A Plastic Micropump Constructed with Conventional Techniques and Materials," *Sensors and Actuators A*, Vol. 77, 1999, pp. 223–228.

[41] Meng, E., et al., "A Check-Valved Silicone Diaphragm Pump," *Proceedings of MEMS'00, 13th IEEE International Workshop Micro Electromechanical System*, Miyazaci, Japan, Jan. 23–27, 2000, pp. 23–27.

[42] Shoji, S., Nakafawa, S., and Esashi, M., "Micropump and Sample-Injector for Integrated Chemical Analyzing Systems," *Sensors and Actuators A*, Vol. 21–23, 1990, pp. 189–192.

[43] Linnemann, R., Richter, M., and Woias, P., "A Self Priming and Bubble Tolerant Silicon Micropump for Space Research," *Proceedings 2nd Round Table on Micro/Nano Technologies for Space (ESTEC)*, Noordwijk, The Netherlands, Oct. 15–17, 1997, pp. 83–90.

[44] Maillefer, D., et al., "A High-Performance Silicon Micropump for an Implantable Drug Delivery System," *Proceedings of MEMS'99, 12th IEEE International Workshop Micro Electromechanical System*, Orlando, FL, Jan. 17-21, 1999, pp. 541–546.

[45] Gass, V., et al., "Integrated Flow-Regulated Silicon Micropump," *Sensors and Actuators A*, Vol. 43, 1994, pp. 335–338.

[46] Lammerink, T. S. J., Elwenspoek, M., and Fluitman, J. H. J., "Integrated Micro-Liquid Dosing System," *Proceedings of MEMS'93, 6th IEEE International Workshop Micro Electromechanical System*, San Diego, CA, Jan. 25–28,1993, pp. 254–259.

[47] Li, H. Q., et al., "A High Frequency High Flow Rate Piezoelectrically Driven MEMS Micropump," *Tech. Dig. Solid-State Sensor and Actuator Workshop*, Hilton Head, SC, 2000, pp. 69–72.

[48] Dario, P., et al., "A Fluid Handling System for a Chemical Microanalyzer," *Journal of Micromechanics and Microengineering*, Vol. 6, 1996, pp. 95–98.

[49] Guo, S., et al., "Design and Experiments of Micro Pump Using ICPF Actuator," *Proc. of Micro Mechatronics and Human Science 96*, 1996, pp. 235–240.

[50] Guo, S., et al., "A New Type of Capsule Micropump Using ICPF Actuator," *Proc. of Micro Mechatronics and Human Science 98*, 1998, pp. 255–260.

[51] Accoto, D., et al., "Theoretical Analysis and Experimental Testing of Miniature Piezoelectric Pump," *Proc. of Micro Mechatronics and Human Science 98*, 1998, pp. 261–268.

[52] Benard, W. L., et al., "Thin Film Shape-Memory Alloy Actuated Micropumps," *Journal of Micro Electromechanical Systems*, Vol. 7, No. 2, 1998, pp. 245–251.

[53] Smits, J. G., "Piezoelectric Micropump with Three Valves Working Peristaltically," *Sensors and Actuators*, Vol. 15, 1988, pp. 153–167.

[54] Judy, J. W., Tamagawa, T., and Polla, D. L., "Surface-Machined Micromechanical Membrane Pump," *Proceedings of MEMS'91, 3rd IEEE International Workshop Micro Electromechanical System*, Nara, Japan, Jan. 30–Feb. 4, 1991, pp. 182–186.

[55] Folta, J. A., Raley, N. F., and Hee, E. W., "Design Fabrication and Testing of a Miniature Peristaltic Membrane Pump," *Technical Digest of the IEEE Solid State Sensor and Actuator Workshop*, Hilton Head Island, SC, June 22–25, 1992, pp. 186–189.

[56] Mizoguchi, H., et al., "Design and Fabrication of Light Driven Micropump," *Proceedings of MEMS'92, 5th IEEE International Workshop Micro Electromechanical System*, Travemuende, Germany, Jan. 25–28, 1992, pp. 31–36.

[57] , Shaegh, S. A. M., et al., "Plug-and-Play Microvalve and Micropump for Rapid Integration with Microfluidic Chips," *Microfluidics and Nanofluidics*, Vol. 19, No. 3, 2015, pp. 557–564.

[58] Unger, M. A., et al., "Monolithic Microfabricated Valves and Pumps by Multilayer Soft Lithography," *Science*, Vol. 228, pp. 113–116.

[59] Shaegh, S. A. M., et al., "Rapid Prototyping of Whole-Thermoplastic Microfluidics with Built-in Microvalves Using Laser Ablation and Thermal Fusion Bonding," *Sensors and Actuators B: Chemical*, Vol. 255, 2018, pp. 100–109.

[60] Grosjean, C., and Tai, Y. C., "A Thermopneumatic Peristaltic Micropump," *Proceedings of Transducers '99, 10th International Conference on Solid-State Sensors and Actuators*, Sendai, Japan, June 7–10, 1999, pp. 1776–1779.

[61] Cabuz, C., et al., "Mesoscopic Sampler Based on 3D Array of Electrostatically Activated Diaphragms," *Proceedings of Transducers '99, 10th International Conference on Solid-State Sensors and Actuators*, Sendai, Japan, June 7–10, 1999, pp. 1890–1891.

[62] Cabuz, C., et al., "The Dual Diaphragm Pump," *Proceedings of MEMS'01, 14th IEEE International Workshop Micro Electromechanical System*, Interlaken, Switzerland, Jan. 21–25, 2001, pp. 519–522.

[63] Shinohara, J., et al., "A High Pressure Resistance Micropump Using Active and Normally Closed Valves," *Proceedings of MEMS'00, 13th IEEE International Workshop Micro Electromechanical System*, Miyazaci, Japan, Jan. 23–27, 2000, pp. 86–91.

[64] Tsai J. H., and Lin, L., "A Thermal Bubble Actuated Micro Nozzle-Diffuser Pump," *Proceedings of MEMS'01, 14th IEEE International Workshop Micro Electromechanical System*, Interlaken, Switzerland, Jan. 21–25, 2001, pp. 409–412.

[65] Gerlach, T., and Wurmus, H., "Working Principle and Performance of the Dynamic Micropump," *Sensors and Actuators A*, Vol. 50, 1995, pp. 135–140.

[66] Jeong, O. C., and Yang, S. S., "Fabrication and Test of a Thermopneumatic Micropump with a Corrugated p+ Diaphragm," *Sensors and Actuators A*, Vol. 83, 2000, pp. 249–255.

[67] Quandt, E., and Ludwig, A., "Magnetostrictive Actuation in Microsystems," *Sensors and Actuators A*, Vol. 81, 2000, pp. 275–280.

[68] Zhao, B., et al., "A Controllable and Integrated Pump-Enabled Microfluidic Chip and Its Application in Droplets Generating," *Scientific Reports*, Vol. 7, 2017, pp. 1–8.

[69] Forster, F. K., et al., "Design, Fabrication and Testing of Fixed-Valve Micro-Pumps," *Proc. of ASME Fluids Engineering Division*, IMECE'95, Vol. 234, 1995, pp. 39–44.

[70] Bardell, R. L., et al., "Designing High-Performance Micro-Pumps Based on No-Moving-Parts Valves," *Proc. of Microelectromechanical Systems (MEMS) ASME*, DSC-Vol. 62/ HTD-Vol. 354, 1997, pp. 47–53.

[71] Tesla, N., *Valvular Conduit*, U.S. patent 1 329 559, 1920.

[72] Matsumoto, S., Klein, A., and Maeda, R., "Development of Bi-Directional Valve-Less Micropump for Liquid," *Proceedings of MEMS'99, 12th IEEE International Workshop Micro Electromechanical System*, Orlando, FL, Jan. 17–21, 1999, pp. 141–146.

[73] Evans, J., Liepmann, D., and Pisano, A., "Planar Laminar Mixer," *Proceedings of MEMS'97, 10th IEEE International Workshop Micro Electromechanical System*, Nagoya, Japan, Jan. 26–30, 1997, pp. 96–101.

[74] Stehr, M., et al., "A New Micropump with Bidirectional Fluid Transport and Selfblocking Effect," *Proceedings of MEMS'96, 9th IEEE International Workshop Micro Electromechanical System*, San Diego, CA, Feb. 11-15, 1996, pp. 485–490.

[75] Nguyen, N. T., et al., "Hybrid-Assembled Micro Dosing System Using Silicon-Based Micropump/Valve and Mass Flow Sensor," *Sensors and Actuators A*, Vol. 69, 1998, pp. 85–91.

[76] Terray, A., Oakey, J., and Marr, D. W. M., "Microfluidic Control Using Colloidal Devices," *Science*, Vol. 296, 2002, pp. 1841–1844.

[77] Jatch, A., et al., "A Ferrofluidic Magnetic Micropump," *Journal of Microelectromechanical Systems*, Vol. 10, No. 2, 2001, pp. 215–221.

[78] Ahn, C. H., and Allen, M. G., "Fluid Micropumps Based on Rotary Magnetic Actuators," *Proceedings of MEMS'95, 8th IEEE International Workshop Micro Electromechanical System*, Amsterdam, The Netherlands, Jan. 29–Feb. 2, 1995, pp. 408–412.

[79] Caton, P. F., and White, R. M., "MEMS Microfilter with Acoustic Cleaning," *Proceedings of MEMS'01, 14th IEEE International Workshop Micro Electromechanical System*, Interlaken, Switzerland, Jan. 21–25, 2001, pp. 479–482.

[80] Meng, A. H., Nguyen, N. T., and White R. M.. "Focused Flow Micropump Using Ultrasonic Flexural Plate Waves," *Biomedical Microdevices Journal*, Vol. 2, No. 3, 2000, pp. 169–174.

[81] Fuhr, G., et al., "Pumping of Water Solutions in Microfabricated Electrohydrodynamic Systems," *Proceedings of MEMS'92, 5th IEEE International Workshop Micro Electromechanical System*, Travemuende, Germany, Feb. 4–7, 1992, pp. 25–30.

[82] Fuhr, G., "From Micro Field Cages for Living Cells to Brownian Pumps for Submicron Particles," *Proc. of IEEE Micro Mechatronics and Human Science 97*, 1997, pp. 1–4.

[83] Ahn, S. H., and Kim, Y. K., "Fabrication and Experiment of a Planar Micro Ion Drag Pump," *Sensors and Actuators A*, Vol. 70, 1998, pp. 1–5.

[84] Richter, A., et al., "A Micromachined Electrohydrodynamic (EHD) Pump," *Sensors and Actuators A*, Vol. 29, 1991, pp. 159–168.

[85] Furuya, A., et al., "Fabrication of Fluorinated Polyimide Microgrids Using Magnetically Controlled Reactive Ion Etching (MC-RIE) and Their Applications to an Ion Drag Integrated Micropump," *Journal Micromechanics Microengineering*, Vol. 6, 1996, pp. 310–319.

[86] Fuhr, G., et al., "Microfabricated Electrohydrodynamic (EHD) Pumps for Liquids of Higher Conductivity," *Journal of Micro Electromechanical Systems*, Vol. 1, No. 3, 1992, pp. 141–145.

[87] Gunji, M., Jones, T. B., and Washizu, M., "DEP Microactuators of Liquids," *Proceedings of MEMS'01, 14th IEEE International Workshop Micro Electromechanical System*, Interlaken, Switzerland, Jan. 21–25, 2001, pp. 385–388.

[88] Harrison, D. J., et al., "Chemical Analysis and Electrophoresis Systems Integrated on Glass and Silicon Chips," *Technical Digest of the IEEE Solid State Sensor and Actuator Workshop*, Hilton Head Island, SC, June 22–25, 1992, pp. 110–113.

[89] Chen, C. H., et al., "Development of a Planar Electrokinetic Micropump," *2000 ASME International Mechanical Engineering Congress and Exposition*, Orlando, FL, 2000, pp. 523–528.

[90] Morf, W. E., Guenat, O. T., and de Rooij, N. F., "Partial Electroosmotic Pumping in Complex Capillary Systems. Part 1: Principles and General Theoretical Approach," *Sensors and Actuators B*, Vol. 72, 2001, pp. 266–272.

[91] Guenat, O. T., et al., "Partial Electroosmotic Pumping in Complex Capillary Systems. Part 2: Fabrication and Application of a Micro Total Analysis System (mTAS) Suited for Continuous Volumetric Nanotitrations," *Sensors and Actuators B*, Vol. 72, 2001, pp. 273–282.

[92] Tseng F. G., et al., "A Surface-Tension-Driven Fluidic Network for Precise Enzyme Batch-Dispensing and Glucose Detection," *Sensors and Actuators A*, Vol. 111, 2004, pp. 107–117.

[93] Wang, X., et al., "Paper Pump for Passive and Programmable Transport," *Biomicrofluidics*, Vol. 7, No. 1, 2013, pp. 1–11.

[94] Takagi, H., et al., "Phase Transformation Type Micropump," *Proc. of Micro Mechatronics and Human Sciences 94*, 1994, pp. 199–202.

[95] Ozaki, K., "Pumping Mechanism Using Periodic Phase Changes of a Fluid," *Proceedings of MEMS'95, 8th IEEE International Workshop Micro Electromechanical System*, Amsterdam, The Netherlands, Jan. 29–Feb. 2, 1995, pp. 31–36.

[96] Jun, T. K., and Kim, C.-J., "Valveless Pumping Using Traversing Vapor Bubbles in Microchannels," *J. Applied Physics*, Vol. 83, No. 11, 1998, pp. 5658–5664.

[97] Song, Y. J., and Zhao, T. S., "Modelling and Test of a Thermally-Driven Phase-Change Nonmechanical Micropump," *Journal of Micromechanics and Microengineering*, Vol. 11, 2001, pp. 713–719.

[98] Geng X., et al., "Bubble-Based Micropump for Electrically Conducting Liquids," *Journal of Micromechanics and Microengineering*, Vol. 11, 2001, pp. 270–276.

[99] Yokoyama, Y., et al., "Thermal Micro Pumps for a Loop-Type Micro Channel," *Sensors and Actuators A*, Vol. 111, 2004, pp. 123–128.

[100] Darhuber, A. A., et al., "Thermocapillary Actuation of Droplets on Chemically Patterned Surfaces by Programmable Microheater Arrays," *Journal of Microelectromechanical Systems*, Vol. 12, 2003, pp. 873–879.

[101] Tseng, Y. T., et al., "Fundamental Studies on Micro-Droplet Movement by Marangoni and Capillary Effects," *Sensors and Actuators A*, Vol. 114, 2004, pp. 292–301.

[102] Matsumoto, H., and Colgate, J. E., "Preliminary Investigation of Micropumping Based on Electrical Control of Interfacial Tension," *Proceedings of MEMS'90, 2nd IEEE International Workshop Micro Electromechanical System*, Napa Valley, CA, Feb. 11–14, 1990, pp. 105–110.

[103] Lee, J., and Kim, C. J., "Liquid Micromotor Driven by Continuous Electrowetting," *Proceedings of MEMS'98, 11th IEEE International Workshop Micro Electromechanical System*, Heidelberg, Germany, Jan. 25–29, 1998, pp. 538–543.

[104] Ren, H., Fair, R. B., and Pollack, M. G., "Automated On-Chip Droplet Dispensing with Volume Control by Electro-Wetting Actuation and Capacitance Metering," *Sensors and Actuators A*, Vol. 98, 2004, pp. 319–327.

[105] Lee, J., et al., "Electrowetting and Electrowetting-on-Dieclectric for Microscale Liquid Handling," *Sensors and Actuators A*, Vol. 95, 2002, pp. 259–268.

[106] Zhou, C., et al., "Chemistry Pumps: a Review of Chemically Powered Micropumps," *Lab Chip*, Vol. 16, 2016, pp. 1797–1811.

[107] Sengupta, S., et al., "Self-Powered Enzyme Micropump," *Nature Chemistry*, Vol. 6, 2016, pp. 415–422.

[108] Böhm, S., et al., "A Closed-Loop Controlled Electrochemically Actuated Micro-Dosing System," *Journal of Micromechanics and Microengineering*, Vol. 10, 2000, pp. 498–504.

[109] Furdui, V. I., Kariuki, J. K., and Harrison, D. J., "Microfabricated Electrolysis Pump System for Isolating Rare Cells in Blood," *Journal of Micromechanics and Microengineering*, Vol. 13, 2003, pp. S164–S170.

[110] Ateya, D. A., Shah, A. A., and Hua S. Z., "An Electrolytically Actuated Micropump," *Review of Scientific Instruments*, Vol. 75, 2004, pp. 915–920.

[111] Hur, J. I., and Kim. C. -J., "Miniature Fuel-Cell System Complete with On-Demand Fuel and Oxidant Supply," *Journal of Power Sources*, Vol. 274, 2015, pp. 916–921.

[112] Lemoff, A. V., and Lee, A. P., "An AC Magnetohydrodynamic Micropump," *Sensors and Actuators B*, Vol. 63, 2000, pp. 178–185.

[113] Jang, J. S., and Lee, S. S., "Theoretical and Experimental Study of MHD (Magnetohydrodynamic) Micropump," *Sensors and Actuators A*, Vol. 80, 2000, pp. 84–89.

[114] Eijkel, J. C. T., "A Circular AC Magnetohydrodyanmic Micropump for Chromatographic Applications," *Sensors and Actuators B*, Vol. 92, 2003, pp. 215–222.

[115] Iwai, K., et al., "Finger-Powered Microfluidic Systems Using Multilayer Soft Lithography and Injection Molding Processes," *Lab Chip*, Vol. 14, 2014, pp. 3790–3799.

# Chapter 8

# Microfluidics for Internal Flow Control: Microflow Sensors

## 8.1 DESIGN CONSIDERATIONS

### 8.1.1 Design Parameters

Microflow sensors are another important component for controlling fluid flow in a microfluidic system. Microflow sensors could complement microvalves and micropumps in a control loop. The function of advanced microfluidic systems depends on the precise monitoring and control of flow rates such as monitoring the flow rate of circulating cell culture medium in microfluidic bioreactors and organ-on-chip devices. The flow rate has a direct impact on flow-induced shear stress over the cultured cells in a bioreactor, as well as on the delivery of oxygen and nutrients to the cells. For such applications, microflow sensors should present a robust performance for continuing in situ measurements and easy integration within the microfluidic chips.

Since flow measurement is a classical field of measurement technology, the sensing principles include almost all fields of physics (Figure 8.1). Similar to other microfluidic components, the first approach of designing microflow sensors follows conventional concepts, which readily exist in large scales. Governed by their small geometry, the biggest advantages of microflow sensors are the low energy consumption, the small device footprint, and the ability to measure very small flow rates [1], on the order of nanoliters per minute to microliters per minute.

Different parameters determine the performance of a microflow sensor: the dynamic range of operation, sensitivity, response time, power consumption, biocompatibility, and chemical compatibility. The range of operation is limited by the maximum flow velocity or flow rate that the sensor can measure. The flow range can be adjusted by geometrical parameters, sensor materials, and the sensing principle. The sensitivity is defined as the derivative of the sensor signal $V$ with respect to the flow rate $\dot{Q}$ or flow velocity $u$:

$$S = \frac{\mathrm{d}V}{\mathrm{d}\dot{Q}} \text{ or } S = \frac{\mathrm{d}V}{\mathrm{d}u} \quad (8.1)$$

For nonlinear characteristics, one can use the zero-sensitivity $S_0$ at zero flow rate:

$$S_0 = \left.\frac{\mathrm{d}V}{\mathrm{d}\dot{Q}}\right|_{\dot{Q}\to 0} \text{ or } \left.\frac{\mathrm{d}V}{\mathrm{d}u}\right|_{u\to 0} \quad (8.2)$$

The response time is defined as the time that the sensor signal requires to stabilize after a change in flow rate. The power consumption depends on the sensor principle. Because of their tiny size,

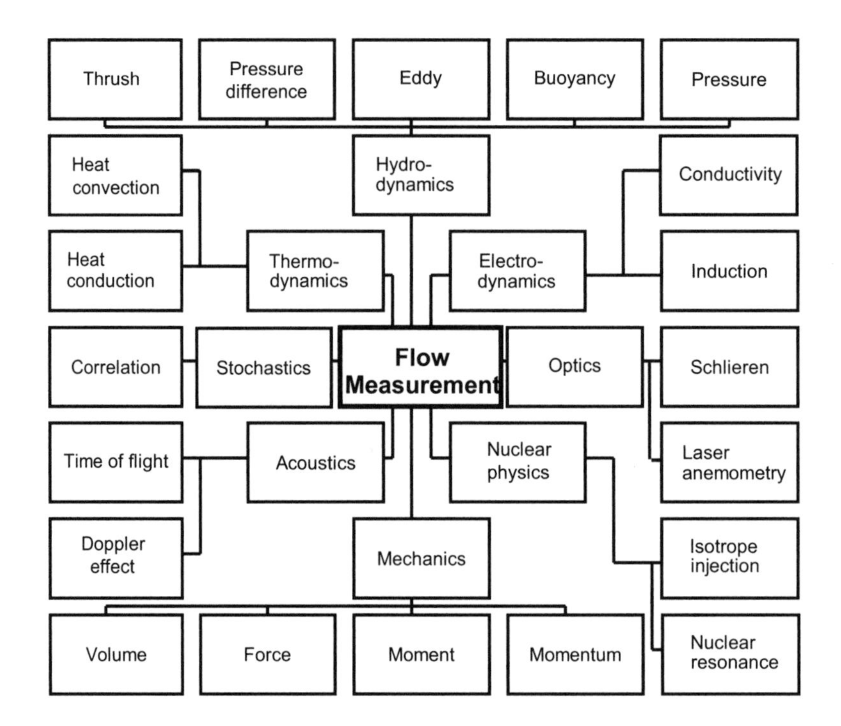


**Figure 8.1** Physical principles of flow measurement.

micromachined thermal flow sensors often consume an electrical power on the order of several milliwatts.

Almost all physical effects are temperature-dependent. Since thermal transfer accompanies the mass transfer of a fluid flow, flow sensors based on thermal principles are most common. Thermal transducers have a wide range of detecting effects. They are simple to implement in silicon technology and represent the majority of developed microflow sensors. Microflow sensors are categorized in this chapter as nonthermal and thermal flow sensors. Micromachined nonthermal or mechanical flow sensors are mainly based on the following principles: differential pressure in microchannels or orifices, drag force, lift force, Coriolis force, and electrohydrodynamics. Thermal flow sensors are categorized as hot film, hot wire, calorimetric sensor, and time-of-flight sensor. Each type can be further categorized by their thermal sensing principles, as thermoresistive, thermocapacitive, thermoelectric, thermoelectronic, and pyroelectric.

### 8.1.2 Nonthermal Flow Sensors

#### 8.1.2.1 Differential Pressure Flow Sensors

Because of the small channel size and, consequently, the small hydraulic diameter as well as a small Reynolds number, flows in microchannels are in most cases laminar. The pressure loss $\Delta p$ along a microchannel is therefore a linear function of the linear flow velocity $u$ (see Section 2.2.4.1):

$$\Delta p = \mathrm{Re} f \frac{\eta L}{2D_{\mathrm{h}}^2} u \tag{8.3}$$

where Re is the Reynolds number, $f$ is the friction coefficient (Re$f = 64$ for circular cross section, Re$f = 50...60$ for rectangular microchannels), $L$ is the channel length, $\eta$ is the dynamic viscosity of the fluid, and $D_{\mathrm{h}}$ is the hydraulic diameter (see Chapter 2).

According to (8.3), the pressure loss is also a linear function of the flow rate, which is the product of the flow velocity and the channel cross section. In (8.3), the dynamic viscosity is the only parameter that is strongly temperature-dependent. Thus, a microflow sensor based on the pressure loss should have integrated sensors for temperature compensation.

Designing a differential pressure flow sensor actually involves designing pressure sensors. The pressure differences are evaluated by the deformation of a sensing element such as a thin membrane. The deformation produces displacements, stresses, and strains, which are evaluated capacitively or piezoresistively to give a calibrated measurement of the pressure.

*Capacitive Sensors.* A capacitive pressure sensor consists of a membrane and a counterelectrode. Pressure load causes the membrane to deform. The deformation leads to a change of the distance between the membrane and its counterelectrode. If the membrane serves as an electrode of a capacitor, the initial value of the capacitor is:

$$C = \frac{\varepsilon_0 \varepsilon_r A}{d} \tag{8.4}$$

where $\varepsilon_0$ and $\varepsilon_r$ are the permittivity of vacuum and the relative permittivity of the material between two electrodes, respectively, $A$ is the overlapping electrode surface, and $d$ is the gap between them. The counterelectrode is stationary and is typically formed by a deposited metal layer. The change of capacitance $\Delta C$ in terms of change in the gap width $\Delta d$ is:

$$\Delta C = \frac{\mathrm{d}C}{\mathrm{d}d} \Delta d = \frac{-\varepsilon_0 \varepsilon_r A \Delta d}{d^2} \tag{8.5}$$

**Example 8.1: Designing a Capacitive Differential Pressure Flow Sensor**

A differential pressure flow sensor is to be developed after the concept shown below. Each of the pressure sensors is made of a circular silicon membrane with a radius of 2 mm. The membrane is 20 μm thick. The initial gap between the sensor electrode is 20 μm. The flow channel has a cross-section area of 100 × 100 μm. The sensor should measure air flow rate up to 100 mL/min. Dynamic viscosity and density of air are $1.82 \times 10^{-5}$ N·sec/m$^2$ and 1 kg/m$^3$, respectively. Determine: (a) the maximum capacitance change of the pressure sensor, and (b) the length of the flow channel.

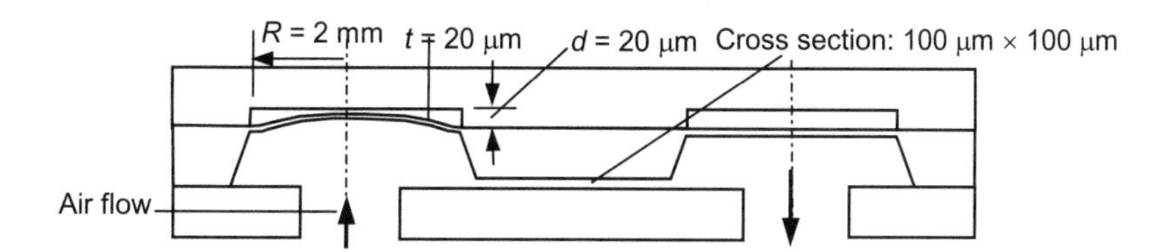


**Solution.** If the deflection of the membrane is smaller than its thickness, the deflection follows the function (see Appendix D):

$$y(r) = \tfrac{2(1-\nu^2)}{16Et^3}(R^2 - r^2)^2 p \text{ (a)}$$

Thus, the maximum deflection at membrane center is:

$$y_0 = y(0) = \tfrac{2(1-\nu^2)}{16Et^3} R^4 p \text{ (b)}$$

The maximum pressure that the sensor can work with is:

$$p_{\max} = \frac{y_0 16Et^3}{3(1-nu^2)R^4} = \frac{20\times10^{-6}\times16\times170\times10^9\times(20\times10^{-6})^3}{3(1-0.25^2)(2\times10^{-3})^4} = 9,671 \text{ Pa}$$

Combining (8.5) with function (a) and integrating the capacitance function across the radial axis $x$ as well as over 360°, the capacitance change can be derived as a function of pressure:

$$\frac{\Delta C}{C} = \frac{(1-\nu^2)R^4}{16Edt^3}p \text{ (c)}$$

Thus, the maximum capacitance change is:

$$\begin{aligned}\frac{\Delta C_{\max}}{C_0} &= \frac{(1-\nu^2)R^4}{16Edt^3}p_{\max}\\ &= \frac{(1-0.25^2)(2\times10^{-3})^4}{16\times170\times10^9\times20\times10^{-6}\times(20\times10^{-6})^3}\times9,671 = 0.33 = 33\%\end{aligned}$$

The maximum mean velocity in the channel is:

$$u_{\max} = \frac{\dot{Q}_{\max}}{A} = \frac{100\times10^{-6}/60}{(100\times10^{-6})^2} = 16.7 \text{ m/s}$$

The Reynolds number at this velocity is (hydraulic diameter $D_{\mathrm{h}} = 100$ μm):

$$\mathrm{Re}_{\max} = \frac{u_{\max}D_{\mathrm{h}}\rho}{\eta} = \frac{16.7\times10^{-4}\times1}{1.82\times10^{-5}} = 918$$

Therefore, it is expected that the flow in the sensor channel is laminar. The required channel length is derived as ($\mathrm{Re}f = 55$ for rectangular microchannels):

$$\begin{aligned}\Delta p = \mathrm{Re}f\frac{\eta uL}{2D_{\mathrm{h}}^2} \rightarrow L &= \frac{1}{\mathrm{Re}f}\frac{2\Delta p_{\max}D_{\mathrm{h}}^2}{\eta u_{\max}}\\ &= \frac{2\times9.671\times(10^{-4})^2}{55\times1.82\times10^{-5}\times167} = 1.16\times10^{-3} \text{ m} = 1.16 \text{ mm}\end{aligned}$$

*Piezoresistive Sensors.* Piezoresistive sensors are the most common types of pressure sensors. The piezoresistive effect is a change in the electric resistance $\Delta R/R$ of a material when stresses or strains $\varphi$ are applied:

$$\frac{\Delta R}{R} = \alpha\varphi \tag{8.6}$$

where $\alpha$ is the piezoresistive coefficient or the gauge factor. If the strain is parallel to the direction of the current flow, $\alpha$ is the longitudinal gauge factor. If the strain is perpendicular to the direction of the current flow, $\alpha$ is the transverse gauge factor. Typical longitudinal gauge factors are on the order of 2 for metals, 10 to 40 for polycrystalline silicon, and 50 to 150 for single-crystalline silicon.

**Example 8.2: Designing a Piezoresistive Differential Pressure Flow Sensor**

If the piezoresistive concept is used for the same geometry of Example 8.1, is the signal detectable?

**Solution.** The stress along the radial axis of the circular can be estimated for small deflection as:

$$\sigma r = \frac{3R^2}{8t^2}\left[(3+\nu)\frac{x^2}{r^2} - (1+\nu)\right]p$$

Thus, the maximum stress on the membrane edge at the pressure of 9,671 Pa is:

$$\sigma_{\max} = \frac{3R^2}{4t^2}p = \frac{3 \times 2,000^2}{4 \times 20^2} \times 9,671 = 7,253.25 \times 10^4 \text{ Pa}$$

The strain at the membrane edge is estimated as:

$$\varphi_{\max} = \frac{\sigma_{\max}}{E/(1-\nu)} = \frac{7,253.25 \times 10^4}{170 \times 10^9/(1-0.25)} = 32 \times 10^{-5}$$

With a piezoresistive coefficient on the order of 100, the maximum change of resistance would be:

$$\frac{\Delta R_{\max}}{R} = \alpha \times \varphi_{\max} = 100 \times 32 \times 10^{-5} = 3.2 \times 10^{-2} = 3.2\%$$

The change is not comparable to the capacitive concept in Example 8.1, but is still good for detection. Attention should be paid to the temperature dependence of the resistors. The design should include measures for temperature compensation such as the integration of temperature sensors.

#### 8.1.2.2 Drag Force Flow Sensors

The flow can be measured indirectly by the drag force using an obstacle such as a rectangular plate in the flow. In laminar flow conditions, with a small Reynolds number, the drag force parallel to the flow direction is given by:

$$F_1 = C_1 L u \eta \tag{8.7}$$

where $F_1$ is the drag force, $C_1$ is a constant depending on the form of the cantilever, $L$ is the dimension of the cantilever, $u$ is the flow velocity, and $\eta$ is the dynamic viscosity of the fluid. The flow rate $\dot{Q}$ and the gap area around the obstacle $A_g$ determine the average velocity $u$. Thus, the drag force is a linear function of the flow rate $\dot{Q}$, and (8.7) becomes:

$$F_1 = \frac{C_1 L \eta \dot{Q}}{A_g} \tag{8.8}$$

The pressure loss in the gap causes an additional force on the obstacle:

$$F_2 = \frac{C_2 \rho \dot{Q}^2}{A_0 A_g^2} \tag{8.9}$$

where $A_0$ is the surface of the obstacle. The total force, which can be evaluated for detecting the flow, is the sum of the two forces $F_1$ and $F_2$:

$$F = F_1 + F_2 = \frac{C_1 L \eta \dot{Q}}{A_g} + \frac{C_2 \rho}{A_0 A_g^2}\dot{Q}^2 \tag{8.10}$$

For very low flow rates, the linear term dominates and the sensor has linear characteristics. For higher flow rates, the quadratic term dominates, and the sensor has quadratic characteristics.

**Example 8.3: Designing a Piezoresistive Drag Force Flow Sensor**

A drag force flow sensor consists of a 1-mm × 2-mm × 10-μm rectangular beam and integrated piezoresistive sensors. Water flow passes through a gap on the tip of the beam. The gap has a hydraulic diameter of 100 μm. Water is fed into the sensor by a circular tube with diameter of 400 μm. Determine the resistive change of the sensor at a volume flow rate of 1 μL/min. Is the change detectable? If the noise-signal ratio is on the order of 0.1%, what is the resolution of the sensor ($\eta_{\text{water}} = 10^{-3}$ kg/m-s)?

**Solution.** The average velocity at the beam tip is:

$$u = \frac{4\dot{Q}}{\pi D_{\text{h}}^2} = \frac{4 \times 10^{-9}/60}{\pi (10^{-4})^2} = 2.1 \times 10^{-3} \text{ m/s}$$

At this low velocity and low dimension, the Reynolds number is small. Thus, the quadratic term of (8.10) is negligible. Using (8.8) and assuming an obstacle dimension of 400 μm, for a circular obstacle ($C_1 = 16$) the force at the tip of the beam is:

$$F = C_1 L \eta u = 16 \times 400 \times 10^{-6} \times 10^{-3} \times 2.1 \times 10^{-3} = 13.44 \times 10^{-9} \text{ N}$$

The stress on the beam surface is estimated as:

$$\sigma = \frac{6FL_{\text{beam}}}{w_{\text{beam}} t_{\text{beam}}^2} = \frac{6 \times 13.44 \times 10^{-9} \times 2 \times 10^{-3}}{10^{-3}(10^{-5})^2} = 1,612.8 \text{ Pa}$$

With a piezoresistive coefficient on the order of 100, the change of resistance would be:

$$\frac{\Delta R}{R} = \alpha\varphi = \frac{\alpha\sigma}{E} = \frac{100 \times 1,612.8}{170 \times 10^9} = 0.95 \times 10^{-6} \approx 10^{-5}$$

The change on the order of $10^{-5}$ is too small and is under the noise level. With the noise-signal ratio of $10^{-3}$, the resolution of the sensor is:

$$\dot{Q}_{\text{max}} = \frac{1 \times 10^{-3}}{10^{-5}} = 100 \text{ μL/min} = 0.1 \text{ mL/min}$$

8.1.2.3 Lift Force Flow Sensors

In contrast to the drag force, the lift force acts perpendicularly to the flow direction. The general expression of the lift force is (see Section 5.3.1):

$$F_{\text{L}} = \frac{C_{\text{L}} A \rho u^2}{2} \tag{8.11}$$

where $C$ is the lift force coefficient, $A$ is the surface area of the airfoil structure, $\rho$ is the fluid density, and $u$ is the free stream velocity. It is obvious that this sensor type has quadratic characteristics.

8.1.2.4 Coriolis Flow Sensors

A true mass flow sensor can be achieved with the Coriolis concept. The basic concept is shown in Figure 8.2. The fluid flows through a U-shaped tube. The tube is fixed at the two ends and oscillates up and down. The twisting Coriolis force that acts on the tube can be written as:

$$F_{\text{C}} = 2\dot{m} \times \omega \tag{8.12}$$

where $F_{\text{C}}$ is the force per unit tube length, $\dot{m}$ is the mass flow rate in the tube, and $\omega$ is the angular velocity of the oscillating tube (Figure 8.2).

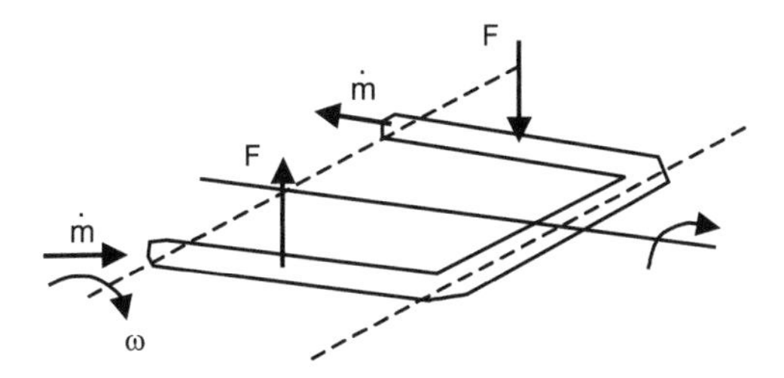


**Figure 8.2** Concept of a Coriolis flow sensor.

### Example 8.4: Designing a Coriolis Flow Sensor

A Coriolis flow sensor is designed based on the concept shown in Figure 8.2. The channel dimension is 1 × 1 mm. The oscillating frequency is 100 Hz. Determine the force on each side of the U-shaped tube at a water flow rate of 10 mL/min.

**Solution.** Assuming a density of 1,000 kg/m$^3$, the mass flow rate in the sensor is:

$$\dot{m} = \dot{Q}\rho = \frac{10 \times 10^{-6}}{60} \times 1,000 = 1.67 \times 10^{-4} \text{ kg/s}$$

The bending force on each side of the U-shaped tube is (the mass flow rate and the vibration axis are perpendicular):

$$F_{\text{C}} = 2\dot{m}\omega = 4\pi\dot{m}f_{\text{osc.}} = 4\pi \times 1.67 \times 10^{-4} \times 100 = 0.21\text{N}$$

### 8.1.3 Thermal Flow Sensors

In Section 8.1.2, a mechanical sensor converts the flow energy directly to mechanical variables such as forces and pressures. Transducers such as piezoresistors convert the mechanical variables into electrical signals. In contrast, thermal flow sensors convert the flow energy indirectly over heat transfer into electrical signals. Because of their structural and electronic simplicity, thermal sensors can be easily integrated into the micromachining fabrication process.

A thermal flow sensor generally consists of a heater and one or more temperature sensors. The effects caused by heat transfer are evaluated in different ways, which correspond to the operating modes of a thermal flow sensor. Measuring the heating power or heater temperature and feeding back to heating current allow the heater to be controlled in two modes: constant heating power and constant heater temperature. The flow velocity is evaluated by measuring heating power, heater temperature, fluid temperature, and time of flight of a heat pulse. Two heater control modes and three evaluation modes result in six operating modes shown in Table 8.1 and three types of thermal flow sensors: hot-wire (hot-film) type, calorimetric type, and time-of-flight type. The following analytical models explain the concept of these operation modes in detail.

#### 8.1.3.1 Sensor Models

A 2-D model of the temperature distribution allows the evaluation of both flow velocity and flow direction. Figure 8.3 shows the 2-D model of a thermal flow sensor. The sensor consists of a circular heater. The heater can be modeled as a linear source with a heat rate of $q'$ (W/m) [Figure 8.3(a)] or as a constant temperature $T_{\text{h}}$ at the heater circumference [Figure 8.3(b)]. For simplification, the flow direction is assumed to be $\theta = 0$. The fluid flow is assumed to have a uniform velocity of $u$. The inlet temperature is the reference ambient temperature $T_0$. Boundary layer effects and hydrodynamic effects behind the heater are neglected in this model.

**Table 8.1**

Operational Modes of Thermal Mass Flow Sensors

| *Heater Controls* | *Constant Heating Power* | | *Constant Heater Temperature* | |
|---|---|---|---|---|
| *Evaluation* | *Heater temperature* | *Temperature difference* | *Heating power* | *Temperature difference* |
| *Operational modes* | Hot-wire and hot-film type | Calorimetric type | Hot-wire and hot-film type | Calorimetric type |
| | Time-of-flight type | | Time-of-flight type | |

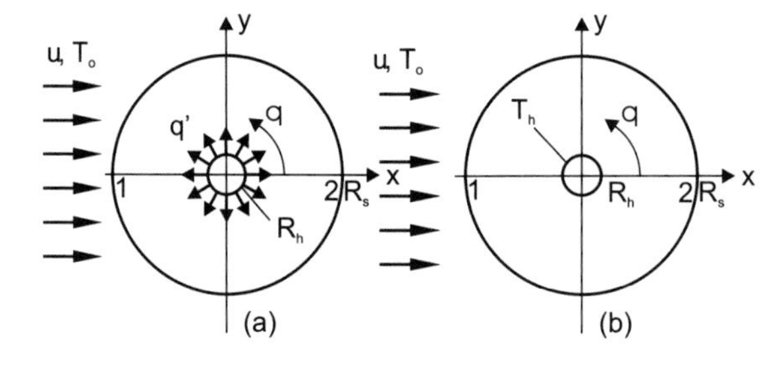


**Figure 8.3** A 2-D model of temperature distribution around a heater: (a) constant heat rate; and (b) constant heater temperature.

Ignoring the source term for a homogenous form, the 2-D governing equation for the temperature field of the model in Figure 8.3 is:

$$\frac{\partial^2 T}{\partial x^2} + \frac{\partial^2 T}{\partial y^2} = \frac{v}{a}\frac{\partial T}{\partial x} \tag{8.13}$$

where $a$ is the thermal diffusivity of the fluid. The temperature difference $\Delta T$ is a product of a velocity-dependent part and a symmetrical part $\Psi$:

$$\Delta T = T - T_0 = \exp\left(\frac{vx}{2\alpha}\right)\Psi(x,y) \tag{8.14}$$

Substitution of:

$$T = T_0 + \Delta T = T_0 + \exp\left(\frac{vx}{2\alpha}\right)\Psi(x,y) \tag{8.15}$$

in (8.13) results to:

$$\frac{\partial^2 \Psi}{\partial x^2} + \frac{\partial^2 \Psi}{\partial y^2} = \left(\frac{v}{2\alpha}\right)\Psi \tag{8.16}$$

with the boundary condition:

$$\left.\frac{\partial \Psi}{\partial x}\right|_{x=\pm\infty} = 0\,, \left.\frac{\partial \Psi}{\partial y}\right|_{y=\pm\infty} = 0$$

Considering the radius $r = \sqrt{x^2 + y^2}$ around the center of the system, the boundary condition for a linear heat source is:

$$\left.\frac{\partial T}{\partial r}\right|_{r=R_{\mathrm{h}}} = -\frac{q'}{2\pi R_{\mathrm{h}} k}$$

In the case of a constant heater temperature, the boundary condition is:

$$T\Big|_{r=R_\mathrm{h}} = T_\mathrm{h}$$

Since the boundary conditions and $\Psi$ are symmetric with respect to $r$, (8.16) can be formulated in the cylindrical coordinate system as:

$$\frac{d^2\Psi}{dr^2} + \frac{1}{r}\frac{d\Psi}{dr} - \left(\frac{v}{2\alpha}\right)^2 \Psi = 0 \tag{8.17}$$

The solution of (8.17) is the modified Bessel function of the second kind and zero order:

$$\Psi = K_0[vr/(2\alpha)] \tag{8.18}$$

In the following analysis, the results are nondimentionalized by introducing the Peclet number $\mathrm{Pe} = 2vR_\mathrm{h}/\alpha$ with the diameter of the heater $2R_\mathrm{h}$ as the characteristic length. The radial variable is nondimentionalized by $R_\mathrm{h}$: $r^* = r/R_\mathrm{h}$.

*Temperature Field with a Constant Heating Power.* Considering the constant linear heat source $q'$ (in W/m), the solution for the temperature difference is:

$$\Delta T(r,\theta) = \frac{q'\alpha}{\pi k R_\mathrm{h}} v^{-1} \frac{K_0[vr/(2\alpha)]}{K_1[vR_\mathrm{h}/(2\alpha)] - K_0[vR_\mathrm{h}/(2\alpha)]\cos\theta} \frac{\exp[vr\cos\theta/(2\alpha)]}{\exp[vR_\mathrm{h}\cos\theta/(2\alpha)]} \tag{8.19}$$

If the dimensionless temperature difference is defined as

$$\Delta T^* = \frac{\Delta T}{2q'/(\pi k)}$$

the dimensionless solution of the temperature field at a constant heating power is:

$$\Delta T^*(r^*,\theta) = \frac{K_0(\mathrm{Pe}r^*/4)/\mathrm{Pe}}{K_1(\mathrm{Pe}/4) - K_0(\mathrm{Pe}/4)\cos\theta}\{\exp[\mathrm{Pe}(r^*-1)/4]\}^{\cos\theta} \tag{8.20}$$

where $K_1$ is the Bessel function of the second kind and first order. The heater temperature is a function of $\theta$ and can be expressed in the dimensionless form as:

$$\Delta T_h^*(1,\theta) = \frac{K_0(\mathrm{Pe}/4)/\mathrm{Pe}}{K_1(\mathrm{Pe}/4) - K_0(\mathrm{Pe}/4)\cos\theta} \tag{8.21}$$

The temperature difference $\Delta T_s^*$ between the two positions 1 and 2 on the evaluation ring across the heater in flow direction is calculated as:

$$\begin{aligned}\Delta T_s^* &= \Delta T_2^* - \Delta T_1^* = \Delta T^*(R_s^*,0) - \Delta T^*(R_s^*,\pi) \\ &= \frac{K_0(\mathrm{Pe}R_s^*/4)}{\mathrm{Pe}}\left\{\frac{\exp[\mathrm{Pe}(R_s^*-1)/4]}{K_1(\mathrm{Pe}/4)-K_0(\mathrm{Pe}/4)} - \frac{\exp[-\mathrm{Pe}(R_s^*-1)/4]}{K_1(\mathrm{Pe}/4)+K_0(\mathrm{Pe}/4)}\right\}\end{aligned} \tag{8.22}$$

Since $\Delta T_h^*$ and $\Delta T_s^*$ are functions of Peclet numbers or the velocities of a given fluid, they are used for measuring the one-dimensional flow. The corresponding methods are called the hot-wire concept and calorimetric concept, respectively.

*Temperature Field with a Constant Heater Temperature.* Considering the constant temperature boundary condition, the solution for the temperature difference is:

$$\Delta T(r,\theta) = \Delta T_{\mathrm{h}} \frac{K_0[vr/(2\alpha)]}{K_0[vR_{\mathrm{h}}/(2\alpha)]} \frac{\exp\left[vr\cos\theta/(2\alpha)\right]}{\exp\left[vR_{\mathrm{h}}\cos\theta/(2\alpha)\right]} \tag{8.23}$$

By introducing the dimensionless temperature difference:

$$\Delta T^* = \Delta T/\Delta T_{\mathrm{h}}$$

the dimensionless temperature field at a constant heater temperature has the form:

$$\Delta T^*(r^*,\theta) = \frac{K_0(\mathrm{Pe}r^*/4)}{K_0(\mathrm{Pe}/4)} \{\exp[\mathrm{Pe}(r^*-1)/4]\}^{\cos\theta} \tag{8.24}$$

For the case of the flow sensor presented in this section, the heat rate can be represented by the dimensionless Nusselt number (Nu). Assuming a linear heat source $q'$ and a characteristic length of $2R_{\mathrm{h}}$, the Nusselt number can be defined as:

$$\mathrm{Nu} = \frac{q'}{\pi k \Delta T_{\mathrm{h}}} \tag{8.25}$$

Using the condition at the heater circumference:

$$\frac{q'(\theta)}{2\pi R_{\mathrm{h}} k} = -\left.\frac{dT(r,\theta)}{dr}\right|_{r=R_{\mathrm{h}}} \tag{8.26}$$

the heat rate at the heater circumference is:

$$q'(\theta) = \frac{\pi k \Delta T_{\mathrm{h}}}{2}\left[\cos\theta - \frac{K_1(\mathrm{Pe}/4)}{K_0(\mathrm{Pe}/4)}\right] \tag{8.27}$$

From (8.25), the average Nusselt number of the heater is:

$$\mathrm{Nu} = \frac{1}{\pi k \Delta T_{\mathrm{h}}} \frac{1}{2\pi} \int_{-\pi}^{\pi} q'(\theta)d\theta = \frac{1}{2}\mathrm{Pe}\frac{K_1(\mathrm{Pe}/4)}{K_0(\mathrm{Pe}/4)} \tag{8.28}$$

The temperature difference between two positions upstream and downstream at a constant heater temperature is:

$$\Delta T_s^* = \frac{K_0(\mathrm{Pe}R_s^*/4)}{K_0(\mathrm{Pe}/4)} \{\exp[\mathrm{Pe}(R_s^*-1)/4] - \exp[-\mathrm{Pe}(R_s^*-1)/4]\} \tag{8.29}$$

Both (8.28) and (8.29) are used in conventional hot-wire sensors and calorimetric sensors to determine the flow velocity, which is represented by the Peclet number.

#### 8.1.3.2 Hot-Wire and Hot-Film Sensors

The effects observed by the analytical model leads to the operation concept of hot-wire and hot-film sensors. Forced convection causes an increase of the heating power at a constant heater temperature and a decrease of the heater temperature at a constant heating power. Figure 8.4 presents typical characteristics of a hot-film sensor. The heating power $P$ for a heated plate with a surface $A_{\mathrm{h}}$ is

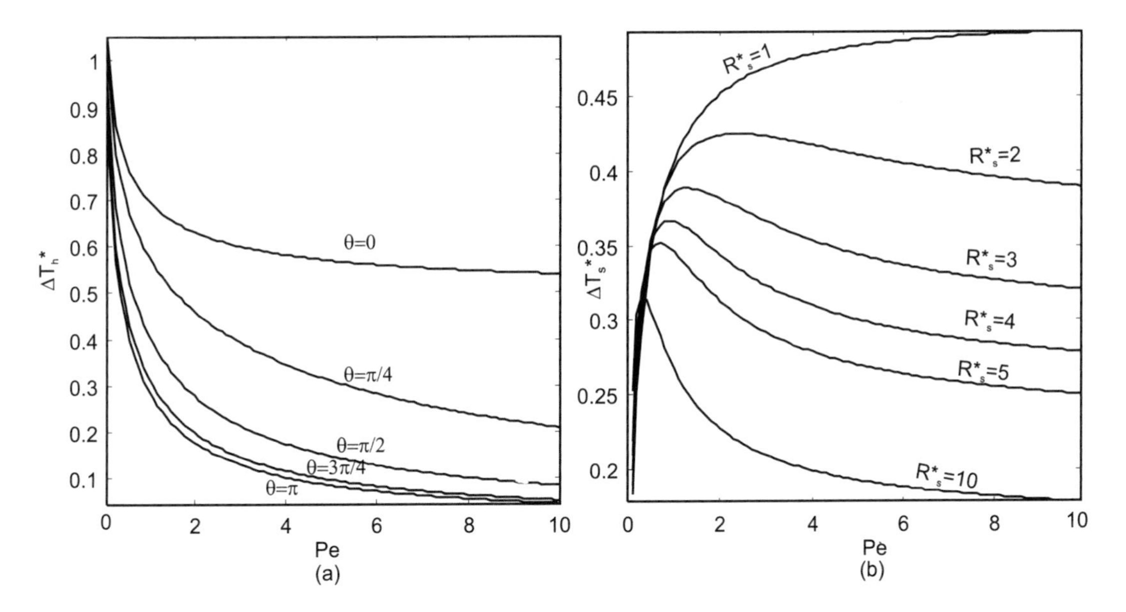


**Figure 8.4** Hot-film and hot-wire sensors: (a) heating power as a function of the flow rate with a constant heater temperature; and (b) heater temperature as a function of the flow rate with a constant heating power.

given as [2]:

$$P = \frac{\mathrm{Nu}\kappa A_{\mathrm{h}}\Delta T}{L} \tag{8.30}$$

where $\kappa$ is the thermal conductivity of the fluid, $A_{\mathrm{h}}$ is the heated area, $\Delta T$ is the temperature difference between the heated body and the ambient, and $L$ is the characteristic length of the body. For a laminar flow, the dimensionless Nusselt number is estimated for flat plates and wires as:

$$\mathrm{Nu} = 0.664\sqrt{\mathrm{Re}}\sqrt[3]{\mathrm{Pr}} \tag{8.31}$$

where Re and Pr are the dimensionless Reynolds number and Prandtl number of the fluid, respectively.

#### 8.1.3.3 Calorimetric Sensors

Calorimetric sensors are thermal flow sensors that measure the asymmetry of temperature profiles around the heater, which is modulated by the fluid flow. Figure 8.5(a) shows the theoretical distribution of the temperature field around a heater. Figure 8.5(a) shows the measured temperature field on a silicon membrane with a heater in the center. In the case of zero flow, the temperature profile is symmetric. When a flow from right to left exists, the temperature field becomes asymmetric. Two temperature sensors upstream and downstream can measure this asymmetry, and, consequently, the flow rate.

**Example 8.5: Characteristics of a Calorimetric Flow Sensor**

The flow sensor described in Figure 8.6 has a distance of 1 mm between the temperature sensor and the heater. The channel height is 0.5 mm. The heater is 0.5 mm wide. If the heater temperature is kept at a constant value of 80K over the fluid temperature, what is the temperature difference between the two temperature sensors at a flow velocity of 100 mm/s? The fluid is assumed to be water with a thermal diffusivity of $0.6 \times 10^{-3}$.

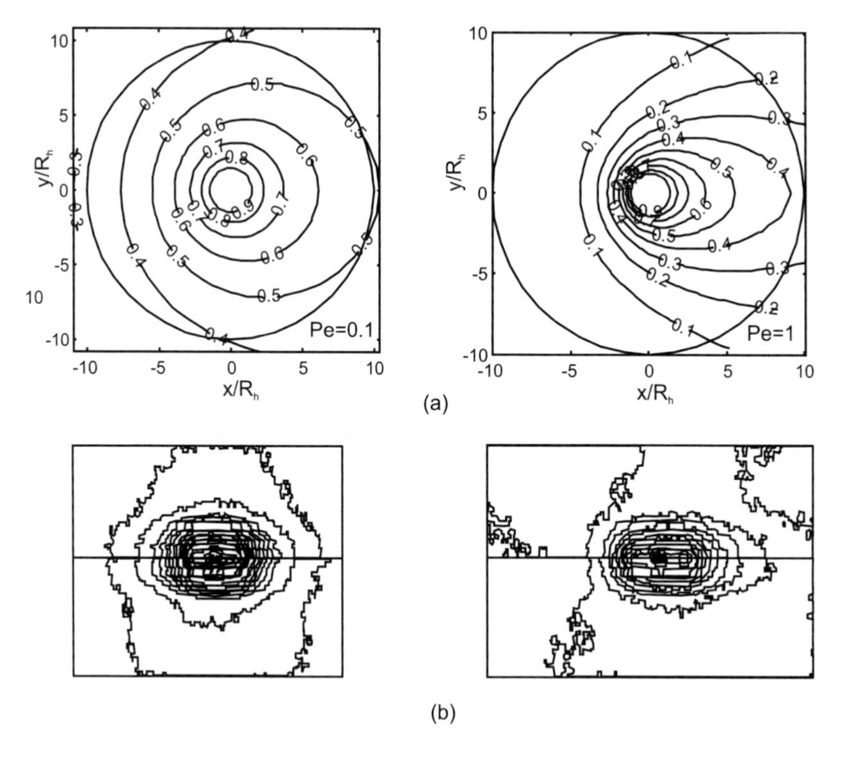


**Figure 8.5** Calorimetric sensors are based on the change of temperature profile around a heater: (a) analytical model; and (b) temperature distribution measured with thermometry around a heater on a silicon membrane (left: zero flow, right: with flow).

**Solution.** For the model depicted in Figure 8.6, the temperature difference between two positions upstream and downstream is calculated as [3]:

$$\Delta T(u) = T_0 \left\{ \exp\left[\gamma_2 \left(l_s - \frac{l_H}{2}\right)\right] - \exp\left[\gamma_1 \left(-l_s + \frac{l_H}{2}\right)\right] \right\} \text{ (a)}$$

with:

$$\gamma_{1,2} = \frac{u \pm \sqrt{u^2 + 8a^2/l_z^2}}{2a} \text{ (b)}$$

and

$$T_0 = \frac{P_H}{8\kappa l_y l_H / l_z + A\kappa(\gamma_1 - \gamma_2)} \text{ (c)} \qquad (8.32)$$

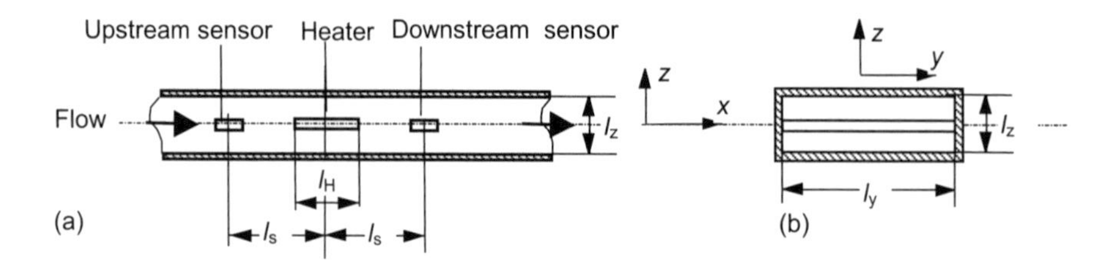


**Figure 8.6** Analytical model for calorimetric sensors: (a) length cut; and (b) cross section.

$a = \kappa/\rho$ is the thermal diffusivity of the fluid. $P_{\mathrm{H}}$ is the heating power. $A = l_{\mathrm{y}}l_{\mathrm{z}}$ is the channel cross section. All other geometrical parameters are defined in Figure 8.6. According to (a), the values for $\gamma_1$ and $\gamma_2$ are:

$$\gamma_1 = \frac{10^{-1} + \sqrt{(10^{-1})^2 + 8(0.6 \times 10^{-3})^2/(0.5 \times 10^{-3})^2}}{2 \times 0.6 \times 10^{-3}} = 2,913$$
$$\gamma_2 = \frac{10^{-1} - \sqrt{(10^{-1})^2 + 8(0.6 \times 10^{-3})^2/(0.5 \times 10^{-3})^2}}{2 \times 0.6 \times 10^{-3}} = -2,746 \quad (8.33)$$

The temperature difference between the downstream and the upstream sensors is:

$$\begin{aligned}\Delta T(u) &= T_0 \left\{\exp\left[\gamma_2 \left(l_{\mathrm{s}} - \tfrac{l_{\mathrm{H}}}{2}\right)\right] - \exp\left[\gamma_1 \left(-l_{\mathrm{s}} + \tfrac{l_{\mathrm{H}}}{2}\right)\right]\right\} \\ &= 80\{\exp[-2,746(10^{-3} - 5 \times 10^{-4}/2)] - \exp[2,913(-10^{-3} + 5 \times 10^{-4}/2)]\} \\ &= 1.2\mathrm{K}\end{aligned}$$

#### 8.1.3.4 Time-of-Flight Sensors

Time-of-flight sensors are thermal flow sensors that measure the time for passage of a heat pulse over a known distance. The transport of the heat generated in a line source through a fluid is governed by the energy equation. The transport of the heat generated in a point source through the fluid is governed by the energy equation:

$$a\frac{\partial^2 \Delta T}{\partial x^2} = u\frac{\partial \Delta T}{\partial x} + \frac{\partial \Delta T}{\partial t} \quad (8.34)$$

With a pulse signal $q'$ (W/m), the solution for (8.34) is [4]:

$$\Delta T(x,t) = \frac{q'}{4\pi kt} \exp\left[-\frac{(x - vt)^2}{4at}\right] \quad (8.35)$$

where $a$ is the thermal diffusivity of the fluid [Figure 8.7]. Measuring the top time $\tau$ at which the signal passes the detection element at ($x = L$) leads to the basic equation of the time of flight of the heat pulse:

$$u = L/t \quad (8.36)$$

For (8.36) to be valid, the term $4at$ in (8.35) must be much smaller than the heater-sensor distance $L$. It assumes that forced convection by the flow is dominating over the diffusive component. In other words, (8.36) is true only at a high Peclet number. When diffusive effect is taken into account, the top time is given by ($u \neq 0$):

$$\tau = \frac{-2a + \sqrt{4a^2 + u^2L^2}}{u^2} \quad (8.37)$$

or for $u = 0$:

$$\tau = \frac{x^2}{4a} \quad (8.38)$$

Figure 8.7 illustrates the time-of-flight concept. The temperature distribution over time along $x$-axis is depicted in Figure 8.7(a), while the sensor characteristics (8.37) and (8.38) are shown in Figure 8.7(b).

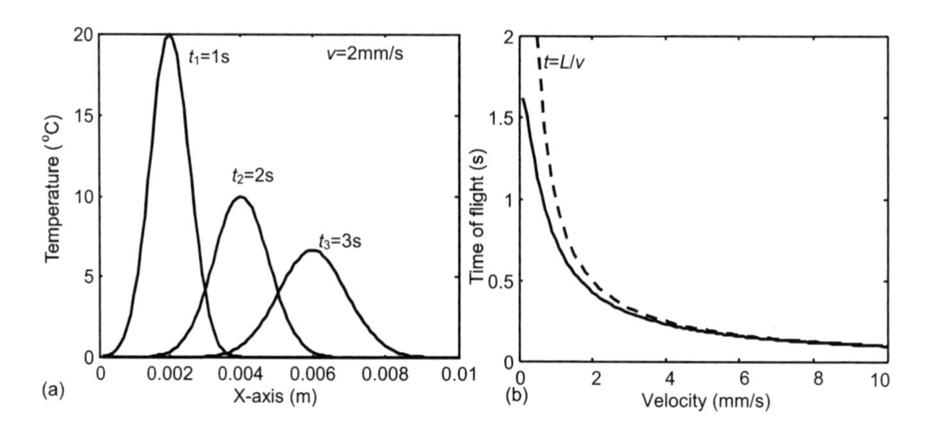


**Figure 8.7** Time-of-flight sensors: (a) temperature profile at different moments; and (b) sensor characteristics (dashed line is the theoretical characteristics; solid line takes the diffusion effect into account).

### Example 8.6: Characteristics of a Time-of-Flight Flow Sensor

The sensor described in Example 8.5 is used as a time-of-flight sensor. Determine the time a pulse signal arrives on the downstream sensor. What is the minimum measurable velocity of this sensor?

**Solution.** Using (8.35), the time needed at $u = 100$ mm/s is:

$$\tau = \frac{-2a+\sqrt{4a^2+u^2L^2}}{u^2}$$

$$= \frac{-2\times0.6\times10^{-3}+\sqrt{4\times0.36\times10^{-6}+0.1^2\times0.001^2}}{0.1^2} = 0.416\times10^{-3}\ \text{sec} = 0.416\ \text{ms}$$

Comparing with the time of thermal diffusion at zero velocity:

$$\tau_0 = \frac{L^2}{4a} = \frac{0.1^2}{4\times0.6\times10^{-3}} \approx 0.417\times10^{-3}\ \text{sec} = 0.417\ \text{ms}$$

we get the approximated minimum measurable velocity of 100 mm/s.

## 8.2 DESIGN EXAMPLES

### 8.2.1 Nonthermal Flow Sensors

#### 8.2.1.1 Differential Pressure Sensors

Figure 8.8(a) shows a flow sensor with two piezoresistive pressure sensors at the two ends of the channel [5]. The sensor used thermoresistive sensors integrated with the same process for temperature compensation. Because the fluid is sucked into the sensor, the flow sensor depicted in Figure 8.8(b) uses only one capacitive pressure sensor to evaluate the pressure loss, because the inlet pressure is equal to the ambient pressure [6]. The sensor shown in Figure 8.8(c) utilizes the same principle as that shown in Figure 8.8(a) [7]. The flow sensor has two capacitive pressure sensors at the channel ends. Table 8.2 compares the different differential pressure flow sensors.

Differential pressure sensors also have been realized through embedding flexible membranes within the liquid microchannel for in situ flow measurements. The pressure increase of liquid results in the membrane deformation that can be quantified by optical microscopy [8] or monitoring the

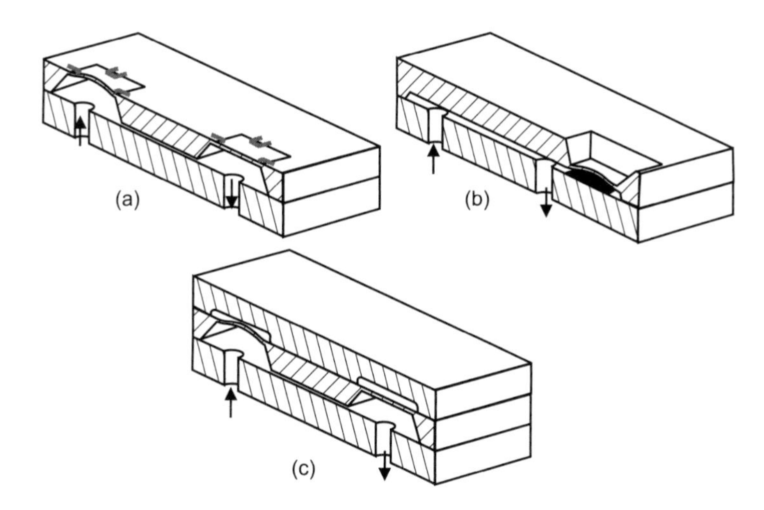


**Figure 8.8** Typical differential pressure flow sensors: (a) piezoresistive; and (b, c) capacitive.

**Table 8.2**
Differential Pressure Flow Sensors
($D_h$: Hydraulic diameter; $L$: Channel length; $\dot{Q}_{max}$: Maximum flow rate; $S_0$: Zero-sensitivity; $\tau$: Response time)

| *Ref.* | $D_h$ *(mm)* | *L (mm)* | $\dot{Q}_{max}$ *(mL/min)* | $\tau$ *(ms)* | *Technology* |
|---|---|---|---|---|---|
| [5] | 33–70 | 1.2–25 | 83 water | $< 2$ | Bulk |
| [6] | 21 | 2.9 | 270 water | $< 0.5$ | Bulk |

membrane electrical resistance [9]. For the latter method, a thin layer of conductive material is coated over the membrane. The electrical resistance of the conductive layer increases once the membrane experiences a deformation (bulge) due to the pressure change (Figure 8.9). Optical microscopy generally requires bulky and expensive accessories that add complexity for integration with microfluidic chips. Monitoring of electrical resistance enables a higher level of miniaturization for microfluidic integration.

#### 8.2.1.2 Drag Force Sensors

The obstacle can be designed in the form of a cantilever or a membrane with an orifice in its center. Figure 8.10(a) depicts a flow sensor with a silicon cantilever as the sensing component. Piezoresistors were integrated into the substrate at the end of the cantilever. The stress caused by the deflection of the cantilever induces a change in the resistors, which can be evaluated with a Wheatstone bridge [10]. The sensor operates in a low flow range and has linear characteristics because the quadratic term in (8.10) can be neglected.

A similar design is reported in [11]. The sensing component has the form of a paddle. The obstacle structure is designed as a square plate. The sensing element is a small cantilever [Figure 8.10(b)]. The design utilizes both drag pressure and differential pressure. The flow range of this sensor is large and the sensor has quadratic characteristics. The sensor depicted in Figure 8.10(c) is actually a piezoresistive pressure sensor with a flow orifice in the middle of the membrane. The mean force in this sensor is caused by the pressure loss at the orifice. Thus, its characteristics are quadratic [12].

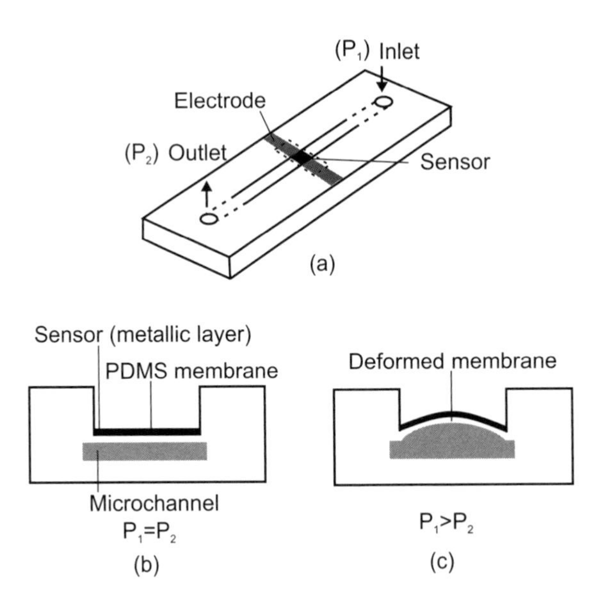


**Figure 8.9** PDMS-based membrane pressure sensing: (a) schematics of the microfluidic chip with the membrane; (b) the cross section of the chip with the flexible thin membrane in static condition (no flow); and (c) deformation the membrane and the thin metallic layer due to pressure change in the liquid channel once liquid flows. $P_1$ and $P_2$ are the inlet and outlet pressures. Membrane deformation due to pressure changes its electrical resistance. (*After:* [9].)

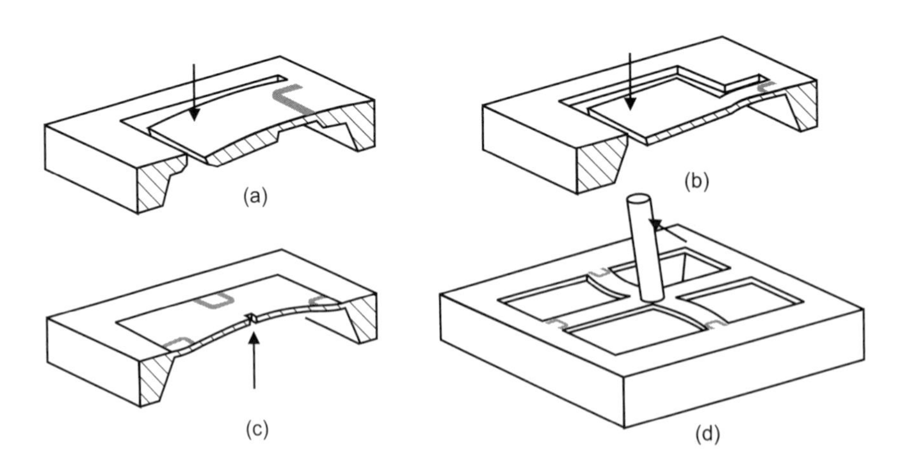


**Figure 8.10** Typical drag force flow sensors: (a) rectangular paddle; (b) square paddle; (c) membrane; and (d) wire.

**Table 8.3**

Typical Parameters of Drag Force Sensors

($W$: Width of the obstacle; $H$: Length or height of the obstacle; $\dot{Q}_{max}$: Maximum flow rate; $S_0$: Zero-sensitivity)

| *Ref.* | *W* (μm) | *H* (μm) | $\dot{Q}_{max}$ *(mL/min)* | $S_0$ | *Sensing Structure* | *Technology* |
|---|---|---|---|---|---|---|
| [10] | 1,000 | 3,000 | 1 water | 4.3 μVmin/μl | Beam | Bulk |
| [11] | 500 | 500 | 200 water | 0.165 Vmin/ml | Beam | Bulk |
| [12] | 28–160 | — | 14 water | 82 Pa min/ml | Membrane | Bulk |
| [13] | — | — | 2 m/s air | 1 mV s/m | Wire | Bulk |

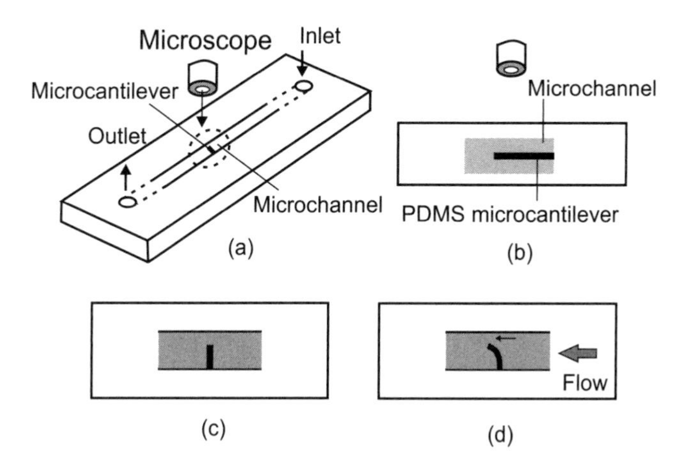


**Figure 8.11** Schematics of microcantilever-based flow sensor: (a) microflow sensor implemented in the microfluidic chip; (b) cross section of the chip showing the cantilever; (c) top view of the cantilever with no liquid flow; and (d) top view of the deformed cantilever due to drag force once exposed to liquid flow. (*After:* [15].)

Figure 8.10(d) shows a design that only evaluates the drag force. The sensing element is an aluminum wire bonded to the center of a crossed beam structure. The beam has piezoresistive sensors at its four ends. The drag force on the aluminum wire bends the beam structure, which converts the stress into an electrical signal. This sensor mimics the wind receptor hairs of insects. Since the sensor is only based on the drag force, its characteristics are linear [13]. Table 8.3 compares the performance of different drag force flow sensors.

Deformable elements such as beams, cantilever, and membranes have been embedded in microfluidic chips, as flow sensors, for direct in-situ flow measurements [14]. Figure 8.11 shows a cantilever directly embedded in a microfluidic chip for flow measurements. The tip deflection of the flexible mircocantilever made in PDMS was measured using optical microscopy [15]. The flow sensor was able to measure flow rates between 200- and 1,300 $\mu$l/min using image processing.

Taking advantage of mechanical deformation of a flexible cantilever once exposed to liquid flow, Lien and Vollmer [16] developed an optofluidic flow sensor with integrated sensing fiber optic at the tip of the cantilever. The fibertip cantilever is aligned with a receiving multi mode fiber once there is no flow rate. Increasing in the flow rate deflects the cantilever and makes it no longer aligned with the receiving fiber [16]. The decrease in transmitted intensity is correlated with the flow rate. The optofluidic sensor allowed for the measurement of flow rate up to 1,500 $\mu$l/min.

Noeth et al. [17] made a SU-8 cantilever of 3.7 $\mu$m in thickness containing microholes for in-channel flow measurement with a resolution of 3.5 nl/min. Flow sensing relied on the determination

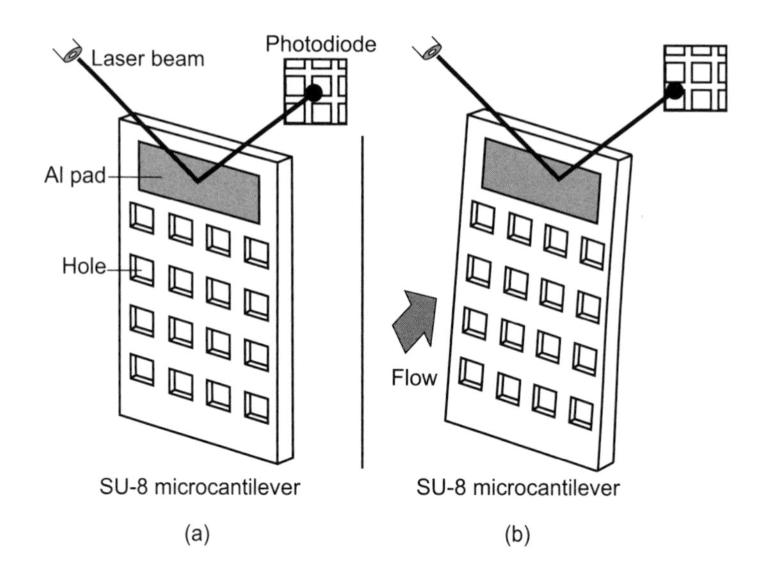


**Figure 8.12** Schematics of the SU-8 cantilever flow sensor: (a) the cantilever with no flow; and (b) deflection of the cantilever once exposed to liquid stream. (*After:* [17].)

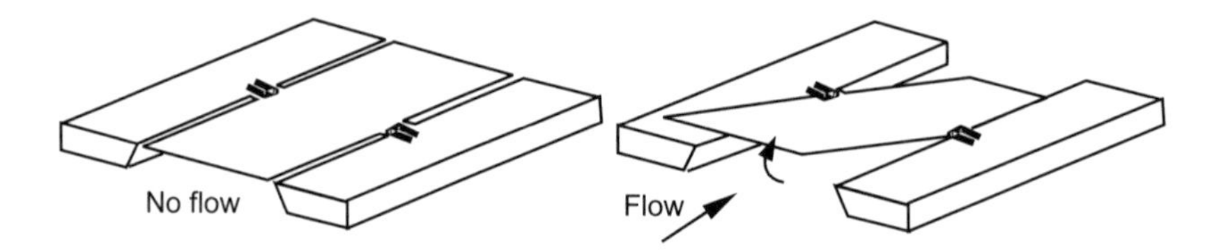


**Figure 8.13** Lift force flow sensors. (*After*: [18].)

of the cantilever deflection using a focused laser beam. An aluminum (Al) pad was implemented at the end of the cantilever to reflect the laser beam into a position sensitive photodiode [Figure 8.12(a)]. The cantilever is placed in the microchannel perpendicular to the flow stream. Thus, the flow-induced deflection of the cantilever changes the position of the laser beam in the photodiode [Figure 8.12(b)]. Sensitivity of the sensor could be tuned by modifying the integrated holes through changing the holes density, their size (from $5 \times 5$ $\mu m^2$ to $20 \times 20$ $\mu m^2$), and hole-to-hole spacing [17]. In this case, the beam second moment of inertia, as well as its resistance to liquid drag force, can be modified to modulate the sensor sensitivity.

### 8.2.1.3 Lift Force Sensors

Figure 8.13 illustrates a design example of lift force sensors. The airfoil structure is a rectangular plate suspended on two beams. The lift force causes a rotational bending of the beams, which have piezoresistors to detect the stress. The airfoil plate is $5 \times 5$ mm large. The sensor is able to measure flow velocities up to 6 m/s [18].

### 8.2.1.4 Coriolis Force Sensor

The Coriolis mass flow sensor presented in [19] consists of two serially connected tube loops located symmetrically in one plane (Figure 8.14). The tube structure is excited electrostatically by an external electrode. The angular amplitudes of the excitation vibration and the Coriolis twisting are detected using a 2-D lateral photodetector and a lock-in amplifier (Figure 8.16). The whole structure

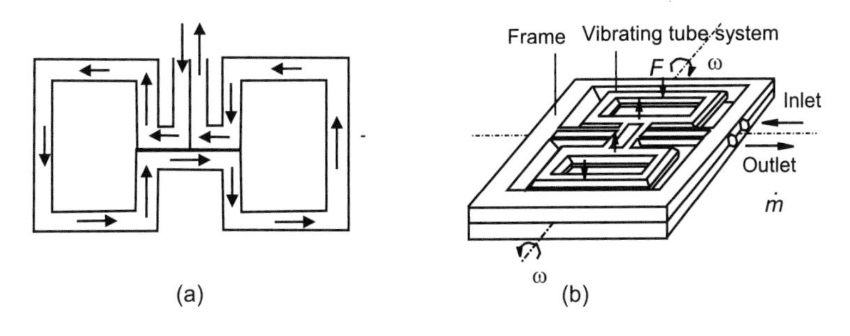


**Figure 8.14** Coriolis flow sensor: (a) flows inside the sensor tube; and (b) design example

**Table 8.4**
Transducer Principles and Realization of Thermal Transducers Using Micromachining Technology

| *Principle* | *Examples* | *Measurement of* |
|---|---|---|
| Thermoresistive | Platinum, polysilicon, silicon, metal alloys (NiCr, NiFe) | Temperature, temperature difference, heating power |
| Thermocapacitive | | |
| Thermoelectrical | pSi-Al (bipolar technology), PolySi-Al (CMOS technology), pPolySi-nPolySi | Temperature, temperature difference |
| Thermoelectronical | Transistors, diodes | Temperature, temperature difference |
| Pyroelectrical | $LiTaO_3$ | Heating power |
| Frequency analog | SAW oscillator, Lamb-wave oscillator | Temperature |

was fabricated using silicon bulk micromachining. The sensor can measure water flow rates up to 30 mL/min in both directions.

The major disadvantage of this sensor type is that the Coriolis force depends on the fluid density. The density in turn depends on the temperature. Temperature compensation is necessary for this sensing principle.

#### 8.2.1.5 Electrohydrodynamic Flow Sensors

A nonmechanical, nonthermal sensing concept is the electrohydrodynamic method, which can be implemented in microtechnology. The working principle is similar to the thermal time-of-flight sensors described in Section 8.1.3.4. This method is based on the charge transport caused by mass transport. First, a charge is injected into the flow by an electrode in the form of a charge pulse. A second electrode downstream detects the charge in the flow. The time of flight of the charge or the time shift between the injected signal and the detected signal is evaluated. The flow velocity is then derived by the known distance between the two electrodes and the time of flight. The sensor presented in [20] can detect flow rates up to 1.6 mL/min with a resolution of 8 μL/min.

### 8.2.2 Thermal Flow Sensors

In the following sections, thermal flow sensors are categorized by their transducers. The most important transducer principles are thermoresistive, thermocapacitive, thermoelectrical, thermoelectronical, pyroelectrical, and frequency analog. Table 8.4 shows an overview of these principles. The indirect transformation over an oscillating mechanical element is referred to here as the frequency analog principle.

**Table 8.5**

Typical Parameters of Thermoresistive Hot-Wire Sensors
($W \times L \times H$: Dimension of the heater; $v_{\max}$: Maximum flow velocity; $\dot{Q}_{\max}$: Maximum volume flow rate; $P_{\max}$: Maximum heating power; $T_{\max}$: Maximum heater temperature; $S_0$: Zero-sensitivity; $\tau$: Response time )

| *Ref.* | $W \times L \times H$ *(μm)* | $v_{\max}$, or $\dot{Q}_{\max}$ | $P_{\max}$ *(mW)* | $\Delta T_{\max}$ *(K)* | $S_0$ | $\tau$ *(ms)* | *Material* |
|---|---|---|---|---|---|---|---|
| [21] | $270 \times 3 \times 1$ | 100 mL/min air | 8 | 230 | 0.3 mV min/ml | 0.5 | Polysilicon |
| [22] | $20 \times 2,000 \times 2$ | 600 nL/min water | 0.14 | 33 | 0.0263 K nL/min | 0.14 | Polysilicon |
| [23] | $2 \times 40 \times 0.45$ | — | 15 | 280 | — | 0.35 | Polysilicon |
| [24] | $290 \times 290 \times 0.5$ | 200 mL/min water | 2,400 | 50 | ≈0.2 V min/ml | 1 | Polysilicon |
| [25] | $200 \times 200 \times 11.8$ | 4 m/s air | 1,600 | 180 | ≈1.7Vs/m | — | Silicon |
| [26] | — | 60 m/s air | 2,400 | 160 | ≈0.5 V s/m | — | Nickel |

#### 8.2.2.1 Thermoresistive Flow Sensors

The approximated linear relation between the resistance $R$ and temperature $T$ is:

$$R = R_0[1 + \alpha(T - T_0)] \tag{8.39}$$

where $R_0$ is the resistance at the temperature $T_0$, and $\alpha$ is the temperature coefficient of resistance (TCR). Since the absolute values of integrated resistors vary widely ($\pm 20\%$), they should be evaluated relatively in a bridge circuitry such as the Wheatstone bridge. The ratio of two resistances fabricated in the same batch is usually accurate with a tolerance less than $1\%$.

*Hot-Wire and Hot-Film Type.* Hot-wire and hot-film sensors generally need one resistor as a heater, and in some cases a second thermoresistor as temperature sensor for the ambient. Microtechnology allows the fabrication of thermoresistive sensors from different materials: polycrystalline silicon, single-crystalline silicon, metals, metal alloys, and temperature-resistant materials. Table 8.5 gives an overview of the characteristics of this sensor type reported in the literature [14, 19].

*Polysilicon.* Polysilicon has a relatively high TCR compared to single-crystalline silicon. The only disadvantage is that the TCR of polysilicon strongly depends on technology parameters such as deposition temperature and annealing temperature. It is also difficult to reproduce the exact TCR in different process batches. The flow sensor reported in [21] has a hot wire in the form of a heavily doped polycrystalline bridge [Figure 8.15(a)]. The bridge is supported on a silicon oxide mesa. The flow channel is etched in silicon. The sensor is able to measure air flow rate up to 100 mL/min. The same hot wire shown in Figure 8.15(b) is integrated in a silicon nitride microchannel [22]. This sensor can resolve water flow rates as low as 10 nL/min. The hot-film sensor is fabricated on a silicon nitride membrane [23] [Figure 8.15(c)] or on a silicon membrane [24] [Figure 8.15(d)].

*Single-Crystalline Silicon.* Single-crystalline silicon can also be used as a heater. However, the resistor is integrated directly in the bulk material, which is thermally conductive. It is therefore difficult to assure thermal isolation. Furthermore, the presence of a p-n junction limits the operating range of the sensor at higher temperatures. The sensor described in [25] has spreading resistors as heaters and temperature sensors.

*Metals and Alloys.* Metals and metal alloys are stable materials for designing heaters and temperature sensors. They have advantages over single-crystalline silicon due to the absence of p-n junctions and over polysilicon due to the low pressure dependence. However, the good electrical conductivity requires a long structure or a large meander form, which represents a hurdle for

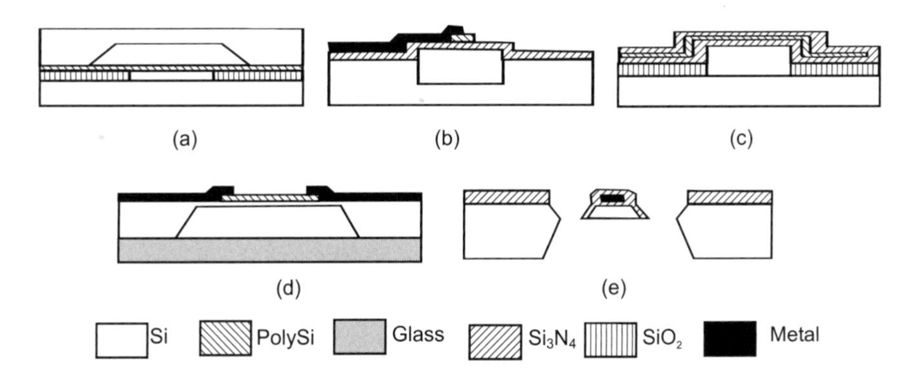


**Figure 8.15** Hot-wire and hot-film sensor based on thermoresistive polysilion and metal: (a) bridge inside a microchannel; (b) embedded in a silicon-nitride microchannel; (c) on a silicon nitride membrane; (d) on a silicon membrane; and (e) metal resistors on a silicon beam.

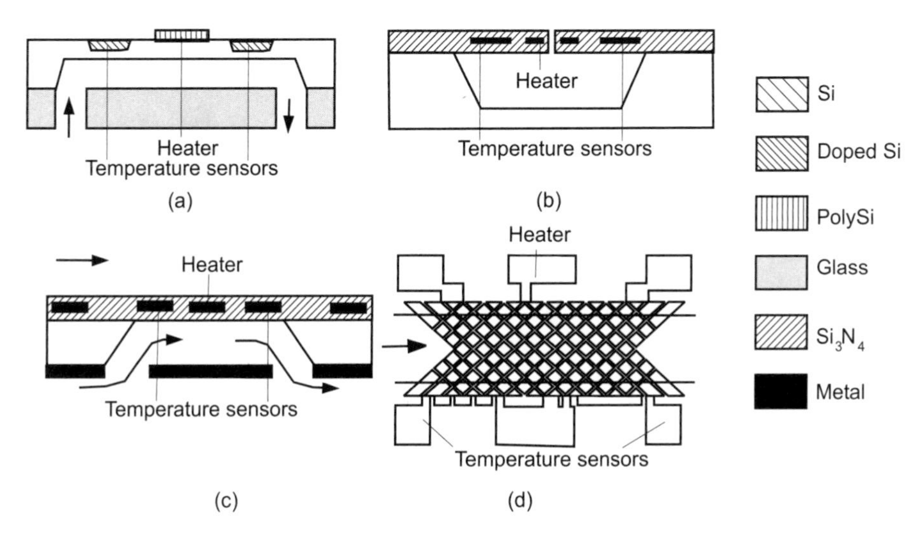


**Figure 8.16** Thermoresistive flow sensors of calorimetric type: (a) silicon; and (b, c, d) metals.

miniaturization. The sensor presented in [26] has a nickel heater embedded in silicon nitride and silicon oxide. The heater is suspended on a silicon beam and works as a hot wire [Figure 8.15(e)]. Materials such as silicon carbide are interesting for high-temperature applications. Thin-film silicon carbide exhibits thermal conductivity on the same order of single-crystalline silicon and has a fast thermal response.

*Calorimetric Type.* Thermoresistive sensors are best suited for the calorimetric principle. The temperature sensors upstream and downstream of the heater can be connected to a Wheatstone bridge, which provides directly the differential signal and does not depend on fabrication tolerance. Similar to hot-wire and hot-film types, a variety of materials can be used.

Single-crystalline silicon is used in [27] to form a full Wheatstone bridge. A second bridge compensates the offset and improves the stability of the sensor. Three operational amplifiers are monolithically integrated on the chip. The sensor reported in [28] has single-crystalline silicon resistors as temperature sensors. The heaters and sensors are located on a membrane out of the flow, which makes measurement of corrosive fluid possible [Figure 8.16(a)]. The drawback of this design is the heat loss through conduction in the silicon membrane.

**Table 8.6**
Typical Parameters of Thermoresistive Calorimetric Sensors
($W \times L$: Dimension of the heater; $v_{\text{max}}$: Maximum flow velocity; $\dot{Q}_{\text{max}}$: Maximum volume flow rate; $P_{\text{max}}$: Maximum heating power; $T_{\text{max}}$: Maximum heater temperature; $S_0$: Zero-sensitivity; $\tau$: Response time)

| *Ref.* | $W \times L$ *(μm)* | $v_{\text{max}}$, or $\dot{Q}_{\text{max}}$ | $P_{\text{max}}$ *(mW)* | $\Delta T_{\text{max}}$ *(K)* | $S_0$ | $\tau$ *(ms)* | *Material* |
|---|---|---|---|---|---|---|---|
| [27] | $4,000 \times 4,000$ | 0.33 m/s air | 250 | 168 | 780 mV s/m | 0.5 | Single crystalline |
| [28] | — | — | — | — | — | — | Single crystalline |
| [29] | $400 \times 500$ | 100 mL/min air | — | 160 | 0.03V min/ml | — | Permalloy |
| [30] | $4,000 \times 6,000$ | 2 m/s | 7.7 | 55 | — | $< 150$ | Gold |
| [3] | $500 \times 1,000$ | 2 m/s air, 0.002 m/s water | 10 | — | 30K s/m air, 2,000K s/m water | — | Au/Cr |
| [32] | $3,000 \times 3,000$ | 3.5 m/s air | 240 | — | 400 mV s/m | 2.5 | SiC |

Nickel-iron alloy (Permalloy) is used for temperature sensors in [29]. The heater and sensors are laminated within a 1-μm-thin thermal isolating silicon nitride layer that are suspended in the form of two bridges over an etched pit in the silicon [Figure 8.16(b)]. The development reported in [23] is similar. A heat sink and flow guide integrated on the backside of the flow sensor are used to achieve an extended measurement range [Figure 8.16(c)].

The flow sensor presented in [3] has suspended CrAu-resistors as heater and temperature sensors. The resistors are carried by a silicon nitride grid. The flow channel is fabricated by using silicon etching and anodic bonding [Figure 8.16(d)]. The newer version of this sensor has high dynamics, which allows the sensor to measure acoustic flow and can be used as a microphone [31].

In the sensor reported in [32], doped silicon carbide film was used as heater and temperature sensors. The silicon carbide resistors form several bridges above a flow channel etched in silicon. Due to the high melting point of silicon carbide (about 2,800°C), high heater temperature is possible. Thermal cleaning can be achieved by elevating the heater temperature to above 1,000°C, at which contaminants are completely burnt. Table 8.6 compares the performance of the above calorimetric sensors.

#### 8.2.2.2 Thermocapacitive Flow Sensors

The temperature dependence of the dielectric constant allows the realization of temperature sensors based on the thermocapacitive principle. The sensor described in [33] has tantalum oxide $Ta_2O_5$ as dielectric material in the sensing capacitor. The sensor works in the hot-film mode. The heater and the sensing capacitor are integrated on a cantilever suspended above a trench etched in silicon [Figure 8.17(a)]. The circuit schematic of the sensor is shown in Figure 8.17(b). The sensor consumes a maximum power of 57 mW. The maximum range of operation is 20 m/s air flow, and the sensitivity is 4.25 mVs/m.

#### 8.2.2.3 Thermoelectric Flow Sensors

The thermoelectric effect, or the Seebeck effect, can be used for measuring temperature differences. The effect yields a voltage between two ends of a semiconductor or metal with a temperature difference of $\Delta T$:

$$\Delta V = \alpha_S \Delta T \tag{8.40}$$

where $\alpha_S$ is the thermoelectric coefficient, or Seebeck coefficient, of the material. When taking two materials with different Seebeck coefficients at the hot junction, the voltages are subtracted. An

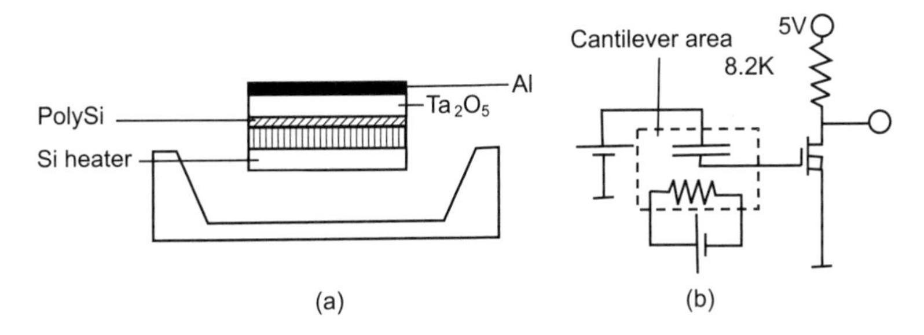


**Figure 8.17** Hot-film sensor based on thermocapacitive principle: (a) design; and (b) evaluation circuit.

**Table 8.7**
Seebeck Coefficients of Different Materials

| *Material* | $\alpha_S$ *(300 K)* μ*V/K* | *Material* | $\alpha_S$ *(300 K)* μ*V/K* |
|---|---|---|---|
| p-Si | 300 to 1,000 | Gold (Au) | 1.94 |
| n-PolySi | −200 to −500 | Copper (Cu) | 1.83 |
| Antimony (Sb) | 43 | Aluminum (Al) | −1.7 |
| Chromium (Cr) | 17.3 | Platinum (Pt) | −5.28 |

effective coefficient remains for the thermocouple. Many thermocouples connected serially form a thermopile, which multiplies the thermoelectric coefficient by the number of thermocouples. Table 8.7 lists the Seebeck coefficient of some common materials [34].

*Hot-Film Type.* Thermopiles are used for measuring the heater temperature in hot-film mode. The cooling of the heater for a constant heating power can be evaluated directly with the thermopile output. The sensor presented in [35] has aluminum/polysilicon thermopiles whose hot junctions are placed next to the heater. The whole structure is suspended on a cantilever [Figure 8.18(a)]. The thermopile output is used in the example of [36] to control the heating current, which is connected to a sigma-delta converter. The sensor delivers a direct digital output. The thermopile in this sensor was made of p-Si and aluminum. The example of [37] has two thermopiles to measure the heater temperature [Figure 8.18(b)]. The thermopile was made of GaAs and AlGaAs, which is compatible with GaAs technology and suitable for high temperature applications.

*Calorimetric Type.* For sensing the temperature difference between downstream and upstream positions, two thermopiles are used. The thermopiles can be fabricated with standard bipolar process [38]. Thermopiles made of aluminum and single crystalline silicon show a larger Seebeck coefficient than those with aluminum and polysilicon (Table 8.7). For better thermal isolation, the sensor structures are suspended on a thin silicon membrane or silicon plate.

The other possibility is the use of doped polysilicon. Thermopiles made of n-doped/p-doped polysilicon [39] or aluminum/polysilicon [40] are compatibile with standard CMOS technology [Figure 8.18(c)]. The technology allows integration of the sensor with digital electronics, which provides an A/D converter with direct digital bitstream output. The sensor presented in [41] was designed for detecting small flow rates generated by acoustic streaming—a gentle pumping principle. The sensor in [42] has thermopiles made of gold and doped polysilicon. The arrangement of thermopiles allows the downstream sensor to work in the time-of-flight mode. Alternative materials such as BiSb and Sb can improve the Seebeck coefficient [43]. Table 8.8 lists the parameters of the above thermoelectric sensors.

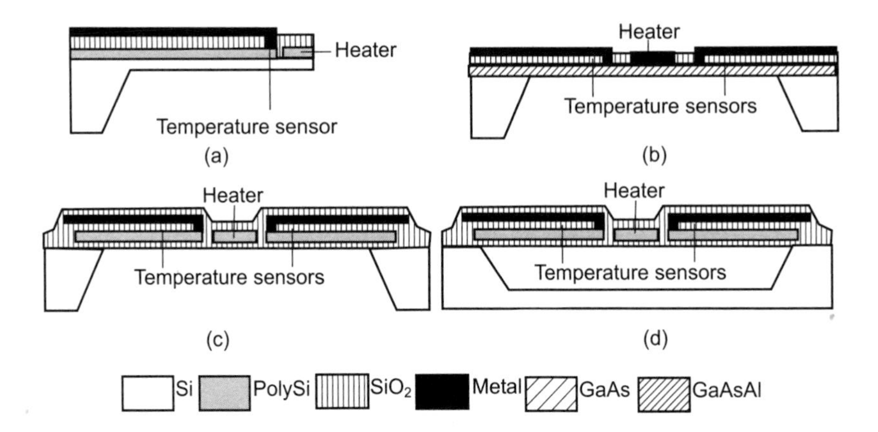


**Figure 8.18** Thermoelectric sensors: (a) aluminum/polysilicon; (b) aluminium/single crystalline silicon; and (c) and (d) aluminum/polysilicon.

### 8.2.2.4 Thermoelectronic Flow Sensors

The thermoelectronic effect is based on the temperature dependence of the pn-junction in diodes or bipolar transistors. The voltage drop over a forward biased base-emitter junction follows the relation:

$$V_{\mathrm{BE}} = V_{\mathrm{G}} + \frac{KT}{e} \ln\left(\frac{I_{\mathrm{C}}}{I_{\mathrm{R}}}\right) \tag{8.41}$$

where $V_{\mathrm{G}}$ is the material bandgap, $K$ is the Boltzmann's constant, $e$ is the elementary charge, $I_{\mathrm{C}}$ the collector current, and $I_{\mathrm{R}}$ is the reference current [44]. Figure 8.19 shows the typical evaluation circuits for thermoelectronic flow sensors.

*Hot-Film Type.* The diode as a temperature sensor can be fabricated by standard microtechniques, described in Chapter 3. For good thermal isolation, the thin single crystalline silicon membrane of an SOI wafer can be used as a substrate for the heater and diode structures [45]. The thermal isolation is achieved with deep trenches, which are filled with thermal oxide. A second identical diode measures the ambient temperature [Figure 8.20(a)]. The analog circuitry for evaluation is shown in Figure 8.19(c).

Using a comparator circuit, the analog output of the diodes can be fed back to the heater [46]. The principle allows the design of a semidigital sensor with pulse width modulated output [46, 47] [Figure 8.19(d)]. The sensor shown in Figure 8.20(b) is thermally isolated by a free-standing cantilever. A trench filled with polyimide separates the two sensing diodes. Furthermore, MOS-FETs can work as both sensor and heater [Figure 8.19(e)] [48].

*Calorimetric Type.* In [49], both heating and sensing elements are diodes that are positioned beneath diamond-shaped silicon pyramids. The pyramids are placed in the flow channel to obtain good thermal contact between fluid flow and the heater as well as sensing elements [Figure 8.20(c)]. The sensing diodes downstream can also work in the time-of-flight mode. Table 8.9 lists the parameters of the above thermoelectronic sensors.

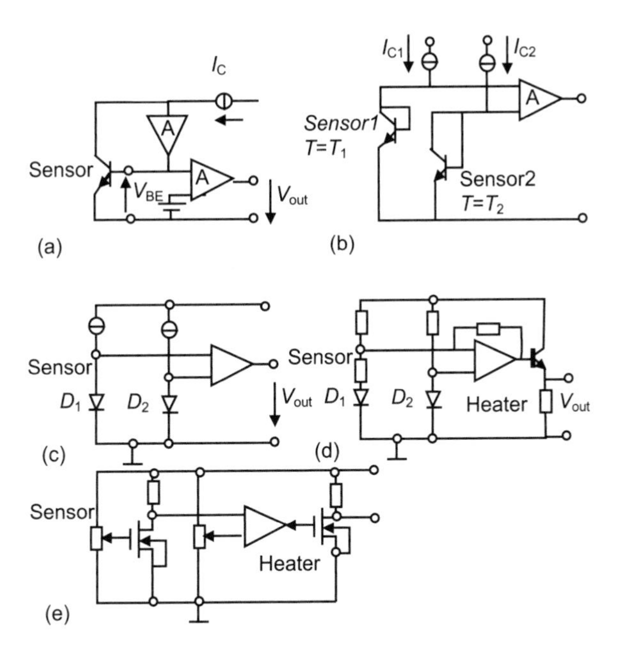


**Figure 8.19** Configuration for evaluation of thermoelectronic flow sensors: (a) hot-film type with transistors; (b) calorimetric type with transistors; (c) hot-film type with diodes; (d) hot-film type with diodes, semidigital output; and (e) hot-film type with field effect transistors.

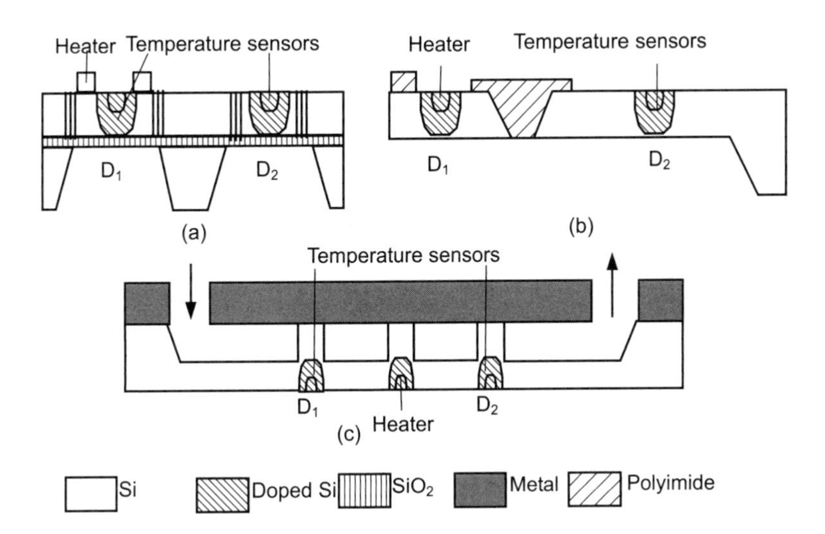


**Figure 8.20** Typical thermoelectronic flow sensors: (a) diodes on a membrane; (b) diodes on a cantilever; and (c) diodes with diamond-shaped silicon pyramids.

**Table 8.8**

Typical Parameters of Thermoelectric Sensors

($W \times L \times H$: Dimension of the heater; $v_{\max}$: Maximum flow velocity; $\dot{Q}_{\max}$: Maximum volume flow rate; $P_{\max}$: Maximum heating power; $T_{\max}$: Maximum heater temperature; $N$: Number of thermocouples; $\alpha_s$: Seebeck coefficient)

| *Ref.* | $W \times L$ *(μm)* | $v_{\max}$, or $\dot{Q}_{\max}$ | $P_{\max}$ *(mW)* | $\Delta T_{\max}$ *(K)* | $N$ | $\alpha_s$ *(ms)* | *Material* |
|---|---|---|---|---|---|---|---|
| | Hot-film type | | | | | | |
| [35] | 250 × 130 | 25 m/s $N_2$ | 10 | 250 | 20 | 58 | n-PolySi/Al |
| [36]] | 5,000 × 5,000 | 50 m/s air | 20 | — | 6 | 833 | p-Si/Al |
| [37] | 1,280 × 2,000 | 30 m/s air | 33 | 300 | 16 | 175 | GaAs/AlGaAs |
| | Calorimetric type | | | | | | |
| [38] | 6,000 × 6,000 | 20 m/s air | 595 | 17 | 25 | 600 | n-Si/Al |
| [39] | 4,000 × 5,000 | 40 m/s air | 3 | — | 20 | 415 | n-PolySi/p-PolySi |
| [40] | 1,000 × 1,000 | 2 m/s air | 3 | 40 | 20 | 108 | n-PolySi/Al |
| [41] | 1,000 × 1,000 | — | 3.5 | — | 10 | ≈100 | n-Poly/Al |
| [42] | 600 × 600 | 10 mm/s water | 50 | — | 20 | — | PolyS/Au |
| [43] | 3,600 × 1,150 | 2.3 m/s air | 1 | 5 | 100 | 135 | BiSb/Sb |

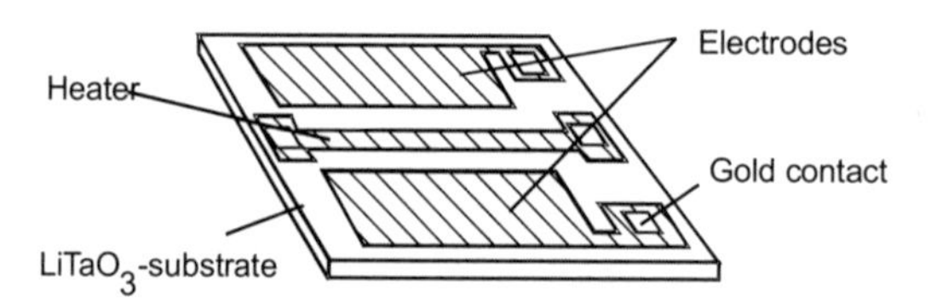


**Figure 8.21** Pyroelectric hot-film flow sensor.

### 8.2.2.5 Pyroelectric Flow Sensors

The pyroelectric effect is based on the dependence of surface charge of a pyroelectric material on the change of the supplied heat. The pyroelectric current is given by [50]:

$$I = C_{\mathrm{p}} A \frac{\mathrm{d}T}{\mathrm{d}t} \tag{8.42}$$

where $C_{\mathrm{p}}$, $A$, and $T$ are the pyroelectric coefficient, the area of the detecting electrode, and temperature of the pyroelectric material, respectively. Therefore, a sinusoidal electric current is supplied to the heater at a frequency $f$ that both heats the pyroelectric substrate to a stationary dc temperature and produces an ac sinusoidal temperature variation at a frequency $2f$.

The heat from the thin-film heater flows through the substrate and is dissipated to the flow (Figure 8.21). The ac temperature variation under the two electrodes includes pyroelectric currents in each of the symmetrically disposed measuring electrodes. The ac pyroelectric currents generated at the two symmetric electrodes produce the sensor signal.

Because lithium tantalate ($LiTaO_3$) has a fairly high pyroelectric coefficient, $C_{\mathrm{p}} = 2.3 \times 10^{-4}$ [$\mathrm{Asm^{-2}K^{-1}}$], which is not affected by ambient temperatures up to about 200°C [50]. The pyroelectric material was used as the substrate material. The electrodes and heater are made from NiCr-alloy or gold thin film [50, 51].

**Table 8.9**
Typical Parameters of Thermoelectronic Flow Sensors
($W \times L$: Dimension of the heater; $v_{\max}$: Maximum flow velocity; $\dot{Q}_{\max}$: Maximum volume flow rate; $P_{\max}$: Maximum heating power; $T_{\max}$: Maximum heater temperature; $\tau$: Response time)

| *Ref.* | $W \times L$ *(μm)* | $v_{\max}$, or $\dot{Q}_{\max}$ | $P_{\max}$ *(mW)* | $\Delta T_{\max}$ *(K)* | $\tau$ *(ms)* | $\tau$ *(ms)* | *Material* |
|---|---|---|---|---|---|---|---|
| | Hot-film type | | | | | | |
| [45] | $4,000 \times 5,000$ | 1 m/s oil | 66 | 55 | — | 50 | p-n Si |
| [46] | $2,000 \times 1,400$ | 30 m/s air | 110 | 130 | 3%/m/s | — | p-nSi |
| [47] | $1,000 \times 1,000$ | 0.7 m/s $N_2$ | — | 25 | 16 mVs/m | 30,000 | Si |
| | Calorimetric type, | Time-of-flight sensor | | | | | |
| [48] | $9,000 \times 5,000$ | 0.1mL/min oil | 60 | — | 3.5 Vmin/ml | 30 | Al/Si/Glass |

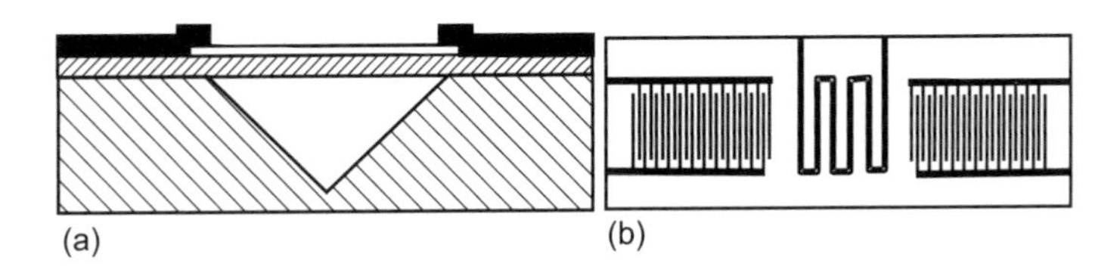


**Figure 8.22** Typical frequency analog flow sensors: (a) resonating bridge; and (b) SAW.

### 8.2.2.6 Frequency Analog Sensors

Frequency analog sensors are realized by using thermal excitation and evaluation of oscillating frequency [52, 53]. The sensing principle is based on the temperature dependence of the oscillating behavior of mechanical elements such as cantilevers and membranes. The change of temperature causes a change of mechanical stress in the element, which leads to a change of its resonant frequency [Figure 8.22(a)]. Further frequency analog sensors are based on the dependence of surface-wave frequency on the temperature [54]:

$$\frac{\Delta f}{f_0} = \frac{\alpha_{\mathrm{f}} A h \Delta T}{G_{\mathrm{eff}}} \tag{8.43}$$

where $\Delta f$ is the frequency shift, $f_0$ is initial oscillator frequency under no gas flow, $\alpha_{\mathrm{f}}$ is the temperature coefficient of frequency for the surface acoustic wave (SAW) substrate, $A$ is the heated surface, $h$ is the heat transfer coefficient, and $G_{\mathrm{eff}}$ is the effective thermal conductance between the substrate and the ambient [Figure 8.22(b)]. A frequency shift of 140 kHz can be reached with a maximum flow rate of 1,000 mL/min. Similar devices can be designed with a Lamb wave [55].

There is a huge interest in developing embedded flow sensors for continuous flow measurements. In particular, recent developments of microfluidic bioreactors and organ-on-chip devices requires the precise control of liquid flow passing through the bioreactors. The flow rate has direct impact on flow-induced shear stress over the cultured cells in a bioreactor, as well as on the delivery of oxygen and nutrients to the cells. To this end, the microflow sensors should present a robust performance for continuing measurements and easy integration within the microfluidic chips.

## 8.3 SUMMARY

This chapter discusses the design and fabrication of microflow sensors. Microflow sensors together with microvalves and micropumps form a feedback controlled system for fluid delivery. Because the majority of flow sensors realized by microtechnology are based on thermal concepts, microflow sensors are categorized in this chapter as nonthermal and thermal types. Most nonthermal flow sensors are based on pressure changes or forces induced by a fluid flow. The design of these flow sensors reduces to designing pressure sensors or force sensors. Thermal flow sensors are further categorized as hot-film/hot-wire sensors, calorimetric sensors, and time-of-flight sensors. All these types are based on heat transfer between a heated body and the fluid flow. The design of thermal sensors typically leads back to designing temperature sensors. Physical phenomena, such as thermoresistive effect, thermocapacitive effect, thermoelectric effect, thermoelectronic effect, and pyroelectric effect, are available for converting temperatures or temperature differences into an electric signal. Thermal-induced stress changes the resonance frequency of a mechanical structure, and thus can also be used for detecting temperature.

### Problems

**8.1** For the sensor of Example 8.4, determine the strain at the ends of the tube. The tube is made of silicon and is 50 μm thick. If piezoresistive effect is used, is the signal detectable at a signal-to-noise ratio of 0.1%?

**8.2** Determine the maximum electric power required for a thermal flow sensor. The sensor consists of a microchannel with a cross section of $500 \times 500$ μm and a heater with a heated surface of $500 \times 200$ μm. The sensor is designed for a maximum water flow rate of 10 μL/min, and the temperature difference between the heater and the fluid is kept constant at 40K.

**8.3** Determine the measured flow range of the sensor in Problem 8.2, working in the electrocaloric mode with a distance of 500 mm between the heater and the temperature sensor. The heater and temperature sensors are suspended in the flow channel; neglect heat loss by conduction to the substrate.

**8.4** The temperature sensor in Problem 8.3 is to be designed as a thermopile with p-type silicon and aluminum ($\alpha_s = 833$ mV/K). Determine the number of thermocouples required for achieving a resolution of 1:50 of the full measurement range. The signal-to-noise ratio of the sensor is 1:1,000, and the maximum output voltage is 5V.

### References

[1] Kuo, J., T., et al., "Micromachined Thermal Flow Sensors-A Review," *Micromachines*, Vol. 2, 2012, pp. 550–573.

[2] Nguyen, N. T., et al., "Investigation of Forced Convection in Micro-Fluid Systems," *Sensors and Actuators A*, Vol. 55, 1996, pp. 49–55.

[3] Lammerink, T. S. J., et al., "Micro-Liquid Flow Sensor," *Sensors and Actuators A*, Vol. 37–38, 1993, pp. 45–50.

[4] van Kuijk, J., et al., "Multiparameter Detection in Fluid Flows," *Sensor and Actuators A*, Vol. 46/47, 1995, pp. 380–384.

[5] Boillant, M. A., et al., "A Differential Pressure Liquid Flow Sensor for Flow Regulation and Dosing Systems," *Proceedings of MEMS'95, 8th IEEE International Workshop Micro Electromechanical System*, Amsterdam, The Netherlands, Jan. 29–Feb. 2, 1995, pp. 350–352.

[6] Cho, S. T., and Wise, K. D., "A High-Performance Microflowmeter with Built-In Self Test," *Sensors and Actuators A*, Vol. 36, 1993, pp. 47–56.

[7] Oosterbroek, R. E., et al., "A Micromachined Pressure/Flow-Sensor," *Sensors and Actuators A*, Vol. 77, 1999, pp. 167–177.

[8] Chung, K., et al., "Multiplex Pressure Measurement in Microsystems Using Volume Displacement of Particle Suspensions," *Lab Chip*, Vol. 9, 2009, pp. 3345–3353.

[9] Wang, L., et al., "Polydimethylsiloxane-Integratable Micropressure Sensor for Microfluidic Chips," *Biomicrofluidics*, Vol. 3, 2009, pp. 1–8.

[10] Gass, V., van der Shoot, B. H., and de Rooij, N. F., "Nanofluid Handling by Micro-Flow-Sensor Based on Drag Force Measurements," *Proceedings of MEMS'93, 6th IEEE International Workshop Micro Electromechanical System*, San Diego, CA, Jan. 25–28, 1993, pp. 167–172.1.

[11] Zhang, L., et al., "A Micromachined Single Crystal Silicon Flow Sensor with a Cantilever Paddle," *International Symposium on Micromechatronics and Human Science*, Nagoya, Japan, Oct. 1997, pp. 225–229.

[12] Richter, M., et al., "A Novel Flow Sensor with High Time Resolution Based on Differential Pressure Principle," *Proceedings of MEMS'99, 12th IEEE International Workshop Micro Electromechanical System*, Orlando, FL, Jan. 17–21, 1999, pp. 118–123.

[13] Ozaki, Y., et al., "An Air Flow Sensor Modeled on Wind Receptor Hairs of Insects," *Proceedings of MEMS'00, 13th IEEE International Workshop Micro Electromechanical System*, Miyazaci, Japan, Jan. 23–27, 2000, pp. 531–536.

[14] Zarifi, M., H., et al., "Noncontact and Nonintrusive Microwave-Microfluidic Flow Sensor for Energy and Biomedical Engineering," *Scientific Reports*, Vol. 8, 2018, pp. 1–10.

[15] Sanati Nezhad, A., et al., "PDMS Microcantilever-Based Flow Sensor Integration for Lab-on-a-Chip," *IEEE Sensors*, Vol. 13, 2013, pp. 601–609.

[16] Lien, V., and Vollmer, F., "Microfluidic Flow Rate Detection Based on Integrated Optical Fiber Cantilever," *Lab Chip*, Vol. 7, 2007, pp. 1352–1356.

[17] Noeth, N., et al., "Fabrication of a Cantilever-Based Microfluidic Flow Meter with nL $\text{min}^{-1}$ Resolution," *J. Micromech. Microeng.*, Vol. 21, 2011, pp. 1–10.

[18] Svedin, N., et al., "A New Silicon Gas-Flow Sensor Based on Lift Force," *Journal of Microelectromechanical Systems*, Vol. 7, No. 3, 1998, pp. 303–308.

[19] Enoksson, P., Stemme, G., and Stemme, E., "A Coriolis Mass Flow Sensor Structure in Silicon," *Proceedings of MEMS'96, 9th IEEE International Workshop Micro Electromechanical System*, San Diego, CA, Feb. 11–15, 1996, pp. 156–161.

[20] Richter, A., et al., "The Electrohydrodynamic Micro Flow Meter," *Proceedings of MEMS'91, 3rd IEEE International Workshop Micro Electromechanical System*, Nara, Japan, Jan. 3–Feb. 4, 1991, pp. 935–938.

[21] Mastrangelo, C. H., and Muller, R. S., "A Constant-Temperature Gas Flowmeter with a Silicon Micromachined Package," *Technical Digest of the IEEE Solid State Sensor and Actuator Workshop*, Hilton Head Island, SC, June 4–7, 1988, pp. 43–46.

[22] Wu, S., et al., "MEMS Flow Sensors for Nano-Fluidic Applications," *Proceedings of MEMS'00, 13th IEEE International Workshop Micro Electromechanical System*, Miyazaci, Japan, Jan. 23–27, 2000, pp. 745–750.

[23] Liu, C., et al., "A Micromachined Flow Shear-Stress Sensor Based on Thermal Transfer Principles," *Journal of Microelectromechanical Systems*, Vol. 8, No. 1, 1999, pp. 90–99.

[24] Nguyen, N. T., and Kiehnscherf, R., "Low-Cost Silicon Sensors for Mass Flow Measurement of Liquids and Gases," *Sensors and Actuators A*, Vol. 49, 1995, pp. 17–20.

[25] Lai, P. T., et al., "Monolithic Integrated Spreading-Resistance Silicon Flow Sensor," *Sensors and Actuators A*, Vol. 58, 1997, pp. 85–88.

[26] Stephan, C. H., and Zanini, M., "A Micromachined, Silicon Mass Air Flow Sensor for Automotive Application," *Proceedings of MEMS'91, 3rd IEEE International Workshop Micro Electromechanical System*, Nara, Japan, Jan. 30–Feb. 4, 1991, pp. 30–33.

[27] van Putten, A. F. P., "An Integrated Silicon Double Bridge Anemometer," *Sensors and Actuators A*, Vol. 4, 1983, pp. 387–396.

[28] Kuttner, H., et al., "Microminiaturized Thermistors Arrays for Temperature Gradient, Flow and Perfusion Measurements," *Sensors and Actuators A*, Vol. 25–27, 1991, pp. 641–645.

[29] Johnson, R. G., and Egashi, R. E., "A Highly Sensitive Silicon Chip Microtransducer for Air Flow and Differential Pressure Sensing Applications," *Sensors and Actuators A*, Vol. 11, 1987, pp. 63–67.

[30] Qiu, L., Obermeier, E., and Schubert, A., "A Microsensor with Integrated Heat Sink and Flow Guide for Gas Flow Sensing Applications," *Proceedings of Transducers '95, 8th International Conference on Solid-State Sensors and Actuators*, Stockholm, Sweden, June 16–19, 1995, pp. 520–523.

[31] de Bree, H. E., et al., "The -FLOWN, A Novel Device Measuring Acoustical Flows," *Proceedings of Transducers '95, 8th International Conference on Solid-State Sensors and Actuators*, Stockholm, Sweden, June 16–19, 1995, pp. 356–359.

[32] Lyons, C., et al., "A High-Speed Mass Flow Sensor with Heated Silicon Carbide Bridges," *Proceedings of MEMS'98, 11th IEEE International Workshop Micro Electromechanical System*, Heidelberg, Germany, Jan. 25–29, 1998, pp. 356–360.

[33] Lin, K. M., Kwok, C. Y., and Huang, R. S., "An Integrated Thermo-Capacitive Type MOS Flow Sensor," *IEEE Electron. Device Letters*, Vol. 17, No. 5, 1996, pp. 247–249.

[34] van Herwaarden A. W., and Sarro, P. M., "Thermal Sensor Based on The Seebeck Effect," *Sensors and Actuators A*, Vol. 10, 1986, pp. 321–346.

[35] Wachutka, G., et al., "Analytical 2D-Model of CMOS Micromachined Gas Flow Sensors," *Proceedings of MEMS'91, 3rd IEEE International Workshop Micro Electromechanical System*, Nara, Japan, Jan. 30–Feb. 4, 1991, pp. 22–25.

[36] Verhoeven, H. J., and Huijsing, J. H., "An Integrated Gas Flow Sensor with High Sensitivity, Low Response Time and Pulse-Rate Output," *Sensors and Actuators A*, Vol. 41–42, 1994, pp. 217–220.

[37] Fricke, K., "Mass-Flow Sensor with Integrated Electronics," *Sensors and Actuators A*, Vol. 45, 1994, pp. 91–94.

[38] van Oudheusden, B. W., "Silicon Thermal Flow Sensors," *Sensors and Actuators A*, Vol. 30, 1992, pp. 5–26.

[39] Moser, D., and Baltes, H., "A High Sensitivity CMOS Gas Flow Sensors on a Thin Dielectric Membrane," *Sensors and Actuators A*, Vol. 37–38, 1993, pp. 33–37.

[40] Robaday, J., Paul, O., and Baltes, H., "Two-Dimensional Integrated Gas Flow Sensors by CMOS IC Technology," *Journal of Micromechanics and Microengineering*, Vol. 5, 1995, pp. 243–250.

[41] Nguyen, N. T., et al., "Integrated Thermal Flow Sensor for In-Situ Measurement and Control of Acoustic Streaming in Flexural-Plate-Wave Pumps," *Sensors and Actuators A*, Vol. 79, No. 2, 2000, pp. 115–121.

[42] Ashauer, M., et al., "Thermal Flow Sensor for Liquids and Gases," *Proceedings of MEMS'98, 11th IEEE International Workshop Micro Electromechanical System*, Heidelberg, Germany, Jan. 25–29, 1998, pp. 351–355.

[43] Dillner, U., et al., "Low Power Consumption Thermal Gas-Flow Sensor Based on Thermopiles of Highly Effective Thermoelectric Materials," *Sensors and Actuators A*, Vol. 60, 1997, pp. 1–4.

[44] Meijer, G. C. M., "Thermal Sensors Based on Transistors," *Sensors and Actuators A*, Vol. 10, pp. 103–125.

[45] Kersjes, R., et al., "An Integrated Sensor for Invasive Blood-Velocity Measurement," *Sensors and Actuators A*, Vol. 37-38, 1993, pp. 674–678.

[46] Stemme, G., "A CMOS Integrated Silicon Gas-Flow Sensor with Pulse-Modulated Output," *Sensors and Actuators A*, Vol. 14, 1988, pp. 293–303.

[47] Pan, Y., and Huijsing, J. H., "New Integrated Gas-Flow Sensor with Duty Cycle Output," *Electronics Letters*, Vol. 24, No. 9, 1988, pp. 542–543.

[48] Tong, Q. Y., and Huang, J. B., "A Novel CMOS Flow Sensor with Constant Chip Temperature (CCT) Operation," *Sensors and Actuators A*, Vol. 12, 1987, pp. 9–12.

[49] Yang, C. Q., and Soeberg, H., "Monolithic Flow Sensor for Measuring Millilitre per Minute Liquid Flow," *Sensors and Actuators A*, Vol. 33, 1992, pp. 143–153.

[50] Yu, D., Hsieh, H. Y., and Zemel, J. N., "Microchannel Pyroelectric Anemometer," *Sensors and Actuators A*, Vol. 39, 1993, pp. 29–35.

[51] Hsieh, H. Y., Bau, H. H., and Zemel, J. N., "Pyroelectric Anemometry," *Sensor and Actuators A*, Vol. 49, 1995, pp. 125–147.

[52] Legtenberg, R., Bouwstra, S., and Fluitman, J. H. J., "Resonating Microbridge Mass Flow Sensor with Low-Temperature Class-Bonded Cap Wafer," *Sensors and Actuators A*, Vol. 25–27, 1991, pp. 723-727.

[53] Geijselaers, H. J. M., and Tijdeman, H., "The Dynamic Mechanical Characteristics of a Resonating Microbridge Mass-Flow Sensor," *Sensors and Actuators A*, Vol. 29, 1991, pp. 37–41.

[54] Joshi, S. G., "Flowsensors Based on Surface Acoustic Waves," *Sensor and Actuators A*, Vol. 44, 1994, pp. 63–72.

[55] Vellekoop, M. J., "A Smart Lamb-Wave Sensor System for the Determination of Fluids Properties," Ph.D. thesis, Delft University of Technology, Delft, Netherlands, 1994.

# Chapter 9

## Microfluidics for Life Sciences and Chemistry: Microneedles

Micromachining technology offers great opportunities for biomedical instrumentation. One of the most common and simplest biomedical instruments is the needle. Traditionally, hypodermic needles are used for delivering drugs and aspirating body fluid through the human skin. The smallest needles fabricated by conventional machining methods have a minimum diameter on the order of 300 μm. These relatively large needles cause pain and are less accurate for targeted delivery in microscale. Furthermore, the advances in biotechnology requires delivery of molecules in nanometer scale with micron precision. Conventional hypodermic needles can not fulfill these requirements.

Micromachining technology is able to fabricate needles smaller than the previous limit of 300 μm. The small size opens new application fields for this simple device. The main applications of microneedles are:

- Painless drug and vaccine delivery through skin (transdermal or intradermal);
- Minimally invasive ocular drug delivery;
- Delivery of active cosmetic ingredients;
- Patient monitoring and diagnosis;
- Closed-loop chemical stimulation of tissues;
- Cell manipulation;
- Sample collection and delivery in chemical and biochemical analysis.

Figure 9.1 shows the structure of human skin. The outer layer is the *stratum corneum* (SC), which is 10 to 15 μm thick and is a dead tissue. The SC layer offers protection for the body. However, this layer poses a barrier and limits the transport rate across the skin in transdermal delivery of drugs. The next layer is viable epidermis (VE) with a thickness of about 50 to 100 μm. This tissue layer consists of living cells, which have blood vessels capable of transporting drugs, but contains very few nerves.

A number of methods, such as injection with needles, chemical/lipid enhancers, iontophoresis, electroporation, acoustic, and photoacoustic effect, have been used for increasing the transport rate across the skin [1]. While direct injection delivers drugs intradermally, the other methods create small pores in the SC layer for improving the transport rate of drug molecules. Most of these methods create pores in the submicron scale. A microneedle (MN) is the another alternative for a transport pathway with pore size on the order of microns, large enough for a high transport rate but still too

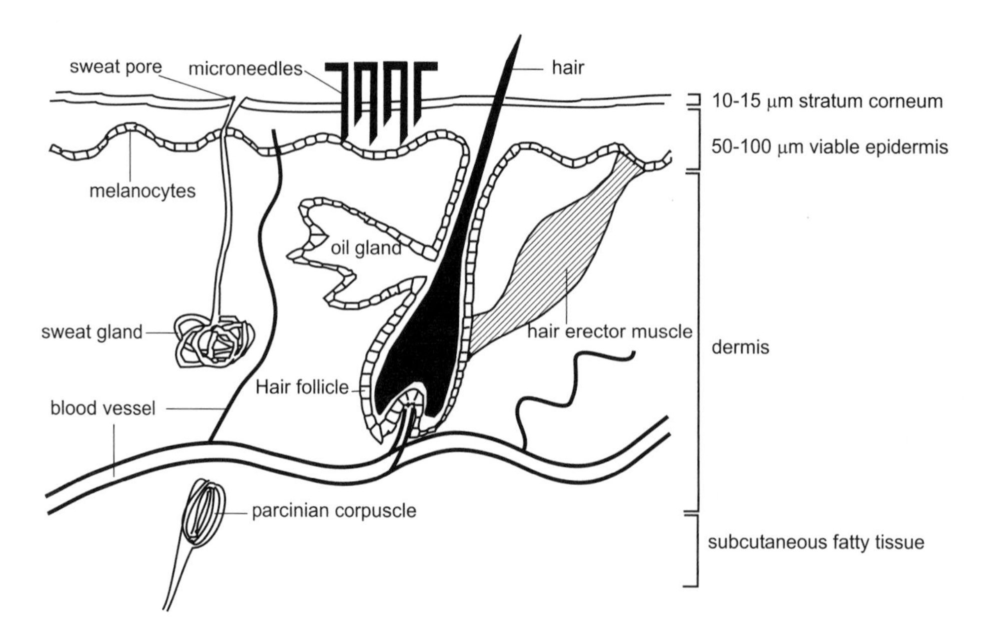


**Figure 9.1** Human skin and microneedle for painless drug delivery.

small to cause damage of clinical significance. Hypodermic needles need medical expertise and are not well suited for controlled delivery of a small amount of drug over a long period of time. Furthermore, microneedles make highly localized, even intracellular delivery possible.

If a microneedle can reach the region of VE, it would be able to deliver drugs without pain. The microneedle should penetrate into the skin from 15 to 100 μm in order to fulfill this function. In transdermal drug delivery, microneedle arrays are needed to puncture the skin. The tiny pores created by the MNs allow faster drug diffusion [2]. MN arrays are considered as minimally invasive drug delivery devices that can bypass the SC barrier to provide drug to the skin microcirculation [3]. This represents a systemic drug delivery via the transdermal route [3]. The microneedles can also remain on the skin after insertion. An array of solid microneedles sandwiched with a gel containing drugs improves the transport rate through the gap between the needle and the punctured skin. Besides the puncturing function, microneedles may also have a delivering function. Higher transport rates can be reached with microneedles coated with drugs. Hollow microneedles allow direct intradermal delivery to the VE region. Drugs are delivered through the bore inside the microneedle by diffusion or pressure-driven flow.

Since micromachining technology is compatible with microelectronics, it is possible to integrate more functions into the needle. The monolithic integration of sensors in microneedles allows measurement of the reaction of tissues and cells. The chemical stimulation function of the drug and the response signal from the sensors make closed-loop controlled cell stimulation possible [4]. Microneedles can manipulate a single cell as well as measure its response. A further application of microneedles is precise fluid sampling for chemical and biochemical analysis. The small size of microneedles can improve the resolution of the dispensed amount.

The following sections first consider the mechanical design of microneedles. Important parameters such as insertion force, buckling force, and fracture force are analyzed. The design should ensure trouble-free insertion into the skin and prevent needle fracture. Mass transport mechanisms such as pressure-driven flow and diffusion with microneedles are discussed. Based on the orientation of insertion direction relative to substrate surface, microneedles can be categorized as in-plane needles and out-of-plane needles. In the following, microneedles are categorized by their function as solid needles and hollow needles. Each type can be further classified by its substrate material, such as silicon, metal, polymer, or glass.

## 9.1 DELIVERY STRATEGIES

Drug delivery through the transdermal route provides the following advantages [3]:

- Inhibition of gastrointestinal degradation;
- Prevention of first-pass hepatic metabolism;
- Prolonged maintenance of relatively constant concentration of drug in blood plasma for up to 7 days using the same patch;
- Avoidance of pain, discomfort, and poor compliance related to injection.

To take advantage of delivery using MN arrays, extensive effort has been paid on exploring various materials, fabrication methods and designs for the development of MN technology toward drug delivery applications [3]. MNs are fabricated in 50 $\mu$m–900$\mu$m height in a density of up to 2,000 MN $cm^{-2}$ [5].

According to the literature, the five MN types as solid, coated, hollow, dissolvable, and hydrogel-forming MNs (Figure 9.2). The five methods are explained as follows [3].

Solid MNs require a two-step drug administration approach. Initially, temporary microchannels are created in SC via MNs. Then a drug formulation is applied [Figure 9.2(a)]. The transport of drug occurs through passive diffusion. The role of the solid MNs is to enhance the permeation of drug formulation. Silicon, metals, and polymers have been used to fabricate solid MNs.

The second method is using drug-coated MNs [Figure 9.2(b)]. Drug formulation is coated over the solid MNs prior to skin application [3]. This is a one-step application process; however, the amount of administered drug is limited to the finite surface area of the MN arrays. Coated MNs have been used for rapid cutaneous delivery of vaccines, proteins, and peptides.

The dissolving MNs are made of biocompatible polymers or sugar via micro moulding process. The MNs contain drug formulation. Dissolution of the MNs occur upon skin insertion of the array and the subsequent exposure of the MNs to skin interstitial fluid [Figure 9.2(c)]. Controlled release of drugs can be obtained by adjusting the fabrication process of MNs or modulating the polymeric composition of the MNs. Water-soluble materials or biodegradable polymers have been explored to make MN arrays. The MNs dissolve or degrade in skin releasing their cargo after insertion.

Hollow MNs enable direct injection of a fluid formulation into the skin through the microchannels inside the needles [Figure 9.2(d)]. Various actuation methods such as diffusion, pressure, or electricity, can be employed for drug injection. In comparison to the other methods, this delivery scheme is able to inject a larger amount of drug. The drug injection is potentially limited because of the clogging of the needle openings. To tackle this issue, the design of the MNs can be modified to accommodate bore-opening at the side of the MN tip. Special attention should be paid to the design of a suitable reservoir to accommodate the liquid drug in a safe and stable condition. Liquid drugs can be very unstable, particularly at elevated temperatures.

The last method is a relatively new approach that uses hydrogel-forming matrices [5]. The MNs are prepared drug-free from a cross-linked polymer. The drug is loaded into a patch-type reservoir located over the array [Figure 9.2(e)]. After insertion of the MNs array into the skin, the tissue interstitial fluid diffuses into the polymer MN enabling drug transport from the reservoir to the tissue through the swollen needles. Hydrogel-forming needles can be removed intact from the

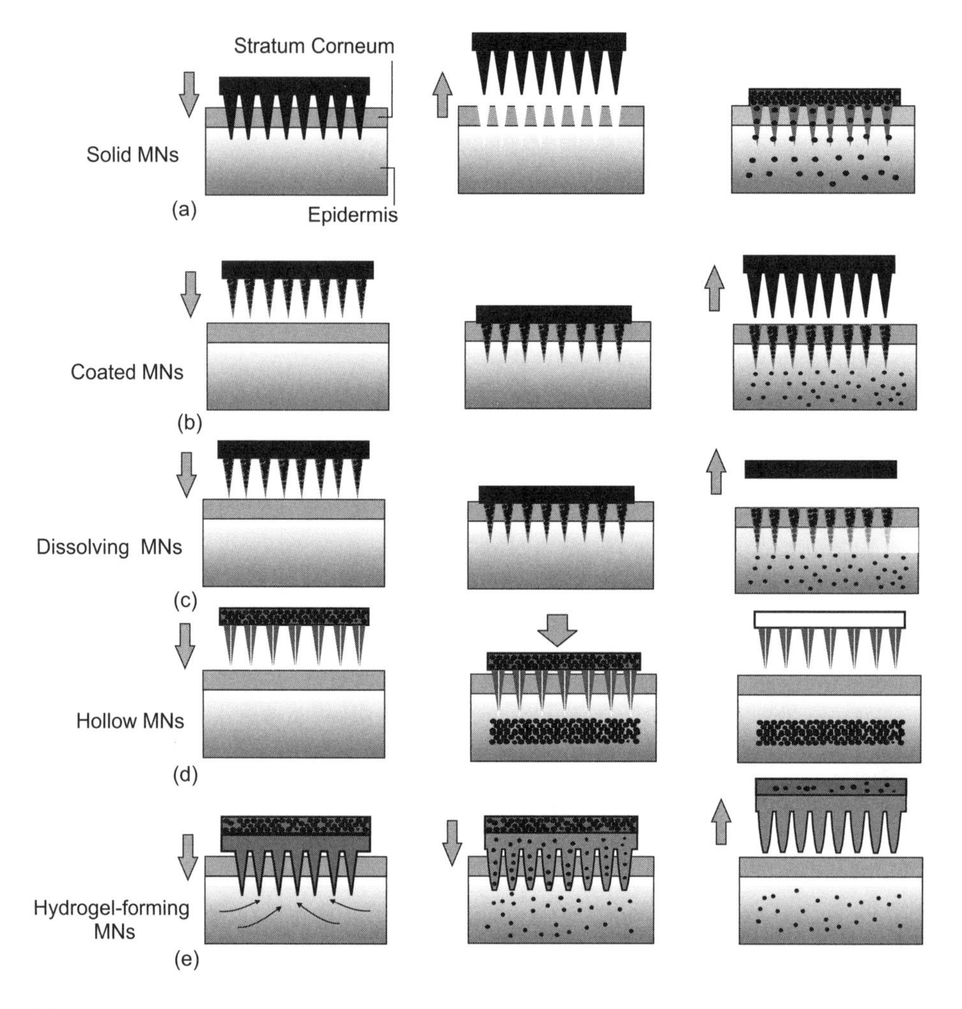


**Figure 9.2** Schematic representing strategies of MN-based transdermal drug delivery: (a) solid MNs; (b) coated MNs; (c) dissolving MNs; (d) hollow MNs for direct infusion or diffusion of drug formulations via needle bores; and (e) hydrogel-forming MNs for direct diffusion of the drug from the local patch to the tissue though the swollen microprojections. (*After* [3].)

skin; however, the reuse of the MN array is not feasible due to the softened needle tips. This feature can decrease the risk of infection transmission.

## 9.2 DESIGN CONSIDERATIONS

The requirements for a microneedle depend on its applications. The general requirements of a microneedle are:

- Sharpness for penetrating tissues;
- Strength against fracture, bending, and buckling;
- Sufficient flow rate in the needle's bore;
- Biocompatibility of needle material;

- Operation considerations for clinical translation.

The first two requirements are characterized by the insertion force $F_{\mathrm{i,max}}$, the buckling force $F_{\mathrm{b}}$, and the fracture force $F_{\mathrm{f}}$. For safety, the insertion force should be less than the buckling force, which in turn should be less then the fracture force ($F_{\mathrm{i,max}} < F_{\mathrm{b}} \ll F_{\mathrm{f}}$). The ratio between the insertion force and the fracture force $F_{\mathrm{i,max}}/F_{\mathrm{f}}$ is called the margin of safety [7].

Since MNs may penetrate into viable layers of the skin, the use of MNs should not cause any infection. Sterilization of the MNs or aseptic manufacturing of MNs should be performed to inhibit the transmission of any microbial load to the body and reduce the infection risks [3]. Required processes should be designed in a way that does not harm the design of the MNs and their cargo. In addition, biocompatibility of the employed materials for the manufacture of the MNs is mandatory to avoid local or systemic reactions. More importantly, the design of the MN-based drug delivery should be optimized to enhance patient and prescriber compliance [3]. MN patches should provide easy-to-use alternative approaches to oral and parenteral delivery methods [3]. To this end, applicators, such as a spring-loaded piston [6], can be used for the insertion of MNs to minimize interindividual variability in skin insertion [3]. For drug delivery, an applicator is placed over the surface of the skin and then pressed to penetrate the MNs into the epidermis.

### 9.2.1 Mechanical Design

#### 9.2.1.1 Insertion Force

In contrast to conventional technologies, micromachining allows fabrication of needle tips with arbitrary shape and sharpness defined by photolithography. Needle tips with a submicron tip radius are possible. The area $A$ of contact surface at the needle tip is considered as the determining factor for the insertion force $F_{\mathrm{i}}$.

The theory of skin insertion is based on the puncture toughness $G_{\mathrm{p}}$, which is the work per area needed to initiate a crack:

$$G_{\mathrm{p}} = \frac{W}{A} \tag{9.1}$$

where $W$ is the total work input and $A$ is the surface of the affected skin area. In case of a hollow needle, the total area determined by the outer perimeter is considered equal the affected skin area. The relation between the work and the insertion force is described as:

$$W = \int_{x=0}^{x=x_{\mathrm{i}}} F_{\mathrm{i}} \mathrm{d}x = G_{\mathrm{p}} A \tag{9.2}$$

where $x_{\mathrm{i}}$ is the the puncture position. The insertion force is assumed to be an exponential function of insertion axis $x$ [7]:

$$F_{\mathrm{i}}(x) = F_0 \exp{(x/\chi)} \tag{9.3}$$

where $F_0$ and $\chi$ are to be determined experimentally by curve fitting. The variable $\chi$ can be considered as the characteristic insertion length. Substituting (9.3) in (9.2) results in:

$$\begin{aligned} W &= \textstyle\int_{x=0}^{x=x_{\mathrm{i}}} F_0 \exp{(x/\chi)} \mathrm{d}x = G_{\mathrm{p}} A \\ F_0 \chi \exp{(x_i/\chi)} &- F_0 \chi = G_{\mathrm{p}} A \\ F_{\mathrm{i,max}} &= F_{\mathrm{i}}(x_{\mathrm{i}}) = F_0 + \tfrac{1}{\chi} G_{\mathrm{p}} A \end{aligned} \tag{9.4}$$

The insertion force in (9.4) is for the static case, where the insertion speed is slow and the kinetic energy of the needle is negligible. Adding kinetic energy into the insertion work in (9.1) will reduce the required insertion force significantly. Yang and Zahn [8] used vibratory actuation and reduced the insertion force to 30% of the value in the static case. The typical vibration frequency is in the kilohertz range.

**Example 9.1: Insertion Force of a Microneedle**

The tip radius of a microneedle is 50 μm. Assume a puncture toughness of the skin of 30 kJ/m$^2$, a characteristic insertion length of 150 μm, and an initial force of $F_0 = 0.1$N. Determine the force required for puncturing the skin.

**Solution.** The affected skin area is:

$$A = \pi r^2 = \pi \times (50 \times 10^{-6})^2 = 7.85 \times 10^{-19}\ \mathrm{m}^2$$

According to (9.4), the insertion force is:

$$F_{\mathrm{i,max}} = F_0 + \frac{1}{\chi} G_{\mathrm{p}} A = 0.1 + \frac{1}{150 \times 10^{-6}} \times 30 \times 10^3 \times 7.85 \times 10^{-9} = 1.67\mathrm{N}$$

9.2.1.2 Buckling Force

*Euler's Model of Microneedle with Constant Cross Section.* If the needle's length is relatively long compared to its width, the first failure mode is buckling. The critical force can be determined based on the simple model of Euler's column [9]. The column is assumed to have the same cross section along its length (Figure 9.3). The deflection curve of the column is governed by the differential equation:

$$EI \frac{\mathrm{d}^2 y}{\mathrm{d}x^2} + M(x) = 0 \tag{9.5}$$

where $E$ is the Young's modulus and $I$ is the moment of inertia of the beam cross section. With a slightly defected position $\delta$ at $x = L$, the bending moment is $M(z) = -F(\delta - y)$. Equation (9.5) has then the form:

$$EI \frac{\mathrm{d}^2 y}{\mathrm{d}x^2} = F(\delta - y) \tag{9.6}$$

Solving (9.6) with the boundary conditions of the fixed end of the needle ($x = 0$):

$$y = 0 \text{ and } \frac{\mathrm{d}y}{\mathrm{d}x} = 0 \tag{9.7}$$

results in:

$$y = \delta(1 - \cos kx) \tag{9.8}$$

where $k^2 = F/EI$. Applying the condition ($y = \delta$) at the insertion point ($x = L$) results in the different buckling modes:

$$kL = (2n - 1)\pi/2 \text{ where } n = 1, 2, 3, ... \tag{9.9}$$

The critical buckling force is considered as the buckling force of the first mode ($n = 1$):

$$F_{\mathrm{b}} = \frac{\pi^2 EI}{4L^2} \tag{9.10}$$

The moments of inertia of different needle cross sections are listed in Table 9.1.

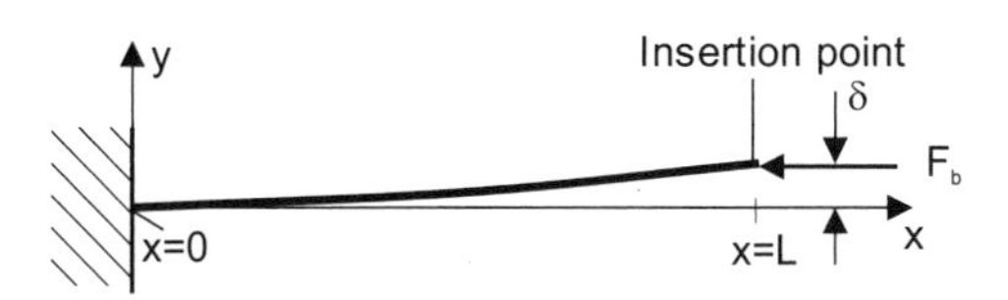


**Figure 9.3** Buckling model of a microneedle.

**Example 9.2: Bending of a Microneedle**

A microneedle is made of parylene C and has channel dimensions of 200 μm × 200 μm × 2 mm. The parylene layer is chemically deposited in 4 hours at a rate of 5 μm/h. Assume that the Young's modulus of parylene C is 3.2 GPa. Determine the tip deflection of the microneedle under a force of 15 mN at the tip.

**Solution.** For a rough estimation, the bending stiffness of the needle can be modeled by the spring constant $k$:

$$k = \frac{3EI}{L^3} = \frac{E}{4L^3}(WH^3 - wh^3)$$

where $E$ is the Young's modulus of the needle material, and $I$ is the needle moment of inertia. The geometry parameters are defined in Table 9.1. The wall thickness of the needle is:

$$5\ \mu\text{m/h} \times 4\ \text{hr} = 20\ \mu\text{m}$$

Thus, the outer dimensions of the needle are:

$$W = 200 + 40 = 240\ \mu\text{m}$$

The bending spring constant of the microneedle is:

$$k = \frac{3E}{12L^3}(W^4 - w^4) = \frac{3 \times 3.2 \times 10^9}{12(2 \times 10^{-3})^3}[(240 \times 10^{-6})^4 - (200 \times 10^{-6})^4] \approx 172\ \text{N/m}$$

The tip deflection is:

$$x = F/k = 0.015/172 = 87 \times 10^{-6} = 87\ \mu\text{m}$$

**Example 9.3: Buckling of a Microneedle**

Determine the critical buckling force for the above microneedle.

**Solution.** The critical buckling force is given by (9.10):

$$F_\text{b} = \frac{\pi^2 E}{4 \times 12 L^2}(W^4 - w^4)$$

$$= \frac{2.25 \times \pi^2 \times 3.2 \times 10^9}{12 \times (2 \times 10^{-3})^2}[(240 \times 10^{-6})^4 - (200 \times 10^{-6})^4] = 2.83 \times 10^{-1}\ \text{N} = 283\ \text{mN}$$

**Example 9.4: Needle Characteristics of Different Materials**

If the above needle is made of polysilicon, what are the tip deflection and the critical buckling force? Assume that all other parameters are the same, and Young's modulus of polysilicon is 150 GPa.

**Table 9.1**

Moments of Inertia for Different Microneedles' Cross Sections. (*After*: [10])

| *Geometry* | *Moment of Inertia* | *Geometry* | *Moment of Inertia* |
|---|---|---|---|
| 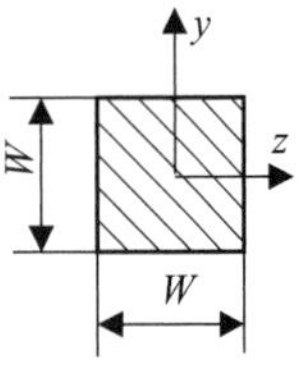 (y, z, W, W) | Solid square $I = \frac{1}{12}W^4$ | 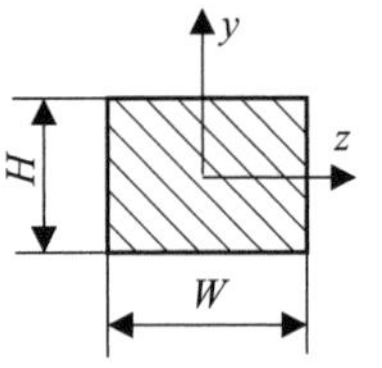 (y, z, H, W) | Solid rectangle $I = \frac{1}{12}WH^3$ |
| 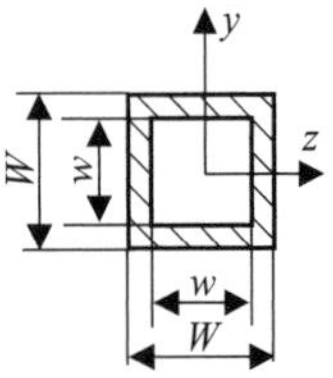 (y, z, W, w, w, W) | Hollow square $I = \frac{1}{12}(W^4 - w^4)$ | 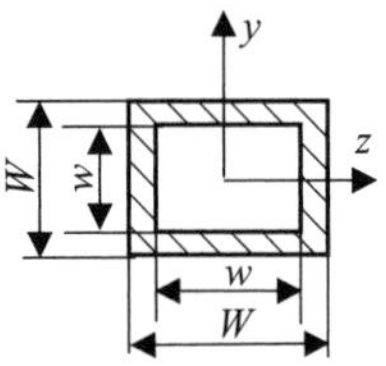 (y, z, W, w, w, W) | Hollow rectangle $I = \frac{1}{12}(WH^3 - wh^3)$ |
| 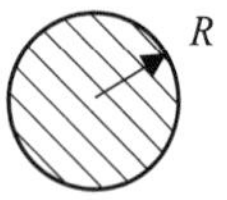 (R) | Solid circle $I = \frac{\pi}{4}R^4$ | 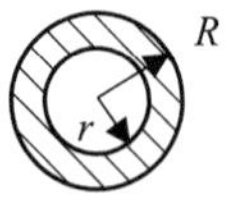 (R, r) | Hollow circle $I = \frac{\pi}{4}(R^4 - r^4)$ |
| 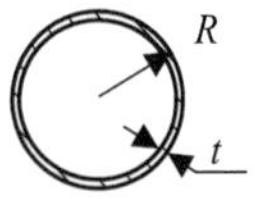 (R, t) | Thin annulus $I = \pi R^3 t$ | | |

**Solution.** The tip deflection of a needle is inversely proportional to its Young's modulus. Thus, the tip deflection of a polysilicon needle is:

$$x_{\text{poly}} = x_{\text{parylene}} \frac{E_{\text{parylene}}}{E_{\text{poly}}} = 87 \times \frac{3.2}{150} = 1.9 \ \mu\text{m}$$

The critical buckling force is proportional to the Young's modulus:

$$F_{\text{poly}} = F_{\text{parylene}} \frac{E_{\text{poly}}}{E_{\text{parylene}}} = 0.283 \times \frac{150}{3.2} = 13.3\text{N}$$

*Microneedle with Continuously Varying Cross Section.* In many practical cases, microneedles are designed with a tapered shape. The sharp tip improves insertion, while the broad base offers mechanical stability. A varying cross section results in a varying moment of inertia. The governing equation of buckling curve has then the form:

$$EI(x)\frac{\mathrm{d}^2 y}{\mathrm{d}x^2} + M(x) = 0 \tag{9.11}$$

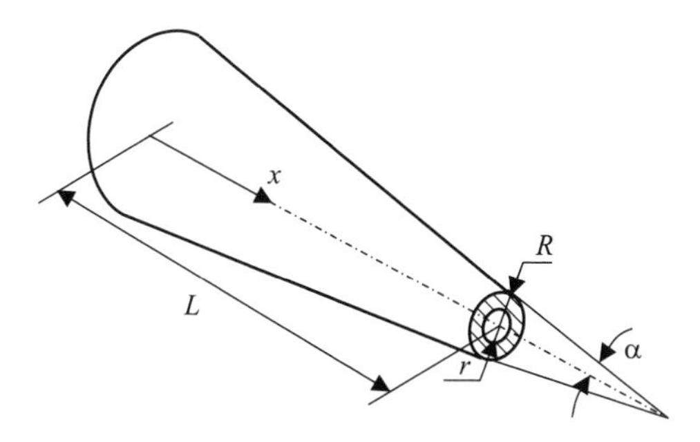


**Figure 9.4** Model of a hollow tapered microneedle.

where the moment of inertia $I(x)$ is a function of $x$-axis. For a solid microneedle, the moment of inertia can be described with the polynomial form:

$$I(x) = \sum_{i=0}^{n} k_i x^n \tag{9.12}$$

The critical buckling force derived from (9.11) has the form [11]:

$$F_\text{b} = \frac{\pi^2 E}{2L^3} \int_0^L \sum_{i=0}^{n} k_i x^n \cos^2\left(\frac{\pi x}{2L}\right) \mathrm{d}x \tag{9.13}$$

Based on this general solution, Kim et al. derived for a microneedle with a shape of a hollow truncated cone (Figure 9.4) of length $L$, angle $\alpha$, inner radius $r$, and outer radius $R$, the following moment of inertia and critical buckling force [12]:

$$I(x) = \frac{\pi}{32}[8(R^4 - r^4) + 32(L-x)\tan\alpha(R^3 - r^3) + 48(L-x)^2\tan^2\alpha(R^2 - r^2) + \\ +32(L-x)^3\tan^3\alpha(R-r)] \tag{9.14}$$

$$F_\text{b} = \frac{E}{40\pi L^2}\Big[\frac{5\pi^4}{2}(R^4 - r^4) + \left(20\pi^2 + 5\pi^4\right)\left(R^3 - r^3\right)L\tan\alpha + \\ +(30\pi^2 + 5\pi^4)(R^2 - r^2)L^2\tan^2\alpha + (-120 + 30\pi^2 + \frac{5}{2}\pi^4)(R-r)L^3\tan^3\alpha\Big] \tag{9.15}$$

With $\alpha = 0$, (9.15) leads to the same form of simple Euler's buckling (9.10) with the moment of inertia given in Table 9.1 for a hollow circle.

#### 9.2.1.3 Fracture Force

The fracture force of a microneedle can be derived from a thin shell model, where the tip radius $r_\text{t}$ should be 10 times larger than the wall thickness $t$ ($r_\text{t} > 10t$) [10]. The fracture force can be estimated as:

$$F_\text{f} = 2\pi r_\text{t} t \sigma_\text{cr} \sin\alpha \tag{9.16}$$

where $r_\text{t}$ is the radius of the needle tip, $\sigma_\text{cr}$ is the critical stress of the needle material, and $\alpha$ is the needle wall angle. For this model, the shear component of the load is neglected due to the relatively large wall angle ($\alpha \to \pi/2$). Since the ratio of tip radius and wall thickness in reality is often smaller than 10, the model overestimates the fracture force. However, it is safe to use this model.

**Example 9.5: Buckling and Fracturing of a Tapered Microneedle**

A microneedle made of electroplated nickel has a wall angle of 3° and a wall thickness of 20 μm. The tip inner diameter is $r = 40$ μm. The needle is 500 μm long. Determine the critical buckling force and the fracture force of the microneedle, if the Young's modulus and yield strength of electroplated nickel are 23 GPa and 830 MPa, respectively.

**Solutions.** Using (9.15) with $r = 40$ μm, $R = r + t = 40 + 20 = 60$ μm, $L = 500$ μm, $\alpha = 3°$, and $E = 23$ GPa, the critical buckling force of the needle is:

$$F_\mathrm{b} = 4.77\mathrm{N}$$

The critical fracture force at the needle tip is:

$$F_\mathrm{f} = 2\pi r_\mathrm{t} t \sigma_\mathrm{cr} \sin\alpha = 2.18 \times 10^{-1}\ \mathrm{N}$$

Thus, the needle will fracture before bending occurs.

### 9.2.2 Delivery Modes

#### 9.2.2.1 Delivery with Hollow Microneedles

The channel inside the needle should be designed as large as possible to minimize the pressure loss along the microneedle and to allow a maximum flow rate at a given pressure. The pressure loss $\Delta p$ along a microneedle can be calculated as (see Section 2.2.4.1):

$$\Delta p = \mathrm{Re}f \frac{\eta L}{2D_\mathrm{h}^2} u \tag{9.17}$$

where $\eta$ is the viscosity of the fluid, $u$ is the average velocity, $L$ is the channel length in the needle, and $D_\mathrm{h}$ is the hydraulic diameter of the channel (see Section 2.2.4.2):

$$D_\mathrm{h} = \frac{4 \times \text{Cross Section Area}}{\text{Wetted Perimeter}} = \frac{4A}{P_\mathrm{wet}} \tag{9.18}$$

The product of the Reynolds number and the factor (Re$f$) is about 64 (circular pipe) and 60 (trapezoidal duct) in conventional, macroscopic fluid dynamics [13]. The friction factor of microchannels is different from that of conventional theories and lies between 50 and 60 [14].

**Example 9.6: Pressure Loss in a Microneedle**

Determine the pressure drop along the needle bore of Example 9.2 if it is used for delivering water at a flow rate of 100 μL/min.

**Solution.** The hydraulic diameter of the needle's bore is:

$$D_\mathrm{h} = \frac{4 \times 200 \times 200}{4 \times 200} = 200\ \mu\mathrm{m}$$

The average velocity of water flow in the needle is (viscosity of water is $\eta = 10^{-3}$ Pa.s):

$$u = \frac{100 \times 10^{-9}/60}{(200 \times 10^{-6})^2} = 41.7 \times 10^{-3}\ \mathrm{m/s} = 41.7\ \mathrm{mm/s}$$

Assuming a factor of Re$f = 55$, the pressure drop becomes:

$$\Delta p = \mathrm{Re}f \frac{\eta u L}{2D_\mathrm{h}^2} = 55 \times \frac{10^{-3} \times 41.7 \times 10^{-3} \times 2 \times 10^{-3}}{2 \times (200 \times 10^{-6})^2} = 57\ \mathrm{Pa}$$

9.2.2.2 Delivery Based on Transdermal Diffusion

Diffusive mass transfer across the skin can be modeled with the mass transfer coefficient $k$:

$$\dot{m} = kA\Delta c \tag{9.19}$$

where $\dot{m}$ is the rate of mass transfer, $A$ is the surface where mass transfer occurs, and $\Delta c$ is the concentration difference. The mass transfer coefficient of skin with punctured holes can be formulated in the simple linear form as [15]:

$$k = f\frac{D}{L} \tag{9.20}$$

where $f = A_{\text{pores}}/A_{\text{skin}}$ is the fractional skin area containing pores from microneedles, $D$ is the diffusion coefficient of the specie in the pore, and $L$ is the pore length. The pore length is approximately the thickness of the epidermis layer (Figure 9.1); thus, $L \approx 50$ µm. If the microneedles are inserted and left on the skin during the diffusion process, the pore can be modeled as an annulus. If the needles are removed after insertion, the pore can be modeled as a cylinder.

**Example 9.7: Transdermal Diffusion of a Drug**

An array with $N = 400$ microneedles is used for transdermal drug delivery. The array occupies an area of $A_{\text{skin}} = 10$ mm$^2$. The drug molecules have a diffusion coefficient of $D = 2 \times 10^{-10}$ m$^2$/s. The tip radius is $R = 10$ µm. Determine the mass transfer coefficient of the drug for two cases: (a) the needles remain on the skin and the annular gap is $w = 250$ nm; and (b) the needles are removed from the skin, leaving pores with a radius of $R_{\text{pore}} = 6$ µm.

**Solution.** For the case with microneedles remaining on skin, the fractional skin area is:

$$\begin{aligned} f_1 &= N\tfrac{\pi[(R+w)^2-R^2]}{A_{\text{skin}}} \\ &= 400 \times \tfrac{\pi[(10\times10^{-6}+250\times10^{-9})^2-(10\times10^{-6})^2]}{10\times10^{-6}} = 6.36 \times 10^{-4} \end{aligned}$$

The corresponding mass transfer coefficient is:

$$k_1 = f_1\frac{D}{L} = 6.36 \times 10^{-4} \times \frac{2 \times 10^{-10}}{50 \times 10^{-6}} = 2.54 \times 10^{-9} \text{ m/s}$$

After removing the microneedles, the fractional skin area is:

$$f_2 = N\frac{\pi R_{\text{pore}}^2}{A_{\text{skin}}} = 400 \times \frac{\pi \times (6 \times 10^{-6})^2}{10 \times 10^{-6}} = 4.52 \times 10^{-3}$$

The mass transfer coefficient of the affected skin area is:

$$k_2 = f_2\frac{D}{L} = 4.52 \times 10^{-3} \times \frac{2 \times 10^{-10}}{50 \times 10^{-6}} = 1.81 \times 10^{-8} \text{ m/s}$$

Removing the microneedles yields a better mass transfer coefficient.

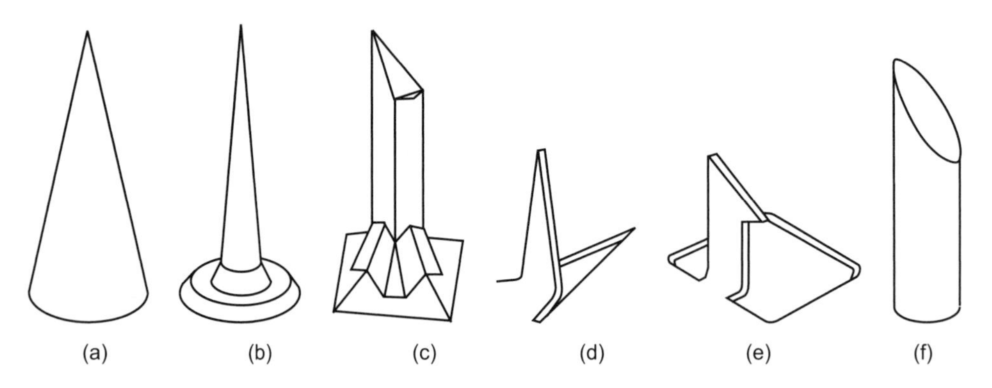


**Figure 9.5** Typical shapes of solid microneedle: (a) silicon, etched in $SF_6$ and $O_2$ plasma [2]; (b) silicon, etched with separated DRIE and RIE [16]; (c) silicon, dicing, and wet etching in KOH [17]; (d) stainless steel, laser machined and bent [18]; (e) titanium, wet etched and bent [19]; and (f) SU-8, photolithography, filled with a sacrificial polymer and etched with a metal mask.

## 9.3 DESIGN EXAMPLES

### 9.3.1 Solid Microneedles

Solid microneedles only have the puncturing function. The design of this needle type is relatively simple. According to (9.4), the needles only need to have a sharp tip to reduce the insertion force and keep this force under the critical buckling force and the fracture force. Solid microneedles can be fabricated by direct micromachining or molding from a micromachined template. Different design examples of solid microneedles are categorized in this section by their materials.

#### 9.3.1.1 Silicon Needles

Single-crystalline silicon is readily available at low cost. Micromachining technologies are well established for silicon (Chapter 3). Thus, the fabrication of silicon microneedles is with a single mask relatively simple. The fabrication process of silicon microneedles consists of two basic steps: forming and sharpening.

Henry et al. [2] combined the two etch gases $SF_6$ and $O_2$ in a reactive ion etcher to shape the needle. While $SF_6$ etches anisotropically and forms the needle body, $O_2$ etches isotropically and sharpens the needle tip [Figure 9.5(a)].

Besides the puncturing function, solid microneedles made of silicon can be used as intracellular sensing probes. Hanein et al. [16] separated the fabrication process into two steps: forming of the needle body and sharpening. First, the needle body and its length are defined by the DRIE process (Chapter 3). Subsequently, the needles are sharpened with RIE in $SF_6$. Protecting pillars that have smaller diameters than the needle were used for forming a strong base for the needle [Figure 9.5(b)].

Shikida et al. [17] proposed a maskless approach, where the needle body was first formed by dicing. Subsequent anisotropic wet etching in KOH sharpens the needles [Figure 9.5(c)]. This technique involves dicing and is not suitable for batch fabrication. Furthermore, the needle size is determined by the geometry of the dicing saw and cannot be adjusted.

#### 9.3.1.2 Needles Made of Metals and Other Materials

Solid microneedles made of metals can be fabricated based on molding techniques. The solid microneedles made of silicon are used as a positive master for casting a polymeric negative mold. The polymeric mold forms the metal replica of the original silicon needles by electroplating.

**Table 9.2**
Typical Parameters of Solid Microneedles

| *Ref.* | *Tip (μm)* | *Base (μm)* | *Length (μm)* | *Material* |
|---|---|---|---|---|
| [2] | 1 | 80 | 200 | Silicon |
| [16] | 0.01 | 70 | 230 | Silicon |
| [17] | 0.01 | 150 | 300 | Silicon |
| [18] | 0.01 | 75 × 200 | 1,000 | Stainless steel |
| [20] | — | 50 | 250–600 | SU-8 |

Laser machining can form microneedles directly from a metal sheet and bend 90° out of the plane of the sheet [18] [Figure 9.5(d)]. A similar approach was reported by Matriano et al. [19]. The needles are etched in-plane on a titanium sheet. They are subsequently bent out of plane [Figure 9.5(e)].

The relation between the critical buckling force and the needle's length (9.10) shows that even a soft material such as a polymer can withstand high axial load if the length is reduced. Thus, a sharp polymer microneedle is able to puncture the skin without buckling. Similar to electroplated metal microneedles, polymer microneedles can be fabricated from a micromachined mold. Polymer replicas of silicon microneedles are fabricated using the negative mold described above. Furthermore, molds for polymeric microneedles can be micromachined directly in silicon, SU-8, or electroplated metal.

Polymer microneedles can directly be formed by photolithography of a thick film resist, such as SU-8 [20]. The sharpening process starts with filling of the gap between the SU-8 cylinders with a sacrificial polymer polyactic-co-glycolic acid (PLGA). The polymer layers (SU-8 and PLGA) was subsequently dry-etched with a metal mask. The mask is positioned asymmetrically to the SU-8 cylinders. Thus, underetching results in a beveled shape of the SU-8 cylinder. Removing the sacrificial polymer with a solvent releases the SU-8 microneedles [Figure 9.5(f)]. Table 9.2 compares typical parameters of the solid microneedles.

### 9.3.2 Hollow Microneedles

In contrast to solid microneedles, the fabrication of hollow microneedles is more complex due to the additional hollow bore. Generally, hollow microneedles can be fabricated either in the bulk substrate, or using surface micromachining with a sacrificial material. In the first approach, the complete hollow microneedles are fabricated directly from a substrate material using bulk micromachining. In the latter approach, a variety of materials can be used as functional layer. First, a functional layer such as polysilicon or electroplated metal is deposited on a sacrificial solid microneedle. Removing the sacrificial material results in a hollow microneedle. Hollow microneedles are categorized according to their substrate materials.

#### 9.3.2.1 Silicon Needles

One of the first approaches to make microneedles with micromachining technology is to fabricate them in silicon. Figure 9.6(a) depicts a solution in single-crystalline silicon reported by Chen et al. [4]. The microchannel is wet-etched by ethylene-diamine pyrocatechol (EDP) (see Section 3.3.1). The highly boron-doped silicon layer is structured as a grid, which allows etch access to the underlying silicon. EDP etching forms the buried channel, because it does not attack the grid covering the channel. The channel is then sealed by thermal oxidation. Deep boron diffusion defines the shape of the needle shank. Wet etching in EDP subsequently releases the needle. Since this

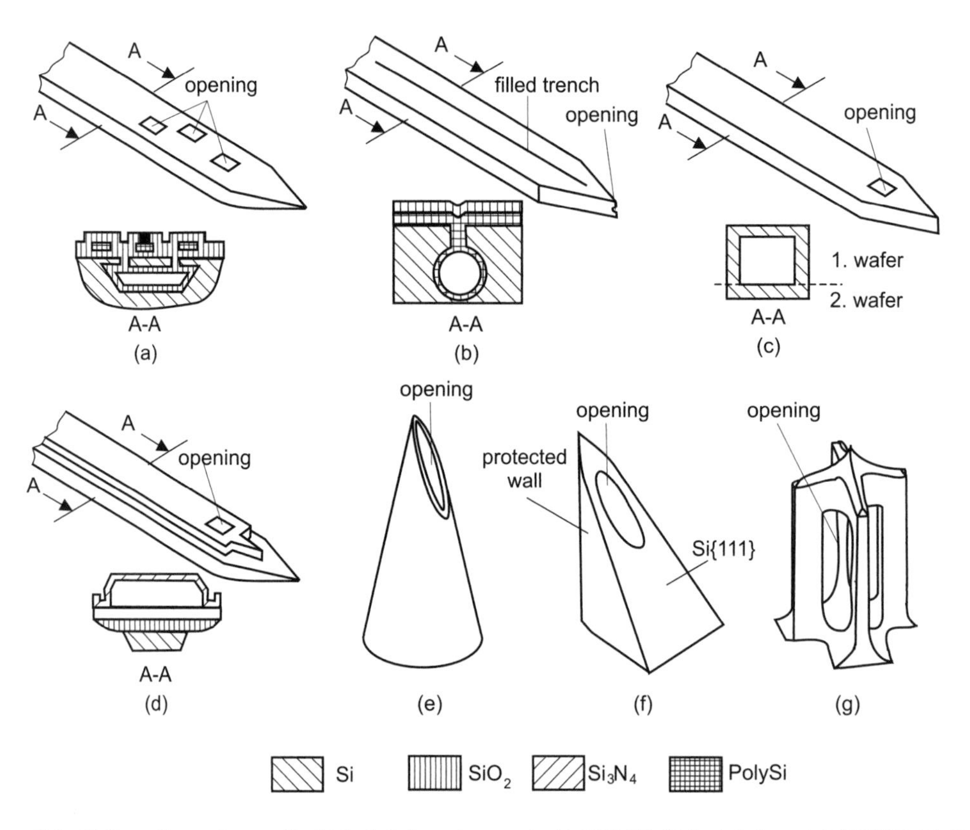


**Figure 9.6** Hollow silicon microneedle: (a) buried channel, wet etched in EDP [4]; (b) buried channel, DRIE [21]; (c) direct bonding, DRIE [22]; (d) surface micromachined channel [23]; (e) out-of-plane, sharpened by dry etching [25]; (f) out-of-plane, sharpened by wet etching [26]; and (g) out-of-plane, sharpened by dry etching with a cross-shaped mask [27].

process is compatible to microelectronics, electrodes can be integrated into the needle for measuring the tissue reaction after injecting drugs. Paik et al. [21] used a similar approach to make a planar array of hollow microneedles. The needles are shaped by DRIE [Figure 9.6(b)]. The advantage over the technology of Chen and Wise is the well-defined needle shape and the opening at the needle's tip.

Since the diffusion profile of boron is not uniform, it is difficult to control the vertical geometry of this needle. The other way of making microneedles in single-crystalline silicon is by using a two-wafer process with DRIE and an SOI wafer (see Section 3.1.2.7). A DRIE process defines the rectangular channel in the first wafer. The channel is covered with an SOI wafer (second wafer). The first wafer is then etched from the back until the trench meets the oxide layer of the SOI wafer. This second DRIE process defines the outer shape of the needle. Etching away the oxide layer releases the microneedle [Figure 9.6(c)] [22].

The example depicted in Figure 9.6(d) is made of highly boron-doped silicon, similar to the example shown in Figure 9.6(a). The channel is made of silicon nitride with surface micromachining. The drawback of this solution is the limited channel height, which is defined by the thickness of the sacrificial layer (on the order of several microns). The access holes for the sacrificial layer are then sealed with a second nitride deposition [23].

For out-of-plane needles that are perpendicular to the wafer surface, the tube structure can be fabricated in single-crystalline silicon using DRIE. The drawback of this solution is that the needle tip is not sharp [24]. Mukerjee et al. [25] used DRIE to etch the bore of the microneedle first. Subsequently, the needle body was defined by dicing. The needles were sharpened by isotropic etching of silicon in a wet etchant composed of hydrofluoric, nitric, and acetic acids [Figure 9.6(e)]. Gardeniers et al. [26] sharpened the out-of-plane hollow microneedles using isentropic etching in KOH. While the needle wall is protected by a nitride layer, one side of the needle is etched isentropically and results in a slanted {111} surface [Figure 9.6(f)]. Griss and Stemme [27] used a cross-shaped mask for the isotropic dry etching process to open the bore on the needle's side wall [Figure 9.6(g)].

The second approach is to make microneedles of polysilicon. The needle is formed by a silicon mold, which has phosphosilicate glass (PSG) as a sacrificial layer [28, 29]. Polysilicon is deposited and annealed in the inner wall of the mold repeatedly, until the desired thickness of 15 to 20 microns is reached. This process results in high yields, but is time-consuming because of the relatively thick deposition layer. The needle shape is similar to that depicted in Figure 9.6(c). The outlet orifice is opened on the side of the needle.

Coating silicon needles with metals, such as titanium, platinum, chromium, gold, or nickel, improves their strength. A 10-μm platinum coating increases the median bending moment of the above polysilicon needle from 0.25 to 0.43 mNm [28]. The fabrication process is compatible with other surface micromachining techniques. Thus, polysilicon needles can be integrated on the same chip with other microfluidic devices, such as micropumps or microvalves [29].

#### 9.3.2.2 Needles Made of Metals and Other Materials

Hollow metal needles can be fabricated by coating a metal layer on sacrificial materials. The sacrificial material works as a mold and defines the shape of the microneedle. The thickness of the metal layer determines the strength of the needle. Because of the required thickness, the metal layer is often electroplated on a thin seed layer, which may be deposited by sputtering.

Figures 9.7(a, b) show the two fabrication methods of hollow microtubes and microneedles using negative mold and positive mold [24]. The negative mold is formed by lithography of a SU-8 layer. Afterward, a seed layer is sputtered on the photoresist wall. Electroplating with the desired metal forms the actual needle. At the end, the sacrificial mold is etched with $O_2/CHF_3$ plasma. Negative molds can only be used for fabrication of microcylinders because of their parallel walls [Figure 9.7(a)].

A needle with a sharp tip needs a positive mold. The positive mold is a solid microneedle, as described in [2]. The negative mold is made of SU-8, which is coated on the positive silicon mold. After etching away silicon, the negative SU-8 mold remains and the process continues as described above. Using this method, needles with 150-μm height, 80-μm base diameter, and 10-μm tip diameter were fabricated [Figure 9.7(b)]. A similar mold can be created by laser machining of a polyethylene terephthalate sheet. After electroplating of nickel, the sacrificial polymer can be removed in boiling NaOH [7].

Figure 9.7(c) shows a solution for in-plane metal microneedles [30, 31, 32]. The needles are fabricated by electroplating on a sacrificial silicon membrane. A thick photoresist sacrificial layer (AZ4620 or AZ9260 [31]) forms the channel. The channel height can be adjusted from 5 to 50 μm by the thickness of the photoresist layer. A second electroplating process with a gold seed layer on the photoresist surface forms the hollow metal needle. After removing the photoresist with acetone, the sacrificial silicon membrane is wet-etched to free the needle structure. With this method, different metals can be used for fabricating the needle.

One of the most important requirements for microneedles is biocompatibility. Silicon or metal needles can be coated with biocompatible materials. While sputtering or evaporation can deposit metal, CVD can deposit other materials, such as parylene or silicon carbide.

Microneedles made of silicon oxide are depicted in Figure 9.7(d) [33]. First, the channel is etched through the entire wafer. Second, the wafer is wet-oxidized. Wet oxidation creates a thick

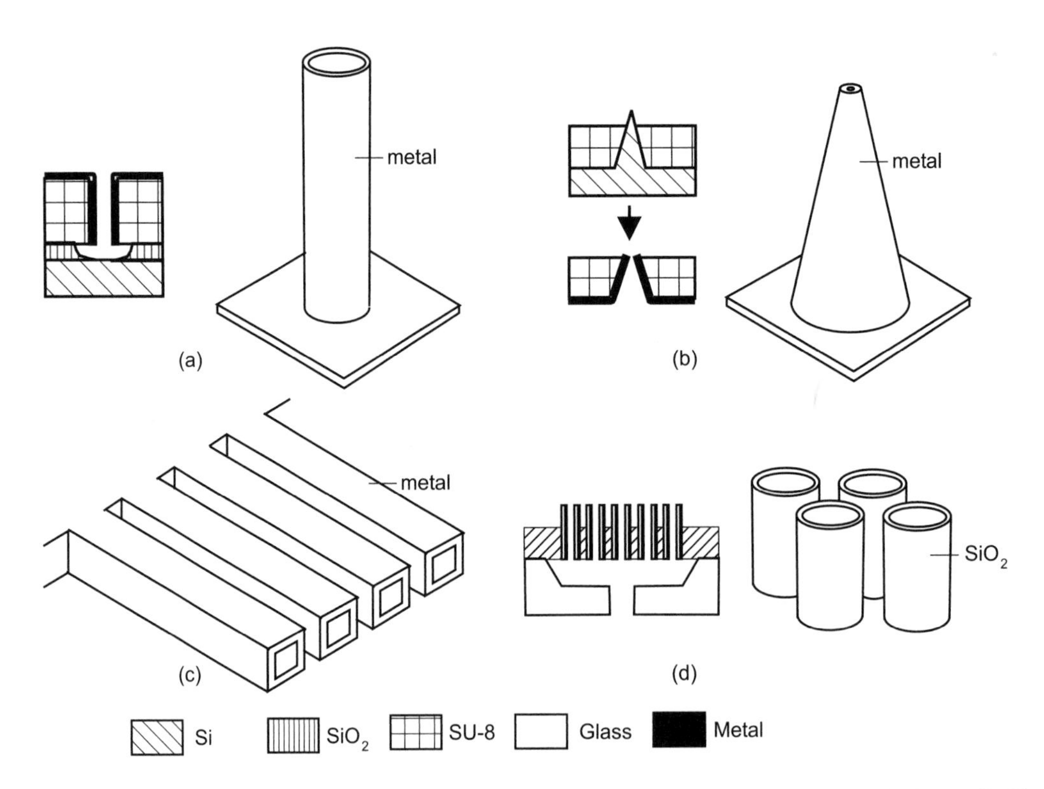


**Figure 9.7** Hollow microneedle made of metals and other materials: (a) microcylinder [24]; (b) tapered microneedle [7]; (c) hollow rectangular needle [30, 31, 32]; and (d) microneedles made of silicon oxide [23].

silicon oxide layer. Etching the silicon wafer from the back side frees the needle. Ohigashi et al. reported a similar process [34]. In this process, a thin fluorocarbon ($CF_x$) layer is deposited by PECVD and works as a protecting layer for the needle. Similar to the process for polysilicon needles in [28], CVD-coated materials such as parylene can be used for fabricating microneedles. Electroplated photoresist works as the sacrificial layer for this process. Parylene needles can also be made with a positive mold. The mold is fabricated from silicon using conventional bulk micromachining. After the conformal deposition and structuring of the parylene layer, etching away silicon releases the needles. Table 9.3 compares the hollow microneedles.

### 9.3.3 Bio-Inspired Microneedles

Natural bio-needles from creatures such as porcupine and mosquitoes have inspired the design of biomimicking microneedles [35]. In particular, the penetration mechanism of mosquito proboscis, an elongated sucking mouthpart, has been studied in details to develop microneedles that require minimum insertion force. Cho et al. has performed a study on American porcupine quill that has a needle-like structure with barbs [36]. The study revealed that the existence of the barbs over the quill results in an increase of the stress distribution and a decrease of the insertion force. Another study, inspired by the the insertion mode of mosquito proboscis in skin, revealed that vibrating MN made of biodegrabel polymer with jagged shape reduce the insertion force [37]. In another research, silicon MN was prepared in an architecture of a central straight channel with two outer jagged needle to mimic the movement of mosquito proboscis. Various movement modes were applied through the PZT actuator. It was revealed that the MN movement mode resembles the mosquito inserting behavior, mode C, required the lowest insertion force compared to mode A and mode B [Figure 9.8].

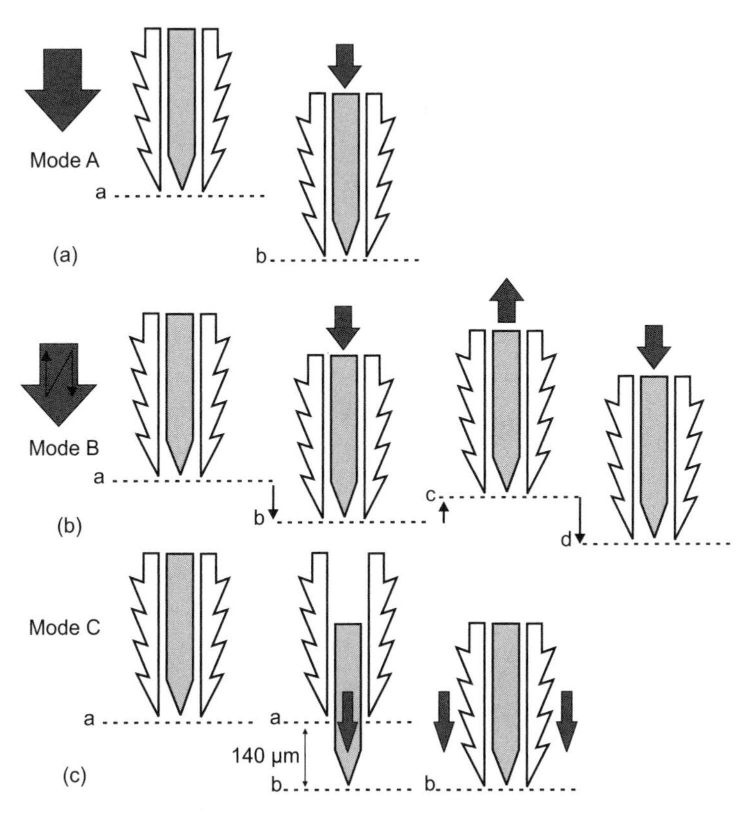


**Figure 9.8** Insertion modes of MNs: (a) simple insertion; (b) vibrating insertion; and (c) insertion of central needle with subsequent movement of the two jagged needles having 180 degrees phase delay [38].

This mode of insertion allowed for the movement of the central needle with subsequent movement of the two jagged needles having 180 degrees phase delay to the central needle [38]. The findings obtained from such studies provide insights for design improvement of MN arrays and the method of needle penetration for painless drug delivery.

### 9.3.4 Polymeric Microneedles

Polymers are mainly employed to fabricate dissolving/biodegradable and hydrogel-forming MNs. Polymeric materials have shown promising characteristics for the fabrication of MNs due to their excellent biocompatibility, biodegradability, low potential of causing toxicity and proper mechanical properties such as strength and toughness [3, 39].

Polysaccharides such as carboxymethlycellulose, amylopectin, dextrin, hydroxypropyl cellulose, alginate and hyaluronic acid are mainly used to form dissolving MN arrays [39]. Such macromolecules provide sufficient mechanical properties for skin piercing.

Biodegradable needles are prepared using poly(lactic acid) (PLA), poly(glycolic acid) (PGA), poly(lactic-co-glycolic acid) (PLGA), and chitosan [39]. Such needles have been explored for the delivery of various therapeutics such as small molecules, macromolecules, and nanoparticles through the skin [39, 40].

Swelling or hydrogel-forming MN arrays are made of polyanhydride type polymers, mixtures of polysaccharides (dextran, gelatin) and poly(vinyl alcohol) (PVA). Conduits are produced once the polymer swell allows for diffusion of drugs from a patch-type reservoir to the dermal circulation [41, 42].

**Table 9.3**
Typical Parameters of Hollow Microneedles

| *Ref.* | *Tip* ($\mu m$) | *Base* ($\mu m$) | *Length* ($\mu m$) | *Wall Thickness* ($\mu m$) | *Material* | *Technology* |
|---|---|---|---|---|---|---|
| [4] | — | $80 \times 9$ | $1,000$–$6,000$ | 1 | Silicon nitride, silicon | Bulk, surface |
| [23] | — | $192 \times 110$ | $7,000$ | 15–20 | Polysilicon, metals | Bulk |
| [25] | $\approx 10$ | 130 | 200 | 3–5 at tip | Silicon | Bulk |
| [27] | — | 100 | 210 | — | Silicon | Bulk |
| [28] | — | $192 \times 110$ | $192 \times 110$ | 15–20 | Polysilicon | Surface |
| [24] | $30 \times 5$ | $30 \times 5$ | 500–10,000 | 20–40 | Metals | Electroplating |
| [7] | 43 | 400 | 500 | 12 | Metals | Electroplating |
| [33] | $5 \times 5$ | $5 \times 5$ | 10–500 | 1–2 | Silicon oxide | Bulk |

## 9.4 SUMMARY

A microneedle is a simple but important component, which directly benefits from the microtechnology. While microneedles can be used for extraction of body fluids and accurate dispensing of liquids, drug delivery remains the most significant application of microneedles [1]. While solid needles improve transdermal diffusion across the punctured skin area, hollow needles allow well-defined diffusion pores or high delivery rate by pressure-driven flow. Solid microneedles were successfully used in combination with an iontophoretic electrode to deliver oligonucleotide [43]. Two orders of improvement were observed with the use of microneedles. Microneedles embedded in the skin allow insulin delivery and significant decreases of blood glucose level [18]. Solid microneedles coated with vaccine show clear antibody response [19].

Polymeric microneedles have gained significant interest for the implementation of dissolvable/biodegradable and hydrogel-forming MN arrays. Polymeric microneedles have superior features of biocompatibility, biodegradability, low toxicity, and suitable mechanical properties. Hydrogel-forming MNs swell once inserted in the skin due to the absorption of interstitial fluid. In this way, the drug can diffuse from the patch-like reservoir into the inserted swollen needle to deliver pharmaceutical cargos.

The mechanical design of microneedles should consider the different critical forces: the insertion force, the buckling force, the fracture force, and the method of needle insertion into the skin. The basic criteria is an insertion force much smaller than the buckling force or the fracture force. Depending on the geometry and material of the needle, fracture or buckling may occur first if the axial load is increased. A thick needle wall can improve the mechanical stability, while the tip radius is the only parameter for a minimum insertion force. Thus, a sharp needle tip is always desirable. MN arrays should be inserted into the skin in a safe and efficacious manner. Unlike conventional patches, the MN arrays need an assistive device such as a piston-like actuator for uniform penetration. For some designs, the MN array is inserted initially into the skin manually and drug delivery is activated upon pushing the MNs into the skin using the assistive device such as a roller [44]. In other methods, the MN arrays are integrated within an applicator and the drug delivery is triggered upon insertion and pushing the MN arrays on the skin using the applicator such as a piston actuator. The applicator provides required force for uniform insertion of the array into the biological tissue at the desired depth [44]. Recently, microneedles have gained attention as a minimally invasive method for ocular delivery of drug formulation. MNs provide a higher precision and accuracy of drug delivery into a target ocular tissue than hypodermic needles [45, 46].

## Problems

**9.1** A microneedle is made of single-crystalline silicon. The channel is 1.5 mm long and has a width of 100 μm and a height of 200 μm. The needle wall is 20 mm thick. Determine: (a) the maximum bending force at the tip of the needle; (b) the critical buckling force; and (c) the water flow rate in the needle if a pressure difference of 100 Pa is applied across the needle.

**9.2** A microneedle is made of single-crystalline silicon. The needle has the form of a cylinder with a radius of 40 μm. The needle wall is 50 μm. Determine the needle length, so that buckling occurs before fracture.

**9.3** If the above needle needs to be 200 μm long, what is the critical needle wall for the case that buckling first occurs?

**9.4** A silicon microneedle has a hollow rectangular cross section. The outer dimension measures $10 \times 50$ μm. Assume a puncture toughness of the skin of 30 kJ/m$^2$, a characteristic insertion length of 150 μm, and an initial force of $F_0 = 0.1$N. Determine the insertion force needed for the needle. If the needle is 150 μm long, what is the safety margin?

**9.5** If the above needle is inserted and remains on the skin for transdermal drug delivery, determine the rate of mass transfer coefficient of a drug with a diffusion coefficient of $1.5 \times 10^{-10}$ m$^2$/s. The gap between the solid needle's wall and the tissue is assumed to be 200 nm.

## References


[1] Prausnitz, M. R., "Microneedles for Transdermal Drug Delivery," *Advanced Drug Delivery Reviews*, Vol. 56, 2004, pp. 581–587.

[2] Henry, S., et al., "Microfabricated Microneedles: a Novel Method to Increase Transdermal Drug Delivery," *J. Pharm. Sci.*, Vol. 87, 1998, pp. 922–925.

[3] Larraneta, E., et al., "Microneedle Arrays as Transdermal and Intradermal Drug Delivery Systems: Materials Science, Manufacture and Commercial Development," *Materials Science and Engineering: R: Reports*, Vol. 104, 2016, pp. 1–32.

[4] Chen, J., et al., "A Multichannel Neural Probe for Selective Chemical Delivery at the Cellular Level," *IEEE Transactions on Biomedical Engineering*, Vol. 44, No. 8, 1997, pp. 760–769.

[5] Donnelly, R. F., et al.,"Hydrogel-Forming Microneedle Arrays for Enhanced Transdermal Drug Delivery," *Adv. Funct. Mater.*, Vol. 22, 2012, pp. 4879–4890.

[6] Lahiji, S. F., et al.,"A Patchless Dissolving Microneedle Delivery System Enabling Rapid and Efficient Transdermal Drug Delivery," *Scientific Reports*, Vol. 5, 2015, pp. 1–7.

[7] Davis, S. P., et al., "Insertion of Microneedles into Skin: Measurement and Prediction of Insertion Force and Needle Fracture Force," *Journal of Biomechanics*, Vol. 37, 2004, pp. 1155–1163.

[8] Yang, M., and Zahn, J. D. "Microneedle Insertion Force Reduction Using Vibratory Actuation," *Biomedical Microdevices*, Vol. 6, No. 3, 2004, pp. 177–182.

[9] Timoshenko, S. P., and Gere, J. M., *Theory of Elastic Stability*, Auckland: McGraw-Hill, 1963.

[10] Young, W. C., *Roark's Formulas for Stress & Strain*, New York: McGraw-Hill, 1989.

[11] Smith, W. G., "Analytic Solution for Tapered Column Buckling," *Computers & Structures*, Vol. 28, 1988, pp. 677–681.

[12] Kim, K., et al., "A Tapered Hollow Metallic Microneedle Array Using Backside Exposure of SU-8," *Journal of Micromechanics and Microengineering*, Vol. 14, 2004, pp. 597–603.

[13] Shah, R. K., and London, A. L., *Laminar Flow Forced Convection in Ducts*, New York: Academic Press, 1978.

[14] Pfahler, J., Harley, J., and Bau, H., "Liquid Transport in Micron and Submicron Channels," *Sensors and Actuators A*, Vol. 21–23, 1990, pp. 431–434.

[15] McAllister D. V., et al., "Microfabricated Needles for Transdermal Delivery of Macromolecules and Nanoparticles: Fabrication Methods and Transport Studies," *PNAS*, Vol. 100, No. 24, pp. 13755–13760.

[16] Hanein, Y., et al., "High-Aspect Ratio Submicron Needles for Intrecellular Applications", *Journal of Micromechanics and Microengineering*, Vol. 13, 2003, pp. 91–95.

[17] Shikida, M., et al., "Non-Photolithographic Pattern Transfer for Fabricating Pen-Shaped Microneedle Structures," *Journal of Micromechanics and Microengineering*, Vol. 14, 2004, pp. 1462–1467.

[18] Martanto, J. A., et al., "Transdermal Delivery of Insulin Using Microneedles In Vivo," *Proceedings of International Symposium on Controlled Release Bioactive Material*, Glasgow, Scotland, 2003, p. 666.

[19] Matriano, J. A., et al., "Macroflux Microprojection Array Patch Technology: a New and Efficient Approach for Intracutaneous Immunization," *Pharmaceutical Research*, Vol. 19, 2002, pp. 63–70.

[20] Park, J. H., et al., "Continuous On-Chip Micropumping Through a Microneedle," *Proceedings of MEMS'03, 16th IEEE International Workshop Micro Electromechanical System*, Kyoto, Japan, Jan. 19–23, 2003, pp. 371–374.

[21] Paik, S. J., et al., "In-Plane Single-Crystal-Silicon Microneedles for Minimally Invasive Microfluid Systems," *Sensors and Actuators A*, Vol. 114, 2004, pp. 276–284.

[22] Sparks, D., and Hubbard, T., "Micromachined Needles and Lancets with Design Adjustable Bevel Angles," *Journal of Micromechanics and Microengineering*, Vol. 14, 2004, pp. 1230–1233.

[23] Lin, L., and Pisano, A. P., "Silicon-Processed Microneedles," *Journal of Microelectromechanical Systems*, Vol. 8, No. 1, 1999, pp. 78–84.

[24] Mc Allister, D. V., et al., "Three-Dimensional Hollow Microneedle and Microtube Arrays," *Proceedings of Transducers '99, 10th International Conference on Solid-State Sensors and Actuators*, Sendai, Japan, June 7–10, 1999, pp. 1098–1107.

[25] Mukerjee, E. V., et al., "Microneedle Array for Transdermal Biological Fluid Extraction and In Situ Analysis," *Sensors and Actuators A*, Vol. 114, 2004, pp. 267–275.

[26] Gardeniers, H. J. G. E., et al., "Silicon Micromachined Hollow Microneedles for Transdermal Liquid Transport," *Journal of Microelectromechanical Systems*, Vol. 12, No. 6, 2003, pp. 855–862.

[27] Griss, P. and Stemme, G., "Side-Opened Out-of-Plane Microneedles for Microfluidic Transdermal Liquid Transfer," *Journal of Microelectromechanical Systems*, Vol. 12, No. 3, 2003, pp. 296–301.

[28] Zahn, J. D., et al., "Microfabricated Polysilicon Microneedles for Minimally Invasive Biomedical Devices," *Journal of Biomedical Microdevices*, Vol. 2, No. 4, 2000, pp. 295–303.

[29] Zahn, J. D., et al., "Continuous On-Chip Micropumping for Microneedle Enhanced Drug Delivery," *Journal of Biomedical Microdevices*, Vol. 6, No. 3, 2004, pp. 183–190.

[30] Papautsky, I., et al., "A Low-Temperature IC-Compatible Process for Fabricating Surface-Micromachined Metallic Microchannels," *IEEE Journal of Microelectromechanical Systems*, Vol. 7, No. 2, 1998, pp. 267–273.

[31] Brazzle, J. D., Papautsky, I., and Frazier, A. B., "Hollow Metallic Micromachined Needle Arrays," *Biomedical Microdevices*, Vol. 2, No. 3, 2000, pp. 197–205.

[32] Papautsky, I., et al., "Micromachined Pipette Arrays," *IEEE Transactions on Biomedical Engineering*, Vol. 47, No. 6, 2000, pp. 812–819.

[33] Chun, K., et al., "An Array of Hollow Microcapillaries for the Controlled Injection of Genetic Materials into Animal/Plant Cells," *Proceedings of MEMS'99, 12th IEEE International Workshop Micro Electromechanical System*, Orlando, FL, Jan. 17–21, 1999, pp. 406–411.

[34] Ohigashi, R., et al., "Micro Capillaries Array Head for Direct Drawing of Fine Patterns," *Proceedings of MEMS'01, 14th IEEE International Workshop Micro Electromechanical Systems*, Interlaken, Switzerland, Jan. 21–25, 2001, pp. 389–392.

[35] Ma, G., and Wu, C., "Microneedle, Bio-Microneedle and Bio-Inspired Microneedle: A review," *Journal of Controlled Release*, Vol. 251, 2017, pp. 11–23.

[36] Cho, W. K., et al.,"Microstructured Barbs on the North American Porcupine Quill Enable Easy Tissue Penetration and Difficult Removal," *Proc. Natl. Acad. Sci.*, Vol. 109, 2012, pp. 21289–21294.

[37] Aoyagi, S., et al., "Biodegradable Polymer Needle with Various Tip Angles and Consideration on Insertion Mechanism of Mosquito's Proboscis," *Sens. Actuators A: Phys*, Vol. 143, 2008, pp. 20–28.

[38] Izumi, H., et al., "Realistic Imitation of Mosquito's Proboscis: Electrochemically Etched Sharp and Jagged Needles and Their Cooperative Inserting Motion," *Sens. Actuators A: Phys.*, Vol. 165, 2011, pp. 115–123.

[39] Hong, X., et al.,"Dissolving and biodegradable microneedle technologies for transdermal sustained delivery of drug and vaccine," *Drug Design, Development and Therapy*, Vol. 7, 2013, pp. 945–952.

[40] Xiaoyun, H., et al., "Hydrogel Microneedle Arrays for Transdermal Drug Delivery," *Nano-Micro Letters*, Vol. 6, 2014, pp. 191–199.

[41] Demir, Y. K., et al.,"Characterization of Polymeric Microneedle Arrays for Transdermal Drug Delivery," *PlosONE*, Vol. 8, 2013, pp. 1-9.

[42] Yang, S., et al.,"A Scalable Fabrication Process of Polymer Microneedles," *International Journal of Nanomedicine*, Vol. 7, 2012, pp. 1415–1422.

[43] Lin, W., et al., "Transdermal Delivery of Antisense Oligonucleotides with Microprojection Path (Macroflux) Technology," *Pharmaceutical Research*, Vol. 18, 2001, pp. 1789–1793.

[44] Donnelly, R. F., et al., "Microneedle-Based Drug Delivery Systems: Microfabrication, Drug Delivery, and Safety," *Drug Delivery*, Vol. 17, 2010, pp. 187–207.

[45] Singh, R. R. T., et al., "Minimally Invasive Microneedles for Ocular Drug Delivery," *Expert Opinion on Drug Delivery*, Vol. 14, 2017, pp. 525–537.

[46] Singh, R. R. T., et al., "Rapidly Dissolving Polymeric Microneedles for Minimally Invasive Intraocular Drug Delivery," *Drug Deliv. Transl. Res.*, Vol. 6, 2016, pp. 800–815.

# Chapter 10

## Microfluidics for Life Sciences and Chemistry: Micromixers

Many miniaturized systems such as on-chip point-of-care diagnostic chips or microfluidic pharmaceutical synthesis require the mixing of two or more aqueous solutions at microscale [1, 2]. While in macroscale, mixing is achieved with turbulence, mixing in microscale relies mainly on diffusion due to the laminar behavior at low Reynolds numbers. The mixing rate is determined by the flux of diffusion $j$:

$$j = -D\frac{\mathrm{d}c}{\mathrm{d}x} \tag{10.1}$$

where $D$ is the diffusion coefficient in meters squared per second and $c$ is the species concentration in kilograms per cubic meter. The relationship in (10.1) is called the Fick's law. The diffusion coefficient is a fluid property, which was derived by Einstein as [3]:

$$D = \frac{RT}{fN_{\mathrm{A}}} \tag{10.2}$$

where $R$ is the gas constant, $T$ is the absolute temperature, $N_{\mathrm{A}} = 6.02 \times 10^{23}$ is the Avogadro number, and $f$ is the friction factor that is proportional to the viscosity $\eta$. At a constant temperature, $D$ is inversely proportional to $\eta$:

$$D = C_{\mathrm{D}}/\eta \tag{10.3}$$

where $C_{\mathrm{D}}$ is the constant incorporating all other factors. Figure 10.1 depicts the range of diffusion coefficients of different materials. Tables 10.1 and 10.2 list some typical values for gases and liquids. Diffusion coefficients of solutions containing large molecules, such as hemoglobin, myosin, or even viruses, are two orders of magnitude lower than those of most liquids [4]. With a constant flux, the diffusive transport is proportional to the contact surface of the species [5]. The average diffusion time $\tau$ over a mixing length $L_{\mathrm{mixing}}$ is characterized by the Fourier number [6]:

$$\mathrm{Fo} = \frac{D\tau}{L^2_{\mathrm{mixing}}} \tag{10.4}$$

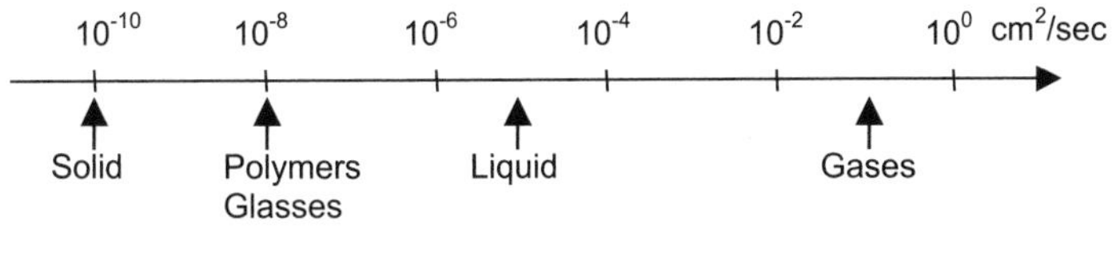

10^-10 10^-8 10^-6 10^-4 10^-2 10^0 cm^2/sec
Solid
Polymers
Glasses
Liquid
Gases

**Figure 10.1** Diffusion coefficient range.

**Table 10.1**
Diffusion Coefficients in Gases at 1 atm [6]

| *Gas Pair* | *T (K)* | *D (cm²/s)* | *Gas Pair* | *T (K)* | *D (cm²/s)* |
|---|---|---|---|---|---|
| Air-$CH_4$ | 273 | 0.196 | $CO_2$-$H_2O$ | 307.5 | 0.202 |
| Air-$H_2$ | 273 | 0.611 | $H_2$-$H_2O$ | 307.1 | 0.915 |
| Air-$H_2O$ | 298.2 | 0.260 | $H_2$-acetone | 296 | 0.424 |
| Air-benzene | 298.2 | 0.096 | $H_2$-ethane | 298.0 | 0.537 |
| Air-butanol | 299.1 | 0.087 | $H_2$-benzene | 311.3 | 0.404 |
| $CH_4$-$H_2$ | 298.0 | 0.726 | $O_2$-$H_2O$ | 308.1 | 0.282 |
| CO-$H_2$ | 295.6 | 0.7430 | $O_2$-benzene | 311.3 | 0.101 |
| $CO_2$-$H_2$ | 298.0 | 0.6460 | Ethylene-$H_2O$ | 307.8 | 0.204 |

Equation (10.4) shows that the diffusion time or the mixing time is proportional to the square of the characteristic mixing length $L_{\mathrm{mixing}}$, which is, for instance, the channel width in the case of two streams coflowing in a microchannel. Because of their small sizes, micromixers decrease the diffusion time significantly. In general, fast mixing can be achieved with smaller mixing path and larger contact surface. If the channel geometry is very small, the fluid molecules collide most often with the channel wall and not with other molecules. In this case, the diffusion process is called Knudsen diffusion [6]. The ratio between the distance of molecules and the channel size is characterized by the dimensionless Knudsen number (Kn):

$$\mathrm{Kn} = \frac{\lambda}{D_{\mathrm{h}}} \tag{10.5}$$

where $\lambda$ is the mean free path and $D_{\mathrm{h}}$ is hydraulic diameter of the channel structure. The mean free path for gases is given by (see Section 2.2.1.1):

$$\lambda = \frac{KT}{\sqrt{2}\pi d_{\mathrm{m}}^2 p} \tag{10.6}$$

where $K$ is the Boltzmann constant ($K = 1.38066 \times 10^{-23}$ J/K), $T$ is the absolute temperature, $p$ is the pressure, and $d_{\mathrm{m}}$ is the molecular diameter of the diffusing species. The Knudsen number for liquid is small, because the mean free path of liquid is on the order of a few angstroms. Thus, Knudsen diffusion may occur only in pores with nanometer sizes. In gases, the mean free path is on the order of a hundred nanometers to microns. For example, at room condition, the mean free path of air is 0.06 μm and that of hydrogen is 0.2 μm. Thus, Knudsen diffusion may occur in microchannels with diameters on the order of a few microns. The Knudsen diffusion coefficient $D_{\mathrm{Kn}}$ (in cm²/s) is given for gases as [7]:

$$D_{\mathrm{Kn}} = 4,850 \times D_{\mathrm{h}}\sqrt{T/M} \tag{10.7}$$

where $D_{\mathrm{h}}$ is the hydraulic diameter of the channel in centimeters, $T$ is the absolute temperature, and $M$ is the molecular weight. $D_{\mathrm{Kn}}$ is independent of pressure.

In the following sections, micromixers are categorized as passive mixers and active mixers. Passive mixers do not have moving parts. Micropumps or microvalves used to deliver fluids to the mixing area are not considered part of the mixer. In active mixers, moving parts are involved. Moving parts are used to manipulate or control the pressure gradients in the mixing area. Because of the nature of the mixing phenomena, the two mixer types are also called static and dynamic mixers.

Because of their simple implementation, passive mixers are a favorable solution for microfluidic systems. The next section presents analytical models describing and discussing the basic passive mixing concepts.

**Table 10.2**
Diffusion Coefficients in Water at 25°C [6]

| *Solute* | *D* ($\times 10^{-5} cm^2/s$) | *Solute* | *D* ($\times 10^{-5} cm^2/s$) |
|---|---|---|---|
| Air | 2.00 | Ammonia | 1.64 |
| Carbon dioxide | 1.92 | Benzene | 1.02 |
| Chlorine | 1.25 | Sulfuric acid | 1.73 |
| Ethane | 1.20 | Nitric acid | 2.60 |
| Ethylene | 1.87 | Acetylene | 0.88 |
| Hydrogen | 4.50 | Methanol | 0.84 |
| Methane | 1.49 | Ethanol | 0.84 |
| Nitrogen | 1.88 | Formic acid | 1.50 |
| Oxygen | 2.10 | Acetic acid | 1.21 |
| Propane | 0.97 | Propionic acid | 1.06 |
| Glycine | 1.06 | Benzoic acid | 1.00 |
| Valine | 0.83 | Acetone | 1.16 |
| Ovalbumin | 0.078 | Urease | 0.035 |
| Hemoglobin | 0.069 | Fibrinogen | 0.020 |

## 10.1 DESIGN CONSIDERATIONS

Conventionally, turbulent flows and mechanical agitation make rapid mixing possible by segregating the fluid in small domains, which increase the contact surface and decrease the mixing path. Since the Reynolds numbers in microfluidic devices are on the order of 1 or less, far below the critical Reynolds number, turbulence is not achievable in microscale. All micromixers work in laminar regime and rely entirely on diffusion. General design requirements for micromixers are:

- Fast mixing time;
- Small device area;
- Integration ability in a more complex system.

Fast mixing time can be achieved by decreasing the mixing path and increasing the interfacial area. The basic concepts for decreasing the mixing path used in passive micromixers are parallel lamination, sequential lamination, sequential segmentation, segmentation based on injection, and focusing. The diffusive/convective transport of the solute in these concepts is governed by the transport equation:

$$\frac{\partial c}{\partial t} + u\frac{\partial c}{\partial x} + v\frac{\partial c}{\partial y} + w\frac{\partial c}{\partial z} = D\Big(\frac{\partial^2 c}{\partial x^2} + \frac{\partial^2 c}{\partial y^2} + \frac{\partial^2 c}{\partial z^2}\Big) + q \tag{10.8}$$

where $c$, $D$, $u, v, w$ are the concentration of the solute, the diffusion coefficient, and the velocity components in $x$-, $y$-, and $z$-axes, respectively. The source term $q$ represents the generation of the solute.

### 10.1.1 Parallel Lamination

Equation (10.4) clearly indicates that the smaller the mixing channel, the faster the mixing process. However, the desired high throughput, particulate in the fluid, and the high driven pressures do not tolerate channels that are too small. The first solution for this problem is making each stream split into $n$ substreams and rejoining them again in a single stream. The mixing time will decrease with a

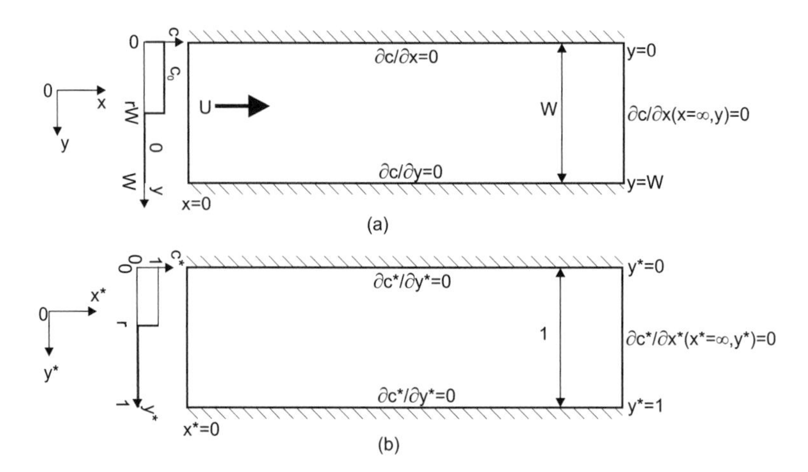


**Figure 10.2** Concentration distribution in a parallel lamination micromixer: (a) the 2-D model; and (b) the dimensionless 2-D model.

factor of $n^2$ [8]. This mixing method is described here as *parallel lamination*. This design concept is simple and is suitable for planar fluidic systems. The drawback is the large lateral area required for the parallel fluid channels. Following a simple model of two streams with an arbitrary mixing ratio is considered. This model can be extended to mixing of multiple streams.

#### 10.1.1.1 Mixing with Two Streams [9]

For simplification, the mixing model assumes a flat microchannel, mixing streams of the same viscosity, and a uniform velocity $u$ across the microchannel. Since the model is two-dimensional, no diffusion flux exists perpendicularly to the considered plane. The 2-D model is depicted in Figure 10.2(a). Basically, the model consists of a long channel of a width $W$, two inlets, and one outlet. One inlet stream is the solute with a concentration of $c = c_0$ and a mass flow rate of $\dot{m}_1$, and the other inlet stream is the solvent with a concentration of $c = 0$ and a mass flow rate of $\dot{m}_2$. The transport equation (10.8) is then reduced to the steady-state 2-D form:

$$u\frac{\partial c}{\partial x} = D\left(\frac{\partial^2 c}{\partial x^2} + \frac{\partial^2 c}{\partial y^2}\right) \tag{10.9}$$

With the same viscosity and fluid density, the dimensionless interface location $r$ is equal to the mass fraction of the solvent in the final mixture $\alpha = \dot{m}_1/(\dot{m}_1 + \dot{m}_2)$ $(0 \leq \alpha \leq 1)$. Thus, the mixing ratio of the solute and the solvent is $\alpha : (1 - \alpha)$. By introducing the dimensionless variables for the coordinates system $x^* = x/W$, $y^* = y/W$, the dimensionless concentration $c^* = c/c_0$, and the Peclet number $\mathrm{Pe} = uW/D$, (10.9) has the dimensionless form:

$$\mathrm{Pe}\frac{\partial c^*}{\partial x^*} = \frac{\partial^2 c^*}{\partial x^{*2}} + \frac{\partial^2 c^*}{\partial y^{*2}} \tag{10.10}$$

The corresponding boundary conditions for the inlets are:

$$f(y^*) = \begin{cases} c^*\Big|_{(x^*=0, 0 \leq y^* < r)} = 1 \\ c^*\Big|_{(x^*=0, r \leq y^* \leq 1)} = 0 \end{cases} \tag{10.11}$$

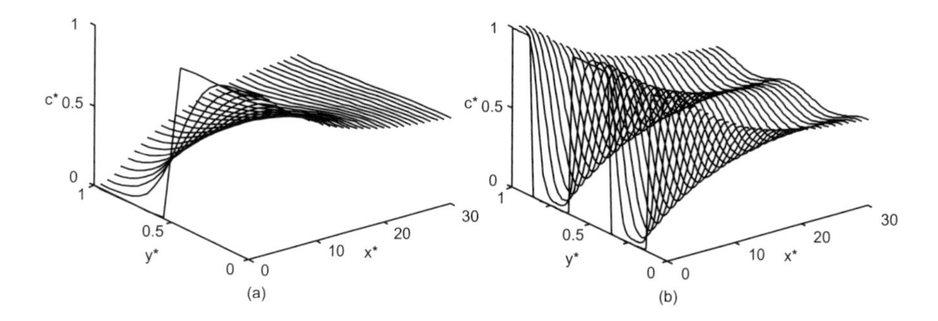


**Figure 10.3** Dimensionless concentration distribution in a microchannel with parallel lamination (Pe $=100$, mixing ratio 1:1 or $\alpha=r=0.5$): (a) two streams; and (b) multiple streams.

Full mixing is assumed for the mixer outlet. Thus, the boundary condition for ($x^*=\infty$) is:

$$\left.\frac{\partial c^*}{\partial x^*}\right|_{(x^*=\infty,0\leq y^*\leq 1)}=0 \tag{10.12}$$

The channel wall is impermeable and the corresponding boundary condition is:

$$\left.\frac{\partial c^*}{\partial y^*}\right|_{y^*=0,1}=0 \tag{10.13}$$

Equation (10.13) is also the symmetry condition for mixing with multiple streams, which is discussed later in the next section.

Separating the variables in (10.10) and applying the corresponding boundary conditions (10.11), (10.12), and (10.13), results in the dimensionless concentration distribution in the mixing channel:

$$c^*(x^*,y^*)=\alpha+\frac{2}{\pi}\sum_{n=1}^{\infty}\frac{\sin\alpha\pi n}{n}\cos(n\pi y^*)\exp\left(-\frac{2n^2\pi^2}{\mathrm{Pe}+\sqrt{\mathrm{Pe}^2+4n^2\pi^2}}x^*\right)\quad n=1,2,3\ldots \tag{10.14}$$

Equation (10.14) consists of three parts: the modulating coefficients, the cosine function, and the exponential function. The decay of concentration profile along flow direction is determined by the exponential function with the Peclet number as a variable. A small Peclet number leads to fast mixing. At a constant diffusion coefficient $D$ and a constant flow velocity $u$, a small Peclet number can be achieved by decreasing the mixing path $W$. Figure 10.3(a) depicts the typical dimensionless concentration distribution for two streams with a mixing ratio of $1:1$.

#### 10.1.1.2 Mixing with Multiple Streams

The solution of (10.14) for mixing with two streams can be extended to the case of multiple mixing streams. The wall boundary condition is identical to the symmetry condition in the middle of each stream. Thus, (10.14) can be extended periodically along the channel width direction. Considering the distance between two neighboring concentration extrema $W_{\mathrm{min,max}}$, (10.14) also represents the case with multiple streams. In this case, the mixing ratio can also be arbitrary, and the Peclet number is evaluated as $\mathrm{Pe}=UW_{\mathrm{min,max}}/D$. The typical concentration distribution of mixing with multiple stream is shown in Figure 10.3(b).

**Example 10.1: Designing a Y-Mixer**

Mix ethanol completely with water in a parallel micromixer with two inlets (Y-mixer) at room temperature. The flow rates of both ethanol and water are 10 μL/min. The Fourier number for this case is assumed to be 0.5. Determine the required length of the mixing channel if the channel cross section has a dimension of 100 μm × 100 μm.

**Solution.** According to Table 10.2, the diffusion coefficient of ethanol in water at room temperature (25°C) is $0.84 \times 10^{-5}$ cm²/s. The characteristic mixing length is the channel width $L_{\text{mixing}} = W = 100$ μm. The required mixing time is:

$$\text{Fo} = \frac{D\tau}{L^2_{\text{mixing}}} \rightarrow \tau = \text{Fo}\frac{L^2_{\text{mixing}}}{D} = \text{Fo}\frac{W^2}{D}$$
$$\tau = 0.5\frac{W^2}{D} = \frac{(100\times10^{-6})^2}{2\times0.84\times10^{-5}\times10^{-4}} = 5.95 \text{ sec}$$

The average velocity of the mixed liquid is:

$$u = \frac{\dot{Q}_{\text{water}} + \dot{Q}_{\text{ethanol}}}{A} = \frac{2\times 10\times 10^{-9}/60}{(100\times 10^{-6})^2} = 33.33\times 10^{-3} \text{ m/s}$$

Thus, the required length of the mixing channel is:

$$L = u\tau = 33.33\times 10^{-3}\times 5.95 = 0.198 \text{ m} = 198 \text{ mm}$$

**Example 10.2: Designing a Long Mixing Channel**

The above mixing channel is to be designed with a meander shape to save lateral device surface. If the channel structure is to be placed inside a square area, determine the dimension of this area. Determine the number of turns.

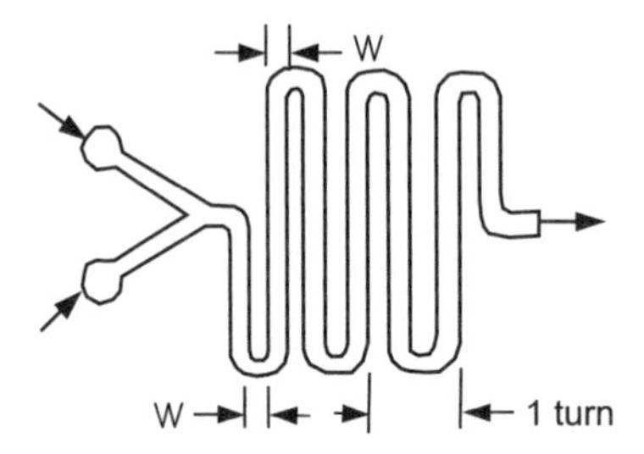


**Solution.** We assume that the channel walls consume the same amount of area as the channel itself. The total surface area required for the mixing channel is:

$$A = 2WL = 2\times 100\times 198\times 10^3 = 3.96\times 10^7 \text{ μm}^2$$

The dimension of the square area is:

$$a = \sqrt{A} = \sqrt{39.6}\times 10^3 \text{ μm} \approx 6,293 \text{ μm}$$

Each turn consumes 2 × 100 μm for walls and 2 × 100 μm for channel width. The total number of turns is:

$$N = \frac{6,293}{4\times 100} \approx 16$$

**Example 10.3: Designing a Parallel Lamination Mixer**

The above mixer has to be redesigned with more lamination layers. In the new design, the channel length should be 1 mm. In how many layers should each stream be separated?

**Solution.** Since the average velocity remains the same as in the mixer of Example 10.2, the new mixing time is proportional to the new length:

$$\tau_{\text{new}} = \tau_{\text{old}} \frac{L_{\text{new}}}{L_{\text{old}}} = 5.95 \times \frac{1}{198} = 3 \times 10^{-2} \text{ sec}$$

Assuming that each stream is separated into $n$ layers, the characteristic mixing length is then $W/n$. The new mixing time is:

$$\tau_{\text{new}} = 0.5 \frac{W^2}{n^2 D}$$

Rearranging the above equation for $n$, we get:

$$n = \sqrt{0.5 \frac{W^2}{\tau_{\text{new}} D}} = \frac{W}{\sqrt{2\tau_{\text{new}} D}} = \frac{100 \times 10^{-6}}{\sqrt{2 \times 3 \times 10^{-2} \times 0.84 \times 10^{-9}}} \approx 14$$

By splitting each stream into 14 layers we can design a mixing channel, which is 200 times shorter than the original Y-mixer. Even if the original channel is coiled in a meander shape, the total dimension is still 4.5 times larger than the new design.

### 10.1.2 Sequential Lamination

Sequential lamination segregates the joined stream into two channels, and rejoins them in the next stage. Using $n$ such splitting stages in serial, one is able to laminate $2^n$ layers, which causes $4^{(n-1)}$ times faster mixing [10]. Figure 10.4 illustrates this mixing concept, with three stages [Figure 10.4(a)] and two stages [Figure 10.4(b)].

The arrangement described in Figure 10.4 is referred to as sequential lamination with vertical lamination and horizontal splitting. Alternatively, sequential lamination can have horizontal lamination and vertical splitting. This mixing concept requires relatively complicated 3-D fluidic structures that can only be made in a bulk system. For the same device area consumed by a channel (parallel concept) or by a stage (sequential concept), the sequential concept results in much faster mixing [Figure 10.4(c)].

**Example 10.4: Designing a Sequential Lamination Mixer**

For the same mixing time of the design in Example 10.3, determine the number of stages required if we decide to make a sequential lamination mixer.

**Solution.** The improvement of mixing time in a parallel lamination mixer with $n$ layers is:

$$\tau_{\text{new}}/\tau = 1/n^2$$

The improvement of mixing time in a sequential lamination mixer with $m$ stages is:

$$\tau_{\text{new}}/\tau = 1/4^{(m-1)}$$

For the same improvement, we get:

$$1/4^{(m-1)} = 1/n^2$$

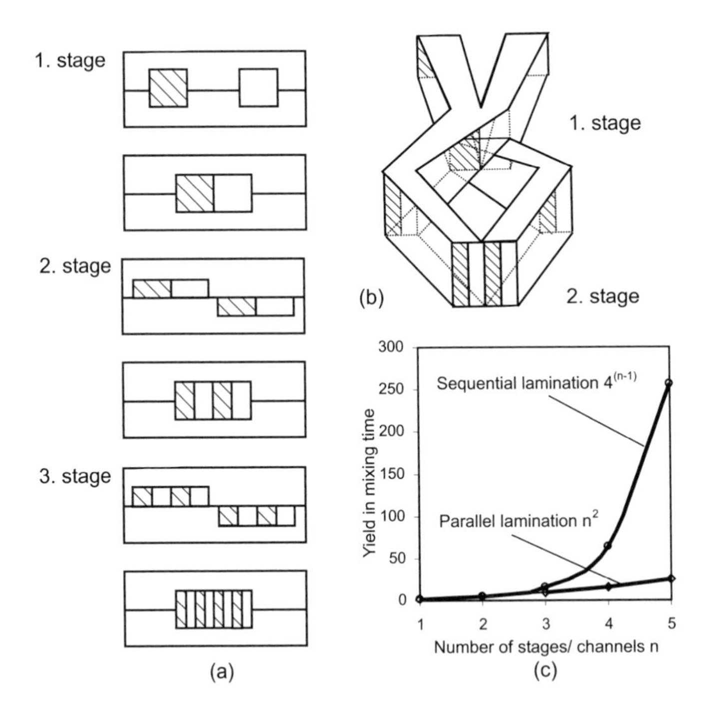


**Figure 10.4** Concept of sequential lamination (vertical lamination and horizontal splitting): (a) $n = 3$; (b) $n = 2$; and (c) comparison between parallel and sequential lamination.

Rearranging the above equation, the required stage number for the sequential lamination mixer is:

$$m = 1 + 2\ln n/\ln 4 = 1 + 2\ln 10/\ln 4 \approx 5$$

That means that if the entire mixing channel is constrained by 1 mm, each mixing stage of the new design only occupies 200 μm.

### 10.1.3 Sequential Segmentation

Sequential segmentation divides the solvent and solute into segments, that occupy the whole channel width. Mixing occurs through diffusion in flow direction. Sequential segmentation can be achieved by alternate switching of the inlet flows, using controlled valves for instance [Figure 10.5(a)]. The mixing ratio, can be adjusted by the switching ratio as depicted for the concentration at the inlet in Figure 10.5(b).

A simple 1-D model is assumed for sequential segmentation. Pressure-driven flow in microchannels with low aspect ratio ($W \gg H$) can be reduced to the model of two parallel plates. The model assumes a flat velocity profile in the channel width direction ($z$-axis) and a Poiseuille velocity profile in the channel height ($y$-axis):

$$u(y) = 6U\left(1 - \frac{y}{H}\right)\frac{y}{H} \tag{10.15}$$

where $U$ is the mean velocity in the flow direction along the $x$-axis. The parabolic velocity profile causes axial dispersion, which can be described by an effective diffusion coefficient $D^*$. Using the Taylor-Aris approach and the velocity profile (10.15), the effective diffusion coefficient can be determined as [11]:

$$D^* = D + \frac{H^2U^2}{210D} \tag{10.16}$$

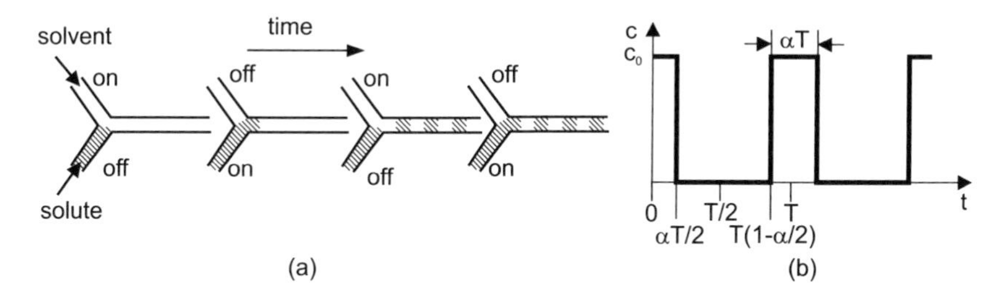


**Figure 10.5** Sequential segmentation: (a) concept of sequential switching; and (b) time signal of the concentration at the inlet for a mixing ration of $\alpha$

where $D$ is the molecular diffusion coefficient of the solute in the solvent, and $H$ is the channel height. In the case of a circular channel of radius $R$, the effective diffusion coefficient is [11]:

$$D^* = D + \frac{R^2U^2}{48D} \tag{10.17}$$

The concentration distribution across the channel width and height is assumed to be uniform; thus, only the concentration profile along the flow direction $x$ is considered. At a flow velocity $u$ and a switching period $T$, the characteristic segment length is defined as $L = uT$. The general transport equation (10.8) can be then reduced to the transient 1-D form:

$$\frac{\partial c}{\partial t} + u\frac{\partial c}{\partial x} = D^*\frac{\partial^2 c}{\partial x^2} \tag{10.18}$$

Figure 10.6 depicts this 1-D model with the boundary condition at the inlet. Figure 10.5(b) already depicts the periodic boundary condition at the inlet ($x = 0$):

$$c(t,0) = \begin{cases} c_0 \;\; 0 \le t \le \alpha T/2 \\ 0 \;\; \alpha T/2 < t \le T - \alpha T/2 \\ c_0 \;\; T - \alpha T/2 < t \le T \end{cases} \tag{10.19}$$

where $c_0$, $T$, and $\alpha$ are the initial concentration of the solute, the period of the segmentation, and the mixing ratio, respectively.

Introducing the dimensionless variables $c^* = c/c_0$, $x^* = x/L$, and $t^* = t/T$, (10.18) has the dimensionless form:

$$\frac{\partial c^*}{\partial t^*} = \frac{1}{\text{Pe}}\frac{\partial^2 c^*}{\partial x^{*2}} - \frac{\partial c^*}{\partial x^*} \tag{10.20}$$

where the Peclet number is defined as $\text{Pe} = uL/D^*$. The dimensionless boundary condition of (10.20) is then:

$$c^*(t^*,0) = \begin{cases} 1 \;\; 0 \le t^* \le \alpha/2 \\ 0 \;\; \alpha < t^* \le 1/2 \end{cases} \tag{10.21}$$

Solving (10.20) with (10.21) and $c^*(\infty) = \alpha$ results in the transient behavior of the concentration profile along the mixing channel as:

$$c^*(x^*,t^*) = \Re\left\langle \alpha + \sum_1^\infty \frac{2\sin(\alpha\pi n)}{\pi n}\left\{\exp\left[\frac{1}{2}\left(\text{Pe} - \sqrt{\text{Pe}^2 + 8\pi n \text{Pe} i}\right)x^*\right] \times \exp(2\pi t^* i)\right\}\right\rangle \tag{10.22}$$

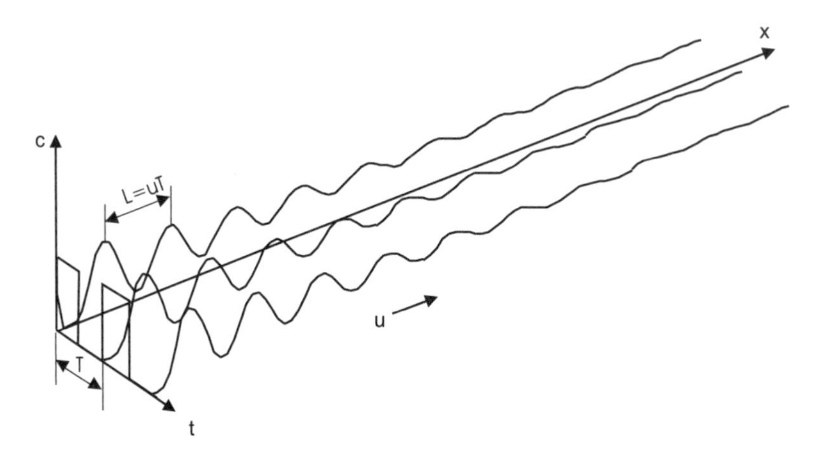


**Figure 10.6** Transient 1-D model for a micromixer with sequential segmentation.

where $i$ is the imaginary unit and $\Re$ indicates the real component of a complex number. The concentration distributions along the mixing channel are depicted in Figure 10.7. Figure 10.7(a) shows clearly that only a short mixing channel is required if the Peclet number is small; that is, either a low flow velocity $u$, a short characteristic segment length $L$, or a short switching time $T$. Sequential segmentation can achieve different final concentrations simply by adjusting the switching ratio $\alpha$ [Figure 10.7(b)].

### 10.1.4 Segmentation Based on Injection

Segmentation based on injection is another passive method for reducing the mixing path and increasing the contact surface between the solvent and the solute. The concept divides the solute flow into many streams, and injects them through a nozzle array into the solute flow. Figure 10.8(a) illustrates this concept with four circular injection nozzles.

For simplification, the analytical model of this mixing concept only considers a single nozzle. The model assumes a circular nozzle with a radius of $R$. The solute enters the solvent flow with a mass flow rate of $\dot{m}_2$ (kilograms per second) and an inlet concentration of $c_0$. The solvent flow is assumed to have a uniform velocity of $u$. The solvent flow has a mass flow rate of $\dot{m}_1$ and an inlet concentration of $c = 0$.

Considering a steady-state, 2-D problem and neglecting the source term, the transport equation (10.8) reduces to the form:

$$\frac{u}{D}\frac{\partial c}{\partial x} = \frac{\partial^2 c}{\partial x^2} + \frac{\partial^2 c}{\partial y^2} \tag{10.23}$$

The solution of (10.23) could be described as a product of a velocity-dependent term and a symmetric term $\Psi$:

$$c = \exp\left(\frac{ux}{2D}\right)\Psi(x, y) \tag{10.24}$$

Substituting (10.24) in (10.23) results in the partial differential equation for the symmetric term $\Psi$:

$$\left(\frac{u}{2D}\right)\Psi = \frac{\partial^2 \Psi}{\partial x^2} + \frac{\partial^2 \Psi}{\partial y^2} \tag{10.25}$$

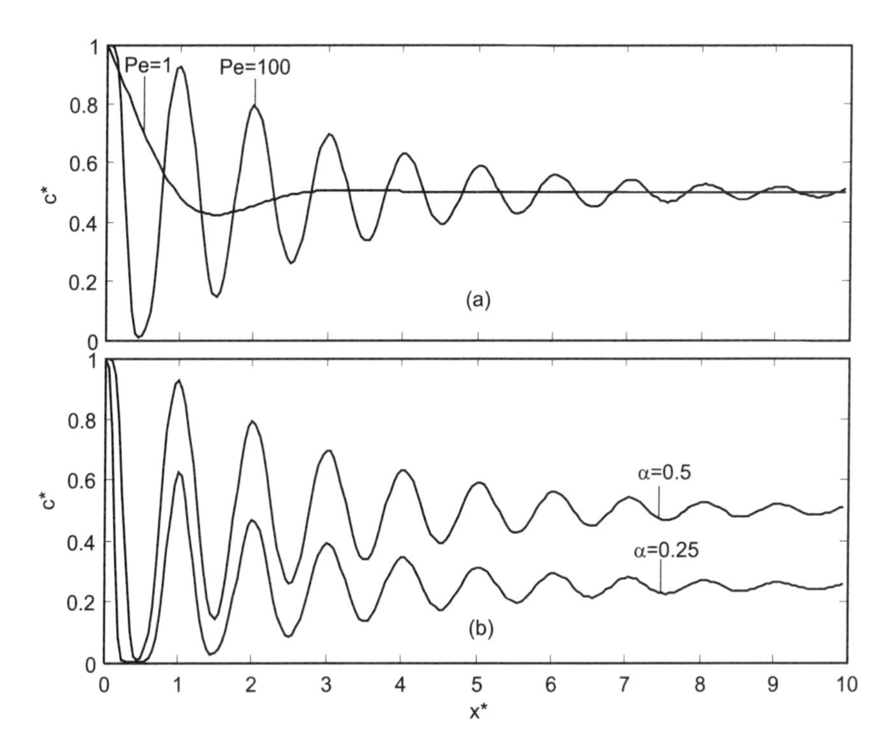


**Figure 10.7** Concentration profile along the flow direction: (a) with different Peclet numbers ($\alpha = 0.5$); and (b) with different mixing ratios (Pe = 100).

Defining the variable $r = \sqrt{x^2 + y^2}$ in the polar coordinate system shown in Figure 10.8(b), the boundary condition of the solute flow is after the Fick's law (10.1):

$$\left.\frac{\mathrm{d}c}{\mathrm{d}r}\right|_{r=R} = -\frac{j}{D} = -\frac{\dot{m}_2}{2\pi RHD} \tag{10.26}$$

where $H$ is the height of the mixing chamber above the injection nozzle. Assuming a small mixing ratio ($\dot{m}_2 \ll \dot{m}_1$), the following boundary condition is acceptable for (10.25):

$$\left.\frac{\mathrm{d}\Psi}{\mathrm{d}}r\right|_{r=\pm\infty} = 0 \tag{10.27}$$

Equation (10.25) can be rewritten for the polar coordinate system as:

$$\frac{\mathrm{d}^2\Psi}{\mathrm{d}r^2} + \frac{1}{r}\frac{\mathrm{d}\Psi}{\mathrm{d}r} - \left(\frac{u}{2D}\right)^2 \Psi = 0 \tag{10.28}$$

The solution of (10.28) is the modified Bessel function of the second kind and zero order:

$$\Psi = K_0[ur/(2D)] \tag{10.29}$$

The solution of (10.23) with the above-mentioned boundary conditions is:

$$c(r,\theta) = \frac{\dot{Q}_2 D}{\pi RH}u^{-1}\frac{K_0[ur/(2D)]}{K_1[uR/(2D)] - K_0[uR/(2D)]\cos\theta}\frac{\exp[ur\cos\theta/(2D)]}{\exp[uR\cos\theta/(2D)]} \tag{10.30}$$

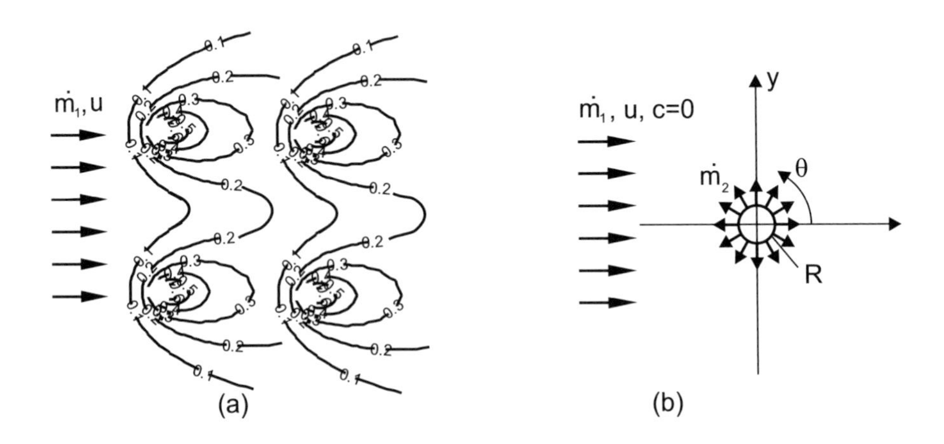


**Figure 10.8** Injection mixer: (a) typical dimensionless concentration distribution of a microplume (Pe = 1); and (b) 2-D model.

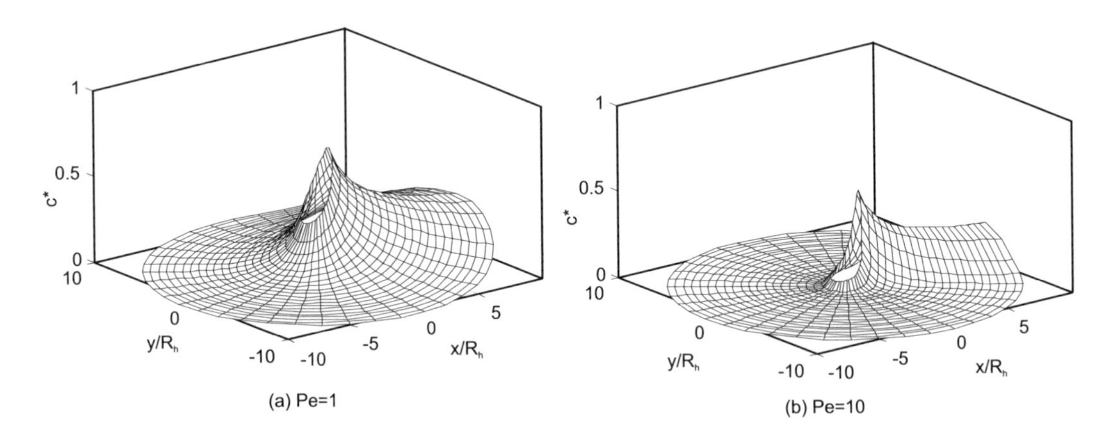


**Figure 10.9** Dimensionless concentration distribution around the injection nozzle: (a) Pe = 1; and (b) Pe = 10.

where $\theta$ is the angular variable of the polar coordinate system.

Introducing the Peclet number $\mathrm{Pe} = 2uR/D$, the dimensionless radial variable $r^* = r/R$, and the dimensionless concentration:

$$c^* = \frac{c}{2\dot{m}_2/(\pi H)} \tag{10.31}$$

the dimensionless form of (10.30) is:

$$c^*(r^*, \theta) = \frac{K_0(\mathrm{Pe}r^*/4)/\mathrm{Pe}}{K_1(\mathrm{Pe}/4) - K_0(\mathrm{Pe}/4)\cos\theta} \{\exp[\mathrm{Pe}(r^* - 1)/4]\}^{\cos\theta} \tag{10.32}$$

where $K_1$ is the modified Bessel function of the second kind and first order. Figure 10.9 shows the typical dimensionless concentration distribution around a single injection nozzle at different Peclet numbers.

### 10.1.5 Focusing of Mixing Streams

Another solution for fast mixing is the focusing of the mixing streams to achieve a smaller stream and thinner lamination width. This concept can be realized by means of geometric focusing [Figure 10.10(a)], or by hydrodynamic focusing [Figure 10.10(b)]. The velocity is inversely proportional to the stream width, while the mixing time according to (10.4) is proportional to the square of the stream width. As a result, the mixing time or the residence time in the mixing channel is proportional to the stream width. Decreasing the stream width causes faster mixing. The following analytical model analyzes the effect of hydrodynamic focusing for reducing the width of the mixing streams [12].

The model of two-phase focusing consists of three streams: the sample stream sandwiched between two identical sheath streams (Figure 10.11). The sample stream and the sheath stream are assumed to be immiscible. In many cases of microfluidics, the flow channel has a rectangular cross section. Figure 10.11(a) shows the actual geometry of the channel cross section with the two phases. The width and height of the channel are $2W$ and $H$, respectively. The position of the interface is $rW$. Since the model is symmetrical regarding $y$-axis and $z$-axis, only one-fourth of the cross section needs to be considered.

The velocity distribution $u_1$ and $u_2$ in the channel can be described by the Navier-Stokes equations:

$$\begin{cases} \dfrac{\partial^2 u_1}{\partial y^2} + \dfrac{\partial^2 u_1}{\partial z^2} = \dfrac{1}{\eta_1}\dfrac{\partial p}{\partial x} \\ \\ \dfrac{\partial^2 u_2}{\partial y^2} + \dfrac{\partial^2 u_2}{\partial z^2} = \dfrac{1}{\eta_2}\dfrac{\partial p}{\partial x} \end{cases} \tag{10.33}$$

where indices 1 and 2 describe the sample flow and the sheath flow, respectively. In (10.23) $\eta_1$ and $\eta_2$ are the viscosities of the sample fluid and of the sheath fluid. Nondimensionalizing the velocity by a reference velocity $u_0$, and the coordinates by $W$ leads to the dimensionless model [Figure 10.11(b)]:

$$\begin{cases} \dfrac{\partial^2 u_1^*}{\partial y^{*2}} + \dfrac{\partial^2 u_1^*}{\partial z^{*2}} = P' \\ \\ \dfrac{\partial^2 u_2^*}{\partial y^{*2}} + \dfrac{\partial^2 u_2^*}{\partial z^{*2}} = P' \end{cases} \tag{10.34}$$

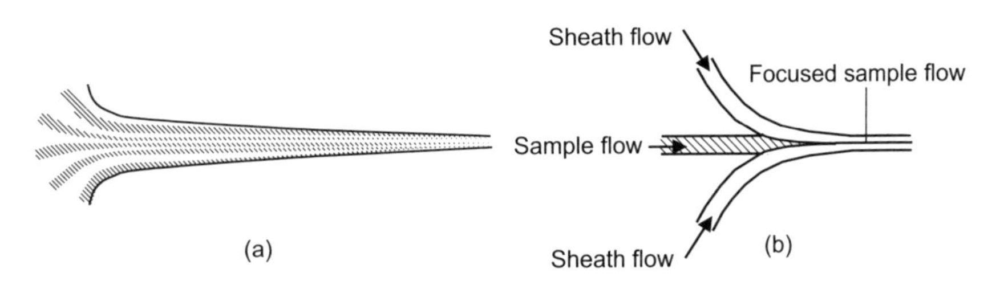


**Figure 10.10** Focusing concepts: (a) geometric focusing; and (b) hydrodynamic focusing.

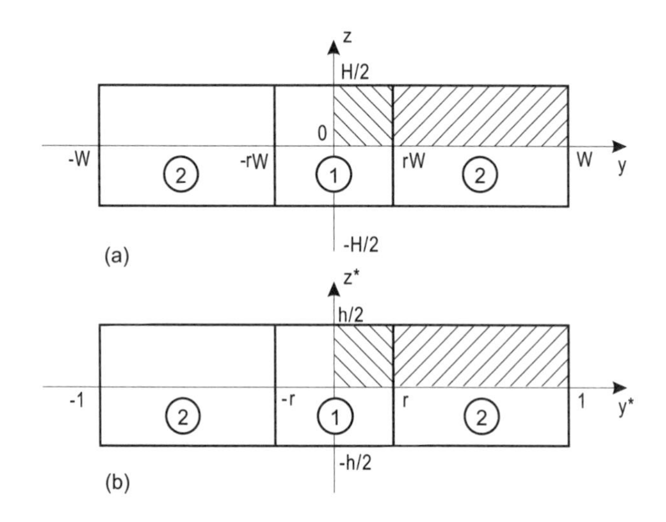


**Figure 10.11** Analytical model for two phases focusing: (a) actual geometry; and (b) the dimensionless model.

With $P' = \dfrac{W}{\eta_1 u_0} \dfrac{\partial p}{\partial x^*}$, $\beta = \eta_2/\eta_1$, and $\theta = (2n-1)\pi/h$, the solutions of (10.34) have the forms ($0 < y^* < 1, 0 < z^* < h/2$):

$$\begin{cases} u_1^*(y^*, z^*) = P' \left[ \dfrac{z^{*2} - h^2/4}{2} + \displaystyle\sum_{n=1}^{\infty} \cos\theta z^* (A_1 \cosh\theta y^* + B_1 \sinh\theta y^*) \right] \\ u_2^*(y^*, z^*) = \dfrac{P'}{\beta} \left[ \dfrac{z^{*2} - h^2/4}{2} + \displaystyle\sum_{n=1}^{\infty} \cos\theta z^* (A_2 \cosh\theta y^* + B_2 \sinh\theta y^*) \right] \end{cases} \tag{10.35}$$

The nonslip conditions at the wall are

$$u_2^*(1, z^*) = 0. \tag{10.36}$$

The symmetry condition at the $z^*$-axis is:

$$\left. \frac{\partial u_1^*}{\partial y^*} \right|_{y^*=0} = 0 \tag{10.37}$$

At the interface between the sample flow and the sheath flow, the velocity and the shear rate are continuous:

$$\begin{cases} u_2^*(r, z^*) &= u_1^*(r, z^*) \\ \left. \dfrac{\partial u_1^*}{\partial y^*} \right|_{y^*=r} &= \beta \left. \dfrac{\partial u_2^*}{\partial y^*} \right|_{y^*=r} \end{cases} \tag{10.38}$$

For a flat channel ($h \ll 1$), the position of the interface can be estimated as:

$$r = \frac{1}{1 + 2\beta\kappa} \tag{10.39}$$

where $\kappa = \dot{m}_2/\dot{m}_1$ is the flow rate ratio between the sheath streams and the mixing streams.

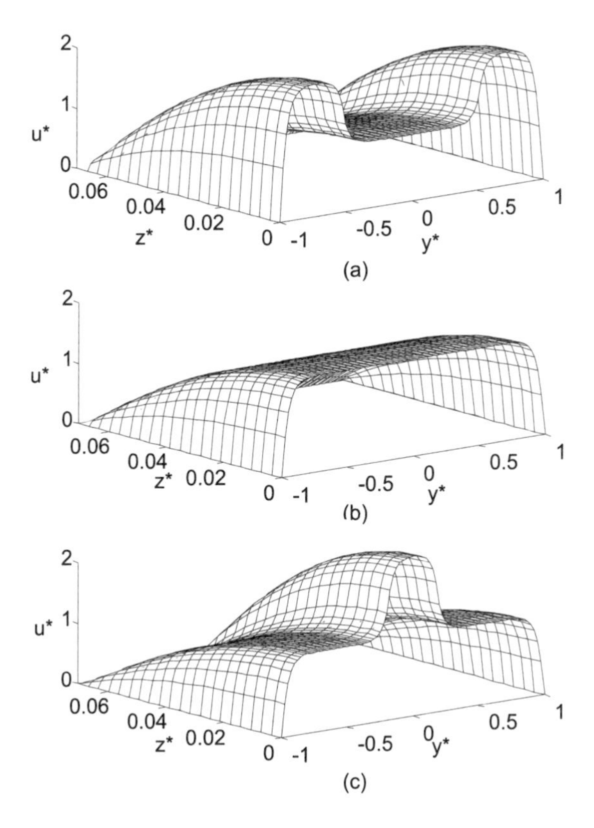


**Figure 10.12** The dimensionless velocity profile ($h = 0.14$, $\kappa = 1$): (a) $\beta = 0.5$; (b) $\beta = 1$; and (c) $\beta = 2$.

A Fourier analysis with the above boundary conditions results in the coefficients of (10.35):

$$\begin{aligned}
A_1 =\ & D\big[\beta \sinh\theta r \cosh^2\theta r \cosh\theta - \sinh^3\theta r \cosh\theta - (\beta-1)\sinh^2\theta r \sinh\theta \cosh\theta - \\
& (\beta-1)\sinh^2\theta r(\cosh\theta - \cosh\theta r) + (\beta-1)\sinh\theta r \cosh\theta r \cosh\theta(\cosh\theta - \cosh\theta r)\big]/ \\
& \big[\beta \cosh^2\theta r \sinh\theta r \cosh^2\theta - \sinh^3\theta r \cosh^2\theta - (\beta-1)\sinh^2\theta r \cosh\theta r \sinh\theta \cosh\theta\big] \\
A_2 =\ & D\big[\beta \cosh^2\theta r \cosh\theta - \sinh^2\theta r \cosh\theta - (\beta-1)\sinh\theta r \cosh\theta r \sinh\theta - \\
& (\beta-1)\sinh\theta r(\cosh\theta - \cosh\theta r)\big]/\big[\beta \cosh^2\theta r \sinh^2\theta - \sinh^2\theta r \cosh^2\theta - \\
& (\beta-1)\sinh\theta r \cosh\theta r \sinh\theta \cosh\theta\big] \\
B_1 =\ & 0 \\
B_2 =\ & D\big[(\beta-1)\sinh\theta r(\cosh\theta - \cosh\theta r)\big]/\big[\beta \cosh^2\theta r \cosh\theta - \sinh^2\theta r \cosh\theta - \\
& (\beta-1)\sinh\theta r \cosh\theta r \sinh\theta\big]
\end{aligned} \tag{10.40}$$

where $D = (-1)^{n+1}\dfrac{4h^2}{(2n-1)^3\pi^3}$. Figure 10.12 shows the typical velocity distribution inside the flow channel for the same flow rate in all streams ($\kappa = 1$). The velocity of the sample flow is lower, if the sample flow is more viscous than the sheath flow ($\beta < 1$) [Figure 10.12(a)]. If the viscosities are equal ($\beta = 1$), the flows behave as a single phase [Figure 10.12(b)]. If the sheath flows are more viscous ($\beta > 1$), the sample flow is faster [Figure 10.12(c)].

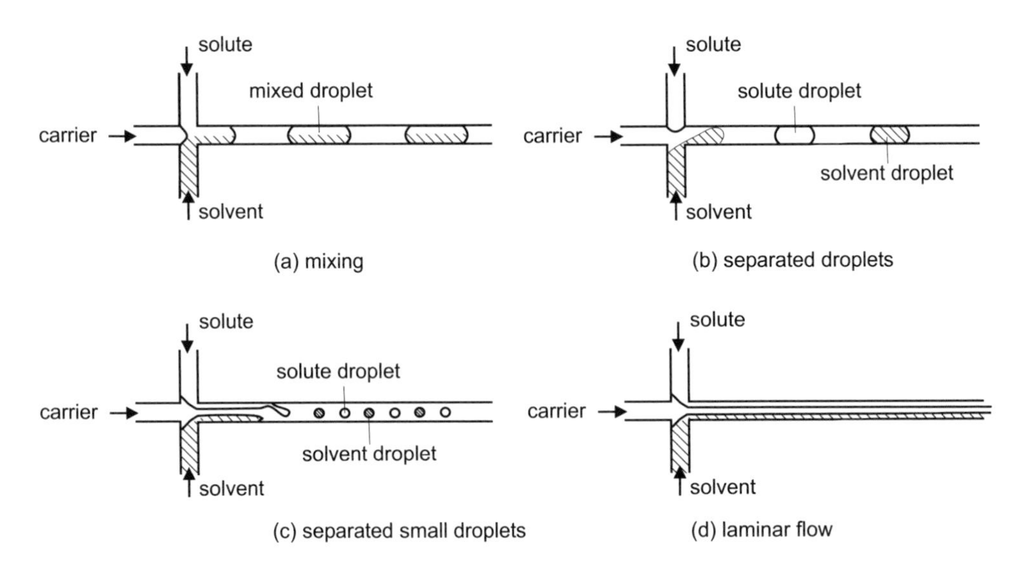


**Figure 10.13** Formation of droplets: (a) merging and mixing of solvent and solute; (b) formation of separated droplets; (c) formation of small separated droplets; and (d) laminar flow. (*After*: [13])

### 10.1.6 Formation of Droplets and Chaotic Advection

Smaller mixing path and possible chaotic advection can be achieved by forming droplets of solvent and solute. The basic configuration is similar to that of hydrodynamic focusing shown in Figure 10.10(b). For the formation of droplets, the middle inlet is used for the carrier fluid, which is immiscible to both solvent and solute. The flows of solvent and solute enter from the two sides. The formation behavior of droplets depend on the capillary number $\mathrm{Ca} = u\nu/\sigma$ and the sample fraction $r = (\dot{Q}_{\mathrm{solvent}} + \dot{Q}_{\mathrm{solute}})/(\dot{Q}_{\mathrm{solvent}} + \dot{Q}_{\mathrm{solute}} + \dot{Q}_{\mathrm{carrier}})$, where $u$ is the average velocity, $\nu$ is the viscosity, and $\dot{Q}_{\mathrm{solvent}}$, $\dot{Q}_{\mathrm{solute}}$, and $\dot{Q}_{\mathrm{carrier}}$ are the flow rates of solvent, solute, and carrier fluid. Section 11.1.2.1 will discuss more about droplet formation in a multiphase system.

At a low capillary number, the solvent and solute can merge into a sample droplet and mix rapidly due to chaotic advection inside the droplet [Figure 10.13(a)]. Increasing the capillary number at the same fraction $r$, the droplets form separately and are not able to merge and mix [Figure 10.13(b)]. Increasing further the capillary number, the alternate droplets become smaller and unstable [Figure 10.13(c)]. At a high capillary number, the three streams flow side by side as in the case of missile fluids [Figure 10.13(d)]. Figure 10.14 depicts the typical phase diagram of droplet formation based on the results reported by Zheng et al. [13].

The droplet train formed in the configuration described above can be stored over a long time, because the carrier fluid, such as an oil, can protect the aqueous sample from evaporation. The long-term stability of the sample allows protein crystallization in the microscale [14].

The flow patterns measured with μPIV inside a droplet moving in a rectangular microchannel are depicted in Figures 10.15(a, b). If both solute and solvent merge and mix, these flow patterns inside the mixed droplet could make chaotic advection possible. However, if the microchannel is straight, as depicted in Figure 10.15(c), the flow pattern inside the droplet is steady. The two vortices containing the solvent and solute exist in each half of the droplet and mixing is not improved significantly. If the microchannel has turns as depicted in Figure 10.15(d), the flow pattern is periodical. Each side of the droplet takes a turn to have a faster vortex. Thus, chaotic advection and consequently faster mixing are possible in this case [15]. Furthermore, the larger vortex may dominate the smaller vortex at each turn, which resembles the concept of reorientation, stretching, and folding in serial lamination discussed previously. Since the number of cycles depends on the

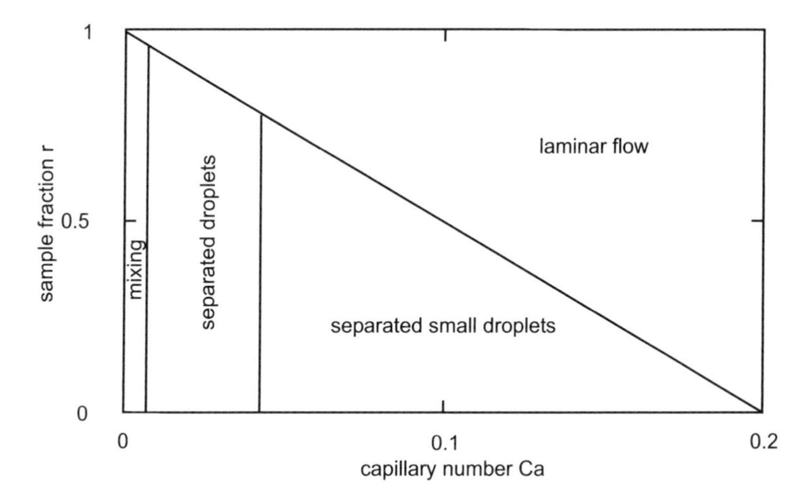


**Figure 10.14** Typical phase diagram of droplet formation in a multiphase system. (*After*: [13].)

number of turns $n$, the improvement of mixing time compared to the case of a straight channel is $4^{(n-1)}$. Compared to diffusive mixing with parallel lamination, the improvement of mixing time can be described as [15]:

$$\frac{t_{\text{chaotic}}}{t_{\text{diffusive}}} = \frac{\text{Pe}}{L^{*} \log(\text{Pe})} \tag{10.41}$$

where $\text{Pe} = uW/D$ is the Peclet number, and $L^{*} = L/W$ is the dimensionless droplet length normalized by the channel width $W$.

## 10.2 DESIGN EXAMPLES

As mentioned in the introduction of this chapter, micromixers can be categorized as passive micromixers and active micromixers. Passive micromixers do not require external disturbance to improve mixing. The passive mixing process relies entirely on diffusion and chaotic advection. Based on the arrangement of the mixed phases, passive mixing concepts can be further categorized as parallel lamination, serial lamination, injection, chaotic advection, and droplet mixing. Active micromixers use external disturbance for accelerating the mixing process. Based on the types of disturbance, active mixing can be categorized in pressure-driven, temperature-induced, electrohydrodynamic, dielectrophoretic, electrokinetic, magnetohydrodynamic, and acoustic concepts. Because of the integrated components and external power supply for the generation of disturbance fields, the design of active micromixers is often complicated and requires a complex fabrication process. The integration of active mixers in a microfluidic system is therefore both challenging and expensive. The major advantage of passive micromixers is the lack of actuators. The simple passive structures are robust, stable in operation, and easy to be integrated. Figure 10.16 illustrates the systematic overview of different micromixer types.

### 10.2.1 Passive Micromixers

The theoretical background of passive mixing was already discussed in Section 10.1. The theories are based on the diffusive/convective transport equation, and show that mixing in passive micromixers relies mainly on molecular diffusion and advection. Increasing the contact surface between the different fluids and decreasing the diffusion path between them could improve molecular diffusion.

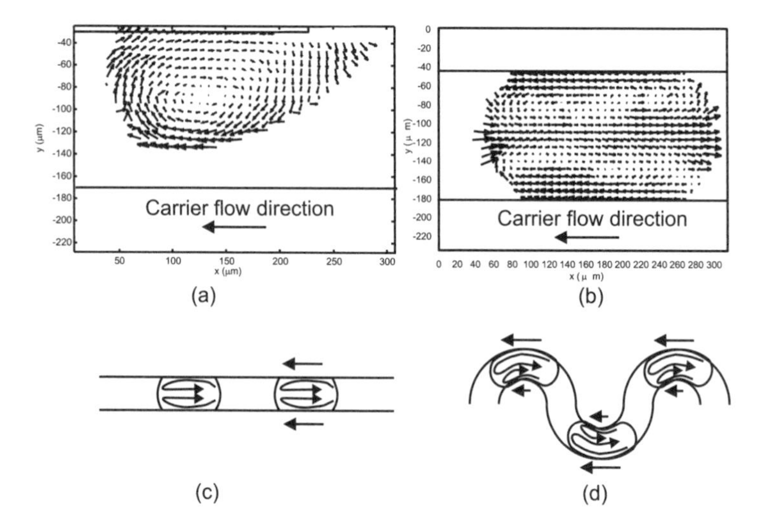


**Figure 10.15** Flow pattern in a microdroplet: (a) measured flow pattern during formation process; (b) measured flow pattern during passage in a straight rectangular channel; (c) schematic flow patterns inside a droplet in a straight channel; and (d) in a channel with turns.

Chaotic advection can be induced by manipulating the laminar flow in microchannels. The resulting flow pattern shortens the diffusion path and thus improves mixing. The following sections discuss the different designs of passive micromixers as categorized in Figure 10.16.

#### 10.2.1.1 Parallel Lamination Micromixers

As discussed in Section 10.1.1, fast mixing can be achieved by decreasing the mixing path and increasing the contact surface between the solvent and the solute. The basic design of a parallel lamination micromixer is a long microchannel with two inlets. According to their shapes, these designs are often called the T-mixer or the Y-mixer [16, 17, 18] [Figures 10.17(a, b)]. The inlet streams of a T-mixer can be twisted and laminated as two thin liquid sheets to reduce the mixing path [19]. As a basic design, a T-mixer is suitable for investigations of basic transport phenomena in microscale, such as scaling law, Taylor's dispersion [16, 18], and other nonlinear effects [20]. According to the analytical model (10.14), a long mixing channel is needed for a high Peclet number, if the channel width and the diffusion coefficient are fixed. A shorter mixing channel can achieve full mixing at extremely high Reynolds numbers [21, 22]. At these high Reynolds numbers, chaotic advection and even turbulent flows can be expected. Wong et al. [22] reported a T-mixer fabricated in glass and silicon. This mixer utilizes Reynolds numbers up to 500, where flow velocity can be as high as 7.6 m/s at a driven pressure up to 7 bars. High shear rate at fluid interfaces under extremely high Reynolds numbers can generate fast vortices, which in turn can be used for mixing. Lim et al. reported fast vortices and fast mixing of fluorescent dyes inside a diamond-shaped cavity close to a straight microchannel [23]. At a flow velocity of 45 m/s, the Reynolds number reaches a value of $Re = 245$. Böhm et al. [24] reported a micromixer with multiple inlet streams focused in a circular chamber to generate fast vortices. In the above-mentioned micromixers, the high velocities on the order of 1 m/s, 10 m/s [24], or even higher (e.g., 45 m/s [23]), cause problems in designing the fluidic interconnects. Such high flow rates require pressures on the order of bars. The high pressures (1.0 to 5.5 bars in [22], 15 bars in [24]) are serious challenges for bonding and interconnection technologies. Instead of a straight design, mixing channels with turns and geometrical obstacles could cause instability in the flow at moderate Reynolds numbers. A Y-mixer made of cofired ceramic tapes with a 90° bend in the mixing channel can generate vortices at Reynolds numbers

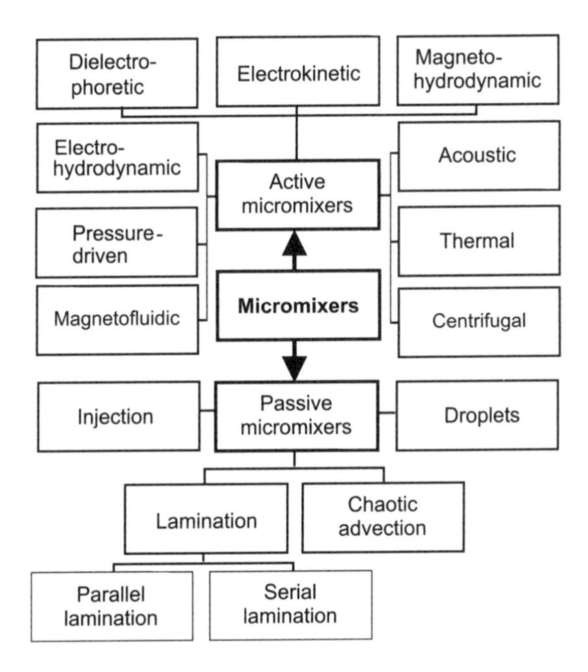


**Figure 10.16** Overview of different micromixer types.

above 10 [21]. At Reynolds numbers higher than 30, mixing is achieved right after the bend. Geometrical obstacles in the basic T-design can be implemented by roughening the channel wall [25] or throttling the channel entrance [26]. Other types of obstacles for generating vortices and chaotic advection are discussed in Section 10.2.1.4. According to Section 10.1.1, a simple method to reduce the mixing path is narrowing mixing streams. Narrow mixing streams are realized with the use of microchannels [27]. Furthermore, smaller streams can be achieved by splitting the solvent and the solute in multiple streams and rejoining them [28, 29, 30] [Figure 10.17(c)] or by interdigitating them three-dimensionally [31]. Bessoth et al. reported a parallel lamination mixer with 32 streams that can achieve full mixing after a few milliseconds [32, 33]. The flow in micromixers based on parallel lamination are usually driven by pressure. However, electro-osmosis can also be used as an alternative for fluid delivery [34, 35, 36].

As analyzed in Section 10.1.5, another concept of reducing the mixing path for parallel lamination micromixers is hydrodynamic focusing [37]. The basic configuration for hydrodynamic focusing is a long microchannel with three inlets. The middle inlet is used for the focused sample flow, while the the sheath flows join though the other two inlets [Figure 10.17(d)]. Knight et al. [37] reported a prototype with a mixing channel of $10 \times 10$ μm cross section. The width of the sample stream was focused by adjusting the pressure ratio between the sample flow and the sheath flow. In the reported experiments, the mixing time can be reduced to a few microseconds [38]. The same configuration of hydrodynamic focusing and mixing were used for cell infection [39]. Wu and Nguyen reported a configuration with two sample streams. The sample streams are focused by two sheath streams. The narrow width improves mixing between the two streams significantly [12]. Table 10.3 summarizes the typical parameters of the above-mentioned parallel lamination micromixers.

#### 10.2.1.2 Serial Lamination Micromixers

Similar to parallel lamination micromixers, serial lamination micromixers also enhance mixing through splitting and later joining the streams [Figure 10.18(a)] [8, 38–40]. The inlet streams are

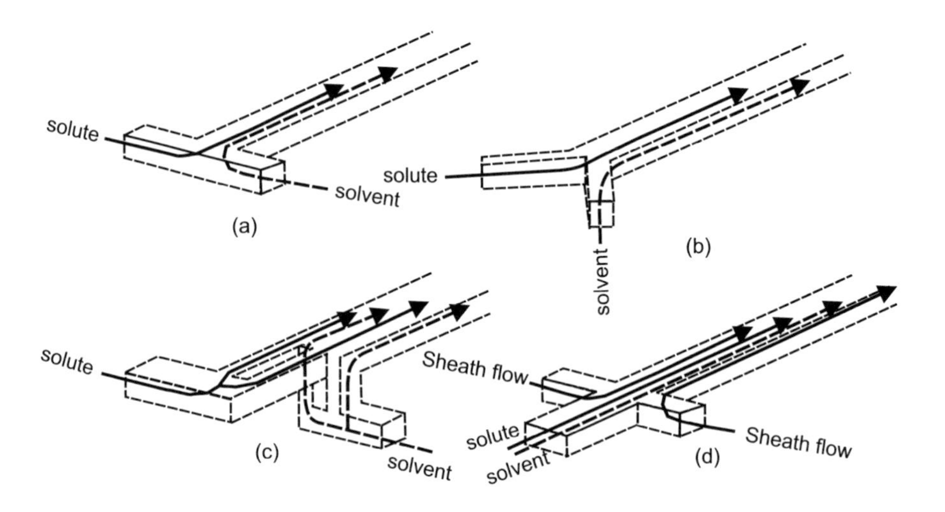


**Figure 10.17** Parallel lamination micromixer: (a) T-mixer; (b) Y-mixer; (c) parallel lamination with multiple streams; and (d) hydraulic focusing.

first joined horizontally and then in the next stage vertically. After $n$ splitting and joining stages, $2^n$ liquid layers can be laminated. The process leads to $4^{n-1}$ times improvement in mixing time. Due to their complex geometry, most of the reported serial lamination mixers [Figure 10.18(b)] were fabricated in silicon, using bulk micromachining technologies such as wet etching in KOH [10, 40] or DRIE technique [41]. Polymeric micromachining is another alternative for making serial lamination micromixers. Lamination of multiple polymer layers also allows making complex 3-D channel structures [42] [Figure 10.18(c)].

The concept of serial lamination can be extended to electrokinetic flows as reported by He et al. [43] [Figure 10.18(d)]. Serial lamination was realized through electro-osmosis flows between the multiple intersecting microchannels. A similar design for a pressure-driven flow was reported by Melin et al. [44]. However, this design only works for a plug of the two mixed liquids. Table 10.4 lists the most important parameters of serial lamination micromixers.

#### 10.2.1.3 Injection Micromixers

As explained in Section 10.1.4, the concept of injection mixer [45, 46, 47, 48, 49] is based on the general concept of decreasing mixing path and increasing contact surface between the solute and the solvent. Instead of splitting both inlet flows, these injection micromixers only split the solute flow into many streams and inject them through a nozzle array into the solvent flow. The nozzle array creates a number of microplumes of the solute [Figure 10.8(a)]. These plumes increase the contact surface and decrease the mixing path. Mixing efficiency can be improved significantly.

The micromixers reported by Miyake et al. [45, 46] have 400 nozzles arranged in a square array. The nozzle array is located in a mixing chamber, which is fabricated in silicon using DRIE. Larsen et al. [47] reported a similar concept with a different nozzle shape and array arrangement. Seidel et al. [48] and Voldman et al. [49] utilized capillary forces for generating microplumes. These mixers use a passive microvalve for releasing one of the two mixed fluids. Table 10.5 compares the parameters of the above-mentioned injection micromixers.

**Table 10.3**

Typical Parameters of Parallel Lamination Micromixers (NA: not applicable)

| *Ref.* | *Type* | *Channel Width (μm)* | *Channel Height (μm)* | *Typical Velocity (mm/s)* | *Re* | *Pe* | *Materials* |
|---|---|---|---|---|---|---|---|
| [16, 17] | T-mixer | 550 | 25 | 6 | 0.3 | 725 | Silicon-glass |
| [18] | Y-mixer | 90 | 90 | 7 | 0.4 | 240 | PDMS-glass |
| [19] | Y-mixer | 1,000 | 20 | 83 | 1.7 | 830 | $CaF_2$-SU8-metal-glass |
| [20] | Y-mixer | 900 | 50 | 0.27 | 0.02 | 150 | PMMA |
| [21] | Y-mixer | 200 | 200 | 50–200 | 80 | 80,000 | Ceramic |
| [22] | T-mixer | 100 | 50 | 7,000 | 500 | 700,000 | Silicon-glass |
| [24] | Vortex | 20 | 200 | 10,000 | 200 | 200,000 | Silicon-glass |
| [25] | Cross-shaped | 30 | 40 | 5,000-10,000 | 170–340 | 150,000 | Ceramic |
| [26] | T-mixer | 500 | 300 | 0.3 | 0.1 | 150 | NA |
| [27] | T-mixer | 100 | 200 | 0.17 | 0.023 | 170 | Silicon-glass |
| [30] | Parallel lamination | 85 | 5 | 0.7 | 0.0035 | 60 | Silicon-glass |
| [32] | Parallel lamination | 20 | 50 | 1.5 | 0.07 | 60 | Glass |
| [35] | T-mixer | 35 | 9 | 1 | 0.014 | 35 | Glass |
| [37] | Focusing | 10 | 10 | 50 | 0.5 | 500 | Silicon-PDMS-glass |
| [39] | Focusing | 200–1,000 | 150 | 1 | 0.15 | 200 | PDMS-glass |
| [12] | Focusing | 900 | 50 | 1 | 0.05 | 100 | PMMA |

#### 10.2.1.4 Micromixers Based on Chaotic Advection

The governing equation (10.8) shows that, besides diffusion, advection is another important form of mass transfer, especially in flows with a low Reynolds number. For all cases discussed in Section 10.1, advection is parallel to the main flow direction and is not useful for the transversal transport of species. Thus, advection in other directions, the chaotic advection, could improve mixing significantly. Chaotic advection can be induced by special geometries in the mixing channel or excited by an external disturbance. While the first type is a passive micromixer, the second type belongs to the active category and will be discussed later in Section 10.2.2.

The basic design concept for generation of advection is the modification of the channel shape for splitting, stretching, folding, and breaking of the laminar flow. Depending on the geometrical configuration, chaotic advection was observed at different Reynolds numbers. Thus, the design examples based on chaotic advection are categorized by their operation range of Reynolds numbers. Although there is no fixed range for a particular design, a range of $\mathrm{Re} > 100$ is considered in this section as high, while the ranges of $10 < \mathrm{Re} < 100$ and of $\mathrm{Re} < 10$ are regarded as intermediate and low, respectively.

*Chaotic Advection at High Reynolds Numbers.* The simplest way to induce chaotic advection is to insert obstacles in the mixing channel. The obstacles can be placed on the wall [Figure 10.19(a)]. Wang et al. numerically investigated the role of obstacles in a mixing channel at high Reynolds numbers [51]. The simulated mixing channel is 300 μm in width, 100 μm in depth, and 1.2 to 2 mm in length; the diameter of the obstacle is 60 μm [Figure 10.19(b)]. The simulation results show that obstacles in a microchannel at low Reynolds numbers cannot generate instabilities or recirculations. However, the obstacles can improve mixing at high Reynolds numbers. Under high Reynolds number conditions, the asymmetric arrangement of obstacles alters the flow directions

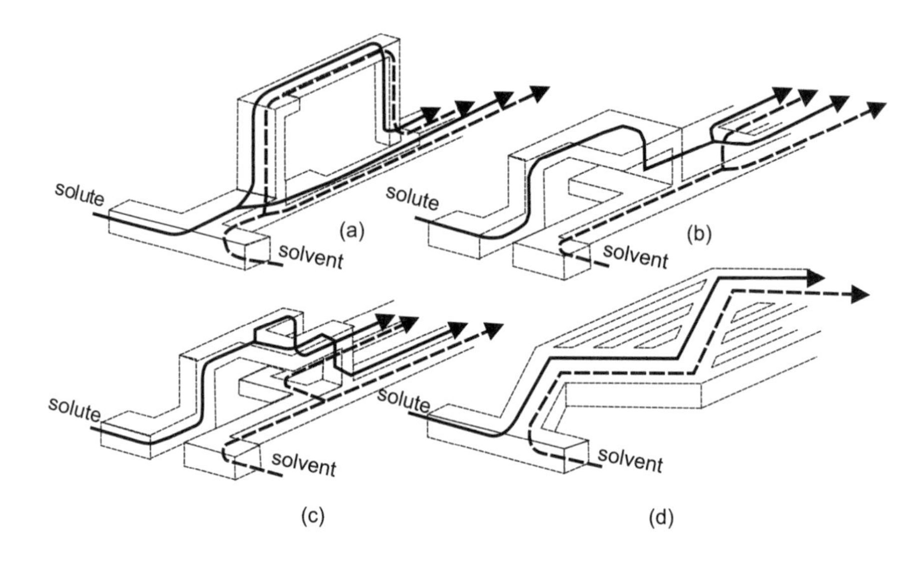


**Figure 10.18** Serial lamination mixer: (a) join-split-join; (b) split-join [10]; (c) split-split-join [42]; and (d) multiple intersecting microchannels [43].

**Table 10.4**

Typical Parameters of Serial Lamination Micromixers (NR: not reported)

| *Ref.* | *Number of Stages* | *Channel Width (μm)* | *Channel Height (μm)* | *Typical Velocity (mm/s)* | *Re* | *Pe* | *Materials* |
|---|---|---|---|---|---|---|---|
| [10] | 3 | 300 | 30 | 1–22 | 0.03-0.66 | 15–330 | Silicon-glass |
| [40] | 5–20 | 400 | 400 | 1.8 | 0.072 | 72 | Silicon-glass |
| [41] | 6 | 200 | 100 | NR | NR | NR | Silicon-glass |
| [42] | 6 | 600 | 100 | 0.5 | 0.05 | 50 | Mylar |
| [43] | 1 | 100 | 10 | 0.25 | 0.0025 | 25 | Quartz |
| [44] | 16 | 50 | 50 | 2 | 0.1 | 14 | Silicon-PDMS |

and creates transversal mass transport. Lin et al. [52] used cylinders placed in a narrow channel to enhance mixing [Figure 10.19(b)]. The 50 × 100 × 100 μm mixing chamber was fabricated in silicon. Seven cylinders of 10-μm diameter were arranged in the mixing chamber. The micromixer worked with Reynolds numbers ranging from 200 to 2,000 and achieved a mixing time of 50 μs.

Another method to generate chaotic advection at high Reynolds numbers is generating recirculation around the turns of a zigzagging mixing channel [Figure 10.19(c)]. Based on a numerical investigation, Mengeaud et al. [53] used the zigzagging period as the optimization parameter. The micromixers were fabricated using excimer laser on polyethylene terephthalate (PET) substrate. The microchannel has a width of 100 μm, a depth of 48 μm, and a length of 2 mm. A critical Reynolds number of 80 was observed. Below this number, the mixing process relied entirely on molecular diffusion. Above this number, fast mixing through recirculations at the turns was observed.

*Chaotic Advection at Intermediate Reynolds Numbers.* At intermediate Reynolds numbers, chaotic advection can be induced by special channel designs. Hong et al. [54] demonstrated an in-plane

**Table 10.5**
Typical Parameters of Injection Micromixers (NR: not reported)

| *Ref.* | *Number of Nozzles* | *Channel Width (μm)* | *Nozzle Size (μm)* | *Channel Height (μm)* | *Typical Velocity (mm/s)* | *Re* | *Pe* | *Materials* |
|---|---|---|---|---|---|---|---|---|
| [45, 46] | 400 | 2,000 | 330 | 15×15 | 1.2 | 0.018 | 18 | Silicon-glass |
| [47] | 10–20 | NR | 100 | 50 | 1 | 0.1 | 100 | Silicon-glass |
| [48] | 1 | 280–600 | 135–175 | 20–43 | NR | NR | NR | Silicon-glass |
| [49] | 1 | 820 | 7 | 70 | 15 | 0.1 | 105 | Silicon-glass |

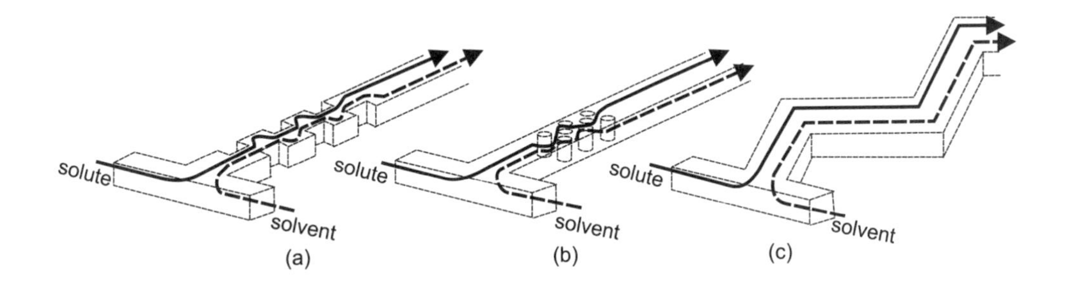


**Figure 10.19** Planar designs for mixing with chaotic advection at high Reynolds numbers: (a) obstacles on walls [25]; (b) obstacles in the channel [50, 52]; and (c) zigzagging channel [53].

micromixer with modified Tesla structures [Figure 10.20(a)]. One branch of the Tesla structure has the form of a nozzle. The flows in the two branches cause the Coanda effect, which leads to chaotic advection and improves mixing significantly. The mixer was made of cyclic olefin copolymer (COC) by hot embossing and thermal direct bonding. The mixer works well at higher Reynolds numbers ($Re > 5$).

Three-dimensionally twisted channels also induce chaotic advection. Liu et al. [55] reported a 3-D serpentine mixing channel fabricated in silicon and glass. The channel was constructed as a series of C-shaped segments positioned in perpendicular planes [Figure 10.20(b)]. The micromixer consists of two inlets joined in a T-junction, a 7.5-mm-long straight channel, and a sequence of six mixing segments. The total mixing length was about 20 mm. Fast mixing caused by chaotic advection was observed at relatively high Reynolds numbers from 25 to 70.

A similar micromixer with a 3-D serpentine mixing channel fabricated in PDMS was reported by Vijiayendran et al. [56]. The channel was designed as a series of L-shaped segments in perpendicular planes [Figure 10.20(c)]. The channel has a width of 1 mm and a depth of 300 μm. The total length of the mixing channel is about 30 mm. The mixer was tested at the Reynolds numbers of 1, 5, and 20. The reported experimental results also indicated that better mixing was achieved at higher Reynolds numbers. Another complex design on PDMS was reported by Chen and Meiners [57]. The mixing unit, called "flow-folding topological structure," by the authors, is formed by two connected out-of-plane L-shapes and measures about 400×300 μm [Figure 10.20(d)]. The microchannel has a width of 100 μm and a depth of 70 μm. With this design, effective mixing can be achieved with a purely laminar flow at Reynolds numbers ranging from 0.1 to 2.

Park et al. reported a more complex 3-D micromixer [58]. The channel rotates and separates the two fluids by partitioned walls [Figure 10.20(e)]. The channel structure was fabricated in PDMS and then bonded to glass. Jen et al. proposed other designs of twisted microchannels for mixing of gases [59]. These designs were only tested numerically. The channel has a width and height of 500

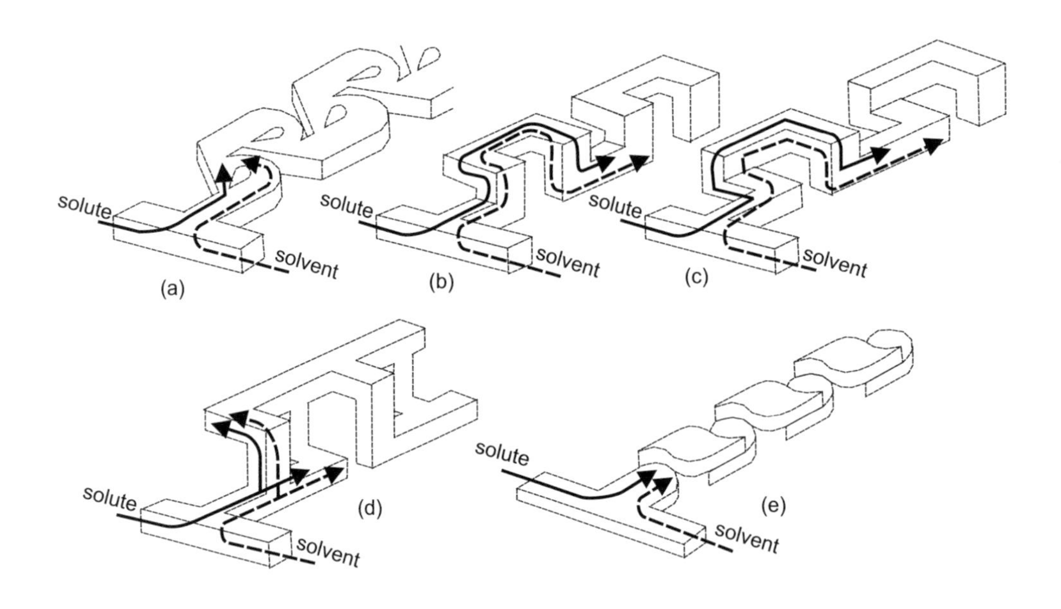


**Figure 10.20** Micromixer designs for mixing with chaotic advection at intermediate Reynolds numbers: (a) modified Tesla structure; (b) C-shape [55]; (c) L-shape [56]; (d) connected out-of-plane L-shapes [57]; and (e) twisted microchannel [58].

and 300 μm, respectively. Mixing of methanol and oxygen at different velocities ranging from 0.5 to 2.5 m/s was considered in the simulation.

Chaotic advection in a mixing chamber was generated actively by a planar pulsed source-sink system [60] as reported by Evans et al. [61]. The mixer was fabricated in silicon on a 1-cm$^2$ area. The mixing chamber measures $1,500 \times 600$ μm with a height of 100 μm.

Micromixers based on unbalanced collision have also been reported for intermediate Reynolds number [62, 63] (Figure 10.21). Mixing occurs based on the asymmetric structure of the channel that produce two different flow rates. Frequent flow splitting and unbalanced collision of the flows improves the mixing. In curved channels [Figure 10.21(a)], mixing is more enhanced due to the presence of Dean vortices combined with the unbalanced collision of the flows [65].

Another method to enhance mixing is using convergent-divergent structures that can generate expansion vortices and enhance the contact area [65]. Micromixer with multiple periods of sinusoidal walls was reported by Afzal et al. [66] that could produce a mixing efficiency of 92% once combined with pulsatile flow [Figure 10.22]. Similar to micromixers with unbalanced collisions, convergent-divergent micromixers with embedded obstacles to produce split-and-recombine flow patterns have been also proposed [67] [Figure 10.22(b)]. Mixing efficiencies of up to 95% have been reported for such mixers due to the presence of secondary flows [65].

Spiral microchannels also have been explored for micromixing. Spiral micromixers benefit from the presence of vortices induced by the Dean effect induced by centrifugal forces in a curved channel [68, 69] [Figures 10.23(b) and (d)]. Thus, high mixing efficiencies of more than 90% can be achieved in low and intermediate Reynolds numbers. Spiral mixers have been realized in 2-D planar and 3-D designs [65]. Some designs of 3-D spiral micromixers have been realized using additive manufacturing (3-D printing) method [70].

*Chaotic Advection at Low Reynolds Numbers.* Similar to their macroscale counterparts, rips [Figure 10.24(a)] or grooves [Figures 10.24(b, c)] on the channel wall can twist the flow and generate chaotic advection. Johnson et al. [71] ablated slanted grooves on the bottom wall of the mixing channel using laser machining. This structure twists an electrokinetic flow into helical stream lines and allows

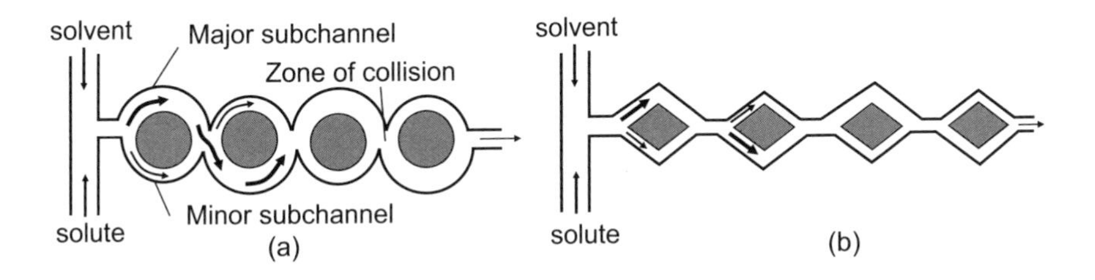


**Figure 10.21** Schematics of mixing based on unbalanced collision at intermediate Reynolds numbers: (a) curved channels; and (b) rhombus channels. (*After:* [64].)

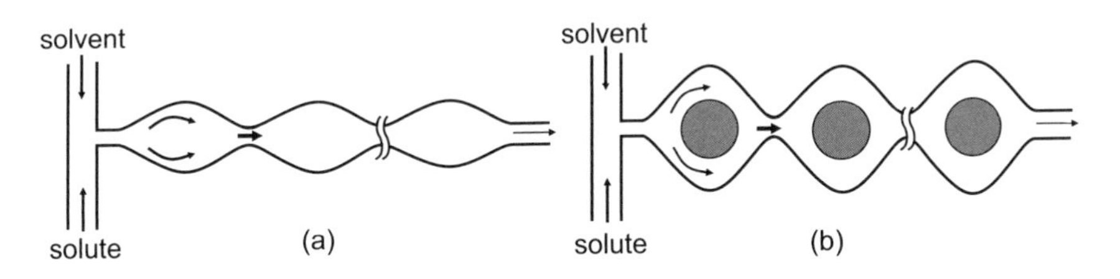


**Figure 10.22** Schematics of mixing based convergent-divergent flow pattern at intermediate Reynolds numbers: (a) sinusoidal walls, [66]; and (b) with embedded obstacles, [67].

mixing at a relatively low velocity of 300 μm/s. The micromixers were fabricated by excimer laser ablation on a polycarbonate sheet (PC). The substrate is bonded to a cover made of polyethylene terephthalate glycol (PETG). The mixing channel was 72 μm wide at the top, 28 μm wide at the bottom, and 31 μm in depth. The width of an ablated groove was 14 μm, the center-to-center spacing between the grooves was 35 μm. The length of the region occupied by the wells from the T-junction was 178 μm.

The same configuration was investigated later by Stroock et al. [72]. Two different groove patterns were considered [Figure 10.24(b, c)]. The staggered herringbone mixer [Figure 10.24(c)] can work well at Reynolds numbers ranging from 1 to 100. This concept can be applied to electrokinetic flow by modifying the surface charge [73]. The effect of chaotic advection with the ripped channel was numerically investigated by Wang et al. [50]. The length, width, and depth of the channels were 5 mm, 200 μm, and 100 μm, respectively. The mean velocity ranges from 100 μm/s to 50 mm/s. The grooves were also ablated on the PDMS substrate by laser [74]. Electrokinetic mixing [75] with only patterned surface modification [76] can also enhance mixing [Figure 10.24 (d–f)]. Compared to the conventional T-configuration, Biddiss et al. reported an improvement of mixing efficiencies from 22 to 68% at Peclet numbers ranging from 190 to 1500.

Kim et al. [77] improved the design of Stroock et al. [72] with embedded barriers parallel to the flow direction. This embedded barrier forces the flow to change the original elliptic mixing pattern [72] to a hyperbolic pattern [77]. The mixing channel of this design is 240 μm in width, 60 μm in depth, and 21 mm in length. The barriers have a cross section of 40 μm × 30 μm.

Bertsch et al. reported a miniaturized version of a conventional mixer with helical flow-twisting elements [78]. The design concept is to modify the 3-D inner wall of a cylindrical mixing channel. Two designs were tested for this concept. The first design was formed by four mixing elements, which was made of 24 rectangular bars placed at 45°. The four mixing elements were arranged at 45° in the channel. The second design consists of right-handed and left-handed helical elements containing six small-helix structures. Because of the complex geometry, the micromixer was fabricated by stereo microphotography (Section 3.4.4), which builds up the complex structure

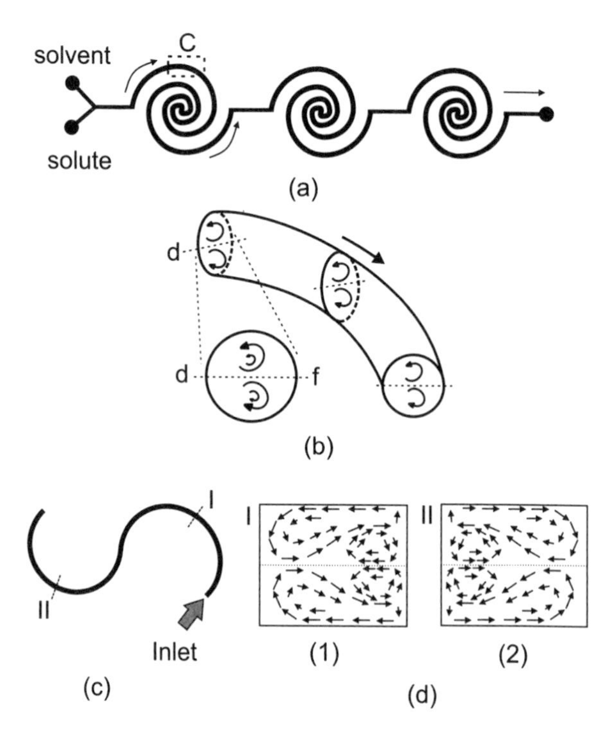


**Figure 10.23** Spiral mixing at intermediate Reynolds numbers: (a) double spiral micromixer [69]; (b) schematic of Dean flow effect in the curved channel with circular cross section shown in the box (C) of (a) [69]; (c) curved channel with rectangular cross section [68]; and (d) the corresponding flow directions at cross sections of I and II shown in (c) [68].

layer by layer. Table 10.6 summarizes the most important parameters of the above micromixers based on chaotic advection.

In addition, placing baffles and barriers in various geometries inside the flow channel has been implemented for micromixers working at various Reynolds numbers. Wang et al. [79], fabricated a micromixer with 64 group of triangle baffles to improve the mixing efficiency within the Reynolds number range of 0.1 to 500 by varying the size, the number and the spacing of the baffles [Figure 10.25(a)].

Tsai and Wua [80] developed a micromixer with two radial barriers having 40-μm and 97.5-μm thickness and length in a curved microchannel to enhance mixing efficiency by creating multidirectional vortices [Figure 10.25(b)]. Dean vortices, produced inside the curved channel, and their expansion induced by the barriers improve the mixing efficiently.

#### 10.2.1.5 Droplet Micromixers

A tiny droplet containing both solvent and solute will reduce the total amount of samples significantly. The small size of the droplet also fulfills the requirement for a short mixing path. If the droplet is set in motion, the flow field inside the droplet further improves the mixing effect. The internal flow field depends on the droplet motion; thus, the mixing pattern can be controlled externally by the motion pattern of the droplet.

In general, droplets can be generated and transported individually using pressure [81] or capillary effects such as thermocapillary [82] and electrowetting [83]. Furthermore, droplets are formed in microchannels with immiscible phases, such as oil/water or water/gas, due to the large difference of surface forces [84]. Hosokawa et al. [81] reported the earliest droplet micromixer,

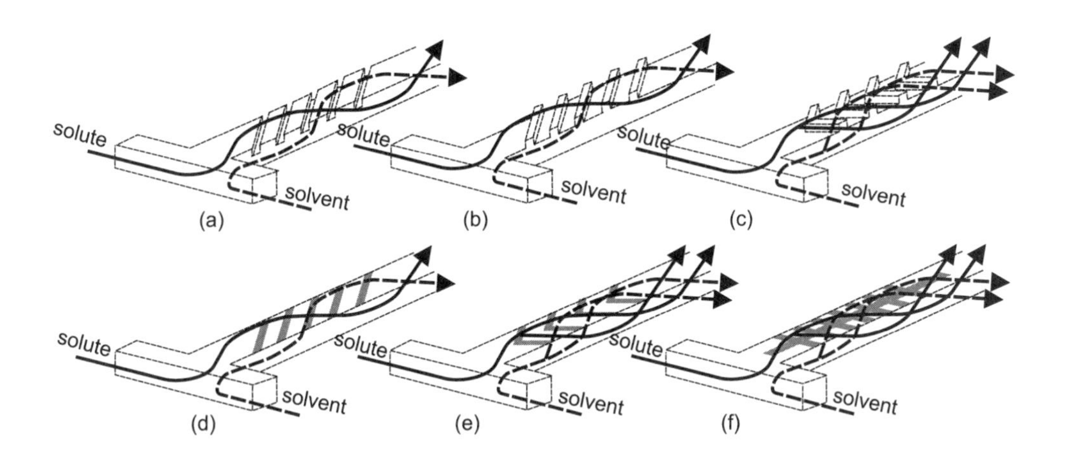


**Figure 10.24** Modification of mixing channel for chaotic advection at low Reynolds numbers: (a) slanted ribs; (b) slanted grooves [72, 73]; (c) staggered herringbone grooves [72, 73]; and (d–f) patterns for surface modification in a micromixer with electrokinetic flows [75].

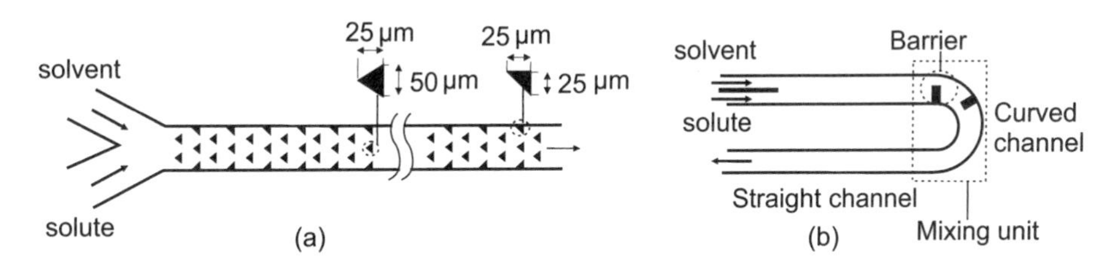


**Figure 10.25** Micromixer designs with baffles and barriers in channel for mixing with chaotic advection at low Reynolds numbers: (a) triangle baffles in channel [79]; (b) barriers in curved channel [80].

which was fabricated in PDMS and covered by a PMMA sheet. The concept utilized a hydrophobic microcapillary vent, which join the solute with the solvent droplet. By simplifying the mass transport equation and introducing an effective dispersion coefficient for a rectangular channel, Handique and Burns reported an analytical model for droplet mixing actuated by thermocapillary [82]. The mixing droplet can also be generated and transported by electrowetting (Section 7.2.2). Paik et al. [83] reported different mixing schemes with the electrowetting concept. Droplets can be merged and split repeatedly to generate the mixing pattern. The merged droplet can then be transported with different motion patterns to induce mixing.

Flow instability between two immiscible liquids can form droplets, which can be used for mixing [84, 85]. Using a carrier liquid such as oil, droplets of the aqueous samples are formed in a microchannel. While moving along the microchannel, the shear force between the carrier liquid and the sample induces an internal flow field and accelerates the mixing process in the droplet. Table 10.7 lists the typical parameters of the above droplet micromixers.

### 10.2.2 Active Micromixers

#### 10.2.2.1 Pressure-Driven Disturbance

Disturbance caused by pressure was used in one of the earliest active micromixers. Deshmukh et al. [86] reported a T-mixer with pressure disturbance [Figure 10.26(a)]. The mixing concept used in this mixer was categorized previously in Section 10.1.3 as a passive concept with sequential

**Table 10.6**

Typical Parameters of Chaotic Advection Micromixers (NR: not reported; NA: not applicable)

| *Ref.* | *Type* | *Channel Width(μm)* | *Channel Height (μm)* | *Typical Velocity (mm/s)* | *Re* | *Pe* | *Materials* |
|---|---|---|---|---|---|---|---|
| [51] | Cyl. obstacles | 300 | 100 | 0.17 | 0.25 | 51 | NA |
| [52] | Cyl. obstacles | 10 | 100 | 20 | 0.2 | 200 | Silicon glass |
| [53] | Zigzag shaped | 100 | 48 | 1.3–40 | 0.26-267 | 130–4,000 | Mylar |
| [54] | 2-D Tesla | 200 | 90 | 5 | 6.2 | $10^4$ | COC |
| [55] | 3-D serpentine | 300 | 150 | 30–350 | 6–70 | 9,000–$10^4$ | Silicon glass |
| [56] | 3-D serpentine | 1,000 | 300 | 2–40 | 1–20 | 2,000–$4\times10^4$ | PDMS |
| [57] | 3-D serpentine | 100 | 70 | 1–20 | 0.1–2 | 10–200 | PDMS |
| [58] | 3-D serpentine | 100 | 50 | NR | 1–50 | 0.015–0.7 | PDMS |
| [59] | 3-D serpentine | 500 | 300 | 2,000 | 48 | 0.36 | NA |
| [61] | Source-sink | $1{,}500\times600$ | 100 | NR | NR | NR | Silicon glass |
| [71] | Patterned wall | 72 | 31 | 0.6 | 0.024 | 15 | PC-PETG |
| [72, 73] | Patterned wall | 200 | 70 | 15 | 0.01 | 3,000 | PDMS |
| [50] | Patterned wall | 200 | 100 | 0.1–50 | 0.0013–6.65 | 20–$10^4$ | PDMS |
| [75] | Patterned wall | 200 | 8 | 0.01–0.09 | 0.08-0.7 | 190–1,500 | PDMS |
| [77] | Patterned wall | 240 | 60 | 11.6 | 0.5 | 2,784 | PDMS |

segmentation. However, considering the integrated pumps, the segmentation can be referred to as a pressure-driven disturbance. The mixer is integrated in a microfluidic system, which is fabricated in silicon using DRIE. An integrated planar micropump drives and stops the flow in the mixing channel causing sequential segmentation, as discussed in Section 10.1.3 [Figure 10.26(a)] [87]. This segmentation concept can also be realized by an external micropump [88].

An alternative method to pressure disturbance is the generation of pulsing velocity [89, 90] [Figure 10.26(b)]. Glasgow and Aubry [89] demonstrated a simple T-mixer and its simulation with a pulsed side flow at a small Reynolds number of 0.3. Niu and Lee [90] generated pressure disturbance using a computer controlled source-sink system. This concept is partly similar to that of Evans et al. [61]. The performance of the mixing process is determined by the pulse frequency and the number of mixing units. Okkels and Tabeling [91] reported a numerical analysis of the mixing pattern in a chamber disturbed by transient pressure field.

Disturbance can be caused directly by moving parts in the mixing channel. Suzuki and Ho [92] reported a micromixer with integrated conductors. The concept is similar to a commercial macroscale magnetic stirrer. The electrical conductors generate a magnetic field, which in turn moves magnetic beads of 1 to 10 μm in diameter. The chaotic movement induced by the magnetic beads improves mixing significantly. An integrated magnetic microstirrer was reported by Lu et al.

**Table 10.7**

Typical Parameters of Droplet Micromixers (NR: not reported)

| *Ref.* | *Transport Type* | *Droplet Size (nL)* | *Channel Width (μm)* | *Channel Height (μm)* | *Materials* |
|---|---|---|---|---|---|
| [81] | Pressure-driven | 10 | 100 | 150 | PDMS/PMMA |
| [83] | Electrowetting | 1,600 | 2,480 | 600–1000 | Glass |
| [84] | Multiple phases | 75–150 | 20–100 | NR | PDMS |

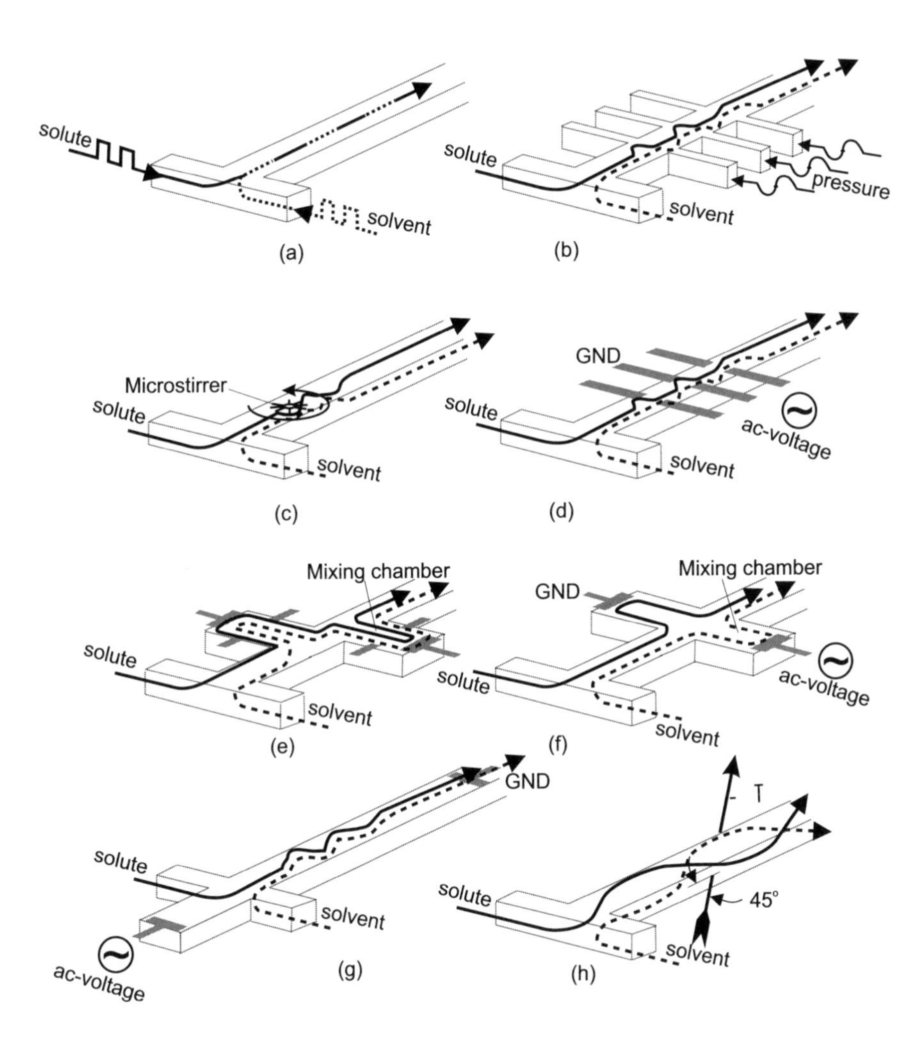


**Figure 10.26** Active micromixers: (a) serial segmentation; (b) pressure disturbance along mixing channel; (c) integrated microstirrer in the mixing channel; (d) electrohydrodynamic disturbance; (e) dielectrophoretic disturbance; (f) electrokinetic disturbance in a mixing chamber; (g) electrokinetic disturbance in a mixing channel; and (h) disturbance caused by thermocapillary convection induced by a tranverse temperature gradient.

[93] [Figure 10.26(c)]. The micromachined stirrer is placed at the interface between two liquids in a T-mixer. An external magnetic field drives the stirrer at speeds ranging between 100 and 600 rpm.

### 10.2.2.2 Electrohydrodynamic Disturbance

Electrohydrodynamic effect (Section 7.1.2) can be used to disturb the laminar streams. The structure of the micromixer with eletrohydrodynamic disturbance reported by El Moctar et al. [94] is similar to the concept reported by Niu and Lee [90] for pressure-driven disturbance. Instead of pressure sources, electrodes are placed along the mixing channel [Figure 10.26(d)]. The mixing channel is 30 mm long, 250 μm wide, and 250 μm deep. A number of titanium electrodes are placed in the direction perpendicular to the mixing channel. By changing the voltage and frequency on the electrodes, good mixing was achieved after less than 0.1 second at a Reynolds number as low as 0.02.

#### 10.2.2.3 Dielectrophoretic Disturbance

Dielectrophoresis (DEP) is the polarization of a particle relative to its surrounding medium in a nonuniform electrical field (Section 7.1.2). This effect causes the particle to move to and from an electrode. Micromixers based on this active concept were reported by Deval et al. [95] and Lee et al. [96]. Chaotic advection was generated by the motion of embedded particles, due to a combination of electrical actuation and local geometry channel variation [Figure 10.26(e)].

#### 10.2.2.4 Electrokinetic Disturbance

In passive micromixers, electro-osmosis can be used to transport liquid as an alternative to pressure-driven flow. Jacobson et al. [36] reported electrokinectically driven mixing in a conventional T-mixer. Lettieri et al. used electro-osmosis to disturb the pressure-driven flow in the mixing channel [97]. In another case [98], oscillating electro-osmotic flow in a mixing chamber is induced by an ac voltage. The pressure-driven flow becomes unstable in a mixing chamber [Figure 10.26(f)] or in a mixing channel [Figure 10.26(g)] and improves mixing through advection.

Similar to the pressure-driven sequential segmentation [86], Tang et al. [99] also utilized switching of electrokinetic flow to generate short segments of solvent and solute in the mixing channel. This flow modulation scheme was capable of injecting reproducible and stable fluid segments into microchannels at a frequency between 0.01 and 1 Hz.

#### 10.2.2.5 Magnetohydrodynamic Disturbance

Transversal transport of species in microchannels can be improved with magnetohydrodynamic effect [100]. In the presence of an external magnetic field, an applied dc voltage on the electrodes generates Lorentz forces, which in turn induce transversal movement in the mixing channel. Depending on the control modes, the Lorentz force can roll and fold the liquids in the mixing channel. However, active mixing based on magnetohydrodynamics only works with an electrolyte solution. The mixer of Bau et al. [100] was fabricated from cofired ceramic tapes. The electrodes are printed with a gold paste.

#### 10.2.2.6 Magnetofluidic Disturbance

Magnetofluidic mixing relies on the interactions occur in a microchannel between a magnetic field (ferrofluid) and a magnetic fluid [101]. To achieve such mixing, water-based streams with different magnetization properties are introduced into the mixing chamber in a core-cladding configuration [Figures 10.27(a, b)]. Ferrofluid can form either the core or the cladding stream. Once exposed to a magnetic field, instabilities at the liquid-liquid interface occurs due to the magnetization mismatch of the two liquid flows. The instabilities can alter the velocity profile within the mixing chamber. Depending on the strength of the magnetic field, diffusive or rapid bulk mixing is feasible. Bulk mixing or stirring occurs at high magnetic fields that is because of the magnetically induced secondary flow in the mixing chamber [101] [Figures 10.27(c, d)]. At low magnetic fields, magnetic particles from the ferrofluid are pulled into the adjacent liquid improving the mixing between the core and the cladding streams.

#### 10.2.2.7 Acoustic Disturbance

Similar to their macroscale counterparts, micromixers can utilize acoustic waves to stir fluids and induce mixing. The proof of concept for acoustic mixing was reported by Moroney et al. [102] with an FPW device. Zhu and Kim [103] reported active mixing with focused acoustic waves in a mixing chamber. The mixing chamber was fabricated in silicon and measures 1mm $\times$ 1mm $\times$ 10 $\mu$m. A zinc oxide membrane is located at the bottom of the mixing chamber. Changing the frequency and

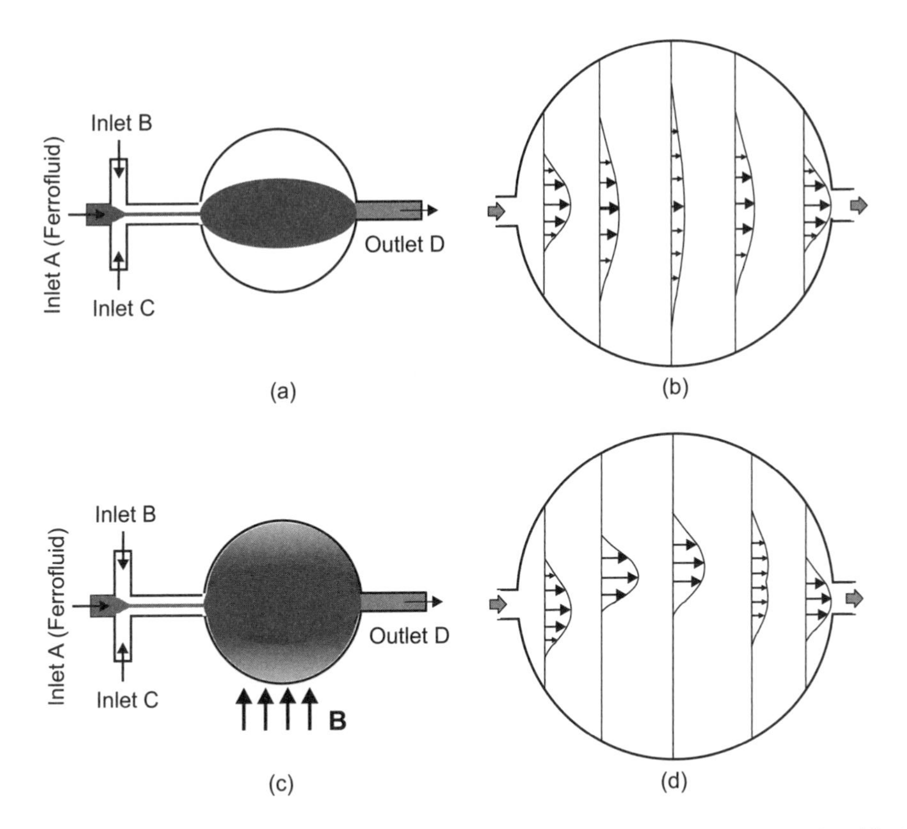


**Figure 10.27** Schematics of magnatofluidic mixing in a uniform magnetic field for two water-based liquids. (a) liquid-liquid interface with ferrofluid core flow at the absence of the magnetic field; (b) velocity profile of the hydrodynamic configuration shown in (a); (c) liquid-liquid interface with the ferrofluid core flow once magnetic field applied; and (d) velocity profile of the the hydrodynamic configuration shown in (c). Inlets B and C have the same liquid flow rate and the same liquid type. (*After*: [101].)

the voltage of the input signal determines the induced acoustic field inside the mixing chamber. The concept of acoustically induced flow, or acoustic streaming, was also used as an active mixing scheme [104]. Focused acoustic streaming with different electrode patterns was used for mixing [105]. Besides the integrated design, stirring at high frequency can also be realized by an external pump [106].

Ultrasonic mixing may have complications for biological analysis. Temperature rise caused by acoustic energy is the first concern, because many biological fluids are highly sensitive to temperature. Secondly, ultrasonic waves around 50 kHz are harmful to biological samples because of the possible cavitations. The acoustic micromixer reported by Yasuda [107] uses loosely focused acoustic waves to generate stirring movements. The wave is generated by a thin zinc oxide film. The actuator was driven by sinusoidal waves with frequencies corresponding to the thickness-mode resonance (e.g., 240 and 480 MHz) of the piezoelectric film. The mixer operated without any significant temperature increase and could be used for temperature-sensitive fluid. Yang et al. [108, 109] reported further acoustic devices for mixing water and ethanol, as well as water and uranine.

Liu et al. [110, 111] utilized acoustic streaming induced around an air bubble for mixing. In this mixer, air pockets with 500-μm diameter and 500-μm depth were used for trapping the air bubbles. Acoustic streaming was induced by the field, generated by an integrated PZT actuator. Yaralioglu et al. [112] also utilized acoustic streaming to disturb the flow in a conventional Y-mixer.

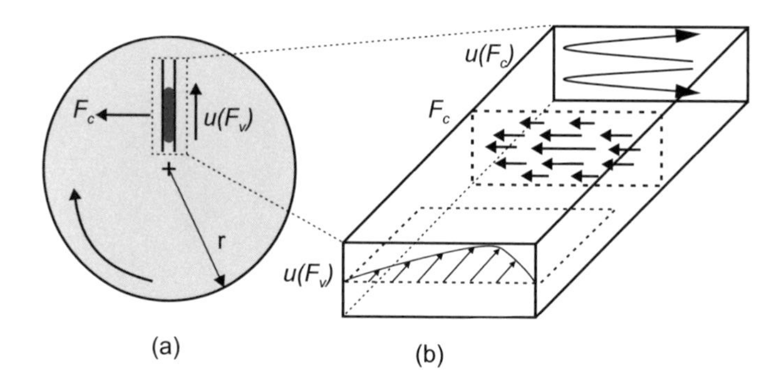


**Figure 10.28** Centrifugal micromixing on a rotating disk: (a) schematics of a rotating disk with embedded microchannel; (b) flows induced in a rotating microchannel. (*After*: [115].)

While the channel is made of PDMS, the acoustic actuator is integrated into the cover quartz wafer. An 8-μm thick zinc oxide layer with gold electrodes works as the actuator.

#### 10.2.2.8 Thermal Disturbance

Since the diffusion coefficient depends highly on temperature, thermal energy also can be used to enhance mixing [113, 114]. Darhuber et al. [113] proposed and numerically analyzed the thermocapillary convection induced by transverse temperature gradient [Figure 10.26(h)]. The temperature gradient induces helical streamlines, which significantly increases the interfacial area between the solvent and the solute. The concept of Tsai and Lin. [114] utilized a thermal bubble to generate disturbance in a mixing channel. Table 10.8 summarizes typical parameters of the above active micromixers.

#### 10.2.2.9 Centrifugal Disturbance

Micromxers can employ Coriolis forces in a network of rotating microchannels. Centrifugal volume forces are produced by rotating the channel network generating continuous flow to drive and mix fluids [115] (Figure 10.28). Coriolis force, $F_c$, is created once the liquid plug starts to move radially through the channel on the rotating disk. Radial movement of the liquid plug, $u(F_v)$, is due to the centrifugal force, $F_v$. $F_c$ is perpendicular to the flow direction. A transversal convection, $u(F_c)$, is induced as a result of the Coriolis force, which travels backward at the upper and lower sections of the microchannel [115]. The combination of the transversal convection and the radial movement of the liquid creates stirring that enhances the contact interface of the liquid solutions within the channel [115]. This results in improved mixing. In addition, barriers can be implemented within the channel to enhance the mixing efficiency. Centrifugal mixing is easily feasible for lab-on-disk microfluidic systems.

## 10.3 SUMMARY

The operation conditions of micromixers can be determined by the characteristic dimensionless numbers, such as Reynolds number (Re) and Peclet number (Pe). From the definitions, the relation between Pe and Re can be derived as:

$$\frac{\mathrm{Pe}}{\mathrm{Re}} = \frac{uL/D}{uD_\mathrm{h}/\nu} = \frac{L}{D_\mathrm{h}}\frac{\nu}{D} \tag{10.42}$$

**Table 10.8**

Typical Parameters of Active Micromixers (NR: not reported; NA: not applicable)

| *Ref.* | *Disturbance* | *Channel Width (μm)* | *Channel Height (μm)* | *Typical Velocity (mm/s)* | *Frequency (Hz)* | *Re* | *Pe* | *St* | *Materials* |
|---|---|---|---|---|---|---|---|---|---|
| [87] | Pressure | 400 | 78 | 0.09 | 1 | 0.01 | 36 | 4.4 | Silicon glass |
| [88] | Pressure | 150 | 150 | 0.9 | 100 | 0.13 | 133 | 17 | PDMS |
| [89] | Pressure | 200 | 120 | 2 | 0.3 | 0.3 | 400 | 0.03 | NA |
| [91] | Pressure | 200 | 26 | 1.6 | 0.85 | 0.04 | 321 | 0.11 | PDMS |
| [92] | Pressure | 160 | 35 | 0.3 | 0.02 | 0.05 | 48 | 4 | Silicon glass |
| [93] | Pressure | 750 | 70 | 0.14 | 5 | 0.01 | 105 | NA | PDMS glass |
| [94] | Electro-hydrodynamic | 250 | 250 | 4.2 | 0.5 | 0.02 | 1,050 | 0.03 | NA |
| [95] | Dielectrophoretic | 50 | 25 | 0.5 | 1 | 0.02 | 25 | 0.1 | Silicon SU-8 glass |
| [96] | Electrokinetic | 200 | 25 | 0.5 | 1 | 0.01 | 100 | 0.4 | NA |
| [98] | Electrokinetic | 1,000 | 300 | 0.5 | 10 | 0.15 | 1,050 | 20 | PDMS-glass |
| [99] | Electrokinetic | 500 | 35 | 1 | 0.17 | 0.04 | 509 | 0.09 | PDMS glass |
| [100] | Magneto-hydrodynamic | 4,700 | 1,000 | NR | NR | NR | NR | NR | Ceramic |
| [102] | Acoustic | 1,000 | 400 | 0.5 | 10 | 0.15 | 1,050 | 20 | Silicon glass |
| [104] | Acoustic | 1,600 | 1,600 | 1 | NR | 1.6 | 1,600 | NR | NR |
| [107] | Acoustic | 2,000 | 2,000 | 6.4 | NR | 12.8 | 12,800 | NR | Silicon glass |
| [109] | Acoustic | 6,000 | 60 | 0.5 | NR | 0.03 | 30 | NR | Silicon glass |
| [110] | Acoustic | 15,000 | 300 | 5 | NR | 1.5 | 1,500 | NR | Silicon glass |
| [112] | Acoustic | 300 | 50 | 1 | NR | 0.86 | 300 | NR | PDMS quartz |

where $u$, $D$, and $\nu$ are the bulk velocity, the diffusion coefficient, and the kinematic viscosity, respectively. The hydraulic diameter $D_{\mathrm{h}}$ and the mixing path $L$ are usually on the same order; therefore, we can assume $L/D_{\mathrm{h}} \approx 1$. The kinematic viscosity and the diffusion coefficient of liquids are on the order of $\nu = 10^{-6}$ m$^2$/s and $D = 10^{-9}$ m$^2$/s, respectively. Thus, (10.42) leads to an estimated relation between Peclet number and Reynolds number of ($\mathrm{Pe} \approx 1000\mathrm{Re}$) for liquids. On a Pr-Re diagram, the relation $\mathrm{Pe} = 1,000\mathrm{Re}$ represents a straight line (Figure 10.29). Operation points of micromixers for liquids are expected to be around this line.

The typical kinematic viscosity and a diffusion coefficient of gases are on the orders $\nu = 10^{-5}$ m$^2$/s and $D = 10^{-5}$ m$^2$/s. The operation points of gases can be expected around the line of Pe = Re (Figure 10.29).

Accurate quantification of the extent of mixing is important for evaluation of performance as well as design optimization of micromixers. While the measurement of concentration can be carried out with full-field methods as described in Chapter 4, the valuation of the 2-D fluorescent images is still a big challenge.

As mentioned in Chapter 4, the common quantification technique is using intensity of fluorescent dye to determine the extent of mixing. The intensity image can then be recorded and evaluated using a fluorescent microscope with a corresponding filter set. Since the concentration of the dye is proportional to the intensity of the recorded image, the uniformity of the concentration image can be quantified by determining the standard deviation of the pixel intensity values [55, 72]. If the standard deviation of intensity values cannot resolve the differences between regions in the image, spatial probability density functions (PDF) of intensities integrated over a finite region can be used to quantify mixing [98]. Furthermore, a 2-D power spectrum of the intensity image can also be considered as another quantification method [98, 90]. Because these techniques are statistical, they

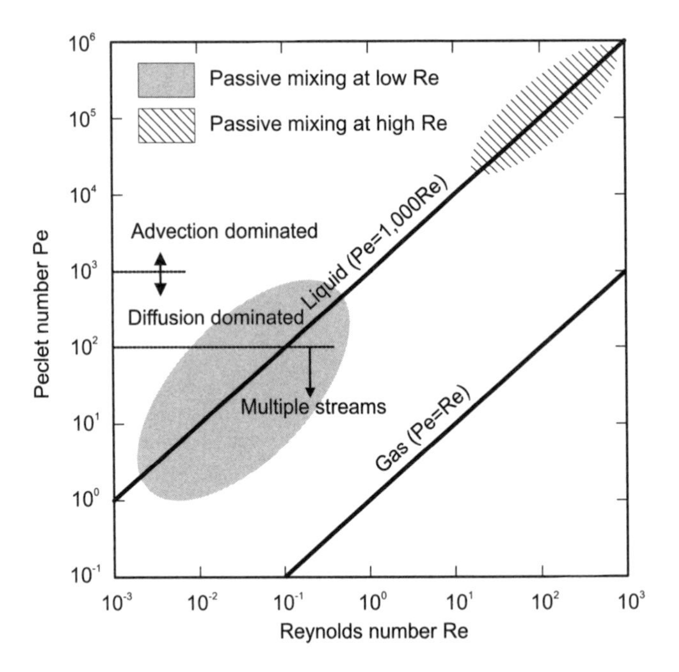


**Figure 10.29** The Pe-Re diagram of micromixers.

rely on the orientation of the mixed fluids relative to the imaging direction. If the imaging direction is perpendicular to the fluid layers, as in the case of the mixer reported by Hinsmann et al. [19], the two layers appear to be completely mixed even at the entrance of the mixing channel. For such cases, a 3-D imaging system with a confocal microscope is necessary for the spacial distribution of the concentration field [18, 37].

While the measurements with fluorescent dyes assume no chemical reaction during the mixing process, measuring the fluorescent product of a chemical reaction [16] can be used to quantify the extent of mixing. Typically, this process is an acid-base reaction, with a dye having a fluorescence quantum yield that is pH-dependent.

Munson and Yager [42] reported a pH-dependent concept for the quantification of mixing. The method relies on the increase of intensity of fluorescein at basic pH. In this method, both liquids are diluted with fluorescein. They only have different buffers with different pH values. The increase in fluorescence in the initially acidic solution overwhelms the small decrease in fluorescence of the other solution. The total fluorescence increases by a factor of 2 and can work as a measure for the extent of mixing [42].

As mentioned in the introduction of this chapter, micromixers are widely used in chemical and biomedical analysis. Almost every chemical assay requires mixing of reagents with a sample. The basic T-mixer was used by Kamholz et al. [16] for measurement of analyte concentrations of a continuous flow. The concentration of a target analyte is measured with the fluorescence intensity of the region where the analyte and a fluorescent indicator have diffused [16, 18]. The micromixer reported by Hinsmann et al. [19] was used for the study of rapid chemical reactions in solution with stopped-flow, time-resolved Fourier transform infrared (TR-FTIR) spectroscopy. Wu et al. [20] used a Y-mixer for investigating nonlinear diffusive behavior of a fluorescein. Micromixers can be used as sensors in environmental monitoring, such as detection of ammonia in aqueous solutions [27]. The fast mixing time in a micromixer benefits time-resolved measurement of reaction kinetics using nuclear magnetic resonance (NMR) [33]. Fluri et al. [34] combined CE separation with a

T-mixer as a postcolumn reactor. Hadd et al. [35] used an electrokinetically driven T-mixer for performing enzyme assays. Walker et al. utilized the short mixing length of hydrodynamic focusing to infect a cell with virus [39]. Fast mixing with a micromixer was used for the freeze-quenching technique, which is useful for trapping metastable intermediates populated during fast chemical or biochemical reactions [52]. Vijiayendran et al. used micromixers for sample preparation of a surface-based biosensor [56].

Pharmaceutical synthesis, in particular nanopharmaceuticals and nanoparticulate drug delivery systems, hold great potential to take advantage of microfluidic systems for clinical translation and manufacturing of nanomedicine [116]. Micromixers are inevitable components of such systems for preparation of nanoparticle precursors and other steps of the synthesis or screening assays. To enhance the quantity of nanoparticles synthesis, 3-D printed vortex mixers have been realized. The scalability of microfluidic systems, and in particular the development of high-throughput micromixers, is necessary for continuous-flow and industrial-scale production of nanocomplexes [117].

Besides the sensing and analysis applications discussed above, micomixers were used as tools for dispersing immiscible liquids and forming microdroplets [31]. Adjusting the inlet conditions of the two immiscible phases makes different droplet sizes possible. Generation of droplets combined with polymerization has a big potential in making nanoparticles, which have a wide range of applications. Furthermore, micromixers work as a separator for particles based on their different diffusion coefficients [118, 119] or as a generator of concentration gradients [120–124].

## Problems

**10.1** Mix water with nitric acid at room temperature in a micro-Y-mixer. Design the mixing channel if the flow rate of water and nitric acid are both 5 mL/min. The channel is 50 μm in width and 150 μm in depth. The channel should be fitted into a rectangular area with edge ratio of 3:2.

**10.2** With the same conditions in Problem 10.1, design a parallel lamination mixer so that the length of the mixing channel cannot exceed 2 mm.

**10.3** How many mixing stages does a sequential mixer require for the same time performance of the mixer designed in Problem 10.2?

**10.4** If the inlet condition of the concentration in the model of sequential segmentation (Section 10.1.3) is $c = c_0 \sin 2\pi t/T$, determine the transient concentration profile along the mixing channel.

## References

[1] Lee, C.-Y., and Fu, L.-M., "Recent Advances and Applications of Micromixers," *Sensors and Actuators B: Chemical*, Vol. 259, 2018, pp. 677–702.

[2] Liu, D., et al., "Current Developments and Applications of Microfluidic Technology Toward Clinical Translation of Nanomedicines," *Advanced Drug Delivery Reviews*, Vol. 128, 2018, pp. 54–83.

[3] Einstein, A., *Investigations on the Theory of the Brownian Movements*, New York: Dover Publications, 1956.

[4] Brody, J. B., and Yager, P., "Low Reynolds Number Microfluidic Devices," *Technical Digest of the IEEE Solid State Sensor and Actuator Workshop*, Hilton Head Island, SC, June 3–6, 1996, pp. 105–108.

[5] Brodky, R. S., *Turbulence in Mixing Operations*, New York: Academic Press, 1975.

[6] Cussler, E. L., *Diffusion Mass Transfer in Fluid Systems*, 2nd ed., New York: Cambridge University Press, 1996.

[7] Cunningham, R. E., and William, R. J. J., *Diffusion in Gases and Porous Media*, New York: Plenum Press, 1980.

[8] Erbacher, C., et al., "Towards Integrated Continuous-Flow Chemical Reactors," *Mikrochimica Acta*, Vol. 131, 1999, pp. 19–24.

[9] Wu, Z., and Nguyen N. T., "Convective-Diffusive Transport in Parallel Lamination Micromixers," *Microfluidics Nanofluidics*, Vol. 1, 2005, pp. 208–217

[10] Branebjerg, J., et al., "Fast Mixing by Lamination," *Proceedings of MEMS'96, 9th IEEE International Workshop Micro Electromechanical System*, San Diego, CA, Feb. 11–15, 1996, pp. 441–446.

[11] Brenner, H., and Edwards, D. A., *Macrotransport Processes*, Boston: Butterworth-Heinemann, 1993.

[12] Wu, Z., and Nguyen N. T., "Rapid Mixing Using Two-Phase Hydraulic Focusing in Microchannels," *Biomedical Microdevices*, Vol. 7, 2005, pp. 13–20

[13] Zheng, B., Tice J. D., and Ismagilov, R. F., "Formation of Droplets of Alternating Composition in Microfluidic Channels and Applications to Indexing of Concentrations in Droplet-Based Assays," *Analytical Chemistry*, Vol. 76, 2004, pp. 4977–4982.

[14] Zheng, B., et al., "A Droplet-Based, Composite PDMS/Glass Capillary Microfluidic System for Evaluating Protein Crystallization Conditions by Microbatch and Vapor-Diffusion Methods with On-Chip X-Ray Diffraction," *Angewandte Chemie-International Edition*, Vol. 43, 2004, pp. 2508–2511.

[15] Bringer, M., et al., "Microfluidic Systems for Chemical Kinetics That Rely on Chaotic Mixing in Droplets," *Phil. Trans. R. Soc. Lond. A*, Vol. 362, 2004, pp. 1087–1104.

[16] Kamholz, A. E. et al., "Quantitative Analysis of Molecular Interaction in Microfluidic Channel: The T-Sensor," *Analytical Chemistry*, Vol. 71, 1999, pp. 5340–5347.

[17] Kamholz, A. E. and Yager, P., "Molecular Diffusive Scaling Laws in Pressure-Driven Microfluidic Channels: Deviation from One-Dimensional Einstein Approximations," *Sensor and Actuators B*, Vol. 82, 2002, pp. 117–121.

[18] Ismagilov, R. F., et al., "Experimental and Theoretical Scaling Laws for Transverse Diffusive Broadening in Two-Phase Laminar Flows in Microchannels," *Applied Physics Letters*, Vol. 76, pp. 2376–2378.

[19] Hinsmann, P., et al., "Design, Simulation and Application of a New Micromixing Device for Time Resolved Infrared Spectroscopy of Chemical Reactions in Solutions," *Lab on a Chip*, Vol. 1, 2001, pp. 16–21.

[20] Wu, Z., Nguyen, N. T., and Huang, X. Y., "Non-Linear Diffusive Mixing in Microchannels: Theory and Experiments," *Journal of Micromechanics and Microengineering*, Vol. 14, 2004, pp. 604–611.

[21] Yi, M., and Bau, H. H., "The Kinematics of Bend-Induced Mixing in Micro-Conduits," *International Journal of Heat and Fluid Flow*, Vol. 24, 2003, pp. 645–656.

[22] Wong, S. H., Ward, M. C. L., and Wharton, C. W., "Micro T-Mixer as a Rapid Mixing Micromixer," *Sensors and Actuators B*, Vol. 100, 2004, pp. 365–385.

[23] Lim, D. S. W., et al., "Dynamic Formation of Ring-shaped Patterns of Colloidal Particles in Microfluidic Systems," *Applied Physics Letters*, Vol. 83, 2003, pp. 1145–1147.

[24] Böhm, S., et al., "A Rapid Vortex Micromixer for Studying High-Speed Chemical Reactions," *Technical Proceedings of Micro Total Analysis Systems MicroTAS 2001*, Monterey, CA, 2001, Oct. 21–25, 2001, pp. 25–27.

[25] Wong, S. H., et al., "Investigation of Mixing in a Cross-Shaped Micromixer with Static Mixing Elements for Reaction Kinetics Studies," *Sensors and Actuators B*, Vol. 95, 2003, pp. 414–424.

[26] Gobby, D., Angeli, P., and Gavriilidis, A., "Mixing Characteristics of T-Type Microfluidic Mixers," *Journal of Micromechanics and Microengineering*, Vol. 11, 2001, pp. 126–132.

[27] Veenstra, T. T., "Characterization Method for a New Diffusion Mixer Applicable in Micro Flow Injection Analysis Systems," *Journal of Micromechanics and Microengineering*, Vol. 9, 1999, pp. 199–202.

[28] Jackman, R. J., et al., "Microfluidic Systems with On-Line UV Detection Fabricated in Photodefineable Epoxy," *Journal of Micromechanics and Microengineering*, Vol. 11, 2001, pp. 263–269.

[29] Möbius et al., "A Sensor Controlled Processes in Chemical Microreactors," *Proceedings of Transducers '95, 8th International Conference on Solid-State Sensors and Actuators*, Stockholm, Sweden, June 16-19, 1995, pp. 775–778.

[30] Koch, M., et al., "Improved Characterization Technique for Micromixers," *Journal of Micromechanics and Microengineering*, Vol. 9, 1999, pp. 156–158.

[31] Haverkamp, V., et al., "The Potential of Micromixers for Contacting of Disperse Liquid Phases," *Fresenius Journal of Analytical Chemistry*, Vol. 364, 1999, pp. 617–624.

[32] Bessoth, F. G., de Mello, A. J., and Manz, A., "Microstructure for Efficient Continuous Flow Mixing," *Anal. Comm.*, Vol. 36, 1999, pp. 213–215.

[33] Kakuta, M., et al., "Micromixer-Based Time-Resolved NMR: Applications to Ubiquitin Protein Conformation," *Analytical Chemistry*, Vol. 75, 2003, pp. 956–960.

[34] Fluri, K., et al., "Integrated Capillary Electrophoresis Devices with an Efficient Postcolumn Reactor in Planar Quartz and Glass Chips," *Analytical Chemistry*, Vol. 68, 1996, pp. 4285–4290.

[35] Hadd, A. G., et al., "Microchip Device for Performing Enzyme Assays," *Analytical Chemistry*, Vol. 69, pp. 3407–3412.

[36] Jacobson, S. C., McKnight, T. E., and Ramsey, J. M., "Microfluidic Devices for Electrokinematically Driven Parallel and Serial Mixing," *Analytical Chemistry*, Vol. 71, 1999, pp. 4455–4459.

[37] Knight, J. B., Vishwanath, A., Brody J. P., and Austin, R. H., "Hydrodynamic Focusing on a Silicon Chip: Mixing Nanoliters in Microseconds," *Physical Review Letters*, Vol. 80, 1998, pp. 3863–3866.

[38] Jensen, K., "Chemical Kinetics: Smaller, Faster Chemistry," *Nature*, Vol. 393, 1998, pp. 735–736.

[39] Walker, G. M., Ozers, M. S., and Beebe, D. J., "Cell Infection Within a Microfluidic Device Using Virus Gradients," *Sensors and Actuators B*, Vol. 98, 2004, pp. 347–355.

[40] Schwesinger, N., et al., "A Modular Microfluid System with an Integrated Micromixer," *Journal of Micromechanics and Microengineering*, Vol. 6, 1996, pp. 99–102.

[41] Gray, B. L., et al., "Novel Interconnection Technologies for Integrated Microfluidic Systems," *Sensors and Actuators A*, Vol. 77, 1999, pp. 57–65.

[42] Munson, M. S., Yager, P., "Simple Quantitative Optical Method for Monitoring the Extent of Mixing Applied to a Novel Microfluidic Mixer," *Analytica Chimica Acta*, Vol. 507, 2004, pp. 63–71.

[43] He, B., et al., "A Picoliter-Volume Mixer for Microfluidic Analytical Systems," *Analytical Chemistry*, Vol. 73, 2001, pp. 1942–1947.

[44] Melin, J., et al., "A Fast Passive and Planar Liquid Sample Micromixer," *Lab on a Chip*, Vol. 3, 2004, pp. 214–219.

[45] Miyake, R., et al., "Micro Mixer with Fast Diffusion," *Proceedings of MEMS'93, 6th IEEE International Workshop Micro Electromechanical System*, San Diego, CA, Jan. 25–28, 1993, pp. 248–253.

[46] Miyake, R., et al., "A Highly Sensitive and Small Flow-Type Chemical Analysis System with Integrated Absorbtionmetric Micro-Flowcell," *Proceedings of MEMS'97, 10th IEEE International Workshop Micro Electromechanical System*, Nagoya, Japan, Jan. 26–30, 1997, pp. 102–107.

[47] Larsen, U. D., Rong, W., and Telleman, P., "Design of Rapid Micromixers Using CFD," *Proceedings of Transducers '99, 10th International Conference on Solid-State Sensors and Actuators*, Sendai, Japan, June 7–10, 1999, pp. 200–203

[48] Seidel, R. U., et al., "Capilary Force Mixing Device as Sampling Module for Chemical Analysis," *Proceedings of Transducers '99, 10th International Conference on Solid-State Sensors and Actuators*, Sendai, Japan, June 7–10, 1999, pp. 438–441.

[49] Voldman, J., Gray, M. L., and Schmidt, M. A., "An Integrated Liquid Mixer/Valve," *Journal of Microelectromechanical Systems*, Vol. 9, 2000, pp. 295–302.

[50] Wang, H., et al., "Numerical Investigation of Mixing in Microchannels with Patterned Grooves," *Journal of Micromechanics and Microengineering*, Vol. 13, 2003, pp. 801–808.

[51] Wang, H., et al., "Optimizing Layout of Obstacles for Enhanced Mixing in Microchannels," *Smart Materials and Structures* Vol. 11, 2002, pp. 662–667.

[52] Lin, Y., et al., "Ultrafast Microfluidic Mixer and Freeze-Quenching Device," *Analytical Chemistry*, Vol. 75, 2003, pp. 5381–5386.

[53] Mengeaud, V., Josserand, J., and Girault, H. H., "Mixing Processes in a Zigzag Microchannel: Finite Element Simulation and Optical Study," *Analytical Chemistry*, Vol. 74, 2002, pp. 4279–4286.

[54] Hong, C. C., Choi, J. W., and Ahn, C. H., "A Novel In-Plane Microfluidic Mixer with Modified Tesla Structures," *Lab on a Chip*, Vol. 4, 2004, pp. 109–113.

[55] Liu, R. H., et al., "Passive Mixing in a Three-Dimensional Serpentine Microchannel," *Journal of Microelectromechanical Systems*, Vol. 9, 2000, pp. 190–197.

[56] Vijiayendran, R. A. et al., "Evaluation of a Three-Dimensional Micromixer in a Surface-Based Biosensor," *Langmuir*, Vol. 19, 2003, pp. 1824–1828.

[57] Chen, H., and Meiners, J. C., "Topologic Mixing on a Microfluidic Chip," *Applied Physics Letters*, Vol. 84, 2004, pp. 2193–2195.

[58] Park, S. J., et al., "Rapid Three-Dimensional Passive Rotation Micromixer Using the Brackup Process," *Journal of Micromechanics and Microengineering*, Vol. 14, 2004, pp. 6–14.

[59] Jen, C. P., et al., "Design and Simulation of the Micromixer with Chaotic Advection in Twisted Microchannels," *Lab on a Chip* Vol. 3, 2003, pp. 77–81.

[60] Johnes, S., and Aref, H., "Chaotic Advection in Pulsed Source-Sink Systems," *Physics of Fluids*, Vol. 31, 1988, pp. 469–485

[61] Evans, J., Liepmann, D., and Pisano, A. P., "Planar Laminar Mixer," *Proceedings of MEMS'97, 10th IEEE International Workshop Micro Electromechanical System*, Nagoya, Japan, Jan. 26–30, 1997, pp. 96–101.

[62] Li, J., et al., "Numerical and Experimental Analyses of Planar Asymmetric Split-and-Recombine Micromixer with Dislocation Sub-Channels," *Journal of Chemical Technology and Biotechnology*, Vol. 88, 2013, pp. 1757–1765.

[63] Ansari, M. A., et al., "A Novel Passive Micromixer Based on Unbalanced Splits and Collisions of Fluid Streams," *J. Micromech. Microeng.*, Vol. 20, 2010, pp. 1–10.

[64] Ansari, M. A., and Kim, K. -Y., "Mixing Performance of Unbalanced Split and Recombine Micomixers with Circular and Rhombic Sub-channels," *Chemical Engineering Journal*, Vol. 162, 2010, pp. 760–767.

[65] Cai, G., et al., "A Review on Micromixers," *Micromachines*, Vol. 8, 2017, pp. 1–27.

[66] Afzal, A., and Kim, K. -Y., "Convergent-Divergent Micromixer Coupled with Pulsatile Flow," *Sensors and Actuators B: Chemical*, Vol. 211, 2015, pp. 198–205.

[67] Afzal, A., and Kim, K. -Y., "Passive Split and Recombination Micromixer with Convergent-Divergent Walls," *Chemical Engineering Journal*, Vol. 203, 2012, pp. 182–192.

[68] Schnfeld, F., and Hardt, S., "Simulation of Helical Flows in Microchannels," *AIChE Journal*, Vol. 50, 2004, pp. 771–778.

[69] Sudarsan, A. P., and Ugaz, V. M., "Fluid Mixing in Planar Spiral Microchannels," *AIChE Journal*, Vol. 6, 2006, pp. 74–82.

[70] Rafeie, M., et al., "An Easily Fabricated Three-Dimensional Threaded Lemniscate-Shaped Micromixer for a Wide Range of Flow Rates," *Biomicrofluidics*, Vol. 11, 2017, pp. 1–15.

[71] Johnson, T. J., Ross, D., and Locascio, L. E., "Rapid Microfluidic Mixing," *Analytical Chemistry*, Vol. 74, 2002, pp. 45–51.

[72] Stroock, A. D., et al., "Chaotic Mixer for Microchannels," *Science*, Vol. 295, 2002, pp. 647–451.

[73] Stroock, A. D., and Whitesides, G. M., "Controlling Flows in Microchannels with Parterned Surface Charge and Topography," *Accounts of Chemical Research*, Vol. 36, 2003, pp. 597–604.

[74] Lim, D., et al., "Fabrication of Microfluidic Mixers and Artificial Vasculatures Using a High-Brightness Diode-Pumped Nd:YAG Laser Direct Write Method," *Lab on a Chip*, Vol. 3, 2003, pp. 318–323.

[75] Biddiss, E., Erickson, D., and Li, D., "Heterogeneous Surface Charge Enhanced Micromixing for Electrokinetic Flows," *Analytical Chemistry*, Vol. 76, 2004, pp. 3208–3213.

[76] Hau, W. I., et al., "Surface-Chemistry Technology for Microfluidics," *Journal Micromechanics and Microengineering*, Vol. 13, 2003, pp. 272–278.

[77] Kim, D. S., et al., "A Barrier Embedded Chaotic Micromixer," *Journal of Micromechanics and Microengineering*, Vol. 14, 2004, pp. 798–805.

[78] Bertsch, A., et al., "Static Micromixers Based on Large-Scale Industrial Mixer Geometry," *Lab on a Chip*, Vol. 1, 2001, pp. 56–60.

[79] Wang, L., et al.,"Mixing Enhancement of a Passive Microfluidic Mixer Containing Triangle Baffles," *Asia-Pac. J. Chem. Eng.*, Vol. 9, 2014, pp. 877–885.

[80] Tsai, R. -T., and Wua, C. -Y., "An Efficient Micromixer Based on Multidirectional Vortices Due to Baffles and Channel Curvature," *Biomicrofluidics*, Vol. 5, 2011, pp. 1–13.

[81] Hosokawa, K., Fujii, T., and Endo, I., "Droplet-Based Nano/Picoliter Mixer Using Hydrophobic Microcapillary Vent," *Proceedings of the IEEE International Workshop Micro Electromechanical System*, Piscataway, NJ, 1999, pp. 388–393.

[82] Handique, K., and Burns, M. A., "Mathematical Modeling of Drop Mixing in a Slit-Type Microchannel," *Journal of Micromechanics and Microengineering*, Vol. 11, 2001, pp. 548–554.

[83] Paik, P., Pamula, V. K., and Fair, R. B., "Rapid Droplet Mixers for Digital Microfluidic Systems," *Lab on a Chip*, Vol. 3, 2003, pp. 253–259.

[84] Song, H. et al., "Experimental Test of Scaling of Mixing by Chaotic Advection in Droplets Moving Through Microfluidic Channels," *Applied Physics Letters*, Vol. 83, 2003, pp. 4664–4666.

[85] Tice, J. D., Lyon, A. D., and Ismagilov R. F., "Effects of Viscosity on Droplet Formation and Mixing in Microfluidic Channels," *Analytica Chimica Acta*, Vol. 507, 2003, pp. 73–77.

[86] Deshmukh, A. A., Liepmann, D., and Pisano, A. P., "Continuous Micromixer with Pulsatile Micropumps," *Technical Digest of the IEEE Solid State Sensor and Actuator Workshop*, Hilton Head Island, SC, June 4-8, 2000, pp. 73–76.

[87] Deshmukh, A. A., Liepmann, D., and Pisano, A. P., "Characterization of a Micro-Mixing, Pumping, and Valving System," *Proceedings of Transducers '01, 11th International Conference on Solid-State Sensors and Actuators*, Munich, Germany, June 6-7, 2001, pp. 779–782.

[88] Fujii, T., et al., "A Plug and Play Microfluidic Device," *Lab on a Chip*, Vol. 3, 2003, pp. 193–197.

[89] Glasgow, I., and Aubry, N., "Enhancement of Microfluidic Mixing Using Time Pulsing," *Lab on a Chip*, Vol. 3, 2003, pp. 114–120.

[90] Niu, X. Z., and Lee, Y. K., "Efficient Spatial-Temporal Chaotic Mixing in Microchannels," *Journal of Micromechanics and Microengineering*, Vol. 13, 2003, pp. 454–462.

[91] Okkels, F., and Tabeling, P., "Spatiotemporal Resonances in Mixing of Open Viscous Fluidis," *Physical Review Letters*, Vol. 92, 2004, pp. 228–301.

[92] Suzuki, H., and Ho, C. M., "A Magnetic Force Driven Chaotic Micro-Mixer," *Proceedings of MEMS'02, 15th IEEE International Workshop Micro Electromechanical System*, Las Vegas, NV, Jan. 20–24, 2002, pp. 40–43.

[93] Lu, L. H., Ryu, K. S., and Liu, C., "A Magnetic Microstirrer and Array for Microfluidic Mixing," *Journal of Microelectromechanical Systems*, Vol. 11, 2002, pp. 462–469.

[94] El Moctar, A. O., Aubry, N., and Batton, J., "Electro-Hydrodynamic Micro-Fluidic Mixer," *Lab on a Chip*, Vol. 3, 2003, pp. 273–280.

[95] Deval, J., Tabeling, P., and Ho, C. M., "A Dielectrophoretic Chaotic Mixer," *Proceedings of MEMS'02, 15th IEEE International Workshop Micro Electromechanical System*, Las Vegas, NV, Jan. 20–24, 2002, pp. 36–39.

[96] Lee, Y. K., et al., "Chaotic Mixing in Electrokinetically and Pressure Driven Micro Flows," *Proceedings of MEMS'01, 14th IEEE International Workshop Micro Electromechanical System*, Interlaken, Switzerland, Jan. 21–25, 2001, pp. 483–486.

[97] Letteri, G. L., et al., "Consequences of Opposing Electrokinetically and Pressure-Induced Flows in Microchannels of Varying Geometries," *Proc. Micro Total Analysis Systems*,Enschede, Netherlands, May 14–18, 2000, pp. 351–354.

[98] Oddy, M. H., Santiago, J. G., and Mikkelsen, J. C., "Electrokinetic Instability Micromixing," *Analytical Chemistry*, Vol. 73, 2001, pp. 5822–5832.

[99] Tang, Z., et al., "Electrokinetic Flow Control for Composition Modulation in a Microchannel," *Journal of Micromechanics and Microengineering*, Vol. 12, 2002, pp. 870–877.

[100] Bau, H. H., Zhong, J., and Yi, M., "A Minute Magneto Hydro Dynamic (MHD) Mixer," *Sensors and Actuators B*, Vol. 79, 2001, pp. 207–215.

[101] Zhu, G. -P., and Nguyen, N. -T., "Rapid Magnetofluidic Mixing in a Uniform Magnetic Field," *Lab Chip*, Vol. 12, 2012, pp. 4772–4780.

[102] Moroney, R. M., White, R. M., and Howe, R. T., "Ultrasonically Induced Microtransport," *Proceedings of MEMS'91, 3th IEEE International Workshop Micro Electromechanical System*, Nara, Japan, Jan. 30–Feb. 4, 1991, pp. 277–282.

[103] Zhu, X., and Kim, E. S., "Acoustic-Wave Liquid Mixer," *Microelectromechanical Systems (MEMS) American Society of Mechanical Engineers, Dynamic Systems and Control Division (Publication) DSC*, ASME, Fairfield, NJ, Vol. 62, 1997, pp. 35–38.

[104] Rife, J. C., et al., "Miniature Valveless Ultrasonic Pumps and Mixers," *Sensors and Actuators A*, Vol. 86, 2000, pp. 135–140.

[105] Vivek, V., Zeng, Y., and Kim, E. S., "Novel Acoustic-Wave Micromixer," *Proceedings of MEMS'00, 13th IEEE International Workshop Micro Electromechanical System*, Miyazaci, Japan, Jan. 23–27, 2000, pp. 668–673.

[106] Woias, P., Hauser, K., and Yacoub-George, E., "An Active Silicon Micromixer for mTAS Applications," *Micro Total Analysis Systems 2000*, A. van den Berg et al. (eds.), Boston: Kluwer Academic Publishers, 2000, pp. 277–282.

[107] Yasuda, K., "Non-Destructive, Non-Contact Handling Method for Biomaterials in Micro-Chamber by Ultrasound," *Sensors and Actuators B*, Vol. 64, 2000, pp. 128–135.

[108] Yang, Z., et al., "Active Micromixer for Microfluidic Systems Using Lead-Zirconate-Titanate (PZT)-Generated Ultrasonic Vibration," *Electrophoresis*, Vol. 21, 2000, pp. 116–119.

[109] Yang, Z., et al., "Ultrasonic Micromixer for Microfluidic Systems," *Sensors and Actuators A*, Vol. 93, 2001, pp. 266–272.

[110] Liu, R. H., et al., "Bubble-Induced Acoustic Micromixing," *Lab on a Chip*, Vol. 2, 2002, pp. 151–157.

[111] Liu, R. H., et al., "Hybridization Enhancement Using Cavication Microstreaming," *Analytical Chemistry*, Vol. 75, 2003, pp. 1911–1917.

[112] Yaralioglu, G. G., et al., "Ultrasonic Mixing in Microfluidic Channels Using Integrated Transducers," *Analytical Chemistry*, Vol. 76, 2004, pp. 3694–3698.

[113] Darhuber, A., A., "Microfluidic Actuation by Modulation of Surface Stresses," *Applied Physics Letters*, Vol. 82, 2003, pp. 657–659.

[114] Tsai. J, H, and Lin, L., "Active Microfluidic Mixer and Gas Bubble Filter Drivern by Thermal Bubble Pump," *Sensors and Actuators A*, Vol. 97–98, 2002, pp. 665–671.

[115] Haeberle, S., et al., "Centrifugal Micromixer," *Chem. Eng. Technol.*, Vol. 28, 2005, pp. 613–616.

[116] Colombo, S., et al., "Transforming Nanomedicine Manufacturing Toward Quality by Design and microfluidics," *Advanced Drug Delivery Reviews*, Vol. 128, 2018, pp. 115–131.

[117] Bohr, A., et al., "High-Throughput Fabrication of Nanocomplexes Using 3D-Printed Micromixers," *J. Pharm. Sci.*, Vol. 106, 2017, pp. 835-842.

[118] Brody, J. P., and Yager, P., "Diffusion-Based Extraction in a Microfabricated Device," *Sensors and Actuators A*, Vol. 58, 1997, pp. 13–18.

[119] Weigl, B., and Yager, P., "Microfluidic Diffusion-Based Separation and Detection," *Science*, Vol. 283, 1999, pp. 346–347.

[120] Burke, B. J., and Regnier, F. E., "Stopped-Flow Enzyme Assays on a Chip Using a Microfabricated Mixer," *Analytical Chemistry*, Vol. 75, 2003, pp. 1786–1791.

[121] Dertinger, S. K. W., et al., "Generation of Gradients Having Complex Shapes Using Microfluidic Networks," *Analytical Chemistry*, Vol. 73, 2001, pp. 1240–1246.

[122] Holden, M. A., et al., "Generating Fixed Concentration Array in a Microfluidic Device," *Sensors and Actuators B*, Vol. 92, 2003, pp. 199–207.

[123] Holden, M. A., et al., "Microfluidic Diffusion Diluter: Bulging of PDMS Microchannels Under Pressure-Driven Flow," *Journal of Micromechanics and Microengineering*, Vol. 13, 2003, pp. 412–418.

[124] Yang, M., et al., "Generation of Concentration Gradient by Controlled Flow Distribution and Diffusive Mixing in a Microfluidic Chip," *Lab on a Chip*, Vol. 2, 2002, pp. 158–163.

# Chapter 11

## Microfluidics for Life Sciences and Chemistry: Microdispensers

Microdispensers are important devices for chemical and biomedical analysis. The precise amount of liquid is often required for dosing and analysis applications. Microdispensers can be seen as microdosing systems, microinjectors, or micropipettes. In general, microdispensers are categorized as closed-loop controlled dispensers or open-loop controlled dispensers.

*Closed-loop controlled dispensers* are complex systems with a fluidic actuator and a flow sensor. The fluidic actuator is a micropump or microvalve. Chapters 6 and 7 discuss these components in detail. In a closed-loop controlled dispenser, the flow rate signal is fed back to a controller where it is compared with the setpoint. The control signal from the controller is connected to the actuator of the pump or the valve to adjust the set-flow rate [1, 2, 3, 4]. This type of dispenser is suitable for continuous flow systems. For a fixed amount of fluid, errors can occur in the starting and stopping phases of the system.

*Open-loop controlled dispensers* are single devices which can be divided into droplet dispensers and in-channel dispensers. The first type generates external droplets with a constant volume for applications such as inkjet printing. The in-channel dispensers prepare droplets for further use on the same chip or in the other systems. In-channel dispensers are currently considered as a subsection of continuous-flow droplet-based microfluidics that contains various methods of in-channel droplet generation mainly for chemical, material, and pharmaceutical sciences. In general, dispensers can be categorized by the actuating principle and the breakup mode for droplet formation (Figure 11.1). In-channel droplet formation can be achieved through passive and active methods. In passive microfluidic devices, fluid instabilities for droplet formation are created through introduction of an immiscible fluid, known as dispersed fluid, into the continuous fluid [5]. In active methods, the use of an external energy source modifies the interfacial instabilities for droplet formation [5].

Common actuating methods for microdispensing, that is, droplet formation, are thermopneumatic (thermal bubble) [6, 7, 8, 9], thermomechanic [10], electrostatic [11, 12], piezoelectric [13, 14], pneumatic [15, 16], electrochemical [17], and electrokinetic. Droplet dispensers have the advantage of massive parallel dosing, which is common in biochemical application. In-channel dispensers are simple and have the advantage of monolithic integration [18]. In the microfluidic literature, droplet-based microfluidics also includes manipulation of discrete droplets on planar surfaces using electrowetting or dielectrophoresis that is known as Digital Microfluidics. This chapter only discusses open-loop controlled dispensers employing channel-based systems.

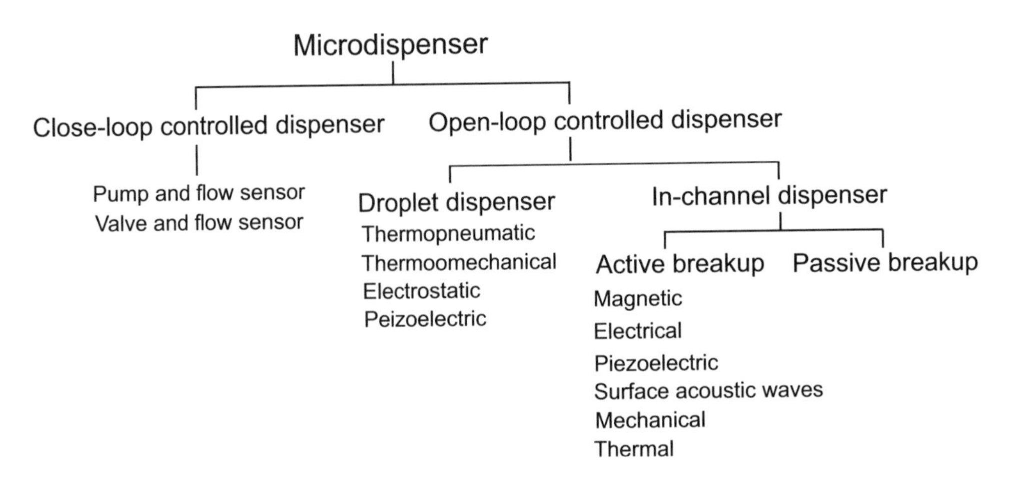


**Figure 11.1** Classification of microdispensers.

## 11.1 DESIGN CONSIDERATIONS

### 11.1.1 Droplet Dispensers

#### 11.1.1.1 Kinetics of a Droplet

Microdroplets have the advantage of small stored energy. The small energy leads to a clean droplet formation. The total energy $E_{\mathrm{d}}$ of a droplet can be described as the sum of surface energy $E_{\mathrm{s}}$ and kinetic energy $E_{\mathrm{k}}$:

$$E_{\mathrm{d}} = E_{\mathrm{s}} + E_{\mathrm{k}} \tag{11.1}$$

The surface energy $E_{\mathrm{s}}$ is calculated as:

$$E_{\mathrm{s}} = \pi d_0^2 \sigma \tag{11.2}$$

where $d_0$ is the droplet diameter and $\sigma$ is the surface tension. The kinetic energy is given by:

$$E_{\mathrm{k}} = \frac{1}{2} m u^2 = \frac{1}{12} \pi \rho d_0^3 u^2 \tag{11.3}$$

where $\rho$ is the density of the liquid and $u$ is the droplet velocity. While the surface energy is proportional to the surface area of the droplet, the kinetic energy is proportional to its volume. According to scaling laws, the surface-volume ratio, or the ratio between surface energy and kinetic energy, increases with miniaturization. As a result, the surface energy dominates in the microscale. The droplet is clean because there is not enough kinetic energy for forming new surfaces, avoiding splashing and formation of satellite droplets after an impact. Comparing the kinetic energy (11.3) to the surface energy (11.2) results in the ratio:

$$\frac{E_{\mathrm{k}}}{E_{\mathrm{s}}} = \frac{1/12 \times \pi \rho d_0^3 u^2}{\pi d_0^2 \sigma} = \frac{1}{12} \frac{\rho u^2 d_0}{\sigma} \tag{11.4}$$

The second term on the right-hand side of (11.4) is known as the Weber number (We):

$$\mathrm{We} = \frac{\rho u^2 L_{\mathrm{ch}}}{\sigma} \tag{11.5}$$

where $L_{\text{ch}}$ is the characteristic length. With the droplet diameter as the characteristic length, the ratio between kinetic energy and surface energy can be described as:

$$\frac{E_{\text{k}}}{E_{\text{s}}} = \frac{1}{12}\text{We} \tag{11.6}$$

Equations (11.5) and (11.6) make clear that the Weber number is proportional to the droplet size. Thus, the kinetic energy is much smaller than the surface energy. The Weber number is therefore a key parameter determining the droplet behavior.

**Example 11.1: Energy Relation in a Microdroplet**

Determine the Weber number and the ratio between the kinetic energy and the surface energy of a water droplet of 50 μm diameter moving with a velocity of 1 m/s. The surface tension and density of water are assumed to be $72\times10^{-3}$ N/m, and 1,000 kg/m$^3$, respectively.

**Solution.** According to (11.5) and $L_{\text{ch}} = d_0$, the Weber number for this case is:

$$\text{We} = \frac{\rho u^2 L_{\text{ch}}}{\sigma} = \frac{1{,}000 \times 1^2 \times 50 \times 10^{-6}}{72 \times 10^{-3}} = 0.694$$

From (11.6), the ratio between kinetic energy and surface energy is:

$$\frac{E_{\text{k}}}{E_{\text{s}}} = \frac{1}{12}\text{We} = \frac{0.694}{12} = 5.79 \times 10^{-2}$$

**Example 11.2: Critical Injection Velocity for Breaking Up a Droplet**

Determine the required minimum velocity of the droplet if the water droplet in Example 11.1 breaks up in two identical droplets. The energy loss due to friction is to be neglected.

**Solution.** If the droplet breaks up in two identical droplets, the new droplet should have one-half of the initial volumes. Thus, the new droplet diameter is:

$$d' = d_0/\sqrt[3]{2}$$

In order to make the breakup possible, the initial kinetic energy should be greater than the surface energies of the new droplets:

$$\begin{aligned}
&\tfrac{1}{12}\pi\rho d_0^3 u^2 \geq 2\pi d'^2\sigma && \rightarrow \tfrac{1}{12}\pi\rho d_0^3 u^2 \geq 2\pi d_0^2 \sqrt[3]{4}\sigma \rightarrow \\
&u \geq \sqrt{\tfrac{24}{\sqrt[3]{4}}\tfrac{\sigma}{\rho d_0}} && \rightarrow u \geq 3.89\sqrt{\tfrac{\sigma}{\rho d_0}} \quad \rightarrow \\
&u \geq 3.89\sqrt{\tfrac{7\times10^{-3}}{1{,}000\times50\times10^{-3}}} && \rightarrow u \geq 4.67 \text{ m/s}
\end{aligned}$$

The required minimum velocity for the breakup is 4.67 m/s.

**Example 11.3: Critical Droplet Size for Breaking Up**

If the droplet velocity is fixed at 1 m/s, determine the smallest droplet size for breaking up into two identical smaller droplets. The energy loss due to friction is to be neglected.

**Solution.** Rearranging the inequality in Example 11.2 for the initial diameter results in:

$$d_0 \geq \frac{24}{\sqrt[3]{4}}\frac{\sigma}{\rho u^2} \rightarrow d_0 \geq 15.1\frac{\sigma}{\rho u^2} \rightarrow$$

$$d_0 \geq 15.1\frac{7\times10^{-3}}{1000\times1^2} = 1.09\times10^{-3}\ \text{m} \rightarrow d_0 \geq 1.09\ \text{mm}$$

At a velocity of 1 m/s, the water droplet should be approximately larger than 1 mm to break up in two identical smaller droplets.

Examples 11.2 and 11.3 show that the impact behavior of a microdroplet on a solid surface can be controlled by the injection velocity and the droplet size. The kinetic energy should be kept smaller than the surface energy to avoid splashing. Thus, the impact of a droplet of a Newtonian fluid can be well characterized by the Reynolds number and the Weber number. However, the impact behavior of the droplet on a solid surface also depends on surface properties, such as roughness and temperature. Furthermore, the kinetic energy can be absorbed by the potential energy of large polymer molecules in a droplet of a non-Newtonian fluid, which has a viscoelastic behavior [19].

Since the Reynolds number characterizes the ratio between the kinetic energy and the friction force between the liquid droplet and surrounding air, the Reynolds number for the droplet is determined as:

$$\text{Re} = \frac{ud_0}{\nu_\text{a}} \tag{11.7}$$

where $\nu_\text{a}$ is the kinematic viscosity of air. The drag coefficient of the droplet depends on the ranges of Reynolds numbers: Stokes regime ($\text{Re} \leq 2$), Allen regime ($2 < \text{Re} \leq 500$), and Newton regime ($500 < \text{Re} \leq 10^5$) [20]. Because of the small size and the relatively slow velocity, droplet microdispensers mainly work in the Stokes regime and the Allen regime.

**Example 11.4: Reynolds Number of a Microdroplet**

Determine the Reynolds number for the droplet in Example 11.1. Kinematic viscosity of air is assumed to be $1.55\times10^{-5}\ \text{m}^2/\text{s}$.

**Solution.** The Reynolds number for a 50-μm droplet at 1 m/s is:

$$\text{Re} = \frac{ud_0}{\nu_\text{a}} = \frac{1\times50\times10^{-6}}{1.55\times10^{-5}} = 3.23$$

The droplet is in the Allen regime.

#### 11.1.1.2 Dynamics of a Droplet

The dynamics of a microdroplet is important for designing the dispensing system. Applying the scaling law for kinetic energy and friction force of a microdroplet, the friction force will dominate the dynamics of the droplet in microscale. Even if the droplet is injected with a high initial velocity, the kinetic energy will be quickly dissipated due to the friction loss. At the end, the droplet only moves with its potential energy stored in the gravitational field. The terminal velocity is determined by the equilibrium between the friction force and the gravitation force. The terminal motion is vertical or in gravitation direction.

Following, the kinetics of a microdroplet is considered for two cases: vertical injection and horizontal injection. The corresponding models are depicted in Figure 11.2. In the first injection mode, the droplet is assumed to be injected into the gravitation direction. The dynamics of the droplet is governed by the acceleration force, the gravitation force, and the friction force. Due to dissipation through friction, the droplet reaches a constant terminal velocity. In the second injection

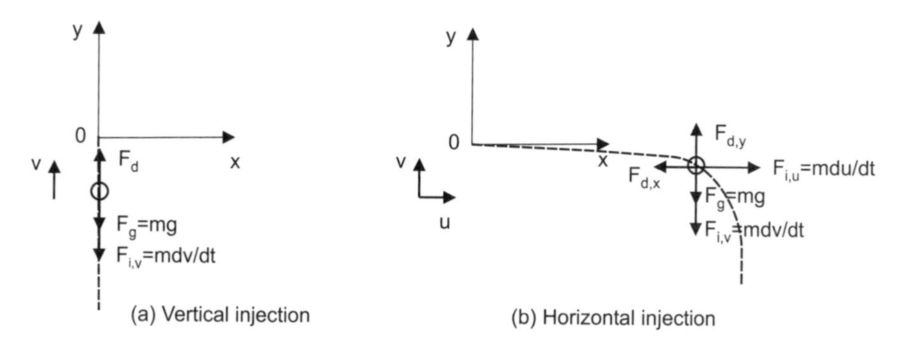


**Figure 11.2** Injection models of a droplet microdispenser: (a) vertical injection, and (b) horizontal injection.

mode, the droplet first moves horizontally due to the initial kinetic energy and then ends up moving vertically with the same terminal velocity as in the first mode. Both the terminal velocity and the trajectory are important parameters for designing the dispensing system.

*Vertical Injection.* In many cases, microdispensers are designed for vertical injection. The dynamics of the droplet can be described by the force balance equation. Figure 11.2(a) shows the three force components acting on the droplet: gravitational force $F_g$, vertical inertial force $F_{i,v}$, and drag force $F_d$. Assuming that Stokes regime for the drag force ($F_d = 3\pi\eta_a d_0 v$), the governing equation of the spherical droplet is:

$$\begin{aligned} F_g + F_d + F_{i,v} &= 0 \\ m\tfrac{dv}{dt} - 3\pi\eta_a d_0 v + mg &= 0 \end{aligned} \tag{11.8}$$

where $\eta_a$ is the dynamic viscosity of air, $m$ is the mass of the droplet, and $g$ is the acceleration of gravitation. The initial velocity at the nozzle exit is $v(0) = -v_0$, and the terminal velocity is $v(\infty) = -v_\infty = -mg/(3\pi\eta d_0) = -\rho g d_0^2/18\eta_a$. When acceleration force is in equilibrium with friction force, the solution for the spherical droplet velocity with the mass of $m = \pi\rho d_0^3/6$ is:

$$v(t) = -v_\infty - (v_0 - v_\infty)\exp\left(\frac{-t}{\rho d_0^2/18\eta_a}\right) \tag{11.9}$$

where $\rho$ is the density of the droplet's liquid. From (11.9), the time constant $\tau_i = \rho d_0^2/18\eta_a$ can be considered as the characteristic time for the acceleration or deceleration of the droplet.

**Example 11.5: Vertical Injection of a Droplet**

If the droplet in Example 11.1 is rejected vertically in the gravitation direction, determine the terminal velocity and the characteristic time constant. The dynamic viscosity of air is assumed to be $1.83 \times 10^{-5}$ Ns/m$^2$. (a) Discuss the characteristics of velocities and traveling distances with different droplet diameters. (b) Discuss the characteristics with different initial injection velocities.

**Solution.** From the above analysis, the terminal velocity of the droplet is:

$$v_\infty = \frac{\rho g d_0^2}{18\eta_a} = \frac{1,000 \times (50 \times 10^{-6})^2}{18 \times 1.83 \times 10^{-5}} = 74.4 \times 10^{-3}\ \text{m/s} = 74.4\ \text{mm/s}$$

The characteristic time constant is:

$$\tau_i = \frac{\rho d_0^2}{18\eta_a} = \frac{1,000 \times (50 \times 10^{-6})^2}{18 \times 1.83 \times 10^{-5}} = 7.59 \times 10^{-3}\ \text{sec} = 7.59\ \text{ms}$$

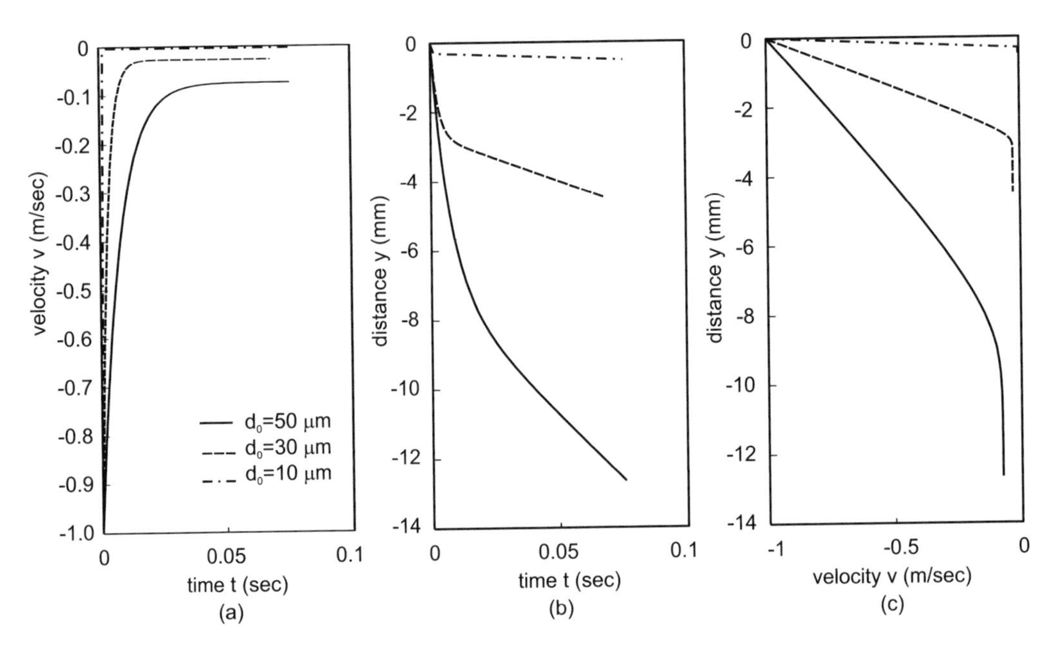


**Figure 11.3** Characteristics of vertically injected microdroplets with different diameters: (a) velocity versus time; (b) distance versus time; and (c) distance versus velocity.

Both terminal velocity and characteristic time constant do not depend on the initial velocity.

From (11.9), the flying distance can be derived as:

$$y = \int_0^t v(\tau)\mathrm{d}\tau = -v_\infty t + (v_0 - v_\infty)\tau_\mathrm{i}[\exp(-t/\tau_\mathrm{i} - 1)]$$

Figure 11.3 plots the characteristics of microdroplets with different diameters 10, 30, and 50 μm. All droplets have an initial velocity of 1 m/s. Since the terminal velocity is proportional to the square of the droplet's diameter, the smaller the droplet, the slower is the terminal velocity [Figure 11.3(a)]. Thus, small droplets may have problems with interference of ambient air flow. In a dispensing system, the receiving part should therefore be placed at a distance where the kinetic energy is still dominant. For instance, with a droplet of 50 μm in diameter, the receiving part should be placed at a distance less than 10 mm from the injection nozzle. For a 10-μm droplet, the allowed distance is only 400 μm [Figure 11.3(b, c)].

As analyzed above, increasing initial velocities has no effect on the terminal velocity and the characteristic time constant. Thus, droplets with different initial velocities reach the terminal velocity at the same time [Figure 11.4(a)]. The initial velocity only affects the flying distance of the droplet. With a higher initial velocity, droplets can travel further away from the nozzle before reaching the terminal velocity. Thus, the distance between the dispensing nozzle and the receiving part can be adjusted by the initial injection velocity [Figure 11.4(b, c)].

*Horizontal Injection.* If the droplet is injected horizontally [Figure 11.2(b)], the dynamics of the droplet is governed by the force balances in the $x$ and $y$ axes. In the $x$ axis, the drag force $F_{\mathrm{d,x}}$ is in balance with the inertial force $F_{\mathrm{i,u}}$. In the $y$ axis, the drag force $F_{\mathrm{d,y}}$ is in balance with the sum of

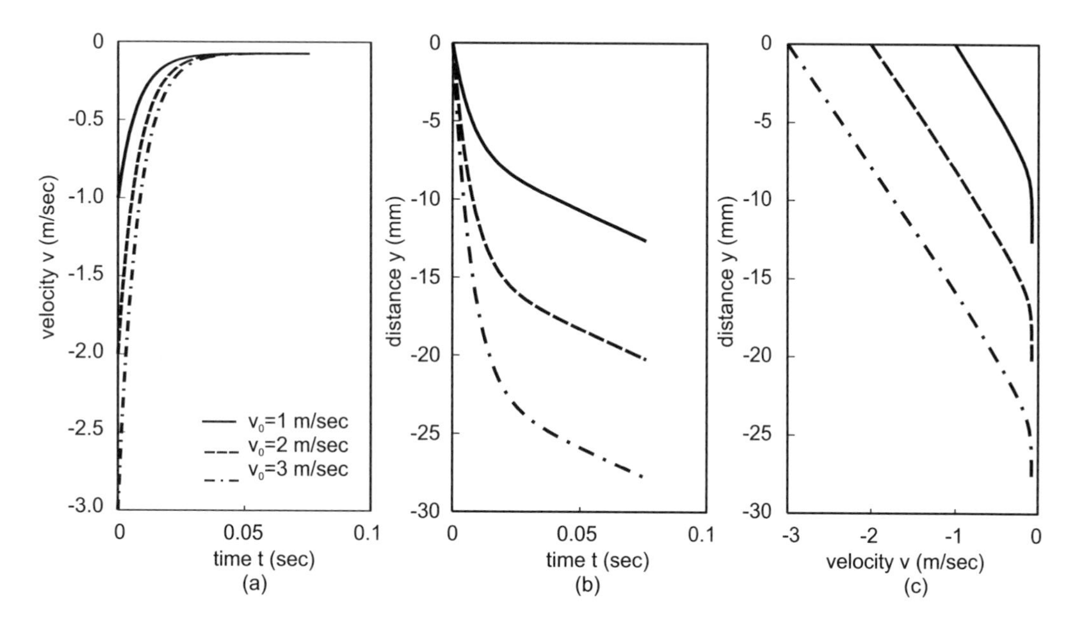


**Figure 11.4** Characteristics of a vertically injected microdroplet with different initial injection velocities: (a) velocity versus time; (b) distance versus time; and (c) distance versus velocity.

the inertial force $F_{\mathrm{i,v}}$ and the gravitational force $F_{\mathrm{g}}$. The force balances result in:

$$\begin{cases} F_{\mathrm{i,u}} + F_{\mathrm{d,x}} & = 0 \\ F_{\mathrm{i,v}} + F_{\mathrm{d,y}} + F_{\mathrm{g}} & = 0 \end{cases} \rightarrow \begin{cases} m\frac{\mathrm{d}u}{\mathrm{d}t} - 3\pi\eta d_0 u & = 0 \\ m\frac{\mathrm{d}v}{\mathrm{d}t} - 3\pi\eta d_0 v + mg & = 0 \end{cases} \quad (11.10)$$

Applying the conditions of the injection velocity $u(0) = u_0$ and the terminal velocity $v(\infty) = -v_\infty$, the results of the velocity components are:

$$\begin{cases} u(t) & = u_0 \exp\left(\frac{-t}{\tau_{\mathrm{i}}}\right) \\ v(t) & = -v_\infty \left[1 - \exp\left(\frac{-t}{\tau_{\mathrm{i}}}\right)\right] \end{cases} \quad (11.11)$$

where $v_\infty$ and $\tau_{\mathrm{i}}$ are the terminal velocity and the characteristic time constant, as defined in the case for vertical injection. Integrating (11.11) over times results in the trajectory of the microdroplet:

$$\begin{cases} x(t) & = u_0\tau_{\mathrm{i}} \left[1 - \exp\left(\frac{-t}{\tau_{\mathrm{i}}}\right)\right] \\ y(t) & = -v_\infty \left\{t - \tau_{\mathrm{i}}\left[1 - \exp\left(\frac{-t}{\tau_{\mathrm{i}}}\right)\right]\right\} \end{cases} \quad (11.12)$$

Equation (11.12) shows that the horizontal flying distance $x$ is proportional to the initial velocity $u_0$.

**Example 11.6: Horizontal Injection of a Droplet**

The droplet in Example 11.1 is rejected horizontally. Discuss the effect of size on the kinetic energy and the trajectory of the droplets. Discuss the effect of initial velocity on the trajectory. The dynamic viscosity of air is assumed to be $1.83 \times 10^{-5}$ Ns/m$^2$.

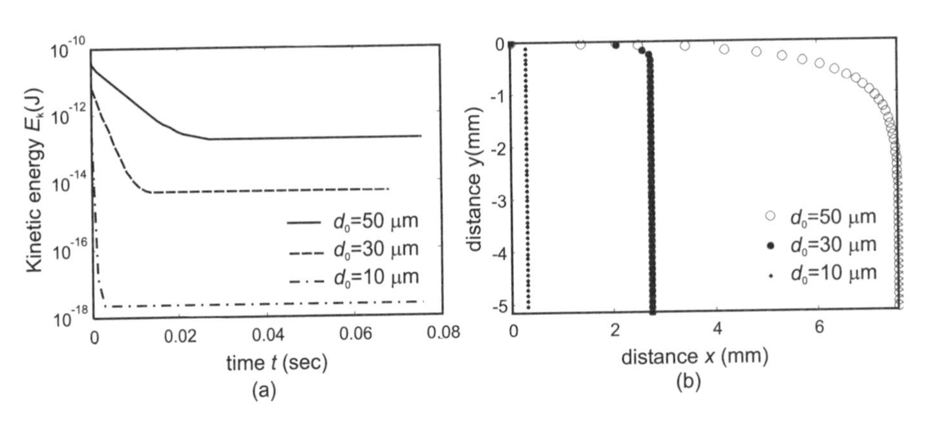


**Figure 11.5** (a) Kinetic energy and (b) trajectory of horizontally injected microdroplets with different diameters.

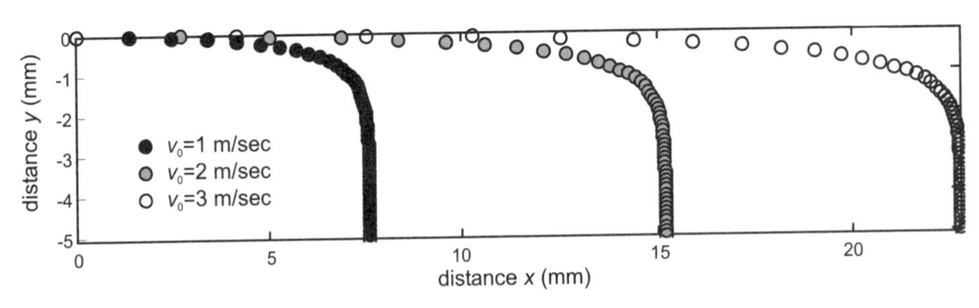


**Figure 11.6** Trajectory of horizontally injected microdroplet with different initial injection velocities.

**Solution.** From (11.11), the droplet velocity can be calculated as:

$$u_\mathrm{d}(t) = \sqrt{u_0^2 \exp^2\left(\frac{-t}{\tau_\mathrm{i}}\right) + v_\infty^2 \left[1 - \exp\left(\frac{-t}{\tau_\mathrm{i}}\right)\right]^2}$$

Thus, the total kinetic energy of the droplet is:

$$E_\mathrm{k}(t) = \frac{mu_\mathrm{d}^2}{2} = \frac{1}{12}\pi\rho d_0^3 \left\{u_0^2 \exp^2\left(\frac{-t}{\tau_\mathrm{i}}\right) + v_\infty^2 \left[1 - \exp\left(\frac{-t}{\tau_\mathrm{i}}\right)\right]^2\right\}$$

Figure 11.5(a) plots the time functions of kinetic energy of droplets with the same initial velocity but different sizes. For all cases, the final kinetic energy is determined by the size of the droplet. Due to the dominant surface effect, the kinetic energy of smaller droplets decay faster than the terminal kinetic energy. Figure 11.5(b) compares the trajectories of the different droplets based on (11.12). Smaller droplets cannot keep a long horizontal trajectory because of the small kinetic energy and the fast dissipation.

Figure 11.6 compares the trajectory of the same droplet at different initial velocities. The trajectory shows clearly that the horizontal flying distance is proportional to the initial velocity.

#### 11.1.1.3 Injection Nozzle

The injection nozzle is an important part of a droplet microdispenser, because the nozzle size is one of the parameters determining the size of the droplet. In addition, the properties of the nozzle

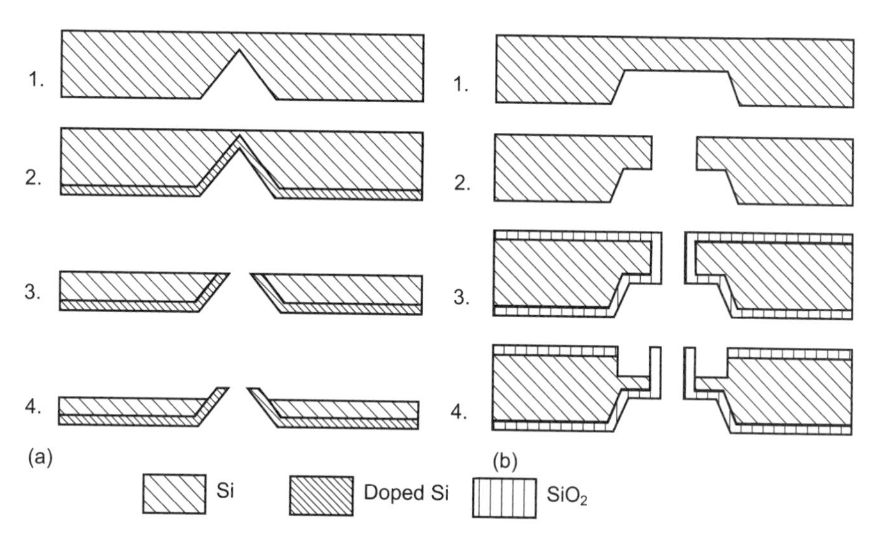


**Figure 11.7** Methods for making a ring-shaped nozzle: (a) 1: wet etching of the pyramid-shaped nozzle, 2: boron-doping the inverted pyramid surface, 3: blanket etching from the front to expose the nozzle orifice, and 4: selective wet etching from the front to shape the nozzle ring (*After*: [12]); and (b) 1: wet etching of the silicon recess, 2: DRIE from the front to form the nozzle structure, 3: selective oxidation to form the nozzle shape, and 4: selective dry etching to free the $SiO_2$-nozzle. (*After*: [21].)

front surface strongly affect the direction of dispensed droplets. There are two ways to minimize this effect: making a ring-shaped nozzle with minimum front surface area, or coating the outer surface with a hydrophilic layer.

Fabrication methods for microneedles described in Section 9.1 can be used for making nozzles with a small front surface. Figure 11.7 shows some typical methods for making ring-shaped nozzles with a small front surface. The method used in Figure 11.7(a) [14] utilizes a highly boron-doped layer as an etch stop for making the nozzle ring. The wet-etching process is selective to the highly doped layer. This layer forms the nozzle structure [Figure 11.7(a), part 4]. The depth of the doped silicon layer determines the ring surface area.

The method depicted in Figure 11.7(b) makes a ring-shaped nozzle of silicon oxide. The ring structure is fabricated using a selective dry-etching process. Silicon oxide remains intact and forms the nozzle [22]. The internal nozzle diameter can be adjusted in both cases by additional deposition of silicon oxide. For better and straight droplet trajectory, the outer surface of the nozzle can be coated with a hydrophobic layer such as silane [14].

#### 11.1.1.4 Actuation Concepts

At ultrasonic driving frequencies, the droplet diameter $d_0$ is a function of the frequency $f$ [23]:

$$d_0 = 0.34 \left( \frac{8\pi\sigma}{\rho f^2} \right)^{\frac{1}{3}} \tag{11.13}$$

where $\sigma$ is the surface tension and $\rho$ is the density of the fluid. At a low frequency or at a single droplet generation, the droplet diameter is assumed to be the same as the nozzle diameter. However, the droplet size also depends on other parameters, such as actuation stroke and viscosity. The droplet speed can be estimated from (11.1), (11.2), and (11.3). The total energy of a droplet $U_d$, or the work done by the actuator per stroke $W_a$, can be estimated by knowing the displaced fluid volume $V_d$ and the energy density $E'_a$ of the actuator, as discussed in Section 6.1.1.1:

$$d_0 = 0.34\left(\frac{8\pi\sigma}{\rho f^2}\right)^{\frac{1}{3}} \quad (11.14)$$

**Example 11.7: Thermal Efficiency of a Thermopneumatic Droplet Dispenser**

A thermopneumatic droplet dispenser has a heater resistance of 57Ω. The ejection of a droplet follows a voltage pulse of 15V and 0.8 μs. The droplet has a volume of 3.2 pL and an initial velocity of 9.6 m/s. Determine the thermal efficiency of this dispenser (data from [8]).

**Solution.** The total electrical energy of the heater is:

$$q_{\text{in}} = \frac{V^2}{R}\Delta t = \frac{15^2}{57} \times 0.8 \times 10^{-6} = 3.16 \times 10^{-6}\ \text{J}$$

According to (11.1), the diameter of a spherical droplet is:

$$V_{\text{d}} = \frac{\pi d^3}{6} \rightarrow d_0 = \sqrt{3}\frac{6V_{\text{d}}}{\pi} = \sqrt{3}\frac{6 \times 3.2 \times 10^{-15}}{\pi} = 18.3 \times 10^{-6}\ \text{m} \approx 18\ \mu\text{m}$$

The energy of the droplet is:

$$U_{\text{d}} = \pi d^2\sigma + \pi\rho d^3 v^2/12$$
$$= \pi(18.28 \times 10^{-6})^2 \times 72 \times 10^{-3} + \pi\times$$
$$1,000 \times (18.28 \times 10^{-6})^3 \times 9.6^2/12$$
$$= 2.23 \times 10^{-10}\ \text{J}$$

Thus, the thermal efficiency of the device is:

$$\eta = \frac{E_{\text{d}}}{q_{\text{in}}} = \frac{2.23 \times 10^{-10}}{3.16 \times 10^{-6}} = 0.7 \times 10^{-4} = 0.007\%$$

### 11.1.2 In-Channel Dispensers

#### 11.1.2.1 Passive In-Channel Droplet Formation

Advancement of droplet formation in multiphasesystems has let to the emerge of a subcategory of microfluidics known as Droplet-based microfluidics that mainly deals with the formation of droplets for (1) sampling purposes in analytical systems or (2) synthesis of microcapsules and microparticles for pharmaceutical and materials sciences. In addition, Chapter 10 already discussed the concept of droplet formation for mixing in microchannel. This concept can be applied for controlled dispensing of droplets. Droplets of sample fluids can be dispensed in microchannels using the instability between two immiscible phases. Thus, the basic condition for droplet formation is that the sample fluid and the carrier fluid are immiscible. The behavior of droplet formation and the size of the dispensed droplet are determined by two parameters: the capillary number (Ca) and the fraction of the sample flow.

The capillary number is defined as:

$$\text{Ca} = \frac{u\eta}{\sigma} \quad (11.15)$$

where $u$ is the average flow velocity, $\eta$ is the dynamic viscosity, and $\sigma$ is the surface tension between the two phases. The capillary number represents the ratio between the friction force and the surface tension in a two-phase system. In microscale, the capillary number is small and surface effects are dominant. Thus, stable and reproducible droplet formation can be achieved. The fraction of the sample flow is defined as the ratio between the volumetric flow rate of the aqueous sample fluid and the total volumetric flow rate of both phases:

$$r = \frac{\dot{Q}_{\text{sample}}}{\dot{Q}_{\text{sample}} + \dot{Q}_{\text{carrier}}} \tag{11.16}$$

The configuration of the microfluidic device impact the droplet formation. Figure 11.8 shows typical geometries employed for droplet generation [5]. Hydrodynamic architectures of cross-flow, co-flow, and flow-focusing are mainly employed to produce viscous shear forces for droplet breakup [5]. Also, channel confinements such as step emulsification, microchannel emulsification, and membrane emulsification are used for droplet formation [5].

The various breakup modes for shear-based droplet formation are squeezing, dripping, jetting, tip-streaming and tip-multibreaking [5] (Figure 11.9). Tip-streaming and tip-multibreaking have not been observed for cross-flow architecture.

Squeezing breakup mode occurs once the viscous forces gives way to the confinement of channel walls. This happens at low capillary numbers, that is, $Ca_c$<O($10^{-2}$) [5]. This mode occurs in cross-flow, co-flow, and flow-focusing configuration, where the junction region is obstructed by the dispersed fluid protrusion as it grows. Since the the continuous flow around the enlarging protrusion is restricted, a pressure gradient in the continuous flow across the developing protrusion is built up. In the other words, the pressure augmentation in the continuous fluid is the main reason for droplet formation. The droplet is formed through squeezing once the pressure gradient is sufficiently high to overcome the pressure inside the dispersed droplet. The squeezed droplet is usually confined by the channel walls, having a plug-type geometry.

Transition from squeezing breakup mode to dripping breakup mode occurs once the capillary number ($Ca_c$) increases. In this case, the viscous shear forces is sufficiently large to break up the droplet before it grows to obstruct the junction. Thus, the droplet diameter is smaller than the channel dimensions. Dripping breakup mode occurs at cross-flow, co-flow and focusing orifice of flow-focusing geometries. In such geometries, the viscous force dominates the interfacial tension that is responsible to stabilize the developing droplet against breakup [5].

Jetting breakup mode occurs once an extended liquid jet is emitted from the dispersed-fluid flow and finally breaks up into droplets because of Rayleigh-Plateau instability.

Tip-streaming breakup is characterized by the steady formation of the cone-jet structure of the dispersed fluid [5]. When dispersed liquid is formed into a conical shape, small droplets are separated from the tip of the cone forming a thin jet. The diameter of injection nozzle is two to three orders of the magnitude larger than the diameter of dispersed jet. This results in the generation of droplets in micrometer or submicron size.

Tip-multibreakup mode is the consequence of the formation of an unsteady tip-streaming cone-jet structure. In this case, the intermittent generation of the droplets form the apex of the cone occurs due to Rayleigh-Plateau instabilities. Because of the oscillating conical meniscus, periodical droplet generation with nonuniform size distribution is formed.

Figure 11.10 shows breakup modes of droplet formation in a cross-flow device once both fluid flows have the same viscosity [24]. At low capillary numbers, the droplets of the dispersed fluid are formed at the junction through the squeezing mode. The size of the droplets and the distance between them are reproducible. For the same fraction $r$, increasing the capillary number decreases the droplet size (dripping mode). Increasing the capillary number further, a laminar side-by-side flow

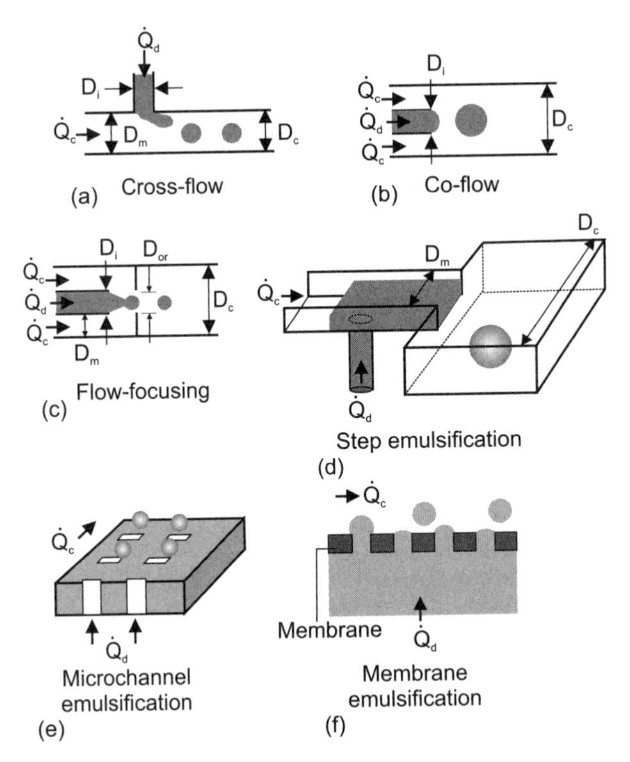


**Figure 11.8** Schematics of typical microfluidic devices for droplet formation: (a) cross-flow (T-junction geometry is shown); (b) co-flow (Quasi-2D planar co-flow is shown); (c) flow-focusing (planar flow-focusing); (d) step emulsification (vertical step is shown); (e) microchannel emulsification (straight-through microchannel is shown); and (f) membrane emulsification (direct membrane is shown). (*After:* [5].)

occurs beyond the junction before separating in droplets. In this case, jetting breakup mode occurs. Figure 11.10 represents quantitatively the phase diagram for a system with the same viscosity [24].

### Example 11.8: Droplet Formation and Breakup in Microchannels

Figure 11.11 depicts a simple model of the formation process of a liquid droplet in another immiscible carrier fluid. The following model only serves the purpose of understanding the relations between key parameters, such as droplet size, formation frequency, flow rates, and, most importantly, the interfacial tension between the two liquid phases. The model assumes a fixed flow rate ratio between the aqueous liquid and carrier liquid ($\alpha = \dot{Q}_{\mathrm{d}}/\dot{Q}_{\mathrm{c}}$). We further assume that the droplet size is small ($\alpha \ll 1$). Since the droplets are formed in microscale and the flows are in steady state, mass-related forces such as inertial force, momentum force, and buoyancy force are neglected in this model. If the aqueous liquid contains a surfactant, the surfactant concentration at the droplet surface is not uniformly distributed during the process of droplet growth. The distributed surfactant concentration leads to a gradient of interfacial tension on the droplet surface. This interfacial tension gradient in turn induces a Marangoni force on the droplet. If the surfactant solution is diluted, the Marangoni force is assumed to be small and negligible. The injection channel and the carrier channel are both assumed to be cylindrical. Determine the droplet sizes and the formation frequency of the droplets.

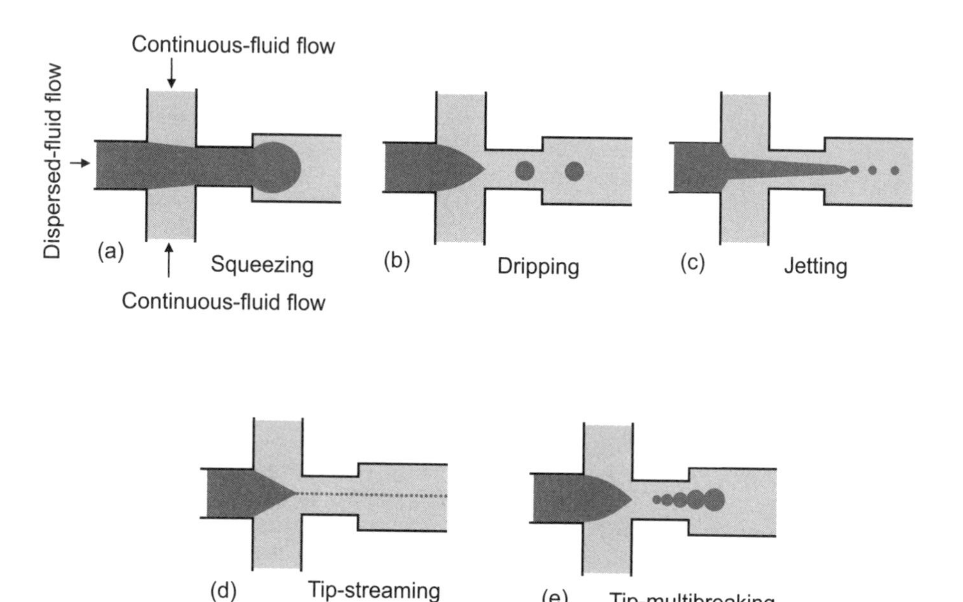


**Figure 11.9** Schematics of droplet breakup modes: (a) squeezing mode; (b) dripping mode; (c) jetting mode; (d) tip-streaming mode; and (e) tip-multibreaking mode. (*After:* [5].)

**Solution.** Considering all the above assumptions, the force balance includes only the drag force of the carrier flow and the interfacial tension at the injection port:

$$F_{\text{drag}} = F_{\text{interfacial tension}}$$

$$\frac{1}{2} C_{\text{D}} \rho U_{\text{c}}^2 A_{\text{D}} = C_{\text{S}} \pi D_{\text{i}} \sigma$$

where $\rho_{\text{c}}$, $U_{\text{c}}$, $A_{\text{D}}$, $D_{\text{i}}$, and $\sigma$ are the density of the carrier fluid, the average velocity of the carrier flow, the effective drag surface, the diameter of the injection port, and the interfacial tension, respectively. In addition, $C_{\text{D}}$ and $C_{\text{S}}$ are the drag coefficient and the coefficient for the interfacial tension. The coefficient $C_{\text{S}}$ depends on the contact angle and the shape of the injection port. In this model, $C_{\text{S}}$ is assumed to be constant. We assume for $C_{\text{D}}$ the drag coefficient of a hard sphere at a low Reynolds number Re:

$$C_{\text{D}} = \frac{24}{\text{Re}}$$

The effective drag interfacial $A_{\text{D}}$ grows with the droplet. Assuming that the droplet is a sphere, the effective drag surface at the detachment moment is:

$$A_{\text{D}} = \frac{\pi D_{\text{d}}^2}{2}$$

where $D_{\text{d}}$ is the diameter of the generated droplet. Initially, the interfacial tension is large enough to keep the small droplet at the injection port. At the detachment moment, the continuous droplet growth makes the drag force large enough to release the droplet. Combining the above equations results in the droplet diameter:

$$D_{\text{d}} = 2\sqrt{\frac{C_{\text{S}}}{C_{\text{D}}} D_{\text{i}} \frac{\sigma}{\rho_{\text{c}} U_{\text{c}}^2}}$$

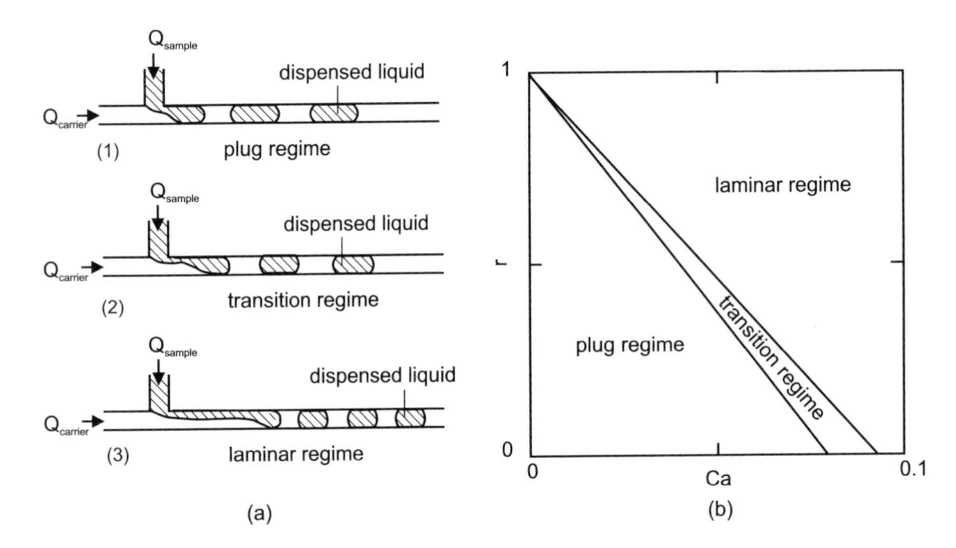


**Figure 11.10** Concept of droplet formation in a cross-flow device: (a) different formation regimes; and (b) phase diagram of the droplet formation for liquids with a same viscosity. (*After:* [24].)

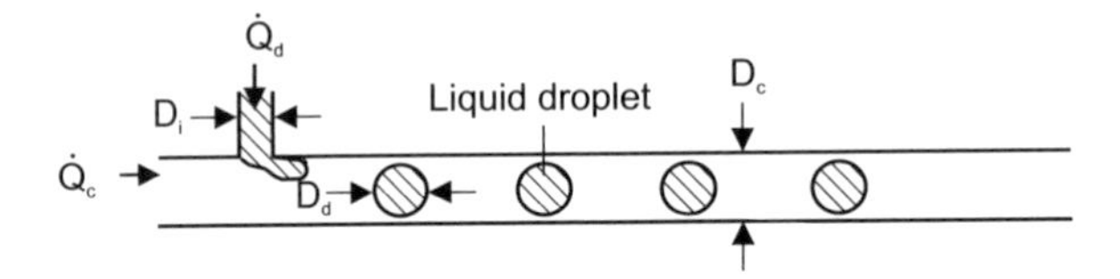


**Figure 11.11** Concept of formation and detection of liquid droplets in a cross-flow device.

The formation frequency can be estimated from the flow rate of the aqueous liquid $\dot{Q}_\mathrm{d}$ and the droplet volume $V_\mathrm{d}$ as:

$$f = \frac{\dot{Q}_\mathrm{d}}{V_\mathrm{d}}$$

Using the droplet diameter $D_\mathrm{d}$ and the relation $\dot{Q}_\mathrm{d} = \alpha\dot{Q}_\mathrm{c}$, the formation frequency in can be expressed as:

$$f = \frac{3\alpha D_\mathrm{c}^2}{16\left(\frac{C_\mathrm{S}}{C_\mathrm{D}} D_\mathrm{i}\right)^{\frac{3}{2}}} \frac{\rho_\mathrm{c}^{\frac{3}{2}} U_\mathrm{c}^4}{\sigma^{\frac{3}{2}}}$$

where $D_\mathrm{c}$ is the diameter of the carrier channel. We can observe a nonlinear relation between the formation frequency and the average carrier's velocity ($f \propto U_\mathrm{c}^4$) or flow rate ($f \propto \dot{Q}_\mathrm{c}^4$).

#### 11.1.2.2 Active In-Channel Droplet Formation

Active droplet formation using an external energy input provides high flexibility in controlling the droplet size and the generation rate. Importantly, active droplet generation also enables the independent control of droplet size and the production frequency. The use of additional energy source modifies the force balance on the interface that influence the interfacial instabilities [5].

Active control of interfacial force balance is achieved through: (1) employing external inputs such as electrical, magnetic, and centrifugal, and also (2) manipulating viscous, inertial, and

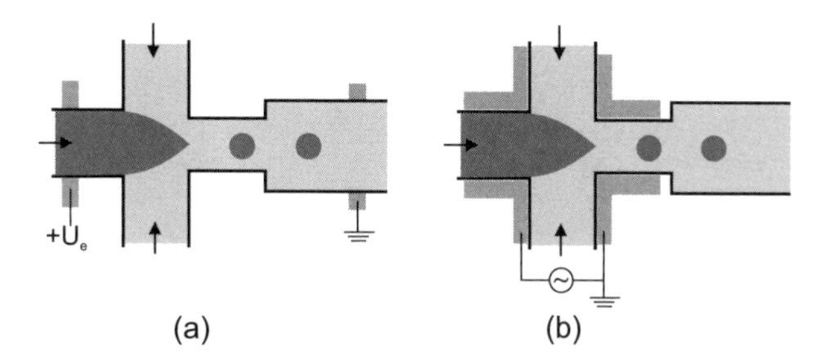


**Figure 11.12** Schematics of two examples of device arrangement for electrical control of droplet formation: (a) direct current (DC); and (b) alternating current (AC). (*After:* [5].)

capillary force. The later ones are realized by modifying the fluids velocity and material properties such as interfacial tension, channel wettability, fluid density, and fluid viscosity [5].

Electrical control can be applied through applying high voltage to the fluidic channel by the implemented electrodes (Figure 11.12). Application of the electrical field stimulates the accumulation of charges on the liquid-liquid interface that facilitates the control of droplet formation. Electrical control can be achieved in direct current (DC), constant or pulsatile voltage, and alternating current (AC) in the operating modes of low frequency and high frequency on the order of kilohertz.

Magnetic control of droplet formation is feasible for ferrofluids containing magnetic nanoparticles (Figure 11.13). Droplet formation can be controlled through modulating the induced magnetic forces to the ferrofluid. Modifying the magnetization, $M$, or the gradient of the magnetic field strength, $\nabla B$, can alter the magnetic forces induced to a ferrofluid, as discussed in Section 2.5.1. To this end, control of magnetic forces can be obtained through changing the location and type of a magnet, polarity, and the uniformity of a magnetic field [5, 25]. Depending on the location of the magnet (direction of the magnetic field) and the device geometry, the size of the droplets and their size variation are changed. As an example, once a magnet is located on the upstream of the junction, magnetic forces pull the droplet back that postpones the droplet formation. It results in the generation of the larger droplets.

Changing the hydraulic pressure and flow resistance to control the fluid velocity and destabilize the liquid-liquid interface provide flexible approaches for active control of droplet formation.

Modification of channel geometry for droplet generation also can be obtained through the use of piezoelectric actuators [Figure 11.14(a)]. Piezoelectric pulsation induces deformation in the flexible channel made in PDMS for on-demand droplet generation or to facilitate the droplet generation. Size of the droplets and their dimension variations can be affected by the profile of the piezoelectric pulse signal. Surface acoustic waves (SAW) also represent another external energy input to destabilize the liquid-liquid interface. Microchannels are implemented over an interdigitated transducer (IDT) that allows for spreading the one liquid into another one [Figure 11.14(b)]. Droplet size can be modulated by the channel geometry, flow rate of the continuous liquid, and pulse power and its duration [5].

Flow manipulation for droplet formation can also be realized with the aid of an on-chip valve. Normally open and normally closed microvalves can be implemented within the channel of the dispersed fluid. The deformation of the valve membrane induces perturbations in the fluid velocity. Figure 11.14(c) shows an on-chip valve for upstream actuation of the dispersed-fluid stream. Microvalves can be implemented at the downstream of the junction to chop a preformed droplet or flow-focused dispersed stream into daughter droplets.

Figure 11.14(d) shows on-demand droplet formation through rapid spot heating of the dispersed fluid. The droplet is produced due to the expansion of the cavitation bubble induced by the laser beam pushing the liquid into the continuous stream.

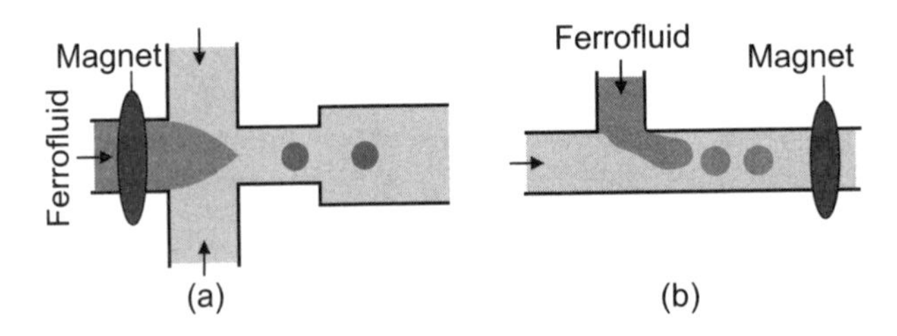


**Figure 11.13** Schematics of two examples of device arrangement for magnetic control of droplet formation: (a) magnet at the upstream; and (b) magnet at the downstream. (*After:* [5].)

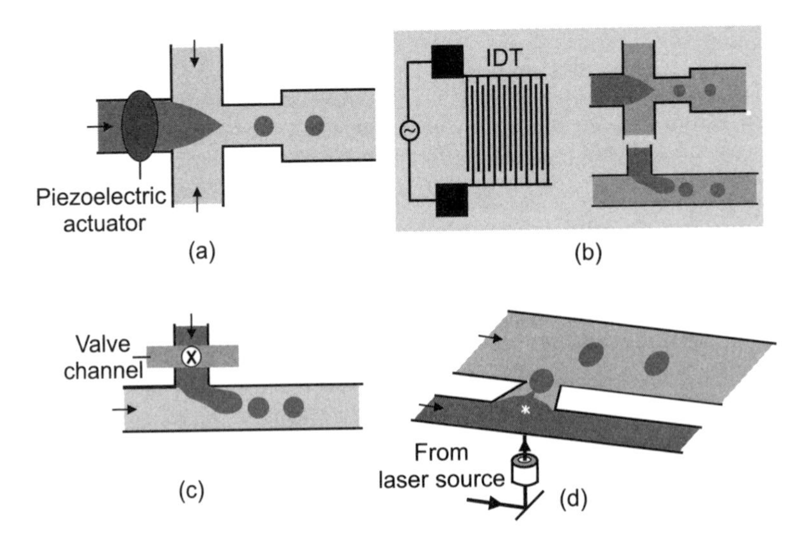


**Figure 11.14** Examples of device arrangements for modulating liquid velocity for active control of droplet formation: (a) piezoelectric pulsation at the upstream of the junction; (b) surface acoustic waves induced by interdigitated transducer; (c) on-chip microvalve actuation for flow manipulation at the upstream of the junction; and (d) pulse laser-induced cavitation for on-demand liquid pumping and droplet formation. (*After:* [5].)

### 11.1.3 Applications of In-Channel Dispensers

Major applications for in-channel droplet formation revolve around cell culturing, microparticle fabrication of drug carriers, and high-throughput single cell screening and drug testing [26, 27].

Droplets with picoliter volumes provide suitable microenvironment for cell culturing with precise control over the cell count and cell type [26]. Cells trapped in droplets can be analyzed and manipulated, and even be cultured if enough oxygen transport through the device and the droplet is provided [Figure 11.15(a)]. The cultures can be sorted using the expression of fluorescent protein or the number of the cells [28]. Lysis of the cells within droplets also could be achieved through various methods such as electroporation, and the introduction of volumes of lysis buffer to the cell droplets during the droplet formation process [26].

In-channel droplet production has enabled the precise fabrication of monodisperse particles containing drug loads for sustained dug release [Figure 11.15(b)]. Drug release profile from the particles can be tuned through modifying the internal structure of the the droplets using double or multiple emulsions [30]. Particle size that determines the drug load and the release time can be controlled by tuning the flow rates, surface tension, and the viscosity of the solutions [30].

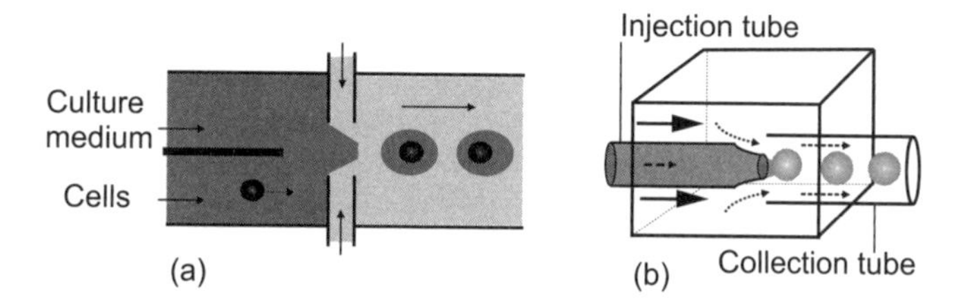


**Figure 11.15** Schematics of two examples of cell encapsulation and synthesis of drug carriers using in-channel droplet production: (a) cell encapsulation in droplets for culturing and screening (*After:* [27]); and (b) production of open-celled poly(N-isopropylacrylamide) (PNIPAM) microgels in a flow-focusing device (*After:* [29]).

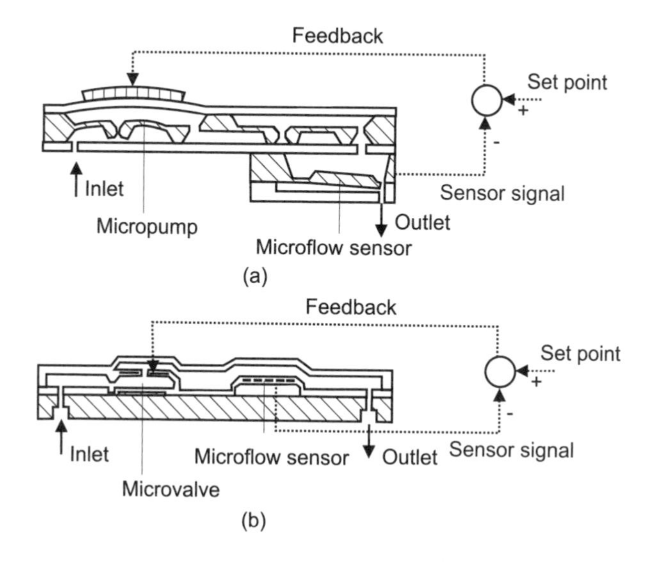


**Figure 11.16** Closed-loop controlled microdispensers: (a) with a pump and a flow sensor [2]; and (b) with a microvalve and a microflow sensor [33].

### 11.1.3.1 Metering Dispenser

In-channel dispensers utilize common actuators used in micropumps for displacing fluids. In closed-loop control, fluid amounts on the order of 1 nL can be dispensed. Such a closed-loop system consists of a valve or a pump and a flow sensor (Figure 11.16). One of the earlier systems with integrated valve and sensor was reported by Gass et al. [2]. The system consists of a piezoelectric microcheck-valve pump and a flow sensor based on drag force. The concepts of the pump and the flow sensor are discussed in Chapters 7 and 8. The device was able to control a flow rate on the order of 100 μL/min. The same flow rate was achieved with a similar closed-loop concept reported by Waibel et al. [31]. Instead of a flow rate sensor, a liquid level sensor was used for metering the liquid amount. The signal from the sensor is used for controlling a piezoelectric valve. Cabuz et al. [32] used a system with a hybrid-assembled microvalve and a thermal flow sensor to control the sheath flows and the sample flow for hydrodynamic focusing. A flow rate on the order of 50 μL/min can be controlled with this system. An integrated solution for microvalves and microflow sensors was reported by Xie et al. [33]. The electrostatic valve and the thermal flow sensor were placed in a 100-μm-wide microchannel. The system can control flow rates on the order of 1 to 10 μL/min. Since the design examples of this dispenser type are those of micropumps, microvalves, and microflow sensors discussed in previous chapters, the following sections will not cover more examples of closed-loop microdispensers.

If only a fixed amount of liquid is required for further analysis, this amount can be measured passively and then delivered by common actuating schemes. The two steps of a metering process are:

- Measuring the fluid amount needed, where the fluid amount is bordered by a stopper on one end and by the outlet of a delivering actuator on the other end;
- Pushing the measured amount to a desired position using a common actuating principle.

There are different ways to realize the stopper structure. In general, it is a one-way valve, which opens if the driving pressure overcomes a certain critical pressure of the valve. It is convenient to use surface tension to realize such a passive valve in microchannels. Looking at the pressure generated by surface tension in a circular channel:

$$\Delta p = \frac{2\sigma}{R} \tag{11.17}$$

it is obvious that both parameters, surface tension $\sigma$ and radius of curvature $R$, can be used for manipulating the pressure of the stopper.

Figure 11.17(a) depicts a hydrophobic patch, which works as the stopper, due to its higher surface tension with the liquid [34, 35]. Figure 11.17(b) shows the second method of narrowing the channel, which also increases the stopping pressure. In both cases, the liquid fills the metering volume using a driving pressure [36]. This pressure should be smaller than the stopping pressure, so that the liquid automatically stops at the desired position. If the channel is small enough, a capillary force is sufficient to drive the liquid. Subsequently, a separating fluid, often the gas phase of the liquid or an air bubble, pushes the measured liquid amount to the desired position. The driving pressure of the second process should be slightly larger than the stopping pressure in (11.17). This dispensing method is convenient for planar microfluidic systems, especially the CD platform described in Section 7.1.1.7, where different driving pressures are available by changing the rotation speed.

Figure 11.17(c) describes another concept of in-channel dispensing [18]. Electrokinetic flow injects the liquid into the measured volume. By switching the injection potential to the dispensing flow, the measured liquid amount is pushed to the desired position. In most cases, the components in the measured liquid undergo a subsequent electrophoretic separation. This method is suitable for conducting liquids and requires glass as a device material or special treatment for the channel surface because of the double layer needed for electrokinetic pumping.

## 11.2 DESIGN EXAMPLES

### 11.2.1 Droplet Dispensers

#### 11.2.1.1 Thermopneumatic

Droplet dispensers were originally designed for printing applications. With an increasing need for massively parallel analysis, the same device can be used for printing the assay matrix of reagents. Figure 11.18(a) describes the simplified structure of the dispenser reported in [6]. The device is an inkjet print head, which has integrated CMOS control circuits. Therefore, its fabrication should be compatible to the CMOS technology.

The fabrication of the nozzles starts after completing the CMOS process. The droplet is injected thermopneumatically by a thermal bubble. The heater is made of hafiniumdiboride. An electroplated Ni/Au layer on top of the device avoids overheating and ensures quick heat spreading after each injection. This top layer is used as the mask for the final dry etching process of the nozzle.

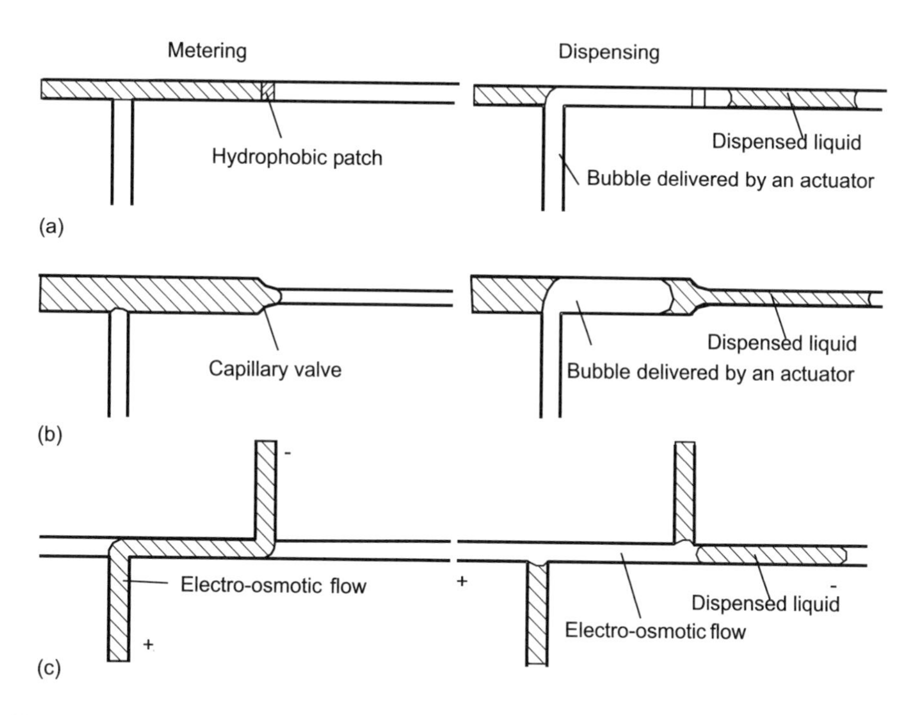


**Figure 11.17** Process steps of in-channel microdispenser: (a) with hydrophobic valve; (b) with capillary valve; and (c) with electro-osmotic flow.

The dispenser depicted in Figure 11.18(b) is fabricated using a combination of bulk machining and surface machining [7]. The heater is made of platinum. In contrast to other designs, this dispenser utilizes two thermopneumatic actuators to control the droplet injection. Beside a large heater structure for droplet injection, a second smaller heater works as a bubble valve, which uses the surface tension to control the refill process. As a result, clean droplets without satellites can be generated.

The dispenser shown in Figure 11.18(c) [22] has a nozzle made of silicon oxide. The fabrication process combines both bulk machining and surface machining. The ring-shaped polysilicon heater creates a doughnut-shaped bubble, which pushes the droplet out of the nozzle. The device chamber is filled automatically by capillary force.

All three dispensers described above have bubbles, which grow in the opposite direction of the droplet. This principle is called *backshooter*. The other principle, called *sideshooter*, ejects the droplet vertically [9] [Figure 11.18(d)]. The sideshooter dispenser utilizes the same technology described in [9] for making microneedles. The heater is made of polysilicon. The nozzles are arranged horizontally and are exposed by sawing through the flow channel.

### 11.2.1.2 Thermomechanical

In contrast to the thermopneumatically driven dispensers, the microdispenser depicted in Figure 11.18(e) utilizes the buckling energy of a thermomechanic actuator to dispense a droplet [10]. The actuator is a circular nickel membrane. The nickel heater is embedded in the electrically insulated silicon oxide layer. If the stress induced by the high temperature of the heater is greater than the critical stress of the membrane, the membrane buckles rapidly. The rapid acceleration adds energy into the droplet and ejects it out of the nozzle. The dispenser is fabricated by bulk micromachining and surface micromachining using aluminum as a sacrificial layer.

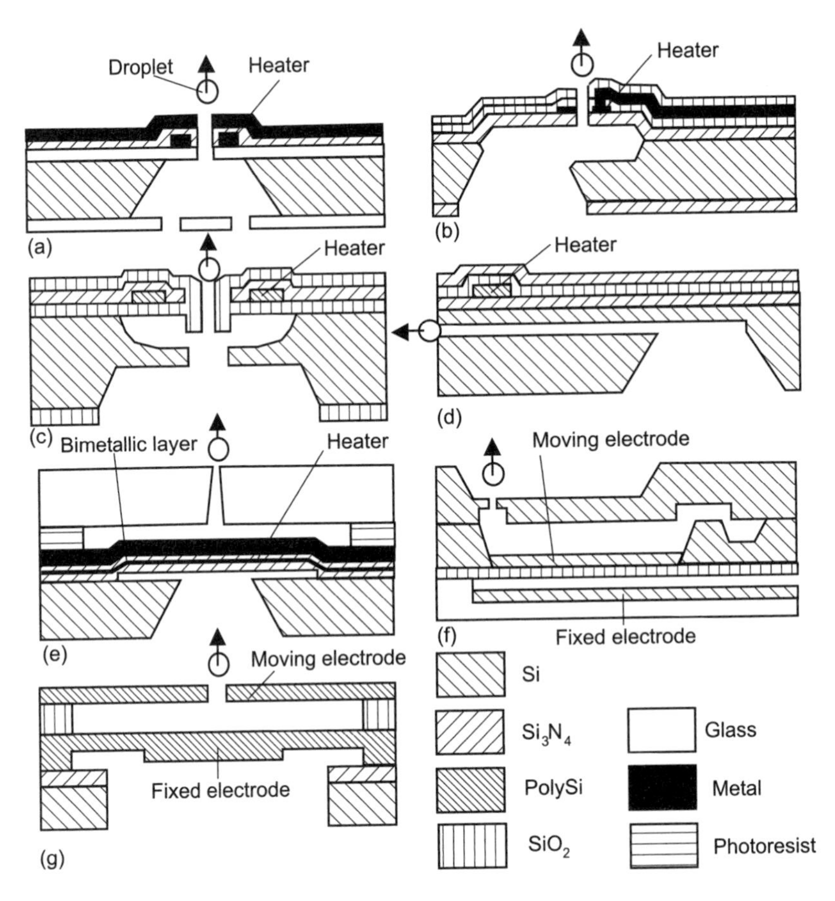


**Figure 11.18** Microdispensers: (a–d) thermopneumatic; (e) thermomechanical; and (f) and (g) electrostatic.

### 11.2.1.3 Electrostatic

All dispensers described above operate at high temperatures. The high temperatures are not critical for the conventional ink-printing application, but can be fatal for printing biochemical reagents. For applications in life sciences, low-temperature actuators, such as the electrostatic actuator and the piezoelectric actuator, should be considered. An electrostatic actuator [11] generates the dispensing pressure in the device depicted in Figure 11.18(f). The silicon membrane opposite the nozzle acts as an electrode of the actuator. The other electrode is made of transparent indium tin oxide (ITO), which is deposited on a glass substrate. Because of the extremely small gap of 0.2 mm, the actuator only needs 26.5V for its operation.

### 11.2.1.4 Piezoelectric

The microdispenser reported in [12] is fabricated entirely with surface micromachining [Figure 11.18(g)]. The nozzle plate and a moveable piston plate are the two electrodes of an electrostatic actuator. The gap between the electrode is about 5 μm. In contrast to the example shown in Figure 11.18(f), the actuating field is applied directly through the dispensed liquid. That means the dispenser can only be used for a dielectric liquid, and the field strength should be below the breakdown strength of the liquid. The other drawback is the possibility of electrolysis, which generates unwanted bubbles and leads to malfunction of the dispenser.

The other type of low-temperature driving mechanism suitable for biochemical application is the piezoelectric actuator. Piezoelectric actuators have the advantage of high drive pressure and fast dynamics. An external stack-type piezoelectric actuator is able to drive several nozzles concurrently. Therefore, passive parallel dispensing is possible with a single actuator. The drive pressure is programmable with the drive voltages. Fast dispensing process and slow reversing avoid air aspiration and allow refilling. This feature and the high dynamics of piezoelectric actuators make fast and clean dispensing possible.

The dispenser depicted in Figure 11.19(a) has a piezobimorph cantilever as an actuator [13]. The driving scheme of this dispenser is similar to that of all other piezoelectric dispensers. The cantilever retracts slowly, allowing the liquid to be filled in the device chamber, while air aspiration is avoided, due to the surface tension of the meniscus at the nozzle outlet. In the dispensing process, the tension is released quickly. The energy released in this fast process allows a droplet to be formed and ejected out of the nozzle. Large energy is required to form the surface of the droplet and to convert into kinetic energy. In this dispenser, only the nozzle plate is made of silicon using micromachining. The piezobimorph beam is mounted on a membrane in the dispenser shown in Figure 11.19(b) [14]. The beam is glued on three cylindrical PMMA stands to generate a large displacement. The ring-shaped nozzle is fabricated with the process described previously in Figure 11.7(a). This design has a fill chamber of 8.8 μL and a droplet volume on the order of 300 nL to 500 nL. Before assembly, the fill chamber and channel surfaces are oxidized to get a thick and stable silicon oxide layer. The hydrophobic characteristics of silicon oxide improve the priming process of the dispenser. The ring-shaped nozzle and the treatment of its outer surface with a hydrophobic layer allow clean droplet generation and accurate trajectory.

The device depicted in Figure 11.19(c) integrates a nozzle on the outlet of a check-valve micropump [37]. A piezodisc drives the pump. This design is more complex than the others and has slow dynamics because of the large amount of liquid to be filled into the pump chamber. The outer surface of the nozzle is treated with a hydrophobic layer. The ring-shaped nozzle is formed by wet etching with a timed stop.

The design described in Figure 11.19(d) has a nozzle in the form of a needle [38]. Two capacitive sensors are integrated into the device for the measurement of displacement and pressure in the fluid chamber. The device works as a pipette. In the aspiration mode, the needle is dipped into the liquid reservoir and the actuator sucks the liquid into the needle.

In the dispensing mode, the actuator pushes the liquid amount out of the needle. The dispenser reported in [22] has a piezodisc as actuator [Figure 11.19(e)]. The extremely small nozzle, on the order of 1 μm, is made of silicon oxide with the process described in Figure 11.7(b). The device works in burst mode with ultrasonic frequency (0.5 MHz). The extremely small nozzle and the high frequency eject droplets on the order of several femtoliters ($10^{-15}$ L).

The device described in [39] has 24 nozzles arranged in a 4×6 array. The nozzles are connected to 24 reservoirs, which contain different liquids. Between the piezoelectric actuator and the nozzles is an air gap. The system allows dispensing of 24 different liquids at once and is used for printing of reagents on a DNA chip.

A simple piezoelectric dispenser is described in Figure 11.19(f) [40]. The acceleration of the piezoelectric bimorph disc transmits the required energy to the droplet. However, the device cannot meet the accuracy and controllability of micromachined devices.

According to the analysis in Section 11.1.1, the two important parameters for the dispensing process are the droplet size and the initial droplet velocity. While the initial droplet velocity is directly proportional to the actuation velocity and can be controlled precisely [39], it is not easy to have a controlled reproducible droplet size. All the above droplet dispensers have a large liquid reservoir. The size of the droplet depends on the actuation stroke and the viscosity of the liquid. For the same liquid, the droplet size is proportional to the actuation stroke [39] and the nozzle diameter

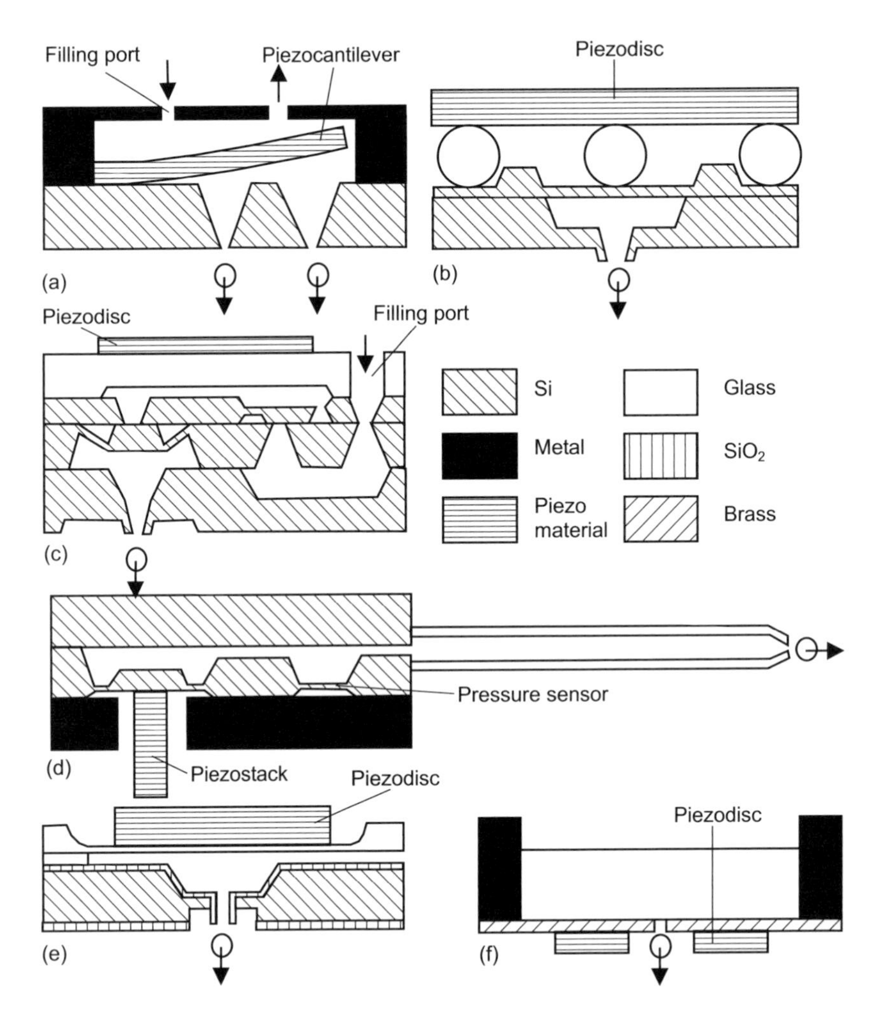


**Figure 11.19** Piezoelectric microdispensers: (a, b) with piezobimorph cantilever; (c) with piezodisc; (d) with piezostack; (e) with piezodisc on a glass membrane; and (f) with piezodisc on a brass membrane.

[41]. For the same actuation condition and nozzle size, the droplet is also proportional to viscosity [42]. In order to achieve a controllable droplet size, Koltay et al. utilized the concept of metering in an in-channel dispenser to accurately dispense the required liquid amount. The metering chamber is positioned directly above the nozzle. A microchannel connects this chamber with the larger reservoir [Figure 11.20(a)]. The microchannel works as the supply channel to fill the metering chamber by using capillary force [Figure 11.20(b, c)]. During injection, both reservoir and metering chamber have the same pressure. Thus, no additional flow crosses the microchannel [Figure 11.20(d)]. The droplet size can be determined precisely by the predesigned volume of the metering chamber [43] [Figure 11.20(e)]. After dispensing, the metering chamber is refilled by capillary force at the microchannel [Figure 11.20(f)]. Table 11.1 gives an overview of the typical parameters of droplet microdispensers.

### 11.2.2 In-Channel Dispensers

Because of their simplicity, in-channel microdispensers are implemented in complex systems, such as a DNA chip [34, 35], and commercial products [36]. Figure 11.21 shows a number of design

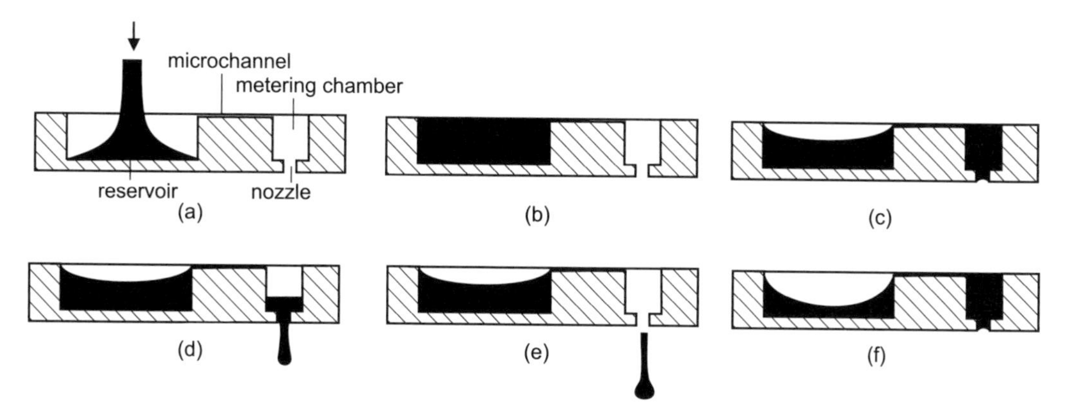


**Figure 11.20** The concept of metering and dispensing: (a) filling the reservoir; (b) filling the metering chamber with capillary force; (c) ready for dispensing; (d) dispensing with a pressure pulse; (e) complete depletion of metering chamber; and (f) refilling of metering chamber. (*After*: [43].)

**Table 11.1**
Typical Parameters of Droplet Microdispensers

| *Ref.* | *Nozzle Size (μm)* | *Array Size* | *Droplet Volume (pL)* | *Droplet Speed (m/s)* | *Actuator* |
|---|---|---|---|---|---|
| [6] | 30×40 | 50×1 | 310 | 10–15 | Thermopneumatic |
| [7] | 40 | 150×1 | 113–382 | 10 | Thermopneumatic |
| [8] | 20–28 | 28×2 | 3.2–9.4 | 10 | Thermopneumatic |
| [9] | 21 | 16×1 | 34 | — | Thermopneumatic |
| [10] | 35 | 1 | 1,150 | 8 | Thermomechanic |
| [11] | 28 | 64×2 | 22.5 | 6 | Electrostatic |
| [12] | 20 | 13×1 | 2 | 10 | Electrostatic |
| [13] | 50 | 67×1 | 268–4,200 | — | Piezoelectric |
| [14] | 60 | 1 | 268–524 | 3.5 | Piezoelectric |
| [21] | 400 | 1 | 140,000 | 0.8 | Piezoelectric |
| [38] | 1.4 | 1 | 0.0034 | — | Piezoelectric |
| [39] | 100 | 4×6 | 4,200 | — | Piezoelectric |
| [40] | 60 | 1 | 125 | 1.64 | Piezoelectric |
| [43] | 50–200 | 8×12 | 50 | — | Piezoelectric |

examples of in-channel dispensers. Both droplet formation and metering in microchannels are based on a multiphase system. While the formation of droplets requires a liquid/liquid system, metering and dispensing a liquid plug requires a gas/liquid system.

The basic configuration of an in-channel droplet dispenser shown in Figure 11.10 can be modified to combine mixing with dispensing. In this case, the sample inlet is designed as a simple Y-mixer, which first merges the solute and solvent before dispensing them with the carrier fluid. If mixing should be avoided before entering the carrying stream, a third stream between the two sample streams can be used to separate them [44].

The example depicted in Figure 11.21(a) uses hydrophobic valves made of SAM films [34]. The hydrophobic patches are patterned on silicon and glass substrates using a lift-off technique. The dispensed volume is determined by the volume of the channel between the outlet of the thermopneumatic actuator and the hydrophobic valve. An overflow channel prevents the driven pressure from exceeding the surface tension at the hydrophobic valve. The hydrophobic vent on the right is used to keep the dispensed liquid at the desired position. Air coming from the

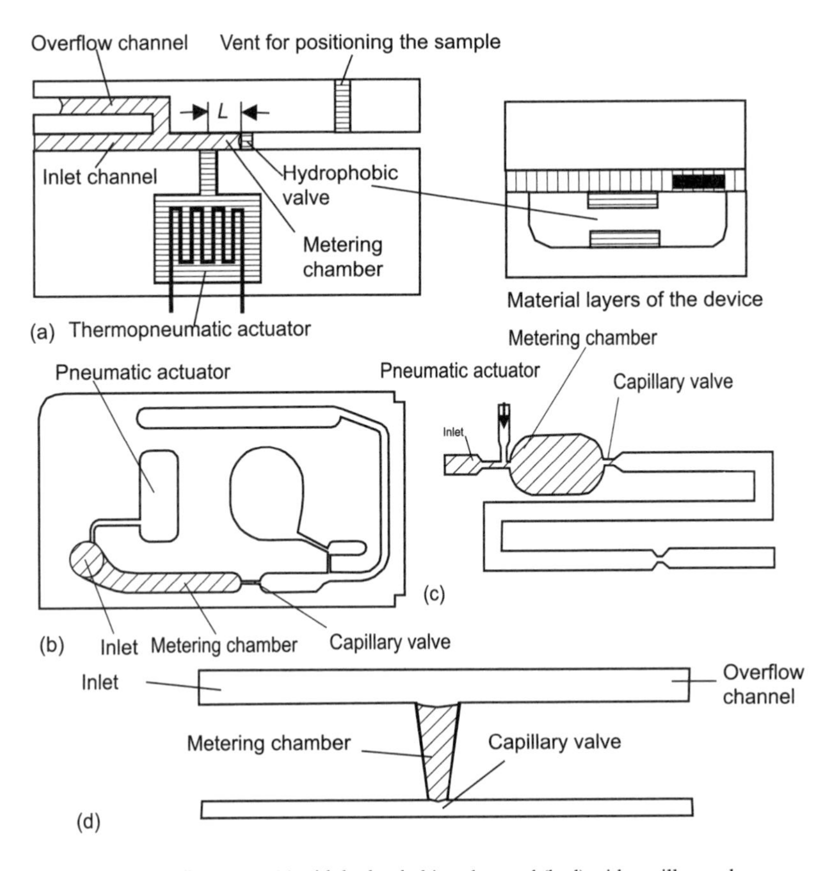


**Figure 11.21** In-channel microdispensers: (a) with hydrophobic valve; and (b–d) with capillary valve.

thermopneumatic actuator exits the dispensing channel at the vent and lets the driven pressure drop down to the initial value. This design is successfully implemented in a DNA analysis system [35].

The dispenser in Figure 11.21(b) is a part of the analysis cartridge, commercially available from i-STAT (i-STAT Inc., Ontario, Canada) [36]. The device is made of plastic. A capillary valve is used to stop the liquid. The measured amount is then pushed to the desired position by a pneumatic actuator. The use of a capillary valve keeps the fabrication of the device simple and cost-effective.

The same concept of a capillary valve is shown in Figure 11.21(c) [45]. The device is made of PDMS using soft-lithography technology; see Section 3.4.5. The volume of the measurement chamber determines the amount of dispensed liquid. Different volumes of 50, 100, and 150 nL were realized. Two capillary valves keep the liquid in the chamber until the external pneumatic actuator pushes out the measured liquid.

The examples shown in Figure 11.21(d) look different, but work with the same concept of in-channel metering and dispensing [46]. The tapered shape of the measurement chamber makes self-filling possible, when the liquid flows into the large inlet channel. The liquid stops at the narrow end of the chamber. The liquid remains in the chamber after withdrawing or pushing away excessive liquid. A quick pressure pulse ejects the measured amount into the smaller channel. The device is made of PDMS and is able to dispense 10 nL. The example reported in [18] uses the conventional method of capillary separation. The measured liquid amount is delivered and separated by electrokinetic pumping. The concept shown in Figure 11.21(d) was used by Yamada and Seki [36] for accurate dosing and mixing. The device was fabricated based on molding of PDMS. The device combines two in-channel dispensers to dispense and mix 3.5-pL droplets. The solvent and

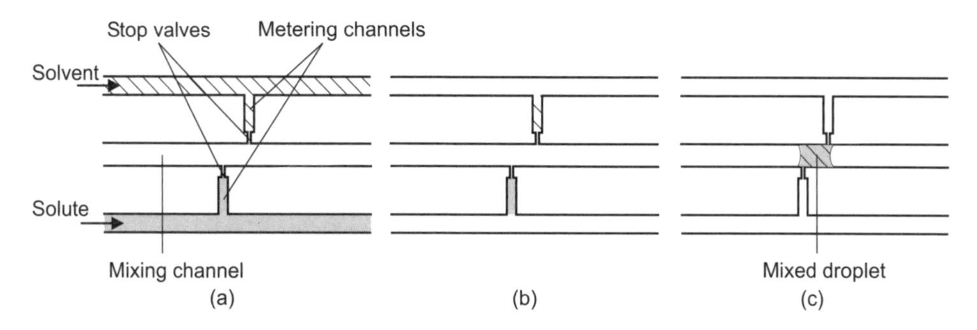


**Figure 11.22** Application of in-channel microdispensers for micromixing. (a) Solute and solvent are introduced into the metering channel. (b) The rest of the liquids are flushed away. (c) Metered solute and solvent are dispensed to form the mixed droplet. (*After*: [36].)

solute are first introduced into the metering channels [Figure 11.22(a)]. Both liquids stop at the capillary valves. Subsequently, the rest of the liquids are flushed away by air flow [Figure 11.22(b)]. Applying a higher pressure than the capillary pressure of the valves dispenses the liquids into the mixing channel [Figure 11.22(c)]. The distance between the two capillary valves is to be carefully determined, so that each dispensed droplet does not block the other.

## 11.3 SUMMARY

This chapter discusses the design consideration and design examples of different microdispenser concepts. In most cases, the dispensed liquid amount is confined by the interface between two or more phases. Injected droplets in air are defined by the liquid/air interface. Droplets or plugs in microchannel are defined by the solid/liquid/liquid or solid/liquid/gas interfaces.

The existence of the interfaces leads to dimensionless numbers that can describe the behavior of the dispensed liquid amount. For the liquid/air system in droplet dispensers, the Weber number (We) describes the ratio between kinetic energy and the surface tension. The Reynolds number based on the air viscosity represents the ratio between the inertia force and the friction force. Thus, a We-Re diagram can be used to describe the behavior of the dispensed droplet.

For the liquid/liquid system in the case of in-channel droplet dispenser, the capillary number (Ca) represents the ratio between the friction force and the surface tension of the droplet. The capillary number is proportional to the Reynolds number. Thus, the other parameter for describing this system is the volumetric fraction $r$ of the droplet liquid. An $r$-Ca diagram can describe well the behavior of the dispensed droplet.

The metering concept uses the volume determined by the geometry to dispense the required liquid amount. Combining this concept with a droplet dispenser can generate droplet sizes, which are independent from other parameters, such as actuation stroke and viscosity.

### Problems

**11.1** A water droplet is rejected from a nozzle with a speed of 2 m/s. Determine the critical diameter of the droplet so that no satellite droplet can occur. The droplet is assumed to be stable if the kinetic energy is smaller than the surface energy. Surface tension and density of water are assumed to be $72\times10^{-3}$ N/m and 1,000 kg/m$^3$, respectively.

**11.2** Determine the minimum gap between the electrodes of an electrostatic actuator, which is used for dispensing a droplet. The actuating voltage is 200V. The droplet has a size of 2 pL and moves with an initial velocity of 10 m/s. [Hint: The electrostatic force is $F = 1/2(\varepsilon\varepsilon_0)A(V^2/x^2)$, which gives the energy to overcome the squeezing force $F_{\text{sf}} = (3w^2\eta A/2)x^3\text{d}x/\text{d}t$, and to form and to eject the droplet. $A$ and $w$ are the area and the lateral dimension of the moving plate, $x$ is the gap, $V$ is the applied voltage, $\eta$ and $\varepsilon$ are the viscosity and relative permittivity of the liquid, and $\varepsilon_0 = 8.854 \times 10^{-12}$ F/m is the permittivity of free space.

**11.3** A piezoelectric stack actuator is used for parallel dispensing of 96 water droplets. The dispensing process occurs during 5 ms. The actuating stroke is 10 mm. Determine the required force if all droplets have a size of 20 pL and an initial velocity of 15 m/s.

**11.4** Is the force in the above parallel dispenser large enough to aspirate air into the nozzles? Assume that the nozzle diameter is equal to the droplet diameter.

**11.5** A droplet of 100 nL is ejected on a glass plate. The droplet meets the plate at 10 m/s. Is it possible that this droplet is split into two satellite droplets after the impact?

**11.6** Design an in-channel microdispenser with capillary valves and thermopneumatic actuators. The amount to be dispensed is 200 nL, and the air temperature of the actuator cannot exceed 50°C.

## References

[1] Lammerink, T. S. J., Elwenspoek, and M., Fluitman, H. J., "Integrated Micro-Liquid Dosing System,“ *Proceedings of MEMS'93, 6th IEEE International Workshop Micro Electromechanical System*, San Diego, CA, Jan. 25–28, 1993, pp. 254–259.

[2] Gass, V., et al., "Integrated Flow-Regulated Silicon Micropump," *Sensors and Actuators A*, Vol. 43, 1994, pp. 335–338.

[3] Boillat, M. A., et al., "A Differential Pressure Liquid Flow Sensor for Flow Regulation and Dosing Systems," *Proceedings of MEMS'95, 8th IEEE International Workshop Micro Electromechanical System*, Amsterdam, The Netherlands, Jan. 29-Feb. 2, 1995, pp. 350–352.

[4] Rossberg, R., and Sandmaier, H., "Portable Micro Liquid Dosing System," *Proceedings of MEMS'98, 11th IEEE International Workshop Micro Electromechanical System*, Heidelberg, Germany, Jan. 25–29, 1998, pp. 526–531.

[5] Zhu, P., and Wang, L.,"Passive an Active Droplet Generation with Microfluidics: A Review," *Lab Chip*, Vol. 17, 2017, pp. 34–75.

[6] Krause, P., Obermeier, E., and Wehl, W., "Backshooter—A New Smart Micromachined Single-Chip Inkjet Printhead," *Proceedings of Transducers '95, 8th International Conference on Solid-State Sensors and Actuators*, Stockholm, Sweden, June 16–19, 1995, pp. 520–523.

[7] Tseng, F. G., Kim, C. J., and Ho, C. M., "A Novel Microinjector with Virtual Chamber Neck," *Proceedings of MEMS'98, 11th IEEE International Workshop Micro Electromechanical System*, Heidelberg, Germany, Jan. 25–29, 1998, pp. 57–62.

[8] Lee, C. S., et al., "A Micromachined Monolithic Inkjet Print Head with Dome Shape Chamber," *Proceedings of Transducers '01, 11th International Conference on Solid-State Sensors and Actuators*, Munich, Germany, June 10–14, 2001, pp. 902–905.

[9] Chen, J., and Wise, K. D., "A High-Resolution Silicon Monolithic Nozzle Array for Inkjet Printing," *IEEE Transactions on Electron Devices*, Vol. 44, No. 9, 1997, pp. 1401–1409.

[10] Hirata, S., "An Inkjet Head Using Diaphragm Microactuator," *Proceedings of MEMS'96, 9th IEEE International Workshop Micro Electromechanical System*, San Diego, CA, Feb. 11–15, 1996, pp. 418–423.

[11] Kamisuki, S., et al., "A High Resolution, Electrostatically Driven Commercial Inkjet Head," *Proceedings of MEMS'00, 13th IEEE International Workshop Micro Electromechanical System*, Miyazaci, Japan, Jan. 23–27, 2000, pp. 793–798.

[12] Galambos, P., et al., "A Surface Micromachined Electrostatic Drop Ejector," *Proceedings of Transducers '01, 11th International Conference on Solid-State Sensors and Actuators*, Munich, Germany, June 10–14, 2001, pp. 906–909.

[13] Ederer, I., Grasegger, J., and Tille, C., "Droplet Generator with Extraordinary High Flow Rate and Wide Operating Range," *Proceedings of Transducers '97, 9th International Conference on Solid-State Sensors and Actuators*, Chicago, IL, June 16–19, 1997, pp. 809–812.

[14] Laurell, T., Wallman, L., and Nilsson, J., "Design and Development of a Silicon Microfabricated Flow-Through Dispenser for On-Line Picolitre Sample Handling," *Journal of Micromechanics and Microengineering*, Vol. 9, 1999, pp. 369–376.

[15] Lee, S. W., Jeong, O. C., and Yang, S. S., "The Fabrication of a Micro Injector Actuated by Boiling and/or Electrolysis," *Proceedings of MEMS'98, 11th IEEE International Workshop Micro Electromechanical System*, Heidelberg, Germany, Jan. 25–29, 1998, pp. 526–531.

[16] Lee, S. W., Sim, W. Y., and Yang, S. S., "Fabrication and In Vitro Test of a Microsyringe," *Sensors and Actuators A*, Vol. 83, 2000, pp. 17–23.

[17] Böhm, S., Olthuis, W., and Bergveld, P., "An Integrated Micromachined Electrochemical Pump and Dosing System," *Journal of Biomedical Microdevices*, Vol. 1, No. 2, 1999, pp. 121–130.

[18] Effenhauser, C. S., Manz, A., and Widmer, H. M., "Glass Chips for High-Speed Capillary Electrophoresis Separations with Submicron Plate Heights," *Analytical Chemistry*, Vol. 65, No. 19, 1993, pp. 2637–2642.

[19] Vance, B., et al., "Controlling Droplet Deposition with Polymer Additives," *Nature*, Vol. 405, No. 6788, 2000, pp. 772–775.

[20] Douglas, J. F., Gasiorek, J. M., and Swaffield, J. A., *Fluid Mechanics*, 4th ed., London: Longman, 1995.

[21] Koide, A., et al., "Micromachined Dispenser with High Flow Rate and High Resolution," *Proceedings of MEMS'00, 13th IEEE International Workshop Micro Electromechanical System*, Miyazaci, Japan, Jan. 23–27, 2000, pp. 424–428.

[22] Luginbuhl, P., et al., "Micromachined Injector for DNA Mass Spectrometry," *Proceedings of Transducers '99, 10th International Conference on Solid-State Sensors and Actuators*, Sendai, Japan, June 7–10, 1999, pp. 1130–1133.

[23] Lang, R. J., "Ultrasonic Atomization of Liquids," *Journal of the Acoustical Society of America*, Vol. 34, No. 1, 1962, pp. 6–8.

[24] Tice, J. D., Lyon, A. D., and Ismagilov R. F., "Effects of Viscosity on Droplet Formation and Mixing in Microfluidic Channels," *Analytica Chimica Acta*, Vol. 507, 2004, pp. 73–77.

[25] Chong, Z. Z., et al., "Active Droplet Generation in Microfluidics," *Lab Chip*, Vol. 16, 2016, pp. 35–58.

[26] Mashaghi, S., et al.,"Droplet Microfluidics: A tool for Biology, Chemistry and Nanotechnology," *TrAC Trends in Analytical Chemistry*, Vol. 82, 2016, pp. 118–125.

[27] Wang, B. L., et al.,"Microfluidic high-throughput culturing of single cells for selection based on extracellular metabolite production or consumption," *Nature Biotechnology*, Vol. 32, 2017, pp. 473–478.

[28] Zang, E., et al.,"Real-Time Image Processing for Label-Free Enrichment of Actinobacteria Cultivated in Picolitre Droplets," *Lab Chip*, Vol. 13, 2013, pp. 3707–3713.

[29] Mou, C. -L., et al.,"Monodisperse and Fast-Responsive Poly(N-isopropylacrylamide) Microgels with Open-Celled Porous Structure," *Langmuir*, Vol. 30, 2014, pp. 1455–1464.

[30] Riahi, R., et al.,"Microfluidics for Advanced Drug Delivery Systems," *Current Opinion in Chemical Engineering*, Vol. 7, 2015, pp. 101–112.

[31] Waibel, G., et al., "Highly Integrated Autonomous Microdosage System," *Sensors and Actuators A*, Vol. 103, 2003, pp. 225–230.

[32] Cabuz, E., et al., "MEMS-Based Flow Controller for Flow Cytometry," *Solid-State Sensor and Actuator Workshop*. Hilton Head Island, SC, June. 2–6, 2002, pp. 110–111.

[33] Xie, J., Shih, J., and Tai, Y. C., "Integrated Surface-Micromachined Mass Flow Controller," *Proceedings of MEMS'03, 16th IEEE International Workshop Micro Electromechanical System*, Kyoto, Japan, Jan. 19–23, 2003, pp. 20–23.

[34] Handique, K., et al., "Microfluidic Flow Control Using Selective Hydrophobic Patterning," *Proceedings of SPIE Conference on Micromachined Devices*, Austin, TX, Sep. 29, 1997, pp. 185–195.

[35] Handique, K., et al., "Nanoliter-Volume Discrete Drop Injection and Pumping in Microfabricated Analysis Systems," *Technical Digest of the IEEE Solid State Sensor and Actuator Workshop*, Hilton Head Island, SC, June 8–11, 1998, pp. 346–349.

[36] Glavina, P. G., "The I-STAT System: Biomedical Application of Microsensor Technology," *Proceedings of the International MEMS Workshop 2001 iMEMS*, Singapore, July 4–6, 2001, pp. 15–20.

[37] Hey, N., et al., "A New Device for Multifunctional Dosage of Liquids by a Free Jet," *Proceedings of MEMS'98, 11th IEEE International Workshop Micro Electromechanical System*, Heidelberg, Germany, Jan. 25–29, 1998, pp. 429–431.

[38] Szita, N., et al., "A Fast and Low-Volume Pipetor with Integrated Sensors for High Precision," *Proceedings of MEMS'00, 13th IEEE International Workshop Micro Electromechanical System*, Miyazaci, Japan, Jan. 23–27, 2000, pp. 409–413.

[39] de Heij, B., et al., "A Tuneable and Highly-Parallel Picolitre-Dispenser Based on Direct Liquid Displacement," *Sensors and Actuators A*, Vol. 103, 2003, pp. 88–92.

[40] Perçin, G., et al., "Controlled Ink-Jet Printing and Deposition of Organic Polymers and Solid Particles," *Applied Physics Letters*, Vol. 73, No. 16, 1998, pp. 2375–2377.

[41] Steinert, C. P., et al., "A Highly Parallel Picoliter Dispenser with an Integrated Novel Capillary Channel Structure," *Sensors and Actuators A*, Vol. 116, 2004, pp. 171–177.

[42] Gutmann, O., et al., "Impact of Medium Properties on Droplet Release in a Highly Parallel Nanoliter Dispenser," *Sensors and Actuators A*, Vol. 116, 2004, pp. 187–194.

[43] Koltay, P., et al., "The Dispensing Well Plate: A Novel Nanodispenser for the Multiparallel Delivery of Liquids (DWP Part I)," *Sensors and Actuators A*, Vol. 116, 2004, pp. 483–491.

[44] Shestopalov, I., Tice, J. D., and Ismagilov, R. F., "Multi-Step Synthesis of Nanoparticles Performed on Millisecond Time Scale in a Microfluidic Droplet-Based System," *Lab on a Chip*, Vol. 4, 2004, pp. 316–321.

[45] Puntambekar, A., et al., "A New Fixed-Volume Metering Microdispenser Module Based on sPROMs Technology," *Proceedings of Transducers '01, 11th International Conference on Solid-State Sensors and Actuators*, Munich, Germany, June 10–14, 2001, pp. 1240–1243.

[46] Aoyama, R., et al., "Novel Liquid Injection Method with Wedge-Shaped Microchannel on a PDMS Microchip System for Diagnostic Analysis," *Proceedings of Transducers '01, 11th International Conference on Solid-State Sensors and Actuators*, Munich, Germany, June 10–14, 2001, pp. 1232–1235.

[47] Yamada, M., and Seki, M., "Nanoliter-Sized Liquid Dispenser Array for Multiple Biochemical Analysis in Microfluidic Devices," *Analytical Chemistry*, Vol. 76, 2004, pp. 895–899.

# Chapter 12

## Microfluidics for Life Sciences and Chemistry: Microfilters and Microseparators

Microfluidics has emerged as a new approach for improving performance and functionality of biochemical and medical analysis. Miniaturization and new effects in microscale promise completely new system solutions in these fields. Dimension reduction results in faster processes and less reagent consumption. The small size also allows parallel processing, in which more compounds can be produced and analyzed. In addition, emerging needs in biotechnology and diagnostics require the development of supporting tools and products. This chapter discusses devices for separation application. The main applications of separation are filtration and sorting. Devices of the first application are considered here as microfilters, while the second types are called micro-separators. In contrast to microfilters as passive devices, microseparators can be designed as either passive or active components that require external actuation. This chapter discusses these two device types.

### 12.1 MICROFILTERS

Microfilters have two unique functions: filtration and collection. Microfilters separate particles for applications that require a clean fluid for further processes. Microfilters are used for washing out waste products of the precedent process. In packed-bed microreactors, filters are used for keeping beads in reaction chambers. In other applications, the filtered particles are of interest. Microfilters separate particles from the mean flow and send them to further processing stages.

Based on the size range of filtered particles, filtration can be categorized as conventional filtration, ultrafiltration, and reverse osmosis. Figure 12.1 depicts the typical sizes of filtered particles. Conventional filtration covers particle sizes above 10 microns. Ultrafiltration can handle macromolecules with molecular size ranging from 10 nanometers to 10 microns. Filtration type for molecules or ions smaller than 10 nanometers is called reverse osmosis.

Based on their fabrication technology and, consequently, their typical pore size, microfilters are categorized as membrane filters and gap filters. The basic structure of the first type is a thin membrane that contains filtering pores. Generally, such membranes are called semipermeable membranes, which allow only one component to pass or to move through with faster velocity, resulting in selective transport across the membrane. The pores can be etched through the membrane with a mask defined by photolithography. Because of the limited resolution of the lithography process, membrane filters have relatively large pores, on the order of microns.

The flux across a filter membrane can be driven by pressure, concentration gradient, or electrical potential. The transport through the membrane depends on the size of the pores. In general, the transport rate is proportional to the pore size, while the selectivity of filtration is inversely

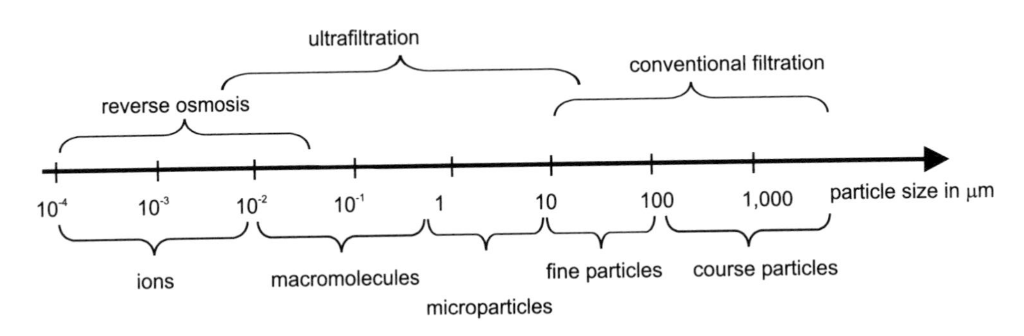


**Figure 12.1** Ranges of typical particle sizes and the corresponding filtration types.

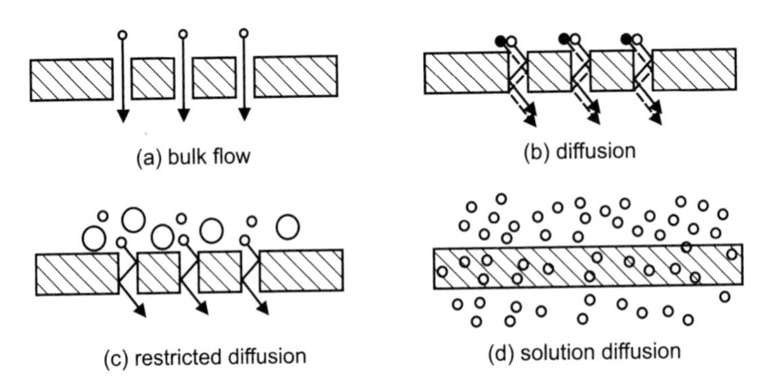


**Figure 12.2** Different transport mechanisms through membranes: (a) bulk flow through pores; (b) diffusion through pores; (c) restricted diffusion or molecular sieving; and (d) solution diffusion.

proportional to the pore size. Micromachined pores are relatively large, larger than the mean free part of the species. The transport through such large pores is governed by the bulk flow, where molecules can pass straight through the pores [Figure 12.2(a)]. Because of the small pore size, the flow can be assumed to be laminar. For circular pores with a diameter of $D_{\mathrm{pore}}$ and a pore length of $L_{\mathrm{pore}}$, the bulk velocity $U$ across the pore is according to Hagen-Poiseuille flow:

$$U = \frac{D_{\mathrm{pore}}^2}{32\eta L_{\mathrm{pore}}}\Delta p \tag{12.1}$$

where $\eta$ is the dynamic viscosity of the fluid, and $\Delta p$ is the pressure difference across the filter membrane. The void fraction $\varepsilon$, also called the porosity, is defined as the ratio between the pore area to the total membrane area $A$:

$$\varepsilon = \frac{A_{\mathrm{opening}}}{A} \tag{12.2}$$

The rate of mass transfer can then be derived as:

$$\dot{m} = u\varepsilon\rho \tag{12.3}$$

where $\rho$ is the density of the species.

If the pores are small compared to the mean free path but large enough for diffusion, the transport mechanism through pores is diffusion. Molecules are passing through the pore in a random manner [Figure 12.2(b)]. This random movement or Brownian motion is actually a diffusion process. Diffusion through pores was briefly discussed in Chapter 9 on microneedles. The rate of mass transport can be derived from Fick's law:

$$\dot{m} = \frac{D_{\text{eff}}}{L_{\text{pore}}} \Delta c \tag{12.4}$$

where $D_{\text{eff}}$ is the effective diffusion coefficient of the species, and $\Delta c$ is the concentration difference across the membrane. The effective diffusion coefficient is a function of the ordinary diffusion coefficient $D$, the porosity $\varepsilon$, and the tortuosity $\tau$:

$$D_{\text{eff}} = \varepsilon \frac{D}{\tau} \tag{12.5}$$

The tortuosity takes into account the longer distance traversed in the pores. Typical tortuosity ranges from 2 to 6.

Restricted diffusion occurs if the pores are large enough only for some species, so that only they can diffuse across the membrane. This transport mechanism is also called molecular sieving [Figure 12.2(c)]. With pore sizes on the order 0.1 to 1 μm, larger particles are restricted from entering the pores. This transport mechanism is common in ultrafiltration. If the pore size is on the order of molecular sizes, for instance, in most polymers, species diffuse through the bulk membrane [Figure 12.2(d)]. The transport mechanism is called solution diffusion or diffusion solubility. The transport rate is determined by:

$$\dot{m} = DH \frac{\Delta c}{L_{\text{membrane}}} \tag{12.6}$$

where $H$ is the partition coefficient or solubility. $D$ and $L$ are the diffusion coefficient and the membrane thickness, respectively. The product $DH$ is called the permeability. Separation for analytical applications often utilizes solution diffusion to distinguish different macromolecules. The next section on microseparators will discuss more about these applications.

Gap filters use the gap between two structural layers as the filtering pore. The gap can be fabricated with surface micromachining or bulk micromachining. In surface micromachining, for example, the gap height or the pore size is determined by the thickness of silicon oxide as a sacrificial layer, which can be controlled precisely by thermal oxidation. In bulk micromachining, controlled etching processes determine the gap. A combination of both micromachining techniques can be used for greater design flexibility.

### 12.1.1 Design Considerations

Microfilters have different requirements:

- Small variation of pore size distribution;
- Low pressure loss;
- Mechanical strength.

In contrast to industry applications, many biomedical applications require an extremely high selectivity ratio of filtration. While a ratio of $1{:}10^2$ is sufficient for industry applications, biomedical applications may need a ratio of $1{:}10^4$ or higher. This high selectivity only allows a small variation of pore size. With precise micromachining technology, such as lithography and controlled growth of sacrificial layers, microfilters could meet these requirements.

Since a microfilter represents an obstacle in the flow and causes pressure loss, a minimum pressure drop over the filter is required. The most important parameter of a microfilter, which has a big impact on pressure loss, is its porosity $\varepsilon$ (12.2). For a maximum porosity, the pores in the membrane should be made as large and as numerous as possible. Another solution for low-pressure losses is replacing the membrane design by a lateral design. The lateral design is a gap filter, which represents a short and wide channel with low fluidic resistance.

The requirement for large porosity conflicts with the desired robustness. Larger holes will decrease the mechanical strength of the filter. The influence of pore geometry on the pressure loss should therefore be considered. Robust filters are realized by a gap design, which is made in bulk silicon and can withstand high inlet pressure. Gap filters can therefore meet both requirements of low-pressure loss and high mechanical strength.

The following examples analyze the effect of pressure drop on the mechanical strength of a membrane filter, as well as the design of gap filters.

**Example 12.1: Stress on Membrane Filter**

Determine the maximum stress on a silicon nitride membrane filter. The membrane has a dimension of $1 \times 1 \times 1$ mm. The Young's modulus for silicon nitride is $3 \times 10^{11}$ Pa, and its Poisson's ratio is 0.25. The porosity is 0.25. The working fluid is air at a flow rate of 100 mL/min. The density and the viscosity of air at room temperature are 1 kg/m$^3$ and $1.82 \times 10^{-5}$ Pa-sec, respectively. Would the filter work if the fluid were water?

**Solution.** The total opening area of the filter is:

$$A_{\text{opening}} = \varepsilon A = 0.25 \times (10^{-3})^2 = 0.25 \times 10^{-6} \text{ m}^2$$

The drag pressure can be estimated as (assuming a flow number $\mu_{\text{flow}} = 0.75$) [1]:

$$\Delta p = \frac{1}{2\mu_{\text{flow}}^2} \frac{\rho \dot{Q}^2}{A_{\text{opening}}^2}$$

$$\Delta p = \frac{1}{2 \times 0.75^2} \frac{1(100/60 \times 10^{-6})^2}{(0.25 \times 10^{-6})^2} = 39.5 \text{ Pa}$$

The maximum stress at the edge of the filter membrane is estimated as (with a Poisson's ratio of 0.25) [1]:

$$\sigma_{\text{total}} = 1.44\sqrt{3}\frac{\Delta p^2 l^2 E}{h^2}$$

$$\sigma_{\text{total}} = 1.44\sqrt{3}\frac{39.5^2 \times (10^{-3})^2 \times 3 \times 10^{11}}{(10^{-6})^2} = 11.2 \text{ MPa}$$

The maximum stress is much smaller than the fracture limit of silicon nitride, which is on the order of $10^4$ MPa. If the fluid is water, the viscosity is 50 times larger, and the density is 1,000 times larger. The drag pressure is more significant, and the total stress will increase with a factor of about 100, which is still under the fracture limit. The filter will work for water at the same flow rate.

**Example 12.2: Adjusting Gap Size in the Gap Filter**

Thermal oxidation is used for adjusting the gap size of a gap filter. Gaps of 2 μm are etched in silicon using DRIE. After thermal oxidation, the oxide thickness is 500 nm. Determine the final gap size of the filter.

**Solution.** Since the thermal oxidation consumes some of the silicon thickness, only 54% of the final oxide thickness is the net increase of the gap surface (see Section 3.1). The thickness gain in each side of the gap is:

$$T_{\text{net}} = 0.54 \times 0.5 = 0.27 \ \mu\text{m}$$

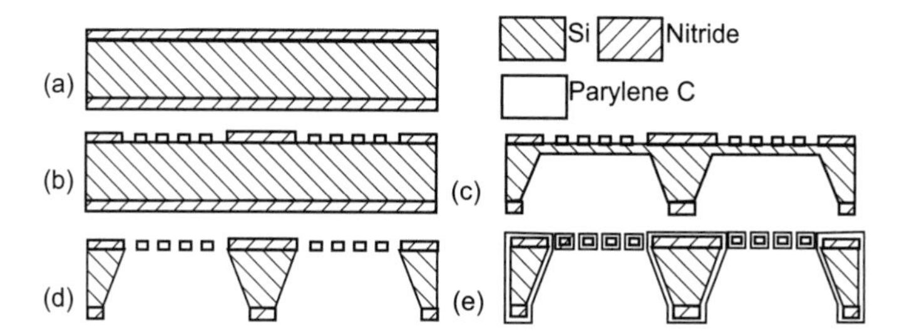


**Figure 12.3** Design examples of microfilters: (a) deposition of low-stress silicon nitride; (b) RIE of the silicon nitride layer; (c) wet etching of the back of the wafer; (d) releasing etch of the membrane; and (e) coating with parylene.

Thus, the final gap size is:

$$D_{\text{final}} = D_{\text{init}} - 2T_{\text{net}} = 2 - 2 \times 0.27 = 1.46\ \mu\text{m}$$

### 12.1.2 Design Examples

#### 12.1.2.1 Membrane Filters

Figure 12.3 illustrates the most common process steps for fabricating membrane microfilters. The material is low-stress silicon nitride. Because of the strength requirement, relatively thick nitride layers (about 1 μm) [2, 3] should be deposited. The low stress is achieved with LPCVD of $SiH_2Cl_2$ and $NH_3$ at 850°C [Figure 12.3(a)]. To have a silicon-rich nitride, the gas flow ratios between $SiH_2Cl_2$ and $NH_3$ are 70:18 [2] or 4:1 [4]. In the next step, silicon nitride is etched by $CHF_3/O_2$ RIE [Figure 12.3(b)]. The membrane is freed by KOH wet-etching from the backside. A thin silicon membrane can be left before the dicing process to avoid damaging the nitride membrane [Figure 12.3(c)] [4]. After dicing, the nitride layer is released with a final etch [Figure 12.3(d)]. A conformal deposition of parylene C can improve the strength of the filter membrane [Figure 12.3(e)] [4]. In addition, varying the thickness of the parylene layer can adjust the pore size.

The relatively thick nitride layer allows membrane sizes from 1 × 1 mm [2, 3] to 8 × 8 mm [4]. The pore sizes range from 4 to 10 μm. In order to go beyond the resolution of lithography, which is limited by the exposure wavelength, a maskless method with interference pattern of laser light can be used for making pore sizes ranging from 65 to 500 nm [5]. However, this method makes the pore distribution and the gaps between them depend on the wavelength of the laser. Figure 12.4 shows typical pore designs that can be formed by photolithography [5]. Circular pores are more robust because of the low stress concentration around them. Hexagonal and rectangular pores can result in higher stress concentrations at the sharp corners.

The membrane filter described in [6] has an integrated acoustic micropump (FPW pump; see Section 7.2.1.5). The device can actively transport particles to the separation area (Figure 12.5).

#### 12.1.2.2 Gap Filters

Gap filters have the advantage of small and precise pore size. Depending on the fabrication method, a gap filter can be categorized as a subtractive gap filter or an additive gap filter.

*Subtractive Gap Filters.* In subtractive gap filters, the etched depth in the substrate defines the filtering gap. The depth depends on etch rates and etching time. The drawback of this filter type is the low reproducibility and the high-pressure loss caused by the filtering channel. Figure 12.6(a) shows the typical subtractive gap filter [7].

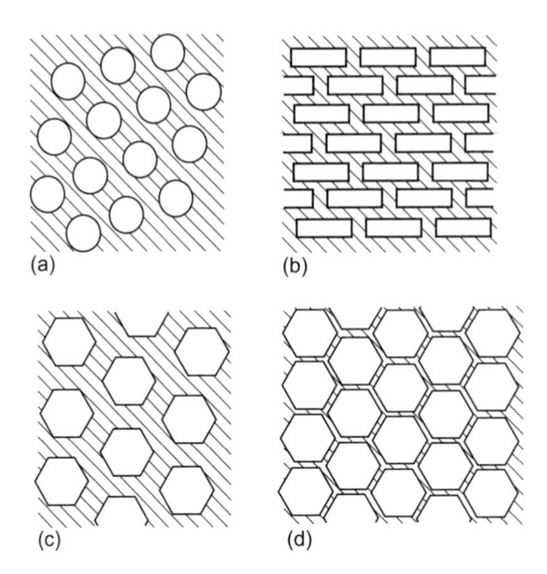


**Figure 12.4** Design of filter pores: (a) circular; (b) rectangular; and (c) and (d) hexagonal. (*After*: [5].)

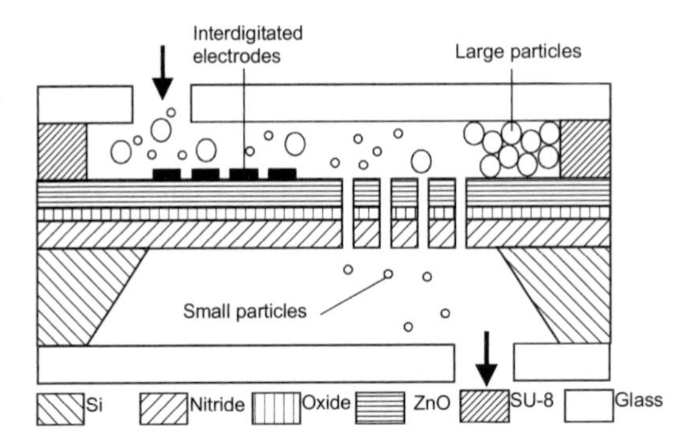


**Figure 12.5** Active microfilter with acoustic cleaning (not to scale). (*After*: [6].)

*Additive Gap Filters.* Additive gap filters have a defining spacer layer. Theoretically, this layer can be any material which can be deposited on the substrate with a controllable growth rate. Thermal oxide is often used as spacer material because of its slow growth rate, which is controlled precisely by time (see Section 3.1). Based on their arrangement between the gap and the substrate surface, additive gap filters are divided into vertical and horizontal types.

*Vertical Filters.* Figure 12.6(b) illustrates the fabrication steps of a vertical gap filter. The vertical gap filter has the same form as the membrane filter described above. The membrane is made of 6-μm highly boron-doped silicon, which is not attacked during wet etching. After a diffusion of boron, large pores are etched into silicon by DRIE [Figure 12.6(b), part 1]. The oxide layer, which defines the actual pore size, is grown by dry thermal oxidation [Figure 12.6(b), part 2]. A layer thickness from 20 to 100 nm corresponds to the pore size on the membrane after sacrificial etching [Figure 12.6(b), part 2]. The large pore is then filled with polysilicon [Figure 12.6(b), part 3]. After opening an etch access on top of the polysilicon layer, the gap previously filled with silicon oxide is freed and works as filtering pores [Figure 12.6(b), part 4] [8].

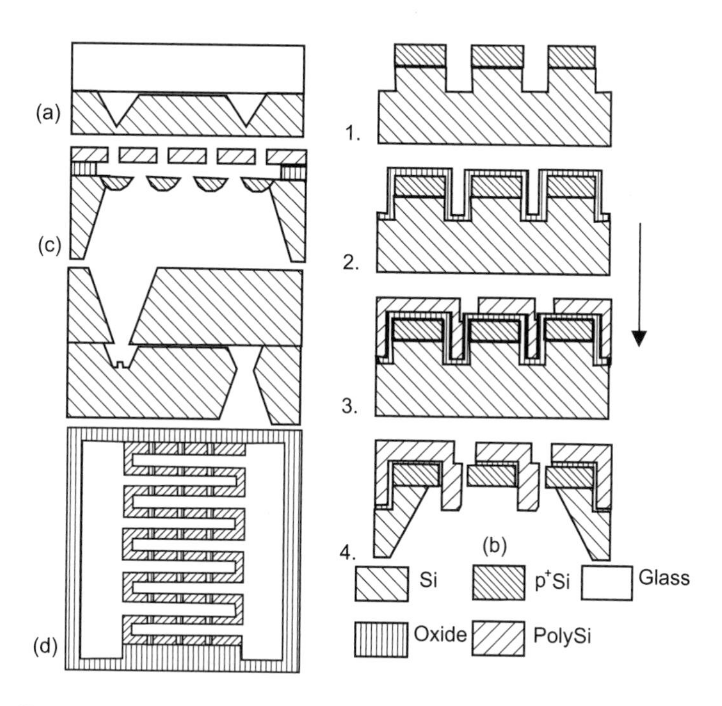


**Figure 12.6** Gap filters: (a) subtractive type; (b) and (c) additive type (vertical); and (d) additive type (horizontal).

**Table 12.1**
Typical Parameters of Microfilters

| *Refs.* | *Pore Size (mm)* | *Maximum Flow Rate (mL/min/cm$^2$)* | *Maximum Pressure (Pa)* | *Technology* |
|---|---|---|---|---|
| [2, 3] | 4 | 1,000 water | $3\times10^5$ | Bulk |
| [5] | 3–20 | 90,000 air | $3\times10^3$ | Bulk |
| [9] | 0.01–0.05 | — | — | Bulk, surface |
| [10] | 0.04 | 1,800 nitrogen | 3.105 | Bulk, surface |

*Horizontal Filters.* Figure 12.6(c) shows another membrane gap filter. The gap is arranged parallel to the substrate surface [9]. A highly boron-doped silicon layer defines the filter membrane. The fabrication process of this filter is similar to the process depicted in Figure 12.6(b).

A robust design is depicted in Figure 12.6(d) [10]. The filtering gaps are placed parallel to the wafer surface. Using bulk micromachining and direct wafer bonding, the filter can withstand several hundred kilopascals of pressure. The filtering gap is designed as a meander structure to minimize the pressure loss across the filter. The meander shape offers a maximum channel width at a constant channel length while occupying a minimum surface area. Table 12.1 lists the most important parameters of the above-mentioned microfilters.

## 12.2 MICROSEPARATORS FOR CELL AND PARTICLE SORTING

Sorters are divided into active and passive, Figure 12.7. Active sorters employ an external source of energy to separate desired cells and particles. The separation of desired cells/particles can be performed in label-based or label-free formats. In label-based methods, the desired cells for

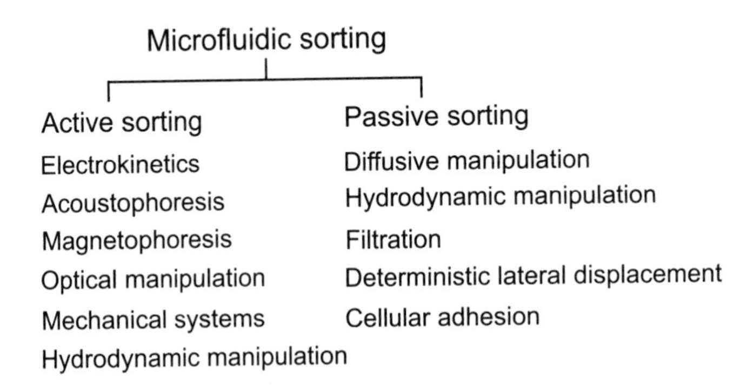


**Figure 12.7** Schemes of microfluidic cell and particle sorting.

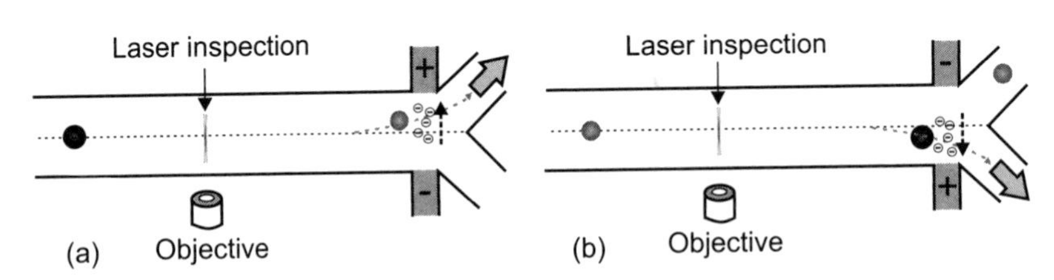


**Figure 12.8** Schematics of sorting using direct current (DC) electro-osmotic flow as a method of flouroscent-based sorting: (a) particle fractionation at Time 0; and (b) particle fractionation at Time 1. (*After:* [11].)

separation are initially labeled with a marker such as fluorophores or beads. The markers are attached to the cell surface proteins. In these cases, fluorescent label-based sorting and bead-based sorting approaches have been achieved. The label-free sorting is another sorting approach that relies on difference in physical properties of cell such as size, elasticity, magnetic susceptibility, shape, and polarizability [11].

The operation of fluorescent label-based sorting has been established upon the identification of cell types using the fluorescent probes. Identification is performed by a laser inspection system. Initially, cells are labeled with a fluorescent marker. The laser beam excites the fluorescent label, which emits light at a longer wavelength. The signal is detected with a epi-fluorescent filter in the microscope to recognize the cell type. Then the separation of the desired cell type(s) from the solution carrying the cells occurs through the use of an active method including electrophoresis, dielectrophoresis, electro-osmosis, acoustophoresis, optical manipulations, and rotary and gating valves. Figure 12.8 shows an example of active sorting using fluorescent detection and electrokinetic actuation. After cell identification through laser inspection, electrically induced flow drags cells to the desired outlet.

In bead-based cell sorting, beads of a particular size, material, and surface binding affinity are bound to desired cells [11]. Then the bound complexes (cell beads) are separated with the aid of an external field. The forces induced to the bound cells is different from unbound cells. Magnetotophoresis, acoustophoresis, and electrokinetic mechanisms can be employed for separation of the bound cells. Figure 12.9 shows the separation of circulating tumor cells (CTCs) from white blood cells (WBC) using an external magnetic field [12]. The CTCs were initially conjugated with magnetic microbeads. Once exposed to the magnetic field after inertial focusing in the curved channel, the bound CTCs are deflected from the main stream and directed to the collecting outlet through magnetophoresis.

Label-free sorting methods do not require any label to identify cell type. Thus, such methods requires a simpler sample preparation process, making them very attractive for microfluidic sorting

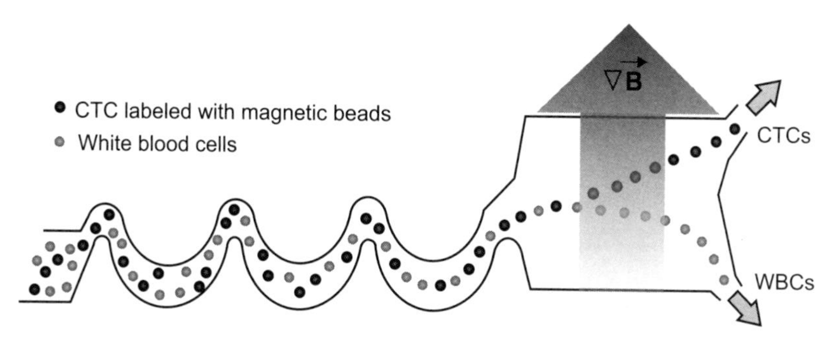


**Figure 12.9** Schematics of sorting using magnetophoresis as a method of bead-based sorting. (*After:* [12].)

purposes [11]. The separation relies on the difference in inherent properties of target cells such as size or deformability once exposed to an external energy source such as acoustics, electrical and magnetic fields, and optical manipulations. Importantly, passive approaches without any need to apply an external energy source have been realized for label-free sorting. Diffusion, inertial forces, hydrodynamic spreading, filtration, deterministic lateral displacement, and cellular adhesion have been widely studied for passive label-free cell sorting.

In the following sections, major active and passive mechanisms developed for sorting purposes are explained.

### 12.2.1 Active Sorting

#### 12.2.1.1 Dielectrophoretic Sorting

Cell trapping and sorting can be realized based on a number of electric concepts: electrophoresis of charged particles in a dc electric field, DEP of polarizable particles in a nonuniform ac field, combination of electrophoresis, and dielectrophoresis. Electrophoresis for separation and analysis will be discussed in Section 12.4. However, most biological cells have similar electrophoretic mobilities. Thus, electrophoresis in a dc electric field cannot be used for cell sorting. DEP is the best candidate for cell sorting using an electric field. If the particle is moving toward regions with higher field strength, the effect is called positive dielectrophoresis. Negative dielectrophoresis is the effect of particles moving to regions with lower field strength; particles are repelled from the electrodes. Negative DEP is more relevant for cell sorting and switching because of the gentle conditions.

The complex electric field around an electrode can be described as [13]:

$$\mathbf{E}(\mathbf{r},t) = \Re\left[\mathbf{E}^{\text{re}}(\mathbf{r}) + \imath\mathbf{E}^{\text{im}}(\mathbf{r})\exp\left(\imath\omega t\right)\right] \tag{12.7}$$

where $\imath$ and $\Re$ represent the imaginary unit and the real component of the complex number, respectively. **E**, **r**, and $\omega$ are the field vector, the space vector, and the radial frequency, respectively. The time-averaged dielectrophoretic force acting on a particle of radius $R$ is:

$$\langle\mathbf{F}\rangle = 2\pi\varepsilon_1\varepsilon_0 R^3 \times \left\{\begin{array}{l} \Re(f_{\text{CM}})\nabla\sum_i\left[(E_i^{\text{re}})^2 + (E_i^{\text{im}})^2\right] + \\ \Im(f_{\text{CM}})\sum_{i,j}\left[\left(E_i^{\text{re}}\frac{\partial E_i^{\text{im}}}{\partial j} - E_j^{\text{im}}\frac{\partial E_i^{\text{re}}}{\partial j}\right)\mathbf{e}_j\right] \end{array}\right\} \tag{12.8}$$

where $i,j = x,y,z$ are indices for the Cartesian coordinates. $\Re$ and $\Im$ indicate the real and imaginary parts of a complex number. The Clausius-Mosotti factor $f_{\text{CM}}$ is determined as:

$$f_{\text{CM}} = \frac{\tilde{\varepsilon_{\text{p}}} - \tilde{\varepsilon_1}}{\tilde{\varepsilon_{\text{p}}} + 2\tilde{\varepsilon_1}} \tag{12.9}$$

where $\tilde{\varepsilon} = \sigma + \imath\omega\varepsilon_0\varepsilon$ is the complex permittivity, $\sigma$ is the electric conductivity, and indices p and l denote the particle and the liquid. In (12.8), the first term is proportional to the rms value of the field strength, while the second term is proportional to the gradient of the field phase. For an ac field with no gradient in the phase, the condition for negative dielectrophoresis is:

$$\Re(f_{\mathrm{CM}}) < 0 \tag{12.10}$$

Thus, the value of the complex permittivity of the particle is lower than that of the liquid. Because of the high permittivity of water, most living cells show negative DEP.

For an idealized electrode, the maximum repelling force perpendicular to the boundary of a capacitor is calculated as [13]:

$$F_{\mathrm{DEP}} = \frac{27\pi^2 V_{\mathrm{rms}}^2}{32}\left(\frac{R}{a}\right)^3 \varepsilon_{\mathrm{l}}\varepsilon_0\Re(f_{\mathrm{CM}}) \tag{12.11}$$

where $R$, $a$, and $V_{\mathrm{rms}}$ are the particle radius, the distance to the electrode, and the rms voltage, respectively. In a continuous-flow system, the DEP force is on the order of the Stokes's drag force. Thus, DEP can be used for trapping and manipulating particles suspended in a hydrodynamic flow.

Fiedler et al. [13] used electrodes made of platinum/titanium and ITO on glass to trap and switch particles and cells. Square-wave signals with a frequency of 1 MHz were applied on the electrodes. The device was able to trap and switch particles suspended in a water flow up to 10 mm/s. Huang et al. [105] reported a 5 × 5 array of circular platinum electrodes. The electrodes are 80 μm in diameter and placed at a 200-μm distance. The device successfully concentrated *Escherichia coli* from a diluted solution, as well as white blood cells from whole blood.

#### 12.2.1.2 Ultrasonic Sorting

Nonlinear effects in an ultrasonic field cause the acoustic streaming. The effect of acoustic streaming was discussed in Section 7.1.1.8. The interesting characteristic caused by an FPW is the high-flow velocity near the actuating membrane. Since this fast-moving layer is only 5 μm [14, 15], which is in the same size order of cells, acoustic streaming on a FPW device can be used for sorting particles. Different particles suspended on the FPW membrane experience different drag forces depending on their size. A typical FPW membrane consists of a nitride membrane, interdigitated aluminum electrodes, piezoelectric zinc oxide, and an aluminium layer as a ground electrode [16]. Using curved electrodes, the acoustic energy and logically acoustic streaming can be focused. A focused FPW device can be used for particle concentration. Figure 12.10(a) shows the arrangement of the curved electrodes and the generated acoustic streaming field. The measurement of the wave amplitudes indicates clearly a focal point [Figure 12.10(b)].

Acoustic streaming can also be focused using the cavity effect. Marmottant and Hilgenfeldt [17] used a 15-μm microbubble to attract cells. The microbubble is driven near its own resonance with a frequency of 180 kHz. Acoustic streaming attracts cells to the microbubble and drives them in a defined trajectory.

Acoustic standing waves can separate particles with different acoustic impedance [18]. Harris et al. [19] used a cavity fabricated in silicon and glass to generate a standing wave. The acoustic wave is generated by a PZT layer deposited on silicon. The opposite glass plate works as the reflector. The standing wave generates a low-pressure node in the middle of the gap between silicon surface and reflector surface. Particles can be trapped in this low-pressure region.

#### 12.2.1.3 Magnetic Sorting

Magnetic force can also be used for sorting particles. In a continuous flow, particles can be deflected from their hydrodynamic trajectory by a magnetic field perpendicular to the flow. The magnitude of

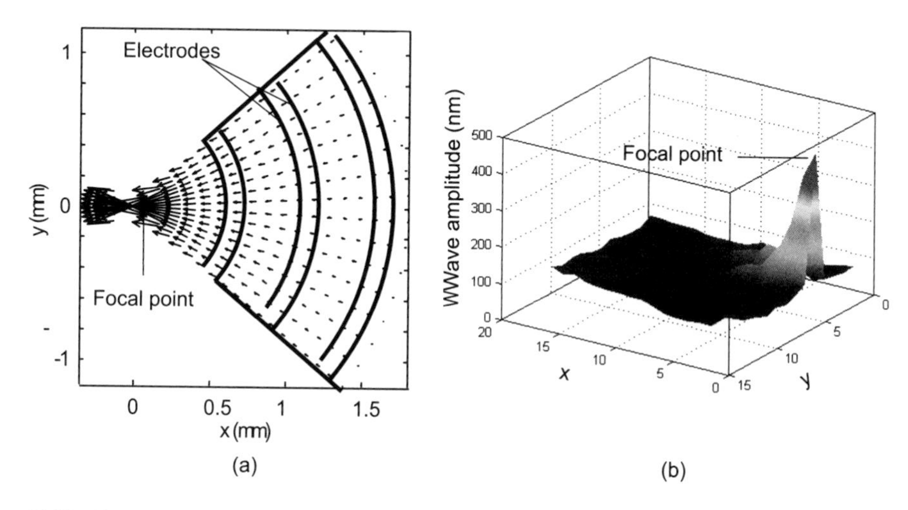


**Figure 12.10** Focused flexural plate wave device for particle concentration: (a) electrode arrangement and the typical acoustic streaming field; and (b) the measured wave amplitudes.

the magnetic force depends on the size and the magnetic property of the particles. Thus, magnetic particles can be separated from each other or from nonmagnetic materials. In a magnetic field, the magnetic force acting on a particle with a volume $V_p$ is [20]:

$$\mathbf{F} = \frac{\chi V_p}{\mu_0}(\nabla \cdot \mathbf{B})\mathbf{B} \tag{12.12}$$

where $\chi$ is the difference in susceptibility between the particle and the fluid, $V_p$ is the volume of the particle, $\mu_0 = 4\pi \times 10^{-7}$ Wb/A-m is the permeability of vacuum, and $\mathbf{B}$ (T) is the flux density field. The susceptibility is the ratio between the magnetization $\mathbf{M}$ (A/m) and the field strength $\mathbf{H}$ (A/m):

$$\chi = \frac{\mathbf{M}}{\mathbf{H}} \tag{12.13}$$

The magnetization of the particle is calculated as:

$$\mathbf{M} = \frac{\mathbf{m}}{V_p} \tag{12.14}$$

where $\mathbf{m}$ is the magnetic moment. The general balance equation of a magnetic system is:

$$\mathbf{B} = \mu_0(\mathbf{H} + \mathbf{M}) \tag{12.15}$$

From (12.12), the magnetic field should be inhomogeneous ($\nabla \cdot \mathbf{B} \neq 0$), so that the magnetic particle can experience a force.

Pamme and Manz [21] demonstrated a simple microfluidic device made of glass. Separation of magnetic particles suspended in the continuous flow was achieved with a permanent magnet. The magnetic sorting concept can be applied for analysis based on magnetic beads or separation of magnetically labeled cells [22].

#### 12.2.1.4 Optical Sorting

A dielectric particle can be trapped in a tightly focused beam. This concept is well known and called optical tweezers [23]. MacDonald et al. demonstrated an optical sorter for microparticles with a 3-D

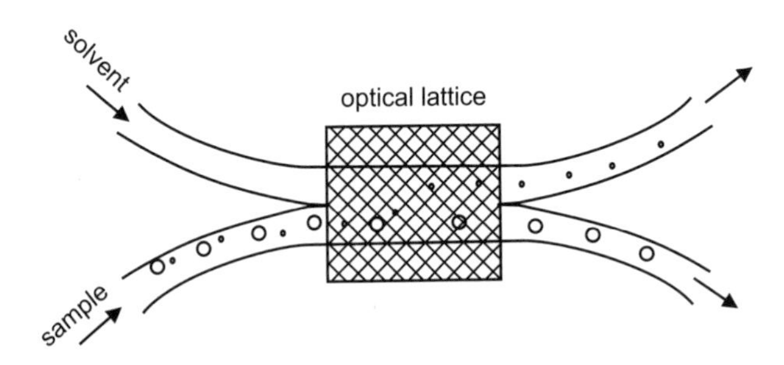


**Figure 12.11** Operation concept of an optical sorter.

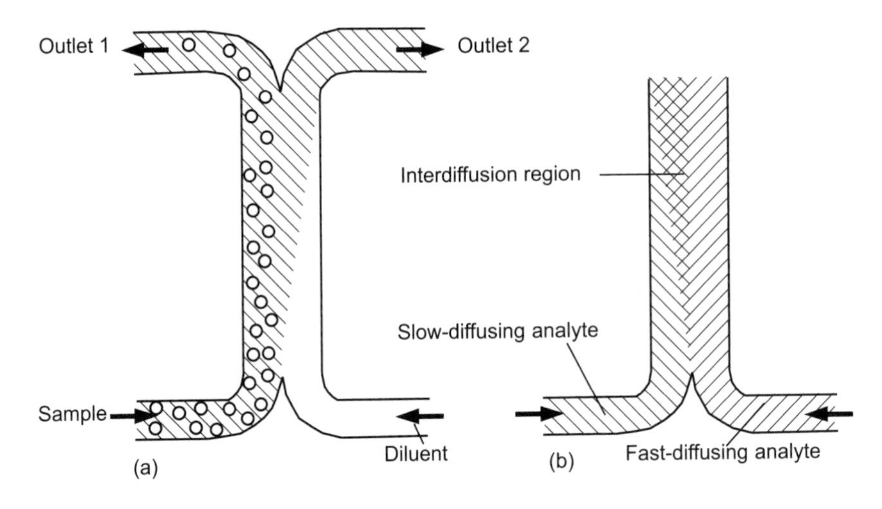


**Figure 12.12** Microseparator based on diffusion: (a) microseparator; and (b) T-sensor.

optical lattice. The optical lattices are created by interferometric patterns of light. Dielectric particles follow defined paths through an optical lattice and can be sorted from a continuous flow (Figure 12.11). The device has the form of the H-filter depicted in Figure 12.12(a). The basic H-configuration can only separate small molecules, because of the required high-diffusion coefficient. Large subjects such as cells cannot be separated. In the optical sorter, an optical lattice is generated by a five-beam interference pattern and positioned at the separation microchannel (Figure 12.11). When the sample flow with different cells passes through the optical lattice, selected cells are deflected to the solvent side, while the others can pass straight. Since the optical force caused by the effect is smaller than the dielectrophoretic force described above, the hydrodynamic Stokes's drag force should be kept small. For instance, 2-μm protein microcapsules were sorted by a total laser power of 530 mW ($\lambda = 1,070$ nm) at a flow speed of only 20 μm/s [24]. Based on this concept, integrating a number of optically assembled, driven, and controlled components on a single microfluidic chip can make an all-optical lab-on-a-microscope possible [25].

### 12.2.1.5 Hydrodynamic Sorting

Active microseparators use feedback control with external components such as valves and pumps to selectively sort particles in a continuous-flow system. The separation concept consists of the three basic steps: focusing, detecting, and switching (Figure 12.13). First, the sample with particles such as cells is focused into a narrow stream, so that each cell can be individually detected. The concept of hydrodynamic focusing was already discussed analytically in Section 10.1.5. Two sheath streams are

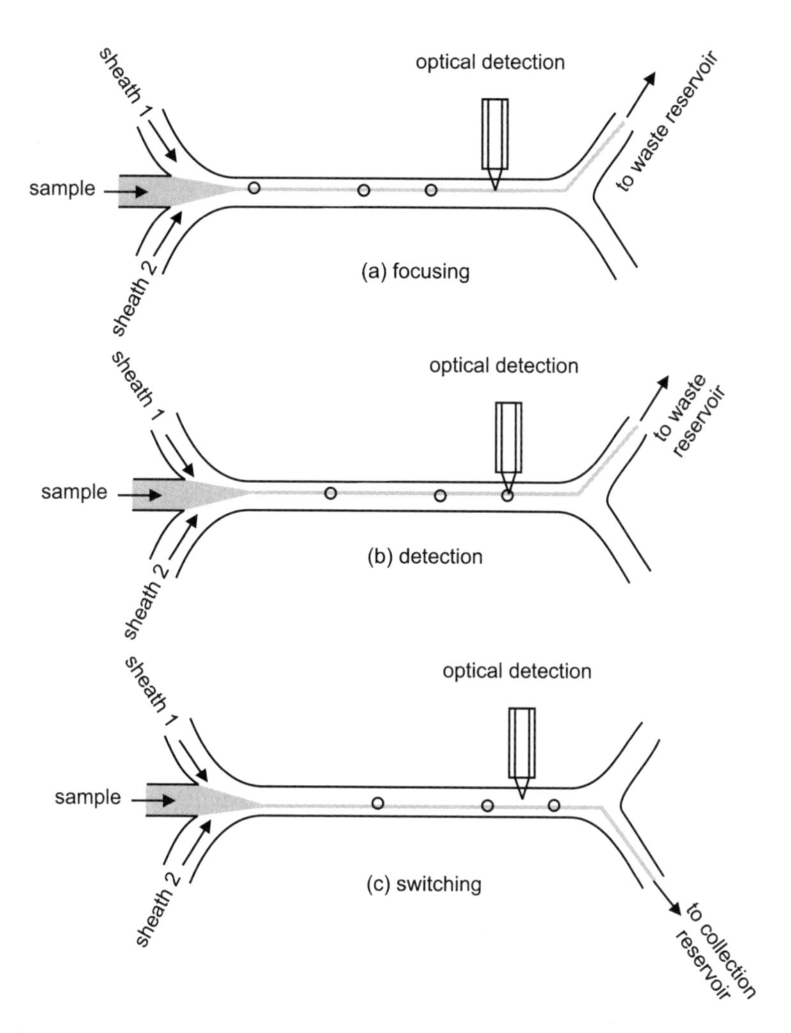


**Figure 12.13** Operation concept of a particle sorter: (a) hydrodynamic focusing; (b) detecting the particle; and (c) switching the focused sample flow to collection reservoir.

used to narrow the sample stream [Figure 12.13(a)]. After being focused, the particles are detected with an optical system, which gives the feedback signal to trigger the flow switch [Figure 12.13(b)]. The switching steps can be followed hydrodynamically by changing the flow rates of the sheath flow and diverting the sample flow into the collecting reservoir [Figure 12.13(c)]. The sample flow can also be switched using a valve, which just blocks or opens the gate to the collecting reservoir.

Krüger et al. developed a microfluidic device for activated cell sorting [26]. The microchannels are formed by SU-8 on a silicon substrate. The channel structure is covered by a glass plate. The switching step was realized by both changing the sheath fluid ratio and external valves. Wolff et al. [27] developed a cell-sorting device in silicon. The microchannels are etched with DRIE and covered by anodic bonding to a glass wafer. Switching was realized by an external valve. This device has integrated waveguides for excitation and detection of the cells. Lin and Lee [28] presented a similar device in glass. The waveguides are glass fibers embedded in the device. Detection and counting were demonstrated, but no switching was reported. Switching by adjusting the sheath flow ratio was demonstrated by Lee et al. [29]. The microchannel was formed by hot embossing of a PMMA substrate.

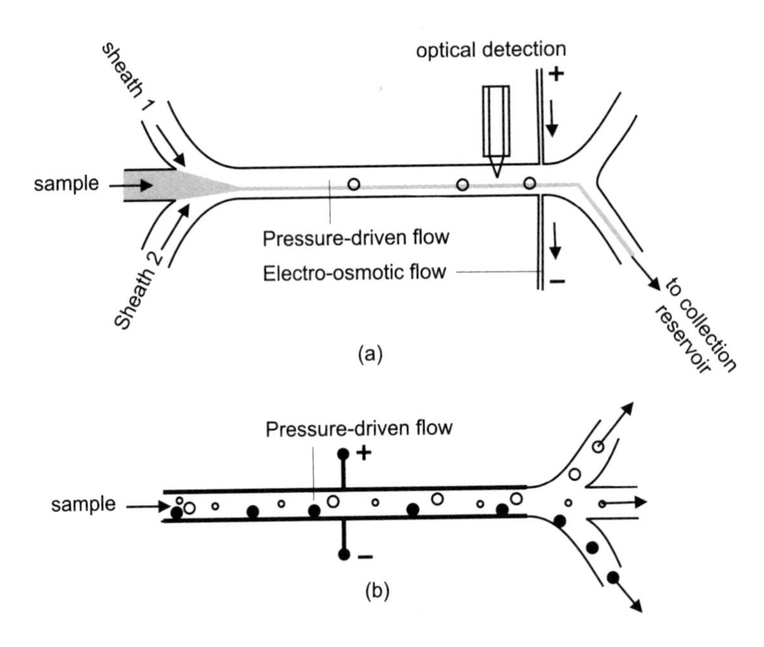


**Figure 12.14** Sorting with electrokinetic forces: (a) electro-osmotic switching; and (b) electrophoretic switching.

Switching can also be achieved by manipulating the exit condition. Dittrich and Schwille [30] used electro-osmotic flow at the exit of a pressure-driven hydrodynamic focusing device to collect the sorted particles [Figure 12.14(a)]. Lu et al. utilized electrophoretic forces to separate particles and divert them into different exit channels [31]. After cell lysing, subcellular organelles were introduced into a flow channel by pressure. Electrodes are positioned on both sides of the channel. The applied electric field separates the particles due to their different charges [Figure 12.14(b)]. This free-flow separation concept is discussed later in Section 12.4.

### 12.2.2 Passive Sorting

#### 12.2.2.1 Diffusive Sorting

In a low Reynolds number regime, a diluent stream such as water is forced into a channel with a sample stream. The same situation occurs in the Y-mixer, if the sample stream contains different species, that have different diffusion coefficients. Smaller species usually have higher diffusion coefficients and quickly diffuse into the diluent stream. Kamholz et al. implemented this concept in a microfluidic device [32]. The channels are etched in silicon and are 10 μm deep. The device is tested by filtering fluorescent dye carboxyfluorescein from 0.6-μm-diameter fluorescent polystyrene spheres. Both species are mixed in one fluid stream. The first one is a relatively small molecule, and thus has a higher diffusion coefficient. Small molecules are extracted from the large particles and sent to the outlet. The configuration of a diffusion filter shown in Figure 12.12(a) is also called the H-filter. The same principle is used for monitoring quickly diffusing molecules. This T-sensor can measure sample concentrations in a continuous-flow system [Figure 12.12(b)] [32].

#### 12.2.2.2 Hydrodynamic Sorting

Hydrodynamic label-free cell sorting relies on forces induced to the particles by the flow. Inertial focusing in curved channel, pinched flow fractionation (PFF), and hydrophoretic cell focusing are major approaches for hydrodynamic particles manipulation.

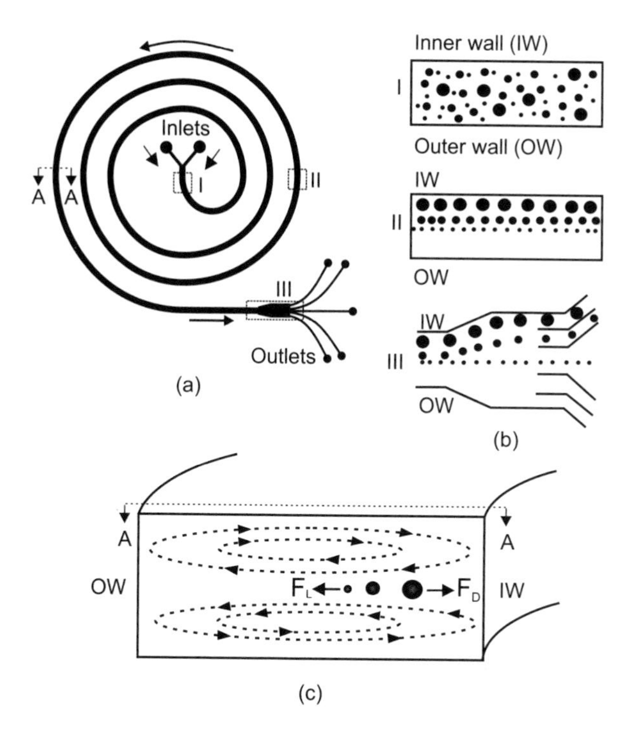


**Figure 12.15** Inertial microfluidics in spiral microchannel: (a) schematics of spiral microfluidics for particle separation; (b) particle fractionation at different sections of the channel; and (c) illustration of forces acting on particles. (*After:* [33].)

Inertial forces in curved microchannels have been used to separate two or more types of particles or cells based on their size (Figure 12.15). In a curvilinear microchannel, a combination of Dean drag, $F_D$, and inertial lift forces, $F_L$, exists that push the particles/cells to the inner wall of the microchannel [33]. The equilibrium position of the particles are determined by the ratio of these forces. Inertial microfluidics in a spiral microchannel has enabled high-throughput separation of rare cells and in particular CTCs from blood cells [34, 35].

In the PFF method, a stream of cell mixture and a cell-free stream are introduced into a Y-shape channel [36]. At the pinched segment, cells are pushed to the vicinity of the sidewall by controlling the flow in the microchhanel (Figure 12.16). Upon entering the cells from the pinch segment to the expanded channel, the smaller cells are more deflected to the sidewall, while the larger particles experience a force directing them to the center of the channel. Thus, two lines of separated particles are formed.

As discussed in detail in Section 1.1, fractionation of particles using isoporous microfabricated membranes and filters [37] is one the earliest methods for size-based separation. Various designs of filter have been explored including Weir filter, pillar filter, and cross-flow filter. The filtration method has been widely used for separation of blood particles. Cross-flow filters were mainly developed to decrease the possibility of clogging (Figure 12.17).

In deterministic lateral displacement (DLD) method [38], particles larger than a critical size, $d_c$, are determined to follow a pathway through an arrays of microposts implemented in the microchannel (Figure12.18). The smaller particles below the critical diameter can follow the streamlines through the micropost arrays. The critical diameter can be considered as 1/3 or 1/4 of the gap between the two pillars ($w$) [39]. In this case, multiple streams of particles based on their size is formed within the channel during sorting process.

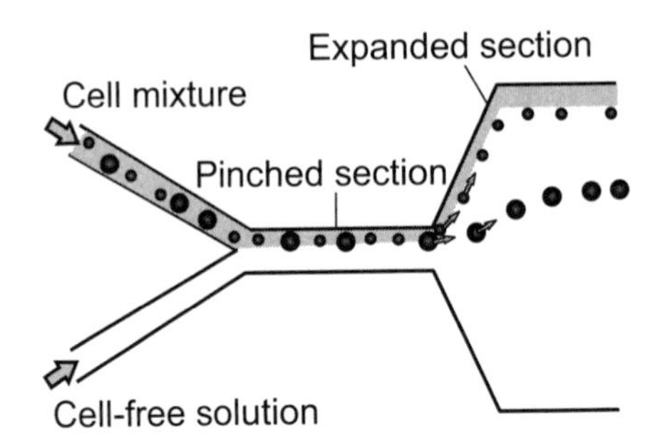


**Figure 12.16** Schematics of PFF in a pinched-expanded flow channel. (*After:* [36].)

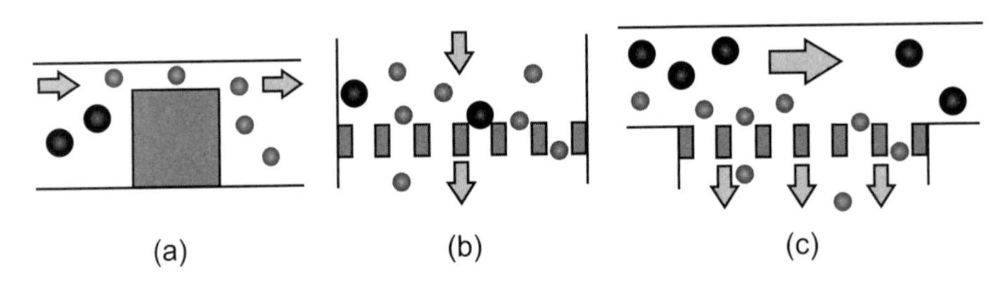


**Figure 12.17** Filtration in microfluidics: (a) schematics of a Weir filter; (b) pillar filter; and (c) cross-flow filter. (*After:* [11].)

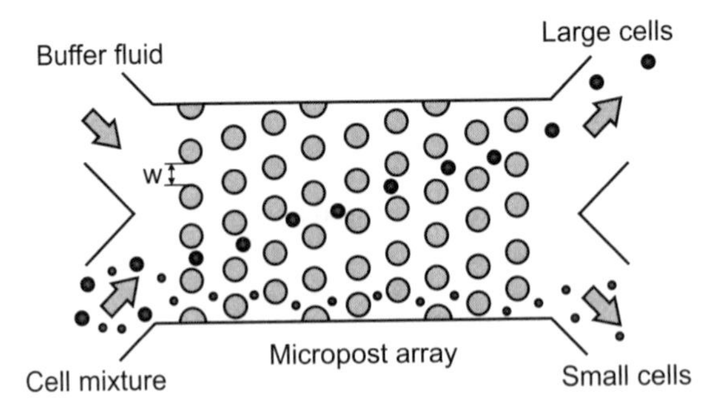


**Figure 12.18** Deterministic lateral displacement (DLD) sorting using micropost arrays. (*After:* [11].)

Sorting based on cellular adhesion mainly relies on binding cells to the molecules bound on the surface of a microchannel. In the other words, the sorting method is based on the interactions occur between cells and a functionalized channel surface. The adhesion can be transient or nontransient. Some non-transient methods, known as cellular immobilization, employed antibodies and aptamers for cell separation from blood. Investigations on the development of microfluidic systems for cell-specific isolation of CTCs and other rare cells from blood have revealed promising results [40, 41].

Hydrodynamic passive cell sorting methods provide unique features for the development of point-of-care microfluidic systems to isolate target cells from bodily fluids using liquid biopsy. Such systems enable the processing of biosamples with minimal preparation effort by unskilled users. Importantly, the isolation/separation processes can be very fast with high throughput in a reliable and cost-effective fashion for widespread screening applications and telemedicine.

The following sections focus on microseparators for analytical applications. The basic concept of these microseparators is based on the different flow velocities of sample molecules in a microchannel.

## 12.3 CHROMATOGRAPHY

### 12.3.1 Design Considerations

In microfluidic systems, microseparators are extremely useful tools for injection analysis. This section focuses on microseparators for analysis applications. Based on the different separation concepts, the devices are categorized as gas chromatography, liquid chromatography, and electrophoresis. Electrophoresis is further divided into capillary electrophoresis, gel electrophoresis, and free-flow electrophoresis.

Chromatography is a separation technique, which is based on the affinity differences between a stationary phase and the components of a mobile phase [42]. The stationary phase, which can be solid or liquid, interacts with the components in the mobile phase and causes their different flow velocities. The mobile phase consists of the sample to be investigated and a carrier medium. If the carrier medium is a gas, the technique is called gas chromatography. If it is a buffer solution, the separation method is called liquid chromatography. The results of chromatography are presented in a chromatogram, which is the time diagram of the concentration of different species in the sample. Each peak in the chromatograph represents a component. The surface area under the peak represents the amount of this component.

*Analysis Time.* The analysis time or the separation time is defined as the time that the slowest component needs to reach the sensor at the end of the separation channel:

$$t = L_{\text{channel}}/U_{\text{min}} \tag{12.16}$$

where $L_{\text{channel}}$ is the channel length and $U_{\text{min}}$ is the speed of the slowest component.

*Separation Speed.* The number of theoretical plates $N$ characterizes the separation efficiency:

$$N = L^2_{\text{channel}}/\sigma^2 \tag{12.17}$$

where $\sigma^2$ is the variance contribution of longitudinal diffusion. This definition indicates that the smaller the signal peak is compared to channel length, the higher is the number of plates and the better the separation. The variance contribution of longitudinal diffusion $\sigma^2$ represents the broadening of the peaks. The variance contribution $\sigma^2$ should be kept as small as possible. Since the broadening caused by the separation channel is fixed, the broadening caused by dead volumes should be minimized. The design of the channel systems should avoid dead volumes.

The separation efficiency can also be characterized by the plate height $H$:

$$H = L_{\text{channel}}/N = \sigma^2/L_{\text{channel}} \tag{12.18}$$

In contrast to micromixers, microseparators should avoid diffusion to gain a better separation. Considering the Taylor-Aris dispersion with the initial pulse condition $c(0) = c_0\delta(x)$, the concentration distribution of a component along the flow axis $x$ after a time $t$ is:

$$c(x,t) = \frac{c_0}{\sqrt{4\pi D^*\tau}} \exp\left[-\frac{(x-Ut)^2}{4D^*t}\right] \tag{12.19}$$

where $D^*$ is the effective diffusion coefficient (see Section 10.1.4). The separation speed or the number of theoretical plates per unit time is inversely proportional to the lateral diffusion time. The relation between the separation time $t$ and the dimension of separation channels $D_{\text{channel}}$ is [43]:

$$N/t \propto 1/\tau \propto D^*/D^2_{\text{channel}} \tag{12.20}$$

*Baseline Resolution.* The resolution between two close peaks is defined as:

$$R = 2\Delta x/(w_1 + w_2) \tag{12.21}$$

where $\Delta x$ is the distance between the peaks, and $w_1$, and $w_2$ are their widths. The baseline resolution is reached if $R$ is equal to 1.5.

*Flow Velocity.* Considering the longitudinal diffusion, which causes the dispersion in the separation results, the optimum flow velocity $U$ in separation channels is proportional to the ratio of effective diffusion coefficient and the channel dimension [43]:

$$U \propto D^*/D_{\text{channel}} \tag{12.22}$$

Thus, miniaturization of separation channels requires a faster flow velocity, which in turn needs higher drive pressure. The problem is apparent in pressure-driven liquid chromatography.

*Self-Heating Effect.* In electrokinetic separation, the channel with the electrolyte represents a resistor, which causes self-heating with an electric current. Since the diffusion coefficient depends strongly on temperature, self-heating causes large dispersion in the separation results. Miniaturization increases the surface-to-volume ratio, which allows better heat conduction and minimizes the self-heating effect.

### 12.3.2 Gas Chromatography

A gas chromatograph consists of a flow controller, a sample injection port, a channel with solid phase, and a detector. The flow controller keeps the flow rate of the carrier gas constant. Thus, a steady-state condition for the analysis is warranted. The injection occurs at a valve, which allows the injection of a defined amount of the sample to be examined. Therefore, the injection port also has the function of a dispenser, as discussed in the previous section. The separation channel is actually a gas-phase microreactor with a catalyst support. In this case, the catalyst is the stationary phase, which can be coated on the channel wall or on packed beads. The reactor chamber remains at high

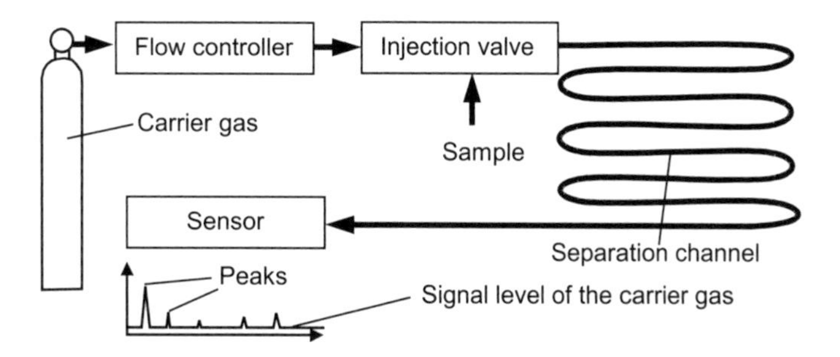


**Figure 12.19** Concept of a gas chromatograph.

temperatures, which keeps the sample in the gas phase. A detector is placed at the end of this system for measuring the time signal of passing components. The concept is depicted in Figure 12.19.

The first microfluidic gas chromatograph is reported in [42]. The device consists of a 1.5-m-long capillary channel, an electromagnetic valve, and a thermal conductivity sensor. The flow channel is 200 μm in width and 40 μm in depth. The channel structure is etched in silicon and covered by a glass plate. The microvalve uses an external solenoid actuator.

The chromatography process works as follows. First, the stationary phase is primed into the channel system. Second, the system is supplied with an inert carrier gas such as nitrogen or helium. The carrier gas flow rate is kept constant. Third, a small sample gas is injected into the channel by quick opening of the valve. After closing the valve, the sample amount is pushed through the long channel. Different molecules have different speeds and are separated at the channel end. A sensor at the channel end detects the change in thermal conductivity of the passing gas. Since different components in the sample have different conductivity values, the sensor signal represents each of the components in its peaks.

### Example 12.3: Characteristics of a Gas Chromatograph

The channel of a gas chromatograph is 1.5m long and has a cross section of 200 × 40 μm. Nitrogen is used as a carrier gas. The pressure drop across the channel is 200 kPa. How long will it take to get the first peak in the sensor signal after a sample injection? The viscosity of nitrogen at room temperature is $1.78 \times 10^{-6}$ Pa-sec.

**Solution.** The pressure drop across the channel is estimated as:

$$\Delta p = \eta L \mathrm{Re} f U/(2D^2)$$

The hydraulic diameter of the channel is calculated as:

$$D = 4\frac{200 \times 40}{2(200 + 40)} \times 10^{-6} = 66.67 \times 10^{-6} \text{ m}$$

Assuming the product Re$f$ is 55 for rectangular microchannels, the average velocity of the carrier gas is:

$$U = \frac{2\Delta p D^2}{\mu L \mathrm{Re} f} = \frac{2 \times 200 \times 10^3 \times (66.67 \times 10^{-6})^2}{1.78 \times 10^{-6} \times 1.5 \times 55} = 12.1 \text{ m/s}$$

The fastest component of the sample will reach the sensor after a time of:

$$t = L/U = 1.5/0.121 = 0.124 \text{ sec}$$

### 12.3.3 Liquid Chromatography

Liquid chromatography is similar to gas chromatography. The mobile phase in this case is a liquid. This separation method is categorized as liquid-liquid chromatography and liquid-solid chromatography. The first type is used for separation of organic compounds with silica gel holding water as stationary phase. The second type is in fact similar to a packed-bead reactor. The stationary phase is the surface of spheres, packed in the separation channel.

Manz et al. presented a micromachined liquid chromatograph [44]. The silicon chip is $5 \times 5$ mm in size and has a separation channel that is 15 cm long and has a cross section of $6 \times 2$ µm. The sensor at the end of the channel measures the electrical conductivity of the passing liquid using platinum electrodes. This work did not demonstrate the separation function.

The first separation experiments in a micromachined liquid chromatograph were reported by Ocvirk et al. [45]. The separation channel is a packed-bead chamber. The exit of the separation chamber is equipped with a 2-µm gap filter, which keeps the 5-µm beads in the chamber. Fluorescent detection is used for the evaluation of the separation results. Because of the higher viscosity of the liquid phase and the pressure drag at the beads, extremely high pressure up to 1 to 15 bars is needed for this liquid chromatograph. This high pressure can only be achieved with a powerful external pump and with special design considerations for the fluidic interconnects. The device can reach a maximum number of 200 theoretical plates.

**Example 12.4: Characteristics of a Liquid Chromatograph**

The channel of a liquid chromatograph is 20 mm long and has a cross section of $200 \times 100$ µm. Water is used as a carrier liquid. The flow rate of the carrier liquid is 800 nL/min. The separation chamber is packed with 5-µm particles, which take about 70% of the total volume of the chamber. Determine the pressure difference required for driving the liquid. How long will it take to get the first peak in the sensor signal after sample injection? Viscosity and density of water at room temperature are $10^{-3}$ Pa-sec and $10^3$ kg/m$^3$, respectively.

**Solution.** The superficial velocity in the separation chamber is:

$$U_{\mathrm{s}} = \frac{\dot{Q}}{wh} = \frac{800 \times 10^{-12}}{60 \times (200 \times 100 \times 10^{-12})} = 0.67 \times 10^{-3} \text{ m/s}$$

The void fraction in the separation chamber is (see Chapter 13)

$$\varepsilon = 1 - 0.7 = 0.3$$

According to (13.28), the Reynolds number in the separation channel is:

$$\mathrm{Re} = \frac{2d_0 U_{\mathrm{s}} \rho}{3\mu(1-\varepsilon)} = \frac{2 \times 5 \times 10^{-6} \times 0.67 \times 10^{-3} \times 1{,}000}{3 \times 10^{-3}(1-0.3)} = 3.2 \times 10^{-3}$$

At this low Reynolds number, we can use the Kozeny-Carman equation (13.28) to get the pressure drop required for the channel:

$$\begin{aligned}\Delta p &= 150\mu U_{\mathrm{s}}(1-\varepsilon)^2 L/(d_0^2\varepsilon^3) \\ &= 150 \times 10^{-3} \times 0.67 \times 10^{-3} \times (1-0.2)^2 \times 20 \times 10^{-3}/(5 \times 10^{-6})^2/0.2^3 \\ &= 14.5 \times 10^5 \text{ Pa} = 14.5 \text{ bar}\end{aligned}$$

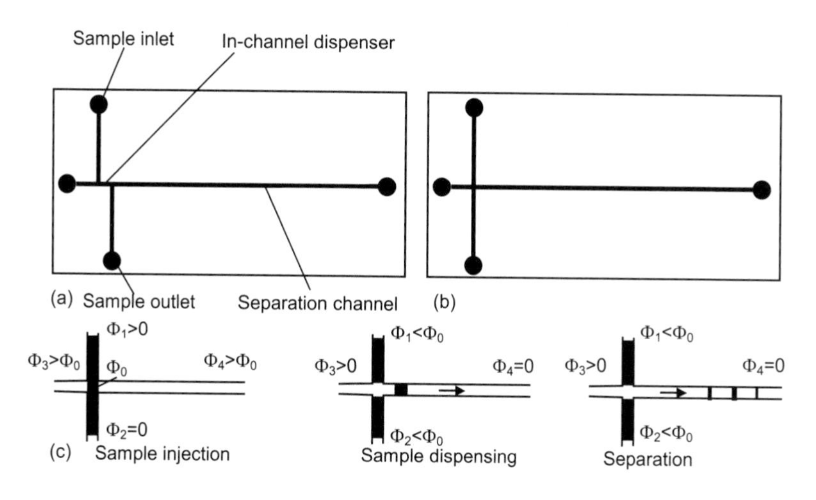


**Figure 12.20** Concept of capillary electrophoresis with free buffer solution: (a, b) separation devices; and (c) separation concept.

The fastest component of the sample moves through the channel with the velocity [46]:

$$U_i = U_\mathrm{s}/\varepsilon = 0.67 \times 10^{-3}/0.3 = 2.22 \times 10^{-3}\ \mathrm{m/s}$$

The flow passes through the beads in a zigzag path with an average angle of $45^\circ$ to the actual path. Thus, the actual flow path is $\sqrt{2}$ times longer than the separation channel. The time that the fastest component of the sample needs to reach the sensor is:

$$t = \sqrt{2}L/U_\mathrm{f} = \sqrt{2} \times 20 \times 10^{-3}/2.23 \times 10^{-3} \approx 12.7\ \mathrm{sec}$$

## 12.4 ELECTROPHORESIS

In contrast to gas chromatography and liquid chromatography, electrophoresis separation uses electrokinetic effects to pump the sample as well as the carrier liquid through the separation channel. Electrokinetic pumping, including electrophoresis and electro-osmosis, are discussed in Chapter 7.

*Capillary Electrophoresis.* In capillary electrophoresis, the liquid is driven by electro-osmosis, and the sample components are separated by electrophoresis. Electrophoretic separations can be further categorized as [43]: separations in free-buffer solution and separations in polymer-sieving media.

Figure 12.20(a) depicts the layout of a microseparator with a free-buffer solution. The channel system is etched in glass [47]. The buffer is primed into the separation channel by electro-osmosis. Subsequently, an in-channel microdispenser injects the sample into the separation channel. The separation occurs with a combination of electrophoresis and electro-osmosis. The separation process is shown in the sequences of Figure 12.20(c). $\Phi_1$ to $\Phi_4$ are the electric potential at the ports, and $\Phi_0$ is the potential at the junction. The dispensed sample volume in the device shown in Figure 12.20(b) has the minimum volume defined by the cross area of the sample channel and the separation channel [48, 49]. The device is fabricated in silicon by wet etching. A cover plate is anodically bonded to the silicon chip. Since silicon is a good conductor, the channel surface should be covered with a silicon oxide insulator. However, the breakdown field strength of silicon oxide limits the range of applied voltage. Due to this problem, glass is a better material for capillary electrophoresis chips [50]. Shallow channels of 12-μm depth and 50-μm width can be etched in glass. The shallow

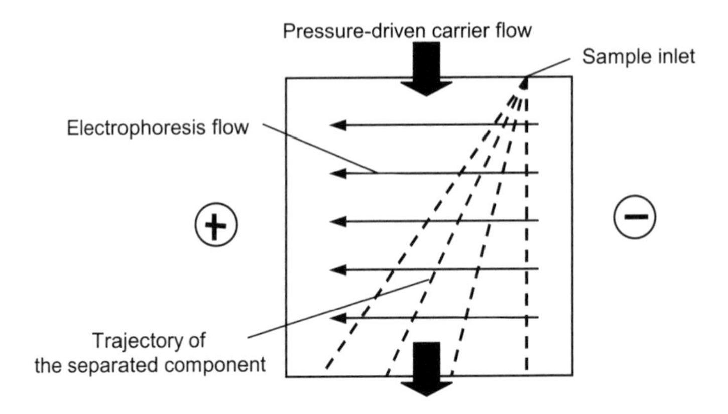


**Figure 12.21** Concept of free-flow electrophoresis with pressure-driven carrier flow.

design allows a better heat dissipation. The channels are covered by a second glass plate by direct bonding at 620°C. Polymer solutions are divided into three types based on their concentration: dilute, semidilute, and entangled. In a dilute solution, the molecules are isolated. With the high concentration of an entangled solution, the polymer molecules are close and interact with each other [51]. In a sieving media, the mobility of a molecule is inversely proportional to its length. Therefore, larger molecules will move more slowly if an electrical field is applied across the sieving media.

The device described in [52] uses polyacrylamide gel as a sieving media. Phosphorothioate oligonucleotides with 10 to 25 bases are successfully separated. A relatively high field strength of 2,300 V/cm is used.

The separation of long double-stranded DNAs was first demonstrated in [53]. An entangled solution of hydroxyethylcellulose is the sieving media. The separation channel is 3.5 cm long. Separation is achieved after 120 seconds. Nowadays, polymer solutions virtually replace polyacrylamide as sieving media.

*Free-Flow Electrophoresis.* Free-flow electrophoresis is a 2-D separation method, which combines hydrodynamic flow with electrophoresis. Figure 12.21 describes this concept. The sample is continuously injected into a carrier flow. The carrier flow and the separation electrical field are perpendicular to each other. Because of different mobility, the components follow different trajectories in the flow channel.

The free-flow electrophoresis device reported by Raymond et al. [54] is made of silicon. A separation channel of 50 μm in depth is etched in silicon. The 100-nm silicon oxide and 600-nm silicon nitride are deposited on the channel surface. The breakdown voltage of this layer is about 200V to 300V. A syringe pump drives the carrier flow, and the electrophoresis separation occurs under the applied voltage between the channel sides. This device is able to separate different biochemical objects such as DNAs, proteins, viruses, bacteria, and cells. Zhang and Manz [55] fabricated the separation channel in PDMS. The separation area consists of an array of interconnecting channels. The small channels increase the electric resistance of the system and thus allow the use of high voltages.

## 12.5 OTHER SEPARATION CONCEPTS

A sieving structure with pore size or gap size on the order of nanometers can replace the gel matrix needed in electrophoretic separation. Such a structure can be categorized as nanofilter. Turner et al.

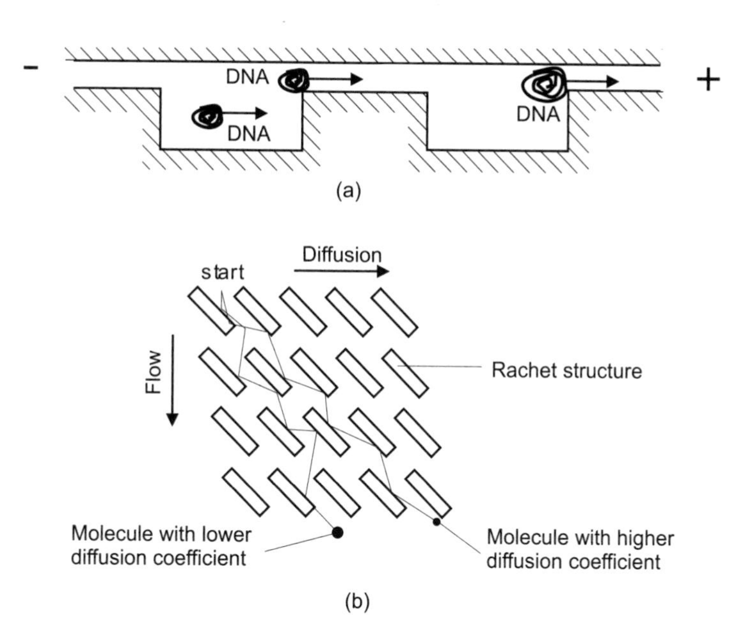


**Figure 12.22** Other separation concepts: (a) separation with entropic traps; and (b) separation with Brownian rachets.

[56] reported a nanofiler consisting of 100-nm pillars, which are arranged at distances from 150 nm to 200 nm. These nanopillars were fabricated with electron beam lithography. Although the sieving structures are in the nanoscale, the gap is still too large compared to molecular size, and thus only suitable for coarse separation. The separation of relatively large DNA molecules of 7.2 and 43 kbp have been reported.

Nanostructures similar to the gap filters described in Section 12.1.2.2 can be used for separation purposes. The gaps work as the entropic traps. The separation channel consists of sections with different gap sizes [Figure 12.22(a)] [57]. If an electric field is applied across the separation channel, DNA molecules move toward the anode under the electrophoretic force. If the gap is much smaller than the radius of gyration of the DNA molecule, the DNA is trapped. The mobility differences of the DNA molecules are determined by their size and the gap size. Han and Craighead [57] reported a separation channel with a width of 30 μm, a length of 15 mm, and alternating depths of from 1.5 to 3 μm and from 75 to 100 nm. The device was able to separate DNA molecules with lengths ranging from 5 to 160 kbp. Nanopores such as ion channels formed by $\alpha$-hemolysin have a size of 5 nm [58]. This size allows the passage of only one DNA fragment at a time. The length of the DNA fragment corresponds to the passage time, which can be measured by monitoring the current through the ion channel. Filters with nanopores can be fabricated with ion beam machining. Li et al. reported the fabrication of nanopores with sizes ranging from 1.8 to 5 nm [59]. In a 1 M KCl solution, with a voltage of 120 mV across a 5-nm pore, an ion current of 1.66 pA was measured. A 500-bp DNA fragment moving through the pore causes a drop of current by 88%.

Brownian ratchets are structures that selectively block the diffusion of particles in one direction. Huang et al. placed a rachet array inside a microchannel. The different diffusion coefficients separate the particles across the channel width. Separation based on Brownian ratchets is similar to free-flow electrophoresis depicted in Figure 12.21. The only difference is that, instead of electrophoresis, Brownian motion causes the separation [Figure 12.22(b)]. Due to the long residence time, the flow should be driven at a very slow speed (e.g., on the order of 1 μ/sec [60]). The drawback of separation based on Brownian ratchets is the relatively long residence time, which may limit the potential use of this concept.

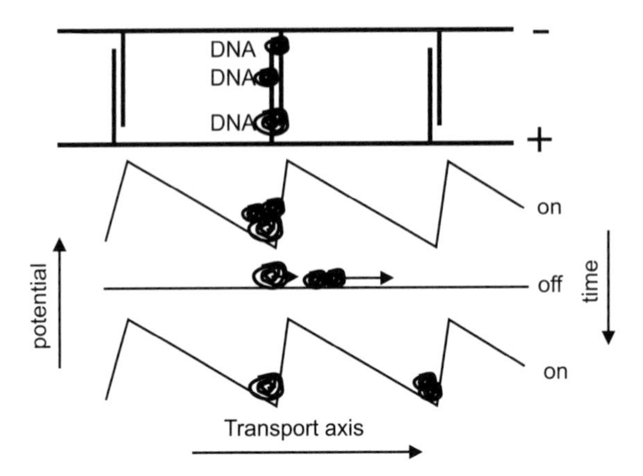


**Figure 12.23** Separation with potential traps and free diffusion.

The asymmetric Brownian ratchet is another trapping concept that can be used for separation. This concept utilizes anisotropic potential fields to trap DNA molecules. The device for this concept consists of asymmetrically interdigitated electrodes (Figure 12.23). If an electric field is applied between the two electrodes, the molecules are trapped in the potential well near the positive electrode. When the potentials are removed, the molecules are allowed to diffuse away [61, 62]. According to Section 10.1, smaller molecules have higher diffusion coefficients. Thus, smaller DNA molecules move further by diffusion and are trapped in the next potential well when the electric field is applied again. Repeating this process separates the DNA molecules based on the combination of diffusion and potential traps. Hammond et al. [61] reported a device with electrodes separated by 0.1 μm. The electrode pair repeats along the transport axis every 1 μm. Separation of small 24-bp DNA fragments from 25-bp fragments was achieved with 12,000 cycles in only 5.4 seconds.

## 12.6 SUMMARY

This chapter discusses two types of separation devices: microfilters and microseparators. Membrane filters fabricated by conventional microtechnologies (Chapter 3) are limited by the resolution of the lithography process or by the wavelength of the exposure light. Submicron filter pores can be achieved by using interference pattern of lasers [5]. However, the resulting pore size is still on the order of hundreds of nanometers. Pore sizes in the molecular range can be achieved. One of the most significant applications of a polymer membrane filter is the microfuel cell, which will be discussed in the next chapter. Gap filters are an alternative to membrane filters to achieve smaller filtering pores. Submicron gaps can be controlled by etching or deposition processes.

While microfilters utilize the different sizes of species for separation, other concepts can be realized based on other properties, such as diffusion coefficient, permittivity, fluorescence, mass, and electric charge. By means of dielectrophoretic force or optical force, particles and cells can be sorted from a sample flow. Labeling the species to be separated, active sorting can be achieved with external actuators. A closed-loop control mechanism can detect the species based on the fluorescent signal and switch them into a collecting reservoir. Microseparators for analytical chemistry utilize the different mobilities of the species. The different mobilities are caused either by the different masses, electric charges, or interactions to the immobile phase on the channel wall. With integrated sensors for detecting the separated species, these microseparators promise a huge market potential for portable and disposable devices for chemical and biochemical analysis.

## Problems

**12.1** A membrane filter is made of single-crystalline silicon. The membrane is 10 μm thick and has the shape of a 1 × 1 mm square. Filter pores are etched in the membrane using DRIE. Each rectangular pore measures 10 × 10 μm. The center distances of the pores are 15 μm. Determine the maximum water flow rate that the filter can withstand.

**12.2** A gap filter is 1 μm high, 100 μm long, and 1 mm wide. Determine the pressure drop across the filter if the flow rate is 100 μL/hr. Assume that the liquid is pure water.

## References

[1] Richter, M., et al., "A Novel Flow Sensor with High Time Resolution Based on Differential Pressure Principle," *Proceedings of MEMS'99, 12th IEEE International Workshop Micro Electromechanical System*, Orlando, FL, Jan. 17–21, 1999, pp. 118–123.

[2] Van Rijin, C. J. M., and Elwenspoek, M. C., "Micro Filtration Membrane Sieve with Silicon Micro Machining for Industrial and Biomedical Applications," *Proceedings of Transducers '95, 8th International Conference on Solid-State Sensors and Actuators*, Stockholm, Sweden, June 16–19, 1995, pp. 83–87.

[3] Farooqui, M. M., and Evans, A. G. R., "Microfabrication of Sub Micron Nozzles in Silicon Nitride," *Proceedings of MEMS'92, 5th IEEE International Workshop Micro Electromechanical System*, Travemuende, Germany, Jan. 25–28, 1992, pp. 150–153.

[4] Yang, X., et al., "Micromachined Membrane Particle Filters," *Sensors and Actuators A*, Vol. 73, 1999, pp. 184–191.

[5] Kuiper, S., "Fabrication of Microsieves with Sub-Micron Pore Size by Laser Interference Lithography," *Journal of Micromechanics and Microengineering*, Vol. 11, No. 1, 2001, pp. 33–37.

[6] Caton, P. F., and White, R. M., "MEMS Microfilter with Acoustic Cleaning," *Proceedings of MEMS'01, 14th IEEE International Workshop Micro Electromechanical System*, Interlaken, Switzerland, Jan. 21–25, 2001, pp. 479–482.

[7] Brody, J. P., et al., "A Planar Microfabricated Fluid Filter," *Proceedings of Transducers '95, 8th International Conference on Solid-State Sensors and Actuators*, Stockholm, Sweden, June 16–19, 1995, pp. 779–782.

[8] Desai, T. A., et al., "Nanopore Technology for Biomedical Applications," *Journal of Biomedical Microdevices*, Vol. 2, No. 1, 1999, pp. 11–40.

[9] Kittilsland, G., et al., "A Sub-Micron Particle Filter in Silicon," *Sensors and Actuators A*, Vol. 23, No. 1–3, 1990, pp. 904–907.

[10] Tu, J. K., et al., "Filtration of Sub-100nm Particles Using a Bulk-Micromachined, Direct-Bonded Silicon Filter," *Journal of Biomedical Microdevices*, Vol. 1, No. 2, 1999, pp. 113–119.

[11] Wyatt Shields IV, C., et al.,"Microfluidic Cell Sorting: A Review of the Advances in the Separation of Cells from Debulking to Rare Cell Isolation," *Lab Chip*, Vol. 15, 2015, pp. 1230–1249.

[12] Ozkumur, E., et al., "Inertial Focusing for Tumor Antigen-Dependent and -Independent Sorting of Rare Circulating Tumor Cells," *Sci. Transl. Med.*, Vol. 5, 2013, pp. 1–11.

[13] Fiedler, S., et al., "Dielectrophoretic Sorting of Particles and Cells in a Microsystem," *Analytical Chemistry*, Vol. 70, 1998, pp. 1909–1915.

[14] Nguyen, N. T., and White, R. M., "Design and Optimization of an Ultrasonic Flexural Plate Wave Micropump Using Numerical Simulation," *Sensors and Actuators A*, Vol. 77, 1999, pp. 229–236.

[15] Nguyen, N. T., et al., "Integrated Flow Sensor for In Situ Measurement and Control of Acoustic Streaming in Flexural Plate Wave Micropumps," *Sensors and Actuators A*, Vol. 79, 2000, pp. 115–121.

[16] Meng, A. H., Nguyen, N. T., and White, R. M., "Focused Flow Micropump Using Ultrasonic Flexural Plate Waves," *Biomedical Microdevices*, Vol. 2, 2000, pp. 169–174.

[17] Marmottant, P., and Hilgenfeldt, S., "Controlled Vesicle Deformation and Lysis by Single Oscillating Bubbles," *Nature*, Vol. 423, 2003, pp. 153–155.

[18] Groschl, M., "Ultrasonic Separation of Suspended Particles. Part I. Fundamentals," *Acustica*, Vol. 3, 1998, pp. 432–447.

[19] Harris, N. R., et al., "A Silicon Microfluidic Ultrasonic Separator," *Sensors and Actuators B*, Vol. 95, 2003, pp. 425–434.

[20] Hatch, G. P., and Stelter, R. E., "Magnetic Design Considerations for Devices and Particles Used for Biological High-Gradient Magnetic Separation (HGMS) Systems," *Journal of Magnetism and Magnetic Materials*, Vol. 225, 2001, pp. 262–276.

[21] Pamme, N., and Manz, A., "On-Chip Free-Flow Magnetophoresis: Continous Flow Separation of Magnetic Particles and Agglomerates," *Analytical Chemistry*, Vol. 76, 2004, pp. 7250–7256.

[22] Furdui, V. I., et al., "Microfabricated Electrolysis Pump System for Isolating Rare Cells in Blood," *Journal of Micromechanics and Microengineering*, Vol. 76, 2004, pp. 7250–7256.

[23] Ashkin, A., et al., "Observation of a Single-Beam Gradient Force Optical Trap for Dielectric Particles," *Optical Letters*, Vol. 11, 1986, pp. 288–290.

[24] MacDonald, M. P., Spalding, G. C., and Dholakia, K., "Microfluidic Sorting in an Optical Lattice," *Nature*, Vol. 426, 2003, pp. 421–424.

[25] Glückstad, J., "Sorting Particles with Light," *Nature Materials*, Vol. 3, 2003, pp. 9–10.

[26] Krüger, J. et al., "Development of a Microfluidic Device for Fluorescence Activated Cell Sorting," *Journal of Micromechanics and Microengineering*, Vol. 12, 2002, pp. 486–494.

[27] Wolff, A., et al., "Integrating Advanced Functionality in a Microfabricated High-Throughput Fluorescent-Activated Cell Sorter," *Lab Chip*, Vol. 3, 2003, pp. 22–27.

[28] Lin, C. H., and Lee, G. B., "Micromachined Flow Cytometers with Embedded Etched Optic Fibers for Optical Detection," *Journal of Micromechanics and Microengineering*, Vol. 13, 2003, pp. 447–453.

[29] Lee, G. B., et al., "Micromachined Pre-Focused 1xN Flow Switches for Continous Sample Injection," *Journal of Micromechanics and Microengineering*, Vol. 11, 2001, pp. 567–573.

[30] Dittrich, P. S., and Schwille, P., "An Integrated Microfluidic System for Reaction, High-Sensitivity Detection, and Sorting of Fluorescent Cells and Particles," *Analytical Chemistry*, Vol. 75, 2003, pp. 5767–5774.

[31] Lu, H., et al., "A Microfabricated Device for Subcellular Organelle Sorting," *Analytical Chemistry*, Vol. 76, 2004, pp. 5705–5712.

[32] Kamholz, A. E., et al., "Quantitative Analysis of Molecular Interaction in a Microfluidic Channel: The T-Sensors," *Analytical Chemistry*, Vol. 71, No. 23, 1999, pp. 5340–5347.

[33] Kuntaegowdanahalli, S. S., et al., "Inertial Microfluidics for Continuous Particle Separation in Spiral Microchannels," *Lab Chip*, Vol. 9, 2009, pp. 2973–2980.

[34] Hou, H. W., et al., "Isolation and Retrieval of Circulating Tumor Cells Using Centrifugal Forces," *Scientific Reports*, Vol. 3, 2013, pp. 1–8.

[35] Warkiani, M. E., et al., "Ultra-Fast, Label-Free Isolation of Circulating Tumor Cells From Blood Using Spiral Microfluidics," *Nature Protocols*, Vol. 11, 2016, pp. 134–148.

[36] Yamada, M., et al., "Pinch Flow Fractionation: Continous Size Separation of Particles Utilizing a Laminar Flow Profile in a Pinched Microchannel," *Anal. Chem.*, Vol. 11, 2016, pp. 134–148.

[37] Warkiani, M. E., et al., "Isoporous Micro/Nanoengineered Membranes," *ACS Nano*, Vol. 7, 2013, pp. 1882–1905.

[38] Huang, L. R., et al., "Continuous Particle Separation Through Deterministic Lateral Displacement," *Proc. Natl. Acad. Sci.*, Vol. 304, 2004, pp. 987–990.

[39] Davis, J. A., et al.,"Deterministic Hydrodynamics: Taking Blood Apart," *Proc. Natl. Acad. Sci.*, Vol. 103, 2006, pp. 14779–14784.

[40] Nagrath, S., et al., "Isolation of rare circulating Tumor Cells in Cancer Patients by Microchip Technology," *Nature*, Vol. 450, 2007, pp. 1235–1239.

[41] Xu, Y., et al.,"Aptamer-Based Microfluidic Device for Enrichment, Sorting, and Detection of Multiple Cancer Cells," *Anal. Chem.*, Vol. 81, 2009, pp. 7436–7442.

[42] Terry, S. C., Jerman, J. H., and Angell, J. B., "A Gas Chromatographic Air Analyzer Fabricated on a Silicon Wafer," *IEEE Transactions on Electron Devices*, Vol. 26, 1979, pp. 1880–1886.

[43] Giddings, J. C., *Unified Separation Science*, New York: Wiley, 1991.

[44] Manz, A., et al., "Design of an Open-Tubular Column Liquid Chromatograph Using Silicon Chip Technology," *Sensors and Actuators B*, Vol. 1, 1990, pp. 249–255.

[45] Ocvirk, G., et al., "Integration of a Micro Liquid Chromatograph onto a Silicon Chip," *Proceedings of Transducers '95, 8th International Conference on Solid-State Sensors and Actuators*, Stockholm, Sweden, June 16–19, 1995, pp. 756–759.

[46] Wilkes, J. O., and Bike, S. G. *Fluid Mechanics for Chemical Engineers*, Upper Saddle River, NJ: Prentice Hall, 1998.

[47] Effenhauser, C. S., "Integrated Chip-Based Microcolumn Separation Systems," in *Microsystem Technology in Chemistry and Life Sciences*, A. Manz and H. Becker (eds.), New York: Springer, 1999, pp. 51–82.

[48] Harrison, D. J., Manz, A., and Glavina, P. G., "Electroosmotic Pumping Within a Chemical Sensor System Integrated on Silicon," *Proceedings of Transducers '91, 6th International Conference on Solid-State Sensors and Actuators*, San Francisco, CA, June 23–27, 1991, pp. 792–795.

[49] Manz, A., et al., "Integrated Electroosmotic Pumps and Flow Manifolds for Total Chemical Analysis Systems," *Proceedings of Transducers '91, 6th International Conference on Solid-State Sensors and Actuators*, San Francisco, CA, June 23–27, 1991, pp. 939–941.

[50] Harrison, D. J., et al., "Micromachining a Miniaturized Capillary Electrophoresis-Based Chemical Analysis System on a Chip," *Science*, Vol. 261, 1993, pp. 895–897.

[51] Grossman, P. D., and Soane, D. S., "Experimental and Theoretical Studies of DNA Separations by Capillary Electrophoresis in Entangled Polymer Solutions," *Biopolymers*, Vol. 31, 1991, pp. 1221–1228.

[52] Effenhauser, C. S., et al., "High-Speed Separation of Antisense Oligonucleotides on a Micromachined Capillary Electrophoresis Device," *Analytical Chemistry*, Vol. 66, 1994, pp. 2949–2953.

[53] Wooly, A. T., and Mathies, R. A., "Ultra-High-Speed DNA Sequencing Using Capillary Electrophoresis Chips," *Analytical Chemistry*, Vol. 67, 1995, pp. 3676–3680.

[54] Raymond, D. E., Manz, A., and Widmer, H. M., "Continuous Separation of High Molecular Weight Compounds Using a Microliter Volume Free-Flow Electrophoresis Microstructure," *Analytical Chemistry*, Vol. 68, 1996, pp. 2515–2522.

[55] Zhang, C. X., and Manz, A., "High-Speed Free-Flow Electrophoresis on Chip," *Analytical Chemistry*, Vol. 75, 2003, pp. 5759–5766.

[56] Turner, S. W., et al., "Monolithic Nanofluid Sieving Structures for DNA Manipulations," *Journal of Vacuum Science and Technology*, Vol. 16, 1998, pp. 3821–3824.

[57] Han, J., and Craighead, H. G., "Separation of Long DNA Molecules in a Microfabricated Entropic Trap Array," *Science*, Vol. 288, 2000, pp. 1026–1029.

[58] Song, L. Z., et al., "Structure of Staphylococcal Alpha-hemolysin, a Heptameric Transmembrane Pore," *Science*, Vol. 274, 1996, pp. 1859–1866.

[59] Li, J., et al., "Ion Beam Sculpting at Nanometer Length Scales," *Nature*, Vol. 412, 2001, pp. 166–169.

[60] Huang, L. R., et al., "Tilted Brownian Rachet for DNA Analysis," *Analytical Chemistry*, Vol. 75, 2003, pp. 6963–6967.

[61] Hammond, R. W., et al., "Differential Transport of DNA by a Rectified Brownian Motion Device," *Electrophoresis*, Vol. 21, 2000, pp. 74–80.

[62] Bader, J. S., et al., "A Brownian-Ratched DNA Pump with Applications to a Single-Nucleotide Polymorhism Genotyping," *Applied Physics A*, Vol. 75, 2002, pp. 275–278.

[63] Huang, Y., et al., "Electric Manipulation of Bioparticles and Macromolecules on Microfabricated electrodes," *Analytical Chemistry*, Vol. 73, 2001, pp. 1549–1559

# Chapter 13

## Microfluidics for Life Sciences and Chemistry: Microreactors

Micromachining technology offers unique opportunities for chemical and biological analysis, as well as chemical production. The miniaturization of reactors and the integration of different components provide new capabilities and functionality, which exceed conventional counterparts. Microreactors offer potentially high throughput and low-cost mass production. Since the requirements and working conditions of microreactors for applications in life sciences are different than those in industrial applications, this chapter considers each type of application accordingly. For industrial applications, the advantages of microreactors can be seen in four main aspects: functionality, safety, cost, and scientific merit.

*Functionality.* The functionality gains of microreactors are small thermal inertia, high gradients of physical properties, uniform temperature, short residence time, and high surface-volume ratio. The small size of the system leads to small thermal inertia of reactors. Since temperature is one of the most important reaction parameters, direct and precise temperature control in microreactors is easier and faster than in conventional reactors. The small size also leads to higher heat and mass transfer rates, which allow reactions carried out under more aggressive conditions than can be achieved in conventional reactors. Furthermore, the high heat transfer rates allow reactions to be performed under more uniform temperature conditions [1]. The small size also results in shorter residence time in reactors. Unstable intermediate products can be transferred quickly to the next process. These characteristics open up new reaction pathways inaccessible for conventional reactions. According to scaling laws, microreactors offer a large surface-to-volume ratio, which leads to the effective suppression of homogeneous side reactions in heterogeneously catalyzed gas phase reactions. Large free surfaces are strong sinks for radical species, which are required for homogeneous reactions. In addition, the reactions become safe because of the suppression of flames and explosions [2], due to the large ratio between surface heat losses and heat generation.

*Safety.* Besides the above-mentioned possibility of carrying out reactions without flames and explosions, the small reactor size also leads to a potentially small quantity of chemicals released accidentally, which does not lead to large-scale hazards and can be contained easily. The possibility of sensor integration increases the safety of microreactors. Failed reactors can be detected, isolated, and replaced. The whole replacement process can be carried out automatically with an array of reactors that are connected redundantly [1].

*Cost.* Rapid screening and high throughput make microreactors more efficient than their counterparts, and that, in turn, makes microreactors cost-effective. The small volume of required reagents also minimizes the amount of expensive reagents. Because microreactors can be fabricated in batch cheaply, numbering-up production can be carried out by replication of reactor units. Numbering-up makes the construction of technical plants less time-consuming because lab-scale results can be transferred directly in commercial production. This aspect is particularly advantageous for fine chemical and pharmaceutical industries with small production amounts per year [1]. The flexibility of the microreactor concept is enormous, because only the number of production units needs to be changed to fit the required production capacity.

*Scientific Merit.* Fluid flows in microreactors are of laminar regime, which is well studied and can be handled by most CFD tools. For submicron structures, the flow can enter the molecular regime and require new computational models. In most practical cases, the laminar model is sufficient for designing microreactors.

## 13.1 DESIGN CONSIDERATIONS

### 13.1.1 Specification Bases for Microreactors

Many microreactors are working on a continuous-flow basis. However, the reagents should have enough time to react. On the other hand, the mean residence time should not be too long to affect the throughput rate of the reactors. The mean *residence time* in a reactor is defined by the typical channel length $L$ and the average velocity $U$:

$$\tau_{\text{residence}} = \frac{L}{U} \tag{13.1}$$

The *reaction time* is determined by characteristic time of the reaction. For heterogeneous reactions, the characteristic reaction time is:

$$\tau_{\text{reaction}} = \frac{D_{\text{h}}^2}{4D\text{Sh}} \tag{13.2}$$

if the reaction is determined by diffusion, and:

$$\tau_{\text{reaction}} = \frac{D_{\text{h}}}{4k_{\text{s}}} \tag{13.3}$$

if the reaction is determined by the reaction rate. For homogenous reaction, the characteristic reaction time is:

$$\tau_{\text{reaction}} = \frac{1}{k_{\text{v}}}. \tag{13.4}$$

In the above equations, $D_{\text{h}}$, $D$, Sh, $k_{\text{s}}$, and $k_{\text{v}}$ are the hydraulic diameter, the diffusion coefficient, the Sherwood number, and the reaction rates for surface and volume, respectively [3]. The *reaction rate* is the flux of the product generated from the reaction. Microreactors can achieve reaction rates on the order of $10^{-6}$ mol/s-m$^2$ [4].

The Sherwood number represents the ratio between overall mass transfer and diffusion:

$$\text{Sh} = \frac{k_{\text{T}} D_{\text{h}}}{D} \tag{13.5}$$

where $K_{\text{T}}$ is the mass transfer coefficient, and $D$ is the diffusion coefficient. Decreasing the channel size may reduce the characteristic reaction time [(13.2) and (13.3)], or not at all (13.4). However, for a fixed throughput, decreasing the channel size also means an increase of two orders in the mean velocity $U$ or a decrease of two orders in the mean residence time $\tau_{\text{residence}} \propto D_{\text{h}}^2$. Thus, in order to match the required throughput with the reaction time, parallel reaction channels should be considered in the design. The parallel channels also help to keep the total pressure drop across the reactor constant.

As mentioned above, one of the advantages of microreactors is safety. Due to the higher surface-to-volume ratio, heat transfer rate is improved and sudden temperature rise can be avoided. Thus, chain reactions and thermal explosions can be avoided. The characteristic time of heat transfer is estimated as [3]:

$$\tau_{\text{thermal}} = \frac{\rho c_p h^2}{4\kappa \text{Nu}}, \tag{13.6}$$

where $\rho$, $c_p$, and $\kappa$ are the density, heat capacity, and thermal conductivity of the fluid in the reactor, respectively. The Nusselt number (Nu) characterizes the heat transfer to the reactor walls:

$$\text{Nu} = \frac{hD_{\text{h}}}{\kappa} \tag{13.7}$$

where $h$ is the heat transfer coefficient. In order to prevent thermal explosion, the reaction time should be shorter than the characteristic time of heat transfer $\tau_{\text{reaction}} \ll \tau_{\text{thermal}}$. *Reaction temperature* determines the activation energy of a reaction. In microreactors, some reactions can be carried out at ambient temperature rather than formerly at cryogenic conditions [5].

It is ideal for a reactor to yield a maximum amount of product in the smallest space and in the shortest time. The performance indication for a microreactor is therefore the space-time yield:

$$\text{STY} = \frac{n}{V_{\text{reactor}}\tau_{\text{residence}}} \tag{13.8}$$

where $n$ (mol) is the amount of the product, and $V_{\text{reactor}}$ is the volume of the reactor. Compared to the macroscopic counterparts, high space-time yield can be expected from microreactors due to their small volume and a shorter residence time. Space-time yields, on the order of $10^4$ to $10^5$ mol/h-m$^3$, are typical for microreactors. This space-time yield is about three orders of magnitude higher than conventional reactors [5].

### 13.1.2 Miniaturization of Chemical Processes

Reducing the size of chemical processes leads to a larger surface-to-volume ratio or the cube-square law. The dominant surface effects are of advantage in reactions with catalysts or multiple phases. Miniaturization also reduces the process path, and increases concentration as well as temperature gradients. Depending on the reaction, microreactors can be divided into heterogeneous and homogeneous types. Heterogeneous reactors involve two different reagent phases. The chemical reactions occur at the interface of these two phases. In homogeneous reactors, the reactions take place in a single phase, and good mixing of reagents is required. In microreactors, some homogeneous reactions may turn into heterogeneous reactions because of the extremely large surface-to-volume ratio.

Reducing the size further to the microns and submicrons scale may not be advantageous for industrial applications. For applications in life sciences and chemical analysis, the microreactors are often disposable; thus, long-term behavior is not a concern while designing these reactors. In contrast, fouling and blocking of small channels could be a serious problem for industrial applications. Furthermore, the small size leads to a shorter residence time, which should be matched with the required reaction time [6]. Heat removal and supply is another important issue for microreactors. Elements such as heaters and heat exchangers may need to be integrated with the microreactor.

For applications in life sciences, miniaturization allows fast heat supply and removal from the reactor. The fast temperature change allows a short thermal cycling period in reactions such as polymerase chain reaction (PCR). Furthermore, the small size allows a more homogeneous temperature distribution in the reactor and thus increases the efficiency of the reaction. However, miniaturization and the consequent high surface-to-volume ratio makes the surface of the reaction chamber one of the key parameters in the design. In contrast to industrial applications with a continuous-flow system, microreactors for life sciences are often a closed system. The volume of the liquid in the reaction chamber could be on the order of picoliters. As already discussed in Section 1.1.1, the total number of molecules available for the reaction also decreases with the small volume. While in macroscale, the reagent molecules in the chamber are numerous enough for the reaction, they could bind to the surface in microscale and inhibit the reaction. Native silicon and silicon nitride, for instance inhibit PCR in the microscale.

The small volume confined in the reaction chamber also poses further problems in fluid handling. In practical applications, the sample liquids need to be prepared, and the reaction results need to be analyzed. Thus, transferring a liquid amount on the order of picoliters between equipment and devices is an almost impossible task. Therefore, designing a microreactor for application in life sciences is often linked to designing integrated sample preconditioning devices and analysis devices. Such a system is complex and requires a more careful design consideration.

### 13.1.3 Functional Elements of a Microreactor

Depending on the type of reaction, microreactors consist of the following elements:

- Microchannels;
- Functional elements for thermal management: heaters, temperature sensors, heat sinks, and heat exchanger;
- Catalyst structures: porous surfaces, catalyst membranes, and catalyst layers;
- Sensing elements: chemical sensors;
- Mixing elements: micromixers.

The designs and operation concepts of different micromixer types were discussed in Chapter 10. The following sections only discuss the remaining four types: microchannels, functional elements for thermal management, catalyst structures, and sensing elements.

#### 13.1.3.1 Microchannels

The most basic elements of a microreactor are microchannels and microreservoirs. Microchannels are used for transport of reagents or work directly as reactors. The fabrication of microchannels based on different technologies and materials was discussed in Chapter 3. From the analysis of the characteristic times in the previous section, the basic steps for designing reactor microchannels are:

- Determining the channel size according to the characteristic reaction time and the characteristic time of heat transfer;
- Determining the required residence time based on the other characteristic times;
- Determining the channel length based on the residence time and the required flow velocity;
- Determining the number of the channels based on the required throughput.

When microchannels are used directly as the reactor and fast heat transfer is needed for the reaction, the size of channel cross section should allow a smaller characteristic time for heat transfer than for the reaction ($\tau_{\text{thermal}} < \tau_{\text{reaction}}$). In contrast, reactions with an explosion hazard should have channel size, which prevents the chain reaction caused by the sudden rise of the temperature ($\tau_{\text{reaction}} \gg \tau_{\text{thermal}}$).

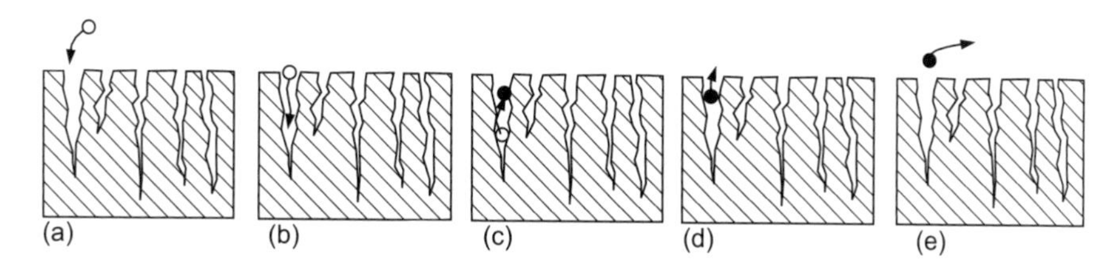


**Figure 13.1** Reaction steps in a microreactor with catalyst support: (a) diffusion to catalyst surface; (b) diffusion in catalyst pores; (c) catalyst reaction; (d) diffusion out of the pores; and (e) transport of reaction products into bulk solution.

With the known size of the channel cross section, the length of the channel is to be determined next. According to (13.1), the length of the channel is calculated from the required residence time and the mean velocity $L = \tau_{\text{residence}} U$. While the residence time can be optimized from either the characteristic times for heat transfer $\tau_{\text{thermal}}$ or reaction $\tau_{\text{reaction}}$, the optimal mean velocity $U$ may be determined by the Peclet number to minimize the effect of axial dispersion (Section 10.1.3). With the known channel cross section $A$ and mean velocity $U$, both flow rate and pressure drop across a channel can be determined. Based on the required throughput or the required total pressure drop, the number of microchannels connected in parallel can be finally calculated.

#### 13.1.3.2 Functional Elements for Thermal Management

Heaters and temperature sensors can be integrated directly in reactors for temperature control. This unique feature allows precise control of temperatures, which is crucial in applications such as PCR for DNA amplification [7]. The design of temperature sensors is well researched, and solutions for integrated thermal sensors are established [8]. Common temperature sensing concepts for microreactors are thermoresistive [7] and thermoelectric [9]. The relatively high thermal conductivity of silicon (157 $\text{Wm}^{-1}\text{K}^{-1}$) makes it a good heat conductor. However, cooling is a more difficult task than heating. For fast thermal control, we cannot rely entirely on heat conduction and free convection. Controllable active cooling is necessary. Common solutions are thermoelectric heat pumps, such as Peltier elements or forced convection. Heat exchangers such as microchannels are fabricated in the same process of reactors.

Integrated heat exchangers can both supply and remove heat to and from the reactor. Heat transfer coefficients on the order of $10^4$ to $10^5$ W/m$^2$-K, and heating rates on the order of $10^6$ K/s, can be achieved with microheat exchangers [10].

#### 13.1.3.3 Catalyst Structures

Many reactions are carried out with catalyst support. The reaction rates in a reactor with catalyst support are proportional to the total reacting surface available. A porous catalyst can increase the reacting surface area. Figure 13.1 describes the main five steps of these reactions [11].

In the first step, reagents diffuse from the bulk solution to the catalyst surface. The mass transfer between the bulk solution and the catalyst controls this step [Figure 13.1(a)]. In the second step, reagents diffuse into the catalyst pores. Diffusion controls this step [Figure 13.1(b)]. For small pores and gaseous reagents, the interactions with the pore wall should be considered. In this case, the Knudsen diffusion coefficient can be used. In the third step, reagents undergo a heterogeneous reaction with the pore walls. The catalyst reaction happens here [Figure 13.1(c)]. In the fourth step, the reaction products diffuse back out of the pore. The process is again controlled by diffusion [Figure 13.1(d)]. In the final step, the products move out of the pore, back into the solution, and can be washed away by the continuous flow. Bulk mass transfer controls this step [Figure 13.1(e)]. Assuming small pores compared to the characteristic size of the catalyst structure, transport equations formulated for a porous catalyst layer can use effective transport coefficients (thermal, diffusion), which are determined by the coefficients of the solid phase and fluid phase.

Most reactors require a catalyst to accelerate the chemical reactions. The catalyst can be implemented in microreactors in the form of packed beads, wires, thin film, or high surface area porous support. While beads [12] and wires [2] are not compatible with batch fabrication, thin-film and porous surface catalysts can be integrated in the fabrication processes of microreactors. Thin film catalysts are deposited by CVD or PVD. Using these techniques, the thin film material is often limited to metals and oxides. In order to increase the surface area, the thin film can be roughened by thermal activation. In contrast to CVD and PVD, wet preparation allows a wider range of catalysts. Using inkjet printing or microplotter [13], a catalyst can be placed precisely in the reactor chamber.

Splinter et al. [14] gave an example of a microreactor with porous silicon. The reaction chamber is comprised of 32 channels. Each channel has a cross section of $50 \times 250$ μm. The channel wall is anodized in a solution of hydrofluoric acid and ethanol. The pore size and the morphology are determined by the current densities. The enzyme is immobilized onto the porous wall. In micromachined reactors, the porous surface can be made of silicon. The silicon surface undergoes an anodization in an electrolyte. The typical composition given in [15] is (50% weight concentration; HF : ethanol = 1:1). A porous layer of 70 μm needs 8 minutes and a current density of 0.25 A/cm$^2$. The pore sizes range from 4 to 8 μm. The resulting surface-to-volume ratio is approximately $10^8$ m$^{-1}$. The advantage of this method is the possibility of making semipermeable membranes, which are often needed in microanalysis systems. Since silicon is not a catalyst for many reactions, the surface of the pores should be covered with an active layer, such as a palladium thin film. In a microreactor, the mixed fluid should be placed next to the catalytic site. Efficient mixing and access to the catalytic site can lead to orders of magnitude enhancement in performance [1].

#### 13.1.3.4 Sensing Elements

*Integrated Sensors.* In situ sensing elements such as temperature sensors and chemical sensors can be implemented in the same fabrication process. These chemical sensors are often based on electrochemical methods. Different methods, such as amperometry, conductimetry, and potentiometry, can be easily implemented with simple integrated microelectrodes [16].

Amperometry is based on the measurement of currents resulting from the oxidation as well as the reduction reactions of analytes at electrodes. The current density at a microelectrode is [17]:

$$i = nFD\frac{c_\infty}{\delta} \tag{13.9}$$

where $n$ is the number of electrons, $F$ is the Faraday constant, $D$ is the diffusion coefficient of reacting species, and $c_\infty$ is the concentration in the bulk of solution. The diffusion layer thickness $\delta$ is time dependent:

$$\delta = \sqrt{\pi D t} \tag{13.10}$$

From (13.9) and (13.10), the current through an electrode area $A$ is:

$$I = nFAc_\infty\sqrt{\frac{D}{\pi t}} \tag{13.11}$$

Equation (13.11) describes the time behavior of the electrode current after switching on the potential. Nonlinear effects at the edges of the microelectrodes with an radius of $r$ at a longer time can be described as:

$$I = nFAc_\infty\left(\sqrt{\frac{D}{\pi t}} + \frac{D}{r}\right) \tag{13.12}$$

In order to increase the sensing current and improve signal-to-noise ratio, an array of electrodes can be used for the measurement.

Conductimetry detects the change in conductivity. This technique is suitable for small ions, which are difficult to be detected by other methods. Potentiometry measures voltages using integrated devices such as ion selective electrodes (ISE) or ion-sensitive field effect transistors (ISFET). These devices are independent from miniaturization and do not benefit the scaling laws.

*Optical Methods.* Compared to electrical elements, a transparent window for optical sensing is simpler and cheaper to implement. Designing a microreactor with glass or plastic automatically provides optical access to the measurement location. Fluorescent detection has been widely used in biochemical applications due to its high selectivity. Amino acids and some biochemically relevant species can be derivatized with a fluorescent tag. The evaluation is achieved with the method of *laser induced fluorescence* (LIF). The species emit fluoresced light if they are excited by a laser beam of a higher wavelength. The source could be a blue argon-ion laser ($\lambda = 488$ nm) [18], a red laser diode ($\lambda = 635$ nm) [19], or a violet laser [20].

Detection based on optical absorption detects the change in light intensity. The optical absorbance $A_{\text{op}}$ can be determined based on the so-called Beer's law:

$$A_{\text{op}} = \varepsilon_{\text{op}} c L \tag{13.13}$$

where $\varepsilon_{\text{op}}$ is the molar absorptivity, $c$ is the molar concentration, and $L$ is the optical path. The molar absorptivity depends on the wavelength; thus, a type of molecule can be detected based on the absorption of its characteristic wavelength. Absorption measurements with wavelengths ranging from UV to visible (vis) spectrum is called UV-vis absorption spectroscopy. According to (13.13), miniaturization results in a short optical path and limits the sensitivity of this method. The sensitivity-of-absorbance method can be improved with multireflection cells to increase the optical path [21].

*Chemiluminescence* is another method that can be detected optically. Some chemical reactions form fluorescent molecules in their excited state. The implementation of chemiluminescence is relatively simple because no light source is required [22].

## 13.2 DESIGN EXAMPLES

### 13.2.1 Gas-Phase Reactors

#### 13.2.1.1 Oxidation Reactors

Common oxidation reactions realized in microscale are oxidation of ammonia, oxidation of ethylene to ethylene oxide, oxidation of 1-butene to maleic anhydride, oxidation of methanol to formaldehyde, oxidation of propene to acrolein, oxidation of isoprene to citraconic anhydride, and oxidation of carbon monoxide to carbon dioxide. For a review of these different oxidation reactions and their realization in microreactors, readers may refer to [5].

*Combustion.* One of the most common gas reactions is oxidation reaction or combustion. Because of the high energy density, the combustion of hydrocarbon or hydrogen has a huge potential in energy generation. An efficient microengine can be realized with a microcombustor, which is capable of producing 10W to 50W of electrical power by consuming only about 7g of jet fuel per hour [23]. The main challenges for a microcombustor are [24]:

- *Shorter residence time for mixing and combustion*: While keeping the same mass flow rate per unit area, miniaturization leads to faster flow velocity and a shorter residence time. With the

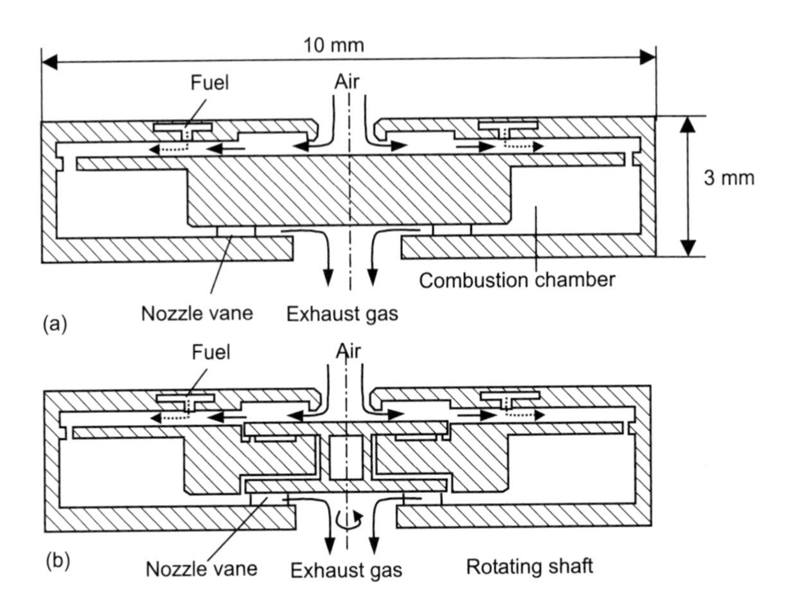


**Figure 13.2** Gas reactors (not to scale): (a) microcombustor; and (b) microengine concept. (*After*: [26])

geometry depicted in Figure 13.2, the residence time will approach the limit of characteristic chemical kinetic time scale for hydrocarbon-air reactions from 0.01 to 0.1 ms.

- *Large heat losses*: The large surface-to-volume ratio and the high temperature gradient lead to heat losses, which is a safety advantage for chemical reactions, as discussed above, but may pose a problem for the flammability of the air/fuel mixture in a microcombustor. The ratio between the surface heat loss rate and the heat generation rate of the combustion should not exceed an upper limit. These ratio scales with the hydraulic diameter of the combustion chamber $D_\mathrm{h}$ as [24]:

$$\frac{\dot{q}_\mathrm{loss}}{\dot{q}_\mathrm{in}} \propto \frac{1}{D_\mathrm{h}^{1.2}} \tag{13.14}$$

  Therefore, there is a critical hydraulic diameter. Below this value, the air/fuel mixture will not be ignited.

- *High temperature*: Combustion temperatures between 1,200K and 1,700K require heat-resistant materials such as ceramics.

Based on the problems discussed above, the design considerations for a microcombustor are [24]:

- Increasing the size of the combustion chamber relative to the size of the device, to increase residence time;
- Premixing of air and fuel before joining the combustion chamber;
- Using special combustion concepts, such as lean burning of hydrogen [25] and catalytic hydrocarbon burning, using platinum or palladium as surface catalysts;
- Using a recirculation jacket, which recovers the heat losses to heat the incoming fuel mixture and allows the compressor discharging air to lower the wall temperatures.

The microcombustor depicted in Figure 13.2(a) is fabricated with deep reactive ion etching in silicon and fusion bonding of a six-wafer stack [26, 27]. Hydrogen combustion is successfully carried out in

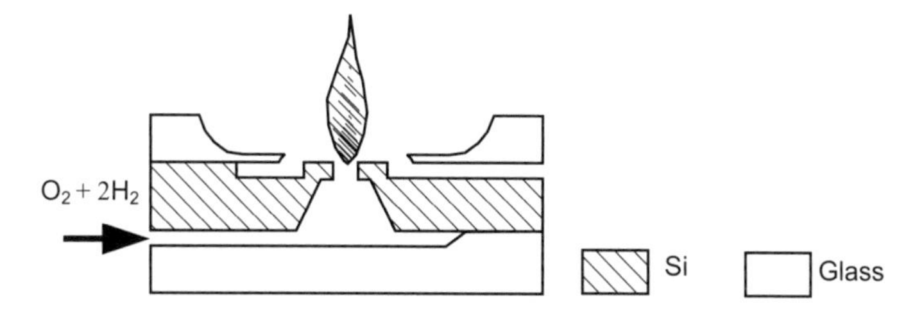


**Figure 13.3** Microburner as a gas-phase reactor.

this microcombustor. A heating wire ignites the combustion reaction. An exit temperature of 1,600K and an efficiency as high as 95% can be reached. Further development uses this microcombustor for a microengine [Figure 13.2(b)]. A rotating shaft connects the compressor and turbine. The electric generator is integrated with the compressor/shaft/turbine system.

The microburner in [28] is able to generate a 1 × 3 mm flame. The burner uses a hydrogen/oxygen mixture provided by an electrolysis cell. The microburner is a part of a microatomic emission flame spectrometer. The device is fabricated in glass and silicon using bulk micromachining (Figure 13.3). The same reaction for the nonflammable regime with a platinum wire as a catalyst is reported in [29].

### Example 13.1: Air-Fuel Ratio in a Microcombustor

A microcombustor uses hydrogen as fuel. The combustor consumes hydrogen with a mass flow rate of 8 mg/s. Assuming that hydrogen burns completely with 100% theoretical air, determine the air-fuel ratio in the combustor.

**Solution.** For air, we assume a molar ratio of oxygen and nitrogen of 1:3.76. The stoichiometric combustion reaction of hydrogen in air is:

$$\mathrm{H_2} + 0.5(\mathrm{O_2} + 3.76\mathrm{N_2}) \rightarrow \mathrm{H_2O} + 1.88\mathrm{N_2}$$

That means that we need:

$$0.5(1 + 3.76) = 2.38 \text{ mol}$$

air for burning 1-mol hydrogen. The mole rate of hydrogen in the combustor is:

$$\dot{n}_{\mathrm{H_2}} = \dot{m}_{\mathrm{H_2}}/M_{\mathrm{H_2}} = 8 \times 10^{-6}/2 = 4 \times 10^{-6} \text{ kmol/s}$$

The mole rate of air is:

$$\dot{n}_{\mathrm{air}} = 2.38\dot{n}_{\mathrm{H_2}} = 2.38 \times 4 \times 10^{-6} = 9.52 \times 10^{-6} \text{ kmol/s}$$

The mass flow rate of air is:

$$\dot{m}_{\mathrm{air}} = \dot{n}_{\mathrm{air}} \times M_{\mathrm{air}} = 9.52 \times 10^{-6} \times 28.97 = 275.8 \times 10^{-6} \text{ kg/s}$$

The air-fuel ratio for the microcombustor is:

$$\dot{m}_{\mathrm{air}}/\dot{m}_{\mathrm{H_2}} = 275.8 \times 10^{-6}/8 \times 10^{-6} = 34.5$$

**Example 13.2: Residence Time in a Microcombustor**

A combustor is designed as shown in Figure 13.2(a). The outer diameter is 10 mm, the inner diameter is 5 mm, and the height is 1 mm. The operating pressure is 4 atm. Estimate the residence time of the air-fuel mixture in the combustor.

**Solution.** Since the volume flow rate is proportional to the molar rate and the volume is inversely proportional to the pressure (for ideal gas), the total volume flow rate of the mixture at the inlet of the combustor chamber is:

$$\dot{Q}_{\text{in}} = \dot{Q}_{\text{air}} \times \frac{1 + 0.5 \times 4.76}{0.5 \times 4.78} \times \frac{1}{4} = 97.92 \times 10^{-6}\ \text{m}^3/\text{s}$$

Thus, the inlet velocity at the outer diameter is:

$$U_{\text{in}} = \frac{\dot{Q}_{\text{in}}}{2\pi r_{\text{out}} h} = \frac{97.92 \times 10^{-6}}{2\pi \times 5 \times 10^{-3} \times 10^{-3}} = 3.12\ \text{m/s}$$

The exhaust gas contains water vapor and remaining nitrogen; thus, the volume flow rate of the exit gas is:

$$\dot{Q}_{\text{exit}} = \dot{Q}_{\text{air}} \times \frac{1 + 1.88}{0.5 \times 4.76} \times \frac{1}{4} = 83.44 \times 10^{-6}\ \text{m}^3/\text{s}$$

The exit velocity at the inner diameter is:

$$U_{\text{exit}} = \frac{\dot{Q}_{\text{exit}}}{2\pi r_{\text{in}} h} = \frac{83.44 \times 10^{-6}}{2\pi \times 2.5 \times 10^{-3} \times 10^{-3}} = 5.31\ \text{m/s}$$

We assume a linear acceleration from inlet to exit, thus, the relation between the pathway and the residence time is:

$$R_{\text{out}} - R_{\text{in}} = a\tau^2/2 = (U_{\text{exit}} - U_{\text{in}})/2$$

Rearranging the above equation, we get the estimated residence time of:

$$\tau = \frac{2(R_{\text{out}} - R_{\text{in}})}{U_{\text{exit}} - U_{\text{in}}} = \frac{2 \times (5 - 2.5) \times 10^{-3}}{5.31 - 3.12} = 2.28 \times 10^{-3}\ \text{sec} = 2.28\ \text{ms}$$

**Example 13.3: Total Energy Released from a Microcombustor**

If air and hydrogen enter the combustor at 25°C and 1 atm, the exhaust gas has an exit temperature of 1,227°C. Determine the heat rate released from the combustion. The related thermodynamic parameters are given in the following table. For further reading on the fundamentals of thermodynamics, see [30].

| | | |
|---|---|---|
| Enthalpy of formation of $H_2$, $N_2$, $O_2$ | $\left(\bar{h}_f^0\right)_{\text{H}_2}, \left(\bar{h}_f^0\right)_{\text{N}_2}, \left(\bar{h}_f^0\right)_{\text{O}_2}$ | 0 |
| Enthalpy of formation of $H_2O$ | $\left(\bar{h}_f^0\right)_{\text{H}_2\text{O}}$ | $-241,820$ kJ/kmol |
| Enthalpy of water vapor at 1,227°C | $\bar{h}_{\text{H}_2\text{O}}(1,227°\text{C})$ | 57,999 kJ/kmol |
| Enthalpy of water vapor at 25°C | $\bar{h}_{\text{H}_2\text{O}}(25°\text{C})$ | 9,904 kJ/kmol |
| Enthalpy of nitrogen at 1,227°C | $\bar{h}_{\text{N}_2}(1,227°\text{C})$ | 57,999 kJ/kmol |

**Solution.** The heat rate released per mole flow rate is determined by the enthalpy difference between the combustion product and combustion reactants:

$$\dot{q}/\dot{n}_{\mathrm{H}_2} = \bar{h}_{\mathrm{p}} - \bar{h}_{\mathrm{r}}$$

Since the enthalpy of formation of the reactants ($H_2$, $N_2$, $O_2$) is zero, and they enter the combustor at reference state (25°C, 1 atm), the enthalpy of the reactants is zero. Thus, the above equation becomes:

$$\begin{aligned}\dot{q}/\dot{n}_{\mathrm{H}_2} &= \bar{h}_{\mathrm{p}} = \left(\bar{h}_f^0 + \Delta h\right)_{\mathrm{H}_2\mathrm{O}} + 1.88\left(\bar{h}_f^0 + \Delta h\right)_{\mathrm{N}_2} \\ &= -241,820 + (57,999 - 9,904) + 1.88(0 + 47,073 - 8,669) \\ &= -121,525.48 \text{ kJ/mol}\end{aligned}$$

The negative sign means that the system rejects heat. With the mole flow rate from Example 13.1, the total heat rate released from the microcombustor is:

$$\dot{Q} = 4 \times 10^{-6} \times 121,525.48 = 0.486 \text{ kW} = 486\text{W}$$

**Example 13.4: Designing a Microengine**

The above microcombustor is used in a microengine. The microengine is a gas turbine engine that follows an air-standard Brayton cycle. Determine the work developed by the turbine, the heat loss of the combustor, and the efficiency of the engine. The related thermodynamic parameters are given in the following table.

| *Temperature (K)* | *Enthalpy h (kJ/kg)* | *Relative Pressure $p_r$* |
|---|---|---|
| 295 | 295.17 | 1.2311 |
| 300 | 300.19 | 1.3860 |
| 440 | 441.61 | 5.332 |
| 450 | 451.80 | 5.775 |
| 1,060 | 1,114.86 | 143.9 |
| 1,080 | 1,137.89 | 155.2 |
| 1,500 | 1,635.97 | 601.9 |

**Solution.** The following figure summarizes the schematic and the temperature-entropy (T-s) diagram of the cycle. We use the air-standard analysis for the microengine.

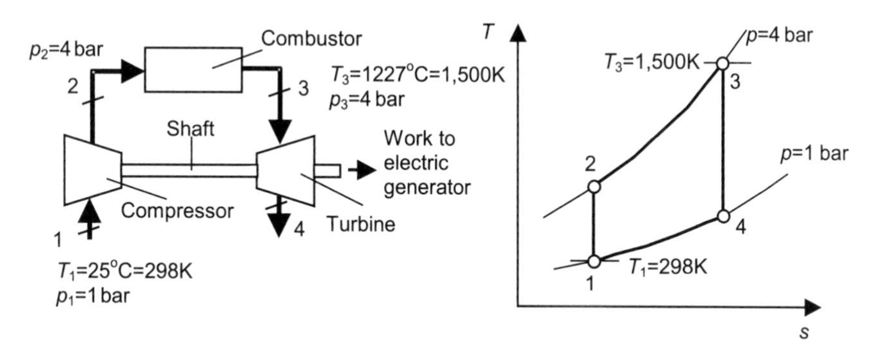

The state points of the cycle are:

- *State 1:* $T_1 = 298\text{K} \rightarrow h_1 = 298.182$ kJ/kg, $p_{\text{r1}} = 1.35432$
- *State 2:* The compression process is isentropic; thus, the relative pressure is:

$$p_{\text{r2}} = (p_2/p_1)p_{\text{r1}} = (4/1) \times 1.35432 = 5.4173$$

  Interpolating from the property table, we get: $T_2 = 442\text{K}$, $h_2 = 443.572$ kJ/kg

- *State 3:* $T_3 = 1,500\text{K} \rightarrow h_3 = 1,635.97$ kJ/kg, $p_{\text{r3}} = 601.9$
- *State 4:* The process across the turbine is isentropic; thus, the relative pressure is:

$$p_{\text{r4}} = (p_4/p_3)p_{\text{r3}} = (1/4) \times 601.9 = 150.475$$

  Interpolating from the property table, we get: $T_4 = 1,066\text{K}$, $h_2 = 1,128.26$ kJ/kg

The calculation is based on the mass flow rate of air from Example 13.2. The work developed by the turbine is:

$$\dot{W}_{\text{t}} = \dot{m}_{\text{air}}(h_3 - h_4) = 275.8 \times 10^{-6} \times (1,635.97 - 1,128.26) = 0.140 \text{ kW} = 140\text{W}$$

The work consumed by the compressor is:

$$\dot{W}_{\text{c}} = \dot{m}_{\text{air}}(h_2 - h_1) = 275.8 \times 10^{-6} \times (443.571 - 298.182) = 0.040 \text{ kW} = 40\text{W}$$

The heat absorbed from the combustor is:

$$\dot{q}_{\text{in}} = \dot{m}_{\text{air}}(h_3 - h_2) = 275.8 \times 10^{-6} \times (1,635.97 - 443.572) = 0.329 \text{ kW} = 329\text{W}$$

Since the total heat released from the combustion is 486W (Example 13.3), the heat loss of the combustor is:

$$\dot{q}_{\text{loss}} = \dot{q}_{\text{total}} - \dot{q}_{\text{in}} = 486 - 329 = 157\text{W}$$

The efficiency of the engine is calculated as the ratio of the work left for the electric generator to the total heat released from the combustion:

$$\eta = (\dot{W}_{\text{t}} - \dot{W}_{\text{c}})/\dot{Q}_{\text{total}} = (140 - 40)/486 = 20.58\%$$

Compare this efficiency with the theoretically maximum efficiency of a power cycle operating between 298K and 1,500K:

$$\eta_{\text{max}} = (T_{\text{H}} - T_{\text{C}})/T_{\text{H}} = (1,500 - 298)/1,500 = 80\%$$

*Microfuel Cells.* In a microengine, energy losses are caused by the combustion reactions. Furthermore, microengines require another step to convert mechanical energy to electrical energy. The high temperature involved in a microengine is a big challenge for material selection and fabrication technology. In contrast to a microengine, the oxidation reaction in fuel cells converts the chemical energy directly into electrical energy at a relatively low temperature, even at room temperature.

A typical fuel cell consists of an electrolyte layer in contact with a porous anode and cathode on either side. Fuel, such as hydrogen, is fed continuously to the anode, and an oxidant, such as oxygen from air, is supplied continuously to the cathode. The electrochemical reactions are activated at the electrodes by catalysts:

$$\begin{aligned} &\text{Anode}: \quad 2\mathrm{H_2} \xrightarrow{\mathrm{Pt}} 4\mathrm{H^+} + 4e^- \\ &\text{Cathode}: \mathrm{O_2} + 4\mathrm{H^+} + 4e^- \xrightarrow{\mathrm{Pt}} 2\mathrm{H_2O} \end{aligned} \tag{13.15}$$

Protons can pass through the electrolyte. The remaining electrons generate a current that can be used by a load. The fuel cell is therefore a true energy converter. Electricity can be continuously produced as long as fuel and oxidant are supplied to the electrodes. The overall reaction of (13.15) is:

$$2\mathrm{H_2} + \mathrm{O_2} \rightarrow 2\mathrm{H_2O} \tag{13.16}$$

Some microfuel cells also use the oxidation reaction of methanol to generate electricity:

$$\begin{aligned} &\text{Anode}: \quad \mathrm{CH_3OH} + \mathrm{H_2O} \xrightarrow{\mathrm{Pt}} \mathrm{CO_2} + 6\mathrm{H^+} + 6e^- \\ &\text{Cathode}: 1.5\mathrm{O_2} + 6\mathrm{H^+} + 6e^- \xrightarrow{\mathrm{Pt}} 3\mathrm{H_2O} \end{aligned} \tag{13.17}$$

The overall reaction of methanol is:

$$\mathrm{CH_3OH} + 1.5\mathrm{O_2} \rightarrow \mathrm{CO_2} + 2\mathrm{H_2O} \tag{13.18}$$

Fuel cells based on methanol are actually liquid/gas reactors. However, this fuel cell type is discussed here for a direct comparison with hydrogen-based fuel cells.

A microfuel cell can be realized by combining thin film materials with the microfluidic technology. Microfuel cells with a power less than 5W can be developed by silicon micromachining or polymeric micromachining (Chapter 3). Electrode structures for collecting the current can be deposited directly on the wall of fuel-delivering microchannels. The key element of a microfuel cell is the membrane electrode assembly (MEA). Nafion membrane (DuPont) is usually used for this purpose, which enables ion (proton) transport from anode to cathode. The electrodes are made of a porous material such as carbon paper. The catalyst layer is coated on the electrode. The catalyst can be made of Nafion-bonded platinum on carbon, which extends the three-phase boundary for electrochemical reaction. This type of fuel cell is called polymer electrolyte membrane fuel cell (PEMFC). Figure 13.4 shows a typical microfuel cell with its characteristics. In all reported works, the membrane electrode assembly needs to be prepared separately. Only the fuel distribution system and current collector can be realized with micromachining technology. Furthermore, fuel storage is still an unsolved problem for a hydrogen-based microfuel cell. The relatively large volume of a high-pressure container for hydrogen is a major concern for safety and suitability in portable applications.

Lee et al. [31] developed a PEMC array with flip-flop interconnection. Hydrogen works as the fuel. Both fuel and air are not delivered by microchannels but distributed by micropillars etched in the substrate. Both glass and silicon were used as substrates. A thin gold layer was sputtered on the structured substrate and acts as current collector. The glass-based microfuel cell had a peak power density of 20 mW/cm$^2$, while the silicon-based device delivered a peak power density of 42 mW/cm$^2$. This microfuel cell works at room temperature. Yu et al. [32] reported a similar device made of silicon with a thicker current collector. The metal system consists of a titanium/tungsten adhesion layer, a thick copper layer, and a gold layer. This microfuel cell achieved a maximum power density of 194.3 mW/cm$^2$ (450 mA/cm$^2$) at room temperature. The good performance is caused by high flow rates of hydrogen and oxygen (50 mL/min). Furthermore, pure oxygen was used as oxidizer.

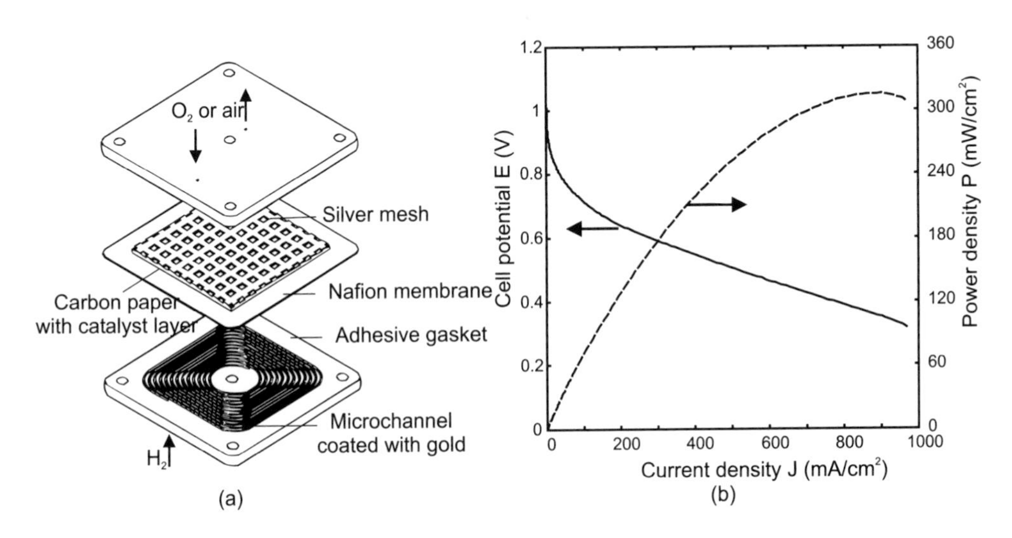


**Figure 13.4** A typical microfuel cell: (a) construction; and (b) characteristics.

As mentioned above, the biggest problem of PEMC using hydrogen is the large fuel storage. Direct methanol fuel cell (DMFC) could solve the problem of fuel storage. Yen et al. [33] fabricated a DMFC based on silicon micromachining. Fuel-delivering microchannels with 400 μm in depth and 700 μm in width were etched using DRIE. A gold layer acting as current collector was sputtered on the silicon substrate with a titanium/copper adhesion layer. This PEMC delivers a peak power density of 47.2 mW/cm$^2$ with 1 M $CH_3OH$ at 60°C. The DMFC reported by Lu et al. [34] used fuel with a higher concentration of 2 M $CH_3OH$. A chromium/copper/gold system was used as a current collector. The peak power density of this fuel cell was 16 mW/cm$^2$ at 23°C and 50 mW/cm$^2$ at 60°C, which is comparable to the device reported by Yen et al.

A microfuel cell can also be fabricated with polymeric technology. Blum et al. [35] developed a water-neutral DMFC made of plastics. The fuel cell worked in stable operation for 900 hours with a peak power density of 12.5 mW/cm$^2$. Shah et al. [36] developed a hydrogen-based microfuel cell made of PDMS. The current collector is a silver grid printed directly on the membrane electrode assembly. The maximum power density achieved was 35 mW/cm$^2$ when the cell was heated to 60°C. Hsieh et al. [37] developed micro fuel cells based on PMMA. Fuel-delivering microchannels were machined by excimer laser. Copper was sputtered on the PMMA substrate and worked as the current collector. The maximum power density achieved was 31 mW/cm$^2$ at room temperature. Chan et al. presented a microfuel cell made of PMMA and gold current collector [38]. The fuel cell can deliver a peak power density of 300 mW/cm$^2$. The superior performance was achieved with the novel design of the fuel-delivering microchannels as well as the special channel shape. The microchannel was micromachined in PMMA with $CO_2$ laser, and has a typical Gaussian shape. This channel shape cannot be achieved with other polymeric micromachining techniques, such as molding [36] or excimer laser machining [37]. The curved channel wall allows sputtering a smooth current collector layer, which decreases the internal electrical resistance of the fuel cell. The construction and the performance characteristics of this microfuel cell are shown in Figure 13.4.

Besides microfuel cells with polymer electrolyte membrane, the implementation of solid oxide fuel cells in microscale has been also reported [39, 40]. The maximum power density could reach 110 mW/cm$^2$ at 570°C [39] and 170 mW/cm$^2$ at 600°C [40]. The reported performance of solid oxide fuel cells is comparable to the polymer fuel cells that operate at room temperature. But these fuel cells work at relatively high temperatures and need a thermal management system for starting up and running.

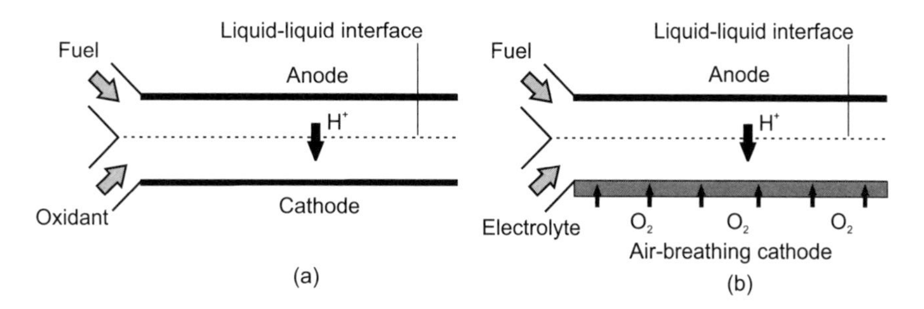


**Figure 13.5** Schematics of microfluidic membraneless fuel cell running on acidic electrolyte. Ion (proton) transport occurs across the channel. Both streams have supporting electrolytes. (a) A microfluidic membraneless fuel cell with flow-over electrodes running of fuel and oxidant containing streams; and (b) a microfluidic membraneless fuel cell with air-breathing cathode to supply oxidant (oxygen) to the cathode. Electrolyte stream is introduced to the channel to avoid fuel crossover to the cathode side. Fuel crossover results in performance degradation of the fuel cell.

*Microfluidic Membraneless Fuel Cells.* Microfluidic membraneless fuel cells, also known as laminar flow-based fuel cells, are considered as a subcategory of microfuel cells. A microfluidic fuel cell is mainly established in a microchannel where two or more streams are stacked to deliver required reagents to the electrodes [41, 42] (Figure 13.5).

In a microfluidic fuel cell, one stream contains fuel while the another stream carries oxidant. Various chemicals can be used as fuel and oxidant. Conventional gaseous chemicals such as hydrogen and oxygen can be dissolved in water to serve fuel and oxidant for oxidation and reduction reactions over the anode and the cathode, respectively. In addition, diluted aqueous solutions of some chemicals such as formic acid and potassium permanganate can be employed as fuel and oxidant, respectively. Due to the laminar nature of the streams, fuel and oxidant are not mixed through convection; their mixing is limited only to a narrow liquid-liquid interdiffusion zone at the middle of the channel. This way eliminates the need for using an ion-exchange membrane to separate the fuel stream from the oxidant stream and to enable ion transport between the electrodes. To perform electrochemical reactions, electrodes are built over the walls of the microchannel through the deposition of electrocatalysists suitable for oxidation and reduction reactions. Similar to membrane-based microfuel cells, the oxidation of the fuel and the reduction of the oxidant are carried out over the catalytic areas. In addition, ions can be produced either over the anode or the cathode. Ion transport between the electrodes occurs through the stacked co-laminar flows by electromigration due to voltage gradient between the electrodes and by ion migration due to the ion concentration gradient between the electrodes [41]. A supporting acidic or alkaline electrolyte is generally added to both fuel and oxidant solutions to facilitate ion transport across the channel. For example, once running a microfluidic fuel cell on dissolved hydrogen and dissolved oxygen, protons are produced at the anode, as indicated in Equation (13.15). The protons migrate from the anode to the cathode side through the co-laminar flows [Figure 13.5(a)].

The operation of any microfluidic fuel cell requires continuous flow of fuel and oxidant that should be supplied from separate reservoirs. Various fuels and oxidants can be employed to perform the electrochemical reactions. In air-breathing microfluidic fuel cells, oxygen from air is supplied to the cathode through a gas permeable hydrophobic membrane, Figure 13.5(b). The catalyst layer is fabricated over the electrode exposed to a stream of supporting electrolyte. In such design, the need to have a reservoir for continuous oxidant supply is eliminated.

In general, the power density of microfluidic fuel cells is lower than that of membrane-based microfuel cells. This partially happens due to flow-over 2-D configuration of the electrodes build on the walls of the microchannel. Flow-over electrodes provide only limited active sites for the electrochemical reactions. In addition, due to the laminar nature of the streams, efficient

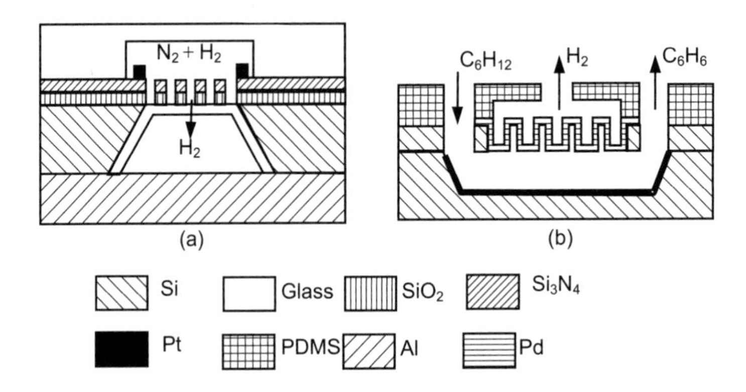


**Figure 13.6** Gas-phase microreactors for (a) hydrogen separation; and (b) dehydrogenation.

replenishment of depletion boundary layers of fuel and oxidant over the electrodes is not feasible. To improve the fuel cell performance, flow-through electrodes have been developed to increase the catalytic active sites and to improve mass transport through the 3-D structure of the electrodes [43].

#### 13.2.1.2 Dehydrogenation Reactors

Thin palladium films have hydrogen-selective characteristics. A palladium membrane is permeable for hydrogen. The reactor depicted in Figure 13.6(a) has a palladium membrane under a silicon oxide/silicon nitride membrane filter. The flow channels are fabricated in PDMS and silicon. An aluminum plate covers the bottom channel. A platinum heater integrated on the membrane filter controls the reactor temperature. The device can separate hydrogen from a nitrogen/hydrogen mixture. The hydrogen flux through the membrane increases with increasing temperature and increasing hydrogen pressure gradient [44]. The palladium membrane can also be used as a catalyst to remove hydrogen in the dehydrogenation reaction of hydrocarbon, such as cyclohexane [45]:

$$3C_6H_12 \stackrel{Pt,Pd}{\longrightarrow} C_6H_6 + 3H_2 \qquad (13.19)$$

#### 13.2.1.3 Hydrogenation Reactors

The conversion of carbon dioxide to methane is an example of a hydrogenation reaction [46]:

$$CO_2 + 4H_2 \rightarrow CH_4 + 2H_2O \qquad (13.20)$$

This reaction is interesting for future manned Mars missions. The Martian carbon dioxide can be converted with terrene hydrogen into fuel and water for astronaut life support systems. On Earth, microreactors for the hydrogenation of carbon dioxide are useful in distributed systems for global carbon dioxide management, which is crucial for reducing global warming [46].

### 13.2.2 Liquid-Phase Reactors

Liquid-phase reactors involve reagents in liquid phases. Most of the reactions are homogeneous; therefore, the reacting species need to be premixed. Unlike gases, liquids have a much smaller diffusion coefficient (see Chapter 10). Designing a fully integrated liquid-phase microreactor is actually designing a micromixer, which was discussed in depth in Chapter 10.

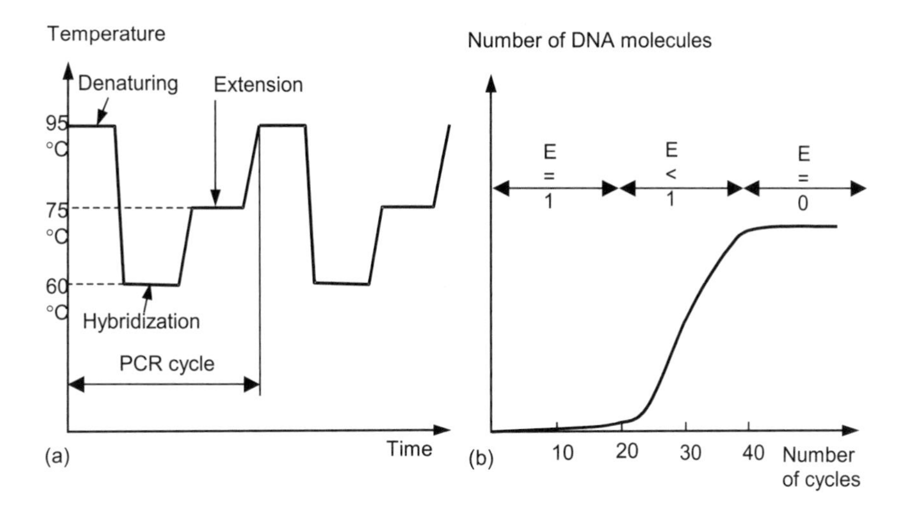


**Figure 13.7** (a) Temperature cycles in the reaction chamber of a PCR reactor; and (b) number of DNA molecules versus number of cycles.

One of the most famous liquid-phase reactors is the device for PCR. The reaction allows amplifying DNA, which stores the genetic information of living species. The technique is necessary for the required concentration of DNA in an analysis (see Section 1.1). This section focuses on the principle and design of PCR reactors.

Genetic information of living species is stored in the structure of a family of molecules called nucleic acids. DNA is a chainlike polymer made of four bases called nucleotides: adenine (A), guanine (G), cytosine (C), and thymine (T). These four bases are complementary in pairs. That means a nucleotide creates a chemical bond with only its complementary partner: adenine with thymine, and guanine with cytosine. In nature, DNA is often present in a double-stranded form. Two complementary strands are chemically bonded together and take up a double helix structure. PCR consists of three different phases: denaturing, hybridization, and extension.

*Denaturing* is a process by which DNA loses its secondary structure. A heating temperature above 90°C breaks a double-stranded DNA molecule into two complementary single-stranded DNA molecules.

*Hybridization*, or annealing, cools the single-stranded DNA molecules at a lower temperature under 60°C. Single DNA strands seek their complementary strands to create double-stranded DNA molecules.

*Extension* is a process by which the incomplete double-stranded DNA molecules are extended with the help of an enzyme called DNA polymerase. The enzyme attaches to the incomplete DNA molecule and replicates the missing complementary bases using available nucleotides called primers. The extension process occurs at a temperature of 70°C. The minimum time of the extension process is limited by the length of the segment to be amplified, the speed of the enzyme itself [30–100 base pairs per second (bp/s)] [47], and the diffusion of primers.

Figure 13.7(a) illustrates the temperature cycles required for the reaction chamber in a PCR process. Miniaturization improves the dynamics of the temperature control, and a faster cycle and faster DNA amplification are possible.

The above three processes complete a cycle called the PCR cycle. Theoretically, repeating the PCR cycle $n$ times amplifies a single double-stranded DNA molecule to $2^n$ folds. In practice, the amplification factor $\Gamma$ is determined by [48]:

$$\Gamma = [1 + E_{\mathrm{PCR}}(n)]^n \tag{13.21}$$

where $E_{\mathrm{PCR}}$ is the efficiency factor, which is a function of the cycle number $n$. For $n < 20$, $E_{\mathrm{PCR}} \approx 1$. For higher cycle numbers ($n > 20$), the efficiency drops [Figure 13.7(b)].

**Example 13.5: Extension Time in PCR**

Determine the extension time for PCR, if the DNA segment to be amplified is 1,000 bp. The enzyme in use is Tag Polymerase, which has an extension speed of 50 bp/s.

**Solution.** The required extension time is the time the enzyme needs for completing the entire segment:

$$\tau = L/U = 1{,}000/50 = 20 \text{ sec} \tag{13.22}$$

#### 13.2.2.1 Design Considerations for PCR Reactors

Although PCR reactors are simple devices, the reaction is sensitive to contamination, especially metal ions. In addition, the chamber surface could adsorb enzyme and DNA and may lead to nonfunctional devices. Tests have been shown that native silicon and silicon nitride can inhibit PCR in microscale. Furthermore, the PCR device should be disposable, because a single DNA strand can contaminate the next sample. Therefore, the substrate material and the machining technology are to be considered carefully.

*Silicon* is used as a device material because silicon technology is established. The high thermal conductivity of silicon allows designing devices with fast thermal response. However, for a fully integrated DNA analysis system, silicon has many disadvantages. First, silicon is electrically conductive and is difficult to incorporate into capillary electrophoresis due to the current required in the fluidic channel. Second, silicon is optically opaque and transparent only to infrared light. Third, DNA tends to stick on native silicon. This sticking effect decreases the number of DNAs available for PCR. It is therefore difficult to use optical detection on silicon devices.

*Glass* is the traditional material in chemical analysis. The low electrical conductivity makes glass the perfect material for capillary electrophoresis. Glass is transparent and has a low native fluorescence, and it is optimal for optical detection. The simple machining technique using wet etching in buffered HF makes glass an attractive material for DNA amplification reactors and analysis devices. However, treatment of substrate surface should be considered.

*Polymers* are the other materials that allow on-chip capillary electrophoresis and optical characterization. However, background fluorescence is an important factor in material choice because many plastics are autofluorescent. Micromachining technology for plastics is still not well established, but it is promising (see Chapter 3).

*Temperature control* is the next important factor in designing PCR reactors. For heating, a heater structure can be integrated into the reaction chamber. Precise temperature measurement is required because the success of the reaction also relies on the accuracy of a thermal cycle. Alternatively, noncontact methods, such as inductive heating [49] or use of an infrared source [50] can simplify the reactor design. For inductive heating, a secondary coil for induced current should be either integrated or placed in close contact with the reactor. For cooling, passive cooling by conduction and natural convection does not require additional design consideration. Fast cooling is desired for a quick cycle. Active cooling using a forced convection or a Peltier element can be incorporated into the device. Kopf-Sill used the Joule-heating effect of an electrolyte solution to realize the temperature cycles for the PCR. An ac voltage was used for heating the liquid to avoid electrolysis and electrophoretic separation of the molecules [51].

Thermal management in a PCR reactor can be categorized as the temporal concept and the spatial concept. With the temporal concept, the temperature changes in time and the fluid remains in the same location. The spatial concept uses continuous flow. The temperature is kept constant in time and the fluid changes its location. The PCR cycle is realized by feeding the solution through three temperature regions, which is repeated in a meander-shaped channel.

**Example 13.6: DNA-Assay Preparation**

If the concentration required for a perfect detection in a DNA assay is $10^5$ copies per mL, how many PCR cycles are needed for an initial sample with a concentration of 100 copies per mL?

**Solution.** The amplification factor required for the reaction is:

$$\Gamma = 10^5/100 = 1,000$$

In (13.4), we assume an efficiency factor of 1, thus:

$$\Gamma = 2^n \rightarrow n = \ln\Gamma/\ln 2 = \ln 1,000/\ln 2 \approx 10$$

**Example 13.7: Designing a Reaction Chamber for PCR**

The chamber of a PCR reactor is made of silicon using DRIE. The chamber is a square with an edge length of $a = 2$ mm and a depth of $d = 200$ μm. Determine the minimum ramping time achievable with this reactor. Assume that the sample liquid is water with a density of $\rho = 1,000$ kg/m$^3$, a specific heat per unit mass of $c = 4,182$ J/kgK, and a thermal conductivity of $\kappa = 0.6$ W/K-m (details on thermal simulation with SIMULINK are given in [52]).

**Solution.** The thermal capacitance or heat capacity of the fluid is estimated as:

$$C_{\text{thermal}} = mc = a^2 d\rho c$$

The thermal resistance from the heating membrane to the top of the chamber is estimated as:

$$R_{\text{thermal}} = d/(\kappa a^2)$$

Thus, the time constant of the liquid in the reaction chamber is:

$$\tau_{\text{thermal}} = R_{\text{thermal}} C_{\text{thermal}} = d^2\rho c/\kappa = (200 \times 10^{-6})^2 \times 10^3 \times 4,182/0.6 = 0.28 \text{ sec}$$

Taking the reactor frame and heat losses to the surroundings, the time constant may be higher than the value above. The ramping time is estimated as:

$$t = 3\tau_{\text{thermal}} = 3 \times 0.28 \text{ sec} = 0.84 \text{ sec}$$

**Example 13.8: Total Preparation Time for DNA Assay**

What is the time required for the preparation of the DNA assay described in Example 13.6 if we use the reactor characterized in Example 13.7 to prepare the sample? The DNA sample and the enzyme for the assay are taken from Example 13.5. The time needed for denaturing and annealing is 5 seconds each.

**Solution.** We assume that the cooling time slope is the same as the ramping time slope. Thus, for one cycle we have three heating and cooling stages. From Example 13.5, the time needed for

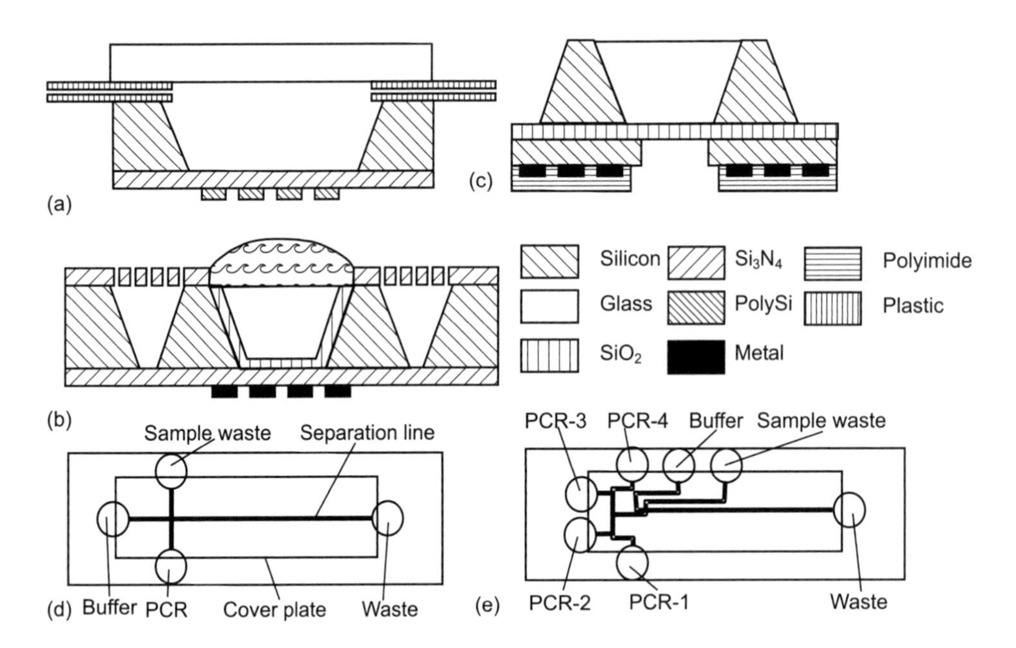


**Figure 13.8** Temporal PCR reactor: (a) silicon/glass; (b) and (c) silicon only; and (d) and (e) glass only.

extension is 20 seconds. The estimated time of a cycle is:

$$\begin{aligned} T &= t_{\text{cooling}} + 2t_{\text{heating}} + t_{\text{denaturing}} + t_{\text{annealing}} + t_{\text{extension}} \\ &= 3 \times 0.84 + 2 \times 5 + 20 = 32.52 \text{ sec} \end{aligned}$$

Ten cycles (Example 13.6) are needed for the required concentration; that means the total preparation time is:

$$t_{\text{total}} = 10 \times T = 10 \times 32.52 \text{ sec} = 325.2 \text{ sec} = 5 \text{ min } 25 \text{ sec}$$

#### 13.2.2.2 Temporal PCR Reactors

Northrup et al. presented for the first time a PCR reactor in silicon [53, 54]. The reactor chamber ($10 \times 10$ mm) is etched in silicon and can contain 50 μL [Figure 13.8(a)]. The temperature control is achieved with a polysilicon heater on the back side of the reactor chip. The reactor chamber is covered with a glass plate using silicone rubber. The small reactor allows a heating rate of 15 K/s and a cycle time of about 1 minute, which is four times faster than conventional PCR cyclers.

The reactor reported by Daniel et al. [55] also uses silicon as the substrate material. Heater and temperature sensors are made of platinum. The reaction chamber is coated with silicon oxide [Figure 13.8(b)]. The chamber is covered with an oil drop to avoid the evaporation of the sample liquid. Surrounding air chambers thermally insulate the chamber. Nitride grids over these chambers avoid thermally conducting liquid getting into the insulating chambers when filling the sample. With a chamber size of 1.5 μL, a thermal time constant of 0.4 second can be achieved.

The reactor array reported by Chaudhari et al. [56] is made of silicon and glass, which are anodically bonded. Temperature control is carried out with an external Peltier element. The Peltier effect is the inverse effect of the thermoelectric effect or Seebeck effect; see Section 8.2. A current passing through metal-metal junctions can control their temperatures. For effective cooling, special thermoelectric junctions such as ($Bi_2Te_3$-CuBr) are required for the Peltier element. These metals

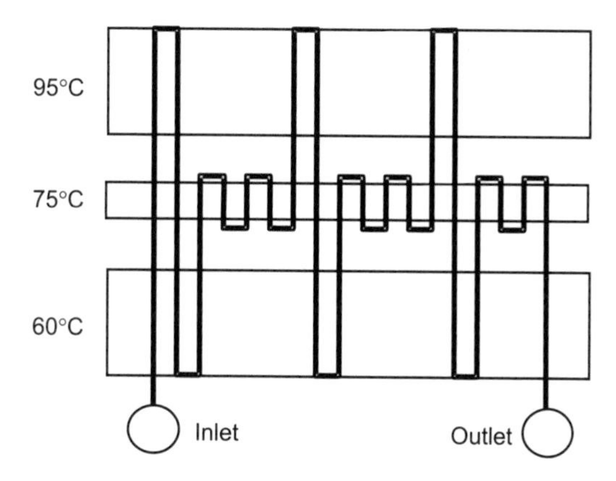


**Figure 13.9** Spatial PCR reactor.

are not compatible to standard silicon processes and are difficult to integrate on the microreactor. Similar reactor arrays are presented in [57]. The chambers are 2 × 2 mm in size and have a volume of 1.4 μL. The device is made in an SOI wafer with integrated titanium heaters [Figure 13.8(c)]. Cheng et al. [58, 59] offered a solution in a silicon-glass chip. The PCR products can be separated directly in a capillary electrophoresis chip.

The PCR reactor presented in [60] is made in 0.55-mm-thick Corning 0211 glass sheets. The reactor chamber is etched with hydrogen fluoride. The glass chip is 15 × 15 mm. The heating and cooling processes of the PCR cycles are carried out by an external Peltier element. Using in situ fluorescence, the PCR efficiency can be detected. The fluorescence density-cycle number characteristics are similar to the typical behavior shown in Figure 13.7(b). A similar solution is presented in [61] using two glass wafers bonded together. The 280-nL reactor chamber is connected to the capillary system for subsequent separation processes [Figure 13.8(d)]. Bulk heating and bulk cooling in this device make parallel analyses with different cycling profiles impossible.

Plastics are used to make sample containers in conventional PCR systems. Polymeric devices could be formed by a sacrificial process with CVD-deposited parylene [48, 62] or hot embossed in bulk PMMA substrate [63].

### 13.2.2.3 Spatial PCR Reactor

In the spatial concept, the sample is forced through three temperature zones with a constant velocity (Figure 13.9). At a constant velocity, the cycle time is proportional to the capillary length. The lengths of the capillary at each temperature zone determine the cycle times. The device can be called a chemical amplifier because the sample concentration at the inlet is amplified at the outlet [64]. The process only depends on the speed of the fluid flow and not on the thermal constant of the system. Therefore, the spatial concept allows a relatively fast cycling.

The channel is fabricated in glass and has a size of 40 μm × 90 μm × 2.2 m. The channel surface is treated with dichlorodimethylsilane to reduce the adsorption of enzyme and DNA. The extremely long and small channel requires high pressure, on the order of bars. This concept of spatial PCR reactor is implemented in a silicon/glass device [65]. The flow channel has a size of 250 μm × 100 μm × 1.512 m. The heaters and temperature sensors are made of thin platinum films on silicon. The heating zones are separated by thermal gaps etched in silicon. The glass wafer and the silicon wafer are bonded anodically. Obeid et al. reported a device based on the same concept [66]. The microchannel was etched in glass and has a cross section of 100 × 55 μm. Outlets are positioned after 20, 25, 30, 35, and 50 cycles. Thus, the number of the total amplification cycles can be selected.

**Example 13.9: Designing a Spatial PCR Reactor**

A spatial PCR reactor has a flow channel etched in glass. The channel height is $H = 40$ μm, and the width is $W = 90$ μm. The flow rates range from 6 to 72 nL/s. What are the dimensions of the reactor depicted in Figure 13.9 if we assume that the three temperature zones have the same width? The time ratio of denaturing, annealing, and extension is 0.5:5:0.5 seconds.

**Solution.** We assume that the sample has properties of water (density 1,000 kg/m$^3$, specific heat per unit mass of 4,182 J/kgK, and a thermal conductivity of 0.6 W/Km).

The thermal capacitance of a channel segment $L$ is given as:

$$C_{\text{thermal}} = mc = LWD\rho c$$

The thermal resistance from the bottom to the top of the channel is estimated as:

$$R_{\text{thermal}} = H/(\kappa a^2)\rho c$$

The time constant for the isothermal condition of the liquid in the channel is:

$$\tau_{\text{thermal}} = R_{\text{thermal}} C_{\text{thermal}} = H^2 \rho c/\kappa = (40 \times 10^{-6})^2 \times 10^3 \times 4,182/0.6 = 0.011 \text{ sec}$$

The smallest design considers the worst case of maximum flow rate or maximum velocity:

$$U = \dot{Q}/(WH) = 72 \times 10^{-12}/(90 \times 10^{-6} \times 40 \times 10^{-6}) = 0.02 \text{ m/s}$$

The time required for passing through the denaturing is:

$$t = 3\tau_{\text{thermal}} + t_{\text{denaturing}} = 3 \times 0.011 + 0.5 = 0.533 \text{ sec}$$

Thus, the channel length in the denaturing zone is:

$$L = Ut = 0.02 \times 0.533 = 10.66 \times 10^{-3} \text{ m} \approx 11 \text{ mm}$$

Since the channel makes one turn in this zone, the zone width will be:

$$W_{\text{zone}} = L/2 = 11/2 = 5.5 \text{ mm}$$

If the extension zone has the same width, the required length is:

$$L_{\text{extension}} = Ut_{\text{extension}} = 0.02 \times (3 \times 0.011 + 5) = 100 \times 10^{-3} \text{ m} = 100 \text{ mm}$$

The ramping time is insignificant in the above equation. The number of turns in the extension zone is:

$$N = L_{\text{extension}}/(2W) = 100/(2 \times 5.5) \approx 9$$

Although buoyancy force is small in microscale, free convection can be used for realizing continuous flow in a closed channel system. The temperature gradients can be used for both driving and realizing the thermal cycles. Krishnan et al. [67] reported a simple configuration with a 35-μL

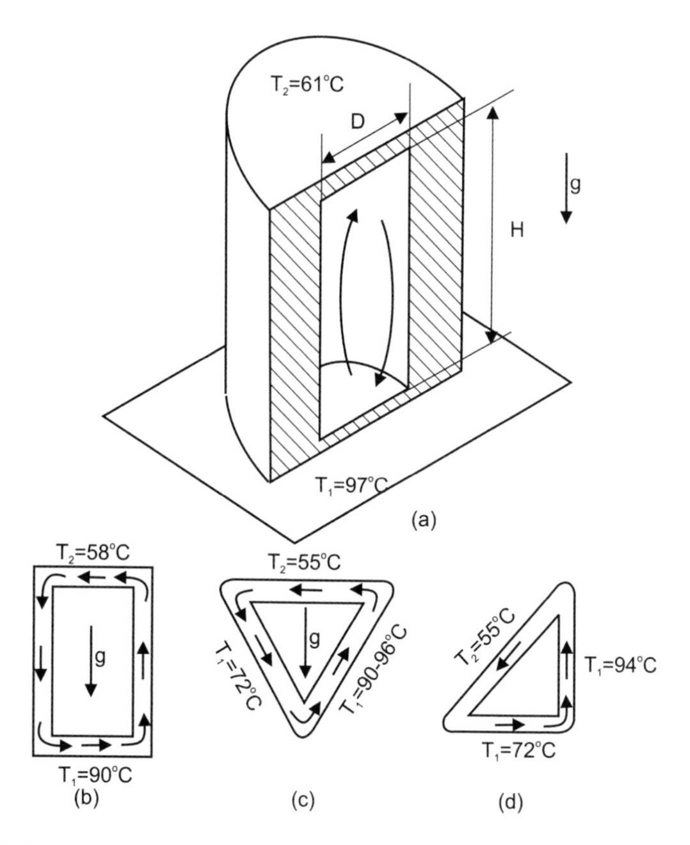


**Figure 13.10** Spatial PCR reactor based on free convection: (a) the basic configuration; (b) rectangular loop; (c) and (d) triangular loops.

cylindrical cavity [Figure 13.10(a)]. The DNA sample was allowed to circulate vertically between 97°C and 61°C. The parameters determining the behavior of the flow are the aspect ratio $(H/D)$ of the chamber and the Rayleigh number:

$$\mathrm{Ra} = \frac{g\alpha\Delta T^3}{H}\nu\kappa \tag{13.23}$$

where $H$ and $D$ are the height and diameter of the reaction chamber, $g$ is the gravitational acceleration, and $\alpha$ is the thermal expansion coefficient of the fluid, $\Delta T$ is the temperature difference, $\nu$ is the kinematic viscosity of the fluid, and $\kappa$ is the thermal conductivity of the fluid. This basic concept can be extended to an array of reaction chambers, a rectangular flow loop [68] [Figure 13.10(b)], a triangular flow loop [69] [Figure 13.10(b, c)], or a semicircular loop [70].

### 13.2.3 Multiphase Reactors

Multiphase reactors discussed in this chapter are reactors for liquid-solid reactions. These reactions are heterogeneous, meaning the reactions occur on the surface of the phases. The reacting surface can be realized with mobile solid phase in the form of microbubbles or microspheres or with a porous substrate material.

*Gas-Liquid Reactors.* In order to increase the reacting surface, gas-liquid reactors can be realized in the form of "trickle-bed" reactors or reactors with bubble columns.

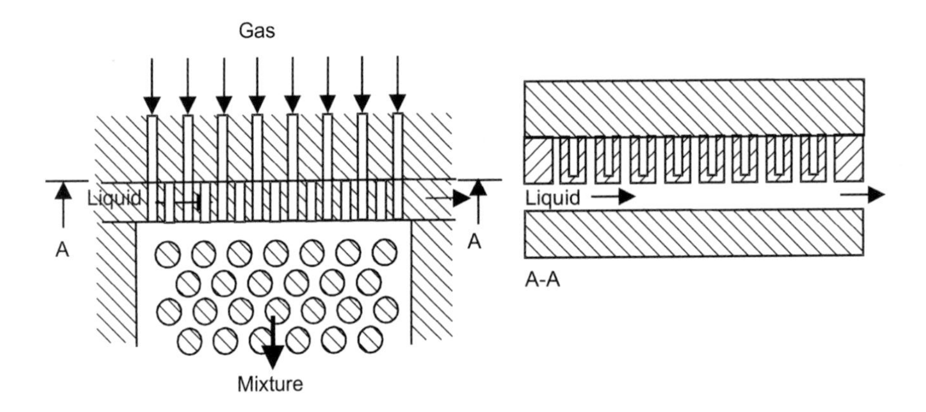


**Figure 13.11** Microreactor for gas-liquid phases. (*After*: [72].)

In the first type, the liquid phase is used to wet the surface of microspheres. The design of such a reactor is similar to reactors with packed beads described below. The wetting efficiency of the beads and the flow distribution of the liquid phase characterize the reaction efficiency.

In the second type, the gas phase is dispersed in the liquid phase in the form of a bubble column. Bubble size and dispersion characterize the reaction efficiency. The dispersion process can be realized with parallel lamination mixers as described in [71]. The mixing effect is further improved with a staggered array of 50-μm-diameter cylinders (Figure 13.11) [72].

*Reactors with Packed Beads.* Reactors with packed beads use the enlarged interfacial area of the beads to improve the mass transfer from reagents coated on the beads to the sample solution, and thus, improve the reaction rates. The amount of reagent transferred is proportional to the concentration difference and the interfacial surface area:

$$\dot{m} = k_{\mathrm{T}} A (c_{\mathrm{i}} - c_{\mathrm{s}}) \tag{13.24}$$

where $k_{\mathrm{T}}$ is the mass transfer coefficient, $A$ is the interfacial area, and $c_{\mathrm{i}}$ and $c_{\mathrm{s}}$ are the concentrations at the interface and the bulk solution. The mass transfer coefficient is a function of flow due to forced convection, while the diffusion coefficient is a property, independent of flow.

The packed beads increase the available surface area, which in turn improves chromatographic separations and immobilizes more reagents for the reaction. The beads are mixed in a solution and fed into the reactor. A microfilter can be integrated in the reactor to contain the microbeads. Gap filters are most appropriate for this application (see Section 12.1). The average velocity across the empty (not packed) chamber is called the superficial velocity $u_{\mathrm{s}}$:

$$U_{\mathrm{s}} = Q / A_{\mathrm{chamber}} \tag{13.25}$$

In a packed-bead reactor, the ratio between the volume left for fluid and the total chamber volume is called the void fraction $\varepsilon$:

$$\varepsilon = (V_{\mathrm{chamber}} - V_{\mathrm{beads}})/V_{\mathrm{chamber}} = (V_{\mathrm{chamber}} - N v_{\mathrm{beads}})/V_{\mathrm{chamber}} \tag{13.26}$$

where $V_{\mathrm{chamber}}$ is the total volume of the reaction chamber, $V_{\mathrm{beads}}$ is the total volume of the beads, and $N$ and $v_{\mathrm{beads}}$ are the number of beads and the volume of a single bead, respectively. The hydraulic diameter in a packed-bead reactor represents the gap between the beads and is defined as:

$$D_{\mathrm{h}} = \frac{4 \times \text{Volume Open to Flow}}{\text{Total Wetted Surface}} = \frac{2\varepsilon}{3(1-\varepsilon)} d_0 \tag{13.27}$$

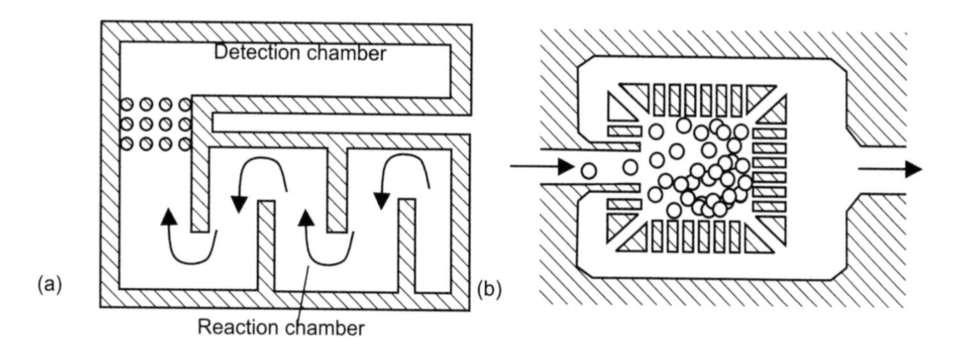


**Figure 13.12** Microreactor with packed beads: (a) made of SU-8; and (b) made of silicon.

where $D$ is the diameter of the beads. Thus, the Reynolds number in a packed bed is calculated with the superficial velocity as:

$$\mathrm{Re} = \frac{2d_0 U_s \rho}{3\eta(1-\varepsilon)} \tag{13.28}$$

where $\rho$ is the density and $\eta$ is the viscosity of the fluid. With a Reynolds number less than about 10, which is often the case in a microreactor, the pressure drop across the reaction chamber with a length $L$ is calculated with the Kozeny-Carman equation [72]:

$$\Delta p = \frac{150\eta U_s (1-\varepsilon)^2 L}{d_0^2 \varepsilon^3} \tag{13.29}$$

The equation clearly indicates that the pressure drop is inversely proportional to the bead diameter.

Figure 13.12(a) shows an example of an enzymatic microreactor [73]. The reactor is integrated monolithically with an electrochemiluminescence (ECL) detector for the evaluation of the reaction results. The reaction chamber is packed with glass beads modified with an immobilized enzyme (glucose oxidase). The active components of the detector are fabricated on the silicon substrate. The reaction and detection chambers are formed with SU-8 and have volumes of 1.5 μL and 2.7 μL, respectively. The gap filter is designed as a group of microcolumns separating the two chambers. The entire device is covered by a PMMA plate with inlet and outlet ports machined in it.

The gap filter of the reactor depicted in Figure 13.12(b) is micromachined in silicon by DRIE. A glass plate is anodically bonded to the silicon structure and seals the device. The volumes of reaction chambers can range from 0.5 to 50 nL [73].

Active filtering of beads is realized with magnetic beads. Magnetic beads are made of magnetite ($Fe_3O_4$) and are coated with the desired enzyme. Magnetic beads can be used for marking cells and DNA molecules. Reference [74] gives an example of integrated electromagnets for trapping magnetic beads. The magnets are integrated as a planar inductive device. Because of the complexity of the integrated magnet, this method is impractical and not attractive for reactors with packed beads.

While the above reactors use a pressure-driven flow to pack microparticles, the packed chamber reported in [75] uses electrokinetic pumping to introduce the beads. The reactor is fabricated in glass by wet etching. The beads are kept in the chamber by two gap filters. Figure 13.13 illustrates the packed chamber. The chamber is a three-way junction, with one inlet, one outlet for sample solution, and one inlet for bead introduction. The total volume of the chamber is 330 pL. The bead introduction channel is designed with a small width to increase its fluidic impedance, and consequently to avoid unpacking the beads. The inlet and the outlet of sample flow are designed with a 1-μm gap filter to retain the beads in the chamber. The bead introduction inlet is designed with a hook structure, which makes the bead flow reverse its direction after filling the chamber [12].

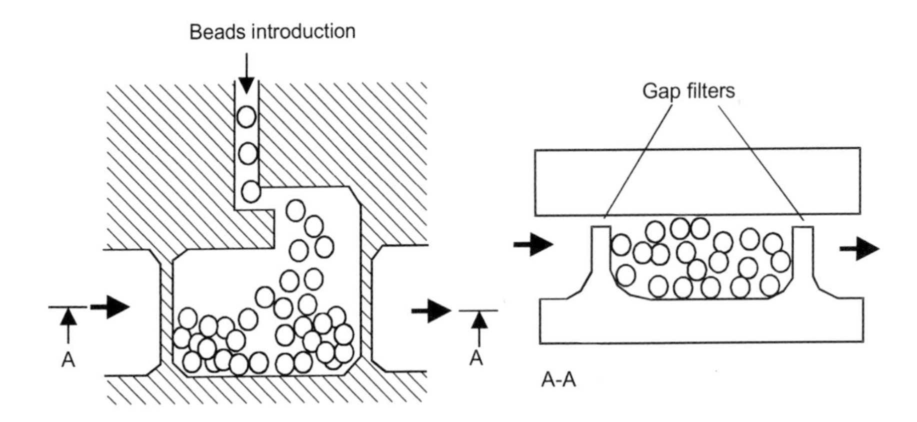


**Figure 13.13** Microreactor with electrokinetic bead introduction. (*After*: [12].)

**Example 13.10: Interfacial Surface**

The reaction chamber of a packed-bead microreactor is etched 200 μm deep in silicon and has a lateral dimension of 500 × 500 μm. The chamber is packed loosely with microspheres, that have a diameter of 10 μm. The microspheres occupy 40% of the chamber volume, and the void fraction is 0.6. Determine the reacting surface and the reacting surface-to-volume ratio.

**Solution.** If the number of packed spheres is $N$, the volume ratio between the spheres and reaction chamber is:

$$\frac{N\pi D^3/6}{WLH} = 40\%$$

Thus, the estimated particle number is:

$$N = 0.4 \times \frac{6WLH}{\pi D^3}$$

$$N = 0.4 \times \frac{6 \times (500 \times 10^{-6})^2 \times 200 \times 10^{-6}}{\pi \times (10 \times 10^{-6})^3} = 38,197$$

The total surface area of the spheres or the reaction surface is:

$$\begin{aligned} A_{\text{total}} &= N\pi D^2 \\ &= 38,197 \times \pi \times (10 \times 10^{-6})^2 = 12 \times 10^{-6}\ \text{m}^2 = 12\ \text{mm}^2 \end{aligned}$$

The ratio between reacting surface and chamber volume is:

$$A_{\text{total}}/V_{\text{total}} = 12 \times 10^{-6}/(500 \times 500 \times 200 \times 10^{-24}) = 2.4 \times 10^{11}\ \text{m}^{-1}$$

**Example 13.11: Mass Transfer in a Packed-Bead Reactor**

The spheres described in Example 13.10 are made of benzoic acid. The mass transfer coefficient at the sphere surface is $10^{-3}$ cm/s. Determine the flow rate of pure water coming into the reaction chamber if the solution at the chamber's exit is 60% saturated with benzoic acid. Assume that water is fed into the chamber laterally. What is the superficial velocity in the reaction chamber?

**Solution.** The flux of the solute at the exit is:

$$\dot{m} = \dot{Q} \times 0.6 \times c_{\text{sat}}$$

where csat is the saturation concentration of the solid sphere. From (13.5), we get:

$$\begin{aligned}
\dot{Q} \times 0.6 \times c_{\text{sat}} &= kA_{\text{total}}(c_{\text{sat}} - 0) \rightarrow \\
\dot{Q} &= kA_{\text{total}}/0.6 \\
\dot{Q} &= 10^{-3} \times 10^{-2} \times 12 \times 10^{-6}/0.6 = 2 \times 10^{-10}\ \text{m}^3/\text{s} = 12\ \mu\text{L/min}
\end{aligned}$$

The superficial velocity is the fluid velocity in the reaction chamber if there are no spheres:

$$U = \dot{Q}/(WH) = 2 \times 10^{-10}/(500 \times 200 \times 10^{-12}) = 2 \times 10^{-3}\ \text{m/s} = 2\ \text{mm/s}$$

### 13.2.4 Microreactors for Cell Culture

The traditional 2-D cell culture significantly differs from the in vivo tissue environment as substrate topography, substrate stiffness, 2-D rather than 3-D cellular structure, and the lack of gradient concentration of various biochemicals such as cytokines and growth factors [76, 77]. Microfluidic-based methods have been developed to address the shortcomings of conventional cell culture approaches. In general, microfluidic-based cell culture occurs in interconnected microfluidic channels or bioreactors with controlled delivery of nutrients and reagents and waste removal [77]. The aim of microfluidic-based cell culture is to present a more realistic in vivo cellular microenvironment under in vitro conditions.

Microfluidic microsystems are able to create a gradient concentration of various biochemicals such as cytokines and growth factors to study the concentration-dependent cellular response. In comparison with the conventional methods for creating a concentration gradient, microfluidic-based gradient generation can provide more flexibility by producing high-resolution drug concentration and the ability of real-time observation of cellular response [78]. Microfluidic gradient generator can be realized either through pure time-evolving diffusion in a microfluidic channel or chamber or steady-state diffusive mixing of laminated parallel streams [78]. Time-evolving gradient generators generally consist of sink and source reservoirs to create a concentration gradient within the connecting channel or chamber, Figure 13.14(a). Since the operation of such gradient generators purely rely on diffusive mixing, the cells do not experience any flow-induced shear stress. The volume of the sink and the source reservoirs are generally limited. Thus, the time-evolving diffusive mixers are mainly suitable for rapid-responding cells [78]. Figure 13.14(b) shows a schematic of a tree-like steady-state gradient generator. The chip consists of two inlets to introduce a drug solution in high concentration and a diluent solution. Various concentrations of drug content are produced in the tree-like section of the microfluidic circuit through serial dilution of the drug-carrying stream. The gradient generator has multiple outlets connected to distinct chambers where single cell type or multiple cell types, to include cell-cell interactions at the presence of a reagent, can be cultured. Unlike time-evolving gradient generators, the steady-state gradient generators may induce flow-induced shear stress to the cultured cells due to the continuous flow passing through the cell culture channels. This concept may induce undesired physical forces that can affect the cellular behavior. Tree-like concentration gradient generators have been employed for high-throughput drug screening applications [79].

In both concentration gradient generators, cells are usually seeded and cultured in a 2-D configuration that may not reflect the proper cellular behavior of 3-D tissues at in vivo conditions. Figure 13.15 shows a 3-D microfluidic chip design developed to create a concentration gradient for 3-D cell culture to enhance cell-matrix interactions. The three channels were created in an agarose membrane sandwiched between a plexiglass cover and a glass slide [80]. Cells with collagen were injected into the central channel. A 3-D cellular matrix was formed in the chamber after polymerization of the collagen. Cell culture media in two different concentrations were pumped to the source and the sink channels and create a concentration gradient at the middle channel

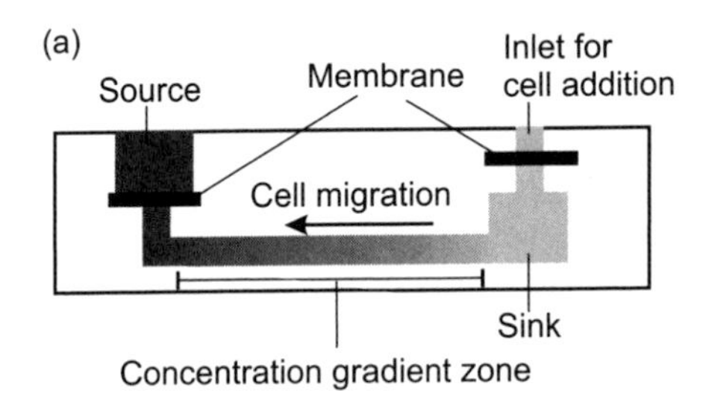


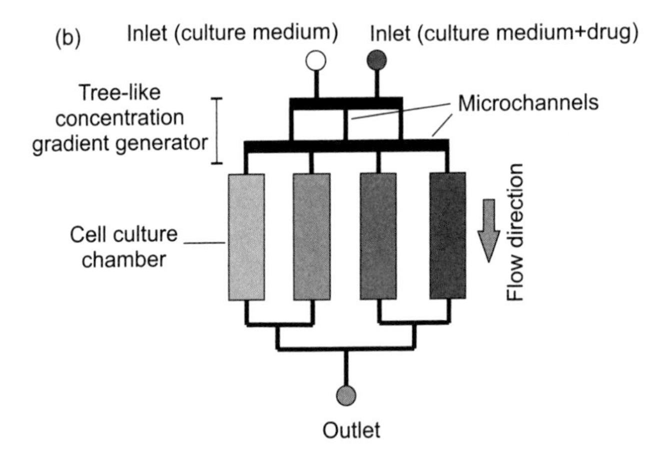


**Figure 13.14** (a) A schematic of a time-evolving chemical gradient generator. Cells are seeded on top of the membrane; some of them start to migrate toward the source that has higher concentration of the chemical. (*After*: [81]). (b) A schematic of a tree-like steady-state concentration gradient generator with cell culture chamber at the downstream. (*After*: [78].)

containing cells [80]. Agarose allows for the the diffusion of nutrients and chemokine from the side channels to the central chamber to maintain cellular viability and to create the concentration gradient. This concept of 3-D microfluidic microdevice with cell-containing hydrogel at the central chamber has been widely used for various studies including modeling of angiogenesis [82] and tumor cell intravasation [83].

Numerous efforts have been carried out to enhance the precision of microfluidic cell culture systems in predicting cellular response at in vitro conditions through mimicking the cell-cell and cell-extracellular matrix interactions. To this end, tissue-engineering methods for cell patterning and modulating matrix stiffness have been employed in advanced 3-D microfluidic systems to develop human organ models [84], known as microfluidic organs-on-chips (OOC) systems [85]. Such a biomimetic microsystem is designed to present the minimum organ-level structural and functional responses of an organ for disease modeling and drug toxicity studies [86]. In general, OOC systems enable the modulation of various parameters including pH, oxygen tension, physical forces such as stretching or flow-induced shear stress, cellular matrix stiffness, and tissue-tissue interfaces. To obtain realistic organ-level responses from an organ-on-chip microsystem, it is essential to present an organ model that mimics the minimum functions of an organ.

Figure 13.16 shows a microfluidic microdevice that reconstitutes the key structural, functional, and mechanical elements of the alveolar-capillary interface of the human lung, known as *lung-on-chip* [86]. The biomimetic lung-on-chip system is composed of two microchannels separated by a flexible and porous membrane with a thickness of 10 $\mu$m made in PDMS. Human alveolar epithelial cells and human pulmonary microvascular endothelial cells were cultured on the opposite

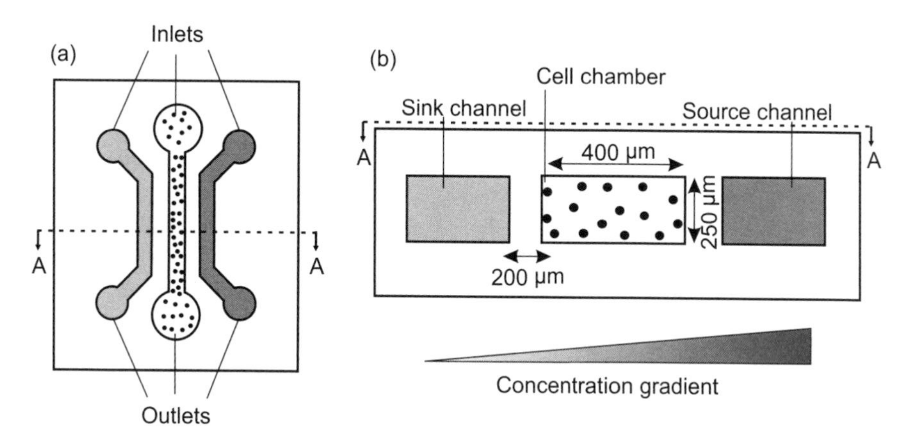


**Figure 13.15** A schematic of a concentration gradient generator with 3-D cell culture chamber at the middle of the device for chemotaxis studies. (a) Top view of the device with two side channels and the middle chamber for cell culture. (b) Cross-sectional view of the microdevice. The side channels are filled with two different concentrations of culture medium, while the cell-containing hydrogel (collagen) is introduced into the middle channel to fill it. The three channels have 400-$\mu$m width and 250-$\mu$m height. Figure was not drawn to scale. The small black circles in the middle chamber are representative of cells (*After*: [80].)

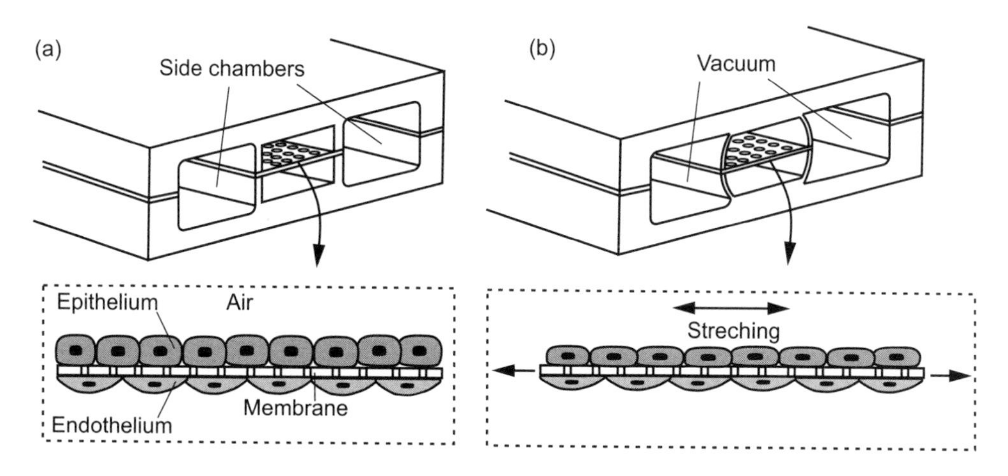


**Figure 13.16** A schematic of a biologically inspired microfluidic lung-on-chip microdevice. (a) The microfabricated chip with distinct chambers to form an alveolar-capillary interface on a thin flexible PDMS membrane. (b) Application of vacuum to the side channels for mechanical deformation of the elastic membrane and the subsequent stretching of the adherent cells at the opposite sides of the membrane. (*After*: [86].)

sides of the membrane [86]. To mimic the expansion of the alveola during inhalation and the induced mechanical forces to the endothelial and the epithelial cells, two lateral microchambers were incorporated into the chip design. The elastic deformation of the flexible membrane and the subsequent stretching of the adherent cells to the membrane is achieved once the side chambers are exposed to vacuum. The membrane and the adherent cells return back to their original size due to the elastic property of the PDMS when the vacuum is removed from the side chambers [86]. The microsystem was employed to reproduce the responses of alveolar-capillary interface to bacteria and inflammatory cytokines presented into the alveolar space. In addition, the microdevice was employed for nanotoxicology studies to investigate the effect of cyclic stretching on epithelial and endothelial uptake of nanoparticles and their transport to the vasculature microchannel. This platform has been widely employed for the development of other organ models such as kidney-on-chip for drug transport and nephrotoxicity assessment [87] and gut-on-chip for modeling the effect of gut peristalsis motion on human intestinal cells [88].

To achieve more realistic responses from OOC systems for toxicity studies, multi-OOC systems may be required to present proper models for pharmacokinetic/pharmacodynamic studies and the cytotoxicity of drug metabolites [89]. To this end, multiple tissue-engineered organ models or organoids are cultured in distinct chambers that are integrated on the same microfluidic chip. A common culture medium, known as blood surrogate, flows through all organ models and mimics the physiological circulation and cross-talk among the different organoids. Continual monitoring of the organoids behavior at various physiological conditions could be achieved through the automated analysis of cell-secreted protein markers [84, 90], in situ microscopy of the organoids in the bioreactors [84], and monitoring of pH and oxygen level of the circulating culture medium (blood surrogate) [91].

### 13.2.5 Microreactors for Cell Treatment

In many applications for biochemical analysis, cells and biological agents often need to be analyzed. The cell itself can be considered here as a confined biological microreactor. Since micromachined structures and cells have the same order of sizes, biochemical and biophysical analysis of a single cell is possible. A biochemical analysis process consists of many steps: sorting and collecting of cells, cell lysis, polymerase chain reaction, and electrophoresis separation. In a biophysical analysis, mechanical [92] or electrical [93] properties of cells can be precisely tested and evaluated. Chapter 12 already discussed the different methods for collecting cells and electrophoresis separation. The design of a PCR microreactor is covered in Section 13.2.2. This section focuses on the different techniques for cell treatment, such as cell lysis, electroporation, and cell fusion.

#### 13.2.5.1 Cell Lysis

Cell lysis breaks the protecting membrane to release DNA. In conventional sample preparation protocols, cell membranes are broken by means of enzymes, chemical lytic agents, heat, mechanical forces, or electric fields.

Cell lysis with enzymes and chemical lytic agents can be carried out in a micromixer as described in Chapter 10. Schilling et al. used a simple T-mixer to dilute lytic agent into the sample flow with cells [94] [Figure 13.17(a)]. After cell lysing, intercellular components with higher diffusion coefficients are separated, as explained earlier in Section 12.2.1. Thermal cell lysis is the simplest technique, that can be realized in microscale. Waters et al. [61] used thermal lysis at 90°C for several minutes to extract DNA from *Escherichia coli*. The DNA was subsequently amplified and analyzed on the same glass chip. For integrated systems, thermal lysis can be achieved with an integrated thermal management system.

In some applications, thermal and chemical treatments are not harsh enough to disrupt the cell membrane. The technique utilizing ultrasonic waves to destroy cell membranes is called

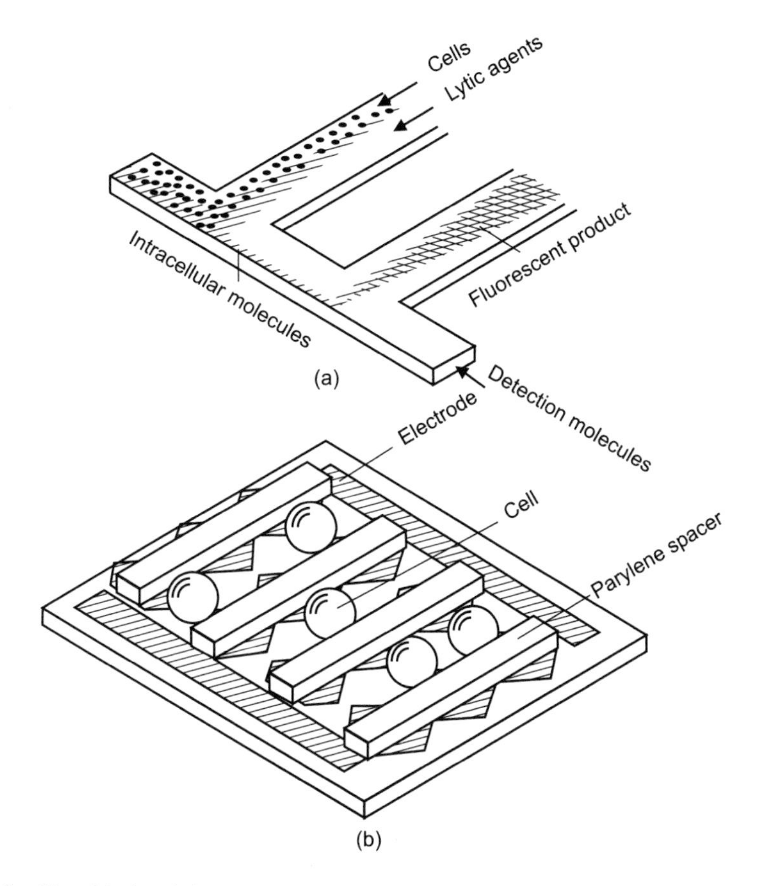


**Figure 13.17** On-chip cell lysis techniques: (a) with chemical lytic agent; and (b) trapping and lysing with electrical fields.

*sonoporation*. Mechanical forces caused by a combination of glass beads and ultrasonic waves deliver a harsher treatment. Belgrader et al. [95] used 106-μm glass beads and ultrasonic waves of 47 kHz to break *Bacillus subtilis* spores. The sonication process used an external sonicator probe and took only 30 seconds. A flexible membrane between the sonicator probe and the sample liquid can avoid cavitation effects [96]. However, cavitation effects in a controlled manner can be used for focusing and lysing cells. Marmottant and Hilgenfeldt [97] used a 15-μm microbubble to attract and disrupt cells. The microbubble is driven with a frequency of 180 kHz, which is near its own resonance. Acoustic streaming attracts the cell to the microbubble and subsequently disrupts the cell due to the focused acoustic energy.

Due to the concentration differences of potassium ions across a cell membrane, an electric field exists across the membrane to keep the potentials on both sides of the membrane in equilibrium. Increasing the field across the cell membrane can make it permeable. If the field strength is higher than a critical value, the effect is irreversible and bursts the membrane. The other effect causing membrane damage is osmosis. The difference in ion concentration caused by a permeable membrane leads further to unbalance in osmotic pressure across the membrane. The osmotic pressure difference causes cell swelling and eventually disrupts the membrane. Lee and Tai [98] used dielecrophoretic forces to trap the cells and position them between two sharp electrodes [Figure 13.17(b)]. The trapping voltage is 6V ac at 2 MHz. After trapping, a short pulse of a higher voltage (20V, 100 μs) bursts the cells open. Yeast cells (*Sacharomyces cerevisiae*) and *Escherichia coli* were successfully lysed with this concept.

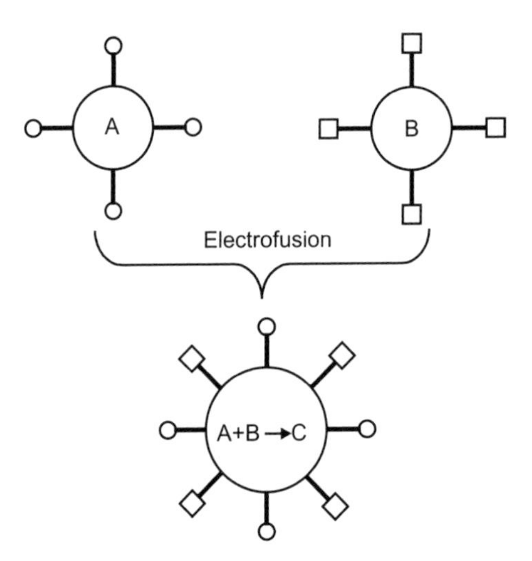


**Figure 13.18** Concept of nanoreactors with liposomes. A liposome containing reactant A is fused with another liposome containing reactant B. Inside the hybrid liposome, A and B react to form C. The circles and squares represent different phospholipods and proteins. (*After*: [106].)

### 13.2.5.2 Electroporation and Cell Fusion

In applications such as gene therapy, effective delivery of genes into living cells is an important task. Conventional gene delivery techniques are virus transfection [99], calcium phosphate-mediated transfection [100], liposome mediated transfection [101], particle bombardment [102], direct injection with microneedle [103], and electroporation [104]. As mentioned above, an impulse of a high electric field makes the cell membrane permeable. The reversibility of permeabilization of the cell membrane depends on the amplitude, length, shape, repetition rate of the voltage pulse, the cell itself, and its development stage. Thus, an electric field can be used for delivering of drug or DNA molecules into a cell. This technique was widely used with commercially available electroporators for protein transfection, drug delivery, and cell fusion.

Commercial electroporators can only treat cells in batch and do not allow the detailed study of electroporation. Microtechnology makes electroporation of a single cell possible. Huang and Rubinsky [105] used a silicon device to trap and to probe a cell. The device immobilizes the cell with hydrodynamic pressure. The cell is positioned on a pore etched in a silicon nitride membrane. With a pair of electrodes made of translucent polysilicon, the electric current across the fluidic chamber filled with an ionic solution and the cell can be measured. The electroporation process can be therefore precisely controlled by monitoring the current across the cell.

The concept of electroporation can be applied to two cells to fuse them together. This process is called *cell fusion*, and is one of the key techniques in biotechnology. Cloning is a typical example of cell fusion, where an adult cell from the mammary gland is fused with an egg cell. In a cell fusion process, the cells are first brought into contact. Techniques such as dielectrophoretic trapping described in Figure 13.17(b) can be used for this purpose. Subsequently, a short voltage pulse causes a high electric field across both cells. This pulse of high electric field fuses together the membrane areas, which are in close contact.

The concept of cell fusion can be well applied to liposomes, which makes *nanoreactors* a reality. *Liposomes* are synthetic lipid-bilayer containers, which are similar to the membranes of natural biological cells. The size of these containers ranges from tens of nanometers to tens of microns [106]. Reagents and molecules in liposomes can mix and react after fusing. Mimicking nature, real biochemical nanoreactors can be realized in this way (Figure 13.18).

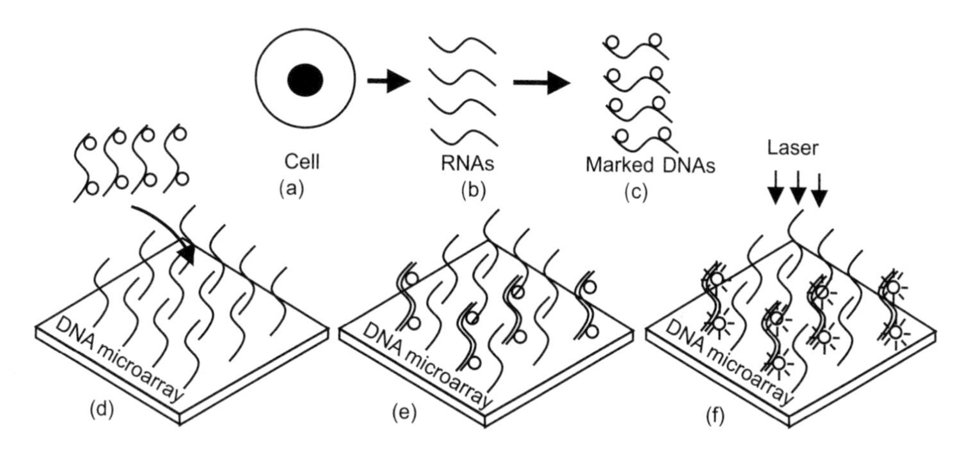


**Figure 13.19** Gene expression using a DNA microarray. (a) Genes of with drug-treated cells are to be compared with those of untreated cells. (b) The genes (DNA) are translated into ribonucleic acid (RNA). (c) The DNA copies of the RNA are tagged with fluorescent dyes. (d) The tagged DNAs are washed over the DNA array. (e) Hybridization reaction occurs with known single-stranded DNA on the array. (f) The DNAs are identified by fluorescent detection.

### 13.2.6 Hybridization Arrays

Hybridization arrays are biochemical microreactors for detection of genetic pathogens, antigens, or antibodies. The massively parallel screening process can be realized on a chip using microtechnology. The concept of these arrays is based on the hybridization binding reaction. Loose hydrogen bonds of complimentary bases on two single-stranded DNA chains attract and form a double-stranded DNA molecule. This concept can detect a small amount of DNA in a solution with a number of unknown molecules. Figure 13.19 shows a typical gene expression process [107]. *Gene expression* indicates a gene's activities in normal cells and in those treated with a drug. In the process of making a protein, the corresponding gene (DNA) is translated into ribonucleic acid (RNA). For gene expression, the RNA is extracted from the cell [Figure 13.19(b)]. The extracted RNA is used to make DNA, which is tagged with a fluorescent dye [Figure 13.19(c)]. The tagged DNA is brought to the array surface, where hybridization reaction occurs [Figure 13.19(d, e)]. The signals of the known DNA on the array are collected with an epifluorescent microscope and a CCD camera [Figure 13.19(f)]. Microarrays for sequence analysis have shorter oligos corresponding to the different disease and gene markers. The mutations of a single base in these DNA strands indicate the disease genes. A sequence analysis array can be used for personalized drug design as well as genetic risk assessments.

A microarray can be fabricated on different substrate materials such as silicon [108], glass [109], or polymer [110]. Microarrays with short DNA strands can be fabricated by the stepwise addition of bases [111]. The array is fabricated layer by layer using conventional lithography techniques. A density of $10^6$ probes/cm$^2$ and a probe size of 8 μm can be achieved [112, 113] (Figure 13.20). Microlithography requires masks for four nucleotides in a single layer. Making long oligos would need a large number of masks. A flexible solution for this problem is the use of a pattern generator. Gao et al. [114] reported the use of a $480 \times 640$ digital micromirror device (DMD) as pattern generator for the microarray. The resolution of this microarray is determined by the digital mirror, which is $30 \times 30$ μm$^2$ in size.

Inkjet concepts can be utilized for printing the bases of a DNA fragment. This in situ printing technique replaces the four colors of the ink (cyan, magenta, yellow, and black) by the

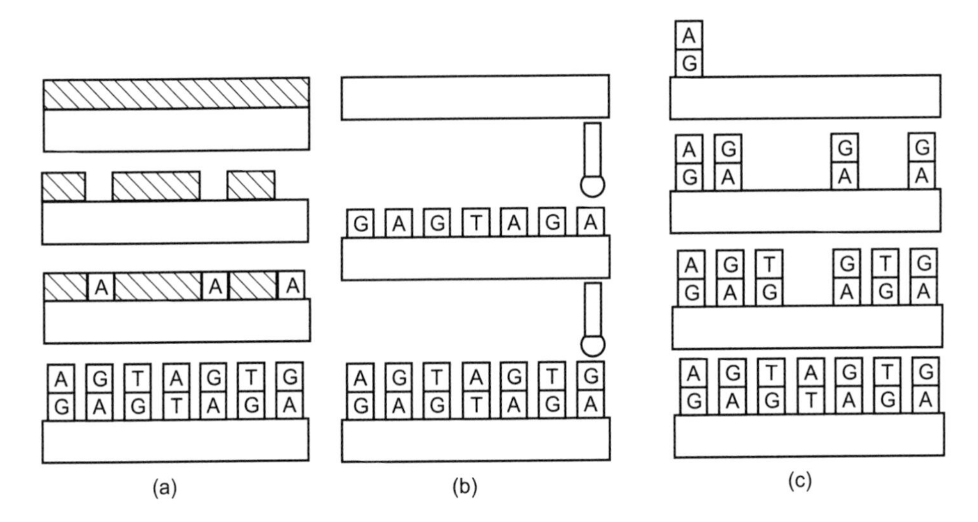


**Figure 13.20** Fabrication techniques for DNA arrays (A: adenine, G: guanine, C: cytosine, T: thymine): (a) microlithography (the hatched area is photoresist); (b) in situ printing with inkjet; and (c) direct printing.

four nucleotides (adenine, guanine, cytosine, and thymine). Layer by layer, this technique can build fragments with up to 60 bases [Figure 13.20(b)] [107].

Inkjet printing can also print presynthesized oligos and gene fragments directly on the substrate, [Figure 13.20(c)] [115]. Inkjet printing has a resolution on the order of 70 to 120 μm and allows making an microarray with about 25,000 spots. Contact printing is another method for bringing DNA fragments to the array surface [116]. A pin is first dipped into the solution with presynthesized DNAs. The pin is subsequently pressed to the array surface and leaves behind a droplet of the solution. In order to avoid repeated dipping, a split tip or a hollow tip can act as a small reservoir for multiple spotting steps. Section 11.1.1 discusses more about droplet dispensers.

A small DNA spot on a flat surface has a limited number of molecules, which in turn causes a weak signal for evaluation. A weak signal also means a higher signal-to-noise ratio. The signal can be improved by increasing the number of immobilized molecules. For a given planar spot size, the number of molecules can only be increased if they are immobilized on all three dimensions. Timofeev et al. [117] immobilized the molecules on a gel matrix, which is coated on the glass substrate. Thus, immobilization occurs in volume, rather than on a surface. However, diffusive transport inside the gel matrix causes different spot sizes and affects hybridization kinetics [118]. In order to avoid this effect, the porous structure can be micromachined into the substrate. Benoit et al. [119] etched tiny pores through the substrate. Each spot occupies a few hundred pores. Molecules are allowed to immobilize on the pore wall. The immobilization rate can be further improved if the sample is allowed to flow through the pores.

## 13.3 SUMMARY

Microreactors find applications in chemical industry, chemical/biochemical analysis, and cell-based studies. The small scale and the high surface-to-volume ratio in microreactors offer a number of advantages for chemical production. The design of a microreactor starts with the requirements from the reaction. Key parameters are the residence time, the reaction time, and the thermal response time. Based on these parameters, microchannels and microchambers for the reaction can be dimensioned. High throughput can be achieved by increasing the number of reactors units.

Microreactors are in favor of the space-time yield, a performance indicator for reactors, because of the small size and short residence time. Microtechnology also allows the integration of other functional elements, such as catalysts, sensors, and heat exchangers in the same device. Microfluidic technology provides numerous advantages for mimicking the cellular microenvironment at in vitro conditions for cell culture. The combination of tissue engineering principles with microfluidics has enabled the development of 3-D organ models known as organ-on-chip systems. Such biomimetic devices are designed to produce some organ-level functions at in vitro conditions. The organ-on-chip devises present unique features suitable for disease modeling and drug toxicity investigations. For chemical and biochemical analysis, scaling down the analysis brings tremendous advantage. Integrating all process steps on a single chip allows handling minute amounts of samples and chemicals. For DNA analysis, for instance, the entire analysis process can be integrated in a single device. While the previous chapters already started with methods and devices for cell sampling, this chapter discussed typical devices for cell lysis, DNA amplification, DNA separation, DNA hybridization, and detection. The same concepts can apply to the analysis of proteins and other molecules.

## Problems

**13.1** A microcombustor uses methane as fuel. The combustor consumes methane ($CH_4$) with a mass flow rate of 50 mg/s. Assuming that methane burns completely with 100% theoretical air, determine the air-fuel ratio and the volume flow rate of air.

**13.2** If the combustor is designed as a diffuser/nozzle structure with an inlet diameter of 2 mm and an outlet diameter of 0.5 mm over a length of 10 mm, the operating pressure is 4 atm. Estimate the residence time of the air/fuel mixture in the combustor.

**13.3** The extension time in a PCR cycle is 50 seconds. The enzyme in use is Tag Polymerase, which has an extension speed of 50 base pairs per second (bp/s). Determine the length of the amplified DNA segment if heating and cooling delays are neglected.

**13.4** Design a reaction chamber in silicon so that the minimum ramping time is 1 second. The silicon wafer is 500 mm thick.

**13.5** Design a spatial PCR reactor. The channel depth is $d = 100$ mm and the width is $w = 200$ mm. The flow rates range from 100 to 500 nL/s. The temperature zones for denaturing and annealing have the same width. The width of the temperature zone for extension is one-half of that of the other two. The time ratio of denaturing, annealing, and extension is 0.5:5:0.5 seconds.

**13.6** Benzoic acid spheres of 5 microns in diameter are packed loosely in a $100 \times 500 \times 500$-μm reaction chamber. The interfacial area-to-volume ratio is about $10^{11}$ $m^{-1}$. The mass transfer coefficient at the sphere surface is $10^{-3}$ cm/s. Determine the flow rate of pure water coming into the reaction chamber if the solution at the chamber exit is 80% saturated with benzoic acid. Assume that water is fed into the chamber laterally. What is the superficial velocity in the reaction chamber?

## References

[1] Jensen, K. F., "The Impact of MEMS on the Chemical and Pharmaceutical Industries," *Technical Digest of the IEEE Solid State Sensor and Actuator Workshop*, Hilton Head Island, SC, June 4–8, 2000, pp. 105–110.

[2] Veser, G., et al., "A Micro Reaction Tool for Heterogeneous Catalytic Gas Phase Reactions," *Proceedings of MEMS'99, 12th IEEE International Workshop Micro Electromechanical System*, Orlando, FL, Jan. 17–21, 1999, pp. 394–399.

[3] Commenge, J. M., *Réacteurs microstructurés: hydrodynamique, thermique, transfert de matiè et application aux procédeés*, Ph.D. Thesis, Institut National Polytechnique de Lorraine, Nancy, 2001.

[4] Kestenbaum, H., et al. "Silver-Catalyzed Oxidation of Ethylene to Ethylene Oxide in a Microreaction System," *Ind. Eng. Chem. Res.*, Vol. 41, 2002, pp. 710–719.

[5] Hessel, V., Hardt, S., and Löwe, H., *Chemical Micro Process Engineering,* Weinheim: Wiley-VCH, 2004.

[6] Wörz, O., et al., "Microreactors—A New Efficient Tool for Reactor Development," *Chemical Engineering Technology*, Vol. 24, 2000, pp. 24–29.

[7] Lao, A. I. K., et al., "Precise Temperature Control of Microfluidic Chamber for Gas and Liquid Phase Reactions," *Sensors and Actuators A*, Vol. 84, 2000, pp. 11–17.

[8] Meijer, G. C. M., and Herwaarden, A. W. (eds), *Thermal Sensors*, Bristol, U.K.: Institute of Physics Publication, 1994.

[9] Poser, S., et al., "Chip Elements for Fast Thermocycling," *Sensors and Actuators A*, Vol. 62, 1997, pp. 672–675.

[10] Schubert, K., et al., "Microstructure Devices for Thermal and Chemical Process Engineering," *Microscale Thermophysical Engineering*, Vol. 5, 2001, pp. 17–39.

[11] Cussler, E. L., *Diffusion Mass Transfer in Fluid Systems,* New York: Cambridge University Press, 1984.

[12] Oleschuk, R. D., et al., "Utilization of Bead Based Reagents in Microfluidic Systems," *Micro Total Analysis Systems 2000*, A. van den Berg, et al., (eds.), Boston: Kluwer Academic Publishers, 2000, pp. 11–14.

[13] Madou, M. J., et al., "A Centrifugal Microfluidic Platform—A Comparison," in *Micro Total Analysis Systems 2000*, A. van den Berg, et al. (eds), Boston: Kluwer Academic Publishers, 2000, pp. 565–570.

[14] Drott, J., et al., "Porous Silicon as the Carrier Matrix in Microstructured Enzyme Reactors Yielding High Enzyme Activities," *Journal of Micromechanics and Microengineering*, Vol. 7, 1997, pp. 14–23.

[15] Splinter, A., et al., "Micro Membrane Reactor: A Flow-Through Membrane for Gas Pre-Combustion," *Proceedings of Transducers '01, 11th International Conference on Solid-State Sensors and Actuators*, Munich, Germany, June 10–14, 2001, pp. 1778–1781.

[16] Schwarz, M. A., and Hauser, P. C., "Recent Developments in Detection Methods for Microfabricated Analytical Devices," *Lab on a Chip*, Vol. 1, 2001, pp. 1–6.

[17] Madou, M. J., and Cubicciotti, R., "Scaling Issues in Chemical and Biological Sensors," *Proceedings of the IEEE*, Vol. 91, 2003, pp. 830–838.

[18] Ocvirk, G., Tang, T., and Harrison, D. J.,"Optimization of Confocal Epifluorescence Microscopy for Microchip-Based Miniaturized Total Analysis Systems," *Analyst*, Vol. 123, 1998, pp. 1429–1434.

[19] Jiang, G., et al.,"Red Diode Laser Induced Fluorescence Detection with a Confocal Microscope on a Microchip for Capillary Electrophoresis," *Biosensors and Bioelectronics*, Vol. 14, 2000, pp. 861–869.

[20] Melanson, J. E., et al.,"Violet (405 nm) Diode Laser for Laser Induced Fluorescence Detection in Capillary Electrophoresis," *Analyst*, Vol. 125, 2000, pp. 1049–1052.

[21] Verpoorte, E., et al.,"A Silicon Flow Cell for Optical Detection in Miniaturized Total Chemical Analysis Systems," *Sensors and Actuators B*, Vol. 6, 1992, pp. 66–70.

[22] Greenway, G. M., et al., "The Use of a Novel Microreactor for High Throughput Continuous Flow Organic Synthesis," *Sensors and Actuators B*, Vol. 63, 2000, pp. 153–158.

[23] Epstein, A. H., et al., "Power MEMS and Microengines," *Proceedings of Transducers '97, 9th International Conference on Solid-State Sensors and Actuators*, Chicago, June 16–19, 1997, pp. 753–756.

[24] Waitz, I. A., et al., "Combustors for Micro-Gas Turbine Engines," *ASME Journal of Fluid Engineering*, Vol. 120, 1998, pp. 109–117.

[25] Mehra, A., and Waitz, I. A., "Development of a Hydrogen Combustor for a Microfabricated Gas Turbine Engine," *Technical Digest of the IEEE Solid State Sensor and Actuator Workshop*, Hilton Head Island, SC, June 2–4, 1998, pp. 144–147.

[26] Mehra, A., et al., "Microfabrication of High Temperature Silicon Devices Using Wafer Bonding and Deep Reactive Ion Etching," *Journal of Microelectromechanical Systems*, Vol. 8, No. 2, 1999, pp. 152–160.

[27] Mehra, A., et al., "A Six-Wafer Cobustion System for a Silicon Micro Gas Turbine Engine," *Journal of Microelectromechanical Systems*, Vol. 9, No. 4, 2000, pp. 517–527.

[28] Zimmermann, S., Wischhusen, S., and Mller, A., "A mTAS—Atomic Emission Flame Spectrometer (AES)," in *Micro Total Analysis Systems 2000*, A. van den Berg, et al. (eds.), Boston: Kluwer Academic Publishers, 2000, pp. 135–138.

[29] Veser, G., et al., "A Micro Reaction Tool for Heterogeneous Catalytic Gas Phase Reactions," in *Micro Total Analysis Systems 2000*, A. van den Berg, et al. (eds.), Boston: Kluwer Academic Publishers, 2000, pp. 394–399.

[30] Moran, M. J., and Shapiro, H. N., *Fundamentals of Engineering Thermodynamics*, 4th ed., New York: Wiley, 2000.

[31] Lee, S. J., et al., "Design and Fabrication of a Micro Fuel Cell Array with Flip-Flop Interconnection," *Journal Power Sources*, Vol. 112, 2002, pp. 410–418.

[32] Yu, J., et al., "Fabrication of Miniature Silicon Wafer Fuel Cells with Improved Performance," *Journal of Power Sources*, Vol. 124, 2003 40–46.

[33] Yen, T. J., et al., "A Micro Methanol Fuel Cell Operating at Near Room Temperature," *Applied Physics Letter*, Vol. 83, 2003, pp. 4056–4058.

[34] Lu, G. Q., et al., "Development and Characterization of a Silicon-Based Micro Direct Methanol Fuel Cell," *Electrochim. Acta*, Vol. 49, 2003, pp. 821–823.

[35] Blum, A., et al., "Water-Neutral Micro Direct-Methanol Fuel Cell (DMFC) for Portable Applications," *Journal of Power Sources*, Vol. 117, 2003, pp. 22–25.

[36] Shah, K., et al., "A PDMS Micro Proton Exchange Membrane Fuel Cell by Conventional and Non-Conventional Microfabrication Techniques," *Sensors and Actuators B*, Vol. 97, 2004, pp. 157–167.

[37] Hsieh, S. S., "A Novel Design and Microfabrication for a Micro PEMFC," *Microsystem Technologies*, Vol. 10, 2004, pp. 121–126.

[38] Chan, S. H., et al., "Development of a Polymeric Micro Fuel Cell Containing Laser-Micromachined Flow Channels," *Journal of Micromechanics and Micro Engineerings*, Vol. 15, 2005, pp. 231–236.

[39] Chen, X., et al., "Thin-Film Heterostructure Solid Oxide Fuel Cells," *Applied Physics Letters*, Vol. 84, 2004, pp. 2700 –2702.

[40] Srikar, V. T., et al., "Structural Design Considerations for Micromachined Solid-Oxide Fuel Cells," *Journal Power Sources*, Vol. 125, 2004, pp. 62–69.

[41] Shaegh, S. A. M., et al., "A Review on Membraneless Laminar Flow-Based Fuel Cells," *International Journal of Hydrogen Energy*, Vol. 36, 2011, pp. 5675–5694.

[42] Kjeang, E., et al., "Microfluidic Fuel Cells: A Review," *Journal of Power Sources*, Vol. 186, 2009, pp. 353–369.

[43] Shaegh, S. A. M., et al., "Air-breathing Membraneless Laminar Flow-Based Fuel Cell With Flow-Through Anode," *International Journal of Hydrogen Energy*, Vol. 37, 2012, pp. 3466–3476.

[44] Franz, A. J., Jensen, K. F., and Schmidt, M. A., "Palladium Based Micromembranes for Hydrogen Separation and Hydrogenation/Dehydrogenation Reactions," *Proceedings of MEMS'99, 12th IEEE International Workshop Micro Electromechanical System*, Orlando, FL, Jan. 17–21, 1999, pp. 382–387.

[45] Cui, T., et al., "Fabrication of Microreactors for Dehydrogenation of Cyclohexane to Benzene," *Sensors and Actuators B*, Vol. 71, 2000, pp. 228–231.

[46] Van der Wiel, D. P., et al., "Carbon Dioxide Conversions in Microreactors," *Proceedings of AlChe 2000 Spring National Meeting*, Atlanta, GA, March 6–9, 2000.

[47] Ehrlich, G. D., and Greenberg, S. J., *PCR-Based Diagnostics in Infectious Disease*, Boston: Blackwell, 1994.

[48] Mastrangelo, C. H., Burns, M. A., and Burke, D. T., "Microfabricated Devices for Genetic Diagnostics," *IEEE Proceedings*, Vol. 86, 1998, pp. 1769–1787.

[49] Debjani, P., and Venkataranman, V., "A Portable Battery-Operated Chip Thermocycler Based on Induction Heating," *Sensors and Actuators A*, Vol. 102, 2002, pp. 151–156.

[50] Hühmer, A. F. R., and Landers, J. P., "Noncontact Infrared-Mediated Thermocycling for Effective Polymerase Chain Reaction Amplification of DNA in Nanoliter Volumes," *Analytical Chemistry*, Vol. 72, 2000, pp. 5507–5512.

[51] Kopf-Sill, A. R., "Microfabricated Devices for Genetic Diagnostics," *Abstracts of Papers of the American Chemical Society*, Vol. 219, 2000, p. 389.

[52] Senturia, S. D., *Microsystem Design*, Boston, MA: Kluwer Academic Publishers, 2001, pp. 617–621.

[53] Northrup, M. A., et al., "DNA Amplification with a Microfabricated Reaction Chamber," *Proceedings of Transducers '93, 7th International Conference on Solid-State Sensors and Actuators*, Yokohama, Japan, June 7–10, 1993, pp. 924–926.

[54] Northrup, M. A., et al., "A MEMS-Based Miniature DNA Analysis System," *Proceedings of Transducers '95, 8th International Conference on Solid-State Sensors and Actuators*, Stockholm, Sweden, June 16–19, 1995, pp. 764–767.

[55] Daniel, J. H., et al., "Silicon Microchambers for DNA Amplification," *Sensors and Actuators A*, Vol. 71, 1998, pp. 81–88.

[56] Chaudhari, A. M., et al., "Transient Liquid Crystal Thermometry of Microfabricated PCR Vessel Arrays," *Journal of Microelectromechanical Systems*, Vol. 7, No. 4, 1998, pp. 345–355.

[57] Akahori, K., et al., "Multi Micro Reactors Consist of Individually Temperature Controlled Silicon Well Arrays Realizing Efficient Biochemical Reactions," in *Micro Total Analysis Systems 2000*, A. van den Berg, et al. (eds.), Boston: Kluwer Academic Publishers, 2000, pp. 493–496.

[58] Cheng, J., et al., "Chip PCR: Surface Passivation of Microfabricated Silicon-Glass Chips," *Nucleic Acids Research*, Vol. 24, No. 2, 1996, pp. 375–379.

[59] Wilding, P., et al., "Integrated Cell Isolation and Polymerase Chain Reaction Analysis Using Silicon Microfilter Chambers," *Anal. Biochem.*, Vol. 257, 1998, pp. 95–100.

[60] Lagally, E. T., et al., "Monolithic Integrated Microfluidic DNA Amplification and Capillary Electrophoresis Analysis System," *Sensors and Actuators B*, Vol. 63, 2000, pp. 138–146.

[61] Waters, L. C., et al., "Microchip Device for Cell Lysis, Multiplex PCR Amplification, and Electrophoretic Sizing," *Analytical Chemistry*, Vol. 70, 1998, pp. 158–162.

[62] Burns, M. A., et al., "An Integrated Nanoliter DNA Analysis Device," *Science*, Vol. 282, 1998, pp. 484–487.

[63] Yu, H., et al., "A Poly-Methylmethacrylate Electrophoresis Microchip with Sample Preconcentrator," *Journal of Micromechanics and Microengineering*, Vol. 11, 2001, pp. 189–194.

[64] Kopp, M. U., de Mello, A. J., and Manz, A., "Chemical Amplification: Continuous-Flow PCR on Chip, *Science*, Vol. 280, 1998, pp. 1046–1048.

[65] Schneegaß, I., Bräutigam, R., and Köhler, J. M., "Miniaturized Flow-Through PCR with Different Template Types in a Silicon Chip Thermocycler," *Lab on a Chip*, Vol. 1, No. 1, 2001, pp. 42–49.

[66] Obeid, P. J., et al., "Microfabricated Device for DNA and RNA Amplification by Continous-Flow Polymerase Chain Reaction and Reserve Transcription-Polymease Chain Reaction with Cycle Number Selection," *Analytical Chemistry*, Vol. 75, 2003, pp. 288–295.

[67] Krishnan, M., Victor, M. U., and Burns, M. A., "PCR in a Rayleigh-Benard Convection Cell," *Science*, Vol. 298, 2002, pp. 793.

[68] Krishnan, M., et al., "Reactions and Fluidics in Miniaturized Natural Convection Systems," *Analytical Chemistry*, Vol. 76, 2004, pp. 6254–6265.

[69] Chen, Z., et al., "Thermosiphon-Based PCR Reactor: Experiment and Modeling," *Analytical Chemistry*, Vol. 76, 2004, pp. 3707–3715.

[70] Wheeler, E. K., et al., "Convectively Driven Polymerase Chain Reaction Thermal Cycler," *Analytical Chemistry*, Vol. 76, 2004, pp. 4011–4016.

[71] Losey, M. W., Schmidt, M. A., and Jensen K. F., "A Micro Packed-Bead Reactor for Chemical Synthesis," *Ind. Eng. Chem. Res.*, Vol. 40, 2001, pp. 2555–2558.

[72] Wilkes, J. O., and Bike, S. G., *Fluid Mechanics for Chemical Engineers*, Upper Saddle River, NJ: Prentice Hall, 1998.

[73] Andersson, H., et al., "Micromachined Flow-Through Filter-Chamber for Chemical Reactions on Beads," *Sensors and Actuators B*, Vol. 67, 2000, pp. 203–208.

[74] Ahn, C. H., et al., "A Fully Integrated Micromachined Magnetic Particle Separator," *Journal of Microelectromechanical Systems*, Vol. 5, No. 3, 1996, pp. 151–158.

[75] Oleschuk, R. D., et al., "Trapping of Bead-Based Reagents Within Microfluidic Systems: On-Chip Solid-Phase Extraction and Electrochromatography," *Analytical Chemistry*, Vol. 72, 2000, pp. 585–590.

[76] Skardal, A., et al., "Multi-Tissue Interactions in an Integrated Three-Tissue Organ-on-Chip Platform," *Scientific Reports*, Vol. 7, 2017, pp. 1–16.

[77] Tehranirokh, M., et al., "Microfluidic Devices for Cell Cultivation and Proliferation," *Biomicrofluidics*, Vol. 7, 2013, pp. 1–32.

[78] Nguyen, N. -T., et al., "Design, Fabrication and Characterization of Drug Delivery Systems Based on Lab-on-a-Chip Technology," *Advanced Drug Delivery Reviews*, Vol. 65, 2013, pp. 1403–1419.

[79] Ye, N., et al., "Cell-Based High Content Screening Using an Integrated Microfluidic Device," *Lab Chip*, Vol. 7, 2007, pp. 1696–1704.

[80] Haessler, U., et al., "An Agarose-Based Microfluidic Platform with a Gradient Buffer for 3D Chemotaxis Studies," *Biomedical Microdevices*, Vol. 11, 2009, pp. 827–835.

[81] Abhyankar, V. V., et al. "Characterization of a Membrane-Based Gradient Generator for Use in Cell-signalling studies," *Lab Chip*, Vol. 6, 2006, pp. 389–393.

[82] Shin, Y., et al., "Microfluidic Assay for Simultaneous Culture of Multiple Cell Types on Surfaces or Within Hydrogels," *Nat. Protoc.*, Vol. 7, 2012, pp. 1247–1259.

[83] Zervantonakis, I. K., et al., "Three-Dimensional Microfluidic Model for Tumor Cell Intravasation and Endothelial Barrier Function," *Proceedings of National Academy of Science*, Vol. 109, 2012, pp. 13515–13520.

[84] Zhang, Y. S., et al., "Multisensor-Integrated Organs-on-Chips Platform for Automated and Continual in situ Monitoring of Organoid Behaviors," *Proceedings of National Academy of Science*, Vol. 114, 2017, pp. 2293–2302.

[85] Bhatia, S. N., and Ingber, D. E., "Microfluidic Organs-on-Chips," *Nature Biotechnology*, Vol. 32, 2014, pp. 760–772.

[86] Huh, D., et al., "Reconstituting Organ-Level Lung Functions on a Chip ," *Science*, Vol. 328, 2010, pp. 1662–1668.

[87] Jang, K. J., et al., "Human Kidney Proximal Tubule-on-Chip for Drug Transport and Nephtotoxicity Assesment," *Integr. Biol.*, Vol. 5, 2013, pp. 1119–1129.

[88] Kim, H. J., and Ingber, D. E., "Gut-on-a-Chip Microenvironment Induces Human Intestinal Cells to Undergo Villus Differentiation," *Integr. Biol.*, Vol. 5, 2013, pp. 1130–1140.

[89] Palaninathan, V., et al., "Multi-organ on a Chip for Personalized Precision Medicine," *MRS Communications*, 2018, pp. 1–16.

[90] Riahi, R., et al., "Automated Microfluidic Platform of Bead-Based Electrochemical Immunosensor Integrated with Bioreactor for Continual Monitoring of Cell Secreted Biomarkers," *Scientific Reports*, Vol. 6, 2016, pp. 1–14.

[91] Shaegh, S. A. M., et al., "A Microfluidic Optical Platform for Real-Time Monitoring of pH and Oxygen in Microfluidic Bioreactors and Organ-on-Chip Devices," *Biomicrofluidics*, Vol. 10, 2016, pp. 1–14.

[92] Tracy, M., et al., "A Microfluidics-Based Instrument for Cytomechanical Studies of Blood," *Proc. IEEEEMBS Conf. Microtech. Med. Bio.*, Vol. 1, pp. 62–67.

[93] Huang, Y., and Rubinsky, B., "Micro-Electroporation: Improving the Efficiency and Understanding of Electrical Permeabilization of Cells," *Biomedical Microdevices*, Vol. 2, 1999, pp. 145–150.

[94] Schilling, E. A., et al., "Cell Lysis and Protein Extraction in a Microfluidic Device with Detection by a Fluorogenic Enzyme Assay," *Analytical Chemistry*, Vol. 74, 2002, pp. 1798–1804.

[95] Belgrader, P., et al., "A Minisonicator To Rapidly Disrupt Bacterial Spores for DNA Analysis," *Analytical Chemistry*, Vol. 71, 1999, pp. 4232–4236.

[96] Taylor, M. T., et al., "Lysing Bacterial Spores by Sonication Through a Flexible Interface in a Microfluidic System," *Analytical Chemistry*, Vol. 73, 2001, pp. 492–496.

[97] Marmottant, P., and Hilgenfeldt, S., "Controlled Vesicle Deformation and Lysis by Single Oscillating Bubbles," *Nature*, Vol. 423, 2003, pp. 153–155.

[98] Lee, S. W., and Tai, Y. C., "A Micro Cell Lysis Device," *Sensors and Actuators A*, Vol. 73, 1999, pp. 74–79.

[99] Yin, L. H., et al., "Results of Retroviral and Adenoviral Approaches to Cancer Gene Therapy," *Stem Cells*, Vol. 16S1, 1998, pp. 247–250.

[100] Lee, J. H., and Welsh, M. J., "Enhancement of Calcium Phosphate-Mediated Transfection by Inclusion of Adenovirus in Coprecipitates," *Gene Tharapy*, Vol. 6, 1999, pp. 676–682.

[101] Schmid, R. M., et al., "Liposome Mediated Gene Transfer into Rat Oesophagus," *Gut*, Vol. 41, 1997, pp. 549–556.

[102] Han, R., et al., "Immunization of Rabbits with Cottontail Rabbit Papillomavirus E1 and E2 Genes: Protective Immunity Induced by Gene Gun-Mediated Intracutaneous Delivery but Not by Intramuscular Injection," *Vaccine*, Vol. 18, 2000, pp. 2937–2944.

[103] Zhang, G., Budker, V., and Wolff, J. A., "High Levels of Foreign Gene Expresion in Hepatocytes After Tail Vein Injection of Naked Plasmid DNA," *Human Gene Therapy*, Vol. 10, 1999, pp. 1735–1737.

[104] Rols, M. P., and Teissie, J., "Electropermeabilization of Mammalian Cells Quantitative Analysis of the Phenomenon," *Biophysics Journal*, Vol. 58, 1990, pp. 1089–1098.

[105] Huang Y., and Rubinsky, B., "Microfabricated Electroporation Chip for Single Cell Membrane Permeabilization," *Sensors and Actuators A*, Vol. 89, 2001, pp. 242–249.

[106] Strömberg, A. et al., "Microfluidic Device for Combinatoral Fusion of Liposomes and Cells," *Analytical Chemistry*, Vol. 73, 2001, pp. 126–130.

[107] Moore, S. K., "Making Chips to Probe Genes," *IEEE Spectrum*, Vol. 38, No. 3, 2001, pp.54–60.

[108] Sosnowski, R. G., et al., "Rapid Determination of Single Base Pair Mutations in DNA Hybrids by Direct Electric Field Control," *Proceeddings of National Academy of Science*, Vol. 94, 1997, pp. 1119–1123.

[109] Lipshutz, R., et al., "Using Oligonucleotide Probe Arrays to Access Genetic Diversity," *Biotechniques*, Vol. 19, 1995, pp. 442–447.

[110] Lipshutz, R. S., et al., "Biopolymer Synthesis on Polypropylene Supports: Oligonucleotide Arrays," *Analytical Biochemistry*, Vol. 224, 1995, pp. 110–116.

[111] Fodor, S. P. A., et al., "Multiplexed Biochemical Assays with Biological Chips," *Nature*, Vol. 364, 1993, pp. 555–556.

[112] McGall, G. H., et al., "Light Directed Synthesis of High Density Oligonucleotide Arrays Using Semiconductor Photoresists," *Proceedings of National Academy of Science*, Vol. 93, 1996, pp. 13555–13560.

[113] Beecher, J. E., McGall, G. H., and Goldberg, M. J., "Chemically Amplified Photolithography for the Fabrication of High Density Oligonucleotide Arrays," *Polymer Material Science and Engineering*, Vol. 76, 1997, pp. 597–598.

[114] Gao, X., et al., "A Flexible Light-Directed DNA Chip Synthesis Gated by Deprotection Using Solution Photogenerated Acids," *Nucleic Acid Research*, Vol. 29, 2001, pp. 4744–4750.

[115] Schena, R. A., et al., "Microarrays: Biotechnology," *Trends in Biotechnology*, Vol. 16, 1998, pp. 301–306.

[116] Shalon, D., Smith, S. J., and Brown, P. O., "A DNA Microarray System for Analysing Complex DNA Samples Using Two Color Fluorescent Probe Hybridization," *Genome Reseacrh*, Vol. 6, 1996, pp. 639–645.

[117] Timofeev, E., et al., "Regionselective Immobilization of Short Oligonucleotides to Acrylic Copolymergels," *Nucleic Acid Reseacrh*, Vol. 24, 1996, pp. 3142–3148.

[118] Lockhart, D., et al., "Expression Monitoring by Hybridization to High Density Oligonucleotides Aarrays," *Nature Biotechnology*, Vol. 14, 1996, pp. 167–168.

[119] Benoit, V., et al., "Evaluation of Three-Dimensional Microchannel Glass Biochip for Multiplexed Nucleic Acid Fluorescence Hybridization Assays," *Analytical Chemistry*, Vol. 73, 2001, pp. 2412–2420.

# Appendix A: List of Symbols

| *Symbol* | *Description* | *Unit* |
|---|---|---|
| $A$ | Thermal diffusivity | $m^2/s$ |
| $A$ | Surface | $m^2$ |
| $B$ | Resolution | m |
| $B$ | Width | m |
| $B$ | Magnetic flux density | Tesla |
| $C$ | Solute concentration | $kg/m^3$ |
| $c^*$ | Exhaust velocity | m/s |
| $c_\mathrm{p}$ | Specific heat at constant pressure | J/kg-K |
| $c_\mathrm{ph}$ | Phase velocity | m/s |
| $c_\mathrm{s}$ | Speed of sound | m/s |
| $c_\mathrm{v}$ | Specific heat at constant volume | J/kg-K |
| $C_\mathrm{valve}$ | Valve capacity | $m^2$ |
| $C$ | Capacitance | F |
| $C_\mathrm{D}$ | Drag coefficient | — |
| $C_\mathrm{F}$ | Thrust coefficient | — |
| $C_\mathrm{L}$ | Lift coefficient | — |
| $C_\mathrm{p}$ | Pyroelectric coefficient | A.sec/K-$m^2$ |
| $D$ | Piezoelectric coefficient | N/C |
| $d, D$ | Diameter | m |
| $D$ | Diffusion coefficient | $m^2/s$ |
| $D_\mathrm{h}$ | Hydraulic diameter | m |
| $e$ | Internal energy | J/kg |
| $e$ | Spacing between pixels | m |
| $E$ | Young's modulus | Pa |
| $E_\mathrm{el}$ | Electric field strength | V/m |
| $E'$ | Energy density | $J/m^3$ |
| $E_\mathrm{c}$ | Eckert number | — |
| $f$ | Frequency | Hz |
| $f$ | Friction factor | — |
| $F$ | Force | N |
| $g$ | Acceleration of gravity | $m^2/s$ |
| $G_\mathrm{eff}$ | Effective thermal conductance | W/K |
| $h$ | Height, pump head | m |
| $h$ | Heat transfer coefficient | — |
| $I$ | Current | A |
| $I_\mathrm{sp}$ | Specific impulse | s |
| $I$ | Moment of inertia | $m^4$ |
| $k$ | Specific heat ratio | — |
| $k$ | Spring constant | N/m |

| *Symbol* | *Description* | *Unit* |
|---|---|---|
| $K_T$ | Mass transfer coefficient | m/sec |
| $K$ | Boltzmann's constant | J/K |
| Kn | Knudsen number | — |
| $l$ | Characteristic length | m |
| $L$ | Length | m |
| $L_e$ | Entrance length | m |
| $L_{valve}$ | Leakage ratio | — |
| $\bar{L}_0$ | Latent heat | kJ/kmol |
| $m$ | Magnetic moment | — |
| $m$ | Mass | kg |
| $m_{gear}$ | Gear module | — |
| $\dot{m}$ | Mass flow rate | kg/sec |
| $M_m$ | Magnetization | A/m |
| $M$ | Molecular weight | kg/kmol |
| $M$ | Magnification of lens | — |
| Ma | Mach number | — |
| $n$ | Rotation speed | rpm |
| $N_A$ | Avogadro number | — |
| NA | Numerical aperture | — |
| Nu | Nusselt number | — |
| $p$ | Pressure | Pa |
| $P$ | Power | W |
| $P_{wet}$ | Perimeter | m |
| $P$ | Polarization vector | $C/m^2$ |
| Pr | Prandtl number | — |
| $q_f$ | Free space charge density | $C/m^3$ |
| $\dot{Q}$ | Volumetric flow rate | $m^3/s$ |
| $R$ | Radius | m |
| $R$ | Wing length | m |
| $R$ | Gas constant | J/kg-K |
| $R$ | Electric resistance | ohm |
| $R$ | Separation resolution | — |
| Re | Reynolds number | — |
| $s$ | Distance | m |
| $S$ | Sensitivity | $Vsec/m^3$ |
| $t$ | Time | sec |
| $T$ | Geometry parameter of a shear stress sensor | m |
| $t$ | Thickness | m |
| $T$ | Absolute temperature | K |
| $u$ | Velocity | m/s |
| $u_\tau$ | Friction velocity | m/s |
| $U$ | Energy | J |
| $v$ | Specific volume | $m^3/kg$ |
| $V$ | Voltage | V |
| $V$ | Volume | $m^3$ |
| $V_{ij}$ | Lennard-Jones potential | J |
| $w$ | Width, distance | m |
| $W$ | Weight | N |
| $W$ | Work | J |
| $x$ | Mass fraction | — |
| $\alpha$ | Temperature coefficient of resistance | ohm/K |
| $\alpha_f$ | Temperature coefficient of frequency | Hz/K |
| $\alpha_S$ | Thermoelectric (Seebeck) coefficient | V/K |
| $\beta$ | Opening factor | — |
| $\Gamma$ | Amplification factor | — |
| $\gamma$ | Thermal expansion coefficient | m/K |

| *Symbol* | *Description* | *Unit* |
|---|---|---|
| $\delta_{\mathrm{a}}$ | Acoustic evanescent length | m |
| $\Delta p$ | Pressure difference | Pa |
| $c$ | Rectification efficiency | — |
| $\Delta T$ | Temperature difference | K |
| $\varepsilon$ | Characteristic energy scale | J |
| $\varepsilon$ | Dielectric constant | F/m |
| $\varepsilon$ | Strain | — |
| $\varepsilon$ | Void fraction | — |
| $\zeta$ | Zeta potential | V |
| $\eta$ | Dynamic viscosity | kg/m-sec |
| $\eta_{\mathrm{S}}$ | Sensor efficiency | — |
| $\eta_{\mathrm{F}}$ | Diodicity | — |
| $\theta$ | Angle | rad |
| $\Theta$ | Compressibility of liquid | $m^2/N$ |
| $\kappa$ | Thermal conductivity | W/K-m |
| $\lambda$ | Wavelength | m |
| $\lambda$ | Mean free path | m |
| $\lambda_{\mathrm{D}}$ | Debye length | m |
| $\mu$ | Permeability | H/m |
| $\mu_{\mathrm{ep}}$ | Electrophoretic mobility | $m^2/s$-V |
| $\mu_{\mathrm{eo}}$ | Electro-osmotic mobility | $m^2/s$-V |
| $\nu$ | Kinematic viscosity | $m^2/s$ |
| $\nu$ | Poisson's ratio | — |
| $\xi$ | Pressure loss coefficient | — |
| $\rho$ | Density | $kg/m^3$ |
| $\sigma$ | Surface tension | Pa |
| $\sigma$ | Characteristic length scale | m |
| $t$ | Time | sec |
| $t_{\mathrm{w}}$ | Wall shear stress | Pa |
| $\phi$ | Gear pressure angle | rad |
| $\psi$ | Compression ratio | — |
| $\Gamma$ | Efficiency factor | — |
| $\phi$ | Dissipation | $J/m^3$ |
| $\Phi$ | Wing beat amplitude | rad |
| $\Phi$ | Mass flux | $kg/m^2$ |
| $\Phi$ | Electrical potential | V |
| $\Omega$ | Aspect ratio | - |
| $\omega$ | Angular velocity | rad/s |

# Appendix B: Resources for Microfluidics Research

## Journals

- *Sensors and Actuators A*: A peer-reviewed journal for sensors and actuators with a lot of papers on silicon microfluidic devices. The journal is published by Elsevier Science, Amsterdam, The Netherlands.

- *Sensors and Actuators B*: A traditional journal for chemical sensors published by Elsevier Science. With the recent issue on micrototal analysis system (mTAS), the journal is also a considerable source for microfluidic devices and applications.

- *Journal of Microelectromechanical Systems*: A peer-reviewed journal for MEMS, published jointly by IEEE and ASME.

- *Journal of Micromechanics and Microengineering*: A peer-reviewed journal published by the Institute of Physics, Bristol, United Kingdom.

- *Journal of Biomedical Microdevices*: A peer-reviewed journal published by Springer. The journal focuses on microdevices for biomedical applications.

- *Lab on a Chip*: A peer-reviewed journal of Royal Society of Chemistry, London, England. The journal is dedicated to applications of microfluidics in life sciences and chemistry. The first issue appeared in September 2001.

- *Microfluidics and Nanofluidics*: A peer-reviewed journal published by Springer. The journal is dedicated to all aspects of microfluidics and nanofluidics. The first issue appeared in November 2004.

- *Biomicrofluidics*: A peer-reviewed journal published by American Institute of Physics (AIP). The journal is dedicated to all aspects of microfluidics and nanofluidics. The first issue published in 2007.

  Different aspects of microfluidics and nanofluidics are also covered in various journals including *Small*, *Biosensors and Bioelectronics*, Nature Publishing Group Journals such as *Scientific Reports*, *Nature Communications*, *Nature Reviews Drug Discovery*, and *Nature Biotechnology*.

## Conferences

- *International Conference on Solid-State Sensors and Actuators (Transducers)*: The conference is held in odd years, rotating between North America, Europe, and Asia.

- *Microelectromechanical Systems Workshop (MEMS)*: The international workshop is held annually, and is sponsored by IEEE.
- *Solid-State Sensor and Actuator Workshop (Hilton Head)*: The workshop is held in even years, focusing on North America, and is sponsored by the Transducers Research Foundation.
- *Micro Total Analysis System (mTAS)*: Since 2000, the conference has been held annually. The conference focuses entirely on mTAS.
- *ASME International Mechanical Engineering Congress & Exposition*: The conference is annual and has technical sessions on microfluidics.
- *AIChe Annual National Spring Meeting, Annual Meeting*: The conferences have technical sessions on microreactors, biochemical applications, and nanosciences.
- *SPIE Micromachining and Microfabrication*: The regular conference is sponsored by SPIE and has a technical session on microfluidics.

# Appendix C: Abbreviations of Different Plastics

| *Abbreviation* | *Plastics* |
|---|---|
| ABS | Acrylonitrile butadiene styrene |
| CA, CAB, CAP | Cellulosics (acetate, butyrates, propionate) |
| COC | Cyclic olefin copolymer |
| ETFE | Ethylene tetrafluoroethylene |
| FEP | Florinated ethylene propylene |
| HDPE | High-density polyethylene |
| HTN | High-temperature nylon (Aromatic) |
| LCP | Liquid crystal polymer |
| LDPE | Low-density polyethylene |
| PA | Polyamide (nylon) |
| PAEK | Polyaryletherketone |
| PBT | Polybutylene terephthalate |
| PC | Polycarbonate |
| PEEK | Polyetheretherketone |
| PEI | Polyetherimide |
| PEKK | Polyetherketoneketone |
| PES | Polyethersulfone |
| PET | Polyethylene terephthalate |
| PETG | Polyethylene terephthalate glycol modified |
| PFA | Perfluoroalkoxy |
| PMMA | Poly methyl methacrylate (acrylic) |
| PMP | Polymethylpentene |
| PP | Polypropylene |
| PPA | Polyphthalamide (hot water moldable) |
| PPE | Modified polyphenylene ether |
| PPS | Polyphenylene sulfide |
| POM | Polyoxymethylene |
| PS | Polystyrene |
| PSU | Polysulfone |
| PTT | Polytrimethylene terephthalate |
| PVDF | Polyvinylidene fluoride |
| RTPU | Rigid thermoplastic polyurethane |
| SAN | Styrene acrylonitrile |
| SPS | Syndiotactic polystyrene |
| TEO | Olefinic thermoplastic elastomer |
| TES | Styrenic thermoplastic elastomer |
| TPI | Thermoplastic polyimide |
| TPU | Polyurethane thermoplastic elastomer |
| TPE | Polyester thermoplastic elastomer |

# Appendix D: Linear Elastic Deflection Models

| *Models* | *Formulas* |
|---|---|
| Cantilever beam with a force at the free end:<br>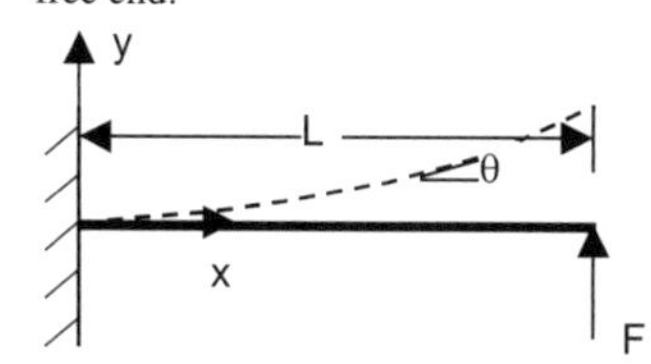<br> | $y = \frac{Fx^2}{6EI}(3L - x),\ y_{\max} = \frac{FL^3}{3EI}$ at $x = L$<br>$\theta = \frac{Fx}{2EI}(2L - x),\ \theta_{\max} = \frac{FL^2}{2EI}$ at $x = L$<br>$M_{\max} = FL$ at $x = 0$ |
| Cantilever beam with a force at a guided end:<br>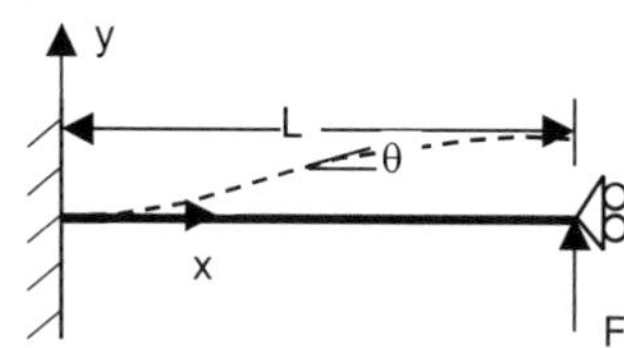<br> | $y = \frac{F}{2EI}\left(\frac{x^3}{3} - \frac{Lx^2}{2}\right),\ y_{\max} = \frac{FL^3}{12EI}$ at $x = L$<br>$\theta = \frac{F}{2EI}(x^2 - Lx),\ \theta_{\max} = \frac{FL^2}{4EI}$ at $x = \frac{L}{2}$<br>$M_{\max} = \frac{FL}{2}$ at $x = 0, L$ |
| Cantilever beam with fixed ends and center load:<br>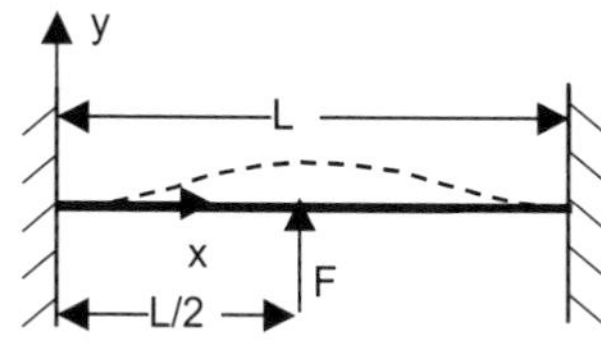<br> | $y = \frac{Fx^2}{48EI}(3L - 4x),\ y_{\max} = \frac{FL^3}{192EI}$ at $x = \frac{L}{2}$<br>$\theta = \frac{Fx}{8EI}(L - 2x)$ at $x \leq \frac{L}{2},\ \theta_{\max} = \frac{FL^2}{64EI}$ at $x = \frac{L}{4}$<br>$M_{\max} = \frac{FL}{8}$ at $x = 0, \frac{L}{2}$ |
| For the above cantilever models: | Function of radius of curvature $r(x)$: $1/r(x) = \mathrm{d}^2x/\mathrm{d}y^2$<br>Strain along the beam $\varphi(x)$: $\varphi(x) = t/2r(x) = t/2\mathrm{d}^2x/\mathrm{d}y^2$ |

| *Models* | *Formulas* |
|---|---|
| Circular membrane with point load: 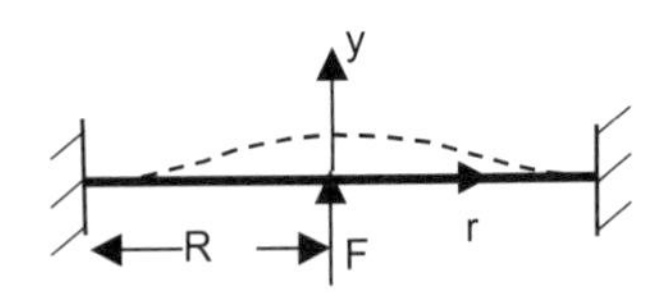  | $$y = \frac{3F(1-nu^2)}{4\pi E t^3}\left[R^2 - r^2\left(1 + 2\ln\frac{R}{r}\right)\right]$$ $$y_{\text{max}} = \frac{3F(1-\nu^2)R^2}{4\pi E t^3} \text{ at } r = 0$$ |
| Circular membrane with distributed load (pressure $p$): 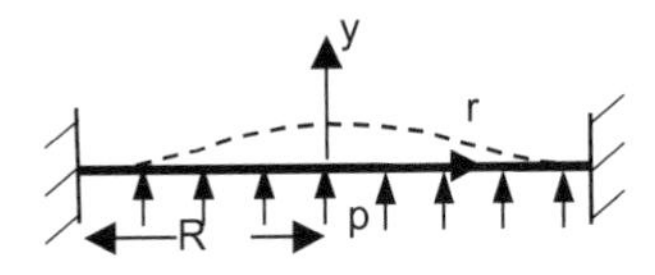  | $$y = \frac{3p(1-\nu^2)}{16Et^3}(R^2 - r^2)^2$$ $$y_{\text{max}} = \frac{3(1-\nu^2)}{16Et^3}R^4 p \text{ at } r = 0$$ $$\sigma = \frac{3R^2}{8t^2}\left[(3+\nu)\frac{x^2}{r^2} - (1+\nu)\right]p$$ $$\sigma_{\text{max}} = \frac{3R^2}{4t^2}p$$ $$\varphi_{\text{max}} = \frac{3R^2(1-\nu)}{4Et^2} \text{ at } r = R$$ |
| | $\theta$: angle of free end, $I$: moment of inertia ($I = bt^3/12$ for rectangular beam with width $b$, and thickness $t$); $\sigma$: stress; $\varphi$: strain; $\nu$: Poisson's ratio; $E$: Young's modulus. |

The above formulas are extracted from the following sources:


- Howell, L. L., *Compliant Mechanisms*, New York: Wiley, 2001.
- Timoshenko, S. P., and Gere, J. M., *Theory of Elastic Stability*, 2nd ed., Auckland, New Zealand: McGraw-Hill, 1961.
- Timoshenko, S. P., and Goodier, J. N., *Theory of Elasticity*, 3rd ed., Auckland, New Zealand: McGraw-Hill, 1970.
- Young, W. C., *Roark's Formulas for Stress & Strain*, 6th ed., New York: McGraw-Hill, 1989.

# About the Authors

**Nam-Trung Nguyen** received his Dip-Ing, Dr Ing and Dr Ing Habil degrees from Chemnitz University of Technology, Germany, in 1993, 1997 and 2004, respectively. In 1998, he was a postdoctoral research engineer in the Berkeley Sensor and Actuator Center (University of California at Berkeley, USA). From 1999 to 2012, he has been a faculty member at Nanyang Technological University in Singapore. Since 2013, he is a Professor and the Director of Queensland Micro- and Nanotechnology Centre at Griffith University, Australia. He is a Fellow of ASME and a Member of IEEE. Nguyen's research is focused on microfluidics, nanofluidics, micro/nanomachining technologies, micro/nanoscale science, and instrumentation for biomedical applications. He published over 370 journal papers and filed 8 patents, of which 3 were granted.

**Steven T. Wereley** Professor Wereley completed his masters and doctoral research at Northwestern University. He joined the Purdue University faculty in August of 1999 after a two-year postdoctoral appointment at the University of California Santa Barbara. During his time at UCSB he worked with a group developing, patenting, and licensing to TSI, Inc., the micro-Particle Image Velocimetry technique. His current research interests include designing and testing microfluidic MEMS devices, investigating biological flows at the cellular level, improving micro-scale laminar mixing, and developing new micro/nano flow diagnostic techniques. Although considerably outside the field of microfluidics, Professor Wereley used his flow measurement expertise to analyze the Deepwater Horizon oil spill in the Gulf of Mexico in 2010, serving on the US government?s Flow Rate Technical Group. His contributions to characterizing the disaster were recognized with the US Geological Survey Director's Award. Professor Wereley is the co-author Particle Image Velocimetry: A Practical Guide, Third Edition (Springer, 2018). He is on the editorial board of Experiments in Fluids and is an Associate Editor of Springer's Microfluidics and Nanofluidics. Professor Wereley has edited Springer's recent Encyclopedia of Microfluidics and Nanofluidics and Kluwer's BioMEMS and Biomedical Nanotechnology.

**Seyed Ali Mousavi Shaegh** received a B.S. and a M.S. in 2004 and 2007 from Yazd University, and Ferdowsi University of Mashhad, Iran, both in mechanical engineering. He received a Ph.D. in 2012 from Nanyang Technological University, Singapore, in mechanical engineering. He joined Harvard University, Brigham and Women's Hospital and Harvard-MIT Division of Health Sciences and Technology (HST), in 2014 as a postdoc fellow after a two-year research appointment at Singapore Institute of Manufacturing Technology (a research institute of Agency for Science, Technology and Research, A*STAR). Currently, he is an assistant professor at Mashhad University of Medical Sciences, in the School of Medicine (Clinical Research Unit and Orthopedic Research Center). Mousavi's research interests are mainly focused on microfluidic systems for diagnostics and drug testing, development of rapid prototyping methods for microfluidic chip fabrication, as well as design, fabrication, and testing of medical devices and implants.

# Index

## Recent Titles in the Artech House Integrated Microsystems Series

*Acoustic Wave and Electromechanical Resonators: Concept to Key Applications,* Humberto Campanella

*Adaptive Cooling of Integrated Circuits Using Digital Microfluidics,* Philip Y. Paik, Krishnendu Chakrabarty, and Vamsee K. Pamula

*Applications of Energy Harvesting Technologies in Buildings,* Joseph W. Matiko and Stephen P. Beeby

*Cost-Driven Design of Smart Microsystems,* Michael Niedermayer

*Design Solutions for Wireless Sensor Networks in Extreme Environments,* Habib F. Rashvand and Ali Abedi

*Fundamentals and Applications of Microfluidics, Third Edition,* Nam-Trung Nguyen, Steven T. Wereley, and Seyed Ali Mousavi Shaegh

*Hermeticity Testing of MEMS and Microelectronic Packages,* Suzanne Costello and Marc P. Y. Desmulliez

*Highly Integrated Microfluidics Design,* Dan E. Angelescu

*Integrated Interconnect Technologies for 3D Nanoelectronic Systems*, Muhannad S. Bakir and James D. Meindl, editors

*Introduction to Microelectromechanical (MEM) Microwave Systems,* Héctor J. De Los Santos

*Integrated Microsystems*, Nam-Trung Nguyen, Steven Wereley, and Seyed Ali Mousavi Shaegh

*An Introduction to Microelectromechanical Systems Engineering,* Nadim Maluf

*Lab-on-a-Chip, Techniques, Circuits, and Biomedical Applications,* Yehya H. Ghallab and Wael Badawy

*MEMS Mechanical Sensors,* Stephen Beeby et al.

*Micro and Nano Manipulations for Biomedical Applications,* Tachung C. Yihllie Talpasanu

*Microfabrication for Microfluidics,* Sang-Joon John Lee and Narayan Sundararajan

*Microfluidics for Biotechnology, Second Edition,* Jean Berthier and Pascal Silberzan

Artech House
685 Canton Street
Norwood, MA 02062
Phone: 781-769-9750
Fax: 781-769-6334
e-mail: artech@artechhouse.com

Artech House
16 Sussex Street
London SW1V 4RW UK
Phone: +44 (0)20 7596-8750
Fax: +44 (0)20 7630-0166
e-mail: artech-uk@artechhouse.com